D0138187

The Basics of Abstract Algebra

The Basics of Abstract Algebra

Paul E. Bland

W. H. Freeman and Company
New York

Acquisitions Editor: Craig Bleyer

Development Editor: Mary Johenk

New Media and Supplements Editor: Mark Santee

Marketing Manager: John Britch

Project Editing and Composition: Publication Services

Cover and Text Design: Cambraia Fernandes

Illustration Coordinator: Bill Page

Illustrations: Publication Services

Photo Researcher: Vikii Wong

Cover Image: Akira Inoue/Photonica

Production Coordinator: Susan Wein

Manufacturing: R R Donnelley & Sons Company

Library of Congress Cataloging-in-Publication Data

Bland, Paul E.
    The basics of abstract algebra / Paul E. Bland.
        p. cm.
    Includes index.
    ISBN 0-7167-4229-2
    1. Algebra, Abstract. I. Title.

QA162 .B585 2002
512'.02--dc21

                                                                          2001023850

© 2002 by W. H. Freeman and Company

No part of this book may be reproduced by any mechanical, photographic, or electronic process, or in the form of a phonographic recording, nor may it be stored in a retrieval system, transmitted, or otherwise copied for public or private use, without the written permission of the publisher.

Printed in the United States of America

First printing 2001

*To my wife, Carole, for her love and support
and to the memory of A. A. McLaughlin,
my first algebra teacher*

Table of Contents

Preface

A New Approach

Students frequently take only one semester of abstract algebra in their undergraduate program in mathematics. More often than not, this one-semester course is composed mainly of group theory or ring theory, depending on how the text chosen for the course begins.

One goal in writing this book was to provide a remedy for this situation by constructing a text that introduces groups, rings, integral domains, and fields early so that students who are exposed to only one semester of abstract algebra will have a better overview of these basic structures. In particular, this approach will aid students who are studying to be secondary teachers of mathematics. Increasingly it is the case that these students must pass a certification examination to qualify as secondary teachers. A complete overview of the basic structures of abstract algebra, as opposed to a course that deals mainly with group or ring theory, will better serve these students.

Another goal was to write a text that presents the development of our number systems from the integers through the rational, real, and complex numbers. Quite often these developments, if they occur at all in the undergraduate curriculum, are split between courses in abstract algebra and real analysis. It is not unusual for the development of the field of real numbers from the rational numbers to be omitted from the undergraduate curriculum. The chapter containing this material has been written so that instructors can choose among the topics depending on their goals for the course. The material on the development of ordered integral domains and the development of the real number system from the field of rational numbers is independent of the remainder of the text and can be omitted if the instructor chooses not to place an emphasis there. In this event, I hope that these sections will be assigned as a reading exercise so that students will have at least a cursory knowledge of how our number systems can be developed.

Students often have difficulty in making the transition from computational mathematics to abstract mathematics and proof. An attempt has been made to write a "student friendly" book, with an informal writing style, that does not sacrifice mathematical content. The book starts slowly, with concepts that students should have seen previously, and builds to concepts that are more abstract and more difficult. The proofs presented in the text are written with more detail than usual in an effort to provide adequate explanation. If an error has been made with regard to the length of explanations, I hope the error will be too much explanation rather than too little. However, the amount of explanation is somewhat reduced as the book proceeds. This is particularly true of the latter chapters of the text, and this is by design: as students gain facility with abstract algebra, they should be required to read and interpret more of the text on their own.

The definitions in the text are often motivated by examples. I believe it is important for a text to motivate definitions by using mathematical entities with which students are already somewhat familiar, though perhaps not in a formal way. The definition of a group is established by pointing out the shared properties of the group of symmetries of a square and the group of units of $\mathbf{Z}_{12}$. Common properties of three classical examples of rings are used to motivate the definition of a ring. Students can compare the abstract definitions of a group and a ring with these examples to see how the definition of a group and a ring might arise. Equivalent techniques are used throughout the text to introduce or illustrate important concepts. Finally, an effort has been made to make the text rich in examples so that students will see concrete examples of the basic structures of abstract algebra.

Structure of the Text

The text contains 12 chapters. The material in Chapter 0 is presented to establish notation and to provide a basis for review. The material on binary operations and congruence relations contained in this chapter may be new to some students, so these concepts should not be overlooked.

Chapter 1 is a study of the basic properties of the integers. The set of positive integers is used to define an order relation on the set of integers, and this construction is used as a model later in the text, when ordered integral domains are developed. Mathematical induction, introduced in this chapter, should be part of every student's mathematical toolbag. Often a student's understanding of induction is "prove a proposition true for 1, assume it's true for k, and prove it's true for $k + 1$." Such students think that this somehow proves the proposition holds for all positive integers. An explanation of how this particular version of induction can be arrived at from a complete statement of the Principle of Mathematical Induction is given in this chapter. The purpose is to give the student a better "picture" of induction. Chapter 1 provides an excellent starting point for students to begin the transition from computational mathematics to proof. Proving statements about properties of the integers provides students with concepts that can be easily understood and, at the same time, presents problems that require at least the beginnings of rigorous proof.

Chapters 2 and 3 introduce the student to the basic structures of abstract algebra: groups, rings, integral domains, division rings, and fields. In these chapters care has been taken to motivate the definitions with examples. These chapters contain basic material on groups, subgroups, normal subgroups, factor groups, group homomorphisms, rings, subrings, ideals, factor rings, and ring homomorphisms. The concept of isomorphism is introduced before homomorphism for pedagogical reasons. Students seem to better understand one-to-one correspondence that preserves the binary operation(s) more quickly than homomorphism in general. This is particularly true when the isomorphic structures are finite and of small order.

Chapter 4 deals with the structure of our number systems. It shows how the field of rational numbers can be developed from the integral domain of integers, how the field of real numbers can be developed from the rational numbers, and how the field of complex numbers can be constructed from the field of real numbers. Including the construction of the real numbers system from the rational numbers in an abstract algebra book has proven to be somewhat controversial. Some argue that this topic does not belong in a book on abstract algebra but more properly belongs in a book on real analysis. However, for the sake of completeness, I have decided to cross this artificial boundary and include it here.

Chapters 5 and 6 present standard topics in abstract algebra. Chapter 5 returns to the study of groups with an investigation of permutation groups, cyclic groups, direct products of groups, and finite abelian groups. Chapter 6 begins with the development of polynomial rings via sequences of ring elements. The Factor Theorem and the Remainder Theorem are discussed in the second section, as are the greatest common divisor and the Euclidean algorithm for polynomials. The last section of this chapter presents a discussion of the Fundamental Theorem of Algebra and the Rational Root Theorem.

Chapter 7 covers standard topics on modular arithmetic in polynomial rings. When F is a field, the similarities between modular arithmetic in the polynomial ring $F[x]$ and modular arithmetic in the ring $\mathbf{Z}$ of integers are pointed out. Euclidean Domains, Principal Ideal Domains, and Unique Factorization Domains are discussed, and the standard relationships among these algebraic structures are developed. Domains of quadratic integers are also investigated in Chapter 7.

Chapter 8 is an introduction to field extensions. Simple field extensions, algebraic field extensions, and the splitting field of a polynomial (whose coefficients lie in a field) are studied. Unfortunately, every topic that one might wish to include in an undergraduate text in abstract algebra cannot be fully developed. This is the case with the algebraic closure of a field. The standard development of the algebraic closure of a field requires the use of concepts that are beyond the scope of this text. For this reason, a complete development of the algebraic closure of a field is not provided. I believe, however, that because the field of complex numbers is the algebraic closure of the field of real numbers, a discussion of algebraic closure should be included. A discussion of the three famous problems of antiquity are also included in this chapter. It is shown that it is impossible to double a cube, to square a circle, and to trisect an angle using only a straightedge and compass. The inclusion of the material on these famous problems gives an interesting application of the theory of algebraic field extensions.

Chapter 9 is an introduction to Galois theory. The material is presented in a historical context in the sense that historical techniques are developed for finding the roots of polynomials of degree 2, 3, and 4. These are followed by a brief introduction to solvable groups. The remainder of the chapter concentrates on showing that, in general, it is not possible to solve polynomials of degree 5 or greater by radicals. The Fundamental Theorem of Galois Theory is presented for fields of characteristic 0 after first being illustrated by several examples. Chapter 9 concludes with a demonstration of the theorem via an example that establishes the Galois correspondence between the set of all subgroups of the Galois group and the intermediate fields of the splitting field of a polymonial.

Chapters 10 and 11 form an introduction to vector spaces and modules and give a standard treatment of these topics. Subspaces, submodules, factor modules, and linear mappings are developed. Inner product spaces are also defined and investigated. Vector spaces are presented first in Chapter 10. Although every vector space is a module, studying these concepts separately demonstrates their similarities and differences more clearly. Every vector space has a basis, but this is not true in general for modules. Examples are given of modules that do not have bases. Moreover, the number of elements in a basis of a vector space is unique—another property that fails to hold for all modules. Chapter 11 concludes with an interesting example of a module that has a basis with one element and a basis with two elements.

Historical Notes

It is enriching for students to learn something of the historical background of the topics they study. For this reason, I have included brief biographies of some of the mathematicians who have made major contributions to the development of abstract algebra. At various places in the text, the contribution of each mathematician is pointed out, and the biography of the mathematician is given at the end of the section in the Historical Notes. If a biography has been presented in the Historical Notes of a previous section, reference to the section is given.

The Use of Technology

Increasingly, technology and computer algebra systems are being used to study some aspects of abstract algebra. One benefit of the use of computer algebra systems in mathematics is that they enable students to embrace experimental mathematics. Through the use of these systems, students can experiment and often answer "What if . . ." questions quickly.

At various places throughout the text, technology problems are provided in the Problem Sets. These exercises are not an integral part of the text, and instructors can use these problems as little or as much as they desire. The problems are usually quite elementary and are intended for students who have little or no experience with a computer algebra system. For the more advanced technology student, a web site has been designed that provides an advanced study of abstract algebra using Mathematica 4.0. This material can be found online at **www.whfreeman.com/bland**.

Solutions Manual

A Solutions Manual containing solutions of all problems in the text is available for instructors. Contact your W. H. Freeman sales representative to receive a copy of this manual.

Acknowledgments

I would like to thank Lowell Abrams, Louis Kolitsch, C. J. Maxon, Michael Neubauer, Emma Previato, Frank Purcell, and Mark Teply, who made meaningful comments while the book was in various stages of development. Your help was very much appreciated.

Paul E. Bland
paul.bland@eku.edu
matbland@prodigy.net

▬0

Preliminaries

Many colleges and universities offer a "transition to advanced mathematics" course intended to bridge the gap between the mainly computational mathematics studied in lower division mathematics courses and the abstract mathematics of more advanced courses. Such courses often contain much of the material presented in Chapter 0 and Chapter 1. For those who have taken such a course, these chapters can be used for a quick review. However, the concept of a congruence relation discussed in Section 0.4 and Theorem 0.4.8 may not have been included in such a transition course. Theorem 0.4.8 will be used several times throughout the text, and it is of particular importance when forming certain algebraic structures. If any portion of the material in Chapters 0 and 1 has not been studied before, then, of course, more time will be required for these chapters.

0.1   SETS

Section Overview. This section establishes notation and gives a brief review of the basic definitions and operations on sets that will be used throughout the text. (See Table 0.1.)

Let Δ be a set and suppose that for each $\alpha \in \Delta$ there is a set A_α. The notation A_α is read A sub-alpha, and the family of sets $\{A_\alpha\}_{\alpha \in \Delta}$ will be referred to as an **indexed family of sets** with indexing set Δ. Defining such families of sets allows us to formulate more general descriptions of set union and set intersection. If $\{A_\alpha\}_{\alpha \in \Delta}$ is an indexed family of sets, then

$$\cup_{\alpha\in\Delta}A_\alpha = \{x \mid x\in A_\alpha \text{ for at least one } \alpha\in\Delta\} \qquad \text{and}$$

$$\cap_{\alpha\in\Delta}A_\alpha = \{x \mid x\in A_\alpha \text{ for all } \alpha\in\Delta\}.$$

We call $\cup_{\alpha\in\Delta}A_\alpha$ the **union** and $\cap_{\alpha\in\Delta}A_\alpha$ the **intersection** of the family of sets $\{A_\alpha\}_{\alpha\in\Delta}$.

You are already quite familiar with the Cartesian product of two and of three sets. For example, if **R** denotes the set of real numbers, you have used $\mathbf{R}\times\mathbf{R}$ as the set of coordinates of points in the plane and $\mathbf{R}\times\mathbf{R}\times\mathbf{R}$ as the set of coordinates of the points that make up 3-space. The Cartesian product, however, can be defined for more than two or three sets. If $A_1, A_2, \ldots, A_n$ are sets, where $n\geq 2$ is an integer, then

$$A_1\times A_2\times\cdots\times A_n = \{(x_1, x_2, \ldots, x_n)\mid x_i\in A_i \text{ for } i = 1, 2, \ldots, n\}.$$

An element $(x_1, x_2, \ldots, x_n)$ of $A_1\times A_2\times\cdots\times A_n$ is referred to as an ordered n-**tuple** and two such n-tuples $(x_1, x_2, \ldots, x_n)$ and $(y_1, y_2, \ldots, y_n)$ are equal if and only if $x_i = y_i$ for $i = 1, 2, \ldots, n$. The Cartesian product of the sets $A_1, A_2, \ldots, A_n$ will often be denoted by $\times_{i=1}^n A_i$, and an element $(x_1, x_2, \ldots, x_n)$ of $\times_{i=1}^n A_i$ may be written simply as (x_i). If $A = A_1 = A_2 = \cdots = A_n$, then $\times_{i=1}^n A_i$ will often be denoted simply by A^n. For example, $\mathbf{R}^4 = \mathbf{R}\times\mathbf{R}\times\mathbf{R}\times\mathbf{R}$.

Table 0.1   **List of Set Notation and Set Operations Used in the Text**

$x\in A$	x is an element of A
$x\notin A$	x is not an element of A
$\varnothing$	The **empty set**
$A\subseteq B$ means that if $x\in A$, then $x\in B$	A is a **subset** of B
$A\subset B$ means $A\subseteq B$ but $A\neq B$	A is a **proper subset** of B
$\mathscr{P}(A)$	The set of all subsets of A; the **power set** of A
$A\setminus B = \mathbf{C}_A B = \{x\mid x\in A \text{ but } x\notin B\}$	The **complement** of B with respect to A
$A\cup B = \{x\mid x\in A \text{ or } x\in B\}$	The **union** of A and B
$A\cap B = \{x\mid x\in A \text{ and } x\in B\}$	The **intersection** of A and B
$A\times B = \{(x, y)\mid x\in A \text{ and } y\in B\}$	The **Cartesian product** of A and B
$A\cap B = \varnothing$	A and B are **disjoint** sets

0.1.1 Example

Let $A_1 = \{1, 2, a, b\}$, $A_2 = \{a, b, c, d\}$, $A_3 = \{3, 4, b, d, e\}$, and $A_4 = \{1, b, d\}$, then the indexing set is $\Delta = \{1, 2, 3, 4\}$.

1. $\bigcup_{i \in \Delta} A_i = A_1 \cup A_2 \cup A_3 \cup A_4$

$$= \{1, 2, a, b\} \cup \{a, b, c, d\} \cup \{3, 4, b, d, e\} \cup \{1, b, d\}$$

$$= \{1, 2, 3, 4, a, b, c, d, e\}.$$

2. $\bigcap_{i \in \Delta} A_i = A_1 \cap A_2 \cap A_3 \cap A_4$

$$= \{1, 2, a, b\} \cap \{a, b, c, d\} \cap \{3, 4, b, d, e\} \cap \{1, b, d$$

$$= \{b\}.$$

3. $\times_{i=1}^{4} A_i = A_1 \times A_2 \times A_3 \times A_4$

$$= \{(x_1, x_2, x_3, x_4) \mid x_i \in A_i, \ i = 1, 2, 3, 4\}$$

An element (x_1, x_2, x_3, x_4) of $A_1 \times A_2 \times A_3 \times A_4$ is an **ordered 4-tuple**. The product $A_1 \times A_2 \times A_3 \times A_4$ actually contains 240 ordered 4-tuples, so it is impractical to list them all. However, $(1, b, e, 1)$ is one element of $A_1 \times A_2 \times A_3 \times A_4$ since $1 \in A_1$, $b \in A_2$, $e \in A_3$, and $1 \in A_4$.

4. Let $A_n = [0, 1 + \frac{1}{n}] \subseteq \mathbf{R}$, where n is a positive integer. Then
$A_1 = [0, 2]$, $A_2 = [0, \frac{3}{2}]$, $A_3 = [0, \frac{4}{3}]$, $A_4 = [0, \frac{5}{4}]$, $A_5 = [0, \frac{6}{5}]$, ...,
and $\bigcap_{n>0} A_n = [0,1]$ and $\bigcup_{n>0} A_n = [0, 2]$.

The following notation will also be used throughout the text:

$\mathbf{N} = \{1, 2, 3, \ldots\}$.

$\mathbf{N}_0 = \{0, 1, 2, 3, \ldots\}$.

$\mathbf{Z}$ is the set of all integers.

$\mathbf{Q}$ is the set of rational numbers.

$\mathbf{R}$ is the set of real numbers.

$\mathbf{C}$ is the set of complex numbers.

Problem Set 0.1

1. If $A = \{a, b, c, d\}$, compute $\mathcal{P}(A)$. That is, form the set $\mathcal{P}(A)$ consisting of all the subsets of A. (Don't forget to include the empty set $\varnothing$.)

2A. Let $A = \{3, 5, 7, 9, 11\}$, $B = \{2, 4, 6, 8, 10\}$, $C = \{2, 4, 5, 7, 8, 9\}$, and $D = \{2, 3, 5, 6, 7, 8, 9, 10\}$. Compute each of the following:

(a) $(D \setminus B) \setminus C$

(b) $(A \cup B) \setminus (A \cap B)$

(c) $(A \cup B) \setminus (C \cap D)$

2B.   For the sets given in 2A, determine if equality holds for each of the following:

(a) Does $(A \cup B) \cap (A \cup C) = A \cap (B \cup C)$?

(b) Does $(A \cup B) \cap (A \cup C) = A \cup (B \cap C)$?

(c) Does $(A \cap B) \cup (A \cap C) = A \cap (B \cap C)$?

Prove 3 through 6 if A, B, C are subsets of X.

3. $A \cup (B \setminus A) = A \cup B$.

4. $A \cap (B \setminus A) = \varnothing$.

5. $A \subseteq B$ if and only if $\mathbf{C}_x B \subseteq \mathbf{C}_x A$.

6.   If $A \subseteq C$, then $A \cup (B \cap C) = (A \cup B) \cap C$.

7.   If $X = \{0, 1, 2, 3, 4, 5, 6, 7, 8, 9\}$, $A = \{2, 3, 4, 5\}$, and $B = \{3, 4, 5, 6\}$, show that $\mathbf{C}_x(A \cup B) = \mathbf{C}_x A \cap \mathbf{C}_x B$ and $\mathbf{C}_x(A \cap B) = \mathbf{C}_x A \cup \mathbf{C}_x B$. These are known as **De Morgan's laws.** De Morgan (1806–1871) was an English logician who attempted to axiomize the basic ideas of algebra.

8.   Prove that $\mathbf{C}_x(A \cup B) = \mathbf{C}_x A \cap \mathbf{C}_x B$ for any three sets A, B, and X. The proof of the other De Morgan law is similar.

9.   If A and B are sets, let $A \Delta B = (A \setminus B) \cup (B \setminus A)$.

(a) If $A = \{1, 2, 3, 4, 5\}$ and $B = \{3, 4, 5, 6, 7\}$, compute $A \Delta B$.

(b) Prove that $A \Delta B = (A \cup B) \setminus (A \cap B)$ for any sets A and B.

10.   For the sets $A = \{1, 2\}$, $B = \{a, b, 2\}$, and $C = \{2, 4\}$, determine if equality holds for each of the following. If equality does hold, can you prove that equality holds for general sets A, B, and C?

(a) Does $A \times (B \cup C) = (A \times B) \cup (A \times C)$?

(b) Does $A \times (C \setminus B) = (A \times C) \setminus (A \times B)$?

11.   Prove or disprove each of the following for sets A and B.

(a) $\mathscr{P}(B \setminus A) = \mathscr{P}(B) \setminus \mathscr{P}(A)$

(b) $\mathscr{P}(A \cup B) = \mathscr{P}(A) \cup \mathscr{P}(B)$

(c) $\mathscr{P}(A \cap B) = \mathscr{P}(A) \cap \mathscr{P}(B)$

12.   If $A_1 = \{1, 2\}$, $A_2 = \{3, 4\}$, and $A_3 = \{5, 6\}$, compute $A_1 \times A_2 \times A_3$ and $A_2 \times A_1 \times A_3$.

13. For $x, y \in \mathbf{R}$, let $[x, y) = \{a \in \mathbf{R} \mid x \le a < y\}$. Compute each of the following:

   (a) $\cup_{n \in \mathbf{N}} [n, n+1)$ (c) $\cap_{n \in \mathbf{N}} [0, \frac{1}{n})$

   (b) $\cup_{n \in \mathbf{N}} [\frac{1}{n}, \sqrt{2} + \frac{1}{n})$ (d) $\cap_{n \in \mathbf{N}} [\frac{1}{n}, \sqrt{2} + \frac{1}{n})$

14. Consider the indexed family of sets $\{A_n\}_{n \in \mathbf{N}}$ defined as follows: for each positive integer n, $A_n = \{n, n+1, n+2\}$. Compute $\cup_{n \in \mathbf{N}} A_n$ and $\cap_{n \in \mathbf{N}} A_n$.

15. For each positive integer n, let $A_n = \mathbf{N} \setminus \{1, 2, 3, \ldots, n\}$. Compute $\cup_{n \in \mathbf{N}} A_n$ and $\cap_{n \in \mathbf{N}} A_n$.

16. Prove that for any indexed family of sets $\{A_\alpha\}_{\alpha \in \Delta}$,
$A \cup (\cap_{\alpha \in \Delta} A_\alpha) = \cap_{\alpha \in \Delta} (A \cup A_\alpha)$.

0.2   RELATIONS

Section Overview. The most important concept presented in this section is that of an equivalence relation. Such relations play a significant role in abstract algebra. One important aspect of an equivalence relation is that it partitions the set. Be sure to take note of the fact that each equivalence relation on a set determines a partition of the set and that each partition of a set determines an equivalence relation.

Relations among the elements of a set are often encountered in mathematics: for example, the relation "x is less than or equal to y" defined on the real numbers and the relation "x is a factor of y" defined on the set of integers. A **relation** R from a set A to a set B is a subset of $A \times B$. If A and B are the same set, then R **is a relation on** A. The set $\{x \in A \mid \text{there is a } y \text{ in } B \text{ such that } (x, y) \in R\}$ is the **domain** of R, while the set $\{y \in B \mid \text{there is an } x \in A \text{ such that } (x, y) \in R\}$ is the **range** of R. **Dom**(R) and **Rng**(R) will denote the domain and the range of R, respectively. Dom(R) is the set of all first elements of the ordered pairs in R and Rng(R) is the set of all second elements. If $(x, y) \in R$, we will often write this as xRy and say that x **is** R **related to** y. Order is important; xRy does not necessarily imply yRx!

In abstract algebra certain kinds of relations play an important role. One relation of fundamental importance is an equivalence relation.

0.2.1   Definition

A relation R on a set A is said to be

1. **Reflexive** if xRx for all $x \in A$
2. **Symmetric** if whenever x and y are in A and such that xRy, then yRx
3. **Transitive** if whenever x, y, and z are in A and such that xRy and yRz, then xRz

A relation E on a set A is said to be an **equivalence relation** if E is reflexive, symmetric, and transitive. If E is an equivalence relation on A, then for any $x \in A$ the set $[x] = \{y \in A \mid xEy\}$ is referred to as the **equivalence class** determined by x.

0.2.2   Examples

1. If the relation E is defined on **Z** by xEy if and only if $x - y$ is a multiple of 3, then E is an equivalence relation on **Z**.

   E is *reflexive*: The relation E is obviously reflexive since if $x \in$ **Z**, $x - x$ is clearly a multiple of 3 since $x - x = 3 \cdot 0$. Hence, xEx for all $x \in$ **Z**.
   E is *symmetric*: If $x, y \in$ **Z** are such that xEy, then $x - y$ is a multiple of 3. If k is an integer such that $x - y = 3k$, then $y - x = 3(-k)$ and thus $y - x$ is a multiple of 3. Therefore, yEx, so E is symmetric.
   E is *transitive*: If $x, y, z \in$ **Z** are such that xEy and yEz, then there are integers k_1 and k_2 such that $x - y = 3k_1$ and $y - z = 3k_2$. Hence, $x - z = (x - y) + (y - z) = 3k_1 + 3k_2 = 3(k_1 + k_2)$, so $x - z$ is a multiple of 3. Consequently, xEy and yEz imply xEz, and so E is transitive.

   There are three distinct equivalence classes determined by the equivalence relation E defined on **Z**, namely,

   $$[0] = \{\ldots, -9, -6, -3, 0, 3, 6, 9, \ldots\}.$$
   $$[1] = \{\ldots, -8, -5, -2, 1, 4, 7, 10, \ldots\}.$$
   $$[2] = \{\ldots, -7, -4, -1, 2, 5, 8, 11, \ldots\}.$$

   Two integers x and y are in the same equivalence class if and only if $x - y$ is a multiple of 3.
2. Define the relation R on **Z** by xRy if and only if x is less than y, that is, if and only if $x < y$. The relation $<$ is neither reflexive nor symmetric, but it is transitive. Thus, $<$ is not an equivalence relation on **Z**. The relation $<$ is sometimes referred to as an **order relation** on **Z**.

Notice in Example 1 of 0.2.2 that $\mathbf{Z} = [0] \cup [1] \cup [2]$ and that $[0] \cap [1] = \varnothing$, $[0] \cap [2] = \varnothing$, and $[1] \cap [2] = \varnothing$. These observations show that $\mathbf{Z}$ is the union of the equivalence classes determined by the equivalence relation E and that any two distinct equivalence classes are disjoint. This is actually true for every equivalence relation on a set A.

0.2.3   Theorem

The following hold for any equivalence relation E defined on a set A.

1. For any $x, y \in A$, either $[x] \cap [y] = \varnothing$ or $[x] = [y]$.
2. For $x, y \in A$, xEy if and only if $[x] = [y]$.
3. $A = \cup_{x \in A} [x]$.

Proof

1. If $[x]$ and $[y]$ have an element in common, we are going to show that every element of $[x]$ is in $[y]$. This will give us $[x] \subseteq [y]$, which is one of the inclusions we need to show $[x] = [y]$. Suppose $[x] \cap [y] \neq \varnothing$. If $z \in [x] \cap [y]$, then $z \in [x]$ and $z \in [y]$ and so xEz and yEz. Since E is symmetric, yEz yields zEy. Hence, xEz and zEy, so transitivity of E gives xEy. If $w \in [x]$, then xEw and thus wEx by symmetry. But then wEx and xEy produce wEy since E is transitive. This in turn gives yEw, again by the symmetry of E. Hence, $w \in [y]$, so $[x] \subseteq [y]$. In a similar fashion it can be shown that $[y] \subseteq [x]$ and, consequently, equality holds.
2. Since this is an "if and only if" statement, we need to show that xEy implies that $[x] = [y]$ and then show that $[x] = [y]$ implies xEy. If xEy, then yEx and so $x \in [y]$. But $x \in [x]$, so $[x] \cap [y] \neq \varnothing$. Therefore, by part 1, $[x] = [y]$. Conversely, suppose that $[x] = [y]$. Since $x \in [x]$, $x \in [y]$ and so yEx, which implies that xEy by the symmetry of E.
3. Clearly $\cup_{x \in A} [x] \subseteq A$, for if $y \in \cup_{x \in A} [x]$, then $y \in [x]$ for some x in A. But $[x] \subseteq A$ and so $y \in A$. Conversely, suppose $y \in A$. Then y is in the equivalence class $[y]$ determined by y. Hence, $y \in [y] \subseteq \cup_{x \in A} [x]$ and therefore $A \subseteq \cup_{x \in A} [x]$. ∎

If E is an equivalence relation on a set A, then every $x \in A$ determines an equivalence class $[x]$. The element x displayed in $[x]$ is said to be the **representative of the equivalence class**. Part 2 of Theorem 0.2.3 shows that if $y \in [x]$, then $[x] = [y]$. Hence, any element of an equivalence class can be chosen as its representative. Theorem 0.2.3 also shows that if E is an equivalence relation on a set A, then $A = \cup_{x \in A}[x]$ and if $[x] \neq [y]$, then $[x] \cap [y] = \varnothing$. This leads to the following definition.

0.2.4   Definition

A family $\mathscr{F}$ of sets is said to be **pairwise disjoint** if every pair of distinct sets in $\mathscr{F}$ has an empty intersection. If $\mathscr{F}$ is a family of pairwise disjoint subsets of a set S such that $S = \cup_{A \in \mathscr{F}} A$ then $\mathscr{F}$ is said to be a **partition** of S.

We have seen in Theorem 0.2.3 that every equivalence relation on a set A determines a partition of A. The converse is also true.

0.2.5   Theorem

Every partition $\mathscr{F}$ of a set S determines an equivalence relation E on S that is defined by xEy if there is a set A in $\mathscr{F}$ such that $x, y \in A$.

0.2.6   Example

The sets $A_1 = \{2, 3, 8\}$, $A_2 = \{4, 6\}$, and $A_3 = \{1, 5, 7\}$ form a partition of the set $A = \{1, 2, 3, 4, 5, 6, 7, 8\}$. These sets induce an equivalence relation E on A that is defined by xEy if and only if there is a set A_i of the partition such that $x, y \in A_i$. This just says that all of the elements of the set A_1 are related, all the elements of A_2 are related, and all of the elements of A_3 are related. Remembering that an equivalence relation on a set must be reflexive, symmetric, and transitive, we see that the ordered pairs in E determined by the elements of A_1 are $(2, 2)$, $(3, 3)$, $(8, 8)$, $(2, 3)$, $(3, 2)$, $(2, 8)$, $(8, 2)$, $(3, 8)$, and $(8, 3)$. Similarly, the sets A_2 and A_3 determine the ordered pairs $(4, 4)$, $(6, 6)$, $(4, 6)$, $(6, 4)$, and $(1, 1)$, $(5, 5)$, $(7, 7)$, $(1, 5)$, $(5, 1)$, $(1, 7)$, $(7, 1)$, $(5, 7)$, $(7, 5)$ in E, respectively. Hence,

$$E = \{(2,2),\ (3,3),\ (8,8),\ (2,3),\ (3,2),\ (2,8),\ (8,2),\ (3,8),$$
$$(8,3),\ (4,4),\ (6,6),\ (4,6),\ (6,4),\ (1,1),\ (5,5),\ (7,7),$$
$$(1,5),\ (5,1),\ (1,7),(7,1),\ (5,7),\ (7,5)\}.$$

Problem Set 0.2

1. Let $A = \{1, 3, 5, 7\}$ and $B = \{2, 4, 6, 8\}$. Give the domain and the range of each of the following relations from A to B.

   (a) $R_1 = \{(3, 2), (5, 2), (7, 4), (3, 6)\}$
   (b) $R_2 = \{(1, 2), (3, 8), (5, 2), (7, 6)\}$
   (c) $R_3 = \{(1, 8), (7, 2), (7, 4), (1, 6)\}$

2. Consider the following relations on the set $\{2, 4, 6\}$. Determine if each is reflexive, symmetric, or transitive. Give reasons for each of your answers.

   (a) $R_1 = \{(2, 2), (4, 4), (4, 6), (6, 4), (6, 6)\}$
   (b) $R_2 = \{(2, 2), (4, 4), (4, 6), (6, 4)\}$
   (c) $R_3 = \{(2, 2), (4, 4), (6, 6), (2, 6), (6, 2), (2, 4), (4, 2), (4, 6), (6, 4)\}$

3. Determine which of the following relations R are equivalence relations on the set **Z**. If R is not an equivalence relation on **Z**, state which of the conditions—reflexive, symmetric, or transitive—fails to hold.

   (a) xRy if and only if $x = -y$.
   (b) xRy if and only if $x = 3y$.
   (c) xRy if and only if x divides y.
   (d) xRy if and only if $xy \geq 0$.
   (e) xRy if and only if $x - y$ is a multiple of 6.

4. In Exercise 3, if a relation R is an equivalence relation, find the equivalence classes determined by R.

5. Consider the set $A = \{1, 2, 3, 4, 5, 6\}$. The collection of sets $A_1 = \{1, 2\}$, $A_2 = \{3, 4\}$, and $A_3 = \{5, 6\}$ is a partition of the set A. Find the equivalence relation E on A determined by this partition. What are the equivalence classes determined by this equivalence relation?

6. Consider the set $A = \{1, 2, 3, 4\}$. Show that the relation E $= \{(1, 1), (2, 2), (3, 3), (4, 4), (2, 3), (3, 2), (1, 4), (4, 1)\}$ is an equivalence relation on A and find the partition of A determined by E.

7. Define the relation R on **Q** by xRy if and only if $x - y \in$ **Z**. Is R an equivalence relation on **Q**?

8. Let S be a finite set and define the relation R on $\mathcal{P}(S)$ by xRy if and only if the sets $x, y \in \mathcal{P}(S)$ have the same number of elements. Show that R is an equivalence relation on $\mathcal{P}(S)$.

9. Define the relation R on $A = $ **Z**$\times($**Z** $\setminus \{0\})$ by (a, b)R(c, d) if and only if $ad = bc$. Prove that R is an equivalence relation on A.

10. Define the relation R on **Z** by xRy if and only if $x - y$ is a multiple of n, where n is a fixed positive integer. Prove that R is an equivalence relation on **Z**.

11. Prove Theorem 0.2.5.

0.3   FUNCTIONS

Section Overview. Section 0.3 establishes the concept of a function. You should master the concepts of an injective function, a surjective function, a bijective function, and the inverse of a function. In particular, pay close attention to what it means for a function to be well-defined. It is important that you understand the difference between how one shows that a function is well-defined and how one shows that a function is injective. These two procedures are different, but similar in appearance, and they often cause confusion.

A relation R from A to B has been defined as a subset of $A \times B$. Relations that satisfy the following conditions play a central role not only in abstract algebra, but in all of mathematics.

0.3.1   Definition

A **function**, or **mapping**, f from A to B, denoted by $f : A \longrightarrow B$, is a relation f from A to B such that for each $x \in A$ there is one and only one $y \in B$ such that $(x, y) \in f$. The set A is the **domain** of f and B is the **codomain**. If $(x, y) \in f$, we write $x \longmapsto y$ and say that y is the **image** of x under f and that x is a **preimage** of y. It is also customary to write this as $y = f(x)$. In this case x is often referred to as the **input** and y is called the **output**. The set $\{y \in B \mid \text{there is an } x \in A \text{ such that } f(x) = y\}$ is the **range** or **image** of f. The range of f will be denoted by **Rng**(f) or **Im**f, and **Dom**(f) will denote the domain of f.

Notation for Functions

For a function $f : A \longrightarrow B$ the correspondence between the elements of A and B may be given by an equation. For example $f : \mathbf{Z} \longrightarrow \mathbf{R}$, where $f(x) = x^2$, is a function from **Z** to **R**. The equation $f(x) = x^2$ is referred to as the **defining**

equation for the function or the **rule of correspondence** between the elements of **Z** and **R**. If a specific integer is substituted for x, say $x = 2$, then $f(2) = 4$ is an element of the codomain **R**. A function such as f is usually represented in one of three different but equivalent ways:

1. $f : \mathbf{Z} \longrightarrow \mathbf{R}$, where $f(x) = x^2$
2. $f : \mathbf{Z} \longrightarrow \mathbf{R}$, where $x \longmapsto x^2$
3. $f : \mathbf{Z} \longrightarrow \mathbf{R} : x \longmapsto x^2$

The first representation for f is the one found in most undergraduate texts in mathematics, and it is the one which we will usually use. However, you may occasionally see 2 or 3.

Definition 0.3.1 states that if $f : A \longrightarrow B$ is a function, then for each $x \in A$ there is one and only one $y \in B$ such that $(x, y) \in f$. For a function f it cannot be the case that there are $x_1, x_2 \in A$ with $x_1 = x_2$ and $f(x_1) \neq f(x_2)$. If f is a relation from A to B and we wish to establish that f is a function from A to B, it must be shown that if we pick two different representations of an element in A, say $x_1 = x_2$, then $f(x_1) = f(x_2)$. This will be referred to as **showing that f is a well-defined function.**

0.3.2   Examples

1. Let $f : \mathbf{Q} \longrightarrow \mathbf{Q}$ be such that $f(x) = x^2 + 1$. Does f define a function from **Q** to **Q**? Note that $\frac{2}{4}$ and $\frac{3}{6}$ are both representations of the same element of **Q** and $\frac{2}{4} = \frac{3}{6} =$ gives $\left(\frac{2}{4}\right)^2 = \left(\frac{3}{6}\right)^2$ since multiplication is well-defined on **Q**. Thus, $\frac{20}{16} = \left(\frac{2}{4}\right)^2 + 1$ and $\left(\frac{3}{6}\right)^2 + 1 = \frac{45}{36}$ are equal in **Q** since addition is well-defined on **Q**. By the same reasoning $x_1 = x_2$ gives $x_1^2 = x_2^2$, so $x_1^2 + 1 = x_2^2 + 1$. Hence, $x_1 = x_2$ implies that $f(x_1) = f(x_2)$ and f is therefore a well-defined function.

2. Consider $f : \mathbf{Q} \longrightarrow \mathbf{Z}$, where $f\left(\frac{p}{q}\right) = pq$. The equation $f\left(\frac{p}{q}\right) = pq$ does *not* define a function from **Q** to **Z**. To see this, pick two different representations of the same element of **Q**, for instance, $\frac{1}{2} = \frac{2}{4}$. Then $f\left(\frac{2}{4}\right) = 1 \cdot 2 = 2$ and $f\left(\frac{2}{4}\right) = 2 \cdot 4 = 8$, so $\frac{1}{2} = \frac{2}{4}$ but $f\left(\frac{1}{2}\right) \neq f\left(\frac{2}{4}\right)$. The ordered pairs $\left(\frac{1}{2}, 2\right), \left(\frac{2}{4}, 8\right) \in f$, so $f\left(\frac{p}{q}\right) = pq$ defines a relation from **Q** to **Z** that is *not* a function.

0.3.3   Definition

A function $f : A \longrightarrow B$ is said to be **injective** (or **one-to-one**) if distinct inputs from A produce distinct outputs in B. That is, if $x_1, x_2 \in A$ are such that $x_1 \neq x_2$, then $f(x_1) \neq f(x_2)$. We say that f is **surjective** (or **onto**) if $\text{Rng}(f) = B$. Thus, a function f is surjective if for each $y \in B$ there is at least one $x \in A$ such that $f(x) = y$. A function that is both injective and surjective is said to be a **bijection** (or **one-to-one correspondence**).

To show that a function is injective, it must be shown that if $x_1 \neq x_2$, then $f(x_1) \neq f(x_2)$. This is equivalent to showing that if $f(x_1) = f(x_2)$, then $x_1 = x_2$. Showing a function to be injective is not to be confused with showing that a function is well-defined, which requires one to show that if $x_1 = x_2$, then $f(x_1) = f(x_2)$. Make sure you understand the differences between what is given in 1 and 2 below.

1. $x_1 = x_2 \Rightarrow f(x_1) = f(x_2)$ shows that f is well-defined.
2. i. $x_1 \neq x_2 \Rightarrow f(x_1) \neq f(x_2)$ shows that f is injective.
    ii. $f(x_1) = f(x_2) \Rightarrow x_1 = x_2$ shows that f is injective.

Statement 2(i) is the **contrapositive** of 2(ii). If S_1 and S_2 are mathematical statements, then not $S_2 \Rightarrow$ not S_1 is the contrapositive statement of $S_1 \Rightarrow S_2$. It is well known that not $S_2 \Rightarrow$ not S_1 if and only if $S_1 \Rightarrow S_2$. Hence, one can show that a function f is injective by showing $x_1 \neq x_2 \Rightarrow f(x_1) \neq f(x_2)$ or by showing $f(x_1) = f(x_2) \Rightarrow x_1 = x_2$.

0.3.4   Examples

The following functions are examples of injective, surjective, and bijective functions.

1. The mapping $f : \mathbf{N} \longrightarrow \mathbf{Z}$ defined by $f(x) = 2x$ is well-defined since, if $x_1 = x_2$, then $2x_1 = 2x_2$, or $f(x_1) = f(x_2)$. If $f(x_1) = f(x_2)$, then $2x_1 = 2x_2$, which implies that $x_1 = x_2$. Hence, f is injective. The range of f is the set $\text{Rng}(f) = \{2, 4, 6, \ldots\}$, and since $\text{Rng}(f) \neq \mathbf{Z}$, f is not surjective.
2. If $f : \mathbf{R} \longrightarrow \mathbf{R}$ is given by $f(x) = x^2$, then f is well-defined and neither injective nor surjective. Since $f(-2) = f(2)$ but $-2 \neq 2$, $f(x_1) = f(x_2)$ does not imply that $x_1 = x_2$ and thus f is not injective. The function f is not surjective since the element -1 has no preimage in the domain of f.

3. If $\mathbf{R}^{\#} = \{x \in \mathbf{R} \mid x \geq 0\}$, then $f : \mathbf{R} \longrightarrow \mathbf{R}^{\#}$, where $f(x) = x^2$, is well-defined and surjective but not injective. If $y \in \mathbf{R}^{\#}$ and $y = 0$, then $x = 0$ is a preimage of y. If $y \neq 0$, then y has two preimages, namely, $x = \pm\sqrt{y}$. This shows that f is surjective but not injective.

Functions can often be combined to form new functions. When the range of a function f is contained in the domain of a function g, **function composition** will produce a new function. If $f : A \longrightarrow B$ and $g : B \longrightarrow C$, then for any $x \in A$, $f(x) \in B$, so $g(f(x)) \in C$. If we let $(g \circ f)(x) = g(f(x))$ for all $x \in A$, then $g \circ f$, which is read "g composed with f," is a new function $g \circ f : A \longrightarrow C$.

0.3.5   Definition

For any set A, the function $1_A : A \longrightarrow A$, where $1_A(x) = x$, is called the **identity function** on A. If $f : A \longrightarrow B$ is a function, then a function $g : B \longrightarrow A$ such that $g \circ f = 1_A$ and $f \circ g = 1_B$ is said to be an **inverse function** for f. When an inverse for f exists, it will often be denoted by f^{-1}.

0.3.6   Examples

1. Let $L : \mathbf{R}^{+} \longrightarrow \mathbf{R}$, where $L(x) = \ln x$, and $E : \mathbf{R} \longrightarrow \mathbf{R}^{+}$, where $E(x) = e^x$, with $\mathbf{R}^{+} = \{x \in \mathbf{R} \mid x > 0\}$. Since $(L \circ E)(x) = \ln e^x = x$, $L \circ E = 1_{\mathbf{R}}$. Also, $(E \circ L)(x) = e^{\ln x} = x$ gives $E \circ L = 1_{\mathbf{R}^+}$. Thus, E is an inverse function for L and L is an inverse function for E.
2. Consider the function $f : \mathbf{R} \longrightarrow \mathbf{R}$ given by $f(x) = x^2$. Does f have an inverse? Suppose that a function $g : \mathbf{R} \longrightarrow \mathbf{R}$ exists such that $g \circ f = 1_{\mathbf{R}}$ and $f \circ g = 1_{\mathbf{R}}$. Then $(f \circ g)(-1) = 1_{\mathbf{R}}(-1) = -1$. But $(f \circ g)(-1) = f(g(-1)) = g(-1)^2 \geq 0$, which is an obvious contradiction. Therefore, an inverse function for f does not exist.

Because of Example 2, you might ask, "When does a function have an inverse?" The following theorem gives a condition that is both necessary and sufficient for a function to have an inverse.

0.3.7   Theorem

A function $f : A \longrightarrow B$ has an inverse function if and only if f is a bijection.

Proof. Let $f : A \longrightarrow B$ have an inverse function $f^{-1} : B \longrightarrow A$. If $x_1, x_2 \in A$ are such that $f(x_1) = f(x_2)$, then $(f^{-1} \circ f)(x_1) = (f^{-1} \circ f)(x_2)$, and thus $1_A(x_1) = $

$1_A(x_2)$ gives $x_1 = x_2$. Hence, f is injective. Also, if $y \in B$, then $f^{-1}(y) \in A$ and $(f \circ f^{-1})(y) = 1_B(y) = y$ and so $f^{-1}(y)$ is a preimage of y. Consequently, f is surjective, so f is a bijection.

Now suppose that f is a bijection. We will use this bijection to define a function $g : B \longrightarrow A$ that will serve as the inverse of f. Since f is a bijection, every element $y \in B$ has a unique preimage in A. Such a preimage exists since f is surjective, and it is unique since f is injective. Call this preimage x_y and define $g : B \longrightarrow A$ by $g(y) = x_y$. It follows that g is well-defined since for every $y \in B$, x_y is unique. Furthermore, $f(g(y)) = f(x_y) = y$ and thus $f \circ g = 1_B$. We also see that if $f(x) = y$, then $x = x_y$ and $g(f(x)) = g(y) = x_y = x$. Hence, $g \circ f = 1_A$ and g is therefore an inverse function for f. ∎

0.3.8   **Definition**

Suppose $A \subseteq B$ and let $f : A \longrightarrow C$ and $g : B \longrightarrow C$ be functions such that $g(x) = f(x)$ for all $x \in A$. Then we say that g **extends** f to B. We also say that f is the **restriction** of g to A. This is denoted by $g|_A = f$ and is read "g restricted to A equals f."

0.3.9   **Example**

Suppose $f : \mathbf{Q} \longrightarrow \mathbf{R}$ is such that $f(x) = x^4 - 2x + 5$. Then f is extended to $\mathbf{R}$ by the function $g : \mathbf{R} \longrightarrow \mathbf{R}$ given by $g(x) = x^4 - 2x + 5$ and f is the restriction of g to $\mathbf{Q}$.

0.3.10   **Definition**

Let $f : A \longrightarrow B$ be a function and suppose that $X \subseteq A$ and $Y \subseteq B$. Then $f(X) = \{f(x) \mid x \in X\}$ and $f^{-1}(Y) = \{x \in A \mid f(x) \in Y\}$. We call $f(X)$ the **image** of X under f, and $f^{-1}(Y)$ is called the **inverse image** of Y under f. Of course, the image of A under f is $\mathrm{Rng}(f) = \mathrm{Im}\, f$.

Don't confuse the notation f^{-1}, for the inverse of a bijective function f, with $f^{-1}(Y)$. In the first case f^{-1} is a function and in the second $f^{-1}(Y)$ is a set! It is the set of all $x \in A$ such that $f(x) \in Y$. These two notations are similar in appearance, but they have very different meanings.

| **0.3.11** | **Examples** |

1. Let $f : \mathbf{R} \longrightarrow \mathbf{R}$ be given by $f(x) = x^2 + 1$. Then $f((-\frac{1}{2},1)) = [1,2)$, and thus

$$f^{-1}\left(f\left(\left(-\tfrac{1}{2},1\right)\right)\right) = f^{-1}([1,2)) = (-1,1).$$

Likewise, $f^{-1}((-2,4]) = [-\sqrt{3}, \sqrt{3}]$, and

$$f(f^{-1}((-2,4])) = f([-\sqrt{3}, \sqrt{3}]) = [1,4].$$

Hence, $\left(-\tfrac{1}{2},1\right) \subset f^{-1}\left(f\left(\left(-\tfrac{1}{2},1\right)\right)\right)$, and

$$f(f^{-1}((-2,4])) \subset (-2,4].$$

Equality does not hold in either case.

2. Let $f : \mathbf{N} \longrightarrow \mathbf{R}$ be such that $f(x) = 2x + 1$. If $X = \{1, 2, 3\}$, then $f(X) = \{3,5,7\}$ and $f^{-1}(f(X)) = X$. It is also easy to see that for $Y = [3,7] \subseteq \mathbf{R}, f^{-1}(Y) = f^{-1}([3,7]) = X$, so $f(f^{-1}(Y)) = f(X) = \{3,5,7\} \subset Y$. Hence,

$$f^{-1}(f(X)) = X \text{ and } f(f^{-1}(Y)) \subset Y.$$

Problem Set 0.3

1. Determine whether or not each of the following equations determines a well-defined function from $\mathbf{Q}$ to $\mathbf{Q}$.

(a) $g\left(\frac{p}{q}\right) = \frac{4p}{q} + 1$

(b) $f\left(\frac{p}{q}\right) = 2p - 3q$

2. Let $A = \{1, 2, 3, 4\}$ and $B = \{1, 2, 3, 4, 5, 6, 7, 8, 9\}$ and suppose that $f : A \longrightarrow B$, where $f(x) = 2x + 1$. Find each of the following:

(a) $\text{Rng}(f)$
(b) $f(X)$, where $X = \{2, 3\}$
(c) $f^{-1}(Y)$, where $Y = \{3, 4, 5, 6, 7\}$

3. Let $f : \mathbf{R} \longrightarrow \mathbf{R}$ be such that $f(x) = \sin x$. If $X = \{x \in \mathbf{R} \mid x \geq 0\}$ and $Y = \{y \in \mathbf{R} \mid y \leq 0\}$, compute $f(X), f^{-1}(Y), f^{-1}(f(X))$, and $f(f^{-1}(Y))$.

4. Consider the following functions and determine which are injective, which are surjective, and which are both or neither.

(a) $f : \mathbf{Z} \longrightarrow \mathbf{Q}$, where $f(x) = \frac{1}{2}x - \frac{4}{5}$

(b) $f: \mathbf{Z} \longrightarrow \mathbf{R}$, where $f(x) = x^2 - 5$
(c) $f: \mathbf{Z} \longrightarrow \mathbf{Z}$, where $f(x) = |x| + x$

5. Show that each of the following functions is a bijection and find the inverse function for each.

(a) $f: \mathbf{R} \longrightarrow \mathbf{R}$, where $f(x) = 3x - 2$
(b) $f: \mathbf{R} \longrightarrow \mathbf{R}$, where $f(x) = x^3 + 1$

6. If $f: A \longrightarrow B$ is an injective function, prove that $X = f^{-1}(f(X))$ for every subset X of A.

7. If $f: A \longrightarrow B$ is a surjective function, prove that $Y = f(f^{-1}(Y))$ for every subset Y of B.

8. If $f: A \longrightarrow B$ is a bijective function with inverse function $f^{-1}: B \longrightarrow A$, prove that f^{-1} is also bijective. What is the inverse function for f^{-1}? Does $(f^{-1})^{-1} = f$?

9. Let $f: A \longrightarrow B$ and $g: B \longrightarrow C$ be injective (surjective) functions. Prove that $g \circ f$ is injective (surjective).

10. If $f: A \longrightarrow A$ is a function, define the relation R on A by aRb if and only if $f(a) = f(b)$. Is R an equivalence relation on A?

11. Let $f: A \longrightarrow B$ and $g: B \longrightarrow C$ be bijective functions. Prove that $f^{-1} \circ g^{-1}: C \longrightarrow A$ is the inverse function for $g \circ f$. Can you conclude from this that $g \circ f$ is a bijection?

12. Let $f: A \longrightarrow B$ and $g: B \longrightarrow C$ be functions. Prove each of the following:

(a) If $g \circ f$ is injective, then f is injective.
(b) If $g \circ f$ is surjective, then g is surjective.

TECHNOLOGY PROBLEMS

Use a computer algebra system to solve each of the following problems.

13. If $f(x) = 3x^{14} + 4x^8 - 6$, find $f(-452.567)$ and $f(516.21315)$.

14. If $f(x) = \frac{3}{2}x^5 + 4x^3 - 8x + \frac{2}{3}$ and $g(x) = \frac{3}{5}x + 3$, compute $(f \circ g)(x)$ and $(g \circ f)(x)$. Instruct your computer algebra system to express each answer as a polynomial in x.

15. If $f(x) = 2x^3 + \frac{4}{5}x - 3$, $g(x) = \frac{1}{2}x^2 + 1$, and $h(x) = \frac{6}{5}x^3 - 1$, compute each of the following. Instruct your computer algebra system to express each answer as a polynomial in x.

(a) $f(g(h(x)))$

(b) $h(f(g(x)))$

(c) $g(h(f(x)))$

16. If $f : [-1, \infty) \longrightarrow [-1, \infty)$ is defined by $f(x) = x^2 + 2x$ and $g : [-1, \infty) \longrightarrow [-1, \infty)$ is given by $g(x) = -1 + \sqrt{x+1}$, show that $g = f^{-1}$.

0.4   BINARY OPERATIONS

Section Overview. Binary operations on sets are the principal topic of this section. Important concepts discussed are associative and commutative binary operations, the identity for a binary operation and the inverse of an element, and what it means for an operation to be closed. Binary operations are used to form the algebraic structures that are the subject of study in abstract algebra. Congruence relations are also introduced in this section. Take special note of Theorem 0.4.8; it will be used several times throughout the text.

Few concepts in mathematics are more basic than that of composition of two elements of a set. This idea is almost inseparable from calculation with whole numbers, where this idea must have originated.

0.4.1   Definition

A **binary operation** $*$ on a set A is a function $* : A \times A \longrightarrow A$. If $(x, y) \in A \times A$, it is usual to denote $*(x, y)$ by $x * y$. The operation $*$ is said to be **associative** if $x * (y * z) = (x * y) * z$ for all x, y, z in A and **commutative** if $x * y = y * x$ for all $x, y \in A$. If $*$ is a binary operation on A and an element $e \in A$ exists such that $x * e = e * x = x$ for all $x \in A$, then e is said to be an **identity element** for $*$. If e is an identity for $*$, we will also say that A has an identity e or that e is an identity for A. If A has an **identity** e and $x \in A$, then if an element $y \in A$ exists such that $x * y = y * x = e$, then y is said to be an **inverse** for x.

A given set may have several binary operations defined on it. In this case it is convenient to give the operations different names and to use a different symbol to denote each operation. If a binary operation on a set A is denoted by xy, it is usually referred to as **multiplication**, and a binary operation on A is called **addition** if it is denoted by $x + y$. These terms will often be used when the set A is not a set of real numbers and the operation has nothing to do with real number arithmetic.

A binary operation $*$ on A always takes its values in A. This is a necessary and sufficient condition for a function with domain $A \times A$ to be a binary operation on A. For example, the operation $-$ of ordinary subtraction is not a binary operation on $\mathbf{N}$ since $3 - 5 \notin \mathbf{N}$. Thus, $-$ does not always take its values in $\mathbf{N}$. When a function with domain $A \times A$ takes its values in A, the operation is referred to as being **closed**. *Therefore, to establish that a function with domain $A \times A$ is a binary operation on A, it is necessary to show that the operation is closed.* If a binary operation on A is designated as multiplication (addition), then an identity for the operation is referred to as a **multiplicative identity** (**additive identity**). Similarly, an inverse for an element $x \in A$, should it exist, is called a **multiplicative inverse** (an **additive inverse**) for x.

0.4.2 | **Examples**

1. The operation of subtraction $-$ is a binary operation on $\mathbf{Z}$. However, as previously observed, when $-$ is restricted to $\mathbf{N}$, $-$ is not a binary operation on $\mathbf{N}$.

2. Consider $M_2(\mathbf{R}) = \left\{ \begin{pmatrix} a & c \\ b & d \end{pmatrix} \middle| a, b, c, d \in \mathbf{R} \right\}$, which is the set of 2×2 matrices with entries from $\mathbf{R}$. (Matrices will be discussed in more detail in the next section.) Define the binary operation of **matrix addition** $+$ on $M_2(\mathbf{R})$ by $\begin{pmatrix} a & c \\ b & d \end{pmatrix} + \begin{pmatrix} e & g \\ f & h \end{pmatrix} = \begin{pmatrix} a+e & c+g \\ b+f & d+h \end{pmatrix}$. The binary operation $+$ on the left in this equation is the operation being defined, whereas the operation $+$ that appears in the matrix on the right is ordinary addition of real numbers. Since addition of real numbers is associative, we see that

$$\left[\begin{pmatrix} a & c \\ b & d \end{pmatrix} + \begin{pmatrix} e & g \\ f & h \end{pmatrix} \right] + \begin{pmatrix} i & k \\ j & l \end{pmatrix} = \begin{pmatrix} a+e & c+g \\ b+f & d+h \end{pmatrix} + \begin{pmatrix} i & k \\ j & l \end{pmatrix}$$

$$= \begin{pmatrix} (a+e)+i & (c+g)+k \\ (b+f)+j & (d+h)+l \end{pmatrix}$$

$$= \begin{pmatrix} a+(e+i) & c+(g+k) \\ b+(f+j) & d+(h+l) \end{pmatrix}$$

$$= \begin{pmatrix} a & c \\ b & d \end{pmatrix} + \begin{pmatrix} e+i & g+k \\ f+j & h+l \end{pmatrix}$$

$$= \begin{pmatrix} a & c \\ b & d \end{pmatrix} + \left[\begin{pmatrix} e & g \\ f & h \end{pmatrix} + \begin{pmatrix} i & k \\ j & l \end{pmatrix} \right],$$

and matrix addition is therefore associative. Inspection of these equations shows that associativity of matrix addition on $M_2(\mathbf{R})$ follows from the fact that addition of real numbers is associative. We say that associativity of + on $M_2(\mathbf{R})$ is **inherited** from the associativity of addition on $\mathbf{R}$. It can also be shown that matrix addition is commutative and that this property is inherited directly from the fact that addition on $\mathbf{R}$ is commutative.

The matrix $\begin{pmatrix} 0 & 0 \\ 0 & 0 \end{pmatrix}$ is an **additive identity** for matrix addition. Every element of $M_2(\mathbf{R})$ also has an **additive inverse**. If $\begin{pmatrix} a & c \\ b & d \end{pmatrix}$ is in $M_2(\mathbf{R})$, then

$\begin{pmatrix} -a & -c \\ -b & -d \end{pmatrix} \in M_2(\mathbf{R})$ is such that $\begin{pmatrix} a & c \\ b & d \end{pmatrix} + \begin{pmatrix} -a & -c \\ -b & -d \end{pmatrix} = \begin{pmatrix} 0 & 0 \\ 0 & 0 \end{pmatrix}.$

3. Similarly, **matrix multiplication** is defined on $M_2(\mathbf{R})$ by

$$\begin{pmatrix} a & c \\ b & d \end{pmatrix}\begin{pmatrix} e & g \\ f & h \end{pmatrix} = \begin{pmatrix} ae+cf & ag+ch \\ be+df & bg+dh \end{pmatrix}.$$

Matrix multiplication is an associative operation with **multiplicative identity** $\begin{pmatrix} 1 & 0 \\ 0 & 1 \end{pmatrix}$. The associativity of matrix multiplication is inherited from properties of addition and multiplication on $\mathbf{R}$. The following example shows that matrix multiplication is not commutative:

$$\begin{pmatrix} 1 & 1 \\ 0 & 0 \end{pmatrix}\begin{pmatrix} 1 & 0 \\ 1 & 0 \end{pmatrix} = \begin{pmatrix} 2 & 0 \\ 0 & 0 \end{pmatrix} \neq \begin{pmatrix} 1 & 1 \\ 1 & 1 \end{pmatrix} = \begin{pmatrix} 1 & 0 \\ 1 & 0 \end{pmatrix}\begin{pmatrix} 1 & 1 \\ 0 & 0 \end{pmatrix}.$$

An element of $M_2(\mathbf{R})$ may not have a **multiplicative inverse**. If $\begin{pmatrix} a & c \\ b & d \end{pmatrix}$ is a matrix in $M_2(\mathbf{R})$ such that $ad - bc \neq 0$, then

$$\begin{pmatrix} \dfrac{d}{ad-bc} & \dfrac{-c}{ad-bc} \\ \dfrac{-b}{ad-bc} & \dfrac{a}{ad-bc} \end{pmatrix}$$

is a multiplicative inverse of $\begin{pmatrix} a & c \\ b & d \end{pmatrix}$. If $ad - bc = 0$, the matrix $\begin{pmatrix} a & c \\ b & d \end{pmatrix}$

does *not* have a multiplicative inverse. For example, $\begin{pmatrix} 1 & 2 \\ 2 & 4 \end{pmatrix}$ fails to have a

multiplicative inverse.

To say that a binary operation $*$ on a set A is associative means that the expression $x * y * z$ is unambiguous. This follows because there are only two ways to perform the operations indicated by $x * y * z$ without commuting the elements. One way is to first "multiply" $x * y$ and then to multiply the result on the right by z, indicated by $(x * y) * z$. The second way is to multiply y by z and then to multiply this result on the left by x, that is, $x * (y * z)$. If the operation is associative, $x * (y * z) = (x * y) * z$.

0.4.3   Example

The binary operation of ordinary subtraction $-$ on **Z** is neither associative nor commutative since $(5 - 4) - 3 \neq 5 - (4 - 3)$ and $8 - 3 \neq 3 - 8$.

0.4.4   Theorem

Let $*$ be a binary operation on A. If A has an identity e, then e is unique.

Proof. Suppose that e' is also an identity for A. Since e is an identity, e has the property that $x * e = x$ for any $x \in A$. In particular, when $x = e'$, $e' * e = e'$. Now e' is also an identity for A, so e' has the property that $e' * x = x$ for any $x \in A$. Hence, if $x = e$, we have $e' * e = e$. Therefore, $e = e' * e = e'$. ∎

0.4.5   Theorem

Let $*$ be an associative binary operation on A and suppose that A has an identity e. If $x \in A$ has an inverse $y \in A$, then y is unique.

Proof. Suppose $y, y' \in A$ are inverses of $x \in A$. Then $y * x = x * y = e$ and $y' * x = x * y' = e$. Hence, $y * x = y' * x$. Multiplying both sides of the last equation on the right by y gives $(y * x) * y = (y' * x) * y$, and thus $y * (x * y) = y' * (x * y)$ since $*$ is associative. Therefore, $y * e = y' * e$ and $y = y'$. ∎

The proof of the following theorem is left as an exercise.

0.4.6   Theorem

If A is a nonempty set and $\mathscr{F}$ is the set of all functions $f : A \longrightarrow A$, then function composition is an associative binary operation on $\mathscr{F}$. Moreover, $\mathscr{F}$ has an identity, and if $f \in \mathscr{F}$ is a bijection, then f has an inverse under this operation.

The multiplicative identity for a binary operation on a set A, should it exist, will be denoted by e, and if there is an additive identity, it will be denoted by 0. Here, 0 may not be the integer zero; 0 is simply the notation chosen for the additive identity. If the operation on A has a multiplicative identity, then the multiplicative inverse of an element $x \in A$, should it exist, will be denoted by x^{-1}. If A has an additive identity and $x \in A$ has an additive inverse, it will be denoted by $-x$.

0.4.7   Definition

Let $*$ be a **binary operation** on a set A and suppose that E is an equivalence relation on A. If xEy and zEw together imply that $(x * z)$E$(y * w)$, then E is said to be **compatible with the binary operation** on A. In such cases we say that the equivalence relation **preserves the binary operation** on A and that the equivalence relation is a **congruence relation** on A. If E is a congruence relation on A and xEy, then x and y are said to be **congruent modulo** E and an equivalence class will be referred to as a **congruence class**.

The next theorem will be used several times throughout the remainder of the text. In particular, it will be used when factor groups and factor rings are formed in later chapters.

0.4.8   Theorem

Let $*$ be a binary operation on A and suppose E is a congruence relation on A. If A/E denotes the set of congruence classes of A determined by E, then the operation $\bullet$ defined on the congruence classes of A/E by $[x] \bullet [y] = [x * y]$ is a well-defined binary operation on A/E.

Proof. To show that the binary operation • is well-defined, it must be shown that $[x] \bullet [z] = [y] \bullet [w]$ when $[x] = [y]$ and $[z] = [w]$. Since $[x] \bullet [z] = [x * z]$ and $[y] \bullet [w] = [y * w]$, it suffices to show that $[x * z] = [y * w]$. If $[x] = [y]$ and $[z] = [w]$, then xEy and zEw, so $(x * z)E(y * w)$ because E is a congruence relation on A. But if $(x * z)E(y * w)$, it follows from Theorem 0.2.3 that $[x * z] = [y * w]$. Since the operation is obviously closed, • is a well-defined binary operation on A/E. ∎

| 0.4.9 | **Example** |

This example is our first construction of an interesting algebraic object. We start with **Z** together with the binary operation of ordinary addition of integers and define the relation E on **Z** by xEy if and only if $x - y$ is a multiple of 3. In Example 1 of 0.2.2 it was shown that E is an equivalence relation on **Z**. The equivalence relation E partitions **Z** into the equivalence classes

$$[0] = \{..., -9, -6, -3, 0, 3, 6, 9, ...\},$$

$$[1] = \{..., -8, -5, -2, 1, 4, 7, 10, ...\},$$

$$[2] = \{..., -7, -4, -1, 2, 5, 8, 11, ...\},$$

so $\mathbf{Z}/E = \{[0], [1], [2]\}$. We claim that E is also a congruence relation on **Z**. Indeed, if xEy and zEw, then there are integers k_1 and k_2 such that $x - y = 3k_1$ and $z - w = 3k_2$. Hence, $(x + z) - (y + w) = (x - y) + (z - w) = 3k_1 + 3k_2 = 3(k_1 + k_2)$, and thus $(x + z)E(y + w)$. Theorem 0.4.8 tells us that the binary operation on $\mathbf{Z}/E$ defined by $[x] + [y] = [x + y]$ is well-defined. (You should notice that two binary operations are involved in the expression $[x] + [y] = [x + y]$. In $[x + y]$ we are adding the integers x and y, and in $[x] + [y]$ we are adding congruence classes.) In our new set of "numbers," $\mathbf{Z}/E = \{[0], [1], [2]\}$, there are only three elements and we have behavior of a sort that we do not see in **Z**. For example, $[1] + [2] = [1 + 2] = [0]$, so $[1]$ is the additive inverse of $[2]$. Hence, $-[2] = [1]$. Note also that $[2] + [2] = [2 + 2] = [1]$.

Problem Set 0.4

*For each of the following, determine if the operation * on the given set is closed.*

1. $x * y = xy$ on $-\mathbf{N} = \{-1, -2, -3, ...\}$

2. $x * y = x + y$ on **N**

3. $x * y = x^2 - y^2$ on **N**

4. $x * y = x$ on $\mathbf{N}$

5. $x * y = x^2y$ on $-\mathbf{N}$

Which of the following binary operations on $\mathbf{R}$ are commutative? Which are associative? Does $\mathbf{R}$ have an identity with respect to the binary operation? If $\mathbf{R}$ has an identity e with respect to the binary operation, which elements of $\mathbf{R}$, other than e, have a multiplicative inverse?

6. $x * y = x^y$, where 0^0 is defined to be 1

7. $x * y = xy + x + y$

8. $x * y = x + xy$

9. $x * y = 3^{xy}$

10. $x * y = x + y - 1$

11. $x * y = |x| + |y|$

For 12 through 14, let $M_2(\mathbf{Q})$ denote the set of 2×2 matrices with entries from $\mathbf{Q}$ together with the binary operations of matrix addition and multiplication as defined in Example 2 of 0.4.2.

12. Show that the identity matrix $I_2 = \begin{pmatrix} 1 & 0 \\ 0 & 1 \end{pmatrix}$ is the multiplicative identity for $M_2(\mathbf{Q})$.

13. Find the multiplicative inverse of each of the following matrices, if it exists.

     (a) $\begin{pmatrix} 1/3 & 2 \\ 5/6 & 1/2 \end{pmatrix}$

     (b) $\begin{pmatrix} -4 & -8 \\ 1/2 & 1 \end{pmatrix}$

     (c) $\begin{pmatrix} 342 & 21 \\ 121 & -342 \end{pmatrix}$

14. Let $A = \begin{pmatrix} 1 & 0 \\ -1 & 2 \end{pmatrix}$, $B = \begin{pmatrix} -1 & 2 \\ 1 & 0 \end{pmatrix}$, and $C = \begin{pmatrix} 1 & -2 \\ 2 & 1 \end{pmatrix}$. Show that

     (a) $A + (B + C) = (A + B) + C$
     (b) $A(BC) = (AB)C$
     (c) $AB \neq BA$
     (d) $A(B + C) = AB + AC$

15. (a) Prove that the relation E defined on $\mathbf{Z}$ by $x\mathrm{E}y$ if and only if $x - y$ is a multiple of 10 is a congruence relation on $\mathbf{Z}$.

     (b) Describe the set $\mathbf{Z}/\mathrm{E}$.

(c) Determine each of the following, where each congruence class is in
Z/E:

[5] + [9]

[6] + [7]

[8] + [8]

(d) What is the additive inverse of each of the following?

[4]

[8]

[3]

(e) Is there a congruence class in **Z**/E that is its own additive inverse?

16. Prove Theorem 0.4.6.

0.5   MATRICES

Section Overview. Matrices arise quite often in mathematics and they can be used to provide many interesting examples in abstract algebra. The important concepts of this section are the definition of addition and multiplication of matrices, equality of matrices, and the identity and zero matrix.

0.5.1   Definition

A matrix

$$A = \begin{pmatrix} a_{11} & a_{12} & \cdots & a_{1n} \\ a_{21} & a_{22} & \cdots & a_{2n} \\ \vdots & \vdots & \vdots & \vdots \\ a_{m1} & a_{m2} & \cdots & a_{mn} \end{pmatrix}$$

is a rectangular array of objects in rows and columns. The objects in a matrix are usually rational, real, or complex numbers, symbols that represent such numbers, or elements from some other algebraic structure. If a matrix A has m rows and n columns, then A is an $m \times n$ matrix, where $m \times n$ is read "m by n." Such a matrix will be denoted by $A = (a_{ij})$ when m and n are understood. The element a_{ij} of a matrix is the (i, j)th entry that lies in the ith row and the jth column of A.

An $n \times n$ matrix is said to be a **square matrix**. The set of $m \times n$ matrices with entries from X will be denoted by $M_{m \times n}(X)$, and $M_n(X)$ will denote the set of $n \times n$ square matrices with entries from X. If A is in $M_{m \times n}(X)$, then A is an $m \times n$ **matrix over** X. If two matrices have the same number of rows and columns, then the matrices are said to be of the **same size**. If (a_{ij}) is a square matrix, then the **diagonal** of (a_{ij}) consists of the entries in (a_{ij}) for which $i = j$. Two matrices (a_{ij}) and (b_{ij}) of the same size are **equal** if $a_{ij} = b_{ij}$ for all i and j. If X is a set together with a binary operation of addition that has an additive identity 0, then a square matrix $(a_{ij}) \in M_n(X)$ is a **diagonal matrix** if all of the off-diagonal entries are 0, that is, if $a_{ij} = 0$ whenever $i \neq j$. If $a_{ij} = 0$ for all i and all j, then (a_{ij}) is a **zero matrix**. If X is also endowed with a binary operation of multiplication and X has an identity e, then a diagonal matrix $I_n = (a_{ij}) \in M_n(X)$ such that $a_{ii} = e$ for $i = 1, 2, \ldots, n$ is an **identity matrix**.

0.5.2   **Examples**

$$\begin{pmatrix} 1 & 5 \\ -2 & 5 \\ 3 & -6 \end{pmatrix} \quad 3 \times 2 \text{ matrix in } M_{3\times2}(\mathbf{Z});$$

$$\begin{pmatrix} \frac{1}{2} & \frac{3}{5} & \frac{2}{3} & -4 \\ -\frac{3}{4} & -\frac{1}{2} & 3 & 0 \\ 5 & -\frac{5}{6} & -4 & 2 \end{pmatrix} \quad 3 \times 4 \text{ matrix in } M_{3\times4}(\mathbf{Q});$$

$$\begin{pmatrix} \sqrt{2} & 3 & -5 \\ -\frac{1}{2} & -\sqrt{23} & \pi \\ 5 & 1 & -12 \end{pmatrix} \quad 3 \times 3 \text{ matrix over } \mathbf{R};$$

$$\begin{pmatrix} -3 & 0 & 0 & 0 \\ 0 & 1 & 0 & 0 \\ 0 & 0 & 2 & 0 \\ 0 & 0 & 0 & 1 \end{pmatrix} \quad \text{Diagonal matrix over } \mathbf{Z};$$

$$\begin{pmatrix} 0 & 0 \\ 0 & 0 \\ 0 & 0 \\ 0 & 0 \end{pmatrix} \quad 4 \times 2 \text{ zero matrix of } M_{4\times2}(\mathbf{Z});$$

$$\begin{pmatrix} 0 & 0 \\ 0 & 0 \end{pmatrix} \qquad 2 \times 2 \text{ zero matrix over } \mathbf{Z};$$

$$I_4 = \begin{pmatrix} 1 & 0 & 0 & 0 \\ 0 & 1 & 0 & 0 \\ 0 & 0 & 1 & 0 \\ 0 & 0 & 0 & 1 \end{pmatrix} \qquad \text{Identity matrix in } M_4(\mathbf{Z});$$

$$I_3 = \begin{pmatrix} 1 & 0 & 0 \\ 0 & 1 & 0 \\ 0 & 0 & 1 \end{pmatrix} \qquad 3 \times 3 \text{ identity matrix over } \mathbf{Z}.$$

When two matrices are of an appropriate size, they can often be added and multiplied if their entries belong to a set on which binary operations of addition and multiplication are defined.

0.5.3   Definition

Let X be a nonempty set on which binary operations of addition and multiplication are defined. If $A = (a_{ij})$ and $B = (b_{ij})$ are matrices in $M_{m \times n}(X)$, then **matrix addition** is defined by $A + B = (a_{ij} + b_{ij})$. The operation of matrix addition is said to be a **componentwise operation**, since corresponding entries of A and B are added to form each entry of the sum of the two matrices. If $A = (a_{ij}) \in M_{m \times r}(X)$ and $B = (b_{ij}) \in M_{r \times n}(X)$, then **matrix multiplication** is defined by $AB = (c_{ij})$, where

$$c_{ij} = \sum_{k=1}^{r} a_{ik} b_{kj} \text{ for } i = 1, 2, \ldots, m \text{ and } j = 1, 2, \ldots, n.$$

If X is as in Definition 0.5.3 and $A = (a_{ij})$ and $B = (b_{ij})$ are matrices in $M_{m \times r}(X)$ and $M_{r \times n}(X)$, respectively, then for each i and j, c_{ij} of the product $AB = (c_{ij})$ is the ith row of A multiplied by the jth column of B:

$$c_{ij} = (a_{i1} \; a_{i2} \; \cdots \; a_{ir}) \begin{pmatrix} b_{1j} \\ b_{2j} \\ \vdots \\ b_{rj} \end{pmatrix} = a_{i1}b_{1j} + a_{i2}b_{2j} + \cdots + a_{ir}b_{rj} \,.$$

The subscripts ij on c_{ij} tell us how to compute c_{ij}. The double subscript ij indicates that we are to multiply the ith row of the first matrix A in the product AB by the jth column of the second matrix B to obtain the (i, j)th entry of (c_{ij}). For example, to compute the c_{12} entry of the product

$$AB = \begin{pmatrix} 1 & 0 & 1 \\ 3 & 1 & 0 \\ 1 & 0 & 0 \\ 2 & 2 & 1 \end{pmatrix} \begin{pmatrix} -1 & 0 \\ 1 & -1 \\ 2 & 1 \end{pmatrix} = \begin{pmatrix} c_{11} & c_{12} \\ c_{21} & c_{22} \\ c_{31} & c_{31} \\ c_{41} & c_{41} \end{pmatrix}$$

multiply the first row of A by the second column of B so that

$$c_{12} = (1\ \ 0\ \ 1) \begin{pmatrix} 0 \\ -1 \\ 1 \end{pmatrix} = 1(0) + 0(-1) + 1(1) = 1, \text{ and}$$

$$(c_{ij}) = \begin{pmatrix} c_{11} & 1 \\ c_{21} & c_{22} \\ c_{31} & c_{31} \\ c_{41} & c_{41} \end{pmatrix}.$$

The other entries of (c_{ij}) are computed in a similar fashion. The following is an example of the sum and product of two matrices. Study the matrix product carefully and make sure that you understand exactly how to compute each of its entries.

0.5.4   Examples

$$\begin{pmatrix} 1 & 2 & 1 \\ 3 & 0 & -1 \\ 0 & 1 & 0 \end{pmatrix} + \begin{pmatrix} -3 & 4 & 2 \\ 2 & -2 & 3 \\ -1 & 0 & -1 \end{pmatrix} = \begin{pmatrix} 1-3 & 2+4 & 1+2 \\ 3+2 & 0-2 & -1+3 \\ 0-1 & 1+0 & 0-1 \end{pmatrix}$$

$$= \begin{pmatrix} -2 & 6 & 3 \\ 5 & -2 & 2 \\ -1 & 1 & -1 \end{pmatrix}.$$

$$\begin{pmatrix} 1 & 2 & 1 \\ 3 & 0 & -1 \\ 0 & 1 & 0 \end{pmatrix} \begin{pmatrix} 1 & 0 \\ 0 & 1 \\ 1 & 0 \end{pmatrix} = \begin{pmatrix} 1(1)+2(0)+1(1) & 1(0)+2(1)+1(0) \\ 3(1)+0(0)+(-1)(1) & 3(0)+0(1)+(-1)(0) \\ 0(1)+1(0)+0(1) & 0(0)+1(1)+0(0) \end{pmatrix}$$

$$= \begin{pmatrix} 2 & 2 \\ 2 & 0 \\ 0 & 1 \end{pmatrix}.$$

The product of two matrices is not always defined, and matrix multiplication is not a commutative operation. In order for a matrix product AB to be defined, the number of columns of A must be equal to the number of rows of B.

Suppose that a set X is endowed with two binary operations: addition and multiplication. When addition on X is associative or commutative, this property is inherited by matrix addition on $M_{m \times n}(X)$. If X has an additive identity, then the $m \times n$ zero matrix is the additive identity for $M_{m \times n}(X)$. If X also has a multiplicative identity, I_n is the multiplicative identity for $M_n(X)$.

0.5.5   **Example**

Consider the matrix

$$A = \begin{pmatrix} 3 & 0 & -1 \\ -1 & 1 & 0 \\ 2 & 1 & 4 \end{pmatrix} \in M_3(\mathbf{Z}).$$

Multiplication by I_3 yields

$$I_3 A = \begin{pmatrix} 1 & 0 & 0 \\ 0 & 1 & 0 \\ 0 & 0 & 1 \end{pmatrix} \begin{pmatrix} 3 & 0 & -1 \\ -1 & 1 & 0 \\ 2 & 1 & 4 \end{pmatrix}$$

$$= \begin{pmatrix} 1(3)+0(-1)+0(2) & 1(0)+0(1)+0(1) & 1(-1)+0(0)+0(4) \\ 0(3)+1(-1)+0(2) & 0(0)+1(1)+0(1) & 0(-1)+1(0)+0(4) \\ 0(3)+0(-1)+1(2) & 0(0)+0(1)+1(1) & 0(-1)+0(0)+1(4) \end{pmatrix}$$

$$= \begin{pmatrix} 3 & 0 & -1 \\ -1 & 1 & 0 \\ 2 & 1 & 4 \end{pmatrix}.$$

Similarly, $AI_3 = A$. This is actually the case for every 3×3 matrix (a_{ij}) in $M_3(\mathbf{Z})$. You can convince yourself of this by careful examination of the example above. If the ith row of (a_{ij}) is multiplied by the jth column of I_3, then the only nonzero entry in the sum is a_{ij}. Hence, $AI_3 = A$. Similar reasoning shows that $I_3A = A$, and thus the identity matrix I_3 is the multiplicative identity of $M_3(\mathbf{Z})$.

Problem Set 0.5

1. Consider the following matrices. Perform each operation indicated below if it is defined. Write DNE (for "does not exist") if an operation cannot be performed.

$$A = \begin{pmatrix} -1 & 0 & 1 \\ 2 & 1 & -1 \\ -3 & 2 & -2 \\ 5 & 3 & 0 \end{pmatrix} \quad B = \begin{pmatrix} 2 & -3 \\ 1 & 4 \\ 3 & 5 \end{pmatrix} \quad C = \begin{pmatrix} 1 & 0 \\ -3 & 1 \end{pmatrix} \quad D = \begin{pmatrix} 0 & 1 & 4 \\ -1 & 0 & 2 \\ -1 & 0 & 3 \\ 1 & -3 & 0 \end{pmatrix}$$

   (a) $A + D$ and $A - D$

   (b) AB and BA

   (c) $A(BC)$ and $(AB)C$

   (d) $C^2 = CC$ and $D^2 = DD$

   (e) $B + C$

2. Find two matrices A and B in $M_3(\mathbf{Z})$ such that $AB \neq BA$.

3. Show that matrix multiplication on $M_2(\mathbf{Q})$ is associative. Identify the property or properties of addition and multiplication on $\mathbf{Q}$ that are responsible for this being true.

4. If $a \in \mathbf{R}$, identify a with the matrix $\begin{pmatrix} a & 0 & 0 \\ 0 & a & 0 \\ 0 & 0 & a \end{pmatrix}$ in $M_3(\mathbf{R})$. Do

   $$\begin{pmatrix} a & 0 & 0 \\ 0 & a & 0 \\ 0 & 0 & a \end{pmatrix} + \begin{pmatrix} b & 0 & 0 \\ 0 & b & 0 \\ 0 & 0 & b \end{pmatrix} \text{ and } \begin{pmatrix} a & 0 & 0 \\ 0 & a & 0 \\ 0 & 0 & a \end{pmatrix}\begin{pmatrix} b & 0 & 0 \\ 0 & b & 0 \\ 0 & 0 & b \end{pmatrix} \text{ correspond to } a + b \text{ and}$$

   ab in $\mathbf{R}$, respectively?

5. If $(a_{ij}) \in M_{m \times n}(X)$, then the **transpose** of (a_{ij}) is defined by $(a_{ij})^t = (a_{ji})$. The transpose of (a_{ij}) is obtained from (a_{ij}) by interchanging rows and columns: row 1 becomes column 1, row 2 becomes column 2, and so on. For example,

$$\begin{pmatrix} 1 & 0 & 5 \\ 3 & 2 & 3 \\ -1 & 4 & 1 \\ 0 & 1 & -2 \end{pmatrix}^t = \begin{pmatrix} 1 & 3 & -1 & 0 \\ 0 & 2 & 4 & 1 \\ 5 & 3 & 1 & -2 \end{pmatrix}.$$

If $A \in M_{m \times n}(X)$, then A^t is in $M_{n \times m}(X)$.

(a) If $A = \begin{pmatrix} 1 & 0 & 5 \\ 3 & 2 & 3 \\ -1 & 4 & 1 \\ 0 & 1 & -2 \end{pmatrix}$ and $B = \begin{pmatrix} 2 & 0 & 9 \\ 5 & 0 & 2 \\ 1 & 1 & 1 \\ 8 & 5 & -6 \end{pmatrix}$, does $(A+B)^t = A^t + B^t$?

(b) If $A = \begin{pmatrix} 1 & 0 & 5 \\ 3 & 2 & 3 \\ -1 & 4 & 1 \end{pmatrix}$ and $B = \begin{pmatrix} 2 & 0 & 9 \\ 5 & 0 & 2 \\ 1 & 1 & 1 \end{pmatrix}$, determine if either of the following is true: $(AB)^t = A^t B^t$ and $(AB)^t = B^t A^t$.

6. Let det: $M_2(\mathbf{R}) \longrightarrow \mathbf{R}$ be given by $\det(A) = ad - bc$, where $A = \begin{pmatrix} a & b \\ c & d \end{pmatrix}$. $\det(A)$ is said to be the **determinant** of A. If $A, B \in M_2(\mathbf{R})$, prove that:

(a) $\det(AB) = \det(A)\det(B)$, but $\det(A + B) \neq \det(A) + \det(B)$.

(b) If $AB = \begin{pmatrix} 0 & 0 \\ 0 & 0 \end{pmatrix}$, then either $\det(A) = 0$ or $\det(B) = 0$.

(c) $\det(A^t) = \det(A)$.

TECHNOLOGY PROBLEMS

Use a computer algebra system to solve each of the following problems.

7. Compute the matrix product

$$
\begin{pmatrix}
2 & \frac{1}{2} & 7 \\
-1 & 0 & -4 \\
0 & \frac{3}{4} & \frac{5}{8} \\
-3 & -1 & 6 \\
4 & 6 & \frac{1}{2}
\end{pmatrix}
\begin{pmatrix}
1 & \frac{1}{4} & 0 & 8 & 5 \\
2 & \frac{1}{2} & 1 & -21 & 6 \\
3 & \frac{8}{7} & 0 & 7 & \frac{2}{3}
\end{pmatrix}.
$$

Instruct your computer algebra system to leave the result in matrix form.

8. Find two 6×6 matrices A and B over **Z** such that $AB \neq BA$.

9. Show that I_4 is the multiplicative identity for $M_4(\mathbf{R})$, where

$$
I_4 =
\begin{pmatrix}
1 & 0 & 0 & 0 \\
0 & 1 & 0 & 0 \\
0 & 0 & 1 & 0 \\
0 & 0 & 0 & 1
\end{pmatrix}.
$$

10. Show that

$$
\det
\begin{pmatrix}
1 & -4 & 8 & \frac{3}{8} \\
3 & 5 & \frac{9}{4} & \frac{1}{4} \\
4 & 12 & 6 & -2 \\
-1 & \frac{2}{3} & 1 & 7
\end{pmatrix}
\neq 0
$$

and find the multiplicative inverse of

$$
\begin{pmatrix}
1 & -4 & 8 & \frac{3}{8} \\
3 & 5 & \frac{9}{4} & \frac{1}{4} \\
4 & 12 & 6 & -2 \\
-1 & \frac{2}{3} & 1 & 7
\end{pmatrix}.
$$

Instruct your computer algebra system to leave the result in matrix form.

11. Solve the following matrix equation for x, y, and z:

$$\begin{pmatrix} 2 & 3 & 5 \\ -1 & 4 & 1 \\ 4 & \frac{1}{3} & 2 \end{pmatrix} \begin{pmatrix} x \\ y \\ z \end{pmatrix} = \begin{pmatrix} 1 \\ \frac{2}{3} \\ 3 \end{pmatrix}$$

12. Show that

$$\det \begin{pmatrix} 1 & -4 & 0 & \frac{3}{8} \\ 3 & 5 & \frac{9}{4} & \frac{1}{4} \\ 4 & 0 & 6 & -2 \\ -1 & \frac{2}{3} & 1 & 7 \end{pmatrix} = \det \begin{pmatrix} 1 & -4 & 0 & \frac{3}{8} \\ 3 & 5 & \frac{9}{4} & \frac{1}{4} \\ 4 & 0 & 6 & -2 \\ -1 & \frac{2}{3} & 1 & 7 \end{pmatrix}^t .$$

1

The Integers

Little in mathematics is more fundamental than the integers. Almost certainly the concept of a binary operation began with addition and multiplication of integers. In this chapter, we will study the fundamental properties of the integers as well as modular arithmetic and proof by mathematical induction. Make sure you master the technique of proof by induction. It is a very powerful technique of proof that you should have in your mathematical tool bag.

1.1   THE INTEGERS

Section Overview. In this section the basic properties of the integers are presented. Important concepts are the definition of order among the integers, the Well-Ordering Axiom, the Division Algorithm, the greatest common divisor, and the Euclidean Algorithm.

It is assumed that you are familiar with the basic properties of the binary operations of addition and multiplication of integers. Both of these operations are associative and commutative, and **Z** has an additive identity 0 and a multiplicative identity 1. Every integer x has an additive inverse $-x$, but only 1 and -1 have multiplicative inverses in **Z**. We begin our study of the integers with the definition of the order relation $<$ on **Z**. (See Example 2 of 0.2.2.) You are already familiar with what it means for one integer x to be less than another integer y, but if you were asked to prove that $x < y$ implies that $x + z < y + z$, where x, y, and z are integers, could you do it? A typical response is to write something like $3 < 7$ and then say, "Well, I can add, say, 2 to both sides of the inequality to get $3 + 2 < 7 + 2$." Or you might write $-4 < 2$ and then add 5 to both sides of this inequality. But this is not a proof! One cannot prove that a mathematical statement is true simply by looking at examples. If we are to prove statements such as "$x < y$ implies that $x + z < y + z$ for integers x, y, and z," what is required is a formal definition of the order relation $<$ on **Z**. Another reason for giving a formal development of this ordering on **Z** is that later in the text we will use the development presented here as a model for defining order on other algebraic structures. Properties of the set **N** will be used to define the order relation $<$ on **Z**.

Properties of the Set N

1. The set **N** is closed under addition and multiplication, so $x + y$ and xy are in **N** whenever $x, y \in$ **N**.
2. For any $x \in$ **Z**, one and only one of the following statements holds:

$$x \in \textbf{N}, \quad x = 0, \quad -x \in \textbf{N}. \quad (\textit{trichotomy property})$$

Now define the ordering of **less than** on **Z** by setting $x < y$ if $y - x$ is in **N**. In this context, $x \leq y$ means that $x < y$ or $x = y$ for any two integers x and y. Similarly, $y \geq x$ means that $x \leq y$. We read $x \leq y$ as "x **is less than or equal to** y," and $y \geq x$ is read "y **is greater than or equal to** x." If $x \in$ **N**, then $x - 0 \in$ **N**, so $x > 0$. Because of this we refer to **N** as the set of **positive elements** of **Z**, and if $x \in$ **N**, we say that x is a **positive integer**. We also see that, because of the trichotomy property, for any $x \in$ **Z** exactly one of the statements $x > 0$, $x = 0$, or $-x > 0$ holds.

This definition of order on **Z** and the properties of the set **N** allow us to prove properties of the order relation $<$ with which you are already quite familiar. As an illustration of the techniques of proof that can be used, we prove four fundamental properties of $<$. The proof of each is actually quite simple. You will be asked to prove addition properties of the relation $<$ in the exercises.

1.1.1   Theorem

Let x, y, and z be integers.

1. If $x > 0$ and $y > 0$, then $xy > 0$.
2. If $x < 0$, then $-x > 0$.
3. If $x < y$, then $x + z < y + z$.
4. If $x < y$ and $z > 0$, then $xz < yz$.

Proof

1. If $x > 0$, then $x = x - 0 \in \mathbf{N}$. Similarly, $y > 0$ implies that $y \in \mathbf{N}$. But $\mathbf{N}$ is closed under multiplication, so $xy \in \mathbf{N}$. Hence, $xy - 0 \in \mathbf{N}$ and thus $xy > 0$.
2. If $x < 0$, then $0 - x \in \mathbf{N}$, and thus $-x - 0 \in \mathbf{N}$. Hence, $-x > 0$.
3. If $x < y$, then $y - x \in \mathbf{N}$. Hence, we see that $(y + z) - (x + z)$ is in $\mathbf{N}$, so $x + z < y + z$.
4. If $x < y$, then $y - x \in \mathbf{N}$. Moreover, $z > 0$, so $z = z - 0 \in \mathbf{N}$. But $\mathbf{N}$ is closed under multiplication, so $yz - xz = (y - x)z \in \mathbf{N}$. Hence, we have $xz < yz$. ∎

Now consider the set $\mathbf{N} = \{1, 2, 3, \ldots\}$ with the ordering $1 < 2 < 3 < \cdots$. It is intuitive that each nonempty subset S of $\mathbf{N}$ contains an element s such that $s \leq x$ for all $x \in S$. We refer to s as the **smallest element** of S. For example, $S_1 = \{2, 4, 6, \ldots\}$ and $S_2 = \{5, 7, 10, 15, 37\}$ are nonempty subsets of $\mathbf{N}$, and each contains a smallest element, namely, 2 and 5, respectively. The fact that every nonempty subset of the set of positive integers has a smallest element is so basic that it cannot be proved without assuming that some other fundamental property of the set $\mathbf{N}$ holds. Because of this, we take this underlying property of the positive integers as a starting point and assume that it holds as an axiom.

1.1.2 Well-Ordering Axiom

Every nonempty subset of the set of positive integers $\mathbf{N}$ contains a smallest element.

The following important theorem about the integers follows from this axiom. Later, when polynomial rings are studied, a similar property will be proved for polynomials.

1.1.3 Division Algorithm

If n is a positive integer, then for any integer x there exist unique integers q and r such that $x = qn + r$, where $0 \leq r < n$. The integer x is called the **dividend**, n the **divisor**, q the **quotient**, and r the **remainder**.

For example, if x is 14 and n is 4, we can find a unique q and r such that $14 = 4q + r$. You have been finding q and r since you learned how to divide: 14 divided by 4 gives 3 (this is q) and leaves a remainder of 2 (this is r). Hence, $14 = 4(3) + 2$. It is not the case that x must be positive. For example, if x is -17 and n is 4, then -17 divided by 4 gives $q = -5$ with a remainder of

$r = 3$. Thus, $-17 = 4(-5) + 3$. However, to put this process on a firm foundation mathematically, we need to provide a proof that such a q and r can always be found. The proof of the theorem depends on the Well-Ordering Axiom and on properties of $<$ with which you are already familiar. Several of the properties used are proved in Theorem 1.1.1.

Proof. Let $S = \{x - yn \mid$ there is a $y \in \mathbf{Z}$ for which $x - yn \geq 0\}$. We are going to show that S has a smallest element and then use this element of S to find q and r. If $x \geq 0$, then $x - 0n \in S$, and if $x < 0$, then $-x > 0$. Thus, $(-x)n \geq -x$ since $n \geq 1$. Hence, $x - xn \geq 0$, so $x - xn \in S$. Consequently, S is nonempty. If $0 \in S$, then 0 is clearly the smallest element of S. If $0 \notin S$, then $S \subseteq \mathbf{N}$, so S contains a smallest element by the Well-Ordering Axiom for $\mathbf{N}$. In any case, S contains a smallest element. If r is the smallest element of S, then $r \geq 0$ and r must be of the form $x - qn$ for some $q \in \mathbf{Z}$. Suppose $q \in \mathbf{Z}$ is such that $x - qn = r$. Then $x = qn + r$ with $r \geq 0$.

Since $x = qn + r$ with $r \geq 0$, the proof will be finished if we can show that $r < n$ and that q and r are unique. Let's see what happens if r is not less than n. If $r \geq n$, then $r - n \geq 0$. But $r - n = (x - qn) - n = x - (q + 1)n$. Hence, $x - (q + 1)n \geq 0$, so $x - (q + 1)n \in S$. If $r_1 = x - (q + 1)n$, then $r_1 = x - (q + 1)n < x - qn = r$, which would contradict the fact that r is the smallest element of S. Therefore, it cannot be the case that $r \geq n$ and thus $r < n$.

Finally, we need to show that q and r are unique. This can be accomplished by assuming that q_1 and r_1 are integers such that $x = q_1 n + r_1$, $0 \leq r_1 < n$, and then showing that $q_1 = q$ and $r_1 = r$. If q_1 and r_1 are such integers, either $r_1 \geq r$ or $r \geq r_1$. Assume that $r_1 \geq r$. Then $0 = x - x = (q_1 n + r_1) - (qn + r) = (q_1 - q)n + (r_1 - r)$ and thus $(q - q_1)n = r_1 - r \geq 0$. Hence, $(q - q_1)n \geq 0$. But $0 \leq r \leq r_1 < n$, so $0 \leq r_1 - r < n - r \leq n$, which indicates that $r_1 - r$ is strictly less than n. Thus, $(q - q_1)n < n$. Since a positive integral multiple of $n > 0$ cannot be strictly less than n, the only possibility is that $q - q_1 = 0$. Therefore, $q_1 = q$, which in turn yields $r_1 = r$. A similar proof holds if $r \geq r_1$, with the roles of r and r_1 interchanged. ∎

In the Division Algorithm if the remainder $r = 0$, then $x = qn$ and the integer x **is a multiple of** n or n **is a factor of** x. This is the same as saying that x **is divisible by** n or that n **divides** x. We write $n \mid x$ to indicate that n divides x, while $n \nmid x$ indicates that n does not divide x.

1.1.4 | **Example**

Since $12 = 3 \cdot 4$, we see that $4 \mid 12$. However, $6 \nmid 15$ since an integer k does not exist such that $15 = 6k$. A divisor can also be negative. For example, $-5 \mid 35$ because $35 = -7(-5)$.

| 1.1.5 | **Definition** |

If x and y are integers, at least one of which is not zero, then an integer d is the **greatest common divisor** of x and y, provided that the following conditions are satisfied:

1. $d > 0$.
2. $d \mid x$ and $d \mid y$.
3. If $c \mid x$ and $c \mid y$, then $c \mid d$.

The greatest common divisor of x and y is denoted by $\gcd(x, y)$. Two integers x and y are said to be **relatively prime** if $\gcd(x, y) = 1$. An expression of the form $ax + by$ is said to be a **linear combination** of x and y.

The greatest common divisor of x and y is unique because if d_1 and d_2 are greatest common divisors of x and y, then $d_1 \mid d_2$ and $d_2 \mid d_1$. Hence, $d_1 = k_2 d_2$ and $d_2 = k_1 d_1$, where k_1 and k_2 are integers. From this it follows that $d_1 = k_2 k_1 d_1$, which implies that $k_1 = k_2 = 1$ and thus $d_1 = d_2$.

| 1.1.6 | **Example** |

It is obvious that $\gcd(45, -30) = 15$, while $\gcd(-12, 0) = 12$. The integers 12 and 47 are relatively prime since $\gcd(12, 47) = 1$.

When $y = 0$ and $x \neq 0$, $\gcd(x, y) = |x|$. Hence, to show that the greatest common divisor of two integers x and y, at least one of which is nonzero, always exists, we need to consider the case where x and y are both nonzero.

| 1.1.7 | **Theorem** |

If x and y are nonzero integers, then $d = \gcd(x, y)$ exists and there are integers a and b such that $d = ax + by$. Furthermore, d is the smallest positive integer that can be written in the form $ax + by$.

Proof. Form the set $S = \{ax + by \mid a, b \in \mathbf{Z}\}$ and let S^+ be the set of all positive integers in S. By the Well-Ordering Axiom, S^+ will have a smallest element provided that it is nonempty. Since $x = 1x + 0y$ and $-x = -1x + 0y$, both x and $-x$ are in S. But $x > 0$ or $-x > 0$, so either x or $-x$ is an element of S^+. Thus, $S^+ \neq \varnothing$.

Let $d = ax + by$ be the smallest element of S^+. We claim that d is the greatest common divisor of x and y. Now $d > 0$ since $d \in S^+$, so the first condition of Definition 1.1.5 is satisfied. By the Division Algorithm 1.1.3 there are integers q and r such that $x = qd + r$, where $0 \le r < d$. If $r \ne 0$, then $r = x - qd = x - q(ax + by) = (1 - qa)x + (-qb)y$ and thus $r \in S^+$. But d is the smallest element in S^+ and $0 \le r < d$, so it must be the case that $r = 0$. Therefore, $x = qd$ and thus $d \mid x$. In a similar manner it can be shown that $d \mid y$, so the second condition of Definition 1.1.5 is satisfied.

Next, suppose that $c \mid x$ and $c \mid y$. Then $x = k_1 c$ and $y = k_2 c$, where k_1 and k_2 are integers, so $d = ax + by = ak_1 c + bk_2 c = (ak_1 + bk_2)c$. Therefore, $c \mid d$, and we have shown that $d = \gcd(x, y)$. Furthermore, d is the smallest positive integer that can be written in the form $ax + by$, where a and b are integers. This follows since any positive integer that can be written as a linear combination of x and y is an element of S^+ and d was chosen to be the smallest element of S^+. ■

It is important to observe that the integers a and b of Theorem 1.1.7 are not unique. This follows because if $d = ax + by$, the integers $a + y$ and $b - x$ are also such that $d = (a + y)x + (b - x)y$.

1.1.8   Corollary

If d is the greatest common divisor of x and y and $x = dm$ and $y = dn$, then $\gcd(m, n) = 1$.

Proof. If d is the greatest common divisor of x and y, then by Theorem 1.1.7 there are integers a and b such that $d = ax + by$. Hence, $d = dma + dnb = d(ma + nb)$, so $1 = ma + nb$. Since 1 is the smallest positive integer that can be written as a linear combination of m and n, it follows from Theorem 1.1.7 that $\gcd(m, n) = 1$. ■

Part 3 of Definition 1.1.5 indicates that if c is a divisor of both x and y, then c is a divisor of $d = \gcd(x, y)$. Hence, $c \le d$, so we are justified in calling d the greatest common divisor of x and y. Theorem 1.1.7 indicates that the greatest common divisor d of two nonzero integers x and y can be written as a linear combination $ax + by$ of x and y, where a and b are integers. This also holds when $y = 0$ and $x \ne 0$. In this case $d = |x|$, so if $x > 0$, $x = 1x + 0y$, and if $x < 0$, $-x = -1x + 0y$. Hence, integers a and b can always be found such that $d = ax + by$. However, Theorem 1.1.7 does not provide a method for finding a and b or for finding $\gcd(x, y)$. One could inspect $S = \{ax + by \mid a, b \in \mathbf{Z}\}$ and try to pick the smallest positive element from this set, but this could prove to be unsatisfactory when the integers x and y are large. In this situation one may have to try many values for a and b before the appropriate integers can be found. The Division Algorithm 1.1.3 can be used to develop a systematic procedure for finding $\gcd(x, y)$. This method, first established by **Euclid of Alexandria** (325 B.C.–265 B.C.), is known as the **Euclidean Algorithm**. As we shall see, the development of a systematic method for finding $\gcd(x,y)$ also produces a systematic method for finding a and b.

The validity of the Euclidean Algorithm depends on the following lemma. (A lemma is a mathematical statement that is used to prove a theorem. Of course, a lemma could be proved within the proof of the theorem itself, but this usually constitutes a diversion from the proof of the theorem.)

1.1.9   **Lemma**

If x, y, q, and r are integers such that $x = qy + r$, then $\gcd(x, y) = \gcd(y, r)$.

Proof. Let A be the set of all common divisors of x and y and B the set of all common divisors of y and r. If $c \in A$, then $c \mid x$ and $c \mid y$, so $x = j_1 c$ and $y = j_2 c$, where j_1 and j_2 are integers. It follows from $x = qy + r$ that $r = (j_1 - qj_2)c$, and so $c \mid r$. Thus, $c \in B$, so $A \subseteq B$. Conversely, let $c \in B$. Then there are integers k_1 and k_2 such that $y = k_1 c$ and $r = k_2 c$. Since $x = qy + r$, $x = (qk_1 + k_2)c$, so $c \mid x$. Hence, $c \in A$, so $B \subseteq A$. Thus, $A = B$, and this proves the lemma since the largest positive integer in A must now coincide with the largest positive integer in B. ∎

1.1.10   **Euclidean Algorithm**

Let x and y be positive integers with $y \leq x$. If $y \mid x$, then $\gcd(x, y) = y$. If $y \nmid x$, apply the Division Algorithm repeatedly as follows:

$$x = q_0 y + r_0 \qquad 0 < r_0 < y$$
$$y = q_1 r_0 + r_1 \qquad 0 < r_1 < r_0$$
$$r_0 = q_2 r_1 + r_2 \qquad 0 < r_2 < r_1$$
$$r_1 = q_3 r_2 + r_3 \qquad 0 < r_3 < r_2$$
$$r_2 = q_4 r_3 + r_4 \qquad 0 < r_4 < r_3$$
$$\vdots \qquad\qquad \vdots$$

Then $r_0 > r_1 > r_2 > \cdots$ is a decreasing sequence of nonnegative integers, and after a finite number of steps a remainder of 0 will be obtained; that is, there is a positive integer k such that

$$\vdots \qquad\qquad\qquad \vdots$$

$$r_{k-2} = q_k r_{k-1} + r_k \qquad 0 < r_k < r_{k-1}$$

$$r_{k-1} = q_{k+1} r_k + 0 \qquad r_{k+1} = 0$$

The last nonzero remainder r_k is the greatest common divisor of x and y.

Proof. Lemma 1.1.9 shows that $\gcd(x, y) = \gcd(y, r_0) = \gcd(r_0, r_1) = \gcd(r_1, r_2)$ $= \cdots = \gcd(r_{k-1}, r_k) = \gcd(r_k, 0) = r_k.$ ∎

1.1.11 | **Example**

Consider the two integers 1224 and 330. Repeated application of the Division Algorithm produces

$$1224 = 3 \cdot 330 + 234$$
$$330 = 1 \cdot 234 + 96$$
$$234 = 2 \cdot 96 + 42$$
$$96 = 2 \cdot 42 + 12$$
$$42 = 3 \cdot 12 + 6$$
$$12 = 2 \cdot 6 + 0$$

The last nonzero remainder is 6, so by the Euclidean Algorithm, $\gcd(1224, 330) = 6$. The preceding information can be used to find integers a and b such that $6 = a(1224) + b(234)$. To find a and b, first solve all but the last of the preceding equations for the remainder and reverse the order of the list:

$$6 = 42 - 3 \cdot 12 \qquad\qquad (1)$$
$$12 = 96 - 2 \cdot 42 \qquad\qquad (2)$$
$$42 = 234 - 2 \cdot 96 \qquad\qquad (3)$$
$$96 = 330 - 1 \cdot 234 \qquad\qquad (4)$$
$$234 = 1224 - 3 \cdot 330 \qquad\qquad (5)$$

Now substitute the right-hand side of equation (2) for **12** in equation (1).

$$6 = 42 - 3(96 - 2 \cdot 42) \quad \text{or}$$

$$6 = 7 \cdot 42 - 3 \cdot 96. \tag{6}$$

Next, substitute the right-hand side of equation (3) for **42** in equation (6).

$$6 = 7(234 - 2 \cdot 96) - 3 \cdot 96 \quad \text{or}$$

$$6 = 7 \cdot 234 - 17 \cdot 96. \tag{7}$$

Continuing, we substitute the right-hand side of equation (4) for **96** in equation (7).

$$6 = 7 \cdot 234 - 17(330 - 1 \cdot 234) \quad \text{or}$$

$$6 = 24 \cdot 234 - 17 \cdot 330. \tag{8}$$

Finally, substitute the right-hand side of equation (5) for **234** in equation (8).

$$6 = 24(1224 - 3 \cdot 330) - 17 \cdot 330 \quad \text{or}$$

$$6 = 24 \cdot 1224 + (-89)330.$$

Hence, $a = 24$ and $b = -89$. When there is a need, this process is referred to as **back-substitution**.

1.1.12   Definition

An integer p is **prime** if $p > 1$ and the only divisors of p are ± 1 and $\pm p$. An integer $x > 1$ that is not prime is said to be a **composite integer**.

1.1.13   Lemma

Every composite integer can be written as a product of primes.

Proof. Let S be the set of all composite integers that cannot be written as a product of primes. If $S = \varnothing$, then there is nothing to prove. Suppose $S \neq \varnothing$. Then, by the Well-Ordering Axiom, S has a smallest element, say x. Since $x \in S$, x is not prime, so x has divisors other than ± 1 and $\pm x$. Suppose $x = ab$, where $1 < a < x$ and $1 < b < x$. Then neither a nor b is in S since x is the smallest element of S. Therefore, a and b both can be written as a product of primes. Let $p_1, p_2, \ldots, p_m$ and $q_1, q_2, \ldots, q_n$ be primes such that $a = p_1 p_2 \cdots p_m$ and $b = q_1 q_2 \cdots q_n$, where it may be the case that $m = 1$ or $n = 1$. Hence, $x = ab = p_1 p_2 \cdots p_m q_1 q_2 \cdots q_n$, so x can be written as a product of primes, and this is a contradiction. Therefore, $S = \varnothing$, and thus every composite integer can be written as a product of primes. ∎

You may have noticed from the proof above that if, say, $m = 1$, then $a = p_1$, so a is the "product" of a single prime. If we extend the definition of a product to allow products with a single factor, Lemma 1.1.13 can be restated to indicate that every integer greater than 1 can be written as a product of primes.

Are there only a finite number of primes? From what you know about the integers, you will probably say "no," but such a conjecture requires a proof. The fact that there are infinitely many primes was first proved by Euclid.

1.1.14   Theorem

There are infinitely many primes.

Proof. Suppose there are only a finite number of primes, p_1, p_2, ..., p_n, and consider $x = p_1 p_2 \cdots p_n + 1$. Now x is either prime or it is not. If x is prime, then x is a new prime since it is greater than any of the primes p_1, p_2, ..., p_n. Hence, there are $n + 1$ primes. But it was assumed that there are only n primes, so x cannot be prime. Since x is not prime, Lemma 1.1.13 shows that x can be written as a product of some of the p_i's. Hence, at least one of the p_i's must divide x. If $p_k \mid x$, then p_k will also divide $x - p_1 p_2 \cdots p_n = 1$, which is an obvious contradiction since $p_k \nmid 1$. Therefore, there must be infinitely many primes. ∎

1.1.15   Lemma

Let x, y, and z be integers. If x and y are relatively prime and $x \mid yz$, then $x \mid z$.

Proof. Since $\gcd(x, y) = 1$ implies that there exist integers a and b such that $1 = ax + by$, we see that $z = axz + byz$. Since $x \mid yz$, there is an integer k such that $yz = kx$, so $z = axz + bkx = (az + bk)x$. Thus, $x \mid z$. ∎

1.1.16   Corollary

If x and y are integers and p is a prime such that $p \mid xy$, then $p \mid x$ or $p \mid y$.

Proof. If $p \mid xy$ and $p \nmid x$, then $\gcd(p, x) = 1$ since p is prime. Hence, by Lemma 1.1.15, $p \mid y$. If $p \mid xy$ and $p \nmid y$, a similar argument shows that $p \mid x$. ∎

We have just seen that every positive integer $x > 1$ can be written as a product of primes. However, such a factorization is not unique in the sense that the order of the factors may be different. For example, 30 can be factored as $2 \cdot 3 \cdot 5$ and as $5 \cdot 3 \cdot 2$. However, the order of the factors in $5 \cdot 3 \cdot 2$ can be rearranged so that this factorization becomes $2 \cdot 3 \cdot 5$. Saying that the factorization of an integer x is **unique up to the order of the factors** means that if $x = p_1 p_2 \cdots p_m$ and $x = q_1 q_2 \cdots q_n$ are two factorizations of x into products of primes, then $m = n$, and after a suitable reindexing of the q_i's, $p_1 = q_1, p_2 = q_2, \ldots, p_n = q_n$. The following theorem is part of almost everyone's basic knowledge of everyday mathematics. It was first proved by Euclid.

1.1.17 Fundamental Theorem of Arithmetic

Every integer $x > 1$ can be written as a product of primes. Moreover, this factorization of x is unique up to the order of the factors.

Proof. If x is a prime, there is nothing to prove, so suppose that x is a composite integer. The fact that every composite integer can be written as a product of primes is Lemma 1.1.13. It remains only to show uniqueness up to the order of the factors. Suppose that $x = p_1 p_2 \cdots p_m$ and $x = q_1 q_2 \cdots q_n$ are factorizations of x into products of primes. Suppose also that $m < n$. Since $p_1 p_2 \cdots p_m = q_1 q_2 \cdots q_n$, $p_1 \mid q_1 q_2 \cdots q_n$. By Corollary 1.1.16, $p_1 \mid q_1$ or $p_1 \mid q_2 \cdots q_n$. If $p_1 \mid q_1$, then $p_1 = q_1$. If $p_1 \nmid q_1$, then $p_1 \mid q_2 q_3 \cdots q_n$, so $p_1 \mid q_2$ or $p_1 \mid (q_3 \cdots q_n)$. If $p_1 \mid q_2$, then $p_1 = q_2$. If $p_1 \nmid q_2$, then $p_1 \mid q_3 \cdots q_n$. If this procedure is continued, then $p_1 = q_i$ for some i, $1 \leq i \leq n$, since there are only a finite number of q_i's. Reindex the q_i's so that $p_1 = q_1 = q_i$ and rearrange the product of the q_i's so that the "new" $q_1 = p_1$ is the leading factor in the product. Next divide both sides of $p_1 p_2 \cdots p_m = q_1 q_2 \cdots q_n$ by p_1 and apply the same procedure again to $p_2 \cdots p_m = q_2 \cdots q_n$. It follows as before that $p_2 = q_k$ for some k, $2 \leq k \leq n$. Reindex again so that $p_2 = q_2 = q_k$, rearrange and divide both sides of $p_2 \cdots p_m = q_2 \cdots q_n$ by p_2. This gives $p_3 \cdots p_m = q_3 \cdots q_n$. Repeating this procedure m times produces $1 = q_{m+1} \cdots q_n$. Since each of $q_{m+1}, \ldots, q_n$ is prime, $q_{m+1} > 1, \ldots, q_n > 1$, so $1 = q_{m+1} \cdots q_n$ is a contradiction. Hence, it cannot be the case that $m < n$, so $m \geq n$. If $n < m$, a repetition of the preceding proof with the roles of the p_i's and q_i's reversed shows that $n \geq m$. Therefore, $m = n$ and $p_1 = q_1, p_2 = q_2, \ldots, p_n = q_n$ after a suitable reindexing of the q_i's. ∎

1.1.18 Example

$660 = 2 \cdot 2 \cdot 3 \cdot 5 \cdot 11 = 11 \cdot 2 \cdot 3 \cdot 2 \cdot 5$. This factorization of 660 into a product of primes is unique except for the order of the prime factors.

Theorem 1.1.17 shows that any composite integer can be written as $x = p_1^{n_1} p_2^{n_2} \cdots p_k^{n_k}$, where $p_1, p_2, \ldots, p_k$ are the unique prime factors of x and $n_1, n_2, \ldots, n_k$ are positive integers that represent the **multiplicity** of each unique prime factor occurring in the factorization of x. For example, the integer $5500 = 2 \cdot 2 \cdot 5 \cdot 5 \cdot 5 \cdot 11$ can be written as $2^2 \cdot 5^3 \cdot 11$. We see that 2 has multiplicity 2, 5 has multiplicity 3, and 11 has multiplicity 1. Expressing two integers x and y as a product of primes according to the multiplicity of the primes gives another method of finding $\gcd(x, y)$. Consider $38500 = 2^2 \cdot 5^3 \cdot 7 \cdot 11$ and $9800 = 2^3 \cdot 5^2 \cdot 7^2$. The prime factors that are common to both factorizations are 2, 5, and 7. Form a product from these common factors, each raised to the lowest power to which it appears in the factorizations of 38,500 and 9800. A little thought shows that $\gcd(38500, 9800) = 2^2 \cdot 5^2 \cdot 7 = 700$.

The fact that the factorization of a composite integer into a product of primes is unique up to rearrangement of prime factors may seem to be something that "everybody knows" and consequently not in need of proof. There are systems, however, where unique factorization into a product of primes in that system does *not* hold. We conclude this section with two examples of such systems.

1.1.19   Examples

1. Consider the set $2\mathbf{Z}$ of even integers, and suppose that we agree to work only with this set of numbers. Define an even integer to be prime in $2\mathbf{Z}$ if it cannot be factored into a product of even integers. For example, the only factorization of 6 is $2 \cdot 3$, but $3 \notin 2\mathbf{Z}$, so 6 is prime in $2\mathbf{Z}$. The integer 8 is composite in $2\mathbf{Z}$ since $8 = 2 \cdot 4$, while 10 is prime in $2\mathbf{Z}$. Each even integer 2, 6, 10, 14, 18, 22, 26, 30, … is prime in $2\mathbf{Z}$. Since $60 = 6 \cdot 10 = 2 \cdot 30$, we see that 60 has two distinct factorizations as a product of primes in $2\mathbf{Z}$. Thus, factorization of an integer in $2\mathbf{Z}$ into a product of primes in $2\mathbf{Z}$ need not be unique.

2. Let $\mathbf{Z}[\sqrt{5}i] = \{x + y\sqrt{5}i \,|\, x, y \in \mathbf{Z}\}$, where $i^2 = -1$, and define addition and multiplication of elements of $\mathbf{Z}[\sqrt{5}i]$ as

$$(x_1 + y_1\sqrt{5}i) + (x_2 + y_2\sqrt{5}i) = (x_1 + x_2) + (y_1 + y_2)\sqrt{5}i \qquad \text{and}$$

$$(x_1 + y_1\sqrt{5}i)(x_2 + y_2\sqrt{5}i) = (x_1x_2 - 5y_1y_2) + (x_1y_2 + x_2y_1)\sqrt{5}i.$$

Every element x of $\mathbf{Z}$ can be written as $x = x + 0(\sqrt{5}i)$, so $\mathbf{Z}$ is a subset of $\mathbf{Z}[\sqrt{5}i]$. Furthermore, the restriction of addition and multiplication on $\mathbf{Z}[\sqrt{5}i]$ to $\mathbf{Z}$ agrees with the usual operations of addition and multiplication of integers. Call an element of $\mathbf{Z}[\sqrt{5}i]$ **irreducible** if it *cannot* be factored nontrivially in $\mathbf{Z}[\sqrt{5}i]$. It can be shown that $1 + \sqrt{5}i$, $1 - \sqrt{5}i$, 2, and 3 are irreducible in $\mathbf{Z}[\sqrt{5}i]$. Since

$$(1 + \sqrt{5}i)(1 - \sqrt{5}i) = 6 = 2 \cdot 3,$$

6 has two different factorizations in $\mathbf{Z}[\sqrt{5}i]$ as a product of irreducible elements. Thus, factorization of an element of $\mathbf{Z}[\sqrt{5}i]$ into a product of irreducible elements need not be unique. A mathematical structure in which every element can be factored into a product of irreducible elements that are in some sense unique is called a **Unique Factorization Domain**. As we have just seen, $\mathbf{Z}[\sqrt{5}i]$ is *not* a Unique Factorization Domain. Unique Factorization Domains are investigated in a subsequent chapter, where these comments are made much more precise.

HISTORICAL NOTES

Euclid of Alexandria (325 B.C.–265 B.C.). Little is known of the life of Euclid, except that he was a teacher of mathematics, but he is one of the most important mathematicians of ancient Greece. Euclid was the head of a group of mathematicians who continued to write books under his name even after his death. He is best known for his work titled *Elements,* which contains 13 sections called books. *Elements* set the standard for the study of geometry for over 2000 years, and it is the oldest surviving work in mathematics that develops mathematics through rigorous procedures. Much of the book deals with geometry, and it is Euclid's geometry that is most often studied in the secondary schools today. *Elements* has been translated into more languages and has appeared in more editions than any book in history with the possible exception of the Bible. *Elements* was a prototype for how "pure" mathematics should be written. It presents a well-thought-out set of axioms, precise definitions, and carefully worded theorems with logically constructed proofs. Euclid proved the Fundamental Theorem of Arithmetic, developed the Euclidean Algorithm, and proved that there are infinitely many primes. He also showed that if x and y are integers and p is a prime such that $p \mid xy$, then $p \mid x$ or $p \mid y$. This is sometimes referred to as *Euclid's lemma.*

Problem Set 1.1

In this set of exercises all variables represent integers.

1. Consider the set $\mathbf{Z}$ together with the ordering $>$ on $\mathbf{Z}$ induced by the set $\mathbf{N}$ of positive integers. Revisit the proof of Theorem 1.1.1, and then use similar techniques to prove each of the following:

    (a) If $x, y \in \mathbf{N}$, then $x + y > 0$.

    (b) If $x \in \mathbf{Z}$, then one and only one of the following holds:
$$x > 0, \qquad x = 0, \qquad -x > 0.$$

    (c) If $x > y$ and $z < 0$, then $xz < yz$.

    (d) If $x > y$ and $y > z$, then $x > z$.

    (e) If $x \in \mathbf{Z}$ and $x \neq 0$, then $x^2 > 0$.

    (f) If $x, y, z, w \in \mathbf{Z}$, $x > y > 0$, and $z > w > 0$, then $xz > yw$.

2. Find the smallest integer in each of the following sets.

    (a) $S^+ = \{x \mid x > 0 \text{ and } x = a4 + b10;\ a, b \in \mathbf{Z}\}$

    (b) $S^+ = \{x \mid x > 0 \text{ and } x = a12 + b30;\ a, b \in \mathbf{Z}\}$

3. Find an example in which $n \mid xy$, but $n \nmid x$ and $n \nmid y$.

4. If $x, y, z \in$ and $x \neq 0$, prove each of the following:

    (a) If $x \mid y$, then $x \mid ay$ for all $a \in \mathbf{Z}$.

    (b) If $x \mid y$ and $x \mid z$, prove that $x \mid (ay + bz)$ for all $a, b \in \mathbf{Z}$.

    (c) If $y \neq 0$, $x \mid y$, and $y \mid z$, then $x \mid z$.

    (d) If $xz \neq 0$ and $xz \mid yz$, then $x \mid y$.

5. Determine which, if any, of the following statements are true. If the statement holds for all integers, give a proof. If the statement is false, find specific integers for which the statement does not hold.

    (a) If p is a prime and $p \mid x$ and $p \mid (x - y)$, then $p \mid y$.

    (b) If p is a prime and $p \mid x$ and $p \mid (x^2 + y^2)$, then $p \mid y$.

    (c) If p is a prime, $p \mid x$, $p \mid (x^2 + y^2)$, and $p \mid (y^2 + z^2)$, then $p \mid (x^2 + z^2)$.

    (d) If p is a prime and $p \mid (x^2 - y^2)$, then $p \mid x$ and $p \mid y$.

6. Prove that the square of an odd integer leaves a remainder of 1 when divided by an integral multiple of 4.

7. Compute $\gcd(x, y)$ for each of the following pairs of integers and use back-substitution to find integers a and b such that $\gcd(x, y) = ax + by$.

    (a) $x = 12{,}345 \qquad y = 5040$

    (b) $x = 85{,}672 \qquad y = 242{,}040$

8. The Euclidean Algorithm provides a method for finding $\gcd(x, y)$ when x and y are positive integers. Show how this method can be adapted to find $\gcd(x, y)$ for any integers x and y, at least one of which is nonzero. Use this method to find $\gcd(x, y)$ for each of the following pairs of integers.

    (a) $x = 756 \qquad y = -552$

    (b) $x = -5088 \qquad y = -1156$

9. Prove that $\gcd(xy, z) = 1$ if and only if $\gcd(x, z) = 1$ and $\gcd(y, z) = 1$.

10. If $x \mid z$, $y \mid z$, and $\gcd(x, y) = 1$, prove that $xy \mid z$.

11. If $n > 0$, prove that $\gcd(nx, ny) = n \cdot \gcd(x, y)$.

12. If x, y, and z are integers such that $x \mid yz$, prove that $x \mid \gcd(x, y)z$.

13. Prove that if the square of an integer is even, then the integer must be even. Hint: Assume that the square of an even integer is odd and arrive at a contradiction.

14. The **least common multiple** of $x \neq 0$ and $y \neq 0$, denoted by $\mathrm{lcm}(x, y)$, is an integer $m > 0$ such that

    (i) $x \mid m$ and $y \mid m$ and
    (ii) if $x \mid c$ and $y \mid c$, then $m \mid c$.
    (a) Prove that a unique $\mathrm{lcm}(x, y)$ exists when $x \neq 0$ and $y \neq 0$.
    (b) If x and y are positive integers, prove that $xy = \gcd(x, y)\,\mathrm{lcm}(x, y)$.

15. Let $\leq$ be the ordering induced on **Z** by the set **N** of positive integers. For any integer x define the **absolute value** of x, denoted $|x|$, by

    $$|x| = \begin{cases} x & \text{if } x \geq 0 \\ -x & \text{if } x < 0 \end{cases}$$

    Prove each of the following:
    (a) $|xy| = |x||y|$
    (b) $-|x| \leq x \leq |x|$
    (c) $|x + y| \leq |x| + |y|$

16. Consider Example 2 of 1.1.19, and define $N(x + y\sqrt{5}i) = x^2 + 5y^2$ for all $x + y\sqrt{5}i \in \mathbf{Z}[\sqrt{5}i]$. If $z_1 = x_1 + y_1\sqrt{5}i$ and $z_2 = x_2 + y_2\sqrt{5}i$, prove that

    (a) $N(z_1 z_2)\,N(z_1)n(z_2)$
    (b) $N(z_1) = 0$ if and only if $z_1 = 0$

17. Use the axiom that **N** is well-ordered to show that for any integer k, the set $\mathbf{N}_k = \{k, k + 1, k + 2, \dots\}$ is well-ordered.

Use a computer algebra system to solve each of the following problems.

18. Factor each of the following into a product of primes.
    (a) 10,313,509,178,944,125
    (b) 16,400,280,126,195,483,918,878,231,607,145,807,700,521,875

19. Find the greatest common divisor of each of the following sets of integers. The greatest common divisor of a finite set of integers can be defined by making the necessary adjustments to Definition 1.1.5. Can you write such a definition?

    (a) 25,875   4,271,484,375
    (b) 2500   8,550,000   992,266,372,125,000
    (c) 2,669,677,734,375   43,439,888,521,963,583,647,921

20. Find the least common multiple of each of the following sets of integers. The least common multiple of a finite set of integers can be defined by making the necessary changes in the definition given in Exercise 14. Can you write such a definition?

    (a) 1944   1296
    (b) 56,924,208   107,811   171,802,697

21. If $x = 69,486,440,625$ and $y = -58,368,610,125$, show that $\gcd(x, y)\operatorname{lcm}(x,y) = |xy|$.

1.2   MATHEMATICAL INDUCTION

Section Overview. Proof by mathematical induction is a method of proof that should be in every student's mathematical tool bag. Section 1.2 presents this method of proof. Proof by induction occurs many times throughout the text, so make sure you master this technique.

Suppose that S is a subset of **N** that has the following two properties: (1) $1 \in S$ and (2) if $k \in S$, then $k + 1 \in S$. Since $1 \in S$ if the second condition is applied to 1, then $1 + 1 = 2 \in S$. If the second condition is now applied to 2, then $2 + 1 = 3 \in S$. If the second condition is applied to 3, then $4 \in S$, and so on. These observations lead one toward the conclusion that S contains every positive integer. This argument is not a proof that $S = \mathbf{N}$, but hopefully it will help convince you that any subset S of **N** that satisfies conditions (1) and (2) must actually contain every positive integer and consequently is equal to **N**. A proof of this fact is actually a consequence of the Well-Ordering Axiom for the positive integers. This axiom allows us to prove the **First Principle of (Mathematical) Induction**. A second form of induction is addressed in the exercises.

1.2.1   First Principle of Induction

Let S be a subset of $\mathbf{N}$ that satisfies the following two conditions:

1. $1 \in S$.
2. If $k \in S$, then $k + 1 \in S$.

Then $S = \mathbf{N}$.

Proof. Let S be a subset of $\mathbf{N}$ that satisfies the two conditions, and suppose $S \neq \mathbf{N}$. Then $S \neq \varnothing$ since by condition 1, $1 \in S$. Now $\mathbf{N} \setminus S$ is a nonempty subset of $\mathbf{N}$, so by the Well-Ordering Axiom $\mathbf{N} \setminus S$ has a smallest element, say k. Since k is the smallest element of $\mathbf{N}$ that is not in S, it must be the case that $k - 1 \in S$. But S satisfies condition 2, so $k = (k - 1) + 1$ is in S. Therefore, $k \in S \cap (\mathbf{N} \setminus S) = \varnothing$, and this is clearly a contradiction. Since the assumption that $S \neq \mathbf{N}$ produces a contradiction, it must be the case that $S = \mathbf{N}$. ∎

The fact that the First Principle of Induction holds for the set $\mathbf{N}$ is a fundamental property of the set of positive integers. We could have assumed the First Principle of Induction as an axiom and then proved that the set $\mathbf{N}$ is well-ordered. Actually, the First Principle of Induction and the Well-Ordering Axiom are equivalent in the sense that each implies the other.

The First Principle of Induction provides a very powerful method of proof. The power of this method results from the fact that a mathematical statement can be shown to be true for an infinite set of integers by utilizing a proof that essentially contains only two steps.

To simplify terminology, the First Principle of Induction will be referred to simply as induction. Proof by induction is demonstrated in the following example.

1.2.2   Example

Prove that for every integer $n \in \mathbf{N}$,

$$\sum_{i=1}^{n} i = \frac{n(n+1)}{2}$$

Proof. Let S be the set of positive integers for which the formula is true. That is, S is the set of all positive integers k for which equality holds in the formula when n is replaced by k. The formula will hold for all positive integers if we can show that $S = \mathbf{N}$. Induction provides the tool for showing that this is the case. Since

$$\sum_{i=1}^{1} i = 1 = \frac{1(1+1)}{2}$$

$1 \in S$. Now suppose that $k \in S$. Then by the definition of S,

$$\sum_{i=1}^{k} i = \frac{k(k+1)}{2}.$$

Adding k + 1 to both sides of this equation gives

$$\left(\sum_{i=1}^{k} i \right) + (k+1) = \frac{k(k+1)}{2} + (k+1).$$

The left-hand side of this last equation produces

$$\sum_{i=1}^{k+1} i = \left(\sum_{i=1}^{k} i \right) + (k+1)$$

and the right side can be summed as

$$\frac{k(k+1)}{2} + (k+1) = \frac{k(k+1)}{2} + \frac{2(k+1)}{2} = \frac{(k+1)(k+2)}{2}.$$

Hence,

$$\sum_{i=1}^{k+1} = \frac{(k+1)(k+2)}{2} = \frac{(k+1)((k+1)+1)}{2}.$$

This shows that the formula holds when $n = k + 1$, so $k + 1 \in S$ whenever $k \in S$. Therefore, by induction, $S = \mathbf{N}$. ∎

It may be the case that a mathematical statement holds only for integers greater than or equal to some fixed positive integer n_0. Often a more general form of induction can be used to prove these statements.

1.2.3   General Form of the First Principle of Induction

For a positive integer n_0, let $\mathbf{N}_{n_0} = \{n_0, n_0 + 1, n_0 + 2, \dots\}$ and suppose that S is a subset of $\mathbf{N}_{n_0}$ that satisfies the following two conditions:

1. $n_0 \in S$.
2. If $k \in S$, then $k + 1 \in S$.

Then $S = \mathbf{N}_{n_0}$.

By letting $n_0 = 1$, we see that the First Principle of Induction is a special case of Theorem 1.2.3. In the General Form of the First Principle of Induction, if $n_0 = 1$, then induction is started at 1; if $n_0 = 2$, induction starts at 2; and if $n_0 = 3$, induction starts at 3, and so on. The proof of Theorem 1.2.3 is very similar to the proof of the First Principle of Induction, so it is left as an exercise.

The use of induction in a proof can be simplified somewhat. If S is the set of positive integers for which a mathematical statement $\mathbf{S}_n$ is true, showing that $1 \in S$ is equivalent to showing that $\mathbf{S}_n$ is true for $n = 1$. Showing that $k \in S$ implies $k + 1 \in S$ is equivalent to assuming that $\mathbf{S}_n$ is true for $n = k$ and, from this, showing that $\mathbf{S}_n$ is true when $n = k + 1$. Hence, to use mathematical induction as a method of proof, one need only complete the following two steps.

Step 1. Show that $\mathbf{S}_n$ is true for $n = 1$.

Step 2. Make the assumption that $\mathbf{S}_n$ is true for an integer $n = k \geq 1$, and from this prove that it is true for $n = k + 1$.

Completing Steps 1 and 2 is sufficient to show that $\mathbf{S}_n$ is true for all positive integers. If we need to show that $\mathbf{S}_n$ is true for all integers greater than or equal to a positive integer n_0, then a rendering of the general form of mathematical induction can be used:

Step 1. Show that $\mathbf{S}_n$ is true for $n = n_0$.

Step 2. Make the assumption that $\mathbf{S}_n$ is true for an integer $n = k \geq n_0$, and from this prove that $\mathbf{S}_n$ is true for $n = k + 1$.

The assumption that a mathematical proposition is true for $n = k$ in the course of proof by induction is called the **induction hypothesis**.

1.2.4   **Example**

Show that for every positive integer n, $4^n - 1$ is divisible by 3.

Proof. Let S_n be the statement that $4^n - 1$ is divisible by 3. Since $4^1 - 1 = 3$ is divisible by 3, S_1 is true. Next, make the induction hypothesis that S_k is true. Then $4^k - 1$ is divisible by 3, so $4(4^k - 1)$ is divisible by 3. If $4(4^k - 1) = 3m$, where m is an integer, then

$$4(4^k - 1) = 3m,$$
$$4^{k+1} - 4 = 3m,$$
$$4^{k+1} - 1 = 3m + 3, \qquad \text{and thus}$$
$$4^{k+1} - 1 = 3(m + 1).$$

Hence, $4^{k+1} - 1$ is divisible by 3, and so S_{k+1} is true. Therefore, S_n is true for all positive integers. In this example, induction started at 1.

1.2.5   **Example**

Show that if n is an integer and $n \geq 4$, then $2^n < n!$.
Proof. Let S_n be the statement that $2^n < n!$. The inequality fails for $n = 1$, 2, and 3, so start induction at 4. Since $16 = 2^4 < 4! = 24$, S_4 is true. Next, make the induction hypothesis that S_k is true where $k \geq 4$. Then $2^k < k!$. But this implies that $2^{k+1} = 2(2^k) < 2k! < (k + 1)k! = (k + 1)!$. Therefore, S_{k+1} is true, so S_n is true for all positive integers greater than or equal to 4. ∎

HISTORICAL NOTES

The use of mathematical induction dates back to approximately 700 A.D. and the Islamic mathematicians of the Middle East. Mathematicians of that period built on the mathematics discovered earlier by the Babylonians and the Greeks. The main contributions of the Islamic mathematicians were in algebra, and these mathematicians used a form of induction to prove mathematical statements for a given value of n. For example, they used a form of induction to show that $1^2 + 2^3 + \cdots + n^3 = (1 + 2 + 3 + \cdots + n)^2$ when $n \leq 10$. The proof given for the formula $1^2 + 2^3 + \cdots + n^3 = (1 + 2 + 3 + \cdots + n)^2$ was first carried out for $n = 10$ and then worked downward to show that it held for $n = 1$. It is generally acknowledged that such proofs demonstrate the earliest use of the two essential steps of induction. However, it was **Blaise Pascal** who first made the process of mathematical induction precise.

Blaise Pascal (1623–1662). Blaise Pascal was a French mathematician whose genius in mathematics showed itself at an early age. Although plagued by poor health and a nervous disposition, Pascal made contributions in many areas of mathematics; his achievements were quite remarkable when one considers that he lived only to the age of 39. At the age of 12, Pascal developed an interest in geometry and discovered for himself that the sum of the angles of a triangle is 180 degrees; he went on to set out many of the propositions of Euclidean geometry on his own. By age 16, he had established many of the properties of projective geometry. His later explorations into mathematical theory helped lay the ground work for probability theory. Pascal is best known today for Pascal's triangle, a triangle of positive integers, each line of which yields the coefficients of the expansion of $(a + b)^n$ for a corresponding value of n. Through the study of the patterns of this triangle, Pascal was the first mathematician to describe mathematical induction precisely.

Problem Set 1.2

Use induction to prove Exercises 1 through 9.

1. For each positive integer n, 4 divides $5^n - 1$.

2. For each positive integer n, 5 divides $8^n - 3^n$.

3. For each positive integer $n \geq 5$, $4^n > n^4$.

4. For each positive integer $n \geq 1$, $\sum_{i=1}^{n} i^2 = \frac{n(n+1)(2n+1)}{6}$.

5. For each positive integer $n \geq 1$, $\sum_{i=1}^{n} i^3 = \left(\sum_{i=1}^{n} i\right)^2$. See Example 1.2.2.

6. For each positive integer $n \geq 1$, $\sum_{i=1}^{n}(3i^2 - 3i + 1) = n^3$.

7. For each positive integer $n \geq 5$, $n^2 < 2^n$.

8. For each positive integer $n \geq 9$, $4^n < n!$.

9. For each positive integer $n \geq 2$, $\sum_{i=1}^{n} \frac{1}{\sqrt{i}} > \sqrt{n}$.

10. The expression $n^2 - n + 41$ yields a prime integer for $n = 1, 2, 3, \ldots, 40$. Show that this is indeed the case for $n = 5, 10, 15,$ and 20. Prove or disprove that this formula produces a prime for each positive integer n.

11. Prove Theorem 1.2.3.

12. Theorem 1.2.3 shows that mathematical induction can be started at any positive integer. This, together with a proof of the following, shows that induction can be started at any $n \in \mathbf{Z}$. If n_0 is an integer less than or equal to 0, let $\mathbf{N}_{n_0} = \{n_0, n_0 + 1, \ldots, 0, 1, 2, \ldots\}$. If S is a subset of $\mathbf{N}_{n_0}$ such that

    (a) $n_0 \in S$ and
    (b) $k \in S$ implies that $k + 1 \in S$,

    prove that $S = \mathbf{N}_{n_0}$.

13. Use induction to show that if S is a finite set with n elements, then the power set $\mathcal{P}(S)$ of S has 2^n elements.

    Exercise 12 shows that induction can be started at any integer. Use this fact to prove Exercises 14 and 15.

14. For all integers $n \geq -5$,

$$\sum_{i=-5}^{n} i = \frac{(n-5)(n+6)}{2}.$$

15. For all integers $n \geq -4$,

$$\sum_{i=-4}^{n} i^2 = \frac{(n+5)(2n^2 - 7n + 36)}{6}.$$

16. If k is a positive integer, let

$$\binom{k}{r} = \frac{k!}{r!(k-r)!},$$

    where r is an integer such that $0 \leq r \leq k$. Prove that

$$\binom{k}{r+1} + \binom{k}{r} = \binom{k+1}{r+1}$$

    for $1 \leq r \leq k$ and that $\binom{k}{r}$ is an integer for each positive integer k.

17. If $x, y \in \mathbf{R}$, prove the **binomial theorem**

$$(x+y)^n = \binom{n}{0}x^n + \binom{n}{1}x^{n-1}y + \binom{n}{2}x^{n-2}y^2 + \cdots$$

$$+ \binom{n}{r}x^{n-r}y^r + \cdots + \binom{n}{n-1}xy^{n-1} + \binom{n}{n}y^n$$

18. The **Second Principle of Mathematical Induction** states that if S is a subset of **N** that satisfies the following two conditions, then $S = $ **N**.
    (i) $1 \in S$.
    (ii) If n is a positive integer and if $k \in S$ for all positive integers $k < n$, then $n \in S$.

    Use the First Principle of Induction to prove the Second Principle.

19. Assume that the Second Principle of Induction holds as an axiom, and prove that the set **N** is well-ordered. Hint: Let S be any subset of **N**, and prove that if S has no smallest element, then S is empty. Then, by the contrapositive of this statement, if S is not empty, then S has a smallest element. To prove that S is empty, use the Second Principle of Induction to prove the statement "If S has no smallest element, then $n \notin S$ for every positive integer n."

20. Explain why the following are equivalent.
    (a) The Well-Ordering Axiom for **N**
    (b) The First Principle of Induction
    (c) The Second Principle of Induction
    See Theorem 1.2.1, Exercise 18, and Exercise 19.

TECHNOLOGY PROBLEMS

Use a computer algebra system to solve each of the following problems.

21. Decide whether or not each of the following holds for the given value(s) of n.
    (a) For $n = 4, 5, 6,$ and 7, $n^2 < 2^n$.
    (b) For $n = 7, 8, 9,$ and 10, $4^n > n!$.
    (c) For $n = 1, 2, 3,$ and 4,

$$\sum_{i=1}^{n} \frac{1}{\sqrt{i}} > \sqrt{n}.$$

    (d) If $n = 1529$, then

$$\sum_{i=1}^{n} i^2 = \frac{n(n-1)(2n-1)}{6}.$$

    (e) If $n = 27{,}561$, then

$$\sum_{i=1}^{n} i^3 = \left[\frac{n(n+1)}{2} \right]^2.$$

(f) If $n = 325$, then

$$\sum_{i=1}^{n} (3i^2 + 3i + 1) = n^3.$$

(g) If $n = 475$, then

$$\sum_{k=-4}^{n} k^2 = \frac{(n+5)(2n^2 - 7n + 36)}{6}.$$

1.3   CONGRUENCE RELATIONS AND MODULAR ARITHMETIC

Section Overview. Linear congruence equations are introduced in this section, and methods for finding solutions of these equations are investigated. The relation modulo n is shown to be a congruence relation, and Theorem 0.4.8 is used to define binary operations of addition and multiplication on the set $\mathbf{Z}_n$ of congruence classes of $\mathbf{Z}$. Properties of addition and multiplication of congruence classes are investigated, and some interesting results emerge.

An equivalence relation E on a set A is a relation on A that is reflexive, symmetric, and transitive. Example 1 of 0.2.2 shows that the relation E defined on $\mathbf{Z}$ by $x \mathrm{E} y$ if and only if $3 \mid (x - y)$ is an equivalence relation on $\mathbf{Z}$. For a positive integer n, we now investigate the relation defined on $\mathbf{Z}$ by the divisibility of $x - y$ by n. It will be shown that this is not only an equivalence relation but also a congruence relation on $\mathbf{Z}$.

1.3.1   Definition

If n is a positive integer, then two integers x and y are said to be **congruent modulo n** if $n \mid (x - y)$. The notation $x \equiv y \pmod{n}$ indicates that the integers x and y are congruent modulo n. The integer n is called the **modulus** of the relation.

1.3.2   Example

Here are a few examples of congruence modulo an integer: $10 \equiv 2 \pmod{4}$ since $4 \mid (10 - 2)$, $-2 \equiv 8 \pmod{5}$ because $5 \mid (-2 - 8)$, and $7 \equiv 7 \pmod{3}$ since $3 \mid (7 - 7)$.

If a, b, and c are integers, certain operations can often be performed on equations such as $ax + b \equiv c \pmod{n}$, and, in fact, under certain conditions such an equation can be solved for x in the integers. Equations such as these are often referred to as **linear congruence equations**. It is also sometimes possible to solve a system of congruence equations such as

$$x \equiv a \pmod{m} \quad \text{and} \quad x \equiv b \pmod{n}.$$

The conditions under which such a set of congruence equations can be solved for x in the integers is known as the **Chinese Remainder Theorem**.

It was **Carl Friedrich Gauss** (1777–1855) who defined the congruence of integers and who established the modern notation $x \equiv y \pmod{n}$. Gauss showed how to solve linear congruences and how to solve the Chinese Remainder Problem.

1.3.3	**Example**

Solutions for simple linear congruence equations can often be found by inspection. For example, $2x \equiv 5 \pmod{7}$ has the integers 6 and -1 as solutions since $7 \mid (2(6) -5)$ and $7 \mid (2(-1) -5)$. Can you find other solutions? You must find integers k such that $2k - 5$ is a multiple of 7.

The set of linear congruence equations $x \equiv 1 \pmod{4}$ and $x \equiv 2 \pmod{3}$ has $x = 5$ as a solution since $4 \mid (5 - 1)$ and $3 \mid (5 - 2)$. Can you find another positive solution? Can you find a negative solution? In the first case, you must find a positive integer k such that $k - 1$ is a multiple of 4 while at the same time $k - 2$ is a multiple of 3. In the second case, you must find a negative integer k with these same properties.

Even though we can sometimes solve a congruence equation by inspection, we would like to go further and, if possible, characterize all solutions of such equations so that we can say "Every solution of the congruence equation looks like ...." When solving equations in the real numbers, we can always add any real number to both sides of the equation, and we can cancel a from $ax = ay$ when $a \neq 0$. Similar operations can sometimes be performed on congruence equations. It is not difficult to show that $x \equiv y \pmod{n}$ if and only if $x + b \equiv y + b \pmod{n}$ for any integer b. It is also the case, and easy to prove, that if $x \equiv y \pmod{n}$, then $ax \equiv ay \pmod{n}$ for any integer a. The converse, however, does not hold. For example, $12(3) \equiv 12(2) \pmod{4}$ since $4 \mid (12(3 - 2))$, yet 3 is not congruent to 2 modulo 4. In general, $ax \equiv ay \pmod{n}$ does not imply $x \equiv y \pmod{n}$ when $n \mid a$. The following lemma clarifies this situation and indicates when a can be canceled from both sides of $ax \equiv ay \pmod{n}$.

1.3.4   Lemma

If x and y are integers such that $x \equiv y \pmod{n}$, then $ax \equiv ay \pmod{n}$ for any integer a. Conversely, if $\gcd(a,n) = 1$ and $ax \equiv ay \pmod{n}$, then $x \equiv y \pmod{n}$.

Proof.   If $x \equiv y \pmod{n}$, then there is an integer k such that $x - y = kn$. Multiplying both sides of this equation by a produces $ax - ay = (ak)n$. Thus, $n \mid (ax - ay)$ and so $ax \equiv ay \pmod{n}$. Conversely, if $\gcd(a,n) = 1$ and $ax \equiv ay \pmod{n}$, then $n \mid a(x - y)$. But since a and n are relatively prime, Lemma 1.1.15 implies that $n \mid (x - y)$. Hence, we have $x \equiv y \pmod{n}$. ∎

Operations That Can Be Performed on a Congruence Equation

1. If $x \equiv y \pmod{n}$, we can add an integer b to both sides of the congruence equation to obtain $x + b \equiv y + b \pmod{n}$.
2. If $x + b \equiv y + b \pmod{n}$, we can cancel b from both sides of the congruence equation, and the result is $x \equiv y \pmod{n}$.
3. If $x \equiv y \pmod{n}$, we can multiply both sides of the congruence equation by any integer a, and this will yield $ax \equiv ay \pmod{n}$.
4. If $ax \equiv ay \pmod{n}$ and $\gcd(a,n) = 1$, then we can cancel a from both sides of the congruence equation to obtain $x \equiv y \pmod{n}$.

The following theorem tells us when we can be assured that an equation of the form $ax + c \equiv d \pmod{n}$ can be solved for x in **Z**. The proof of the theorem also provides a method of finding x. Since $ax + c \equiv d \pmod{n}$ implies that $ax \equiv d - c \pmod{n}$, if we let $b = d - c$, then solving $ax + c \equiv d \pmod{n}$ for x in **Z** amounts to being able to solve $ax \equiv b \pmod{n}$ for x in **Z**. The theorem also characterizes the solutions of $ax \equiv b \pmod{n}$. Before stating and proving the theorem, let's look at an example.

1.3.5   Example

Let's see if we can find all the solutions of the congruence equation $4x \equiv 12 \pmod{10}$. We need to find the integers x such that $10 \mid (4x - 12)$. But if $10 \mid (4x - 12)$, then there is an integer k such that $4x - 12 = 10k$. Now $\gcd(4, 10) = 2$ and $2 \mid 12$, so we can divide both sides of the equation by 2 to obtain $2x - 6 = 5k$. Hence, we are seeking solutions of the equation $2x \equiv 6 \pmod{5}$. Since 2 and 5 are relatively prime, 2 can be canceled from both sides of this equation to obtain $x \equiv 3 \pmod{5}$. Several solutions of this last equation are $\ldots, -7, -2, 3, 8, 13, \ldots$. All the integers in the list are congruent modulo 5, and all are solutions of $4x \equiv 12 \pmod{10}$.

The following theorem is a general statement of what was demonstrated in the preceding example.

1.3.6   Theorem

Let a, b, and $n > 0$ be integers. If $\gcd(a,n) = d$, then $ax \equiv b \pmod{n}$ can be solved for x in the integers if and only if $d \mid b$. Furthermore, if m is an integer such that $n = md$ and y is a solution of $ax \equiv b \pmod{n}$, then y' is also a solution if and only if $y \equiv y' \pmod{m}$.

Proof.   First, we show that if there is a solution of $ax \equiv b \pmod{n}$, then $d \mid b$. Suppose that y is a solution. Then $n \mid (ay - b)$, so there is an integer q such that $ay - b = nq$ and thus $b = ay + n(-q)$. Since $d \mid a$ and $d \mid n$, $d \mid b$.

Conversely, we show that if $d \mid b$, then we can construct a solution. If $d \mid b$, there is an integer q' such that $b = dq'$. Since there are integers s and t such that $d = as + nt$, $dq' = a(sq') + n(tq')$. Hence, $a(sq') - b = n(-tq')$, so $n \mid (a(sq') - b)$. Therefore, $a(sq') \equiv b \pmod{n}$, so $x = sq'$ is a solution in the integers of the equation $ax \equiv b \pmod{n}$.

Finally, we show that if y is a solution of $ax \equiv b \pmod{n}$, then y' is also a solution if and only if $y \equiv y' \pmod{m}$, where m is an integer such that $n = md$. Suppose that y and y' are solutions of $ax \equiv b \pmod{n}$. Then $ay \equiv ay' \pmod{n}$, so $n \mid a(y - y')$. Exercise 12 of Problem Set 1.1 indicates that $n \mid d(y - y')$, so $d(y - y') = kn$ for some integer k. But then $y - y' = km$ and so $m \mid (y - y')$. Therefore, $y \equiv y' \pmod{m}$, and any two solutions of $ax \equiv b \pmod{n}$ are congruent modulo m. Conversely, suppose that y is a solution of $ax \equiv b \pmod{n}$ and that $y \equiv y' \pmod{m}$. Then $y = y' + mk$ for some integer k, so $ay = ay' + amk$. Now y is a solution of $ax \equiv b \pmod{n}$, so there is an integer q such that $ay = b + nq$. Hence, we see that $b + nq = ay' + amk$ or $ay' - b = nq - amk$. Now $d \mid a$, so there is an integer k' such that $a = dk'$. From this it follows that $ay' - b = nq - dmkk' = n(q - kk')$ and thus $n \mid (ay' - b)$. Therefore, y' is a solution of $ax \equiv b \pmod{n}$. ∎

1.3.7   Corollary

If a, b, and $n > 0$ are integers such that $\gcd(a,n) = 1$, then any two solutions of $ax \equiv b \pmod{n}$ are congruent modulo n.

1.3.8   Corollary

If a and $n > 0$ are integers, then $ax \equiv 1 \pmod{n}$ can be solved for x in the integers if and only if $\gcd(a,n) = 1$.

The proof of Theorem 1.3.6 provides a method of finding the solutions of $ax \equiv b \pmod{n}$ whenever $\gcd(a,n) = 1$.

1.3.9   **Example**

1.   Find the solutions of the congruence equation $6x \equiv 17 \pmod 7$.

Solution. If 7 is divided by 6, the remainder is 1, and $7 = 6(1) + 1$. Hence, $1 - 6(-1) = 7$, so $6(-1) \equiv 1 \pmod 7$. If both sides of this congruence equation are multiplied by 17, we get $6(-17) \equiv 17 \pmod 7$. Thus, $x = -17$ is a solution. The smallest positive solution must be congruent to -17 modulo 7. This is $x = 4$, so any solution is congruent to 4 modulo 7.

2.   Find the solutions of congruence equation $20x \equiv 12 \pmod{72}$.

Solution. First find the gcd$(20, 72)$, which is 4. Since $4 \mid 12$, the congruence equation has solutions in the integers. Now divide each integer in the congruence equation by 4 to obtain $5x \equiv 3 \pmod{18}$. Now gcd$(5, 18) = 1$, so we solve this congruence equation for x in the integers. We use the Euclidean Algorithm to find integers s and t such that $1 = 5s + 18t$. This process produces the following sequence of equations: $18 = (3)5 + 3, 5 = 3 + 2, 3 = 2 + 1, 2 = 2 + 0$. The last nonzero remainder is 1, and back-substitution gives $1 = 2(18) + (-7)(5)$. Hence, $5(-7) \equiv 1 \pmod{18}$. If both sides of this congruence equation are multiplied by 3, then $5(-21) \equiv 3 \pmod{18}$. From this we see that a solution of the equation is $x = -21$. The smallest positive solution is $x = 15$. You can check that $x = 15$ is a solution of $20x \equiv 12 \pmod{72}$ and that, in fact, by Theorem 1.3.6, every solution of $20x \equiv 12 \pmod{72}$ is congruent to 15 modulo 18.

3.   Find all solutions of the congruence equation $14x \equiv 21 \pmod 4$.

Solution. Since gcd$(14, 4) = 2$ and 2 does not divide 21, Theorem 1.3.6 indicates that there are no solutions of this congruence equation.

Let's now turn our attention to the Chinese Remainder Problem of finding common solutions of a finite system of linear congruence equations. We restrict ourselves to the case of two equations. However, it is sometimes possible to find common solutions for a finite set of linear congruence equations.

1.3.10   **Chinese Remainder Theorem**

If m and n are positive integers such that gcd$(m, n) = 1$, then the pair of congruence equations

$$x \equiv a \pmod m \quad \text{and} \quad x \equiv b \pmod n$$

have a common solution in the integers, and any two solutions are congruent modulo mn.

Proof. Since $\gcd(m, n) = 1$, there are integers s and t such that $1 = sm + tn$. Hence, $a = asm + atn$, so $atn \equiv a \pmod{m}$. Similarly, $bsm \equiv b \pmod{n}$. If we let $x = bsm + atn$, then $x \equiv atn \pmod{m}$, so $x \equiv a \pmod{m}$. Likewise, $x \equiv b \pmod{n}$. Therefore, $x = bsm + atn$ is a solution of the congruence equations. Adding any multiple of mn to x obviously will produce another solution. Conversely, if x_1 and x_2 are solutions of the congruence equations, then $x_1 \equiv x_2 \pmod{m}$ and $x_1 \equiv x_2 \pmod{n}$. Thus, $x_1 - x_2$ is divisible by m and n and so by Exercise 10 of Section 1.1, $x_1 - x_2$ is divisible by mn. Hence, x_1 and x_2 are congruent modulo mn. ■

The proof of Theorem 1.3.6 produced a method for solving a congruence equation of the form $ax \equiv b \pmod{n}$ when $\gcd(a, n) = 1$. The proof of the Chinese Remainder Theorem 1.3.10 also provides a method for solving a set of congruence equations $x \equiv a \pmod{m}$ and $x \equiv b \pmod{n}$ when $\gcd(m, n) = 1$.

1.3.11	**Example**

Find a solution of $x \equiv 12 \pmod{4}$ and $x \equiv 5 \pmod{7}$.

Solution. A solution exists since $\gcd(4,7) = 1$. To find a solution, we need to complete the equation $x = bsm + atn$ of the proof of Theorem 1.3.10. If we let $m = 4$ and $n = 7$, then since $-5(4) + 3(7) = 1$, $s = -5$ and $t = 3$. Hence, a solution is given by $x = -5(4)5 + 3(7)12 = 152$. Any two solutions are congruent modulo $4(7) = 28$, so the smallest positive solution is $x = 12$ since $12 \equiv 152 \pmod{28}$.

It was pointed out earlier that the relation modulo n is an equivalence relation on **Z**. In fact, you were asked to prove this in Exercise 10 of Problem Set 0.2. However, since the relation modulo n is important to our discussion, we prove it here.

1.3.12	**Theorem**

If n is a positive integer, the relation modulo n defined on **Z** by $x \equiv y \pmod{n}$ if and only if $n \mid (x - y)$ is an equivalence relation on **Z**.

Proof

1. The relation modulo n is obviously reflexive because if $x \in$ **Z**, $x - x$ is divisible by n. Hence, $x \equiv x \pmod{n}$ for all $x \in$ **Z**.

2. If $x, y \in \mathbf{Z}$ are such that $x \equiv y \pmod{n}$, then $n \mid (x - y)$. Thus, there is an integer k such that $x - y = nk$. But then $y - x = n(-k)$, so $n \mid (y - x)$. Therefore, $y \equiv x \pmod{n}$, which shows that the relation modulo n is symmetric.

3. If $x, y, z \in \mathbf{Z}$ are such that $x \equiv y \pmod{n}$ and $y \equiv z \pmod{n}$, then $n \mid (x - y)$ and $n \mid (y - z)$, so there are integers k_1 and k_2 such that $x - y = nk_1$ and $y - z = nk_2$. Hence, $x - z = (x - y) + (y - z) = nk_1 + nk_2 = n(k_1 + k_2)$ and thus $n \mid (x - z)$. Consequently, $x \equiv y \pmod{n}$ and $y \equiv z \pmod{n}$ imply that $x \equiv z \pmod{n}$, so the relation modulo n is transitive. ∎

In the presence of a binary operation on A, it is sometimes the case that an equivalence relation E on A is compatible with the binary operation. In Definition 0.4.7 such an equivalence relation was called a congruence relation on A. It was shown in Theorem 0.4.8 that a binary operation can be defined on the set of equivalence classes A/E determined by such a congruence relation E.

1.3.13   Theorem

The relation modulo n is a congruence relation on $\mathbf{Z}$ with respect to the binary operations of addition and multiplication of integers. That is, if $x \equiv y \pmod{n}$ and $z \equiv w \pmod{n}$, then

1. $x + z \equiv y + w \pmod{n}$ and
2. $xz \equiv yw \pmod{n}$.

Proof.   Theorem 1.3.12 shows that equality modulo n is an equivalence relation on $\mathbf{Z}$, so it remains only to show that the relation modulo n is compatible with the binary operation of addition and multiplication of integers modulo n.

1. Suppose $x \equiv y \pmod{n}$ and $z \equiv w \pmod{n}$. Then there are integers k_1 and k_2 such that $x - y = k_1 n$ and $z - w = k_2 n$. Hence, $(x + z) - (y + w) = (x - y) + (z - w) = k_1 n + k_2 n = (k_1 + k_2)n$. Therefore, $x + z \equiv y + w \pmod{n}$.

2. Again let $x \equiv y \pmod{n}$ and $z \equiv w \pmod{n}$. By Lemma 1.3.4, $xz \equiv yz \pmod{n}$ and $yz \equiv yw \pmod{n}$. Therefore, by transitivity of equality modulo n, $xz \equiv yw \pmod{n}$. ∎

$\mathbf{Z}_n$ will now denote the set of all congruence classes of $\mathbf{Z}$ determined by the congruence relation modulo n on $\mathbf{Z}$.

1.3.14   Corollary

The binary operations defined on $\mathbf{Z}_n$ by $[x] +_n [y] = [x + y]$ and $[x] \bullet_n [y] = [xy]$ are well-defined binary operations on $\mathbf{Z}_n$. (The subscripts on $+_n$ and on $\bullet_n$ serve as a reminder that these are operations on congruence classes.)

Proof. This follows immediately from Theorem 0.4.8 and Theorem 1.3.13. ■

Care must be taken with notation when more than one congruence relation modulo n is under consideration on the integers. For example, $[3] \in \mathbf{Z}_4$ and $[3] \in \mathbf{Z}_5$, but the two congruence classes are quite different. In $\mathbf{Z}_4$, $[3] = \{\ldots, -5, -1, 3, 7, 11, \ldots\}$, and in $\mathbf{Z}_5$, $[3] = \{\ldots, -7, -2, 3, 8, 13, \ldots\}$. When there is a need, such differences are denoted by $[3]_4$ and $[3]_5$, respectively.

At this point you might be wondering how many congruence classes are determined by the congruence relation modulo n on $\mathbf{Z}$. The following theorem shows that there are n distinct congruence classes in $\mathbf{Z}_n$.

1.3.15   Theorem

Let n be a positive integer. If the remainder of the division of an integer x by n is r, then $x \equiv r \pmod{n}$ and $[x] = [r]$.

Proof.   If the quotient of the division of x by n is q and the remainder is r, then $x = qn + r$. Hence, $x - r = qn$, so $x \equiv r \pmod{n}$. Therefore, by Theorem 0.2.3, $[x] = [r]$. ■

Since the division of any integer by n leaves a remainder of $0, 1, 2, \ldots, n - 1$, the congruence classes determined by the congruence relation modulo n are $[0]$, $[1], [2], \ldots, [n - 1]$. It is now a simple matter to determine the congruence class to which an integer belongs. Simply divide the integer by n to obtain the remainder r that satisfies the condition $0 \le r < n$. The integer belongs to the congruence class determined by this remainder. It is the usual practice to represent the congruence classes determined by the relation modulo n by $[0], [1], [2], \ldots, [n - 1]$.

1.3.16   Example

Let $n = 6$ and consider the congruence relation modulo 6 on $\mathbf{Z}$. The congruence classes determined by this relation are

$$[0] = \{..., -18, \; -12, \; -6, \; 0, \; 6, 12, 18, \; ...\},$$
$$[1] = \{..., -17, \; -11, \; -5, \; 1, \; 7, 13, 19, \; ...\},$$
$$[2] = \{..., -16, \; -10, \; -4, \; 2, \; 8, 14, 20, \; ...\},$$
$$[3] = \{..., -15, \; -9, \; -3, \; 3, \; 9, 15, 21, \; ...\},$$
$$[4] = \{..., -14, \; -8, \; -2, \; 4, \; 10, 16, 22, \; ...\}, \quad \text{and}$$
$$[5] = \{..., -13, \; -7, \; -1, \; 5, \; 11, 17, 23, \; ...\}.$$

To what congruence class does 267 belong? The remainder from the division of 267 by 6 is 3, so $267 \in [3]$. Similarly, since the remainder from the division of -7987 by 6 is 5, $-7987 \in [5]$.

Addition of Congruence Classes

Corollary 1.3.14 shows that the binary operation $[x] +_6 [y] = [x + y]$ on $\mathbf{Z}_6$ is well-defined. This operation is independent of the representatives used for the congruence classes. For example, $[3] +_6 [4] = [3 + 4] = [7] = [1]$. If different representatives are chosen for the congruence classes, say $[3] = [-15]$ and $[4] = [10]$, then $[-15] +_6 [10] = [-15 + 10] = [-5] = [1]$, which yields the same result as adding $[3]$ and $[4]$. It should be obvious that $[0]$ is the additive identity for $\mathbf{Z}_6$.

Since ordinary addition of integers is commutative and associative, these properties are inherited by $+_6$. It is also interesting and maybe unexpected that each element of $\mathbf{Z}_6$ has an additive inverse.

$$[0] +_6 [0] = [0 + 0] = [0],$$
$$[1] +_6 [5] = [1 + 5] = [6] = [0],$$
$$[2] +_6 [4] = [2 + 4] = [6] = [0],$$
$$[3] +_6 [3] = [3 + 3] = [6] = [0].$$

If $-[x]$ denotes the additive inverse of $[x]$, then

$$-[0] = [0] = [-0], \qquad -[3] = [3] = [-3],$$
$$-[1] = [5] = [-1], \qquad -[4] = [2] = [-4],$$
$$-[2] = [4] = [-2], \qquad -[5] = [1] = [-5].$$

Hence, $-[x] = [-x]$, so the additive inverse $-x$ of x in $\mathbf{Z}$ determines the additive inverse $-[x]$ of $[x]$ in $\mathbf{Z}_6$.

Multiplication of Congruence Classes

The equivalence relation modulo 6 is a congruence relation on $\mathbf{Z}$ with respect to multiplication of the integers. Hence, $[x] \bullet_6 [y] = [xy]$ is a well-defined binary operation on $\mathbf{Z}_6$. For example, $[2] \bullet_6 [4] = [2 \cdot 4] = [8] = [2]$, and this operation is also independent of the particular representatives chosen for the congruence classes. Since ordinary multiplication of integers is associative and commutative, $\bullet_6$ is associative and commutative. The set $\mathbf{Z}_6$ has a multiplicative identity e, namely, $e = [1]$. However, there are elements of $\mathbf{Z}_6$ that do not have multiplicative inverses. For example, consider

$$[2] \bullet_6 [3] = [2 \cdot 3] = [6] = [0]. \tag{1}$$

If $[2]$ has a multiplicative inverse, then there is an element $[a]$ in $\mathbf{Z}_6$ such that

$$[a] \bullet_6 [2] = [2] \bullet_6 [a] = [1].$$

Multiplying both sides of the equation (1) by $[a]$ produces

$$[a] \bullet_6 ([2] \bullet_6 [3]) = [a] \bullet_6 [0] = [a \cdot 0] = [0].$$

By using associativity, we see that

$$[a] \bullet_6 ([2] \bullet_6 [3]) = ([a] \bullet_6 [2]) \bullet_6 [3] = [1] \bullet_6 [3] = [3].$$

Thus, $[3] = [0]$, which is certainly not the case in $\mathbf{Z}_6$. Therefore, $[2]$ does *not* have a multiplicative inverse in $\mathbf{Z}_6$. However, if x and 6 are relatively prime, then $\gcd(x,6) = 1$, so there are integers a and b such that $1 = ax + b6$. Therefore, $1 - ax = 6b$ and $6 \mid (1 - ax)$. But then $ax \equiv 1 \pmod{6}$, so $[ax] = [1]$ or $[a] \bullet_6 [x] = [1]$. Thus, $[a]$ is the multiplicative inverse of $[x]$ in $\mathbf{Z}_6$. The elements $[x]$ of $\mathbf{Z}_6$ such that x is relatively prime to 6 are $[1]$ and $[5]$. Both have multiplicative inverses, since $[1] \bullet_6 [1] = [1 \cdot 1] = [1]$ and $[5] \bullet_6 [5] = [5 \cdot 5] = [25] = [1]$.

Tables for Addition and Multiplication of Congruence Classes

Tables can be constructed for addition and multiplication of congruence classes in $\mathbf{Z}_6$. These tables, known as Cayley tables, were discovered by **Arthur Cayley** (1821–1895), who was the first mathematician to use them in his work. The following is a Cayley table for $\mathbf{Z}_6$.

$\downarrow$

$+_6$	[0]	[1]	[2]	[3]	[4]	[5]
[0]	[0]	[1]	[2]	[3]	[4]	[5]
[1]	[1]	[2]	[3]	[4]	[5]	[0]
$\rightarrow$ [2]	[2]	[3]	[4]	[5]	[0]	[1]
[3]	[3]	[4]	[5]	[0]	[1]	[2]
[4]	[4]	[5]	[0]	[1]	[2]	[3]
[5]	[5]	[0]	[1]	[2]	[3]	[4]

To compute the sum $[2] +_6 [3]$ using this table, locate $[2]$ in the left-hand column and trace along this row until you are under $[3]$ in the top row. (These positions are indicated by the arrows.) The number in the box at the intersection of this row and column is the sum of $[2]$ and $[3]$. Hence, $[2] +_6 [3] = [5]$. Similarly, $[5] +_6 [4] = [3]$, and so on.

A similar table can be constructed for $•_6$. The product of two congruence classes can be determined using the same method as for addition. For example, from the following table, $[2] •_6 [3] = [0]$, where again the selection of the correct row and column are indicated by the arrows.

$•_6$	[0]	[1]	[2]	[3] ↓	[4]	[5]
[0]	[0]	[0]	[0]	[0]	[0]	[0]
[1]	[0]	[1]	[2]	[3]	[4]	[5]
→ [2]	[0]	[2]	[4]	[0]	[2]	[4]
[3]	[0]	[3]	[0]	[3]	[0]	[3]
[4]	[0]	[4]	[2]	[0]	[4]	[2]
[5]	[0]	[5]	[4]	[3]	[2]	[1]

These observations of the modular arithmetic in $\mathbf{Z}_6$ lead to the following interesting theorem.

1.3.17   Theorem

For any positive integer n, congruence class addition $[x] +_n [y] = [x + y]$ and congruence class multiplication $[x] •_n [y] = [xy]$ defined on $\mathbf{Z}_n$ have the following properties:

1. The operations $+_n$ and $•_n$ are associative and commutative.
2. The additive identity of $\mathbf{Z}_n$ is $[0]$ and the multiplicative identity is $[1]$.
3. Every element $[x]$ of $\mathbf{Z}_n$ has an additive inverse $-[x] = [-x]$.
4. The congruence class $[x] \in \mathbf{Z}_n$ has a multiplicative inverse $[x]^{-1}$ if and only if $\gcd(x, n) = 1$.

Proof. We prove the associativity and commutativity of $+_n$ and leave the proof of the associativity and commutativity of $•_n$ as an exercise. The associativity and commutativity of addition of congruence classes are inherited from the associative and commutative properties of addition of the integers. The following shows that $+_n$ is commutative and associative. For $[x], [y], [z] \in \mathbf{Z}_n$,

$$[x] +_n [y] = [x+y]$$
$$= [y+x]$$
$$= [y] +_n [x];$$
$$[x] +_n ([y] +_n [z]) = [x] +_n [y+z]$$
$$= [x + (y+z)]$$
$$= [(x+y)+z]$$
$$= [x+y] +_n [z]$$
$$= ([x] +_n [y]) +_n [z].$$

The congruence class $[0]$ is clearly the additive identity, and if $[x] \in \mathbf{Z}_n$, then $[x] +_n [-x] = [x + (-x)] = [x - x] = [0]$ and thus $-[x] = [-x]$. The congruence class $[1]$ is the multiplicative identity of $\mathbf{Z}_n$.

Finally, if $[x] \in \mathbf{Z}_n$ and $\gcd(x, n) = 1$, then by Theorem 1.1.7, there are integers a and b such that $1 = ax + bn$, so $1 - ax = bn$. This implies that $n \mid (1 - ax)$ or $ax \equiv 1 \pmod{n}$. Hence, $[ax] = [1]$ or $[a] \bullet_n [x] = 1$, so $[x]^{-1} = [a]$. Conversely, if $[y] \in \mathbf{Z}_n$ is such that $[x] \bullet_n [y] = [1]$, then $[xy] = [1]$, and so $xy \in [1]$. Thus, $n \mid (1 - xy)$, so if $1 - xy = nk$, where k is an integer, then $1 = xy + nk$ and, by Theorem 1.1.7, $\gcd(x, n) = 1$. ■

1.3.18 Corollary

If p is a prime integer, then every nonzero element of $\mathbf{Z}_p$ has a multiplicative inverse.

To simplify notation, we will now write $[x][y]$ for $[x] \bullet_n [y]$ and $[x] + [y]$ for $[x] +_n [y]$. In the equation $[x][y] = [xy]$, the left side of the equation represents multiplication of congruence classes, and the multiplication xy within the brackets on the right side of the equation represents multiplication of integers. Likewise, in the equation $[x] + [y] = [x+y]$, the plus sign on the left of the equation represents the addition of congruence classes, while the sum $x + y$ on the right side of the equation within the brackets represents the addition of integers.

1.3.19 Example

Sometimes it is possible to solve equations in $\mathbf{Z}_n$. For example, when $n = 6$, the equation $[2][x] = [4]$ has a solution for $[x]$ in $\mathbf{Z}_6$. Since there are only a finite number of possibilities for $[x]$, one method of finding a solution is just to try all the possibilities. The possibilities are [0], [1], [2], [3], [4], and [5]. Hence,

$$[2][0] = [2 \cdot 0] = [0] \neq [4],$$
$$[2][1] = [2 \cdot 1] = [2] \neq [4],$$
$$[2][2] = [2 \cdot 2] = [4],$$
$$[2][3] = [2 \cdot 3] = [6] = [0] \neq [4],$$
$$[2][4] = [2 \cdot 4] = [8] = [2] \neq [4],$$
$$[2][5] = [2 \cdot 5] = [10] = [4].$$

Therefore the equation has two solutions in $\mathbf{Z}_6$, namely, $[x] = [2]$ and $[x] = [5]$. It may also be the case that an equation may *not* have a solution. For example, consider the equation $[x]^2 + [1] = [0]$ in $\mathbf{Z}_4$, where $[x]^2 = [x][x]$. Since

$$[0]^2 + [1] = [0] + [1] = [1] \neq [0],$$
$$[1]^2 + [1] = [1] + [1] = [2] \neq [0],$$
$$[2]^2 + [1] = [4] + [1] = [1] \neq [0],$$
$$[3]^2 + [1] = [9] + [1] = [2] \neq [0],$$

the equation does not have a solution in $\mathbf{Z}_4$. When n is large and we are seeking a solution for $[a][x] = [b]$, it may be impractical to try all the possibilities. If z is a solution for x in the equation $ax \equiv b \pmod{n}$, then $[x] = [z]$ is a solution for $[a][x] = [b]$ in $\mathbf{Z}_n$. The proof of Theorem 1.3.6 gives a method for finding solutions of $ax \equiv b \pmod{n}$ when $d \mid b$, where $\gcd(a, n) = d$. This method was demonstrated in Example 2 of 1.3.9. Hence, if a and n are relatively prime, a solution will exist in $\mathbf{Z}_n$ for $[a][x] = [b]$. If p is a prime, $[a][x] = [b]$ always has $[x] = [a]^{-1}[b]$ as a solution in $\mathbf{Z}_p$, provided, of course, that $[a] \neq [0]$.

HISTORICAL NOTES

Arthur Cayley (1821–1895). Arthur Cayley, although born in Richmond, Surrey, England, spent the first eight years of his life in St. Petersburg, Russia, where his father was a merchant involved in Russian trade. In 1829 the family returned to England, and Arthur was educated first at a private school and later at King's College in London. He studied mathematics at Trinity College, Cambridge, where his academic achievements earned him a postgraduate fellowship as a Fellow of Trinity and assistant tutor. His duties were light and gave him the opportunity to publish some 25 papers during the three years he held the post. Because there was no suitable teaching position available, Cayley left Cambridge in 1846 and began to prepare himself for a legal career. He never abandoned mathematics, however, and during his 14 years as a lawyer, he published between 200 and 300 papers in mathematics. In 1863 Cayley was appointed to a newly established professorship in pure mathematics at Cambridge. He was a prolific mathematician and during his life produced nearly 1000 articles dealing with almost every aspect of mathematics. Some of his most important work was in developing the algebra of matrices and his contribution to geometry and group theory. Cayley was the first to define an abstract group and to display multiplication tables for groups. Such tables are now often referred to as Cayley tables.

Carl Friedrich Gauss (1777–1855). Carl Friedrich Gauss's genius developed early, and he is considered to be one of the greatest mathematicians of all time. He was a brilliant child prodigy in mathematics. In elementary school he found the sum of the integers from 1 to 100 by recognizing that the sum was 50 pairs of numbers, with each pair summing to 101, and then multiplying 50 by 101 in his head. Gauss received his Ph.D. from the University of Helmstedt for a dissertation that dealt with the Fundamental Theorem of Algebra. He published several books, the first of which was *Disquisitiones Arithmeticae,* which dealt primarily with number theory. One of his discoveries at the age of 19 was that it is possible to construct a regular 17-gon using only a straightedge and compass. This problem had defied solution for over 2000 years. Gauss published only what he considered to be the most important of his results in mathematics and then only after his results were complete and rigorously done. His motto was "few but ripe." However, he left behind a tremendous amount of mathematical writing in 12 volumes of diaries. Gauss's interest in astronomy led him in 1806 to accept the position of professor of astronomy at the University of Gottingen, where he remained for the rest of his life. In 1809 he published his second book, a work that dealt with the motion of celestial bodies and set the standard for computational and practical astronomy for years to come.

Gauss made contributions to many areas of mathematics. He worked in geometry, and he attempted to derive the axiom of parallels from the other axioms of Euclidean geometry. He seemed to be aware of the existence of non-Euclidean geometries but failed to publish on the subject for fear that his reputation would suffer if he took such a radical stand in mathematics. Gauss's work in statistics led to his developing the method of least squares. He also contributed to differential geometry, and his idea of curvature on a surface (Gaussian curvature) bears his name. The mathematical work of Gauss was so profound that it continues to influence mathematicians to this day. Moritz Cantor, Richard Dedekind, and Bernhard Riemann, who themselves went on to become famous mathematicians, were among his doctoral students.

Problem Set 1.3

1. Perform the following operations in $\mathbf{Z}_8$.

    (a) $[4] + [5]$
    (b) $[4][7]$
    (c) $[2][4] + [4][7]$
    (d) $[5][4] + [7][2]$
    (e) Does $[5]([6] + [3]) = ([5][6]) + ([5][3])$?

2. Make addition and multiplication tables for each of the following.

    (a) $\mathbf{Z}_2$ (c) $\mathbf{Z}_5$
    (b) $\mathbf{Z}_3$ (d) $\mathbf{Z}_6$

3. Show that the square of each element of $\mathbf{Z}_4$ is either $[0]$ or $[1]$.

4. If a is a positive integer and $a = a_0 + a_1 10 + a_2 10^2 + \cdots + a_n 10^n$, prove that $a \equiv a_0 + a_1 + \cdots + a_n \pmod 9$.

5. For each positive integer n, let

    $$n[x] = [x] + [x] + \cdots + [x] \quad \text{where there are } n \text{ summands.}$$

    Show that $n[x] = [n][x]$ and $-n[x] = -[n][x]$ for each positive integer n.

6. Perform the following multiplications and replace $[x][x]$ by $[x]^2$.

    (a) $([3][x] + [4])([2][x] + [5])$ in $\mathbf{Z}_6$
    (b) $([2][x] + [3])([5][x] + [2])$ in $\mathbf{Z}_7$
    (c) $([2][x] + [3])([3][x] + [2])$ in $\mathbf{Z}_6$
    (d) $([2][x] + [3])([3][x] + [2])$ in $\mathbf{Z}_7$

7. Describe the distinct solution(s) modulo n for each of the following equations, if such exist.

    (a) $22x \equiv 4 \pmod 4$ (c) $25x + 12 \equiv 19 \pmod{13}$
    (b) $141x \equiv 45 \pmod{939}$ (d) $68x + 12 \equiv 24 \pmod{14}$

8. Find a common positive solution in the integers of the following pairs of congruence equations, if a solution exists.

    (a) $x \equiv 8 \pmod 5$ and $x \equiv 4 \pmod 3$
    (b) $x \equiv 12 \pmod{15}$ and $x \equiv 11 \pmod{14}$
    (c) $3x \equiv 9 \pmod 2$ and $4x \equiv 12 \pmod 3$

9. If $[x], [y],$ and $[z]$ are elements of $\mathbf{Z}_n$, prove that $[x]([y] + [z]) = [x][y] + [x][z]$. Because of this, we say that (congruence class) **multiplication is distributive over addition from the left**. (Notice that $[x]$, the multiplier, is to the left of the sum.) Show also that multiplication is distributive over addition from the right.

10. For each positive integer k, let

$$[x]^k = [x][x]\cdots[x] \quad \text{where there are } k \text{ factors.}$$
$$= [xx\cdots x]$$
$$= [x^k]$$

If $[a] \in \mathbf{Z}_n$, then an element $[x] \in \mathbf{Z}_n$ such that $[x]^k = [a]$ is said to be a **kth root** of $[a]$. When $k = 2$, $[x]$ is said to be a **square root** of $[a]$, and when $k = 3$, $[x]$ is a **cube root** of $[a]$, and so on. Notice that if $[x]$ is a square root of $[a]$, then $-[x]$ is also a square root of $[a]$, since $(-[x])^2 = [-x]^2 = [-x][-x] = [(-x)(-x)] = [x^2] = [x]^2 = [a]$. For example, in $\mathbf{Z}_8$, $[2]$ is a square root of $[4]$, as is $-[2] = [6]$ since $[6]^2 = [36] = [4]$. Thus, $[x] = \pm[2] = [2], [6]$ are square roots of $[4]$ in $\mathbf{Z}_8$. Find the square roots of each of the following elements, if such exist.

(a) $[1]$ in $\mathbf{Z}_3$ (c) $[8]$ in $\mathbf{Z}_9$ (e) $[0]$ in $\mathbf{Z}_4$
(b) $[2]$ in $\mathbf{Z}_7$ (d) $[9]$ in $\mathbf{Z}_{10}$
(f) Is it possible to find a positive integer n and an element of $\mathbf{Z}_n$ that has more than two square roots?

11. Since $6 = 2 \cdot 3$, we say that 2 and 3 are factors of 6 or that 2 and 3 are divisors of 6. This terminology is carried over to $\mathbf{Z}_n$. For example, in $\mathbf{Z}_8$, $[2][3] = [6]$ and so we say that $[2]$ and $[3]$ are divisors of $[6]$ in $\mathbf{Z}_8$. Since $[2][4] = [0]$, $[2]$ and $[4]$ are divisors of $[0]$ in $\mathbf{Z}_8$. In general, we say that a nonzero element $[x] \in \mathbf{Z}_n$ is a **divisor of zero** if there is a nonzero element $[y] \in \mathbf{Z}_n$ such that $[x][y] = [0]$.

Find all the divisors of zero in each of the following, if such exist.

(a) $\mathbf{Z}_{12}$ (b) $\mathbf{Z}_{35}$ (c) $\mathbf{Z}_{31}$

12. If x and y are integers such that $x \equiv y \pmod{mn}$, prove that $x \equiv y \pmod{m}$ and $x \equiv y \pmod{n}$.

13. Prove property 1 of Theorem 1.3.17 for multiplication $\bullet_n$.

14. An element $[x] \in \mathbf{Z}_n$ is said to be **idempotent** if $[x]^2 = [x]$. Find all the idempotent elements in each of the following:

(a) $\mathbf{Z}_6$ (b) $\mathbf{Z}_{12}$ (c) $\mathbf{Z}_{30}$
(d) Prove that if p is a prime, then $\mathbf{Z}_p$ has no idempotent elements other than $[0]$ and $[1]$.

15. An element $[x] \in \mathbf{Z}_n$ is said to be **nilpotent** if $[x]^k = [0]$ for some positive integer k. Find all the nilpotent elements in each of the following:

(a) $\mathbf{Z}_4$ (b) $\mathbf{Z}_{12}$ (c) $\mathbf{Z}_{27}$

Use a computer algebra system to solve each of the following problems.

16. Determine the congruence class to which each of the following integers belongs.

    (a) 342,543,514,365,498 modulo 23
    (b) –21,324,593,439,368,294 modulo 56
    (c) 32,435,468,182,381,234,934,239,343,546 modulo 3

17. Compute each of the following in $\mathbf{Z}_{274,132}$.

    (a) (54,643)(3245)
    (b) (591,234)(294,935) + (24,356,795)(23,546)

18. Find all solutions of each of the following congruence equations that are greater than or equal to zero but less than the given modulus.

    (a) $2x = 12 \pmod 6$
    (b) $10x = 140 \pmod 8$
    (c) $23x + 10 = 12 \pmod{20}$
    (d) $45x - 128 = 10 \pmod{11}$

19. Find a solution of each of the following sets of congruence equations.

    (a) $x \equiv 132 \pmod 3$ $x \equiv 6576 \pmod 5$
    (b) $x \equiv 1534 \pmod 7$ $x \equiv 13,965,734 \pmod{11}$
    (c) $x \equiv 2314 \pmod 3$ $x \equiv 12,536 \pmod 4$ $x \equiv 54,672 \pmod 7$

20. Show that there are four square roots of [4] in $\mathbf{Z}_{12}$. What are they?

21. How many fifth roots are there of [3] in $\mathbf{Z}_{32}$? What are they?

22. Find the smallest positive integer n such that $\mathbf{Z}_n$ has two square roots of [2].

2

Sets with One Binary Operation: Groups

It is common practice in abstract algebra, as well as in many other areas of mathematics, to pick out the fundamental and shared properties of mathematical structures and then try to find other mathematical structures that satisfy these same properties. If these "abstracted" properties are well-chosen and it turns out that there are other commonly occurring mathematical structures that satisfy these properties, then this process may result in a classification system for these structures. Such classification systems and the accompanying theory often arise quite slowly, and they are usually the end result of dedicated efforts by many mathematicians. Such is the case for a mathematical structure known as a group. Historically, the concept of a group arose quite slowly through the study of several areas of mathematics: geometry, number theory, and the theory of algebraic equations.

2.1   GROUPS

Section Overview. In this section the definition of a group is motivated by picking out the shared properties of two nonempty sets, each of which is endowed with a single binary operation. After a group is defined, several examples of groups are given. Make sure you study these carefully and convince yourself that each is indeed an example of a group. The remainder of the section is devoted to giving properties that hold in every group and to stating definitions of particular features of groups.

As a motivation for the definition of a group, we first investigate an algebraic structure that can be formed from the rotations and reflections of a square, and then we look at certain algebraic properties of a subset of $\mathbf{Z}_{12}$. The first example is actually an example of a group of permutations. The shared properties of these two algebraic structures provide a model for the definition of a group.

2.1.1   Examples

1. A square can be rotated counterclockwise through certain angles or reflected about certain lines, and the square will end up with its original appearance. However, the corners of the original square have actually been moved. If the corners of the square are labeled, the effect of such rotations or reflections becomes apparent. In the following rotation figures, R_θ indicates that the square has been rotated about its center counterclockwise through an angle of θ degrees, whereas in the reflection figures, L_2, for example, indicates that the square has been reflected through the diagonal of the square labeled L_2 in the figure. These rotations and reflections are known as the **symmetries of a square** (Figure 2.1). If a reflection or rotation is followed by a reflection or rotation, the result can be achieved by a single reflection or rotation. For example, if we first reflect the square about the line L_1 and then rotate the result through 270 degrees, the end result is the same as reflecting about the line L_3 (Figure 2.2).We will see later in the text that these operations are actually mappings, so we follow the usual convention of reading the composition of functions from right to left. If this operation is viewed as multiplication of the elements in the set

$$D_4 = \{R_0, R_{90}, R_{180}, R_{270}, L_1, L_2, L_3, L_4\}$$

then the operation is well-defined and closed, since a rotation or reflection followed by a rotation or reflection can be achieved by a single rotation or reflection. Hence, the operation is actually a binary operation on the set D_4. A multiplication table for this binary operation on D_4 is given as Table 2.1.

A little thought should convince you that the binary operation is associative; for example, $R_{270}(L_3L_1) = (R_{270}L_3)L_1$. Inspection of the table shows that $e = R_0$ is a multiplicative identity for D_4 and that every element of D_4 has a multiplicative inverse that is in D_4. For

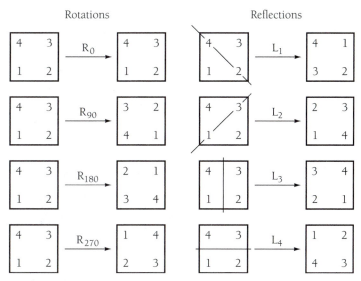

Figure 2.1

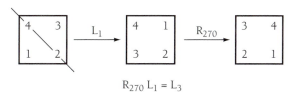

$$R_{270} L_1 = L_3$$

Figure 2.2

Table 2.1

•	R_0	R_{90}	R_{180}	R_{270}	L_1	L_2	L_3	L_4
R_0	R_0	R_{90}	R_{180}	R_{270}	L_1	L_2	L_3	L_4
R_{90}	R_{90}	R_{180}	R_{270}	R_0	L_4	L_3	L_1	L_2
R_{180}	R_{180}	R_{270}	R_0	R_{90}	L_2	L_1	L_4	L_3
R_{270}	R_{270}	R_0	R_{90}	R_{180}	L_3	L_4	L_2	L_1
L_1	L_1	L_3	L_2	L_4	R_0	R_{180}	R_{90}	R_{270}
L_2	L_2	L_4	L_1	L_3	R_{180}	R_0	R_{270}	R_{90}
L_3	L_3	L_2	L_4	L_1	R_{270}	R_{90}	R_0	R_{180}
L_4	L_4	L_1	L_3	L_2	R_{90}	R_{270}	R_{180}	R_0

instance, $R_{90}^{-1} = R_{270}$ since $R_{90}R_{270} = R_{270}R_{90} = R_0 = e$. Since $L_3R_{90} \neq R_{90}L_3$, the binary operation is *not* commutative.

2. Consider the set $U(\mathbf{Z}_{12}) = \{[1], [5], [7], [11]\}$ of elements of $\mathbf{Z}_{12}$ that have multiplicative inverses in $\mathbf{Z}_{12}$ under multiplication modulo 12. Recall that an element $[x] \in \mathbf{Z}_{12}$ has a multiplicative inverse in $\mathbf{Z}_{12}$ when x and 12 are relatively prime. (See Theorem 1.3.17.) Let's construct a multiplication table for $U(\mathbf{Z}_{12})$. (See Table 2.2.)

Table 2.2

·	[1]	[5]	[7]	[11]
[1]	[1]	[5]	[7]	[11]
[5]	[5]	[1]	[11]	[7]
[7]	[7]	[11]	[1]	[5]
[11]	[11]	[7]	[5]	[1]

The operation is closed since the table contains only elements of $U(\mathbf{Z}_{12})$, so the operation is a binary operation on $U(\mathbf{Z}_{12})$. The operation is associative since multiplication of integers is associative. The congruence class [1] is a multiplicative identity for $U(\mathbf{Z}_{12})$, and every element of $U(\mathbf{Z}_{12})$ has a multiplicative inverse in $U(\mathbf{Z}_{12})$. This operation is commutative.

The two preceding examples have the following properties in common: (1) both sets are nonempty, (2) both sets are endowed with an associative binary operation, (3) both sets have a multiplicative identity, and (4) every element of each set has a multiplicative inverse that is in the set. These examples are illustrations of a very fundamental and important algebraic structure known as a group, which we now define.

2.1.2   Definition

A **group** is a nonempty set G together with a binary operation $G \times G \rightarrow G : (x, y) \longmapsto x * y$ such that

1. the binary operation is associative: $x * (y * z) = (x * y) * z$ for all $x, y, z \in G$,
2. there is an element $e \in G$ such that $x * e = e * x = x$ for all $x \in G$, that is, G has an identity, and
3. each element of G has an inverse in G: For each $x \in G$, there is an element $x^{-1} \in G$ such that $x * x^{-1} = x^{-1} * x = e$.

If the binary operation is commutative, G is said to be an abelian group. The **order of a group** G, denoted by $|G|$, is the number of elements in the group. If G has a finite number of elements, then G is said to be a **group of finite order**. If G is not a finite group, then G is a **group of infinite order**. Abelian groups are named after **Niels Abel** (1802–1829), a Norwegian mathematician who made important contributions to group theory.

If the binary operation $*$ in the definition of a group is written as xy, the group is referred to as a **multiplicative group**. For a multiplicative group G, e is said to be a **multiplicative identity** of G, and the element x^{-1} is a **multiplicative inverse** of x. If the binary operation on a group G is denoted by $x + y$, then G is said to be an **additive group**. An identity of an additive group is denoted by 0, and 0 is said to be an **additive identity** of G. In an additive group, the inverse of an element x is written as $-x$ and $-x$ is an **additive inverse** of x.

It is important to point out that additive groups and multiplicative groups are *not* two different types of groups. There is only one definition of a group, and the only difference between an additive group and a multiplicative group is in the notation used for the binary operation. Any given group can be described using either notation. However, one notation often seems more natural for a particular group. It is also important to realize that an additive identity 0 of an additive group need not be the integer 0. The symbol 0 is simply the notation chosen for the identity of an additive group.

2.1.3 Some Familiar Examples of Groups

1. The set $\mathbf{Z}$ together with the usual operation of addition of integers is an additive abelian group. The set $\mathbf{Z}$ together with the usual operation of multiplication is *not* a group. The integer 1 is a multiplicative identity for $\mathbf{Z}$, but $x \in \mathbf{Z}$, $x \neq \pm 1$, does not have a multiplicative inverse in $\mathbf{Z}$.

2. The set $\mathbf{Q}$ of rational numbers together with the usual operation of addition is an additive abelian group. The set $\mathbf{Q}$ together with multiplication is *not* a group. The integer 1 is a multiplicative identity for $\mathbf{Q}$, but the integer $0 \in \mathbf{Q}$ fails to have a multiplicative inverse. However, $\mathbf{Q}^* = \mathbf{Q} \setminus \{0\}$ is an abelian group under ordinary multiplication of rational numbers.

3. The set $\mathbf{R}$ together with the binary operation of ordinary addition of real numbers is an additive abelian group, while $\mathbf{R}^* = \mathbf{R} \setminus \{0\}$ is an abelian group under multiplication of real numbers.

4. The set **C** of complex numbers $a + bi$, where $a, b \in \mathbf{R}$ and $i = \sqrt{-1}$, is an additive abelian group under addition of complex numbers defined by $(a + bi) + (c + di) = (a + c) + (b + d)i$. Two complex numbers $a + bi$ and $c + di$ are equal if and only if $a = c$ and $b = d$. An additive identity is the complex number $0 + 0i$, and an additive inverse of $a + bi$ is $-a + (-b)i$. The zero complex number $0 + 0i$ is usually abbreviated as 0. The set **C*** of nonzero complex numbers $a + bi$ is an abelian group under multiplication if multiplication is defined as

$$(a + bi)(c + di) = (ac - bd) + (ad + bc)i.$$

To compute the product of two complex numbers, multiply in the usual fashion and then replace i^2 by -1. A multiplicative identity is the complex number $1 + 0i$, which can be written as 1. A nonzero complex number $a + bi$ has

$$\frac{a}{a^2 + b^2} + \frac{-b}{a^2 + b^2}i$$

as a multiplicative inverse.

2.1.4   Some Less Familiar Examples of Groups

1. The set $M_2(\mathbf{R})$ of 2×2 matrices with entries from **R** is an additive abelian group under matrix addition, defined in Definition 0.5.3. An additive identity is $\begin{pmatrix} 0 & 0 \\ 0 & 0 \end{pmatrix}$ and an additive inverse of the matrix $\begin{pmatrix} a & b \\ c & d \end{pmatrix}$ is the matrix $-\begin{pmatrix} a & b \\ c & d \end{pmatrix} = \begin{pmatrix} -a & -b \\ -c & -d \end{pmatrix}$. The set of all such nonzero matrices, however, is *not* a group under matrix multiplication as defined in Definition 0.5.3, even though $M_2(\mathbf{R})$ has a multiplicative identity, $\begin{pmatrix} 1 & 0 \\ 0 & 1 \end{pmatrix}$. As shown in Example 2 of 0.4.2, there are nonzero matrices in $M_2(\mathbf{R})$ that fail to have a multiplicative inverse.

2. Let $GL_2(\mathbf{R}) = \left\{ \begin{pmatrix} a & b \\ c & d \end{pmatrix} \mid a, b, c, d \in \mathbf{R} \text{ and } ad - bc \neq 0 \right\}$. The set $GL_2(\mathbf{R})$ is a nonabelian group under matrix multiplication with identity $\begin{pmatrix} 1 & 0 \\ 0 & 1 \end{pmatrix}$.

    A multiplicative inverse of $\begin{pmatrix} a & b \\ c & d \end{pmatrix} \in GL_2(\mathbf{R})$ is $\begin{pmatrix} \dfrac{d}{ad - bc} & \dfrac{-b}{ad - bc} \\ \dfrac{-c}{ad - bc} & \dfrac{a}{ad - bc} \end{pmatrix}$.

    $GL_2(\mathbf{R})$ is the **general linear group** of 2×2 matrices with entries from **R**.

3. The set $G = \{-1, -i, i, 1\}$, where $i = \sqrt{-1}$, is an abelian group under the operation of multiplication of complex numbers. The multiplication table for G follows.

$\bullet$	-1	$-i$	i	1
-1	1	i	$-i$	-1
$-i$	i	-1	1	$-i$
i	$-i$	1	-1	i
1	-1	$-i$	i	1

4. Let $G = \{[0], [1], [2], [3], [4], [5]\}$ denote the set of congruence classes determined by the congruence relation $x \equiv y \pmod 6$ on **Z**. Then G is an additive abelian group under congruence-class addition modulo 6. An additive identity is $[0]$, and an additive inverse of $[x] \in G$ is $-[x] = [-x]$.

5. The set $G = \{[1], [2], [3], [4]\}$ of congruence classes determined by the congruence relation $x \equiv y \pmod 5$ on **Z** is a multiplicative abelian group under congruence-class multiplication modulo 5. A multiplicative identity is $[1]$. Note that $[1][1] = [1]$, $[2][3] = 1$, and $[4][4] = [1]$, so each element of G does indeed have a multiplicative inverse.

6. D_4, the set of all symmetries of a square introduced at the beginning of this chapter, is a nonabelian group. It is known as the **dihedral group** of order 8. A similar group D_3 with six elements can be constructed from the set of all symmetries of an equilateral triangle. One can also form a group D_5 with 10 elements from the symmetries of a regular pentagon. In fact, one can form a group D_n with $2n$ elements for $n \geq 3$ from the symmetries of any regular n-gon. The nonabelian groups D_n are known as the **dihedral groups**. These groups will be visited again in the exercises and when permutation groups are studied.

7. If n is a positive integer and $U(\mathbf{Z}_n)$ is the set of elements of $\mathbf{Z}_n$ that have multiplicative inverses in $\mathbf{Z}_n$, then $U(\mathbf{Z}_n)$ is a multiplicative abelian group under congruence-class multiplication modulo n. It is the **group of units** of $\mathbf{Z}_n$.

We have seen that a group G may be written additively or multiplicatively. Any theorem proven for a multiplicative group remains valid for an additive group when an appropriate change in language and notation is made in the theorem. Conversely, a theorem about additive groups becomes a theorem for multiplicative groups under such a change. When such a translation is made, the result is *not* a new theorem. It is simply the same theorem expressed in a different way. A theorem can be translated by making the changes suggested by the following correspondence, plus other minor changes that may be required to correct the language.

$$\text{Multiplicative} \quad \longleftrightarrow \quad \text{Additive}$$
$$xy \quad \longleftrightarrow \quad x + y$$
$$e \quad \longleftrightarrow \quad 0$$
$$x^{-1} \quad \longleftrightarrow \quad -x$$

It is important that you learn how to translate a theorem about multiplicative groups to one about additive groups and vice versa. Throughout the text, we often make use of results for additive groups that were previously proved for groups that were written multiplicatively. As an illustration of such a translation, consider the following theorem, which was actually proved in Chapter 0, though it was not stated using the concept of a group. (See Theorems 0.4.4 and 0.4.5.)

2.1.5   Theorem

If G is a multiplicative group, then the multiplicative identity e of G is unique. Furthermore, the multiplicative inverse x^{-1} of each $x \in G$ is also unique.

The following is a restatement of Theorem 2.1.5 for additive groups. Compare 2.1.5* with 2.1.5, and make sure you see exactly what changes have been made.

2.1.5*   Theorem

If G is an additive group, then the additive identity 0 of G is unique. Furthermore, the additive inverse $-x$ of each $x \in G$ is also unique.

Because of Theorem 2.1.5, there is only one identity in a group, and each element of a group has exactly one inverse. Hence, we can refer to *the* identity of a group and to *the* inverse of an element of a group. Now let's prove some of the basic properties that hold for every group. Other properties are presented in the exercises.

2.1.6   Theorem

If G is any group, then the following properties hold.

1. **Left-Hand Cancellation Property:** For any $a, x, y \in G$, $ax = ay$ implies that $x = y$.

2. **Right-Hand Cancellation Property**: For any $a, x, y \in G$, $xa = ya$ implies that $x = y$.

Proof. We prove property 1. The proof of property 2 is similar; just switch sides. If $ax = ay$, then $a^{-1}(ax) = a^{-1}(ay)$. But the operation is associative, so $(a^{-1}a)x = (a^{-1}a)y$. Hence, $ex = ey$, which gives $x = y$. ∎

2.1.7   Theorem

The following properties hold in any group G.

1. For any $x \in G$, $(x^{-1})^{-1} = x$.
2. For any $x, y \in G$, $(xy)^{-1} = y^{-1}x^{-1}$.

Proof

1. If $x \in G$, then $x^{-1} \in G$, so x^{-1} has an inverse in G. But $xx^{-1} = x^{-1}x = e$ implies that x is the inverse of x^{-1}. Thus, $(x^{-1})^{-1} = x$ since inverses are unique.
2. Note that $(xy)(y^{-1}x^{-1}) = x(yy^{-1})x^{-1} = xex^{-1} = xx^{-1} = e$. Similarly, $(y^{-1}x^{-1})(xy) = e$, so $y^{-1}x^{-1}$ is an inverse for xy. Because inverses are unique, we see that $(xy)^{-1} = y^{-1}x^{-1}$. ∎

2.1.8   Theorem

If G is a group and $a, b \in G$, then each of the equations $ax = b$ and $xa = b$ has a unique solution for x in G.

Proof. Note that $x = a^{-1}b$ and $x = ba^{-1}$, both of which exist in G, are solutions of $ax = b$ and $xa = b$, respectively. This follows since $a(a^{-1}b) = (aa^{-1})b = eb = b$ and $(ba^{-1})a = b(a^{-1}a) = be = b$. Now let's show uniqueness. If z_1 and z_2 are solutions of $ax = b$, then $az_1 = b$ and $az_2 = b$ implies that $az_1 = az_2$, so $z_1 = z_2$ by Theorem 2.1.6. Hence, the solution $x = a^{-1}b$ for $ax = b$ is unique. Similarly, it can be shown that the solution $x = ba^{-1}$ for $xa = b$ is unique. ∎

Exponents and Multiples in a Group

Let G be a multiplicative group, and suppose that $x \in G$. Then x^2 is defined to be the product xx, x^3 is the product xxx, and, in general, if n is a positive integer, $x^n = xx \cdots x$ with n factors. We also define x^{-n} to be the product $(x^{-1})^n$, that is, $x^{-n} = (x^{-1})^n$. If $n = 0$, we set $x^0 = e$. If G is an additive group, $x \in G$, and n is a positive integer, then $nx = x + x + \cdots + x$, where there are n summands and $(-n)x = (-x) + (-x) + \cdots + (-x)$. For example, $3x = x + x + x$ and $(-2)x = (-x) + (-x)$. Of course $0x = 0$, where the zero in $0x$ is the integer 0 while the zero on the right is the additive identity of the group.

We conclude our introduction to groups with a list of properties of exponents for multiplicative groups and a list of properties of multiples for an additive group. Notice that the Laws of Multiples for Additive Groups are just the translation to additive groups of the Laws of Exponents for Multiplicative Groups, and vice versa. The proof of each part of the laws can be effected by what is essentially a proof by induction.

2.1.9   Laws of Exponents for Multiplicative Groups

If x is any element of a multiplicative group G, then

1. For every integer n, $x^n x^{-n} = x^{-n} x^n = e$.
2. For all integers m and n, $x^m x^n = x^{m+n}$.
3. For all integers m and n, $(x^m)^n = x^{mn}$.
4. If G is an abelian group, then $(xy)^n = x^n y^n$ for every integer n and all $x, y \in G$.

2.1.10   Laws of Multiples for Additive Groups

If G is an additive group and x is any element of G, then

1. For every integer n, $nx + (-n)x = 0$.
2. For all integers m and n, $mx + nx = (m+n)x$.
3. For all integers m and n, $m(nx) = (mn)x$.
4. If G is an abelian group, then $n(x+y) = nx + ny$ for any integer n and all $x, y \in G$.

Finally, note that law 4 of 2.1.9 does not always hold unless the group is abelian. To see this, consider the general linear group $GL_2(\mathbf{R}) = \left\{ \begin{pmatrix} a & b \\ c & d \end{pmatrix} \middle| a, b, \right.$ $c, d \in \mathbf{R}$ and $ad - bc \neq 0 \Big\}$ and of Example 2 of 2.1.4. The group $GL_2(\mathbf{R})$ is a nonabelian group under matrix multiplication. If $A = \begin{pmatrix} 1 & 1 \\ 0 & 1 \end{pmatrix}$ and $B = \begin{pmatrix} 1 & 0 \\ 1 & 1 \end{pmatrix}$, then both A and B are in $GL_2(\mathbf{R})$. We claim that $(AB)^2 \neq A^2 B^2$. Now $AB = \begin{pmatrix} 2 & 1 \\ 1 & 1 \end{pmatrix}$ and $(AB)^2 = \begin{pmatrix} 5 & 3 \\ 3 & 2 \end{pmatrix}$. But $A^2 = \begin{pmatrix} 1 & 2 \\ 0 & 1 \end{pmatrix}$, $B^2 = \begin{pmatrix} 1 & 0 \\ 2 & 1 \end{pmatrix}$, and $A^2 B^2 = \begin{pmatrix} 5 & 2 \\ 2 & 1 \end{pmatrix}$. Hence, $(AB)^2 \neq A^2 B^2$.

HISTORICAL NOTES

   Neils Henrik Abel (1802–1829). Niels Abel was a Norwegian mathematician who made important contributions in many areas of mathematics before his untimely death from tuberculosis at the age of 27. His work expanded our knowledge not only of what eventually became known as abstract algebra, but also of infinite series, elliptic functions and integrals, and *abelian* integrals. (Abel is honored today through the use of the term *abelian group,* even though he did not use the term *group* in his work.) Abel is usually credited as being the first mathematician to prove that it is impossible to find a formula that can be used to solve all 5th-degree polynomials for their roots. This problem had gone unsolved for nearly 300 years, despite the dedicated efforts of many mathematicians. In Chapter 9 we shall see that **Paolo Ruffini** actually presented a proof some years earlier, but his proof was incomplete and he had difficulty in having it verified.

Problem Set 2.1

In exercises 1 through 5 decide if the given set together with the indicated binary operations forms a group.

1. The set $2\mathbf{Z}$ of all even integers together with the operation of ordinary addition of integers

2. The set $\{[0], [2], [4], [6], [8]\}$ under the operation of congruence-class addition modulo 10

3. The set $2\mathbf{Z}$ of all even integers together with the ordinary operation of multiplication of integers

4. The set of all irrational real numbers under the usual operation of addition of real numbers

5. The set $\{[1], [5], [7], [11]\}$ under the operation of congruence-class multiplication modulo 12

6. The set $G = \{1, -1, i, -i, j, -j, k, -k\}$ is a group with identity 1 and multiplication given by

$$(-1)^2 = 1^2 = 1$$
$$i^2 = j^2 = k^2 = -1$$
$$ij = -ji = k$$
$$jk = -kj = i$$
$$ki = -ik = j.$$

Make a multiplication table for G and identify the multiplicative inverse of each element of G.

7. Find elements x and y of a multiplicative group G such that $(xy)^{-1} \neq x^{-1}y^{-1}$. Prove that $(xy)^{-1} = x^{-1}y^{-1}$ for all $x, y \in G$ if and only if G is an abelian group.

8. Theorem 2.1.8 shows that in a multiplicative group G, each of the equations $ax = b$ and $xa = b$ has a unique solution for x for all $a, b \in G$. Formulate and prove a similar property for additive groups.

9. Let G be a group, and suppose that $x \in G$ is such that $x^n = e$ for some positive integer n.

    (a) Show that there is a smallest positive integer n_0 such that $x^{n_0} = e$.
    (b) Show that if m is a positive integer such that $x^m = e$, then $n_0 \mid m$.

10. Prove that $(xy)^2 = x^2y^2$ for all elements x and y of a multiplicative group G if and only if G is abelian.

11. Let $\mathcal{P}(A)$ denote the power set of a set A. Define addition on $\mathcal{P}(A)$ by $X + Y = (X \setminus Y) \cup (Y \setminus X)$ for all $X, Y \in \mathcal{P}(A)$. Assume that addition is associative, and then prove that $\mathcal{P}(A)$ is an abelian group under this operation of addition.

12. Let $A = \{x, y, z\}$. Make an addition table for the group defined in Exercise 11. Identify the additive identity and the additive inverse of each element of $\mathcal{P}(A)$.

13. Let G be a nonempty set together with an associative binary operation of multiplication. An element $e_L \in G$ is said to be a **left identity** for G if $e_L x = x$ for all $x \in G$. If G has left identity and $x \in G$, then an element $x_L^{-1} \in G$ such that $x_L^{-1} x = e_L$ is said to be a **left inverse** for x.

    (a) If G has a left identity and every element of G has a left inverse in G, prove that G is a group.
    (b) If a **right identity** and a **right inverse** of an element of G are defined in a similar fashion, does a similar result hold for G?

14. An element x of a group G is said to be **idempotent** if $x^2 = x$. Prove that every multiplicative group G has exactly one idempotent element.

15. Let G be the set of all ordered pairs (x, y) of real numbers with $x \neq 0$, and define multiplication on G by $(x, y)(z, w) = (xz, yz + w)$. Prove that G is a nonabelian group under this operation.

16. Suppose that G is the set of all rational numbers except -1. Define the operation $*$ on G by $x * y = x + y + xy$. Prove that G together with this binary operation is an abelian group.

17. Let $G = \left\{ \begin{pmatrix} 1 & x \\ 0 & 1 \end{pmatrix} \mid x \in \mathbf{Q}, \ x \neq 0 \right\}$. Show that G is an abelian group under the operation of matrix multiplication.

18. Consider the set $\mathbf{Z}_{24}$ with the binary operation of multiplication modulo 24. Then $\mathbf{Z}_{24}$ is *not* a group since there are elements of $\mathbf{Z}_{24}$ that do not have a multiplicative inverse. Let $U(\mathbf{Z}_{24})$ denote the set of all congruence classes $[x] \in \mathbf{Z}_{24}$ such that $\gcd(x, 24) = 1$. Prove that $U(\mathbf{Z}_{24})$ is a group under multiplication modulo 24. $U(\mathbf{Z}_{24})$ is the **group of units** of $\mathbf{Z}_{24}$. Generalize this for $\mathbf{Z}_n$ and prove your conclusion.

19. Consider the set G of 2×2 matrices $\begin{pmatrix} x & y \\ -y & x \end{pmatrix} \neq \begin{pmatrix} 0 & 0 \\ 0 & 0 \end{pmatrix}$, where $x, y \in \mathbf{R}$. If the operation on G is matrix multiplication, show that G is an abelian group under this operation.

20. Prove that any group with two or three elements must be abelian.

21. Make a multiplication table for the dihedral group D_3 of all symmetries of an equilateral triangle. Show that D_3 is indeed a group that is nonabelian.

22. Suppose that $G = \{x \in \mathbf{R} \mid x > 0 \text{ and } x \neq 1\}$. Define the operation $*$ on G by $x * y = x^{\ln y}$. Prove that G together with this binary operation on G is an abelian group.

23. If G is a group such that $x^2 = e$ for all $x \in G$, prove that G is an abelian group.

24. Prove that a group G has the property that $(xy)^n = x^n y^n$ for all $x, y \in G$ and every integer n if and only if G is abelian.

TECHNOLOGY PROBLEMS

Use a computer algebra system to solve the following problems.

25. Compute an addition table for the additive group $\mathbf{Z}_{15}$.

26. Let $[4]\mathbf{Z}_{60} = \{[4][x] = [4x] \mid [x] \in \mathbf{Z}_{60}\}$ and $[30]\mathbf{Z}_{250} = \{[30][x] = [30x] \mid [x] \in \mathbf{Z}_{250}\}$. Compute the addition table for each of these groups.

27. Use your computer algebra system to show that $[3]\mathbf{Z}_{20} = \mathbf{Z}_{20}$ but that $[4]\mathbf{Z}_{20} \neq \mathbf{Z}_{20}$. See Exercise 26 for the meaning of $[3]\mathbf{Z}_{20}$ and $[4]\mathbf{Z}_{20}$.

28. Find $U(\mathbf{Z}_{100})$, the group of units of $\mathbf{Z}_{100}$.

2.2   SUBGROUPS AND FACTOR GROUPS

Section Overview. This section introduces the notion of a subgroup. Theorem 2.2.3 is important since it provides a method to quickly determine whether or not a nonempty subset is a subgroup. Study this theorem carefully, for it is used frequently in the text. Cosets determined by a subgroup are also introduced, and the invariance of the number of elements in the cosets of a group leads to Lagrange's Theorem. Of particular importance is the concept of a normal subgroup. Such a subgroup induces a congruence relation on the group, and the resulting congruence classes are the cosets determined by the subgroup. The result of defining a binary operation on these cosets is called a factor group. Factor groups are fundamental in the study of groups, so make sure you understand this concept.

There are certain nonempty subsets of a group G that are themselves groups with respect to the same binary operation defined on G. The set $2\mathbf{Z}$ of even integers (or integral multiples of 2) is a group under the binary operation of addition of integers. Addition of even integers is an associative operation since if it was not, addition of integers would not be associative in $\mathbf{Z}$. The set $2\mathbf{Z}$ is closed under addition, for if $x = 2k_1$ and $y = 2k_2$ are even integers, then $x + y = 2k_1 + 2k_2 = 2(k_1 + k_2)$ is an even integer. The even integer $0 = 2 \cdot 0$ is the additive identity of $2\mathbf{Z}$. Finally, if $x = 2k$ is an even integer, then the even integer $-x = 2(-k)$ is the additive inverse of x. Hence, $2\mathbf{Z}$ is indeed a group under the same binary operation of addition that is defined on $\mathbf{Z}$. The group $2\mathbf{Z}$ is known as a **subgroup** of $\mathbf{Z}$. Likewise, $n\mathbf{Z}$, the set of all integral multiples of n, is a subgroup of $\mathbf{Z}$ for any integer n. These observations lead to the following definition of a subgroup.

2.2.1   Definition

A nonempty subset H of a group G is a **subgroup** of G if H is a group when the binary operation on G is restricted to H. A subgroup of G strictly smaller than G is a **proper subgroup** of G. The notation $H \subseteq G$ will indicate that H is a subgroup of G.

The notation $H \subseteq G$ has also been used to indicate that H is a subset of G, so the context of the discussion determines whether $H \subseteq G$ means that H is a subgroup or only a subset of G.

There are many examples of subgroups. Work through each of the following examples carefully, and convince yourself that each is indeed a subgroup of the given group.

2.2.2   Examples

1. If G is a multiplicative group, then G always has at least two subgroups, namely, $\{e\}$ and G itself.

2. If G is a multiplicative group and $x \in G$, then $\langle x \rangle = \{x^n \mid n \in \mathbf{Z}\}$ is a subgroup of G. We refer to $\langle x \rangle$ as the subgroup of G **generated by** x. Any subgroup of G generated in this way by a single element is said to be a **cyclic subgroup** of G. Note that $\langle x \rangle = \{\ldots, x^{-3}, x^{-2}, x^{-1}, e, x, x^2, x^3, \ldots\}$. If there is an integer m such that $x^m = e$, then there is a positive integer n such that $x^n = e$. (Can you show this?) The smallest such positive integer is said to be the **order** of x. The order of x is denoted by $\mathbf{o}(x)$. If there is an element $x \in G$ such that $G = \langle x \rangle$, then G is a **cyclic group**. If a group G is cyclic, then every subgroup of G is cyclic. This is proved in a later chapter where cyclic groups are investigated in more detail, but why don't you try to prove it now?

3. The set of integers $\mathbf{Z} = \{n1 \mid n \in \mathbf{Z}\}$ is an additive cyclic group with generator 1. (-1 also generates $\mathbf{Z}$.) For every integer k, the subgroup $\langle k \rangle = k\mathbf{Z}$ of $\mathbf{Z}$ is a cyclic subgroup of $\mathbf{Z}$.

4. The finite additive group $\mathbf{Z}_{12}$ is a cyclic group of finite order with generator [1]. Each element of $\mathbf{Z}_{12}$ generates a cyclic subgroup of $\mathbf{Z}_{12}$. However, not all of these subgroups are distinct. If $[x] \in \mathbf{Z}_{12}$ and x is relatively prime to 12, the subgroup generated by $[x]$ is $\mathbf{Z}_{12}$. If x is not relatively prime to 12, then $[x]$ generates a proper subgroup of $\mathbf{Z}_{12}$. A quick way to compute the subgroup of $\mathbf{Z}_{12}$ generated by $[x]$ is to add $[x]$ to itself to obtain an element of the subgroup. Now add $[x]$ to this element to obtain another element of the subgroup. Continue until [0] is obtained. The elements generated together with $[x]$ make up the subgroup generated by $[x]$. The following subgroups are the subgroups of $\mathbf{Z}_{12}$ generated by its elements.

$\langle[0]\rangle = \{[0]\}$,
$\langle[1]\rangle = \{[0], [1], [2], [3], [4], [5], [6], [7], [8], [9], [10], [11]\} = \mathbf{Z}_{12}$,
$\langle[2]\rangle = \{[0], [2], [4], [6], [8], [10]\}$,
$\langle[3]\rangle = \{[0], [3], [6], [9]\}$,
$\langle[4]\rangle = \{[0], [4], [8]\}$,
$\langle[5]\rangle = \{[0], [1], [2], [3], [4], [5], [6], [7], [8], [9], [10], [11]\} = \mathbf{Z}_{12}$,
$\langle[6]\rangle = \{[0], [6]\}$,
$\langle[7]\rangle = \{[0], [1], [2], [3], [4], [5], [6], [7], [8], [9], [10], [11]\} = \mathbf{Z}_{12}$,
$\langle[8]\rangle = \{[0], [4], [8]\}$,
$\langle[9]\rangle = \{[0], [3], [6], [9]\}$,
$\langle[10]\rangle = \{[0], [2], [4], [6], [8], [10]\}$,
$\langle[11]\rangle = \{[0], [1], [2], [3], [4], [5], [6], [7], [8], [9], [10], [11]\} = \mathbf{Z}_{12}$.

Hence, we see that the only distinct subgroups of $\mathbf{Z}_{12}$ are $\langle[0]\rangle$, $\langle[2]\rangle$, $\langle[3]\rangle$, $\langle[4]\rangle$, $\langle[6]\rangle$, and $\langle[1]\rangle = \mathbf{Z}_{12}$.

The subgroup structure of $\mathbf{Z}_{12}$ is displayed in the following diagram, which is referred to as the **lattice of subgroups** of $\mathbf{Z}_{12}$.

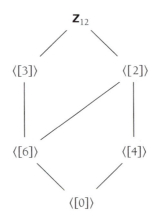

Each group in the lattice is a subgroup of a group that lies above it in the lattice if it is connected to the group by a sequence of line segments. Hence, $\langle[0]\rangle \subseteq \langle[6]\rangle \subseteq \langle[3]\rangle \subseteq \mathbf{Z}_{12}, \langle[0]\rangle \subseteq \langle[6]\rangle \subseteq \langle[2]\rangle \subseteq \mathbf{Z}_{12}$, and $\langle[0]\rangle \subseteq \langle[4]\rangle \subseteq \langle[2]\rangle \subseteq \mathbf{Z}_{12}$ are **chains of subgroups** of $\mathbf{Z}_{12}$.

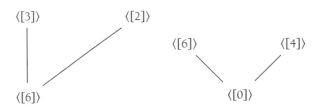

Notice also that $\langle[2]\rangle \cap \langle[3]\rangle = \langle[6]\rangle$ and $\langle[4]\rangle \cap \langle[6]\rangle = \langle[0]\rangle$. We can also see this by inspecting the appropriate sections of the lattice shown earlier. $\langle[6]\rangle$ is the intersection of the two groups that lie directly above it in the lattice. The same is true of $\langle[0]\rangle$. Hence, the intersection of two subgroups is a subgroup, at least for these subgroups. We will prove that this is always true a little later.

5. If D_4 is the nonabelian dihedral group of order 8 given at the beginning of this chapter, then

$$\langle R_{90}\rangle = \{R_0, R_{90}, R_{90}^2 = R_{180}, R_{90}^3 = R_{270}\}$$

is a cyclic subgroup of D_4. Since a cyclic group is abelian, we see that it is possible for a nonabelian group to have nontrivial subgroups that are abelian.

6. Consider the general linear group $GL_2(\mathbf{R})$ of Example 2 of 2.1.4.

$$SL_2(\mathbf{R}) = \left\{ \begin{pmatrix} a & b \\ c & d \end{pmatrix} \middle| \; a, b, c, d \in \mathbf{R} \; \text{and} \; ad - bc = 1 \right\}$$

is a subgroup of $GL_2(\mathbf{R})$. $SL_2(\mathbf{R})$ is the **special linear group** of 2×2 matrices with entries from $\mathbf{R}$.

7. The set $\mathbf{R}^+ = \{x > 0 \mid x \in \mathbf{R}\}$ is a group under multiplication of real numbers. It is a subgroup of the group $\mathbf{R}^*$ of all nonzero real numbers under multiplication.

8. If $M_3(\mathbf{R})$ is the set of all 3×3 matrices with entries from $\mathbf{R}$, then $M_3(\mathbf{R})$ is a group under matrix addition. The set

$$H = \left\{ \begin{pmatrix} a & 0 & 0 \\ 0 & a & 0 \\ 0 & 0 & a \end{pmatrix} \middle| \; a \in \mathbf{R} \right\}$$

under matrix addition is a subgroup of $M_3(\mathbf{R})$.

9. The set $\mathbf{N}$ is a *not* a subgroup of $\mathbf{Z}$ under addition. Addition is indeed a binary operation on $\mathbf{N}$ since the sum of two positive integers is a positive integer. However, $0 \notin \mathbf{N}$, so $\mathbf{N}$ does not have an additive identity. Thus, $\mathbf{N}$ cannot be a subgroup of $\mathbf{Z}$. The elements of $\mathbf{N}$ also fail to have additive inverses because if $x \in \mathbf{N}$, then the additive inverse $-x$ of x is *not* in $\mathbf{N}$.

10. $\mathbf{Z}_2 \times \mathbf{Z}_2$ is an additive abelian group if addition is defined on $\mathbf{Z}_2 \times \mathbf{Z}_2$ by $([x], [y]) + ([z], [w]) = ([x] + [z], [y] + [w])$. It is known as the **Klein 4-group**. The group $\mathbf{Z}_2 \times \mathbf{Z}_2$ has four subgroups, $\{[0]\} \times \{[0]\}$, $\mathbf{Z}_2 \times \{[0]\}$, $\{[0]\} \times \mathbf{Z}_2$, and $\mathbf{Z}_2 \times \mathbf{Z}_2$. The lattice of subgroups of $\mathbf{Z}_2 \times \mathbf{Z}_2$ is given below.

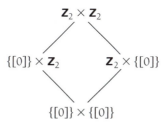

It is an accepted practice when one is proving properties of groups to write the group multiplicatively. We follow this practice. The corresponding results for additive groups can be obtained by the appropriate translation of language and notation to that of an additive group. Several of the following theorems provide you with an opportunity to translate theorems proven for multiplicative groups to equivalent results for additive groups.

If H is a nonempty subset of a multiplicative group G, then multiplication of elements of H is always associative. This follows because if $x, y, z \in H$ and $x(yz) \neq (xy)z$, then $x, y, z \in G$, and this would indicate that multiplication is not associative in G. For this reason, we say that **associativity of multiplication in H is inherited from associativity of multiplication in G.** Hence, when showing that a nonempty subset of a group G is a subgroup of a group G, the associativity of the binary operation on H automatically holds, and we need not prove this fact.

The following two theorems give two ways to show that a subset of a group G is a subgroup of G.

2.2.3   Theorem

If H is a nonempty subset of a group G, then H is a subgroup of G if and only if $xy^{-1} \in H$ whenever $x, y \in H$.

Proof. Let H be a nonempty subset of a group G that is a subgroup of G. If $x, y \in H$, then y has an inverse $y^{-1} \in H$, so $xy^{-1} \in H$ since H is closed under the binary operation on G. Conversely, suppose that H is a nonempty subset of G and $xy^{-1} \in H$ whenever $x, y \in H$. We need to show that H is a group. First, if $x = y$, then $e = xx^{-1} \in H$, so H has an identity. But $e, x \in H$, so $x^{-1} = ex^{-1} \in H$. Thus, H has inverses. Since H has inverses, $x, y^{-1} \in H$, whenever $x, y \in H$, so $xy = x(y^{-1})^{-1}$ is in H. Hence, H is closed under the binary operation on G. Finally, associativity of multiplication of elements of H is inherited from G, so we have shown that H is a group. Therefore, H is a subgroup of G. ∎

2.2.4   Theorem

If H is a subset of a group G, then H is a subgroup of G if and only if the following three conditions hold:

1. $H \neq \varnothing$.
2. $xy \in H$ whenever $x, y \in H$.
3. $x^{-1} \in H$ whenever $x \in H$.

Proof. If H is a subgroup of a group G, it follows immediately from the definition of a subgroup that conditions 1, 2, and 3 hold. Conversely, if the three conditions hold, then H is nonempty, so suppose that $x, y \in H$. Then y^{-1} is in H by condition 3. But condition 2 now indicates that $xy^{-1} \in H$, so by Theorem 2.2.3, H is a subgroup of G. ∎

By translating Theorem 2.2.3 to additive groups, we see that a nonempty subset H of an additive group G is a subgroup of G if and only if whenever $x, y \in H$,

$x - y \in H$, where **subtraction** is defined in G by $x - y = x + (-y)$. The translation of Theorem 2.2.4 to additive groups indicates that H is a subgroup of an additive group G if and only if $H \neq \varnothing$, $x + y \in H$, and $-x \in H$ whenever x and y are in H.

Theorem 2.2.3 can be used to quickly prove that the intersection of any family of subgroups of a group G is a subgroup of G.

2.2.5   Theorem

If $\{H_\alpha\}_{\alpha \in \Delta}$ is a family of subgroups of a group G, then $\cap_{\alpha \in \Delta} H_\alpha$ is a subgroup of G.

Proof. Note first that $\cap_{\alpha \in \Delta} H_\alpha \neq \varnothing$ since $e \in H_\alpha$ for all $\alpha \in \Delta$. Next, if $x, y \in \cap_{\alpha \in \Delta} H_\alpha$, then x and y are in H_α for each $\alpha \in \Delta$. Thus, by Theorem 2.2.3, $xy^{-1} \in H_\alpha$ for each $\alpha \in \Delta$, so $xy^{-1} \in \cap_{\alpha \in \Delta} H_\alpha$. Hence, $\cap_{\alpha \in \Delta} H_\alpha$ is a subgroup of G. ∎

2.2.6   Definition

Let H be a subgroup of a group G. If $x \in G$, let $xH = \{xh \mid h \in H\}$ and $Hx = \{hx \mid h \in H\}$. The subset xH is said to be a **left coset** of H in G (Note that x is on the left of H in xH.) and Hx is a **right coset** of H in G. The element x is said to be the **representative of the coset** xH. The set of all left cosets of H in G is denoted by G/H, as is the set of all right cosets of H in G. The context of the discussion determines which is being considered. If x is an element of an additive group G, then $x + H = \{x + h \mid h \in H\}$ and $H + x = \{h + x \mid h \in H\}$ are left and right cosets of H in G, respectively.

2.2.7   Examples

1. The set $4\mathbf{Z} = \{\ldots, -8, -4, 0, 4, 8, \ldots\}$ is a subgroup of the additive abelian group $\mathbf{Z}$. There are exactly four distinct left cosets of $4\mathbf{Z}$ in $\mathbf{Z}$:

    $$0 + 4\mathbf{Z} = \{\ldots, -8, -4, 0, 4, 8, \ldots\} = 4\mathbf{Z},$$
    $$1 + 4\mathbf{Z} = \{\ldots, -7, -3, 1, 5, 9, \ldots\},$$
    $$2 + 4\mathbf{Z} = \{\ldots, -6, -2, 2, 6, 10, \ldots\},$$
    $$3 + 4\mathbf{Z} = \{\ldots, -5, -1, 3, 7, 11, \ldots\}.$$

    Note that these cosets are pairwise disjoint and that $\mathbf{Z}$ is the union of these cosets.

2. The set $H = \langle [5] \rangle = \{[0], [5], [10], [15]\}$ is a subgroup of the cyclic group $\mathbf{Z}_{20}$. The distinct cosets of H in $\mathbf{Z}_{20}$ are

    $$[0] + H = \{[0], [5], [10], [15]\},$$

$$[1] + H = \{[1], [6], [11], [16]\},$$
$$[2] + H = \{[2], [7], [12], [17]\},$$
$$[3] + H = \{[3], [8], [13], [18]\}, \qquad \text{and}$$
$$[4] + H = \{[4], [9], [14], [19]\}.$$

As in the first example, the cosets are pairwise disjoint, and $\mathbf{Z}_{20}$ is the union of these cosets. It is also the case that the order 4 of H divides the order 20 of $\mathbf{Z}_{20}$ and $20 \div 4 = 5$ is the number of distinct cosets of H in $\mathbf{Z}_{20}$.

3. An interesting example is also provided by the dihedral group

$$D_4 = \{R_0, R_{90}, R_{180}, R_{270}, L_1, L_2, L_3, L_4\},$$

given in the introduction to Section 2.1. It has already been pointed out that $\langle R_{90} \rangle = \{R_0, R_{90}, R_{180}, R_{270}\}$ is a cyclic subgroup of D_4. It is the subgroup of rotations of D_4. The distinct left cosets of $\langle R_{90} \rangle$ in D_4 are $R_0 \langle R_{90} \rangle = \langle R_{90} \rangle$ and $L_1 \langle R_{90} \rangle$. The elements of each coset can be determined by inspection of the multiplication table for D_4:

$$\langle R_{90} \rangle = \{R_0, R_{90}, R_{180}, R_{270}\},$$
$$L_1 \langle R_{90} \rangle = \{L_1, L_2, L_3, L_4\}.$$

Again notice that these cosets are disjoint and that $D_4 = \langle R_{90} \rangle \cup L_1 \langle R_{90} \rangle$. Notice also that the order 4 of $\langle R_{90} \rangle$ divides the order 8 of D_4 and that $8 \div 4 = 2$ is the number of distinct cosets of $\langle R_{90} \rangle$ in D_4.

There are several things you should notice in the preceding examples:

1. A coset need not be a subgroup. In the first example, only the coset $0 + 4\mathbf{Z} = 4\mathbf{Z}$ is a subgroup of $\mathbf{Z}$. The other cosets are not closed under addition and thus cannot be subgroups of $\mathbf{Z}$. For example, 1 and 5 are in $1 + 4\mathbf{Z}$, but $1 + 5 = 6 \notin 1 + 4\mathbf{Z}$. Similar observations can be made with regard to the cosets of the second and third examples.
2. In each of the examples, the set of all cosets forms a partition of the group.
3. When the group is finite, each coset has the same number of elements.
4. When the group is finite, the order of the subgroup divides the order of the group and the quotient of this division is the number of distinct cosets determined by the subgroup.

We now show that the second and third observations hold for every group. The fourth observation that for a finite group G, the order of a subgroup H divides the order of G was first proved by **Joseph Louis Lagrange** (1736–1813), an Italian mathematician of French ancestry who made lasting contributions to mathematics.

2.2.8   **Theorem**

If G is a group with subgroup H, then the following statements are true:

1. If $x \in G$, then $xH = H$ if and only if $x \in H$.
2. If $x, y \in G$, then either $xH = yH$ or $xH \cap yH = \varnothing$ and the set G/H of all left cosets of H in G forms a partition of G.
3. If $x, y \in G$, then $f: xH \to yH$ defined by $f(xh) = yh$ is a bijective function.

Proof

1. Suppose $xH = H$. If $xh \in H$, then $xh = h'$ for some $h' \in H$. But then $x = h'h^{-1} \in H$. Conversely, if $x \in H$, then $xH \subseteq H$ since H is closed under the binary operation on G. Next, let $h \in H$. Since $x \in H$ and H is a subgroup of G, $x^{-1} \in H$. Therefore, $x^{-1}h \in H$, so we see that $h = x(x^{-1}h) \in xH$. Thus, $H \subseteq xH$ and so $xH = H$.

2. Suppose $xH \cap yH \neq \varnothing$, and let $xh' = yh'' \in xH \cap yH$. Then $x = yh''h'^{-1}$. If $xh \in xH$, then $xh = yh''h'^{-1}h \in yH$ since $h''h'^{-1}h \in H$. Hence, $xH \subseteq yH$. A similar argument shows that $yH \subseteq xH$, so $xH = yH$. Therefore, the distinct cosets in G/H are pairwise disjoint. Now let Δ be the set of $x \in G$ such that $\{xH\}_{x \in \Delta}$ is the family of distinct cosets of H in G. To show that $\{xH\}_{x \in \Delta}$ forms a partition of G, it must be shown that G is the union of these cosets. Clearly, $\cup_{x \in \Delta} xH \subset G$, so let y be an arbitrary element of G. We claim that $y \in xH$ for some $x \in \Delta$. Suppose that $y \notin xH$ for all $x \in \Delta$. Then $yH \cap xH = \varnothing$ for all $x \in \Delta$. Thus, yH is a coset that is distinct from all of the cosets in $\{xH\}_{x \in \Delta}$. But this is impossible since every distinct coset of H in G is in $\{xH\}_{x \in \Delta}$. Thus, for an arbitrary element $y \in G$, there is an $x \in \Delta$ such that y is an element of xH. Hence, we have $y \in \cup_{x \in \Delta} H_\alpha$ and so $G \subseteq \cup_{x \in \Delta} xH$.

3. Let $x, y \in G$, and suppose that $f: xH \to yH$ is defined by $f(xh) = yh$. If $xh = xh'$, then $x^{-1}xh = x^{-1}xh'$ and thus $h = h'$. But multiplication is well-defined in G, so $yh = yh'$. Hence, $f(xh) = f(xh')$, so f is well-defined. The function f is clearly surjective because if $yh \in yH$, then $f(xh) = yh$. Now suppose that $f(xh) = f(xh')$. Then $yh = yh'$, so $y^{-1}yh = y^{-1}yh'$, which implies that $h = h'$. Hence, $xh = xh'$ and so f is injective. Therefore, f is a bijective function. ∎

If S is a finite set, the number of elements in S is denoted by $|S|$. If G is a finite group, recall that $|G|$ denotes the **order** of G. Since a bijective function between two finite sets just says that the two sets have the same number of elements, part 3 of Theorem 2.2.8 shows that if H is a finite subgroup of G, then $|xH| = |yH|$ for any x and y in G. We have now developed the machinery sufficient to prove the Theorem of Lagrange mentioned earlier.

2.2.9   Lagrange's Theorem

If H is a subgroup of a finite group G, then the order of H divides the order of G.

Proof. In view of Theorem 2.2.8, the set G/H of all left cosets of H in G forms a partition of G. Furthermore, every pair of left cosets of H in G has the same number of elements, so $|H| = |xH|$ for all $x \in G$. Since G is finite, there can only be a finite number of distinct left cosets of H in G. If $x_1, x_2, \ldots, x_n$ belong to G and are such that $x_1H, x_2H, \ldots, x_nH$ are the disjoint, distinct left cosets of H in G, then $G = x_1H \cup x_2H \cup \cdots \cup x_nH$. Consequently, G is the union of n distinct cosets, each with $|H|$ elements. Thus, $|G| = n|H|$ and so $|H|$ divides $|G|$. ■

You will be asked to prove the following two corollaries to Lagrange's Theorem in the exercises.

2.2.10   Corollary

If G is a finite group and $x \in G$, then $o(x)$ divides the order of G.

2.2.11   Corollary

Any finite group with prime order is cyclic.

If G is a group of order n and k is a positive integer that divides n, then it is possible for G to have a subgroup of order k. Lagrange's Theorem, however, *does not* say that G *must* have a subgroup of order k but only that it is *possible* for G to have a subgroup of order k. Later in the text we shall see that the converse of Lagrange's Theorem holds for cyclic groups. That is, we will see that if G is a finite cyclic group of order n and k is a divisor of n, then G has a subgroup of order k.

2.2.12   Definition

If H is a subgroup of a group G, then the number of distinct left cosets of H in G is called the **index** of H in G. The index of H in G is denoted by $[G : H]$.

If G is a finite group, Lagrange's Theorem shows that the index of H in G is $\frac{|G|}{|H|}$. If G is an infinite group, $\frac{|G|}{|H|}$ does not make sense. However, groups of infinite order may have subgroups that are of finite index. The additive group $\mathbf{Z}$ is such that $[\mathbf{Z} : n\mathbf{Z}] = n$ for each positive integer n, while the index of $\{0\}$ in $\mathbf{Z}$ is infinite. More specifically, Example 1 of 2.2.7 shows that $[\mathbf{Z} : 4\mathbf{Z}] = 4$.

Theorem 2.2.8 shows that when H is a subgroup of a group G, the set of all (distinct) left cosets G/H of H in G is a partition of G. By invoking Theorem 0.2.5, we see that G/H determines an equivalence relation E on G. This equivalence relation is defined by xEy if and only if $x, y \in zH$ for some coset $zH \in G/H$. Thus, $x = zh_1$ and $y = zh_2$, where $h_1, h_2 \in H$. But then $xh_1^{-1} = yh_2^{-1}$, so $y^{-1}x = h_2^{-1}h_1 \in H$. Therefore, xEy if and only if $y^{-1}x \in H$. The following theorem verifies that E is indeed an equivalence relation on G and that the equivalence classes determined by E are the left cosets of H in G.

2.2.13   Theorem

Suppose that H is a subgroup of a group G. If the relation E is defined on G by xEy if and only if $y^{-1}x \in H$, then E is an equivalence relation on G, and the equivalence classes determined by E are the left cosets of H in G.

Proof.   Clearly, xEx for any $x \in G$ since $x^{-1}x = e \in H$. Hence, E is reflexive. If xEy, then $y^{-1}x \in H$. Since H is a group, $x^{-1}y = (y^{-1}x)^{-1} \in H$. Therefore, yEx, so E is symmetric. If xEy and yEz, then $y^{-1}x$ and $z^{-1}y$ are in H. Consequently, $z^{-1}x = (z^{-1}y)(y^{-1}x) \in H$, which indicates that xEz, so E is transitive. Thus, E is an equivalence relation on G. Finally, if $[x]$ is the equivalence class determined by $x \in G$, then $y \in [x]$ if and only if yEx, which in turn is true if and only if $x^{-1}y \in H$. But $x^{-1}y \in H$ if and only if $y = xh$ for some $h \in H$, so $[x] = xH$. ∎

If H is a subgroup of a group G, then one wonders if it is possible to make the set of left cosets G/H of H in G into a group in a nontrivial way. The answer is yes, provided that the subgroup H possesses a property that is sufficient to turn the equivalence relation of Theorem 2.2.13 into a congruence relation on G. These subgroups, defined next, are very closely connected to the concept of a group homomorphism, which is studied in the next section of this chapter.

2.2.14   Definition

If G is a group, then a subgroup H of G is said to be a **normal subgroup** of G if $xH = Hx$ for all $x \in G$.

The following theorem is often quite useful in showing that a subgroup is normal.

2.2.15   Theorem

If H is a subgroup of G, then H is a normal subgroup of G if and only if for each $x \in G$, $xhx^{-1} \in H$ for all $h \in H$.

Proof.   Suppose that H is normal in G, and let $x \in G$ and $h \in H$. Now $xh \in xH$, and since $xH = Hx$, $xh = h'x$ for some $h' \in H$. Hence, $xhx^{-1} = h'$ and thus $xhx^{-1} \in H$. Conversely, suppose that $xhx^{-1} \in H$ for each $x \in G$ and all $h \in H$. To show that H is normal, we must show that $xH = Hx$ for each x in G. Let $x \in G$, and suppose that $y \in xH$. If $y = xh$, then by assumption xhx^{-1} is in H. If $xhx^{-1} = h' \in H$, then $y = xh = h'x \in Hx$. Hence, $xH \subseteq Hx$. Similarly, $Hx \subseteq xH$, so $xH = Hx$. ■

When H is a normal subgroup of G, every left coset is a right coset and every right coset is a left coset, so there is no need to make a distinction between the left and right cosets of H in G.

If H is a normal subgroup of G, then $xH = Hx$ for each $x \in G$. You should not be led to believe, however, that $xH = Hx$ means that $xh = hx$ for each $h \in H$. If $xh \in xH$, the most that can be said is that there is an $h' \in H$ such that $xh = h'x \in Hx$. Similarly, if $hx \in Hx$, there is an $h' \in H$ such that $hx = xh' \in xH$. Of course, if G is an abelian group, then $xh = hx$ for each $h \in H$. In fact, if G is an abelian group, then every subgroup of G is a normal subgroup.

2.2.16   Theorem

If H is a normal subgroup of a group G, then the equivalence relation E of Theorem 2.2.13 is a congruence relation on G. Furthermore, the binary operation defined on G/H by $(xH)(yH) = xyH$ is well-defined, and G/H together with this binary operation is a group.

Proof.   If it can be shown that E is a congruence relation on G, then invoking Theorem 0.4.8 shows that the binary operation is well-defined. So let's show that E is a congruence relation on G. If $x_1 E y_1$ and $x_2 E y_2$, we need to show that $x_1 x_2 E y_1 y_2$. According to the definition of E, this means that we are to show that $(y_1 y_2)^{-1} (x_1 x_2) \in H$. If $x_1 E y_1$ and $x_2 E y_2$, then $y_1^{-1} x_1, y_2^{-1} x_2 \in H$, so $x_1 = y_1 h_1$ and $x_2 = y_2 h_2$, where $h_1, h_2 \in H$. Thus, $x_1 x_2 = y_1 (h_1 y_2) h_2$. But since H is a normal subgroup of G, $h_1 y_2 \in H y_2 = y_2 H$ implies that there is an $h_3 \in H$ such that $h_1 y_2 = y_2 h_3$. Hence, $x_1 x_2 = y_1 (y_2 h_3) h_2 = (y_1 y_2)(h_3 h_2)$, so $(y_1 y_2)^{-1} (x_1 x_2) = h_3 h_2 \in H$. Hence, E is indeed a congruence relation on G. Thus, by Theorem 0.4.8, multiplication of the congruence classes in G (the left cosets of H in G) is a well-defined binary operation on G/H. It now follows easily that G/H is a group. The identity of G/H is eH and the multiplicative inverse of xH is $x^{-1}H$. ■

When H is a normal subgroup of G, the group G/H is said to be the **factor group** of G by H. The group G/H has the cosets of H in G as its elements, and the binary operation on G/H is defined by $(xH)(yH) = xyH$. We say that G/H has been formed by **factoring out** H.

2.2.17 | **Examples**

For the first three of the following four examples of factor groups, we return to the three examples of 2.2.7.

1. The additive abelian group $\mathbf{Z}$ has $4\mathbf{Z}$ as a normal subgroup. Thus, we can form the factor group $\mathbf{Z}/4\mathbf{Z}$ whose elements are

$$0 + 4\mathbf{Z} = \{..., -8, -4, 0, 4, 8, ...\} = 4\mathbf{Z},$$
$$1 + 4\mathbf{Z} = \{..., -7, -3, 1, 5, 9, ...\},$$
$$2 + 4\mathbf{Z} = \{..., -6, -2, 2, 6, 10, ...\}, \quad \text{and}$$
$$3 + 4\mathbf{Z} = \{..., -5, -1, 3, 7, 11, ...\}.$$

The binary operation on this factor group is coset addition. For example, $(2 + 4\mathbf{Z}) + (3 + 4\mathbf{Z}) = (2 + 3) + 4\mathbf{Z} = 5 + 4\mathbf{Z} = 1 + (4 + 4\mathbf{Z}) = 1 + 4\mathbf{Z}$. The additive identity of $\mathbf{Z}/4\mathbf{Z}$ is $0 + 4\mathbf{Z}$.

2. The group $H = \{[0], [5], [10], [15]\}$ is a normal subgroup of the additive abelian group $\mathbf{Z}_{20}$. The distinct cosets in the factor group $\mathbf{Z}_{20}/H$ are

$$[0] + H = \{[0], [5], [10], [15]\} = H,$$
$$[1] + H = \{[1], [6], [11], [16]\},$$
$$[2] + H = \{[2], [7], [12], [17]\},$$
$$[3] + H = \{[3], [8], [13], [18]\}, \quad \text{and}$$
$$[4] + H = \{[4], [9], [14], [19]\}.$$

The binary operation is coset addition. For example, $([3] + H) + ([4] + H) = ([3] + [4]) + H = [3 + 4] + H = [7] + H = [2] + H$. Note that $|H| = 4$, $|\mathbf{Z}_{20}| = 20$, and $4 \,|\, 20 = 5$. The index of H in $\mathbf{Z}_{20}$ is 5, the number of distinct cosets in the factor group $\mathbf{Z}_{20}/H$.

3. The nonabelian dihedral group D_4 has $H = \langle R_{90} \rangle$ as a subgroup. The set D_4/H of distinct left cosets of H in D_4 is the set with the following two elements:

$$H = \{R_0, R_{90}, R_{180}, R_{270}\},$$
$$L_1 H = \{L_1, L_2, L_3, L_4\}.$$

We claim that H is a normal subgroup of D_4. According to Theorem 2.2.15, H is a normal subgroup of D_4 if $xhx^{-1} \in H$ for each $x \in D_4$ and all $h \in H$. To show this directly would involve a lot of cases, so let's try to convince ourselves that H is normal by another method. Think of the square as a piece of paper. A rotation rotates the paper and a reflection flips it over. Now let $x \in G$ and $h \in H$. Since h is always a rotation, there are only two cases to consider: the case when x is a rotation and the case when x is a reflection. If x is a rotation, then x^{-1} is a rotation since H is a subgroup of G. Hence, xhx^{-1} is a product of three rotations, which is a single rotation since H is a subgroup. Hence, $xhx^{-1} \in H$. If x is a reflection, then $x^{-1} = x$ since a reflection turns the paper over face down and a subsequent reflection about the same line returns the paper to its original position face up. Hence, $xhx^{-1} = xhx$. Now x flips the paper over so its face is down, h rotates the paper, and x flips the paper again to leave it face up, so the result is a rotation of the paper. Hence, again we have $xhx^{-1} \in H$. Therefore, H is a normal subgroup of D_4. The following multiplication table for D_4/H shows that the factor group D_4/H is abelian. Hence, we see that it is possible for a nonabelian group to have an abelian factor group.

$\cdot$	H	L_1H
H	H	L_1H
L_1H	L_1H	H

4. Consider the nonabelian group $G = \{1, -1, i, -i, j, -j, k, -k\}$ with identity 1 and multiplication given by

$$(-1)^2 = 1^2 = 1, \qquad ij = -ji = k, \qquad ki = -ik = j,$$
$$i^2 = j^2 = k^2 = -1, \qquad jk = -kj = i.$$

The set $H = \{-1, 1\}$ is a subgroup of G and G/H is the set of left cosets whose elements are H, iH, jH, and kH. Since $xH = \{-x, x\} = Hx$ for any $x \in G$, H is a normal subgroup of G. The multiplication table for the factor group G/H follows.

$\cdot$	H	iH	jH	kH
H	H	iH	jH	kH
iH	iH	H	kH	jH
jH	jH	kH	H	iH
kH	kH	jH	iH	H

The factor group G/H is abelian, so this is another example of a nonabelian group that has an abelian factor group.

We have seen in Theorem 2.2.5 that an intersection of subgroups is a subgroup. A similar result holds for normal subgroups.

2.2.18   Theorem

If $\{H_\alpha\}_{\alpha \in \Delta}$ is a family of normal subgroups of a group G, then $H = \cap_{\alpha \in \Delta} H_\alpha$ is also a normal subgroup of G.

Proof. Theorem 2.2.5 shows that H is a subgroup of G. If $x \in G$ and $h \in H$, then $h \in H_\alpha$ for each $\alpha \in \Delta$. But then xhx^{-1} is in each H_α since each H_α is a normal subgroup of G. Therefore, $xhx^{-1} \in H$, so Theorem 2.2.15 indicates that H is a normal subgroup of G. ■

If H is a normal subgroup of a group G, then elements of G can, in a weak sense, be commuted past elements of H. Indeed, if $x \in G$ and $h \in H$, then $xh \in xH = Hx$, so there is an $h' \in H$ such that $xh = h'x$. Hence, when H is a normal subgroup of G, we have a quasi-commutative condition between the elements of G and those of H. There are instances when elements of two normal subgroups actually commute as shown by the following theorem.

2.2.19 **Theorem**

If H and K are normal subgroups of G and $H \cap K = \{e\}$, then $xy = yx$ for all $x \in H$ and all $y \in K$.

Proof. Consider the element $xyx^{-1}y^{-1}$ of G, where $x \in H$ and $y \in K$, and note that $xyx^{-1}y^{-1} = (xyx^{-1})y^{-1}$ implies that $xyx^{-1}y^{-1} \in K$ since xyx^{-1} and y^{-1} are in K. Likewise, $xyx^{-1}y^{-1} = x(yx^{-1}y^{-1}) \in H$ since x and $yx^{-1}y^{-1}$ are in H. Hence, we have $xyx^{-1}y^{-1} \in H \cap K = \{e\}$, so $xyx^{-1}y^{-1} = e$. Therefore, $xy = yx$. ■

We conclude with the following theorem, which presents some interesting facts about normal subgroups. We prove parts 1 and 2 of the theorem. Parts 3 and 4 have similar proofs.

2.2.20 **Theorem**

Suppose that H and K are subgroups of a group G, and let $HK = \{hk \mid h \in H$ and $k \in K\}$.

1. If either H or K is a normal subgroup of G, then HK is a subgroup of G.
2. If both H and K are normal subgroups of G, then HK is a normal subgroup of G.
3. If K is a normal subgroup of G, then $H \cap K$ is a normal subgroup of H.
4. If K is a normal subgroup of G, then K is a normal subgroup of HK.

Proof

1. Obviously, $e \in HK$ since e is in both H and K. Hence, $HK \neq \varnothing$. Now suppose that H is normal in G, and let $h_1k_1, h_2k_2 \in HK$. Then for some

$h' \in H, (h_1k_1)(h_2k_2) = h_1(k_1h_2)k_2 = h_1(h'k_1)k_2 = (h_1h')(k_1k_2) \in HK$, so HK is closed under multiplication. If $hk \in HK$, then for some $h'' \in H$, $(hk)^{-1} = k^{-1}h^{-1} = h''k^{-1}$ is in HK. Thus, HK has inverses and so, by Theorem 2.2.4, HK is a subgroup of G. A similar proof can be given if K is a normal subgroup of G.

2. Suppose that both H and K are normal in G. By part 1, HK is a subgroup of G. If $hk \in HK$ and $x \in G$, then there exist $h' \in H$ and $k' \in K$ such that $x(hk)x^{-1} = (xh)(kx^{-1}) = (xh)(x^{-1}k') = x(hx^{-1})k' = x(x^{-1}h')k' = (xx^{-1})(h'k') = h'k' \in HK$. Theorem 2.2.15 now shows that HK is a normal subgroup of G. ∎

HISTORICAL NOTES

Joseph Louis Lagrange (1736–1813). Lagrange was born in Turin, Italy, and he developed an early interest in mathematics. Despite his father's encouragement to study law, Lagrange had by the age of 19 accepted a position as professor of mathematics at the Royal Artillery School in Turin. At the age of 30, Lagrange was appointed head of the Berlin Academy, where he enjoyed the patronage of the Prussian king, Frederick the Great. After Frederick's death in 1786, Lagrange moved to Paris as a member of the French Academy, and he remained there for the rest of his life.

Throughout his life, Lagrange made many contributions of fundamental importance to mathematics. He contributed to the development of the theory of equations, differential equations, and number theory, and he proved the theorem in group theory that now bears his name. Moreover, he is responsible, through his methods of solving third- and fourth-degree polynomial equations by radicals, for laying the foundation for what is now known as Galois theory. This work allowed Abel and Galois to develop methods to advance the theory of solutions of polynomial equations. Lagrange also headed a commission charged with devising a new system of weights and measures; from this effort the metric system emerged.

Problem Set 2.2

1. (a) Let $H = \{[0], [3], [6], [9]\}$. How many distinct cosets of H in $\mathbf{Z}_{12}$ are there, and what are they?

    (b) Let $H = \{[0], [2], [4], [6], [81, [10]\}$. How many distinct cosets of H in $\mathbf{Z}_{12}$ are there, and what are they?

2. Suppose that G is a finite group with 54 elements. Is it possible for G to have a subgroup with 27 elements? With 7 elements?

3. Construct the lattice of subgroups for each of the following additive groups.

    (a) $\mathbf{Z}_{15}$ (b) $\mathbf{Z}_{20}$

4. Construct the lattice of subgroups of the multiplicative group G that has the following multiplication table.

$\bullet$	e	x	y	z
e	e	x	y	z
x	x	e	z	y
y	y	z	e	x
z	z	y	x	e

5. If H is a subgroup of a group G and G is a subgroup of a group K, prove that H is a subgroup of K.

6. If H is a subgroup of G, then H and G both have identities e_H and e_G, respectively. Show that $e_H = e_G$.

7. Consider the group $GL_2(\mathbf{R})$ of Example 2 of 2.1.4. Show that each of the following is a subgroup of $GL_2(\mathbf{R})$.

    (a) $H_1 = \left\{ \begin{pmatrix} a & 0 \\ 0 & a \end{pmatrix} \middle| \ a \in \mathbf{R}, \ a \neq 0 \right\}$

    (b) $H_2 = \left\{ \begin{pmatrix} a & b \\ -b & a \end{pmatrix} \middle| \ a, b \in \mathbf{R}, \ a \neq 0 \ \text{or} \ b \neq 0 \right\}$

    (c) $H_3 = \left\{ \begin{pmatrix} a & b \\ c & d \end{pmatrix} \middle| \ a, b, c, d \in \mathbf{R}, \ ad - bc = 1 \right\}$

    (d) Is H_1 a subgroup of H_2? Is H_2 a subgroup of H_3?

8. If G is a group and $x \in G$, prove that $\langle x \rangle = \{x^n \mid n \in \mathbf{Z}\}$ is a subgroup of G.

9. Prove Corollary 2.2.10.

10. Prove parts 3 and 4 of Theorem 2.2.20.

11. If H and K are subgroups of a group G, let $HK = \{hk \mid h \in H, k \in K\}$. Prove that $H^2 = HH = H$ for any subgroup H of G. If H is a finite nonempty subset of G such that $H^2 \subseteq H$, prove that H is a subgroup of G.

12. Let H and K be subgroups of a group G, and define HK as in Exercise 11. Prove that HK is a subgroup of G if and only if $KH \subseteq HK$.

13. Consider the multiplicative group $\mathbf{C}^*$ of nonzero complex numbers. Show that the set H of complex numbers $a + bi$ such that $a^2 + b^2 = 1$ is a subgroup of $\mathbf{C}^*$.

14. Let H be a subgroup of a group G, and let $\mathscr{L}$ and $\mathscr{R}$ denote the set of left and right cosets of H in G, respectively. Define $f : \mathscr{L} \to \mathscr{R}$ by $f(xH) = Hx^{-1}$. Show that f is a well-defined bijection. Conclude that if $[G : H]_L$ and $[G : H]_R$ denote the "number" of left and right cosets of H in G, respectively, then $[G : H]_L < \infty$ if and only if $[G : H]_R < \infty$ and, when this is the case, $[G : H]_L = [G : H]_R$.

15. If G is a finite group with subgroups H_1 and H_2 such that $H_1 \subseteq H_2$, show that $[G : H_1] = [G : H_2][H_2 : H_1]$.

16. Let H be a normal subgroup of G. Define the relation $\equiv$ on G by $x \equiv y$ if and only if $xH \equiv yH$. Prove that $\equiv$ is a congruence relation on G.

17. Let H be a subgroup of a finite group G. If the index of H in G is 2, must H be a normal subgroup of G? If your answer is yes, give a proof. If your answer is no, give a counterexample.

18. If H is a subgroup of a group G, prove that xHx^{-1} is a subgroup of G for any $x \in G$. The group xHx^{-1} is said to be a **conjugate** of H, and H and xHx^{-1} are said to be **conjugate subgroups**. If H is a subgroup of G, prove that H is a normal subgroup of G if and only if every conjugate subgroup of H coincides with H.

19. Suppose that G is a group, and let S be a subset of G. Prove that the intersection of all the subgroups of G that contain S is the smallest subgroup of G that contains S. This group is said to be the **subgroup of G generated by** S and is denoted by $\langle S \rangle$. Describe the group $\langle \varnothing \rangle$. If S is a subset of G, prove that $\langle S \rangle$ is the group described by $\{a_1^{\epsilon_1} a_2^{\epsilon_2} \cdots a_n^{\epsilon_n} \mid n \in \mathbf{Z}, a_i \in S$, and $\epsilon_i = \pm 1$ for each i, where all of the a_i need not be distinct$\}$. That is, $\langle S \rangle$ is the set of all finite products of elements of S whose exponents are 1 or -1.

TECHNOLOGY PROBLEMS

Use a computer algebra system to solve each of the following problems.

20. Compute all of the subgroups of $\mathbf{Z}_{18}$.

21. Determine all of the distinct cosets of the subgroup $H = \{[0], [4], [8], [12], [16], [20], [24], [28]\}$ in the additive group $\mathbf{Z}_{32}$.

22. The sets $[4]\mathbf{Z}_{24}$ and $[6]\mathbf{Z}_{24}$ are subgroups of $\mathbf{Z}_{24}$, so $[4]\mathbf{Z}_{24} + [6]\mathbf{Z}_{24}$ is a subgroup of $\mathbf{Z}_{24}$. (See Theorem 2.2.20.) Use a computer algebra system to list the elements of $[4]\mathbf{Z}_{24} + [6]\mathbf{Z}_{24}$. Determine whether there is a positive integer n such that $[4]\mathbf{Z}_{24} + [6]\mathbf{Z}_{24} = [n]\mathbf{Z}_{24}$. If so, what is the smallest positive integer n for which this is true?

2.3   ALGEBRAICALLY EQUIVALENT GROUPS: GROUP HOMOMORPHISMS

Section Overview. This section introduces the idea of group homomorphism and of group isomorphism. It is shown that the kernels of group homomorphisms are closely connected to normal subgroups and to factor groups. It is shown that kernels are normal subgroups and that normal subgroups are kernels. Three important isomorphism theorems and a correspondence theorem for groups are also presented in this section. Arguably, the most fundamental of these four theorems is Theorem 2.3.11, the First Isomorphism Theorem for Groups.

One of the most important concepts studied in abstract algebra is that of isomorphism. It is sometimes the case that two groups G and H are algebraically equivalent. A one-to-one correspondence can be established between the elements of two such groups in such a way that the algebraic properties of the two groups are preserved by the correspondence. These ideas are illustrated by the following two examples, which provide a model for the definition of a group homomorphism.

2.3.1 | Examples

1. Consider the groups $G = \{-1, -i, i, 1\}$ and $H = \{R_0, R_{90}, R_{180}, R_{270}\}$, where H is the subgroup of rotations of the dihedral group D_4 given in Section 2.1. The multiplication tables for G and H follow.

$\cdot$	1	i	-1	$-i$
1	1	i	-1	$-i$
i	i	-1	$-i$	1
-1	-1	$-i$	1	i
$-i$	$-i$	1	i	-1

G

$*$	R_0	R_{90}	R_{180}	R_{270}
R_0	R_0	R_{90}	R_{180}	R_{270}
R_{90}	R_{90}	R_{180}	R_{270}	R_0
R_{180}	R_{180}	R_{270}	R_0	R_{90}
R_{270}	R_{270}	R_0	R_{90}	R_{180}

H

These two groups are actually algebraically equivalent groups. Consider the following correspondence between the elements of G and H.

CORRESPONDENCE TABLE

$$G \longleftrightarrow H$$
$$1 \longleftrightarrow R_0$$
$$i \longleftrightarrow R_{90}$$
$$-1 \longleftrightarrow R_{180}$$
$$-i \longleftrightarrow R_{270}$$

This establishes a one-to-one correspondence, or bijective function, between the elements of G and H, and the integrity of the binary operations on the two groups is preserved by this correspondence. If one computes the product of two elements in G and then computes the product of the two corresponding elements in H, the results correspond. For example $-i \cdot i = 1$ in G. The elements that correspond to $-i$ and i in H are R_{270} and R_{90}, respectively, and $R_{270} * R_{90} = R_0$. From the correspondence table, we see that 1 and R_0 correspond. As a second example, $-1 \cdot -i = i$ in G. The elements in H that correspond to -1 and $-i$ in G are R_{180} and R_{270}, respectively, and $R_{180} * R_{270} = R_{90}$ in H. The correspondence table shows that i and R_{90} are corresponding elements of G and H, respectively. What is happening can be easily seen if one considers the group G geometrically.

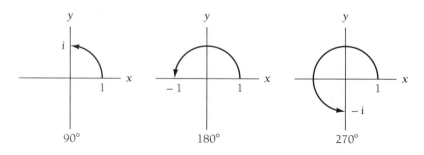

In the complex plane, multiplication of 1 by i represents a rotation of 1 to i through an angle of 90 degrees, as shown above. Multiplication of 1 by -1 represents a rotation of 1 to -1 through an angle of 180 degrees, and multiplication of 1 by $-i$ represents a rotation of 1 to $-i$ through an angle of 270 degrees. With this in mind, the correspondence displayed in the correspondence table seems quite natural. It can be shown that for these two groups, products of corresponding elements always correspond. Hence, the two groups are algebraically equivalent, and the two groups can actually be identified through this correspondence of elements. If the bijective function from G to H given by the correspondence is named f, then we have a function $f : G \rightarrow H$ such that

$$
\begin{array}{ccc}
G & \xleftrightarrow{\quad f \quad} & H \\
x & \longleftrightarrow & f(x) \\
y & \longleftrightarrow & f(y) \\
x \cdot y & \longleftrightarrow & f(x) * f(y)
\end{array}
$$

for all $x, y \in G$. Since $x \cdot y$ and $f(x) * (y)$ always correspond under f,

$$ f(x \cdot y) = f(x) * f(y) \qquad \text{for all } x, y \in G $$

In this situation, we say that the **group operations are preserved** by f.

2. There are mappings between groups that are not bijective functions, yet the group operations are preserved by the mapping. For example, consider the function $f : \mathbf{Z} \rightarrow \mathbf{Z}_4$ that is defined by $f(x) = [x]$. If $x, y \in \mathbf{Z}$, then $f(x + y) = [x + y] = [x] + [y] = f(x) + f(y)$, so the group operations are preserved by f.

The properties of the function f in the first example provides us with a model for the definition of a group isomorphism. The second example, which is more general than the first in the sense that f is not a bijective function, leads us to the definition of a group homomorphism.

2.3.2   Definition

If G and H are groups, a function $f : G \rightarrow H$ such that $f(xy) = f(x)f(y)$ for all $x, y \in G$ is said to be a **group homomorphism**. A group homomorphism that is an injective function is a **group monomorphism**, while a group homomorphism that is a surjective mapping is a **group epimorphism**. If $f : G \rightarrow H$ is a group monomorphism, then we say that G **embeds** in H. When $f : G \rightarrow H$ is a group epimorphism, H is said to be a **homomorphic image** of G. A group homomorphism that is both injective and surjective is a **group isomorphism**. If $f : G \rightarrow H$ is a group isomorphism, then G and H are **isomorphic groups**, and H is said to be a **copy** of G.

Group homomorphisms occur throughout mathematics, and they are an important tool for the study and classification of groups. For example, later in the text we shall see that every cyclic group is isomorphic to $\mathbf{Z}$ or to $\mathbf{Z}_n$ for some positive integer n and that finite abelian groups have a particularly nice structure. Although quite a bit of preliminary work will be carried out in the classification of finite abelian groups, the final step in the classification relies intrinsically on the concept of group isomorphism.

Two examples of group homomorphism were presented in the opening remarks of this section, and additional examples will be given later. Let's now investigate some of the basic properties of group homomorphisms. The following theorem shows that under a group homomorphism the identities of the two groups always correspond, as do inverses.

2.3.3   Theorem

If $f: G \rightarrow H$ is a group homomorphism, then $f(e_G) = e_H$ and $f(x^{-1}) = f(x)^{-1}$ for all $x \in G$.

Proof. We show first that $f(e_G) = e_H$. Since $f(x) = f(e_G x) = f(e_G)f(x)$, we have $f(x) = f(e_G)f(x)$. If we multiply both sides of this equation on the right by $f(x)^{-1}$, we see that $e_H = f(x)f(x)^{-1} = [f(e_G)f(x)]f(x)^{-1} = f(e_G)\,[f(x)f(x)^{-1}] = f(e_G)e_H = f(e_G)$. Therefore, $f(e_G) = e_H$. Finally, let's show that $f(x^{-1}) = f(x)^{-1}$. Since $f(e_G) = e_H$, we see that $e_H = f(e_G) = f(xx^{-1}) = f(x)f(x^{-1})$. Similarly, $f(x^{-1})f(x) = e_H$, so we see that $f(x^{-1})$ is a multiplicative inverse of $f(x)$. Since $f(x)^{-1}$ is unique, it must be the case that $f(x)^{-1} = f(x^{-1})$. ∎

If H is a normal subgroup of a group G, then we can form the factor group G/H. Normal subgroups are fundamental in the study of groups, and they are very closely connected to group homomorphism. The following definition is helpful in showing how this connection can be established.

2.3.4   Definition

If $f: G \rightarrow H$ is a group homomorphism, then the set $K = \{x \in G \mid f(x) = e_H\}$ is the **kernel** of f. The set K is often denoted by ker f.

The first link between group homomorphisms and the normal subgroups of a group is given in part 1 of the following theorem.

2.3.5	**Theorem**

Let $f: G \rightarrow H$ be a group homomorphism.

1. The kernel K of f is a normal subgroup of G.
2. The homomorphism f is a monomorphism if and only if $K = \{e_G\}$.

Proof

1. Since $f(e_G) = e_H$, $e_G \in K$ and thus $K \neq \varnothing$. If $x, y \in K$, then $f(x) = f(y) = e_H$. Hence, by Theorem 2.3.3, $f(y^{-1}) = f(y)^{-1} = e_H^{-1} = e_H$ and so $y^{-1} \in K$. Therefore, $f(xy^{-1}) = f(x)f(y^{-1}) = e_H e_H = e_H$, so $xy^{-1} \in K$ and, by Theorem 2.2.3, K is a subgroup of G. If $x \in G$ and $k \in K$, then $f(xkx^{-1}) = f(x)f(k)f(x^{-1}) = f(x)e_H f(x)^{-1} = f(x)f(x)^{-1} = e_H$. Hence, $xkx^{-1} \in K$ and, by Theorem 2.2.15, K is a normal subgroup of G.

2. Suppose that $K = \{e_G\}$. If $f(x) = f(y)$, then $f(x)f(y)^{-1} = e_H$. Since $f(xy^{-1}) = f(x)f(y)^{-1} = e_H$, $xy^{-1} \in K$. Thus, $xy^{-1} = e_G$, which implies that $x = y$. Therefore, f is a monomorphism. Conversely, if f is a group monomorphism and $x \in K$, then $f(x) = e_H = f(e_G)$. But the fact that f is a monomorphism implies that $x = e_G$, so $K = \{e_G\}$. ∎

2.3.6	**Examples of Group Isomorphisms**

1. Let $\mathcal{C} = \{(a, b) \mid a, b \in \mathbf{R}\}$ and define addition on $\mathcal{C}$ by $(a, b) + (c, d) = (a + c, b + d)$. Then $\mathcal{C}$ is an additive abelian group with additive identity $(0, 0)$. The additive inverse of (a, b) is $-(a, b) = (-a, -b)$. This group is isomorphic to the additive group of complex numbers $\mathbf{C}$. For two complex numbers $a + bi$ and $c + di$, $(a + bi) + (c + di) = (a + c) + (b + d)i$, so it is fairly obvious that these two groups are isomorphic. This should be evident if you compare

$$(a, b) + (c, d) = (a + c, b + d) \qquad \text{with}$$

$$(a + bi) + (c + di) = (a + c) + (b + d)i$$

However, we'll go through a formal proof very carefully to illustrate how to prove that two groups are isomorphic. Sometimes, as in our second example, the connection between the two groups won't be as obvious as it is here. Basically, there are two major steps in showing that a well-defined function is a group isomorphism: show that the function is a bijection and show that the function is a group homomorphism. Now let's prove that $\mathcal{C}$ and $\mathbf{C}$ are isomorphic. Let $f: \mathcal{C} \rightarrow \mathbf{C}$ be defined by $f((a, b)) = a + bi$. We claim that f is a well-defined function that is a group isomorphism. There are several steps involved in showing that this is true:

Step 1. The mapping f is well-defined: pick two different representations of the same element in $\mathcal{C}$, say, $(a, b) = (c, d)$. Then $a = c$ and $b = d$, so $a + bi = c + di$. Hence, $f((a, b)) = f((c, d))$, so f is well-defined.

Step 2. The function f is injective: if $f((a, b)) = f((c, d))$, then by the definition of f, $a + bi = c + di$, so $a = c$ and $b = d$. It follows that $(a, b) = (c, d)$.

Step 3. The function f is surjective: if $a + bi \in \mathbf{C}$, then $(a, b) \in \mathcal{C}$ and $f((a, b)) = a + bi$.

Step 4. The function f is a group homomorphism: suppose $(a, b), (c, d) \in \mathcal{C}$. Then

$$
\begin{aligned}
f((a, b) + (c, d)) &= f((a + c, b + d)) \\
&= (a + c) + (b + d)i \\
&= (a + bi) + (c + di) \\
&= f((a, b)) + f((c, d))
\end{aligned}
$$

Hence, $\mathcal{C}$ and $\mathbf{C}$ are isomorphic groups, and we can say that the additive group of complex numbers $\mathbf{C}$ is a copy of the additive group $\mathcal{C}$.

2. Another example is afforded by the multiplicative group of positive real numbers $\mathbf{R}^+$ and the additive group of real numbers $\mathbf{R}$. The function $L : \mathbf{R}^+ \to \mathbf{R}$, where $L(x) = \ln x$, has an inverse function $E : \mathbf{R} \to \mathbf{R}^+$ defined by $E(x) = e^x$ since $(L \circ E)(x) = L(e^x) = \ln e^x = x$ and $(E \circ L)(x) = E(\ln x) = e^{\ln x} = x$. Hence, by Theorem 0.3.7, L is a bijection. Since $L(xy) = \ln(xy) = \ln x + \ln y = L(x) + L(y)$, the operations are preserved, and L is a group isomorphism. Therefore, the additive group $\mathbf{R}$ and the multiplicative group $\mathbf{R}^+$ are isomorphic. This establishes that the additive algebraic structure on $\mathbf{R}$ is algebraically equivalent to the multiplicative algebraic structure on $\mathbf{R}^+$.

   Notice that we took an "alternate route" in the proof. Rather than directly showing that L is a bijection as we did in Example 1, we demonstrated that L has an inverse function, which is sufficient, because of Theorem 0.3.7, to show that L is a bijection.

It was pointed out in the second example of the opening remarks of this section that not all group homomorphisms are isomorphisms. We now give an example of a group monomorphism and of a group epimorphism, neither of which is a group isomorphism.

Example

1. Consider the additive group $\mathcal{C}$ of Example 1 of 2.3.6, and let $M_2(\mathbf{R})$ be
   the additive group of 2×2 matrices whose entries are real numbers.
   Let $f : \mathcal{C} \to M_2(\mathbf{R})$ be defined by

   $$f((a, b)) = \begin{pmatrix} a & 0 \\ 0 & b \end{pmatrix}$$

   The mapping f is clearly well-defined since if $(a, b) = (c, d)$, then $a = c$
   and $b = d$, so

   $$\begin{pmatrix} a & 0 \\ 0 & b \end{pmatrix} = \begin{pmatrix} c & 0 \\ 0 & d \end{pmatrix}$$

   Hence, $f((a, b)) = f((c, d))$, and f is well-defined as asserted. Now let's
   show that f is a group homomorphism. This follows since

   $$f((a, b) + (c, d)) = f((a + c, b + d))$$

   $$= \begin{pmatrix} a + c & 0 \\ 0 & b + d \end{pmatrix} = \begin{pmatrix} a & 0 \\ 0 & b \end{pmatrix} + \begin{pmatrix} c & 0 \\ 0 & d \end{pmatrix}$$

   $$= f((a, b)) + f((c, d))$$

   We claim that f is an injective function. If

   $$f((a, b)) = \begin{pmatrix} 0 & 0 \\ 0 & 0 \end{pmatrix} \qquad \text{then}$$

   $$\begin{pmatrix} a & 0 \\ 0 & b \end{pmatrix} = \begin{pmatrix} 0 & 0 \\ 0 & 0 \end{pmatrix}$$

   and so $a = 0$ and $b = 0$.

   Therefore, $(a, b) = (0, 0)$, and the kernel of f contains only the additive
   identity $(0, 0)$ of $\mathcal{C}$. Hence, by Theorem 2.3.5, f is a group monomorphism.
   Thus, we see that $\mathcal{C}$ embeds in $M_2(\mathbf{R})$. It is interesting to note that

   $$H = \left\{ \begin{pmatrix} a & 0 \\ 0 & b \end{pmatrix} \middle| (a, b \in \mathbf{R}) \right\}$$

   is a proper subgroup of $M_2(\mathbf{R})$ that is isomorphic to $\mathcal{C}$. Finally, if
   $\begin{pmatrix} a & b \\ c & d \end{pmatrix} \in M_2(\mathbf{R})$ such that $b \neq 0$ or $c \neq 0$, then $\begin{pmatrix} a & b \\ c & d \end{pmatrix}$ has no preimage
   in $\mathcal{C}$, so f is not an isomorphism.

2. The mapping $f : \mathbf{Z} \to \mathbf{Z}_6$ given by $f(x) = [x]$ is a group epimorphism from the additive group of integers $\mathbf{Z}$ to the additive group $\mathbf{Z}_6$. This mapping is not injective. Indeed, $7 \neq 13$, yet $f(7) = [7] = [1]$ and $f(13) = [13] = [1]$. Thus $7 \neq 13$ but $f(7) = f(13)$. Hence, distinct inputs do not produce distinct outputs. The fact that f is a group homomorphism follows easily from $f(x + y) = [x + y] = [x] + [y] = f(x) + f(y)$. The mapping f is surjective because if $[x] \in \mathbf{Z}_6$, then $f(y) = [x]$ for any $y \in [x]$. Thus, $\mathbf{Z}_6$ is a homomorphic image of $\mathbf{Z}$.

In Example 1 of 2.3.7, $f : \mathcal{C} \to M_2(\mathbf{R})$ is a group monomorphism, and $f(\mathcal{C}) = H$ is a subgroup of $M_2(\mathbf{R})$ that is isomorphic to $\mathcal{C}$. When $f : G \to G'$ is a group homomorphism, $f(G)$ is a subgroup of G', but it is not necessarily the case that G and $f(G)$ are isomorphic. In general, under a group homomorphism, subgroups go to subgroups and inverse images of subgroups are subgroups.

2.3.8   Theorem

Let $f : G \to G'$ be a group homomorphism.

1. If H is a subgroup of G, then $f(H)$ is a subgroup of G'.
2. If H' is a subgroup of G', then $f^{-1}(H')$ is a subgroup of G.

Proof

1. If H is a subgroup of G, then $H \neq \varnothing$. If $h \in H$, then $f(h) \in f(H)$ and thus $f(H) \neq \varnothing$. Now suppose that $f(x), f(y) \in f(H)$, where $x, y \in H$. Since $x, y \in H$, $xy^{-1} \in H$ and so $f(xy^{-1}) \in f(H)$. But $f(x)f(y)^{-1} = f(x)f(y^{-1}) = f(xy^{-1})$, so $f(x)f(y)^{-1} \in f(H)$. Therefore, by Theorem 2.2.3, $f(H)$ is a subgroup of G'.
2. Now let's show that inverse images of subgroups are subgroups. Let e and e' denote the identities of G and G', respectively, and suppose that H' is a subgroup of G'. Since $e' \in H'$ and $f(e) = e'$, $e \in f^{-1}(H')$. Hence, $f^{-1}(H') \neq \varnothing$. If $x, y \in f^{-1}(H')$, then $f(x), f(y) \in H'$. Thus, $f(xy^{-1}) = f(x)f(y)^{-1} \in H'$ and so $xy^{-1} \in f^{-1}(H')$. Consequently, $f^{-1}(H')$ is a subgroup of G. ■

It was pointed out earlier that normal subgroups of a group are closely connected to group homomorphisms. The following theorem indicates just how close this connection really is. Loosely speaking, the theorem shows that kernels are normal subgroups and that normal subgroups are kernels.

2.3.9 | Theorem

A subgroup K of a group G is a normal subgroup of G if and only if K is the kernel of a group homomorphism whose domain is G.

Proof. If $f : G \to H$ is a group homomorphism with kernel K, then Theorem 2.3.5 shows that K is a normal subgroup of G.

Conversely, suppose that K is a normal subgroup of G. The mapping $\eta : G \to G/K : x \mapsto xK$ can easily be shown to be a well-defined group homomorphism with kernel K. Therefore, when K is a normal subgroup of G, K is the kernel of a group homomorphism with domain G. ∎

2.3.10 | Examples

1. The function $f : \mathbf{Z} \to \mathbf{Z}_4$, where $f(x) = [x]$, is a well-defined group epimorphism, and the kernel of f is $4\mathbf{Z}$. We claim that $\phi : \mathbf{Z}/4\mathbf{Z} \to \mathbf{Z}_4$, defined by $\phi(x + 4\mathbf{Z}) = [x]$, is a group isomorphism. We leave it as an exercise to show that ϕ is well-defined. The fact that ϕ is a group homomorphism follows immediately since

    $$\begin{aligned}
    \phi((x + 4\mathbf{Z}) + (y + 4\mathbf{Z})) &= \phi((x + y) + 4\mathbf{Z}) \\
    &= [x + y] \\
    &= [x] + [y] \\
    &= \phi(x + 4\mathbf{Z}) + \phi(y + 4\mathbf{Z})
    \end{aligned}$$

    If $[x] \in \mathbf{Z}_4$, then $x + 4\mathbf{Z} \in \mathbf{Z}/4\mathbf{Z}$ and $\phi(x + 4\mathbf{Z}) = [x]$, so ϕ is surjective. Finally, we need to show that ϕ is injective. If $\phi(x + 4\mathbf{Z}) = \phi(y + 4\mathbf{Z})$, then $[x] = y]$, so $x \equiv y \pmod 4$. Thus, $x - y$ is a multiple of 4. If $x - y = 4k$, where k is an integer, then $x + 4\mathbf{Z} = (y + 4k) + 4\mathbf{Z} = y + (4k + 4\mathbf{Z}) = y + 4\mathbf{Z}$. Hence, $\phi(x + 4\mathbf{Z}) = \phi(y + 4\mathbf{Z})$ implies that $x + 4\mathbf{Z} = y + 4\mathbf{Z}$, so ϕ is an injective function. Therefore, ϕ is a group isomorphism, and $\mathbf{Z}/4\mathbf{Z}$ and $\mathbf{Z}_4$ are isomorphic groups.

2. Consider the set H of all 2×2 matrices of the form

    $$\begin{pmatrix} a & 0 \\ 0 & d \end{pmatrix},$$

    where a and d are in $\mathbf{R}$. H is an additive abelian group if addition is defined on H by

    $$\begin{pmatrix} a & 0 \\ 0 & d \end{pmatrix} + \begin{pmatrix} a' & 0 \\ 0 & d' \end{pmatrix} = \begin{pmatrix} a + a' & 0 \\ 0 & d + d' \end{pmatrix}.$$

Now define $f: M_2(\mathbf{R}) \to H$ by

$$f\left(\begin{pmatrix} a & b \\ c & d \end{pmatrix}\right) = \begin{pmatrix} a & 0 \\ 0 & d \end{pmatrix},$$

where we are considering $M_2(\mathbf{R})$ as a group under matrix addition. Then f is a well-defined group epimorphism. If K is the kernel of f, then

$\begin{pmatrix} a & b \\ c & d \end{pmatrix} \in K$ if and only if $f\left(\begin{pmatrix} a & b \\ c & d \end{pmatrix}\right) = \begin{pmatrix} a & 0 \\ 0 & d \end{pmatrix} = \begin{pmatrix} 0 & 0 \\ 0 & 0 \end{pmatrix}$ if and only if

$a = d = 0$. Thus, $K = \{\begin{pmatrix} 0 & b \\ c & 0 \end{pmatrix} \mid b, c \in \mathbf{R}\}$. If we factor out K, the result is

the factor group $M_2(\mathbf{R})/K$. A typical element of $M_2(\mathbf{R})/K$ is of the form

$$\begin{pmatrix} a & b \\ c & d \end{pmatrix} + K = \begin{pmatrix} a & 0 \\ 0 & d \end{pmatrix} + \begin{pmatrix} 0 & b \\ c & 0 \end{pmatrix} + K = \begin{pmatrix} a & 0 \\ 0 & d \end{pmatrix} + K.$$

Hence, every element $\begin{pmatrix} a & b \\ c & d \end{pmatrix} + K$ of $M_2(\mathbf{R})/K$ can be written in the form
$\begin{pmatrix} a & 0 \\ 0 & d \end{pmatrix} + K$. This clearly shows the effect of factoring out K
when forming the factor group $M_2(\mathbf{R})/K$. It is not difficult to show that $M_2(\mathbf{R})/K$ and H are isomorphic under the correspondence

$$M_2(\mathbf{R})/K \longleftrightarrow H \qquad \text{and}$$

$$\begin{pmatrix} a & 0 \\ 0 & d \end{pmatrix} + K \longleftrightarrow \begin{pmatrix} a & 0 \\ 0 & d \end{pmatrix}$$

In the first example, $f: \mathbf{Z} \to \mathbf{Z}_4$ defined by $f(x) = [x]$ is a group epimorphism with kernel $4\mathbf{Z}$, and the factor group $\mathbf{Z}/4\mathbf{Z}$ is isomorphic to $\mathbf{Z}_4$. In the second example, $f: M_2(\mathbf{R}) \to H$, where

$$f\left(\begin{pmatrix} a & b \\ c & d \end{pmatrix}\right) = \begin{pmatrix} a & 0 \\ 0 & d \end{pmatrix},$$

is also a group epimorphism, and $M_2(\mathbf{R})/K$ and H are isomorphic. If $f: G \to H$ is a group epimorphism with kernel K, is it always the case that G/K and H are isomorphic groups? The answer lies in the following theorem.

2.3.11   First Isomorphism Theorem for Groups

If $f: G \to H$ is a group epimorphism with kernel K, then G/K and H are isomorphic groups.

Proof. Let $\phi : G/K \rightarrow H$ be defined by $\phi(xK) = f(x)$, and let's show that ϕ is a group isomorphism. If $xK = yK$, then $x = yk$ for some k in K. But then $f(x) = f(yk) = f(y)f(k) = f(y)e_H = f(y)$. Hence, $\phi(xK) = \phi(yK)$, so ϕ is well-defined. Furthermore, if $xK, yK \in G/K$, then $\phi((xK)(yK)) = \phi(xyK) = f(xy) = f(x)f(y) = \phi(xK)\phi(yK)$ and so ϕ is a group homomorphism. Finally, it must be shown that ϕ is a bijective function. If $\phi(xK) = \phi(yK)$, then $f(x) = f(y)$, so $f(x)f(y)^{-1} = e_H$. Hence, $f(xy^{-1}) = f(x)\, f(y^{-1}) = f(x)f(y)^{-1} = e_H$ and so $xy^{-1} \in K$. If $xy^{-1} = k$, then $x = ky$, so $xK = Kx = Kky = Ky = yK$. Hence, ϕ is injective. If $z \in H$, then since f is an epimorphism, there is an $x \in G$ such that $f(x) = z$. But then $xK \in G/K$ and $\phi(xK) = f(x) = z$. Thus, ϕ is surjective, and the proof is complete. ∎

2.3.12   Corollary

If $f : G \rightarrow H$ is a group homomorphism with kernel K, then G/K and $f(G)$ are isomorphic groups.

Proof. Theorem 2.3.8 shows that $f(G)$ is a subgroup of H, so the mapping $\bar{f} : G \rightarrow f(G)$ defined by $\bar{f}(x) = f(x)$ is a group epimorphism, and $K = \ker \bar{f} = \ker f$. The corollary now follows from Theorem 2.3.11. ∎

2.3.13   Example

The mapping $f : \mathbf{Z}_{24} \rightarrow \mathbf{Z}_{12}$, where $f([x]_{24}) = [4x]_{12}$, is easily shown to be a group homomorphism, and the kernel of f is

$$K = \{[0]_{24}, [3]_{24}, [6]_{24}, [9]_{24}, [12]_{24}, [15]_{24}, [18]_{24}, [21]_{24}\}.$$

By Theorem 2.3.8, $f(\mathbf{Z}_{24}) = \{[0]_{12}, [4]_{12}, [8]_{12}\}$ is a subgroup of $\mathbf{Z}_{12}$, and it is not difficult to show that

$$\mathbf{Z}_{24}/K = \{[0]_{24} + K, [1]_{24} + K, [2]_{24} + K\}.$$

The isomorphism $\phi : \mathbf{Z}_{24}/K \rightarrow f(\mathbf{Z}_{24})$ of Corollary 2.3.12 is given by $\phi([x]_{24} + K) = f([x]_{24}) = [4x]_{12}$, and since both are finite groups of small order, the isomorphism can be displayed explicitly:

$$\mathbf{Z}_{24}/K \longleftrightarrow f(\mathbf{Z}_{24}),$$
$$[0]_{24} + K \longleftrightarrow [0]_{12},$$
$$[1]_{24} + K \longleftrightarrow [4]_{12},$$
$$[2]_{24} + K \longleftrightarrow [8]_{12}.$$

There are two additional isomorphism theorems for groups and a correspondence theorem. We prove the first two of the following three theorems and sketch the proof of the third. You should fill in the details to complete the last proof.

2.3.14   Second Isomorphism Theorem for Groups

If H_1 and H_2 are normal subgroups of a group G with $H_1 \subseteq H_2$, then:

1. H_2/H_1 is a normal subgroup of G/H_1.
2. The factor groups $(G/H_1)/(H_2/H_1)$ and G/H_2 are isomorphic.

Proof

1. To show that H_2/H_1 is a normal subgroup of G/H_1, consider the mapping $f : G/H_1 \rightarrow G/H_2$ given by $f(xH_1) = xH_2$. Then f is well-defined, for if $xH_1 = yH_1$, then $y^{-1}x \in H_1 \subseteq H_2$ and thus it follows that $xH_2 = yH_2$. Hence, $f(xH_1) = f(yH_1)$. Note that $xH_1 \in \ker f$ if and only if $f(xH_1) = eH_2 = H_2$ if and only if $xH_2 = H_2$ if and only if $x \in H_2$. Therefore, $\ker f = H_2/H_1$. Since kernels of group homomorphisms are normal subgroups of their domain, H_2/H_1 is a normal subgroup of G/H_1.
2. Since the group homomorphism of part 1 is an epimorphism, we see that $(G/H_1)/(H_2/H_1)$ and G/H_2 are isomorphic groups by invoking Theorem 2.3.11. ∎

2.3.15   Third Isomorphism Theorem for Groups

If H_1 and H_2 are subgroups of a group G and H_2 is normal in G, then $H_1/(H_1 \cap H_2)$ and H_1H_2/H_2 are isomorphic groups.

Proof. Since H_2 is a normal subgroup of G, Theorem 2.2.20 shows that H_1H_2 is a subgroup of G. Again by Theorem 2.2.20, H_2 is a normal subgroup of H_1H_2, so we can form the factor group H_1H_2/H_2. The mapping $f : H_1 \rightarrow H_1H_2/H_2$ given by $f(x) = xH_2$ is a well-defined group epimorphism with kernel $H_1 \cap H_2$, so the result follows from Theorem 2.3.11. ∎

2.3.16   Correspondence Theorem

If $f : G \rightarrow H$ is a group epimorphism, there is a one-to-one correspondence between the subgroups of G that contain $\ker f$ and the subgroups of H. Moreover, if $G_1 \subseteq G$ and $H_1 \subseteq H$ are subgroups that correspond under f, then G_1 is a normal subgroup of G if and only if H_1 is a normal subgroup of H.

Proof. (Sketched) Let S_G denote the set of subgroups G_1 of G such that $\ker f \subseteq G_1$, and let S_H denote the set of subgroups H_1 of H. Define $\phi : S_G \to S_H$ by $\phi(G_1) = f(G_1)$. Then $\phi(G_1) \in S_H$ because of Theorem 2.3.8. Next, define $\psi : S_H \to S_G$ by $\psi(H_1) = f^{-1}(H_1)$. Then $\psi(H_1) \in S_G$, again by Theorem 2.3.8. Since $\phi \circ \psi = 1_{S_H}$ and $\psi \circ \phi = 1_{S_G}$, ϕ is a bijective function. Let G_1 be a normal subgroup of G, and let $y \in H$. Then there is an $x \in G$ such that $f(x) = y$. Thus, if $y f(g_1) y^{-1} \in y f(G_1) y^{-1}$, where $g_1 \in G_1$, then $y f(g_1) y^{-1} = f(x) f(g_1) f(x)^{-1} = f(x g_1 x^{-1})$ is in $f(G_1)$ since $x g_1 x^{-1} \in G_1$. Therefore, $f(G_1)$ is a normal subgroup of H. A similar proof can be given when H_1 is a normal subgroup of H. ■

2.3.17 | **Example**

Consider the group homomorphism $f : \mathbf{Z} \to \mathbf{Z}_{12}$ defined by $f(x) = [x]$. From Example 4 of 2.2.2, the subgroups of $\mathbf{Z}_{12}$ are

$\langle [1] \rangle = \{[0], [1], [2], [3], [4], [5], [6], [7], [8], [9], [10], [11]\} = \mathbf{Z}_{12}$,

$\langle [2] \rangle = \{[0], [2], [4], [6], [8], [10]\}$,

$\langle [3] \rangle = \{[0], [3], [6], [9]\}$,

$\langle [4] \rangle = \{[0], [4], [8]\}$,

$\langle [6] \rangle = \{[0], [6]\}$,

$\langle [0] \rangle = \{[0]\}$.

If $x \in \ker f$, then $f(x) = [0]$, so $[x] = [0]$ in $\mathbf{Z}_{12}$. But $[x] = [0]$ in $\mathbf{Z}_{12}$ if and only if x is a multiple of 12. Hence, $\ker f = 12\mathbf{Z}$. Now the subgroups of $\mathbf{Z}$ that contain $12\mathbf{Z}$ are the subgroups $2\mathbf{Z}$, $3\mathbf{Z}$, $4\mathbf{Z}$, $6\mathbf{Z}$, and $12\mathbf{Z}$. The lattice of subgroups of $\mathbf{Z}$ that contain $\ker f$ and the lattice of subgroups of $\mathbf{Z}_{12}$ are given below. The one-to-one correspondence between these two sets of subgroups is evident from the lattice diagrams.

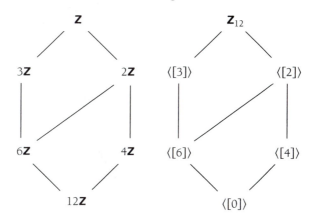

Our visit with groups is intended only as an introduction. There is certainly much more that can be said on this subject, and we return to the study of groups in a subsequent chapter.

Problem Set 2.3

Let **R*** denote the group of all nonzero real numbers under the operation of multiplication of real numbers. Decide which of Exercises 1 through 5 are group homomorphisms from **R*** to **R***.

1. $f(x) = |x|$
2. $g(x) = x^2$
3. $h(x) = 1/x$
4. $k(x) = 2x$
5. $f(x) = x$
6. Consider the additive group $\mathbf{Z}_8$. Prove that $f : \mathbf{Z}_8 \to \mathbf{Z}_8$ where $f([x]) = [2x]$, is a group homomorphism.
7. If $f : G \to H$ and $g : H \to K$ are group homomorphisms, prove that $g \circ f : G \to K$ is a group homomorphism.
8. Show that $f : M_2(\mathbf{R}) \to \mathbf{R}$ defined by

$$f\left(\begin{pmatrix} a & b \\ c & d \end{pmatrix}\right) = \det\begin{pmatrix} a & b \\ c & d \end{pmatrix} = ad - bc$$

is *not* a group homomorphism from $M_2(\mathbf{R})$, the additive group of 2×2 matrices over **R**, to the additive group **R**.

9. Let G and H be multiplicative groups, and suppose that $f : G \to H$ is a group homomorphism. Prove that $f(x^n) = f(x)^n$ for each $n \in \mathbf{Z}$ and every $x \in G$. Use this to explain why the following are true for each $n \in \mathbf{Z}$ and every $x \in G$.

    (a) $f(nx) = nf(x)$ when G and H are additive groups
    (b) $f(nx) = f(x)^n$ when G is additive and H is multiplicative
    (c) $f(x^n) = nf(x)$ when G is multiplicative and H is additive

10. If $f : G \to H$ is a group isomorphism, prove that the inverse function $f^{-1} : H \to G$ is also a group isomorphism.

11. Suppose that G and H are groups each with exactly two elements. Prove that G and H must be isomorphic. Prove that this is also true if G and H each contain exactly three elements.

12. Prove that G is an abelian group if and only if the mapping $f: G \to G$ given by $f(x) = x^{-1}$ is a group isomorphism.

13. Consider the set G of all mappings of the form $f: \mathbf{R} \to \mathbf{R}$ defined by $f(x) = ax + b$, where $a, b \in \mathbf{R}$, $a \neq 0$. Prove that G is a group under the operation of function composition.

14. Let G be a group, and suppose that $f: G \to H$ is a bijective mapping. (H is just a set.) Show that H can be made into a group in a natural way by using the binary operation on G and the function f.

15. Let $\mathbf{Q}^*$ be the multiplicative group of nonzero rational numbers and H the set of rational numbers not equal to 1. Define multiplication on H by $x \circ y = x + y - xy$. Prove that H is a group under the multiplication $\circ$, and show that the mapping $f: \mathbf{Q}^* \to H$, where $f(x) = 1 - x$, is a group isomorphism.

16. Consider the additive group $\mathbf{Z}_{12}$ and the mapping $f: \mathbf{Z}_{12} \to \mathbf{Z}_{12}$ given by $f([x]) = [4x]$. Prove that f is a group homomorphism, and describe the kernel of f.

17. Suppose that G is a multiplicative group, and let x be a fixed element of G. Let $f: \mathbf{Z} \to G$ be such that $f(n) = x^n$, and prove that f is a group homomorphism from the additive group $\mathbf{Z}$ to G. Describe the kernel of f.

18. Let $\mathbf{C}$ be the group of complex numbers under addition. If

$$G = \left\{ \begin{pmatrix} x & -y \\ y & x \end{pmatrix} \middle| \; x, y \in \mathbf{R} \right\},$$

then G is a group under matrix addition. Show that $\mathbf{C}$ and G are isomorphic groups.

19. Let $\mathbf{C}^*$ be the group of nonzero complex numbers under multiplication. If G^* is the set of nonzero matrices of G in Exercise 18, then G^* is a group under matrix multiplication. Prove that $\mathbf{C}^*$ and G^* are isomorphic groups.

20. If $G = \{a + b\sqrt{3} \mid a, b \in \mathbf{Q}\}$ then G is a group under addition. Prove that G is isomorphic to the additive group

$$H = \left\{ \begin{pmatrix} x & 3y \\ y & x \end{pmatrix} \middle| \; x, y \in \mathbf{Q} \right\}.$$

21. Demonstrate the Correspondence Theorem as in Example 2.3.17 for each of the following group epimorphisms.

    (a) $f: \mathbf{Z} \to \mathbf{Z}_{18}$, where $f(x) = [x]$
    (b) $f: \mathbf{Z} \to \mathbf{Z}_{24}$, where $f(x) = [x]$
    (c) $f^*: \mathbf{Z}_{24} \to f(\mathbf{Z}_{24})$, where $f^*([x]) = f([x])$ and $f: \mathbf{Z}_{24} \to \mathbf{Z}_{24}$ is defined by $f([x]) = [4x]$

22. Fill in the details of the proof of Theorem 2.3.16.

Use a computer algebra system to solve each of the following problems.

23. The mapping $f : \mathbf{Z}_{64} \to \mathbf{Z}_{32}$, where $f([x]) = [4x]$, is a group homomorphism. Determine the kernel of f and the cosets of $K = \ker f$ in $\mathbf{Z}_{64}$. Compute the elements of $\mathbf{Z}_{64}/K$. Compute $f(\mathbf{Z}_{24})$. What is the correspondence between $\mathbf{Z}_{64}/K$ and $f(\mathbf{Z}_{64})$ that establishes an isomorphism between these two groups?

24. The mapping $f : \mathbf{Z}_{512} \to \mathbf{Z}_{256}$ defined by $f([x]) = [8x]$ is a group homomorphism. What is the kernel of f? Compute $f(\mathbf{Z}_{512})$. What is the correspondence between the subgroups of $\mathbf{Z}_{512}$ that contain $K = \ker f$ and the subgroups of $f(\mathbf{Z}_{512})$?

3

Sets with Two Binary Operations: Rings

*The algebraic structure known as a ring is not only ubiquitous in mathematics, but it is also one of the fundamental algebraic structures that underpins our number systems. Important work leading to the concept of a ring was conducted by **Carl Friedrich Gauss** (1777–1855), **Ernst Eduard Kummer** (1810–1893), and **Richard Dedekind** (1831–1916). (Information on Krummer and Dedekind can be found in the Historical Notes at the end of Section 3.1, while information on Gauss can be found in the Historical Notes of Section 1.3.)*

3.1 RINGS

Section Overview. The main purpose of this section is to introduce the concept of a ring and to establish some of the fundamental properties that hold in every ring. Make sure that you understand the similarities and differences among each of the following: ring, domain, integral domain, division ring, and field. Several examples are given of each structure. Study each example carefully, and convince yourself that it possesses the attributes described. Polynomial rings are also introduced in this section. These rings are investigated in more detail in a subsequent chapter. Be sure to study the classification scheme for classes of rings given at the end of this section.

In Section 2.1, the definition of group was motivated by considering properties shared by two examples of sets, each of which was endowed with a single binary operation. A similar approach will be taken to introduce the definition of a ring. The properties shared by the following examples of sets with two binary operations will be used as a model for the definition of a ring.

3.1.1 | Examples

1. The first example of a set endowed with two binary operations is one with which you are already quite familiar. It is the set $\mathbf{Z}$ under addition and multiplication of integers. This set, together with these binary operations, has the following properties:
    (i)   The set $\mathbf{Z}$ together with addition of integers forms an additive abelian group. The additive identity is the integer 0, and if $x \in \mathbf{Z}$, the additive inverse of x is the integer $-x$.
    (ii) The operation of multiplication of integers is associative, that is, $x(yz) = (xy)z$ for all $x, y, z \in \mathbf{Z}$.
    (iii) Multiplication is distributive over addition from both the left and the right. If $x, y, z \in \mathbf{Z}$, then $x(y + z) = xy + xz$ and $(y + z)x = yx + zx$.
    (iv) The multiplicative identity of $\mathbf{Z}$ is $e = 1$, and multiplication is commutative.
2. The second example is the set $\mathbf{Z}_n$ under the operations of congruence-class addition and multiplication modulo n.
    (i)   The set $\mathbf{Z}_n$ together with congruence-class addition forms an additive abelian group. The additive identity is $[0]$, and if $[x] \in \mathbf{Z}_n$, the additive inverse of $[x]$ is $-[x] = [-x]$.
    (ii) The operation of multiplication of congruence classes is associative; that is, $[x]([y][z]) = ([x][y])[z]$ for all $[x]$, $[y]$, $[z] \in \mathbf{Z}_n$.
    (iii) Multiplication is distributive over addition from both the left and the right. If $[x]$, $[y]$, $[z] \in \mathbf{Z}_n$, then $[x]([y] + [z]) = [x][y] + [x][z]$ and $([y] + [z])[x] = [y][x] + [z][x]$.
    (iv) The multiplicative identity of $\mathbf{Z}_n$ is $e = [1]$, and multiplication is commutative.

3. Finally, consider the set $M_2(\mathbf{R})$ of all 2×2 matrices over $\mathbf{R}$ under the binary operations of matrix addition and multiplication as defined in Definition 0.5.3. This set together with these two binary operations has the following properties:

   (i) The set $M_2(\mathbf{R})$ together with matrix addition forms an additive abelian group. The additive identity is the zero matrix $\begin{pmatrix} 0 & 0 \\ 0 & 0 \end{pmatrix}$, and $\begin{pmatrix} -a & -b \\ -c & -d \end{pmatrix}$ is the additive inverse of $\begin{pmatrix} a & b \\ c & d \end{pmatrix}$.

   (ii) The operation of matrix multiplication is associative. Hence, for any matrices $A, B, C \in M_2(\mathbf{R})$, $A(BC) = (AB)C$.

   (iii) Matrix multiplication is distributive over addition from both the left and the right. Thus, for any matrices $A, B, C \in M_2(\mathbf{R})$, $A(B + C) = AB + AC$ and $(B + C)A = BA + CA$.

   (iv) Matrix multiplication is *not* commutative, but $M_2(\mathbf{R})$ does have a multiplicative identity, namely, $e = \begin{pmatrix} 1 & 0 \\ 0 & 1 \end{pmatrix}$.

The shared properties given in (i), (ii), and (iii) of the three preceding examples provide a model for the following definition of a ring.

3.1.2 **Definition**

A **ring** is a nonempty set R together with two binary operations

$$R \times R \to R, \text{ where } (x, y) \longmapsto x + y, \quad \text{ and}$$
$$R \times R \to R, \text{ where } (x, y) \longmapsto xy,$$

called addition and multiplication that satisfy the following three conditions:

1. R together with addition forms an additive abelian group.
2. Multiplication is associative.
3. Multiplication is distributive over addition from both the left and the right:

$$x(y + z) = xy + xz \quad \text{ and}$$
$$(y + z)x = yx + zx \quad \text{ for all } x, y, z \in R.$$

If R multiplication is commutative, then R is said to be a **commutative ring**. If R has a multiplicative identity e, R is a **ring with identity**. A ring R always has an additive identity, so when we say that R is a ring with identity, we are indicating that R has a multiplicative identity. **Subtraction** is defined in a ring R by $x - y = x + (-y)$ for all $x, y \in R$.

The Laws of Exponents 2.1.9 and the Laws of Multiples 2.1.10 hold in any ring. There is one additional generalized property for rings that can easily be proved by induction.

3.1.3   The Generalized Distributive Properties

If $x_1, x_2, \ldots, x_n \in R$, then for any $x \in R$,

1. $x(x_1 + x_2 + \cdots + x_n) = xx_1 + xx_2 + \cdots + xx_n$ and
2. $(x_1 + x_2 + \cdots + x_n)x = x_1x + x_2x + \cdots + x_nx$.

In addition to the three examples given in the introduction of this chapter, other examples of rings are often encountered in mathematics. Several of these are given next. You should consider each example carefully and verify that it satisfies all the requirements to be a ring.

3.1.4   Some Familiar Examples of Rings

1. The set $2\mathbf{Z}$ of even integers is a commutative ring under addition and multiplication of integers. The additive identity of $2\mathbf{Z}$ is 0, but $2\mathbf{Z}$ is *not* a ring with identity since $1 \notin 2\mathbf{Z}$. The operations of addition and multiplication of integers are indeed binary operations on $2\mathbf{Z}$ since the sum and product of two even integers is an even integer. Hence, the operations are closed, a requirement if these operations are to be binary operations on $2\mathbf{Z}$. Of course, $\mathbf{Z}$ is a commutative ring with identity under the binary operations of addition and multiplication of integers since $\mathbf{Z}$ was one of our models for the definition of a ring.
2. The set $\mathbf{Q}$ of rational numbers is a commutative ring with identity under the usual operations of addition and multiplication of rational numbers.
3. If I denotes the set of irrational numbers, then I is *not* a ring under the operations of addition and multiplication of real numbers. In fact, neither of these operations is closed on I since, for example, $\sqrt{2}$ and $-\sqrt{2}$ are irrational numbers, yet $\sqrt{2} + (-\sqrt{2}) = 0$ is a rational number. Similarly, $\sqrt{2}\sqrt{2} = 2$ is a rational number.
4. The set $\mathbf{R}$ of real numbers is a commutative ring with identity $e = 1$ under the operations of addition and multiplication of real numbers.
5. The set $\mathbf{C}$ of complex numbers is a commutative ring with identity if addition and multiplication of complex numbers are defined as

$$(a + bi) + (c + di) = (a + c) + (b + d)i \qquad \text{and}$$

$$(a + bi)(c + di) = (ac - bd) + (ad + bc)i.$$

The additive identity for this ring is the complex number $0 + 0i$, and the multiplicative identity is $1 + 0i$. The additive and multiplicative identities are usually written simply as 0 and 1, respectively.

It was shown in Theorem 0.4.4 that the additive identity 0 and the multiplicative identity e (whenever e exists) of a ring R are unique, and Theorem 0.4.5 shows that the additive inverse $-x$ of an element $x \in R$ is unique.

The following properties hold in any ring. We prove two of these properties and leave the others as exercises.

3.1.5   Theorem

The following hold in any ring R.

1. For every $x \in R$, $x0 = 0x = 0$.
2. For all $x, y \in R$, $(-x)(-y) = xy$.
3. For all $x, y \in R$, $(-x)y = x(-y) = -(xy)$.
4. For all $x, y, z \in R$, $x(y - z) = xy - xz$.
5. For all $x, y, z \in R$, $(y - z)x = yx - zx$.

Proof. A proof of properties 1 and 3 is given. You should justify each step of both proofs.

1. Since $0 + 0 = 0$, $x(0 + 0) = x0$. But by the left-hand distributive property, $x(0 + 0) = x0 + x0$. Hence, $x0 + x0 = x0$, so adding $-(x0)$ to both sides of this equation produces

$$-(x0) + [x0 + x0] = -(x0) + x0$$
$$[-(x0) + x0] + x0 = 0$$
$$0 + x0 = 0$$
$$x0 = 0.$$

   Similarly, $0x = 0$.
3. Since $-x + x = 0$, $(-x)y + xy = (-x + x)y = 0y$. But $0y = 0$ by part 1, so

$$(-x)y + xy = 0.$$

We also have

$$-(xy) + xy = 0,$$

so it must be the case that $(-x)y = -(xy)$ since the additive inverse of xy is unique. The fact that $-(xy) = x(-y)$ has a similar proof. ∎

Several other types of rings are now introduced. It is possible for the product of two elements to be zero when neither element is zero. For example, in $\mathbf{Z}_6$, $[2][3] = [6] = [0]$. However, the ring of integers $\mathbf{Z}$ has the property that if $x, y \in \mathbf{Z}$

are such that $xy = 0$, then $x = 0$ or $y = 0$. Other rings share this property with the ring of integers. We give commutative rings with this property a special name.

3.1.6   Definition

If R is a ring, then an element $x \neq 0$ of R is said to be a **zero-divisor** of R if there is an element $y \neq 0$ in R such that either $xy = 0$ or $yx = 0$. A commutative ring with at least two elements that is free of zero-divisors is said to be a **domain**. A domain that has an identity is an **integral domain**. A ring R with identity and at least two elements in which every nonzero element has a multiplicative inverse is said to be a **division ring**. A commutative division ring is a **field**.

The ring of integers $\mathbf{Z}$ is a commutative ring with identity with at least two elements, and $\mathbf{Z}$ does not have zero-divisors. Since $\mathbf{Z}$ has an identity, $\mathbf{Z}$ is an integral domain. The ring of even integers $2\mathbf{Z}$ is a domain that is not an integral domain since it does not have an identity. The rings $\mathbf{Q}$, $\mathbf{R}$, and $\mathbf{C}$ are examples of fields under the usual operations of addition and multiplication. Each is a commutative ring with identity with at least two elements, and each nonzero element of $\mathbf{Q}$, $\mathbf{R}$, and $\mathbf{C}$ has a multiplicative inverse. The multiplicative inverse of a nonzero element $\frac{p}{q} \in \mathbf{Q}$ is $\frac{q}{p}$, and the multiplicative inverse of a nonzero element $r \in \mathbf{R}$ is $\frac{1}{r}$. If $a + bi$ is a nonzero element of $\mathbf{C}$, the multiplicative inverse of $a + bi$ is $\dfrac{a}{a^2 + b^2} - \dfrac{b}{a^2 + b^2} i$.

One might wonder why the condition "has at least two elements" is part of the definitions of a domain, an integral domain, and a division ring and field. If $R = \{0\}$ and addition and multiplication are defined on R by $0 + 0 = 0$ and $0 \cdot 0 = 0$, then R is a commutative ring with additive identity 0 and a multiplicative identity that is also 0. R is the **zero ring**. It is often the practice to eliminate this trivial ring from consideration when discussing rings. This is the reason that the condition "has at least two elements" is part of the definitions given in Definition 3.1.6. You will be asked to show in the exercises that a ring R with identity e contains at least two elements if and only if $e \neq 0$.

3.1.7   Theorem

Every division ring is free of zero-divisors.

Proof. Suppose R is a division ring, and let $x, y \in R$ be such that $xy = 0$. If $x \neq 0$, then x^{-1} exists in R, so, by Theorem 3.1.5, $x^{-1}(xy) = x^{-1}0 = 0$. But $x^{-1}(xy) = (x^{-1}x)y = ey = y$. Hence, $y = 0$. Similarly, if $xy = 0$ and $y \neq 0$, then $x = 0$. Thus, R cannot have zero-divisors. ∎

Theorem 3.1.7 shows that every field is an integral domain. The converse does not hold, since $\mathbf{Z}$ is an integral domain that is not a field. However, under certain conditions, integral domains are fields. The following theorem gives a condition that is sufficient for an integral domain to be a field. The proof hinges on the following property of finite sets: suppose that B is a finite set with, say, four elements. If A is a subset of B and A has four elements, then it must be the case that $A = B$. Once you have convinced yourself that this is true for every finite set, the proof of the theorem should be easy to understand.

3.1.8 Theorem

Any integral domain with a finite number of elements is a field.

Proof. Let $R = \{0, e, x_1, \ldots, x_n\}$ be a finite integral domain, where all the elements of R are distinct. (Note that R has $n + 2$ elements.) To show that R is a field, we need only show that each nonzero element of R has a multiplicative inverse. Consider the set $R^* = \{e, x_1, \ldots, x_n\}$ containing the $n + 1$ nonzero elements of R and let $x \in R^*$. If $x = e$, then x clearly has a multiplicative inverse. If $x \neq e$, consider the set $xR^* = \{x, xx_1, \ldots, xx_n\}$. Since R is an integral domain, $0 \notin xR^*$, so we see that $xR^* \subseteq R^*$. Now R^* has $n + 1$ elements, so if we can show that xR^* also has $n + 1$ elements, it will have to be the case that $xR^* = R^*$. To show that xR^* has $n + 1$ elements, it suffices to show that the elements of xR^* are distinct. If there is an i, $1 \leq i \leq n$, such that $xx_i = x$, then $x(x_i - e) = 0$, so $x_i = e$ since $x \neq 0$ and R is a domain. But this is a contradiction since the elements of R are distinct. If there exist i and j, $1 \leq i, j \leq n$, such that $xx_i = xx_j$, then $x(x_i - x_j) = 0$, so $x_i = x_j$, which again produces a contradiction. Thus, the elements of xR^* are distinct and so $xR^* = R^*$. Since $e \in R^* = xR^*$ and $x \neq e$, there must exist an i, $1 \leq i \leq n$, such that $xx_i = e$. Therefore, x has a multiplicative inverse. Since x was chosen arbitrarily in R^*, every nonzero element of R^* has a multiplicative inverse, so R is a field. ∎

Several familiar examples of rings were given in Examples 3.1.4. You may not be as familiar with the following examples of rings as you were with those of 3.1.4.

3.1.9   Some Less Familiar Examples of Rings

1. Let $q(\mathbf{R})$ be the set of all elements of the form $a + bi + cj + dk$, where $a, b, c, d \in \mathbf{R}$ and i, j, and k are subject to the following rules of multiplication:

$$\begin{aligned}
ij &= k & ik &= -j \\
jk &= i & kj &= -i & i^2 &= j^2 = k^2 = -1 \qquad (*) \\
ki &= j & ji &= -k
\end{aligned}$$

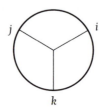

The mixed products of i, j, and k are easily remembered using the figure above and the following rule: counterclockwise, the product of two consecutive letters is equal to the next letter on the circle, and clockwise, the product of two consecutive letters is equal to the negative of the next letter. To add two elements of $q(\mathbf{R})$, simply add using the usual rules of algebra and collect coefficients. For example,

$$\begin{aligned}
(2 + 3i - 2j &+ 5k) + (-1 + 2i - j + \tfrac{1}{2}k) \\
&= (2 - 1) + (3 + 2)i + (-2 - 1)j + (5 + \tfrac{1}{2})k \\
&= 1 + 5i - 3j + \tfrac{11}{2}k.
\end{aligned}$$

To multiply, use the usual rules of algebra, but do not commute products involving i, j, and k: replace the mixed products of i, j, and k by their equivalent expressions using the equations in $(*)$; replace i^2, j^2, and k^2 by -1; and then collect coefficients.

$$\begin{aligned}
(2 + 3i - 2j + 5k)\left(-1 + 2i - j + \tfrac{1}{2}k\right) = \; & 2\left(-1 + 2i - j + \tfrac{1}{2}k\right) \\
& + 3i\left(-1 + 2i - j - \tfrac{1}{2}k\right) \\
& - 2j\left(-1 + 2i - j + \tfrac{1}{2}k\right) \\
& + 5k\left(-1 + 2i - j + \tfrac{1}{2}k\right)
\end{aligned}$$

$$= -2 + 4i - 2j + k$$
$$- 3i + 6i^2 - 3ij + \frac{3}{2}ik$$
$$+ 2j - 4ji + 2j^2 - jk$$
$$- 5k + 10ki - 5kj + \frac{5}{2}k^2$$
$$= -2 + 4(14)i - 2j + k$$
$$- 3i + 6(-1) - 3k + \frac{3}{2}(-j)$$
$$+ 2j - 4(-k) + 2(-1) - i$$
$$- 5k + 10j - 5(-i) + \frac{5}{2}(-1)$$
$$= \left(-2 - 6 - 2 - \frac{5}{2} \right)$$
$$+ (4 - 3 - 1 + 5)i$$
$$+ (-2 - \frac{3}{2} + 2 + 10)j$$
$$+ (1 - 3 + 4 - 5)k$$
$$= -\frac{25}{2} + 5i + \frac{17}{2}j - 3k.$$

The end result of executing these same procedures with two general elements $a_1 + b_1 i + c_1 j + d_1 k$ and $a_2 + b_2 i + c_2 j + d_2 k$ of $q(\mathbf{R})$ gives the formal rules for the definition of the operations of addition and multiplication on $q(\mathbf{R})$:

$$(a_1 + b_1 i + c_1 j + d_1 k) + (a_2 + b_2 i + c_2 j + d_2 k) =$$
$$(a_1 + a_2) + (b_1 + b_2)i + (c_1 + c_2)j + (d_1 + d_2)k$$
$$(a_1 + b_1 i + c_1 j + d_1 k)(a_2 + b_2 i + c_2 j + d_2 k) =$$
$$+ (a_1 a_2 - b_1 b_2 - c_1 c_2 - d_1 d_2)$$
$$+ (a_1 b_2 + a_2 b_1 + c_1 d_2 - c_2 d_1)i$$
$$+ (a_1 c_2 + a_2 c_1 + b_2 d_1 - b_1 d_2)j$$
$$+ (a_1 d_2 + a_2 d_1 + b_1 c_2 - b_2 c_1)k$$

The set $q(\mathbf{R})$ together with these operations is a ring with identity. The additive identity is $0 = 0 + 0i + 0j + 0k$, and the multiplicative identity is $1 = 1 + 0i + 0j + 0k$. Each of i, j, and k is an element of $q(\mathbf{R})$ since, for example, $i = 0 + 1i + 0j + 0k$. If $a + bi + cj + dk \neq 0$, then at least one of a, b, c, or d must be nonzero. Hence, $r = a^2 + b^2 + c^2 + d^2 \neq 0$. If $a + bi + cj + dk \neq 0$ is multiplied by $\frac{a}{r} - \frac{b}{r}i - \frac{c}{r}j - \frac{d}{r}k$ either on the left or the right, the result is 1. Thus, every nonzero element of $q(\mathbf{R})$ has a multiplicative inverse. However, $q(\mathbf{R})$ is not commutative since, for example, $ij \neq ji$. Thus, $q(\mathbf{R})$ is a division ring that is not a field. We call $q(\mathbf{R})$ the division ring of **real quaternions**. The modifier *real* is used in connection with this ring since a, b, c, and d are real numbers. If a, b, c, and d are taken only from $\mathbf{Q}$, then $q(\mathbf{Q})$ is also a division ring called the ring of **rational quaternions**. Quaternions were discovered by **William Hamilton** (1805–1865), an Irish mathematician who set out to justify the use of negatives and imaginaries in algebra. He thought that these concepts had a poor foundation. What Hamilton developed was the division ring of real quaternions. This ring possessed all of the properties Hamilton had sought, except that it was noncommutative.

2. For any positive integer n, $\mathbf{Z}_n$ is a commutative ring with identity [1] under the binary operations of congruence-class addition and multiplication. This was one of our models for the definition of a ring. If n is a positive integer that is not prime and $n \neq 1$, then $\mathbf{Z}_n$ has zero-divisors. Indeed, if $n = n_1 n_2$, where n_1 and n_2 are positive integers greater than 1, then $[n_1][n_2] = [n] = [0]$ with $[n_1] \neq [0]$ and $[n_2] \neq [0]$. When n is a prime integer p, Corollary 1.3.18 shows that every nonzero element of $\mathbf{Z}_p$ has a multiplicative inverse, so $\mathbf{Z}_p$ is a field.

3. The set $\mathcal{C}$ of all 2×2 matrices of the form $\begin{pmatrix} a & b \\ -b & a \end{pmatrix}$, where $a, b \in \mathbf{R}$, is a ring under the operations of matrix addition and matrix multiplication. The operations can easily be shown to be closed so that they are indeed binary operations on $\mathcal{C}$. Since $\begin{pmatrix} 1 & 0 \\ 0 & 1 \end{pmatrix} = \begin{pmatrix} 1 & 0 \\ -0 & 1 \end{pmatrix}$ and $\begin{pmatrix} 0 & 0 \\ 0 & 0 \end{pmatrix} = \begin{pmatrix} 0 & 0 \\ -0 & 0 \end{pmatrix}$, the additive identity and the multiplicative identity of $M_2(\mathbf{R})$ are in $\mathcal{C}$. Multiplication is also commutative, and if $\begin{pmatrix} a & b \\ -b & a \end{pmatrix} \neq \begin{pmatrix} 0 & 0 \\ 0 & 0 \end{pmatrix}$, then at least one of a and b is nonzero, and it can be shown that

$$\begin{pmatrix} a & b \\ -b & a \end{pmatrix}^{-1} = \begin{pmatrix} \dfrac{a}{a^2 + b^2} & \dfrac{-b}{a^2 + b^2} \\[2ex] \dfrac{b}{a^2 + b^2} & \dfrac{a}{a^2 + b^2} \end{pmatrix}$$

Hence, every nonzero element of $\mathcal{C}$ has a multiplicative inverse, so $\mathcal{C}$ is a field. You should verify the details. $\mathcal{C}$ is actually another way of describing the field of complex numbers.

4. Let $\mathbf{Z}[i] = \{x + yi \mid x, y \in \mathbf{Z}\}$, where $i = \sqrt{-1}$, and define addition and multiplication on $\mathbf{Z}[i]$ by

$$(x_1 + y_1 i) + (x_2 + y_2 i) = (x_1 + x_2) + (y_1 + y_2)i \qquad \text{and}$$

$$(x_1 + y_1 i)(x_2 + y_2 i) = (x_1 x_2 - y_1 y_2) + (x_1 y_2 + x_2 y_1)i$$

Then $\mathbf{Z}[i]$ together with these operations is a commutative ring with identity. We refer to $\mathbf{Z}[i]$ as the **ring of Gaussian integers**. This ring is named in honor of **Carl Friedrich Gauss** (1777–1855) who was the first mathematician to discover and investigate its properties, although he did not use the terminology associated with rings. (See the Historical Notes of Section 1.3 for information on Gauss.)

5. For a set X, recall that $\mathcal{P}(X)$ denotes the collection of all subsets of X. If $A, B \in \mathcal{P}(X)$, define addition and multiplication on $\mathcal{P}(X)$ by $A + B = (A \cup B) \setminus (A \cap B)$ and $AB = A \cap B$ for all $A, B \in \mathcal{P}(X)$. Then $\mathcal{P}(X)$ together with these two binary operations is a commutative ring with identity. The additive identity is the empty set $\varnothing$, and the multiplicative identity is the set X. If X has at least two elements, can you show that $\mathcal{P}(X)$ is not an integral domain? Note that $A^2 = A$ for each $A \in \mathcal{P}(X)$. An element x of a ring R such that $x^2 = x$ is said to be an **idempotent element** of R. Every ring has at least one idempotent element, namely, the additive identity 0. If every element of a ring is idempotent, the ring is said to be a **Boolean ring**. Boolean rings are named in honor of **George Boole** (1815–1864), an English logician who was one of the first mathematicians to study this type of ring. Boolean rings, and more generally Boolean algebras, have become quite useful in the design of electrical networks.

3.1.10 Polynomial Rings

One important example of a ring is the ring $R[x]$ of all polynomials over a ring R. A **polynomial** over a ring R is a formal expression of the form $f(x) = a_0 + a_1 x + a_2 x^2 + \cdots + a_n x^n$, where $a_i \in R$ for $i = 0, 1, 2, \ldots, n$. The ring R may not be commutative, and R need not have an identity. We say that $f(x)$ is a polynomial in (the **indeterminate**) x that has its **coefficients** in R. We assume that $ax = xa$ for all $a \in R$. If n is any nonnegative integer, the polynomial $f_0(x) = 0 + 0x + 0x^2 + \cdots + 0x^n$ is the **zero polynomial**, where if $n = 0$, we set $0x^0 = 0$. The zero polynomial has multiple representations: $f_0(x) = 0$,

$f_0(x) = 0 + 0x$, $f_0(x) = 0 + 0x + 0x^2$, and so on are representations of the same zero polynomial. Thus, a polynomial $f(x) = a_0 + a_1x + a_2x^2 + \cdots + a_nx^n$ is the zero polynomial if $a_0 = a_1 = \cdots = a_n = 0$. If $f(x) = a_0 + a_1x + a_2x^2 + \cdots + a_nx^n$ is not the zero polynomial, then the largest nonnegative integer n such that $a_n \neq 0$ is called the **degree of the polynomial**, and a_n is said to be the **leading coefficient** of $f(x)$. The degree of a polynomial is denoted by $\deg(f(x))$. For the zero polynomial, there is no nonnegative integer n such that $a_n \neq 0$, so no degree is assigned to this polynomial. A **constant polynomial** is a polynomial of the form $f(x) = a$, where $a \in R$. Two nonzero polynomials $f(x) = a_0 + a_1x + a_2x^2 + \cdots + a_nx^n$ and $g(x) = b_0 + b_1x + b_2x^2 + \cdots + b_mx^m$ are said to be **equal** if $m = n$ and $a_i = b_i$ for $i = 0, 1, \ldots, n$. This just says that two nonzero polynomials are equal if they have the same degree and their corresponding coefficients are equal. In the polynomials $f(x) = a_0 + a_1x + a_2x^2 + \cdots + a_nx^n$ and $g(x) = b_0 + b_1x + b_2x^2 + \cdots + b_mx^m$, a_0 and b_0, a_1x and b_1x, a_2x^2 and b_2x^2, etc. are said to be **terms that are alike**, or simply **like terms**, of the polynomials. If a_kx^k is a term in a polynomial, a_k is said to be the **coefficient** of x^k.

To add two polynomials in $R[x]$, just add the coefficients of like terms, and to compute the product of two polynomials, simply "multiply out" the product using the distributive property in $R[x]$ and then collect coefficients of like terms. There is one point to keep in mind during this process: if the ring is not commutative, care must be taken not to commute elements of the ring in the multiplication process. If R is a not a commutative ring, then neither is $R[x]$. For example, if $f(x) = a_0 + a_1x$ and $g(x) = b_0 + b_1x$ are polynomials in $R[x]$, where $a_0, b_0 \in R$ are such that $a_0b_0 \neq b_0a_0$, then

$$f(x)g(x) = (a_0 + a_1x)(b_0 + b_2x)$$
$$= a_0(b_0 + b_1x) + a_1x(b_0 + b_1x)$$
$$= a_0b_0 + (a_0b_1 + a_1b_0)x + a_1b_1x^2$$

and

$$g(x)f(x) = (b_0 + b_1x)(a_0 + a_1x)$$
$$= b_0(a_0 + a_1x) + b_1x(a_0 + a_1x)$$
$$= b_0a_0 + (b_0a_1 + b_1a_0)x + b_1a_1x^2.$$

Hence, $f(x)g(x) \neq g(x)f(x)$ since $a_0b_0 \neq b_0a_0$. The set $R[x]$ together with these operations of addition and multiplication of polynomials is a ring. The additive identity is the zero polynomial $f_0(x)$ and the additive inverse of $f(x) = a_0 + a_1x + a_2x^2 + \cdots + a_nx^n$ is $-f(x) = (-a_0) + (-a_1)x + (-a_2)x^2 + \cdots + (-a_n)x^n$, which is usually written as $-f(x) = -a_0 - a_1x - a_2x^2 - \cdots - a_nx^n$. When R has an identity e, $R[x]$ has the constant polynomial $f_e(x) = e$ as an identity. If R is a commutative ring, then $R[x]$ is also commutative. We refer to $R[x]$ as **the ring of polynomials in the indeterminate x that have their coefficients in R**.

| 3.1.11 | **Examples of Polynomial Rings** |

1. Polynomials in the rings $\mathbf{Z}[x]$, $\mathbf{Q}[x]$, $\mathbf{R}[x]$ and $\mathbf{C}[x]$ have their coefficients in the integers, the rational numbers, the real numbers, and the complex numbers, respectively. Each of these polynomial rings is a commutative ring with identity.

2. For any integer $n > 1$, $\mathbf{Z}_n[x]$ is the ring of polynomials in x with their coefficients in $\mathbf{Z}_n$. For example, if $n = 6$ and if $f(x) = [2] + [3]x$ and $g(x) = [5]x + [2]x^2$ are polynomials in $\mathbf{Z}_6[x]$, then

$$f(x) + g(x) = ([2] + [3]x) + ([5]x + [2]x^2)$$
$$= [2] + ([3] + [5])x + [2]x^2$$
$$= [2] + [2]x + [2]x^2$$

and

$$f(x)g(x) = ([2] + [3]x)([5]x + [2]x^2)$$
$$= [2][5]x + [2][2]x^2 + [3][5]x^2 + [3][2]x^3$$
$$= [4]x + [1]x^2.$$

3. For an example of a polynomial ring where the coefficient ring is noncommutative, consider $M_2(\mathbf{Z})[x]$. If $f(x) = \begin{pmatrix} 1 & -1 \\ 0 & 2 \end{pmatrix} + \begin{pmatrix} 2 & 0 \\ 1 & 1 \end{pmatrix}x^2$ and

$g(x) = \begin{pmatrix} -1 & 0 \\ 0 & 1 \end{pmatrix}x^2 + \begin{pmatrix} 1 & 0 \\ 0 & 1 \end{pmatrix}x^3$, then

$$f(x) + g(x) = \begin{pmatrix} 1 & -1 \\ 0 & 2 \end{pmatrix} + \begin{pmatrix} 1 & 0 \\ 1 & 2 \end{pmatrix}x^2 + \begin{pmatrix} 1 & 0 \\ 0 & 1 \end{pmatrix}x^3,\ \text{and}$$

$$f(x)g(x) = \left[\begin{pmatrix} 1 & -1 \\ 0 & 2 \end{pmatrix} + \begin{pmatrix} 2 & 0 \\ 1 & 1 \end{pmatrix}x^2 \right]\left[\begin{pmatrix} -1 & 0 \\ 0 & 1 \end{pmatrix}x^2 + \begin{pmatrix} 1 & 0 \\ 0 & 1 \end{pmatrix}x^3 \right]$$

$$= \begin{pmatrix} 1 & -1 \\ 0 & 2 \end{pmatrix}\begin{pmatrix} -1 & 0 \\ 0 & 1 \end{pmatrix}x^2 + \begin{pmatrix} 1 & -1 \\ 0 & 2 \end{pmatrix}\begin{pmatrix} 1 & 0 \\ 0 & 1 \end{pmatrix}x^3$$

$$+ \begin{pmatrix} 2 & 0 \\ 1 & 1 \end{pmatrix}\begin{pmatrix} -1 & 0 \\ 0 & 1 \end{pmatrix}x^4 + \begin{pmatrix} 2 & 0 \\ 1 & 1 \end{pmatrix}\begin{pmatrix} 1 & 0 \\ 0 & 1 \end{pmatrix}x^5$$

$$= \begin{pmatrix} -1 & -1 \\ 0 & 2 \end{pmatrix}x^2 + \begin{pmatrix} 1 & -1 \\ 0 & 2 \end{pmatrix}x^3 + \begin{pmatrix} -2 & 0 \\ -1 & 1 \end{pmatrix}x^4 + \begin{pmatrix} 2 & 0 \\ 1 & 1 \end{pmatrix}x^5.$$

A very loose construction has been given for polynomial rings $R[x]$, which leaves unanswered some fundamental questions about these rings. For example, exactly what is the indeterminate x, and is x always an element of $R[x]$? If R has an identity e and we set $x = ex$, then $x \in R[x]$. But what if R does not have an identity? And why should we expect that $(a + b)x = ax + bx$? This is certainly true if x were in R, but, in general, $x \notin R$. A more formal development of polynomial rings is presented in a later chapter where these questions are answered.

Classification Scheme for Classes of Rings

In the following lattice of classes of rings, each class of rings is a subclass of a class above it, if the two classes are connected by a sequence of line segments.

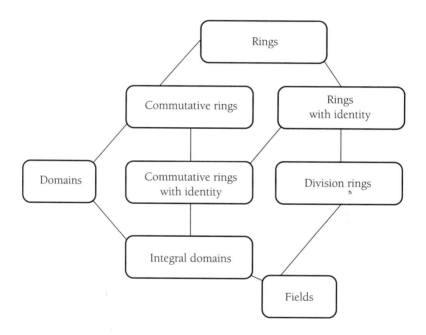

HISTORICAL NOTES

The concept of a ring developed slowly and almost certainly had its roots in number theory and the study of modular arithmetic. Gauss made important contributions to this area of mathematics, and he is credited with introducing the modern definition and notation, $x \equiv y \pmod{n}$, of modular arithmetic. Gauss studied the set of complex numbers **Z**[i] of the form $x + yi$, where x and y are integers. Today we know that **Z**[i] is an integral domain under the operations of addition and multiplication of complex numbers. Gauss, however, did not use the terminology associated with rings since rings were yet to be defined. Gauss defined a prime in $\mathbf{Z}[i]$ and showed that $x + yi \in \mathbf{Z}[i]$ can be factored uniquely in $\mathbf{Z}[i]$ into a product of primes. Factorization is unique up to the order of the factors and the insertion of elements of $\mathbf{Z}[i]$, each of which has a multiplicative inverse in $\mathbf{Z}[i]$. For a time, it was thought that factorization into a product of primes was unique in any domain of the form

$$\mathbf{Z}[\omega] = \{a_0 + a_1\omega + \cdots + a_{p-1}\omega^{p-1} \mid a_i \in \mathbf{Z} \text{ for } i = 0, 1, \ldots, p - 1\},$$

where

$$\omega = \cos\left(\frac{2\pi}{p}\right) + i\sin\left(\frac{2\pi}{p}\right)$$

and p is a prime. However, **Ernst Kummer** (1810–1893) showed that this assumed fact was false, and he went on to consider "ideal" complex numbers in an attempt to correct this difficulty. He was not successful, but his work led **Julius Wilhelm Richard Dedekind** (1831–1916) to create a new concept in an attempt to restore unique factorization. He concluded that there was no need to create a new number such as the ideal complex numbers of Kummer. Dedekind decided that what was required was to consider a system of numbers that already existed, and this led him to define what he called an "ideal set" of numbers. Today we refer to such a set in a ring as an ideal.

Ernst Kummer (1810–1893). Ernst Kummer was born in Sorau, Germany. He enrolled in the University of Halle at the age of 18 and received a Ph.D. in mathematics at the age of 21. He spent the early portion of his life teaching at the high school level and doing research in mathematics. In 1842 Kummer was appointed to a position at Breslau and 13 years later moved to the University of Berlin. Kummer was known as an outstanding teacher of mathematics, and he did his best work on unique factorization domains and on Fermat's Last Theorem, where he laid the foundation for the development of the idea of an ideal in a ring with his concept of an "ideal" number.

Julius Wilhelm Richard Dedekind (1831–1916). Dedekind was born in what is now Brunswick, Germany, and after he was grown dropped his first two names. He was educated at Collegium Carolium in Brunswick and at the University of Göttingen. Dedekind completed his doctorate in two years under Gauss, and he began his teaching career at the Polytechnikum in Zurich. In the meantime, Collegium Carolium was upgraded to the Brunswick Polytechnikum, and Dedekind was appointed to a position there, where he remained for the rest of his life. Dedekind is best known for his work with what are known today as Dedekind cuts. He posited the idea that every real number r divides the set of rational numbers into two sets, those that are less than r and those that are greater than r. Dedekind went on to develop the field of real numbers from the rational numbers through the use of these cuts. It was Dedekind who extended Kummer's idea of an ideal number to the concept of an ideal in a ring.

Problem Set 3.1

1. If R is any ring and $b, c \in R$, prove that the equation $x + b = c$ has a unique solution for x in R. Show, furthermore, that if F is a field, then the equation $ax + b = c$ has a unique solution for x in F, where $a, b, c \in F$ with $a \neq 0$.

2. Consider the set $\mathbf{Z}[\sqrt{2}] = \{x + y\sqrt{2} \mid x, y \in \mathbf{Z}\}$. If elements of $\mathbf{Z}[\sqrt{2}]$ are added and multiplied using the usual rules of algebra, show that $\mathbf{Z}[\sqrt{2}]$ is an integral domain that is not a field. Now replace $\mathbf{Z}$ by $\mathbf{Q}$ and show that $\mathbf{Q}[\sqrt{2}]$ is a field.

3. If R is a ring that has no zero-divisors, prove that the **left (right) cancellation property**, $ax = ay$ $(xa = ya)$ implies that $x = y$, holds for all $a, x, y \in R, a \neq 0$.

4. Complete the proof of Theorem 3.1.5.

5. Define the operations of addition $\oplus$ and multiplication $\otimes$ on $\mathbf{Z}$ by $x \oplus y = x + y - 1$ and $x \otimes y = x + y - xy$. Show that under these operations of addition and multiplication, $\mathbf{Z}$ is a commutative ring with identity. What is the additive identity of this ring? Is this ring an integral domain?

6. Show that the following hold in any ring R. Recall that subtraction is defined in a ring by $x - y = x + (-y)$.

    (a) $-(-x) = x$
    (b) $-(x + y) = -x - y$
    (c) $-(x - y) = -x + y$

7. Apply the procedures in Example 1 of 3.1.9 for adding and multiplying the elements of $q(\mathbf{R})$ to

    $$a_1 + b_1 i + c_1 j + d_1 k \qquad \text{and} \qquad a_2 + b_2 i + c_2 j + d_2 k$$

    to develop the general definition of the sum and product of these elements of $q(\mathbf{R})$. Show that the multiplicative inverse of a nonzero element $a + bi + cj + dk \in q(\mathbf{R})$ is $\frac{a}{z} - \frac{b}{z}i - \frac{c}{z}j - \frac{d}{z}k$, where $z = a^2 + b^2 + c^2 + d^2$.

8. If x and y are elements of a commutative ring R, show that $(x + y)^2 = x^2 + 2xy + y^2$. Find a noncommutative ring R and elements $x, y \in R$ such that $(x + y)^2 \neq x^2 + 2xy + y^2$.

9. Show that neither of the following can be an addition table for a ring R.

+	x	y	z	w
x	x	y	z	w
y	y	z	x	w
z	z	x	w	y
w	y	z	w	x

+	a	b	c	d
a	a	b	c	d
b	b	c	d	a
c	c	d	a	b
d	d	d	b	c

10. Find all the zero-divisors, if any, of each of the following rings. Also, find all the elements of each ring that have multiplicative inverses.

   (a) $\mathbf{Z}_{24}$ (b) $\mathbf{Z}_{35}$ (c) $\mathbf{Z}_7$

11. Prove that every Boolean ring is commutative. Hint: Consider $(x + x)^2$ and then $(x + y)^2$.

12. Show that every Boolean ring with identity and at least three elements has zero-divisors.

13. Prove that a Boolean ring R is a field if and only if $R = \{0, e\}$.

14. Prove that a ring R with identity is a Boolean ring if and only if $(x + y)xy = 0$ for all $x, y \in R$.

15. Let $R = \{x, y, z, w\}$. Complete the multiplication table below so that R will be a ring. Hint: Use the distributive property and the addition table.

+	x	y	z	w
x	x	y	z	w
y	y	z	w	x
z	z	w	x	y
w	w	x	y	z

•	x	y	z	w
x	x	x	x	x
y	x	z		
z	x		x	
w	x		x	z

16. Show that the multiplicative inverse of a nonzero complex number $a + bi$ is

$$(a + bi)^{-1} = \frac{a}{a^2 + b^2} + \frac{-b}{a^2 + b^2} i.$$

17. Prove that multiplication defined for the ring $\mathcal{C}$ of Example 3 in 3.1.9 is commutative. Show also that the multiplicative inverse of a nonzero element $\begin{pmatrix} a & b \\ -b & a \end{pmatrix}$ of $\mathcal{C}$ is

$$\begin{pmatrix} \dfrac{a}{a^2 + b^2} & \dfrac{-b}{a^2 + b^2} \\[2ex] \dfrac{b}{a^2 + b^2} & \dfrac{a}{a^2 + b^2} \end{pmatrix}.$$

18. Let R be a ring with identity e. Prove that R contains at least two elements if and only if $e \neq 0$. Conclude that the requirement that a ring with identity contain at least two elements can be replaced by the condition $e \neq 0$.

19. Let $R = \mathbf{Z} \times \mathbf{Z}$, and define addition and multiplication on R by $(a, b) + (c, d) = (a + c, b + d)$ and $(a, b)(c, d) = (ac + 2bd, ad + bc)$. Prove that R is a commutative ring with identity. Is R an integral domain?

20. Prove that every ring with identity and exactly two elements must be a field.

21. Let $R = \mathbf{Z} \times \mathbf{Z} \times \mathbf{Z}$ be the set of all ordered triples (a, b, c), where a, b, and c are in $\mathbf{Z}$. Define addition and multiplication on R by

$$(a_1, b_1, c_1) + (a_2, b_2, c_2) = (a_1 + a_2, b_1 + b_2, c_1 + c_2) \qquad \text{and}$$

$$(a_1, b_1, c_1)(a_2, b_2, c_2) = (a_1 a_2, b_1 a_2 + c_1 b_2, c_1 c_2).$$

Of course, $(a_1, b_1, c_1) = (a_2, b_2, c_2)$ if and only if $a_1 = a_2$, $b_1 = b_2$, and $c_1 = c_2$. Verify that R is a ring with additive identity $(0, 0, 0)$ and multiplicative identity $(1, 0, 1)$. Is R commutative, and does every nonzero element of R have a multiplicative inverse? Is R a domain?

22. Let R_1 and R_2 be rings with identities e_1 and e_2, respectively, and suppose that $R = R_1 \times R_2$ is the set of ordered pairs (x, y) where $x \in R_1$ and $y \in R_2$. Define addition and multiplication on R componentwise, that is, by $(x_1, y_1) + (x_2, y_2) = (x_1 + x_2, y_1 + y_2)$ and $(x_1, y_1)(x_2, y_2) = (x_1 x_2, y_1 y_2)$. Show that R is a ring with identity under these two operations. Show also that if R_1 and R_2 are commutative, then R is commutative. Show that R is *not* an integral domain even if R_1 and R_2 are integral domains.

23. Compute the sum and product of the following pairs of polynomials in the given polynomial ring.

    (a) $f(x) = \frac{1}{2} - \frac{3}{2}x^2$ and $g(x) = 2 + 4x - 8x^3$ in $\mathbf{Q}[x]$

    (b) $f(x) = [2] + [5]x^2 + [3]x^3$ and $g(x) = [3]x + [2]x^4$ in $\mathbf{Z}_6[x]$

    (c) $f(x) = \begin{pmatrix} 3 & 2 \\ -1 & 1 \end{pmatrix} + \begin{pmatrix} 1 & 0 \\ 1 & 2 \end{pmatrix} x^2$ and

    $g(x) = \begin{pmatrix} -2 & 1 \\ -1 & 1 \end{pmatrix} x + \begin{pmatrix} 0 & -3 \\ 0 & 5 \end{pmatrix} x^2$ in $M_2(\mathbf{Z})[x]$.

TECHNOLOGY PROBLEMS

Use a computer algebra system to solve each of the following problems.

24. Make an addition and a multiplication table for the ring $\mathbf{Z}_{12}$.

25. Instruct your computer algebra system to print out all pairs of integers $\{x, y\}$ such that $[x][y] = [0]$ in $\mathbf{Z}_{27}$, where $1 \leq x, y \leq 26$.

26. Make addition and multiplication tables for $[5]\mathbf{Z}_{60}$. Is $[5]\mathbf{Z}_{60}$ a ring?

27. Compute the product $f(x)g(x)$ if $f(x) = [34]x^4 + [24]x^3 + [16]$, $g(x) = [55]x^5 + [23]x^4 - [10]x^3 + [4]x^2 - [47]x + [45]$, and the coefficients are in $\mathbf{Z}_{56}$.

28. Instruct your computer algebra system to print out all pairs $\{x, y\}$ of integers such that $[x][y] = [1]$ in $\mathbf{Z}_{56}$, where x and y are such that $1 \leq x, y \leq 55$. Is $U(\mathbf{Z}_{56})$ a field under the operations of addition and multiplication of congruence classes modulo 56? Explain why or why not.

3.2   SUBRINGS, IDEALS, AND FACTOR RINGS

Section Overview. This section introduces subrings, left and right ideals, principal ideals, and factor rings. Be sure to take note of Theorem 3.2.3. It can often be used to determine whether a nonempty subset of a ring is a subring. Theorem 0.4.8 is used to show that the set of all cosets in the ring determined by an ideal can be made into a ring. Principal ideal rings are defined, and it is shown that the ring of integers forms a principal ideal ring. Prime ideals and maximal ideals are also defined, and Theorem 3.2.13 connects prime ideals to integral domains and maximal ideals to fields.

There are subsets S of a ring R that are themselves rings when the binary operations on R are restricted to S. The set $2\mathbf{Z}$ of even integers is a ring if the binary operations on $\mathbf{Z}$ are restricted to $2\mathbf{Z}$. In fact, this is true of $n\mathbf{Z}$ for any integer n.

3.2.1   Definition

A nonempty subset S of a ring R is said to be a **subring** of R if S is a ring when the binary operations on R are restricted to S. A subring of R that is a field is a **subfield** of R.

There are many examples of rings that have nontrivial subrings.

3.2.2   Examples

1. A ring R always has at least two subrings, namely, R and the zero ring $R_0 = \{0\}$. These are the obvious subrings of any ring R.

2. The set S of all polynomials in $\mathbf{R}[x]$ that have zero constant term is a subring of $\mathbf{R}[x]$. The ring $\mathbf{R}[x]$ has an identity $f_e(x) = 1$, whereas S does not have an identity.

3. The set $M_3(\mathbf{R})$ of all 3×3 matrices with entries from $\mathbf{R}$ is a noncommutative ring with identity that has zero-divisors. The set S of all 3×3 matrices of the form

$$\begin{pmatrix} a & 0 & 0 \\ 0 & a & 0 \\ 0 & 0 & a \end{pmatrix}$$

where $a \in \mathbf{R}$ is a subring of $M_3(\mathbf{R})$. S is closed under addition and multiplication since

$$\begin{pmatrix} a & 0 & 0 \\ 0 & a & 0 \\ 0 & 0 & a \end{pmatrix} + \begin{pmatrix} b & 0 & 0 \\ 0 & b & 0 \\ 0 & 0 & b \end{pmatrix} = \begin{pmatrix} a+b & 0 & 0 \\ 0 & a+b & 0 \\ 0 & 0 & a+b \end{pmatrix} \quad \text{and}$$

$$\begin{pmatrix} a & 0 & 0 \\ 0 & a & 0 \\ 0 & 0 & a \end{pmatrix}\begin{pmatrix} b & 0 & 0 \\ 0 & b & 0 \\ 0 & 0 & b \end{pmatrix} = \begin{pmatrix} ab & 0 & 0 \\ 0 & ab & 0 \\ 0 & 0 & ab \end{pmatrix}.$$

The ring S is a commutative subring of a noncommutative ring. The identity of $M_3(\mathbf{R})$ is the identity of S, and S is a subring of $M_3(\mathbf{R})$ that is a field.

4. The set $S = \{[0], [3], [6], [9]\}$ is a subring of $\mathbf{Z}_{12}$. The addition and multiplication tables for S are:

+	[0]	[3]	[6]	[9]
[0]	[0]	[3]	[6]	[9]
[3]	[3]	[6]	[9]	[0]
[6]	[6]	[9]	[0]	[3]
[9]	[9]	[0]	[3]	[6]

$\cdot$	[0]	[3]	[6]	[9]
[0]	[0]	[0]	[0]	[0]
[3]	[0]	[9]	[6]	[3]
[6]	[0]	[6]	[0]	[6]
[9]	[0]	[3]	[6]	[9]

5. For this example, we first prove that $\sqrt{2}$ is not a rational number: $\sqrt{2}$ is either a rational number or it is not. Suppose $\sqrt{2}$ is rational. Then there are integers p and $q \neq 0$ such that $\sqrt{2} = \frac{p}{q}$. Assume that $\frac{p}{q}$ has been reduced to lowest terms. Then $p = \sqrt{2}\,q$, so $p^2 = 2q^2$. Therefore, p^2 is an even integer, and thus by Exercise 13 of Problem Set 1.1, p is

even. Let $p = 2k$, where k is an integer. Then $p^2 = 2q^2$ becomes $4k^2 = 2q^2$, which gives $q^2 = 2k^2$. Hence, q^2 is even, so q *is also even.* Therefore, p and q are both even, which contradicts the assumption that $\frac{p}{q}$ was reduced to lowest terms. Thus, $\sqrt{2}$ cannot be a rational number.

Let $\mathbf{Q}[\sqrt{2}] = \{a + b\sqrt{2} \mid a, b \in \mathbf{Q}\}$. We claim that $\mathbf{Q}[\sqrt{2}]$ is a subfield of $\mathbf{R}$. The additive identity of $\mathbf{Q}[\sqrt{2}]$ is $0 + 0\sqrt{2}$, the multiplicative identity is $1 + 0\sqrt{2}$, and the additive inverse of $a + b\sqrt{2}$ is $-a - b\sqrt{2}$. The set $\mathbf{Q}[\sqrt{2}]$ is closed under addition and multiplication since

$$(a + b\sqrt{2}) + (c + d\sqrt{2}) = (a + c) + (b + d)\sqrt{2} \in \mathbf{Q}[\sqrt{2}]$$

and

$$(a + b\sqrt{2})(c + d\sqrt{2}) = (ac + 2bd) + (ad + bc)\sqrt{2} \in \mathbf{Q}[\sqrt{2}].$$

Since $\mathbf{Q}[\sqrt{2}] \subseteq \mathbf{R}$, the other ring properties are inherited from $\mathbf{R}$. That is, if a ring property, say, associativity of multiplication, doesn't hold for elements of $\mathbf{Q}[\sqrt{2}]$, then this ring property won't hold for all elements of $\mathbf{R}$, which would be a contradiction. Therefore, $\mathbf{Q}[\sqrt{2}]$ is a commutative subring of $\mathbf{R}$ that has an identity. To complete the proof that $\mathbf{Q}[\sqrt{2}]$ is a subfield of $\mathbf{R}$, we need to show that every nonzero element of $\mathbf{Q}[\sqrt{2}]$ has a multiplicative inverse. If $a + b\sqrt{2} \neq 0$, then $a \neq 0$ or $b \neq 0$.

Case 1. If $a \neq 0$ and $b = 0$, then $a + b\sqrt{2} = a$, so $(a + b\sqrt{2})^{-1} = a^{-1}$.

Case 2. If $a = 0$ and $b \neq 0$, then $a + b\sqrt{2} = b\sqrt{2}$, and we see that $(a + b\sqrt{2})^{-1} = \frac{1}{2}b^{-1}\sqrt{2}$.

Case 3. If $a \neq 0$ and $b \neq 0$, then

$$\frac{1}{a + b\sqrt{2}} = \frac{a - b\sqrt{2}}{(a + b\sqrt{2})(a - b\sqrt{2})} = \frac{a - b\sqrt{2}}{a^2 - 2b^2}$$

$$= \frac{a}{a^2 - 2b^2} - \frac{b}{a^2 - 2b^2}\sqrt{2},$$

and from this it follows easily that

$$(a + b\sqrt{2})\left(\frac{a}{a^2 - 2b^2} - \frac{b}{a^2 - 2b^2}\sqrt{2}\right) = 1.$$

Hence, $a + b\sqrt{2}$ will have a multiplicative inverse in $\mathbf{Q}[\sqrt{2}]$ provided that $a^2 - 2b^2 \neq 0$. Now we know that $a \neq 0$ and $b \neq 0$, so if $a^2 - 2b^2 = 0$, then

$\sqrt{2} = \pm \frac{a}{b} \in \mathbf{Q}$. But this indicates that $\sqrt{2}$ is a rational number, which has just been shown not to be the case. Hence, if $a \neq 0$ and $b \neq 0$, then $a - 2b^2 \neq 0$.

Therefore, if $a + b\sqrt{2} \neq 0$, then $(a + b\sqrt{2})^{-1}$ exists in $\mathbf{Q}[\sqrt{2}]$, so $\mathbf{Q}[\sqrt{2}]$ is a subfield of $\mathbf{R}$. The field $\mathbf{Q}[\sqrt{2}]$ is known as a **quadratic number field**. If $n > 1$ is a square free integer, it can be shown that $\mathbf{Q}[\sqrt{n}]$ is a subfield of $\mathbf{R}$. A **square free integer** is an integer n different from 0 and 1 that is not divisible by the square of any integer $k \neq \pm 1$. More generally, if n is a square free integer (which may be negative), then $\mathbf{Q}[\sqrt{n}]$ is a subfield of $\mathbf{C}$.

We have previously seen by Theorem 2.2.3 that a nonempty subset H of an additive group G is a subgroup of G if and only if $x - y \in H$ whenever $x, y \in H$. A similar proposition holds for subrings.

3.2.3   Theorem

A nonempty subset S of a ring R is a subring of R if and only if $x - y \in S$ and $xy \in S$ for all $x, y \in S$.

Proof. Let S be a subring of R. Then S is closed under addition and multiplication. If $x, y \in S$, then $-y \in S$ since S is a subgroup of the additive group of R. Hence, $x - y = x + (-y) \in S$. Likewise, $xy \in S$ since S is closed under multiplication.

Conversely, suppose that $x - y$ and xy are in S whenever $x, y \in S$. Then by Theorem 2.2.3, S is a subgroup of the additive group of R. The associativity of multiplication and the distributive property of multiplication over addition are inherited from R. Since S is closed under multiplication, S is a subring of R. ∎

Normal subgroups play an important role in group theory. These subgroups allow us to form factor groups, and they are closely connected to group homomorphisms. In the theory of rings, a parallel situation holds. There are subrings called ideals that can be used to form factor rings, and these subrings have a close connection to ring homomorphisms. Ring homomorphisms are introduced in the next section.

3.2.4   Definition

If I is a nonempty subset of a ring R, then I is said to be an **ideal** of R if I is a subgroup of the additive group of R and xa and ax are in I whenever $x \in I$ and $a \in R$. If I is a subgroup of the additive group of R such that $xa \in I$ $(ax \in I)$ whenever $x \in I$ and $a \in R$, then I is a **right ideal** (**left ideal**) of R. An ideal I of R is a **proper ideal** of R if $I \neq R$.

Since a nonempty subset I of a ring R is a subgroup of the additive group of R if and only if $x - y \in I$ whenever $x, y \in I$, I is an ideal of R if and only if I is closed under subtraction and xa, $ax \in I$ for each $a \in R$ and all $x \in I$. An ideal of a ring can loosely be described as a subgroup that absorbs multiplication by a ring element from the left and from the right. Similar observations hold for left ideals and right ideals of a ring: a left ideal is a subgroup that absorbs multiplication by ring elements from the left, and a right ideal is a subgroup that absorbs multiplication by ring elements from the right.

Of course, an ideal of a ring is both a left and a right ideal and when the ring is commutative, a left or right ideal of the ring is an ideal. *An ideal I of R is a subring of R* (Do you see why?), and if $e \in I$, then $I = R$. This follows easily since if $x \in R$, then $x = ex \in I$, so $R \subseteq I$. Every ring has at least two ideals, namely, the zero ideal $\{0\}$ and R itself. The concept of an ideal was formed from the work of **Ernst Kummer** (1810–1893), a German mathematician who worked on unique factorization domains. Kummer laid the foundation for the idea of an ideal in a ring with his concept of an "ideal" number. (See the Historical Notes at the end of Section 3.1 for additional information on Kummer and the development of the concept of an ideal in a ring.)

| 3.2.5 | **Examples** |

1. If R is any ring and $x \in R$, consider $xR = \{xa \mid a \in R\}$. If $xa, xb \in xR$, then $xa - xb = x(a - b) \in xR$. Also, if $xa \in xR$ and $b \in R$, then $(xa)b = x(ab) \in xR$. Thus, xR is a right ideal of R. Similarly, Rx is a left ideal of R. In particular, if n is any integer, $n\mathbf{Z}$ is an ideal of R since $n\mathbf{Z}$ is a subgroup of $\mathbf{Z}$, and if $nx \in n\mathbf{Z}$ and $k \in \mathbf{Z}$, then $(nx)k = n(xk) \in n\mathbf{Z}$.

2. Consider the ring $\mathbf{Z}_6$ under the operations of addition and multiplication modulo 6. Then the set $I = \{[0], [2], [4]\}$ is an ideal of $\mathbf{Z}_6$. The addition table for I is

+	[0]	[2]	[4]
[0]	[0]	[2]	[4]
[2]	[2]	[4]	[0]
[4]	[4]	[0]	[2]

which shows that I is an additive subgroup of the additive group of $\mathbf{Z}_6$. To complete the demonstration that I is an ideal of $\mathbf{Z}_6$, since $\mathbf{Z}_6$

is a commutative ring, it suffices to show that *I* is closed under multiplication by ring elements from the left.

$$[0]I = \{[0][0], [0][2], [0][4]\} = \{[0], [0], [0]\} \subseteq I$$
$$[1]I = \{[1][0], [1][2], [1][4]\} = \{[0], [2], [4]\} = I$$
$$[2]I = \{[2][0], [2][2], [2][4]\} = \{[0], [4], [2]\} = I$$
$$[3]I = \{[3][0], [3][2], [3][4]\} = \{[0], [0], [0]\} \subseteq I$$
$$[4]I = \{[4][0], [4][2], [4][4]\} = \{[0], [2], [4]\} = I$$
$$[5]I = \{[5][0], [5][2], [5][4]\} = \{[0], [4], [2]\} = I$$

3. Consider the ring $M_2(\mathbf{Z})$ of 2×2 matrices with entries from $\mathbf{Z}$, and let

$$I_1 = \left\{ \begin{pmatrix} x & 0 \\ y & 0 \end{pmatrix} \middle| x, y \in \mathbf{Z} \right\}, \quad I_2 = \left\{ \begin{pmatrix} x & y \\ 0 & 0 \end{pmatrix} \middle| x, y \in \mathbf{Z} \right\},$$

$$I_3 = \left\{ \begin{pmatrix} x & y \\ z & w \end{pmatrix} \middle| x, y, z, w \in 2\mathbf{Z} \right\}, \quad \text{and} \quad I_4 = \left\{ \begin{pmatrix} x & 0 \\ 0 & 0 \end{pmatrix} \middle| x \in \mathbf{Z} \right\}.$$

Each of these nonempty subsets of $M_2(\mathbf{Z})$ is an additive subgroup of the additive group of $M_2(\mathbf{Z})$. If $\begin{pmatrix} a & b \\ c & d \end{pmatrix}$ is an arbitrary element of $M_2(\mathbf{Z})$, then the following hold:

(i) $\begin{pmatrix} a & b \\ c & d \end{pmatrix}\begin{pmatrix} x & 0 \\ y & 0 \end{pmatrix} = \begin{pmatrix} ax + by & 0 \\ cx + dy & 0 \end{pmatrix} \in I_1$, but

$\begin{pmatrix} x & 0 \\ y & 0 \end{pmatrix}\begin{pmatrix} a & b \\ c & d \end{pmatrix} = \begin{pmatrix} xa & xb \\ ya & yb \end{pmatrix} \notin I_1.$

Hence, I_1 is a left ideal but *not* a right ideal of $M_2(\mathbf{Z})$.

(ii) $\begin{pmatrix} a & b \\ c & d \end{pmatrix}\begin{pmatrix} x & y \\ 0 & 0 \end{pmatrix} = \begin{pmatrix} ax & ay \\ cx & cy \end{pmatrix} \notin I_2$, but

$\begin{pmatrix} x & y \\ 0 & 0 \end{pmatrix}\begin{pmatrix} a & b \\ c & d \end{pmatrix} = \begin{pmatrix} xa + yc & xb + yd \\ 0 & 0 \end{pmatrix} \in I_2,$

so I_2 is a right ideal but *not* a left ideal of $M_2(\mathbf{Z})$.

(iii) $\begin{pmatrix} a & b \\ c & d \end{pmatrix}\begin{pmatrix} x & y \\ z & w \end{pmatrix} = \begin{pmatrix} ax + bz & ay + bw \\ cx + dz & cy + dw \end{pmatrix} \in I_3$ and

$$\begin{pmatrix} x & y \\ z & w \end{pmatrix} \begin{pmatrix} a & b \\ c & d \end{pmatrix} = \begin{pmatrix} xa + yc & xb + yd \\ za + wc & zb + wd \end{pmatrix} \in I_3 \text{ since the entries in}$$

the product are even integers when a, b, c, and $d \in \mathbf{Z}$ and x, y, z, $w \in 2\mathbf{Z}$. Hence, I_3 is an ideal of $M_2(\mathbf{Z})$.

(iv) $\begin{pmatrix} x & 0 \\ 0 & 0 \end{pmatrix} \begin{pmatrix} y & 0 \\ 0 & 0 \end{pmatrix} = \begin{pmatrix} xy & 0 \\ 0 & 0 \end{pmatrix} \in I_4$, so I_4 is a subring of $M_2(\mathbf{Z})$.

However,

$$\begin{pmatrix} a & b \\ c & d \end{pmatrix} \begin{pmatrix} x & 0 \\ 0 & 0 \end{pmatrix} = \begin{pmatrix} ax & 0 \\ cx & 0 \end{pmatrix} \notin I_4 \text{ and}$$

$$\begin{pmatrix} x & 0 \\ 0 & 0 \end{pmatrix} \begin{pmatrix} a & b \\ c & d \end{pmatrix} = \begin{pmatrix} xa & xb \\ 0 & 0 \end{pmatrix} \notin I_4.$$

Hence, I_4 is neither a left nor a right ideal of $M_2(\mathbf{Z})$.

4. If R and S are rings, then $R \times S = \{(x, y) \mid x \in R \text{ and } y \in S\}$ is a ring if addition and multiplication are defined on $R \times S$ by $(x, y) + (z, w) = (x + z, y + w)$ and $(x, y)(z, w) = (xz, yw)$. Consider the subset $\{0\} \times S = \{(0, y) \mid y \in S\}$ of $R \times S$. If $(0, y), (0, w) \in \{0\} \times S$, then $(0, y) - (0, w) = (0, y - w) \in \{0\} \times S$, so $\{0\} \times S$ is a subgroup of the additive group of $R \times S$. If $(0, y) \in \{0\} \times S$ and $(a, b) \in R \times S$, then $(0, y)(a, b) = (0a, yb) = (0, yb) \in \{0\} \times S$ and $(a, b)(0, y) = (a0, by) = (0, by) \in \{0\} \times S$. Thus, $\{0\} \times S$ is an ideal of $R \times S$. Similarly, $R \times \{0\}$ is an ideal of $R \times S$.

Since R is an abelian group under addition, if I is an ideal of R, then I is a normal subgroup of the additive group of R, and thus we can consider the set R/I of cosets of I in R. Since R is abelian, $x + I = I + x$ for all $x \in R$, so there is no need to make a distinction between the left and right cosets of I in R. It is also easy to show that $x + I = y + I$ if and only if $x - y \in I$. You should take special notice of this fact since it is used frequently, often without mention.

The following two theorems show that R/I can be made into a ring.

3.2.6   **Theorem**

If I is an ideal of a ring R and a relation E is defined on R by xEy if and only if $x - y \in I$, then E is a congruence relation on R with respect to both the binary operations, addition and multiplication, defined on R. Furthermore, the congruence classes determined by E are the cosets of I in R.

Proof. Since R is an abelian group under addition, I is a normal subgroup of R. It follows from Theorems 2.2.13 and 2.2.16 that E is a congruence relation on R with respect to addition and that the congruence classes determined by E are the cosets of I in R. It remains only to show that E is a congruence relation with respect to multiplication. If xEx' and yEy', then $x - x' \in I$ and $y - y' \in I$. Suppose that $x = x' + i$ and $y = y' + j$, where $i, j \in I$. Then $xy = (x' + i)(y' + j) = x'y' + x'j + iy' + ij$. Hence, $xy - x'y' = x'j + iy' + ij \in I$ since I is an ideal of R. Therefore, $xyEx'y'$, so E is a congruence relation on R with respect to multiplication, and this completes the proof. ■

Notice in the preceding proof that $x'j + iy' + ij \in I$ since I absorbs multiplication by ring elements from the left and the right. It is precisely this absorption property of ideals that turns coset membership into a congruence relation that preserves multiplication. For this reason, multiplication of cosets can be performed by multiplying their representatives in R.

3.2.7   Theorem

If I is an ideal of R, then R/I is a ring under the binary operations of addition and multiplication defined on R/I by

$$(x + I) + (y + I) = (x + y) + I \qquad \text{and}$$
$$(x + I)(y + I) = xy + I.$$

If R is a commutative ring, then R/I is a commutative ring, and if R has an identity e, then R/I is a ring with identity $e + I$.

Proof. It is a consequence of Theorems 3.2.6 and 0.4.8 that the two binary operations are well-defined. If Theorem 2.2.16 is translated to additive notation, it follows immediately that R/I is an additive group under coset addition. The additive identity is $0 + I = I$, and the additive inverse of $x + I$ is $(-x) + I$. Since $(x + I) + (y + I) = (x + y) + I = (y + x) + I = (y + I) + (x + I)$, R/I is an abelian group under coset addition. Hence, it remains only to show that multiplication of cosets is associative and that the left and right distributive properties hold. To show that multiplication of cosets is associative, suppose that $x + I$, $y + I$, and $z + I$ are cosets in R/I. Then

$$(x + I)[(y + I)(z + I)] = (x + I)(yz + I)$$
$$= x(yz) + I$$
$$= (xy)z + I$$
$$= (xy + I)(z + I)$$
$$= [(x + I)(y + I)](z + I).$$

The proof that coset multiplication in R/I is distributive over coset addition from both the left and the right is just as straightforward and is left as an exercise. If the ring R has an identity e, then $e + I$ is clearly an identity for R/I, and if R is a commutative ring, then it is easy to show that R/I is also commutative. ∎

When I is an ideal of R, then we call the ring R/I of Theorem 3.2.7 the **factor ring** of R formed by factoring out I.

| 3.2.8 | **Examples** |

1. The set $5\mathbf{Z}$ of all integral multiples of 5 is an ideal of $\mathbf{Z}$, and $\mathbf{Z}/5\mathbf{Z} = \{0 + 5\mathbf{Z}, 1 + 5\mathbf{Z}, 2 + 5\mathbf{Z}, 3 + 5\mathbf{Z}, 4 + 5\mathbf{Z}\}$. The set $\mathbf{Z}/5\mathbf{Z}$ is a commutative ring with identity under coset addition and coset multiplication. For example, $(2 + 5\mathbf{Z}) + (4 + 5\mathbf{Z}) = (2 + 4) + 5\mathbf{Z} = 6 + 5\mathbf{Z} = 1 + 5 + 5\mathbf{Z} = 1 + 5\mathbf{Z}$ and $(2 + 5\mathbf{Z})(4 + 5\mathbf{Z}) = 2 \cdot 4 + 5\mathbf{Z} = 8 + 5\mathbf{Z} = 3 + 5 + 5\mathbf{Z} = 3 + 5\mathbf{Z}$. The additive identity of $\mathbf{Z}/5\mathbf{Z}$ is $0 + 5\mathbf{Z}$, and the multiplicative identity is $1 + 5\mathbf{Z}$.

2. Consider the ring $\mathbf{Z}_{12}$. Then $I = [3]\mathbf{Z}_{12} = \{[0], [3], [6], [9]\}$ is an ideal of $\mathbf{Z}_{12}$. The distinct cosets of I in $\mathbf{Z}_{12}$ are $[0] + I$, $[1] + I$, and $[2] + I$. (You should verify this.) Hence, $\mathbf{Z}_{12}/I = \{[0] + I, [1] + I, [2] + I\}$, and $\mathbf{Z}_{12}/I$ is a commutative ring with identity under coset addition and multiplication.

3. Consider the polynomial ring $\mathbf{R}[x]$, and let $I = \{f(x) \in \mathbf{R}[x] \mid f(x)$ has zero constant term$\}$. We claim that I is an ideal of $\mathbf{R}[x]$. I is clearly closed under subtraction since the difference of two polynomials with zero constant term will have zero constant term. Moreover, if $f(x) \in I$ and $g(x) \in \mathbf{R}[x]$, then both $f(x)g(x)$ and $g(x)f(x)$ will have zero constant term and so are in I. Thus, I is an ideal of $R[x]$, so we can form the factor ring $R[x]/I$. Now let's see what the cosets of $\mathbf{R}[x]/I$ look like. Consider $f(x) = \sqrt{5} + 2x - \frac{1}{2}x^3$. Then $f(x) = \sqrt{5} + g(x)$, where $g(x) = 2x - \frac{1}{2}x^3$ has zero constant term. It now follows that $f(x) + I = \sqrt{5} + g(x) + I = \sqrt{5} + I$ since $g(x) \in I$. In general, if $f(x) \in R[x]$ has constant term a, then $f(x) = a + g(x)$, where $g(x) \in R[x]$ has zero constant term. Hence, $f(x) + I = a + g(x) + I = a + I$ since $g(x) + I = I$. Thus, the coset $a + I$ consists of all the polynomials of $f(x) \in \mathbf{R}[x]$ that have constant term a. Similarly, $b + I$ is made up of all polynomials $f(x) \in \mathbf{R}[x]$ that have constant term b. Therefore, $\mathbf{R}[x]/I = \{a + I \mid a \in \mathbf{R}\}$.

The following two examples provide the motivation for the definitions of certain types of ideals that play an important role in the theory of commutative rings.

3.2.9   Examples

1. In the ring of integers, if p is a prime integer, then the ideal $p\mathbf{Z}$ has the property that if $xy \in p\mathbf{Z}$, then either $x \in p\mathbf{Z}$ or $y \in p\mathbf{Z}$. This follows since if $xy \in p\mathbf{Z}$, then there is an integer z such that $pz = xy$. But then $p\,|\,xy$, and since p is prime, $p\,|\,x$ or $p\,|\,y$. Hence, either x or y is a multiple of p, so either $x \in p\mathbf{Z}$ or $y \in p\mathbf{Z}$.

2. As a second example of a special type of ideal in a ring, we turn to the polynomial ring $\mathbf{Z}[x]$. Consider $I = \{f(x) \in \mathbf{Z}[x] \,|\, f(x)$ has a constant term that is an even integer$\}$. I is clearly a subgroup of $\mathbf{Z}[x]$, and if $f(x) = 2a_0 + a_1x + \cdots + a_nx^n \in I$ and $g(x) = b_0 + b_1x + \cdots + b_mx^m \in \mathbf{Z}[x]$, then the constant term of the product $f(x)g(x)$ is $2a_0b_0$, which is an even integer. Hence, $f(x)g(x) \in I$, so I is an ideal of $\mathbf{Z}[x]$. Let J be an ideal of $\mathbf{Z}[x]$ such that $I \subseteq J$. We claim that either $I = J$ or $J = \mathbf{Z}[x]$. If $I \neq J$, then J must contain a polynomial of the form

$$f(x) = (2k+1) + a_1x + a_2x^2 + \cdots + a_nx^n,$$

where k is an integer; that is, the constant term of $f(x)$ is an odd integer. But then

$$g(x) = 2k + a_1x + a_2x^2 + \cdots + a_nx^n$$

is in I and consequently in J since $I \subseteq J$. Therefore, $1 = f(x) - g(x) \in J$ because J is closed under subtraction. It was pointed out earlier that if an ideal I contains the identity of the ring R, then $I = R$. Because of this, we have $J = \mathbf{Z}[x]$.

In the first preceding example, $p\mathbf{Z}$ was an ideal of $\mathbf{Z}$ such that if $xy \in p\mathbf{Z}$, then either $x \in p\mathbf{Z}$ or $y \in p\mathbf{Z}$. In the second example, the ideal I had the property that if J was an ideal of $\mathbf{Z}[x]$ such that $I \subseteq J \subseteq \mathbf{Z}[x]$, either $J = I$ or $J = \mathbf{Z}[x]$. These two properties of ideals provide the motivation for the following definition.

3.2.10   Definition

A proper ideal P of a commutative ring R is a **prime ideal** of R if, whenever $xy \in P$, either $x \in P$ or $y \in P$. A proper ideal M of a ring R (not necessarily commutative) is said to be a **maximal ideal** of R if whenever I is an ideal of R

such that $M \subseteq I \subseteq R$, then either $I = M$ or $I = R$. If R is a commutative ring and $x \in R$, then an ideal of the form xR, where $x \in R$, is said to be a **principal ideal** of R. The principal ideal xR is often denoted by (x). A commutative ring in which every ideal is principal is a **principal ideal ring**.

Returning to the two examples of 3.2.9, we see that the ideal $p\mathbf{Z}$ of $\mathbf{Z}$ is a principal ideal that is prime, and the ideal I of $\mathbf{Z}[x]$ is a maximal ideal.

Prime and maximal ideals play an important role in commutative rings. We are now going to prove a sequence of theorems that show how prime and maximal ideals are related and how they are connected to integral domains and fields. You don't yet have a lot of experience with these concepts, so for now concentrate on the statements of the theorems and on the examples. Work through a couple of the proofs to get a feeling for the kinds of arguments involved. We return to these ideas in a later chapter. To establish the connection between prime ideals and integral domains and between maximal ideals and fields, we need the following lemma, the proof of which is left as an exercise.

3.2.11   Lemma

If J and K are ideals (left ideals, right ideals) of a ring R, then $J + K$ is an ideal (a left ideal, a right ideal) of R, where $J + K = \{j + k \mid j \in J$ and $k \in K\}$. Furthermore, $J \subseteq J + K$ and $K \subseteq J + K$.

Before we state and prove the following theorem, let's look at two examples that show the connection between prime ideals and integral domains and between maximal ideals and fields.

3.2.12   Examples

1. Consider the factor ring $\mathbf{Z}/6\mathbf{Z}$. The ideal $6\mathbf{Z}$ is not prime, since $xy \in 6\mathbf{Z}$ does not imply $x \in 6\mathbf{Z}$ or $y \in 6\mathbf{Z}$. For example, $2 \cdot 3 \in 6\mathbf{Z}$, yet $2 \notin 6\mathbf{Z}$ and $3 \notin 6\mathbf{Z}$. It is also the case that the ring $\mathbf{Z}/6\mathbf{Z}$ has zero-divisors since $2 + 6\mathbf{Z} \neq 0$ in $\mathbf{Z}/6\mathbf{Z}$ and $3 + 6\mathbf{Z} \neq 0$ in $\mathbf{Z}/6\mathbf{Z}$, yet $(2 + 6\mathbf{Z})(3 + 6\mathbf{Z}) = 6 + 6\mathbf{Z} = 0$. This shows the close connection between zero-divisors in $\mathbf{Z}/6\mathbf{Z}$ and the failure of $6\mathbf{Z}$ to be a prime ideal.
2. For this example, consider the factor ring $\mathbf{Z}/5\mathbf{Z}$. We claim that $5\mathbf{Z}$ is a maximal ideal of $\mathbf{Z}$. This follows since if I is an ideal of $\mathbf{Z}$ such that $5\mathbf{Z} \subset I \subseteq \mathbf{Z}$, then there is an integer $x \in I$ such that $x \notin 5\mathbf{Z}$. Therefore,

$5 \nmid x$ and so $\gcd(5, x) = 1$. Now we know that there are integers a and b such that $1 = a5 + bx$. But $5 \in 5\mathbf{Z} \subset I$, so we see that $a5 \in I$. Furthermore, $bx \in I$ since $x \in I$, so it follows that $1 = a5 + bx \in I$. Hence, $I = \mathbf{Z}$, which shows that $5\mathbf{Z}$ is a maximal ideal of $\mathbf{Z}$. Finally, the factor ring $\mathbf{Z}/5\mathbf{Z}$ is a field: each of the elements $1 + 5\mathbf{Z}, 4 + 5\mathbf{Z}$ is its own multiplicative inverse and $(2 + 5\mathbf{Z})(3 + 5\mathbf{Z}) = 6 + 5\mathbf{Z} = 1 + 5\mathbf{Z}$.

3.2.13   Theorem

The following hold for any commutative ring R with identity e:

1. An ideal P of R is prime if and only if R/P is an integral domain.
2. An ideal M of R is maximal if and only if R/M is a field.

Proof. Since R is a commutative ring with identity, both R/P and R/M are commutative rings with identities. To prove that R/P is an integral domain, it suffices to prove that R/P has no zero-divisors, and to show that R/M is a field, we must show that every nonzero element of R/M has a multiplicative inverse in R/M.

1. Let P be a prime ideal of R. If $x + P, y + P \in R/P$ are such that $(x + P)(y + P) = 0 + P$, then $xy + P = 0 + P$, which implies that $xy \in P$. But since P is prime, either $x \in P$ or $y \in P$. If $x \in P$, then $x + P = 0 + P$, and if $y \in P$, then $y + P = 0 + P$. Hence, R/P is free of zero-divisors and so is an integral domain. Conversely, if R/P is an integral domain and $xy \in P$, then $(x + P)(y + P) = xy + P = 0 + P$ in R/P. But R/P is an integral domain, so $x + P = 0 + P$ or $y + P = 0 + P$. Hence, $x \in P$ or $y \in P$, and thus P is a prime ideal of R.
2. Let M be a maximal ideal of R. If $x + M$ is a nonzero element of R/M, we need to show that $x + M$ has a multiplicative inverse in R/M. Now xR and M are ideals of R, so $M + xR$ is, by Lemma 3.2.11, an ideal of R and $M \subseteq M + xR$. Since $x + M$ is a nonzero element of R/M, $x \notin M$. But $x = 0 + xe \in M + xR$, so $M \neq M + xR$. But M is a maximal ideal of R and so $M + xR = R$. Therefore, $m + xy = e$ for some $m \in M$ and $y \in R$. Hence, $e - xy = m \in M$, so $xy + M = e + M$ in R/M. But then $e + M = (x + M)(y + M)$, which shows that $x + M$ has a multiplicative inverse in R/M. Hence, R/M is a field.
    Conversely, suppose that R/M is a field, and let I be an ideal of R such that $M \subseteq I \subseteq R$. If $I \neq M$, there is an element $x \in I$ such that $x \notin M$ and so $x + M$ is nonzero in R/M. If $y + M \in R/M$ is such that $(x + M)(y + M) = e + M$, then $xy + M = e + M$, so $xy - e \in M \subseteq I$. If $xy - e = k \in I$, then

$e = xy - k \in I$ since xy and k are in I. But any ideal of **R** that contains e must coincide with **R**. Thus, $I = R$, so M is a maximal ideal of **R** when R/M is a field. ∎

3.2.14 Corollary

In a commutative ring with identity, every maximal ideal is prime.

Proof. If M is a maximal ideal of a commutative ring R with identity, then Theorem 3.2.13 shows that R/M is a field. But any field is, by Theorem 3.1.7, an integral domain. Hence, R/M is an integral domain, and Theorem 3.2.13 now shows that M is a prime ideal of R. ∎

The following two theorems shed light on the ideal structure of the ring **Z** of integers.

3.2.15 Theorem

Z is a principal ideal ring.

Proof. Let I be an ideal of **Z**. If I is the zero ideal, then $I = 0\mathbf{Z}$ and thus I is principal. If I is not the zero ideal, let $x \in I$, $x \neq 0$. Then $-x \in I$, because I is an additive abelian group. Now either x or $-x$ is a positive integer, so the set I^+ of positive integers in I is nonempty. Since **N** is well-ordered, let n be the smallest positive integer in I^+. We claim that $I = n\mathbf{Z}$. Since $n \in I$, it is immediate that $n\mathbf{Z} \subseteq I$, so the proof will be complete if it can be shown that $I \subseteq n\mathbf{Z}$. If $y \in I$, then by the Division Algorithm 1.1.3 there exist unique integers q and r such that $y = qn + r$, where $0 \leq r < n$. Now $y, n \in I$, so $y - qn \in I$. Hence, $r \in I$ and $r < n$, which is a contradiction unless $r = 0$. Therefore, $y = qn \in n\mathbf{Z}$, so we have $I \subseteq n\mathbf{Z}$. Hence, $I = n\mathbf{Z}$. ∎

Two mathematical statements S_1 and S_2 are equivalent if each implies the other, that is, if S_1 implies S_2 and S_2 implies S_1. In the following theorem we are going to prove that three statements, numbered 1, 2, and 3, are equivalent. One method of proving such an equivalence is to use a "loop" proof that involves showing that 1 implies 2, 2 implies 3, and then 3 implies 1. We adopt the notation $A \Rightarrow B$ for A implies B. Showing that $1 \Rightarrow 2 \Rightarrow 3 \Rightarrow 1$ actually proves that any two of the three statements are equivalent. For example, after showing that $1 \Rightarrow 2 \Rightarrow 3 \Rightarrow 1$, we know that 2 is equivalent to 3. This follows since $3 \Rightarrow 1 \Rightarrow 2$ and so $3 \Rightarrow 2$. Also, $2 \Rightarrow 3$ directly. Hence, 2 if and only if 3, or $2 \Leftrightarrow 3$.

3.2.16   Theorem

If $p > 1$ is an integer, the following are equivalent:

1. The integer p is prime.
2. The ideal $p\mathbf{Z}$ is a maximal ideal of $\mathbf{Z}$.
3. The ideal $p\mathbf{Z}$ is a prime ideal of $\mathbf{Z}$.

Proof

1. $\Rightarrow$ 2. Suppose that I is an ideal of $\mathbf{Z}$ such that $p\mathbf{Z} \subseteq I$. If $p\mathbf{Z} \neq I$, then I must contain an integer x that is not a multiple of p. Thus, $\gcd(p, x) = 1$. Let a and b be integers such that $xa + pb = 1$. Then $pb \in p\mathbf{Z}$, so $pb \in I$ since $p\mathbf{Z} \subseteq I$. Since $xa \in I$, $1 = xa + pb \in I$, so $I = \mathbf{Z}$. Therefore, $p\mathbf{Z}$ is a maximal ideal of $\mathbf{Z}$.

2. $\Rightarrow$ 3. This is immediate from Corollary 3.2.14.

3. $\Rightarrow$ 1. Let $p\mathbf{Z}$ be a prime ideal of $\mathbf{Z}$, and let's show that p must be prime. If p is not prime, let $p = n_1 n_2$, where $1 < n_1 < p$ and $1 < n_2 < p$. Then $n_1 n_2 = p \in p\mathbf{Z}$, so either $n_1 \in p\mathbf{Z}$ or $n_2 p\mathbf{Z}$. If $n_1 \in p\mathbf{Z}$, then $n_1 = pk$ for some positive integer k. Hence, $1 < pk < p$, which is a contradiction. Similarly, we get a contradiction if $n_2 \in p\mathbf{Z}$. Therefore, if $p\mathbf{Z}$ is a prime ideal of $\mathbf{Z}$, then p must be a prime integer. ∎

3.2.17   Corollary

The following are equivalent:

1. The integer p is a prime integer.
2. The ring $\mathbf{Z}/p\mathbf{Z}$ is a field.
3. The ring $\mathbf{Z}/p\mathbf{Z}$ is an integral domain.

In the ring of integers $\mathbf{Z}$, the zero ideal $\{0\}$ is a prime ideal since $xy \in \{0\}$ implies $xy = 0$, which in turn implies that $x = 0$ or $y = 0$. Hence, $xy \in \{0\}$ implies $x \in \{0\}$ or $y \in \{0\}$. However, $\{0\}$ is properly contained in the ideal $2\mathbf{Z}$ of $\mathbf{Z}$, so $\{0\}$ is not a maximal ideal of $\mathbf{Z}$. Hence, one cannot say that an ideal of $\mathbf{Z}$ is maximal if and only if it is prime, but only that a *nonzero* ideal of $\mathbf{Z}$ is maximal if and only if it is prime. The following two examples yield additional information about ideals in a ring. The first example demonstrates that a maximal ideal of a ring may not be prime when the ring does not have an identity, and the second example shows that even if a ring has an identity, a nonzero prime ideal need not be maximal.

3.2.18	**Examples**

1. Corollary 3.2.14 shows that in a commutative ring with identity every maximal ideal must be prime. This example shows that this may be false if the ring does not have an identity. Consider the commutative ring $2\mathbf{Z}$ of even integers. Note that $4\mathbf{Z}$ is an ideal of $2\mathbf{Z}$ that is not prime since $2 \cdot 2 \in 4\mathbf{Z}$, but $2 \notin 4\mathbf{Z}$. We claim that $4\mathbf{Z}$ is a maximal ideal of $2\mathbf{Z}$. Let I be an ideal of $2\mathbf{Z}$ such that $4\mathbf{Z} \subseteq I \subseteq 2\mathbf{Z}$. Then I^+, the set of positive elements of I, contains a smallest element, say n. Since $n \in 2\mathbf{Z}$, n is even. Let $n = 2k$, where k is a positive integer. If $x \in I$, then by the Division Algorithm 1.1.3 there are integers q and r such that $x = q(2k) + r$, where $0 \le r < 2k$. Since x and $2k$ are in I, $r = x - q(2k) \in I$, so $r = 0$ since $2k$ is the smallest positive integer in I. Therefore, $x = q(2k)$, so $I \subseteq (2k)\mathbf{Z}$. Since $2k \in I$, $(2k)\mathbf{Z} \subseteq I$, so $I = (2k)\mathbf{Z}$. Thus, $4\mathbf{Z} \subseteq (2k)\mathbf{Z} \subseteq 2\mathbf{Z}$. Since $4 \in 4\mathbf{Z}$, $4 \in (2k)\mathbf{Z}$. Hence, $4 = (2k)m$ for some integer m, and thus it follows that $k = 1$ and $m = 2$ or $k = 2$ and $m = 1$. If $k = 1$, then $(2k)\mathbf{Z} = 2\mathbf{Z}$, and if $k = 2$, then $(2k)\mathbf{Z} = 4\mathbf{Z}$. Therefore, $4\mathbf{Z}$ is a maximal ideal of $2\mathbf{Z}$, and so if a ring does not have an identity, a maximal ideal need not be prime.

2. In this example we show that a prime ideal need not be maximal, even if the ring has an identity. Define the binary operations of addition and multiplication on $\mathbf{Z} \times \mathbf{Z}$ by $(x, y) + (z, w) = (x + z, y + w)$ and $(x, y)(z, w) = (xz, yw)$. Under these operations, $\mathbf{Z} \times \mathbf{Z}$ is a commutative ring with identity $(1, 1)$. The set $\mathbf{Z} \times \{0\} = \{(x, 0) \mid x \in \mathbf{Z}\}$ is an ideal of $\mathbf{Z} \times \mathbf{Z}$. If $(x, y)(z, w)$ is in $\mathbf{Z} \times \{0\}$, then $yw = 0$. But $\mathbf{Z}$ is an integral domain, so $y = 0$ or $w = 0$. Hence, $(x, y) = (x, 0) \in \mathbf{Z} \times \{0\}$ or $(z, w) = (z, 0) \in \mathbf{Z} \times \{0\}$. Thus, $\mathbf{Z} \times \{0\}$ is a prime ideal of $\mathbf{Z} \times \mathbf{Z}$. However, $\mathbf{Z} \times \{0\}$ is not a maximal ideal of $\mathbf{Z} \times \mathbf{Z}$ since the ideal $I = \mathbf{Z} \times 2\mathbf{Z}$ properly contains $\mathbf{Z} \times \{0\}$ and $I \ne \mathbf{Z} \times \mathbf{Z}$. Hence, there exist prime ideals in $\mathbf{Z} \times \mathbf{Z}$ that are not maximal.

Both of the preceding examples are quite different from the ring $\mathbf{Z}$ of integers where a nonzero ideal is prime if and only if it is maximal. The property possessed by the ring of integers, which is sufficient for this property of ideals to hold, is that $\mathbf{Z}$ is a principal ideal domain that has an identity. Although we have not seen an example of a principal ideal domain with identity other than $\mathbf{Z}$, mathematicians are always interested in proving properties of algebraic structures in the most general setting. Principal ideal domains are studied in more detail later in the text where examples of principal ideal domains other than $\mathbf{Z}$ are presented. For now, we will be satisfied to prove the following theorem, which relates prime and maximal ideals in such a domain with identity. Hopefully, this will whet your appetite for things to come.

3.2.19   Theorem

If R is a principal ideal domain that has an identity, then a nonzero ideal I of R is prime if and only if it is maximal.

Proof. Since R is a commutative ring with identity, Corollary 3.2.14 shows that every maximal ideal in R is prime. It remains then only to show that nonzero prime ideals are maximal. Let xR be a nonzero prime ideal of R, and suppose that yR is an ideal of R such that $xR \subseteq yR \subseteq R$. Now $x \in xR \subseteq yR$ and thus $x = ya$ for some $a \in R$. But $ya \in xR$ implies that either $y \in xR$ or $a \in xR$ since xR is prime. If $y \in xR$, it immediately follows that $xR = yR$. If $a \in xR$, then $a = xb$ for some $b \in R$. Hence, $x = ya = yxb$, which implies that $yb = e$ since $x \neq 0$ and R is a domain. Thus, $e \in yR$ and so $yR = R$. Therefore, when $xR \subseteq yR \subseteq R$, either $xR = yR$ or $yR = R$, so xR is a maximal ideal of R. ∎

Problem Set 3.2

1. Define binary operations of addition and multiplication on $\mathbf{Q} \times \mathbf{Q}$ by $(x, y) + (z, w) = (x + z, y + w)$ and $(x, y)(z, w) = (xz, yw)$. Then $\mathbf{Q} \times \mathbf{Q}$ is a commutative ring with identity $(1, 1)$. Show that $\mathbf{Q} \times \{0\}$ and $\{0\} \times \mathbf{Q}$ are both subrings of $\mathbf{Q} \times \mathbf{Q}$. If we define exactly the same binary operations on $\mathbf{R} \times \mathbf{R}$, is $\mathbf{Q} \times \mathbf{Q}$ a subring of $\mathbf{R} \times \mathbf{R}$?

2. If $\mathbf{Z}[\sqrt{3}] = \{x + y\sqrt{3} \mid x, y \in \mathbf{Z}\}$ and elements of $\mathbf{Z}[\sqrt{3}]$ are added and multiplied in the usual way, is $\mathbf{Z}[\sqrt{3}]$ a subring of $\mathbf{R}$? of $\mathbf{Q}$?

3. Let $S = \{[0], [2], [4], [6], [8]\} \subseteq \mathbf{Z}_{10}$. Make an addition and multiplication table for S modulo 10. Is S a subring of $\mathbf{Z}_{10}$? Does S have an identity? Hint: Find an element $[x] \in S$ such that $[2][x] = [2]$, and then compute $[4][x]$, $[6][x]$, and $[8][x]$. Conclude that a ring R with identity e_R can have a subring S with identity e_S such that $e_R \neq e_S$. It will be shown subsequently that this cannot be the case for fields.

4. Consider the ring $M_2(\mathbf{Z})$ of 2×2 matrices with the binary operations of matrix addition and multiplication. If $S = \left\{ \begin{pmatrix} x & y \\ 0 & z \end{pmatrix} \mid x, y, z \in \mathbf{Z} \right\}$, is S a subring of $M_2(\mathbf{Z})$? Does S have an identity? If so, is the identity of S the same as the identity of $M_2(\mathbf{Z})$?

5. If S_1 and S_2 are subrings of a ring R, show that $S_1 \cap S_2$ is a subring of R. Now generalize your proof to show that if $\{S_\alpha\}_{\alpha \in \Delta}$ is a family of subrings of R, then $\cap_{\alpha \in \Delta} S_\alpha$ is a subring of R.

6. Find subrings S_1 and S_2 of **Z** such that $S_1 \cup S_2$ is *not* a subring of **Z**.

7. If $\{F_\alpha\}_{\alpha \in \Delta}$ is a family of subfields of a field F, prove that $\cap_{\alpha \in \Delta} F_\alpha$ is a subfield of F.

8. Consider the ring $\mathbf{Z}[i] = \{x + yi \mid x, y \in \mathbf{Z}\}$ under the binary operations of addition and multiplication of complex numbers. Show that $2\mathbf{Z}[i] = \{x + yi \mid x, y \in 2\mathbf{Z}\}$ is an ideal of $\mathbf{Z}[i]$.

9. If I_1 and I_2 are right ideals (left ideals, ideals) of a ring R, prove that $I_1 \cap I_2$ is a right ideal (a left ideal, an ideal) of R. Formulate and prove a similar statement for a family $\{I_\alpha\}_{\alpha \in \Delta}$ of right ideals of R, for a family of left ideals of R, and for a family of ideals of R.

10. Suppose that R is a ring with identity. Prove that R is a division ring if and only if the only left ideals of R are $\{0\}$ and R. Show that a similar result holds if "left" is replaced by "right."

11. Consider the ring $M_2(\mathbf{R})$ of all 2×2 matrices under the operations of matrix addition and multiplication. Show that $M_2(\mathbf{R})$ is not a division ring and that the only ideals of $M_2(\mathbf{R})$ are $\{0\}$ and $M_2(\mathbf{R})$. Thus, the conclusion of Exercise 10 does not hold if "left ideal" ("right ideal") is replaced by "ideal." A ring with identity **R** in which the only ideals are $\{0\}$ and **R** is said to be a **simple ring**.

12. Prove Lemma 3.2.11.

13. If R is a ring and $x \in R$, prove that $\mathrm{ann}_l(x) = \{a \in R \mid ax = 0\}$ is a left ideal of R and $\mathrm{ann}_r(x) = \{a \in R \mid xa = 0\}$ is a right ideal of R, called the **left** and **right annihilator** of x, respectively.

14. Prove that if I is a left ideal, then $\mathrm{ann}_l(I) = \{a \in R \mid ax = 0 \text{ for all } x \in I\}$ is an ideal of R called the **left annihilator** of I. Make a similar statement concerning the **right annihilator** of a right ideal of R. When the ring is commutative, $\mathrm{ann}_l(I) = \mathrm{ann}_r(I)$, which is written as $\mathrm{ann}(I)$.

15. Let R be a ring without identity, and define addition and multiplication on $S = \mathbf{Z} \times R$ by

    $$(m, x) + (n, y) = (m + n, x + y) \qquad \text{and}$$
    $$(m, x)(n, y) = (mn, my + nx + xy)$$

    for all $(m, x), (n, y) \in S$.

    (a) Show that S is a ring with identity $(1, 0_R)$.
    (b) Show that $S' = \{(0, x) \mid x \in R\}$ is a subring of S.

16. Determine whether $2\mathbf{Z}[x]$ is a subring of the polynomial ring $\mathbf{Z}[x]$.

17. Show that the set $\mathbf{Z}[\sqrt{5}i]^* = \{x + y(\sqrt{5}i) \mid x, y \in \mathbf{Z}, x - y \in 2\mathbf{Z}\}$ is an ideal of $\mathbf{Z}[\sqrt{5}i] = \{x + y(\sqrt{5}i) \mid x, y \in \mathbf{Z}\}$ if addition and multiplication are defined on $\mathbf{Z}[\sqrt{5}i]$ in the usual fashion.

18. If $I_1 \subseteq I_2 \subseteq I_3 \subseteq \cdots \subseteq I_n \subseteq \cdots$ is an ascending chain of ideals (of right ideals, of left ideals) of a ring R, prove that $\cup_{n \geq 1} I_n$ is an ideal (a right ideal, a left ideal) of R.

19. Compute $10\mathbf{Z} \cap 15\mathbf{Z}$ in the ring $\mathbf{Z}$ of integers under the usual operations of addition and multiplication of integers. If $x\mathbf{Z}$ and $y\mathbf{Z}$ are ideals in $\mathbf{Z}$, prove that $x\mathbf{Z} \cap y\mathbf{Z} = m\mathbf{Z}$, where $m = \mathrm{lcm}(x, y)$, the least common multiple of x and y. See Problem Set 1.1, Exercise 14.

20. In the ring of integers $\mathbf{Z}$, prove that $x\mathbf{Z} + y\mathbf{Z}$ is an ideal of $\mathbf{Z}$ and that $x\mathbf{Z} + y\mathbf{Z} = d\mathbf{Z}$, where $d = \gcd(x, y)$.

21. Define addition and multiplication on $\mathbf{Z} \times \mathbf{R}$ componentwise by $(x_1, y_1) + (x_2, y_2) = (x_1 + x_2, y_1 + y_2)$ and $(x_1, y_1)(x_2, y_2) = (x_1 x_2, y_1 y_2)$. Show that $\mathbf{Z} \times \mathbf{R}$ is a commutative ring with identity $(1, 1)$ and that $\mathbf{Z} \times \{0\} = \{(x, 0) \mid x \in \mathbf{Z} \text{ and } 0 \in \mathbf{R}\}$ is an ideal of $\mathbf{Z} \times \mathbf{R}$. Prove that the factor ring $(\mathbf{Z} \times \mathbf{R}) / (\mathbf{Z} \times \{0\})$ is a field.

22. For the ring $\mathbf{Z}_{24}$ under addition and multiplication modulo 24, show that $I = \{[0], [4], [8], [12], [16], [20]\}$ is an ideal of $\mathbf{Z}_{24}$. How many cosets are in $\mathbf{Z}_{24}/I$? Compute the cosets in $\mathbf{Z}_{24}/I$ and make an addition and a multiplication table for this ring.

23. Suppose that $R = \left\{ \dfrac{a}{b} \mid a, b \in \mathbf{Z} \text{ and } 5 \nmid b \right\}$. Prove that R is a commutative ring with identity under the usual operations of addition and multiplication of rational numbers. Show that the set $I = \left\{ \dfrac{a}{b} \in R \mid 5 \mid a \right\}$ is an ideal of R and that the factor ring R/I is a field. Does a similar result hold if 5 is replaced by any prime p?

24. Let I be an ideal of a ring R. Prove or disprove that $I[x]$ is an ideal of the polynomial ring $R[x]$.

25. Prove or disprove that $2\mathbf{Z}[x]$ is an ideal of the polynomial ring $\mathbf{Z}[x]$.

26. Prove or disprove that $O[x]$ is an ideal of the polynomial ring $\mathbf{Z}[x]$, where $O[x]$ is the set of all polynomials in $\mathbf{Z}[x]$ whose coefficients are odd integers.

Use a computer algebra system to solve each of the following problems.

27. Consider the ring $\mathbf{Z}_{20}$. Instruct your computer algebra system to compute an addition table for $K = \{[0], [4], [8], [12], [16]\}$ modulo 20. Does this show that K is a subgroup of the additive group of $\mathbf{Z}_{20}$? Also instruct your computer algebra system to compute a table for $[n]K$ modulo 20 for $n = 0, 1, \ldots, 19$. Explain why this shows that K is an ideal of $\mathbf{Z}_{20}$.

28. Inspect the table for $[n]K$ of Exercise 27 and form ann(K). (See Exercises 13 and 14.) Is ann(K) an ideal of $\mathbf{Z}_{20}$?

29. Compute all of the distinct ideals of $\mathbf{Z}_{18}$.

30. In Exercise 27, $K = \{[0], [4], [8], [12], [16]\}$ was shown to be an ideal of $\mathbf{Z}_{20}$. Compute all the distinct left cosets of K in $\mathbf{Z}_{20}$, and make an addition and multiplication table for $\mathbf{Z}_{20}/K$.

3.3   RINGS THAT ARE ALGEBRAICALLY EQUIVALENT: RING HOMOMORPHISMS

Section Overview. This section presents the ideas of ring homomorphism and ring isomorphism. It is shown that the kernels of ring homomorphisms are closely connected to ideals in rings and to factor rings. It is established that ideals are kernels and that kernels are ideals. Properties of rings that are preserved by ring homomorphisms are investigated. You should study this material carefully. It will give you a feel for the kinds of algebraic properties that are transferred from one ring to another by a ring homomorphism. Three important isomorphism theorems and a correspondence theorem are presented. The most fundamental of these is Theorem 3.3.12, the First Isomorphism Theorem for Rings. Be sure to study the comparative summary of groups and rings given at the end of this section; this summary lists the major similarities of groups and rings presented in Chapters 2 and 3.

As with groups, it is sometimes the case that two rings are algebraically equivalent. If R and S are rings, then R and S are algebraically equivalent if two things can be shown to happen: (1) there is a function $f : R \rightarrow S$ that establishes a one-to-one correspondence between the elements of R and S, and (2) the sum and product of corresponding elements correspond. If $f: R \rightarrow S$ is such a bijective function and $x, y \in R$, then x and $f(x)$, and y and $f(y)$, are corresponding elements of R and S, respectively.

$$R \xleftrightarrow{\quad f \quad} S$$
$$x \longleftrightarrow f(x)$$
$$y \longleftrightarrow f(y).$$

If the sum and product of corresponding elements correspond, then

$$x + y \longleftrightarrow f(x) + f(y)$$
$$xy \longleftrightarrow f(x)f(y),$$

so it must be the case that

$$f(x + y) = f(x) + f(y) \qquad \text{and}$$
$$f(xy) = f(x)f(y) \qquad \text{for all } x, y \in R.$$

When this is the case, we say that the **binary operations are preserved** by f.

3.3.1 | Examples

Consider the field of real numbers **R** and the ring $R = \{(x, 0) \mid x \in \mathbf{R}\}$, where addition and multiplication are defined on R by $(x, 0) + (y, 0) = (x + y, 0)$ and $(x, 0)(y, 0) = (xy, 0)$. The one-to-one correspondence $f: \mathbf{R} \rightarrow R$ defined by $f(x) = (x, 0)$ is such that $f(x + y) = (x + y, 0) = (x, 0) + (y, 0) = f(x) + f(y)$ and $f(xy) = (xy, 0) = (x, 0)(y, 0) = f(x)f(y)$. Hence, the two rings are algebraically equivalent rings.

The preceding example and the observations made in the introduction of this chapter lead to the definition of a ring isomorphism and more generally to the definition of a ring homomorphism. One important use of isomorphism is in the classification of algebraic structures. In particular, the concept of isomorphism leads to meaningful results in the classification of rings. For example, ring isomorphism is used later in the text to show that there is essentially only one ring **Z** of integers in the sense that any integral domain that has a set of positive elements that is well-ordered is ring isomorphic to **Z**. Ring isomorphism is also used to show that, under certain conditions, **Q**, the field of rational numbers; **R**, the field of real numbers; and **C**, the field of complex numbers, are unique up to isomorphism. If you are to appreciate these classifications through the use of isomorphism, it is essential that you understand the concepts of ring homomorphism and ring isomorphism.

| **3.3.2** | **Definition** |

If R and S are rings, then a function $f : R \to S$ is said to be a **ring homomorphism** provided that

1. $f(x + y) = f(x) + f(y)$ for all $x, y \in R$ and
2. $f(xy) = f(x)f(y)$ for all $x, y \in R$.

A ring homomorphism that is injective is said to be a **ring monomorphism**, and a ring homomorphism that is surjective is a **ring epimorphism**. When $f : R \to S$ is a ring monomorphism, R is said to **embed** in S, and if $f : R \to S$ is a ring epimorphism, then S is a **homomorphic image** of R. A ring homomorphism $f : R \to S$ that is a bijective function is a **ring isomorphism**. In this case, R and S are referred to as **isomorphic rings**. If $f : R \to S$ is a ring isomorphism, then S is said to be a **copy** of R.

| **3.3.3** | **Examples of Ring Homomorphisms** |

1. Consider the rings $\mathbf{Z}$ and $\mathbf{Z}_n$, where n is a positive integer. If $f : \mathbf{Z} \to \mathbf{Z}_n$, where $f(x) = [x]$, then f is a ring homomorphism since $f(x + y) = [x + y] = [x] + [y] = f(x) + f(y)$ and $f(xy) = [xy] = [x][y] = f(x)f(y)$. If $[x] \in \mathbf{Z}_n$, then $f(y) = [x]$ for any $y \in \mathbf{Z}$ for which $y \equiv x \pmod{n}$, so f is surjective. Hence, $\mathbf{Z}_n$ is a homomorphic image of $\mathbf{Z}$. However, f is not an injective mapping since if $x \neq y$ but $x \equiv y \pmod{n}$, then $f(x) = [x] = [y] = f(y)$. Thus, $x \neq y$ does not imply $f(x) \neq f(y)$. Intuitively, one knows that $\mathbf{Z}$ and $\mathbf{Z}_n$ cannot be isomorphic rings since a one-to-one correspondence between the elements of $\mathbf{Z}$ and those of $\mathbf{Z}_n$ would require that these two sets have the same number of elements. But $\mathbf{Z}$ is an infinite ring, while $\mathbf{Z}_n$ has n elements.

2. Let $\mathbf{C}$ be the field of complex numbers and $\mathcal{C}$ the field given in Example 3 of 3.1.9. We claim that $\mathbf{C}$ and $\mathcal{C}$ are isomorphic fields. Consider the correspondence given by

$$\mathbf{C} \xleftarrow{\ \ f\ \ } \mathcal{C}$$

$$a + bi \longleftrightarrow \begin{pmatrix} a & b \\ -b & a \end{pmatrix},$$

and let's show that this correspondence satisfies the requirements to be a ring isomorphism.

(i) The mapping f is a well-defined function because if
$a + bi = c + di$, then $a = c$ and $b = d$ and thus $\begin{pmatrix} a & b \\ -b & a \end{pmatrix} = \begin{pmatrix} c & d \\ -d & c \end{pmatrix}$.
Hence, $f(a + bi) = f(c + di)$.

(ii) If $f(a + bi) = f(c + di)$, then $\begin{pmatrix} a & b \\ -b & a \end{pmatrix} = \begin{pmatrix} c & d \\ -d & c \end{pmatrix}$, so $a = c$ and $b = d$.
Thus, $a + bi = c + di$, and so f is injective.

(iii) If $\begin{pmatrix} a & b \\ -b & a \end{pmatrix} \in \mathcal{C}$, then $a, b \in \mathbf{R}$, so $a + bi \in \mathbf{C}$. Furthermore,
$f(a + bi) = \begin{pmatrix} a & b \\ -b & a \end{pmatrix}$, so f is surjective. Hence, f is a bijective

function that establishes a one-to-one correspondence between the elements of $\mathbf{C}$ and those of $\mathcal{C}$. The fields $\mathbf{C}$ and $\mathcal{C}$ are isomorphic if we can show that f preserves addition and multiplication.

(iv) If $a + bi, c + di \in \mathbf{C}$, then

$$f((a + bi) + (c + di)) = f((a + c) + (b + d)i)$$

$$= \begin{pmatrix} a + c & b + d \\ -(b + d) & a + c \end{pmatrix}$$

$$= \begin{pmatrix} a & b \\ -b & a \end{pmatrix} + \begin{pmatrix} c & d \\ -d & c \end{pmatrix}$$

$$= f(a + bi) + f(c + di).$$

and

$$f((a + bi)(c + di)) = f((ac - bd) + (ad + bc)i)$$

$$= \begin{pmatrix} ac - bd & ad + bc \\ -(ad + bc) & ac - bd \end{pmatrix}$$

$$= \begin{pmatrix} a & b \\ -b & a \end{pmatrix}\begin{pmatrix} c & d \\ -d & c \end{pmatrix}$$

$$= f(a + bi)f(c + di).$$

Thus, f preserves addition and multiplication, so $\mathbf{C}$ and $\mathcal{C}$ are isomorphic fields. We began this example with the knowledge that both $\mathbf{C}$ and $\mathcal{C}$ are fields. If F is a field, R is a ring, and $f : F \rightarrow R$ is a ring isomorphism, do you think that R will be a field? Think about your answer; we will consider this question in more detail a little later.

3. The set $M_3(\mathbf{R})$ of all 3×3 matrices with entries from $\mathbf{R}$ is a noncommutative ring with identity that has zero-divisors. The set S of all 3×3 matrices of the form

$$\begin{pmatrix} a & 0 & 0 \\ 0 & a & 0 \\ 0 & 0 & a \end{pmatrix},$$

where $a \in \mathbf{R}$, is a subring of $M_3(\mathbf{R})$. The mapping $f : \mathbf{R} \to S$ defined by

$$f(a) = \begin{pmatrix} a & 0 & 0 \\ 0 & a & 0 \\ 0 & 0 & a \end{pmatrix}$$

is a ring isomorphism. Thus, $M_3(\mathbf{R})$ contains a subring that is isomorphic to $\mathbf{R}$. If we identify each $a \in \mathbf{R}$ with

$$\begin{pmatrix} a & 0 & 0 \\ 0 & a & 0 \\ 0 & 0 & a \end{pmatrix} \in M_3(\mathbf{R}),$$

then we can consider $\mathbf{R}$ to be a subfield of $M_3(\mathbf{R})$. Hence, it is possible for a noncommutative ring that is not even a division ring to contain a subring that is a field.

4. The field $\mathbf{C}$ of complex numbers contains a subring that is isomorphic to the field $\mathbf{R}$ of real numbers. Indeed, $f : \mathbf{R} \to \mathbf{C}$, where $f(a) = a + 0i$, is a ring monomorphism. Hence, $\mathbf{R}$ embeds in $\mathbf{C}$. If each element $a \in \mathbf{R}$ is identified with the element $a + 0i \in \mathbf{C}$, then $\mathbf{R}$ can be considered to be a subfield of $\mathbf{C}$.

5. The division ring of real quaternions $q(\mathbf{R})$ contains a subring that is isomorphic to $\mathbf{C}$. The mapping $f : \mathbf{C} \to q(\mathbf{R})$ that is given by $f(a + bi) = a + bi + 0j + 0k$ is a ring monomorphism. Hence, $\mathbf{C}$ embeds in $q(\mathbf{R})$. Note that $q(\mathbf{R})$ also contains a subring that is isomorphic to $\mathbf{R}$. This follows since the mapping $f : \mathbf{R} \to q(\mathbf{R})$, where $f(a) = a + 0i + 0j + 0k$, is a ring monomorphism. Thus, $q(\mathbf{R})$ contains a copy of $\mathbf{R}$ and of $\mathbf{C}$.

6. Let $R = \{a, b, c, d\}$. Then R is a ring under addition and multiplication as defined in the following addition and multiplication tables.

+	a	b	c	d
a	a	b	c	d
b	b	a	d	c
c	c	d	a	b
d	d	c	b	a

•	a	b	c	d
a	a	a	a	a
b	a	b	c	d
c	a	a	a	a
d	a	b	c	d

Similarly, $S = \{\alpha, \beta, \gamma, \delta\}$ is a ring if addition and multiplication are defined by the following tables.

+	α	β	γ	δ
α	γ	δ	α	β
β	δ	γ	β	α
γ	α	β	γ	δ
δ	β	α	δ	γ

$\bullet$	α	β	γ	δ
α	α	β	γ	δ
β	γ	γ	γ	γ
γ	γ	γ	γ	γ
δ	α	β	γ	δ

The one-to-one correspondence f given by the following correspondence table establishes a ring isomorphism between the rings R and S.

$$R \xleftrightarrow{\ f\ } S$$
$$a \longleftrightarrow \gamma$$
$$b \longleftrightarrow \alpha$$
$$c \longleftrightarrow \beta$$
$$d \longleftrightarrow \delta$$

One can check (although it requires a bit of work) that the various ring properties hold for R and S and that, under this correspondence, sums and products of corresponding elements always correspond, so R and S are isomorphic rings.

7. Suppose that R and S are rings, and consider the set $R \times S = \{(x, y) \mid x \in R \text{ and } y \in S\}$. Define addition and multiplication on $R \times S$ by $(x, y) + (z, w) = (x + z, y + w)$ and $(x, y)(z, w) = (xz, yw)$. Then $R \times S$ is a ring. Consider the subset $\{0\} \times S = \{(0, y) \mid y \in S\}$ of $R \times S$. If $(0, y), (0, w) \in \{0\} \times S$, then $(0, y) - (0, w) = (0, y - w) \in \{0\} \times S$, so $\{0\} \times S$ is closed under subtraction. Thus, $\{0\} \times S$ is a subgroup of the additive group of $R \times S$. If $(0, y), (0, w) \in \{0\} \times S$, then $(0, y)(0, w) = (0, yw) \in \{0\} \times S$. It follows that $\{0\} \times S$ is a subring of $R \times S$. Similarly, $R \times \{0\}$ is a subring of $R \times S$. The functions $f_1 : R \to R \times \{0\}$ defined by $f(x) = (x, 0)$ and $f_2 : S \to \{0\} \times S$, where $f_2(y) = (0, y)$, are ring isomorphisms. Therefore, $R \times S$ contains subrings that are isomorphic to R and S, so $R \times S$ contains a copy of R and of S.

There is a practical benefit from studying ring homomorphisms. The following simple example demonstrates this point. Consider the ring homomorphism $f : \mathbf{Z} \to \mathbf{Z}_4$ given by $f(x) = [x]$. The ring homomorphism f can be used to answer the following question: Does the equation $x^2 - 8x + 1 = 0$ have a solution in $\mathbf{Z}$? If y is an integer that is a solution of this equation, then $y^2 - 8y + 1 = 0$ in $\mathbf{Z}$, and thus we have

$$f(y^2 - 8y + 1) = f(0),$$
$$f(y^2) - f(8y) + f(1) = f(0),$$

$$f(y)f(y) - f(8)f(y) + f(1) = f(0),$$
$$[y][y] - [8][y] + [1] = [0].$$

But $[8] = [0]$ in $\mathbf{Z}_4$, so the last equation reduces to

$$[y]^2 + [1] = [0].$$

Hence, if y is a solution of $x^2 - 8x + 1 = 0$ in $\mathbf{Z}$, then $[y]$ is a solution of the equation $[x]^2 + [1] = [0]$ in $\mathbf{Z}_4$. But $[x]^2 + [1] = [0]$ has no solution in $\mathbf{Z}_4$. Indeed, since $\mathbf{Z}_4$ has only four elements, we can simply test all the elements to show that $[x]^2 + [1] = [0]$ has no solution in $\mathbf{Z}_4$.

$$[0][0] + [1] = [1] \neq [0]$$
$$[1][1] + [1] = [2] \neq [0]$$
$$[2][2] + [1] = [1] \neq [0]$$
$$[3][3] + [1] = [2] \neq [0]$$

Therefore, y cannot be a solution of $x^2 - 8x + 1 = 0$ in $\mathbf{Z}$, so the equation $x^2 - 8x + 1 = 0$ has no solutions in $\mathbf{Z}$. Of course, if we find a solution of an equation in $\mathbf{Z}_4$, it does not mean that the "corresponding" equation in $\mathbf{Z}$ has a solution. For example, $[x] = [1]$ and $[x] = [3]$ are solutions of $[x]^2 + [3] = [0]$ in $\mathbf{Z}_4$, yet $x^2 + 3 = 0$ has no solution in $\mathbf{Z}$.

Because of the preceding example, one wonders what other benefits might be provided by ring homomorphisms, and in particular, one wonders what ring properties are preserved by such mappings. For example, if $f: R \to S$ is a ring homomorphism and 0_R is the additive identity of R, is $f(0_R)$ the additive identity 0_S of S? If R has an identity, must S also be a ring with identity? Are additive inverses and multiplicative inverses (when the latter exist) preserved by ring homomorphisms? If R is a domain (an integral domain), must S also be a domain (an integral domain)? The answer to each of these questions is sometimes yes and sometimes no for a general ring homomorphism. However, the answer is yes when the ring homomorphism is chosen appropriately. The following theorems are devoted to answering these kinds of questions. First, we look at two properties that are preserved by *all* ring homomorphisms.

3.3.4   **Theorem**

If $f: R \to S$ is a ring homomorphism, then:

1. If 0_R is the additive identity of R, then $f(0_R) = 0_S$, the additive identity of S.
2. If $x \in R$, then $f(-x) = -f(x)$.

Proof. Since $f: R \to S$ is a group homomorphism from the additive group of R to the additive group of S, this theorem is just Theorem 2.3.3 with additive notation. ■

3.3.5   Theorem

If $f : R \to S$ is a ring epimorphism, then:

1. If R is a commutative ring, then so is S.
2. If R is a ring with identity e_R, then S is a ring with identity e_S and $e_S = f(e_R)$.
3. If R has an identity and $x \in R$ has a multiplicative inverse in R, then $f(x)$ has a multiplicative inverse in S and $f(x)^{-1} = f(x^{-1})$.

Proof

1. If $y_1, y_2 \in S$, then since f is a surjective mapping, there are $x_1, x_2 \in R$ such that $f(x_1) = y_1$ and $f(x_2) = y_2$. Hence, $y_1 y_2 = f(x_1)f(x_2) = f(x_1 x_2) = f(x_2 x_1) = f(x_2)f(x_1) = y_2 y_1$, so S is a commutative ring.
2. Suppose $y \in S$, and let $x \in R$ be such that $f(x) = y$. Then $f(e_R)y = f(e_R)f(x) = f(e_R x) = f(x) = y$. Similarly, $yf(e_R) = y$. Since y was arbitrarily chosen in S, $f(e_R)$ is an identity for S, so $e_S = f(e_R)$.
3. By part 2, S has an identity $e_S = f(e_R)$. If $x \in R$ has a multiplicative inverse x^{-1} in R, then $f(x)f(x^{-1}) = f(xx^{-1}) = f(e_R) = e_S$. Similarly, $f(x^{-1})f(x) = e_S$. Thus, $f(x^{-1})$ is a multiplicative inverse for $f(x)$ in S and so $f(x)^{-1} = f(x^{-1})$. ∎

3.3.6   Theorem

If $f : R \to S$ is a ring isomorphism and R is a domain (division ring), then so is S.

Proof. Suppose $y_1, y_2 \in S$ are such that $y_1 y_2 = 0_S$. Since f is an isomorphism, f is a surjective mapping, so there are elements $x_1, x_2 \in R$ such that $f(x_1) = y_1$ and $f(x_2) = y_2$. Hence, $f(x_1 x_2) = f(x_1)f(x_2) = y_1 y_2 = 0_S = f(0_R)$. Since f is an injective function, $f(x_1 x_2) = f(0_R)$ implies that $x_1 x_2 = 0_R$. Now R is a domain, so either $x_1 = 0_R$ or $x_2 = 0_R$. If $x_1 = 0_R$, then $y_1 = f(x_1) = f(0_R) = 0_S$. Likewise, if $x_2 = 0_R$, then $y_2 = 0_S$. Part 1 of Theorem 3.3.5 shows that S is commutative, so S is a domain. Parts 2 and 3 of Theorem 3.3.5 show that S is a division ring whenever R enjoys this property. ∎

3.3.7   Corollary

If $f : R \to S$ is a ring isomorphism, then:

1. If R is an integral domain, then S is an integral domain.
2. If R is a field, then S is a field.

Proof. Parts 1 and 2 of Theorem 3.3.5 show that S is a commutative ring with identity when R has this property, so if R is an integral domain, Theorem 3.3.6 shows that S is an integral domain. If R is a field, part 3 of Theorem 3.3.5 shows that S is a field. ∎

In general, Theorem 3.3.6 fails when $f : R \rightarrow S$ is not an isomorphism. The mapping $f : \mathbf{Z} \rightarrow \mathbf{Z}_{12}$ defined by $f(x) = [x]$ is a ring homomorphism that is not an isomorphism, and $\mathbf{Z}$ is a domain while $\mathbf{Z}_{12}$ has zero-divisors.

In several of the examples of 3.3.3, it was pointed out that when $f : R \rightarrow S$ is a ring monomorphism, $f(R)$ is a subring of S. A more general form of this observation holds. In fact, when R' is a subring of R, the image of R' under a ring homomorphism f is a subring of S, and it is even true that the inverse image under f of a subring S' of S is a subring of R.

<table>
<tr><td>3.3.8</td></tr>
</table>

Theorem

Let $f : R \rightarrow S$ be a ring homomorphism.

1. If R' is a subring of R, then $f(R')$ is a subring of S.
2. If S' is a subring of S, then $f^{-1}(S')$ is a subring of R.

Proof

1. If R' is a subring of R, then $R' \neq \varnothing$. Thus, if $x \in R'$, then $f(x) \in f(R')$, so $f(R') \neq \varnothing$. If $f(x), f(y) \in f(R')$, where $x, y \in R'$, then $f(x) - f(y) = f(x - y) \in f(R')$ since $x - y \in R'$. Note also that $f(x)f(y) = f(xy) \in f(R')$ because $xy \in R'$. Hence, by Theorem 3.2.3, $f(R)$ is a subring of S.
2. Let S' be a subring of S. Since $f(0_R) = 0_S \in S'$, then $0_R \in f^{-1}(S')$, so $f^{-1}(S') \neq \varnothing$. If $x, y \in f^{-1}(S')$, then $f(x - y) = f(x) - f(y) \in S'$ since $f(x)$ and $f(y)$ are in S' and S' is a subring of S. Thus, $x - y$ is in $f^{-1}(S')$. Similarly, $f(xy) = f(x)f(y) \in S'$ implies that $xy \in f^{-1}(S')$. Hence, $f^{-1}(S')$ is a subring of R. ∎

When group homomorphisms were investigated in Section 2.3, we saw that the normal subgroups of a group were very closely connected to group homomorphisms. In fact, H is a normal subgroup of a group G if and only if H is the kernel of a group homomorphism whose domain is G. After defining the kernel of a ring homomorphism $f : R \rightarrow S$, it will be shown that the kernel of f is actually an ideal of R and that ideals are very closely connected to ring homomorphisms.

3.3.9   Definition

If $f : R \to S$ is a ring homomorphism, the **kernel** of f is the set $K = \{x \in R \mid f(x) = 0_S\}$. The kernel of f is denoted by $\ker f$.

3.3.10   Theorem

Let $f : R \to S$ be a ring homomorphism.

1. The kernel of f is an ideal of R.
2. The homomorphism f is a ring monomorphism if and only if $\ker f = \{0_R\}$.

Proof. Let $K = \ker f$.

1. Note that $K \neq \emptyset$ since, by Theorem 3.3.4, $f(0_R) = 0_S$. If $x, y \in K$, then $f(x - y) = f(x) - f(y) = 0_S - 0_S = 0_S$. Hence, $x - y \in K$ whenever x and y are in K, so K is an additive subgroup of R. If $x \in K$ and $a \in R$, then $f(xa) = f(x)f(a) = 0_S f(a) = 0_S$. Thus, $xa \in K$. Similarly, $ax \in K$, and so K is an ideal of R.
2. If $K = \{0_R\}$, let $x, y \in K$ be such that $f(x) = f(y)$. Then $0_S = f(x) - f(y) = f(x) + f(-y) = f(x - y)$, so $x - y \in \{0_R\}$. Hence, $x = y$ and f is a monomorphism. Conversely, if f is a monomorphism and $x \in K$, then $f(x) = 0_S = f(0_R)$ implies that $x = 0_R$. Thus, $K = \{0_R\}$. ■

3.3.11   Theorem

A subset K of a ring R is an ideal of R if and only if K is the kernel of a ring homomorphism whose domain is R.

Proof. If K is an ideal of R, the **canonical ring epimorphism** $\eta : R \to R/K$ given by $\eta(x) = x + K$ has kernel K. Conversely, if $f : R \to S$ is a ring homomorphism, then, by Theorem 3.3.10, the kernel of f is an ideal of R. ■

In Section 2.3 three important and very fundamental isomorphism theorems for groups were established, as well as a correspondence theorem. Since ideals in a ring play a role in ring theory that is quite similar to the role played by normal subgroups in group theory, one might suspect that there are similar isomorphism theorems for rings and a correspondence theorem for rings. This is indeed the case.

3.3.12   First Isomorphism Theorem for Rings

If $f : R \to S$ is a ring epimorphism with kernel K, then R/K and S are isomorphic rings.

Proof. Theorem 3.3.10 indicates that K is an ideal of R, so we can form the factor ring R/K. Let $\phi : R/K \to S$ be given by $\phi(x + K) = f(x)$. The mapping ϕ is well-defined because if $x + K = y + K$, then $x - y \in K$, so $f(x) - f(y) = f(x - y) = 0$. Hence, $f(x) = f(y)$ and thus $\phi(x + K) = \phi(y + K)$. If $x + K, y + K \in R/K$, then

$$\phi((x + K) + (y + K)) = \phi((x + y) + K)$$
$$= f(x + y)$$
$$= f(x) + f(y)$$
$$= \phi(x + K) + \phi(y + K)$$

and

$$\phi((x + K)(y + K)) = \phi(xy + K)$$
$$= f(xy)$$
$$= f(x)f(y)$$
$$= \phi(x + K)\phi(y + K)$$

Hence, ϕ is a ring homomorphism. If $\phi(x + K) = \phi(y + K)$, then $f(x) = f(y)$, so $f(x - y) = f(x) - f(y) = 0$. Hence, $x - y \in K$, and so $x + K = y + K$. Therefore, f is an injective mapping. Finally, if $s \in S$, then since f is an epimorphism, there is an $x \in R$ such that $f(x) = s$. But then $\phi(x + K) = f(x) = s$, so ϕ is surjective. This shows that ϕ is a ring isomorphism. ∎

3.3.13   Corollary

If $f : R \to S$ is a ring homomorphism with kernel K, then R/K and $f(R)$ are isomorphic rings.

3.3.14   Examples

1. For any positive integer n, the rings $\mathbf{Z}/n\mathbf{Z}$ and $\mathbf{Z}_n$ are isomorphic.

    Proof. If $f : \mathbf{Z} \to \mathbf{Z}_n$ is defined by $f(x) = [x]$, then we claim that f is a ring epimorphism. The mapping f is obviously well-defined and a ring homomorphism since

$$f(x + y) = [x + y]$$
$$= [x] + [y]$$
$$= f(x) + f(y) \qquad \text{and}$$
$$f(xy) = [xy]$$
$$= [x][y]$$
$$= f(x)f(y)$$

If $[x] \in \mathbf{Z}_n$, then $f(y) = [x]$ for any $y \in \mathbf{Z}$ such that $y \equiv x \pmod{n}$. Hence, f is a ring epimorphism. If K is the kernel of f, then $k \in K$ if and only if $f(k) = [0]$ if and only if $[k] = [0]$ in $\mathbf{Z}_n$. But this is true if and only if $n \mid (k - 0)$, which in turn is true if and only if $k \in n\mathbf{Z}$. Hence, $K = n\mathbf{Z}$, and from Theorem 3.3.12 we see that $\mathbf{Z}/n\mathbf{Z}$ and $\mathbf{Z}_n$ are isomorphic rings. ■

2. If I is an ideal of R, then $I[x]$ is an ideal of $R[x]$, and $R[x]/I[x]$ and $(R/I)[x]$ are isomorphic rings.

Proof. The successful completion of Exercise 24 of Problem Set 3.2 shows that $I[x]$ is an ideal of $R[x]$, so it remains only to show that the rings $R[x]/I[x]$ and $(R/I)[x]$ are isomorphic. Consider the mapping $\phi : R[x] \to (R/I)[x]$ that is defined by $\phi(a_0 + a_1x + \cdots + a_nx^n) = (a_0 + I) + (a_1 + I)x + \cdots + (a_n + I)x^n$. The proof that ϕ is a well-defined ring epimorphism is left as an exercise. Now $a_0 + a_1x + \cdots + a_nx^n \in \ker \phi$ if and only if $(a_0 + I) + (a_1 + I)x + \cdots + (a_n + I)x^n$ is the zero polynomial in $(R/I)[x]$ if and only if $a_i + I = 0 + I$ for $i = 0, 1, 2, \ldots, n$. But this is true if and only if $a_i \in I$ for $i = 0, 1, 2, \ldots, n$. Hence, $a_0 + a_1x + \cdots + a_nx^n$ is a polynomial in $\ker \phi$ if and only if $a_0 + a_1x + \cdots + a_nx^n \in I[x]$. Therefore, $\ker \phi = I[x]$, so by Theorem 3.3.12, $R[x]/I[x]$ and $(R/I)[x]$ are isomorphic rings. In particular, $n\mathbf{Z}$ is an ideal of $\mathbf{Z}$, and so the rings $\mathbf{Z}[x]/n\mathbf{Z}[x]$ and $(\mathbf{Z}/n\mathbf{Z})[x]$ are isomorphic rings. Since $\mathbf{Z}/n\mathbf{Z}$ and $\mathbf{Z}_n$ are isomorphic, $\mathbf{Z}[x]/n\mathbf{Z}[x]$ and $\mathbf{Z}_n[x]$ are isomorphic rings. ■

A sketch is given for a proof of each of the following two isomorphism theorems. You should fill in the details of each proof. The proof of the Correspondence Theorem is left as an exercise.

3.3.15   Second Isomorphism Theorem for Rings

If I_1 and I_2 are ideals of a ring R and $I_2 \subseteq I_1$, then:

1. I_1/I_2 is an ideal of R/I_2.

2. The factor rings $(R/I_2)/(I_1/I_2)$ and R/I_1 are isomorphic.

Proof

1. Consider the mapping $f: R/I_2 \to R/I_1$ defined by $f(x + I_2) = x + I_1$. This map is easily shown to be a well-defined ring homomorphism with kernel I_1/I_2. Hence, by Theorem 3.3.10, I_1/I_2 is an ideal of R.

2. The mapping $f: R/I_2 \to R/I_1$ of part 1 is a well-defined ring epimorphism with kernel I_1/I_2. The result follows from the First Isomorphism Theorem for Rings. ∎

3.3.16   Third Isomorphism Theorem for Rings

If I_1 and I_2 are ideals in a ring R, then $I_1/(I_1 \cap I_2)$ and $(I_1 + I_2)/I_2$ are isomorphic rings.

Proof. The mapping $f: I_1 \to (I_1 + I_2)/I_2$ given by $f(x) = x + I_2$ is a well-defined ring epimorphism with kernel $I_1 \cap I_2$. The result follows from the First Isomorphism Theorem for Rings. ∎

3.3.17   Correspondence Theorem

If $f: R \to S$ is a ring epimorphism, there is a one-to-one correspondence between the subrings of R that contain $K = \ker f$ and the subrings of S. Furthermore, if R' and S' are corresponding subrings of R and S, respectively, then R' is an ideal of R if and only if S' is an ideal of S.

Proof. The proof is left as an exercise. See Theorem 2.3.16. ∎

Comparative Summary of Groups and Rings

GROUPS	RINGS
1. A nonempty set G with one binary operation $G \times G \to G$ $$(x, y) \longmapsto xy$$ that satisfies the requirements to be a group.	1. A nonempty set R with two binary operations $R \times R \to R$ $$(x, y) \longmapsto x + y$$ $$(x, y) \longmapsto xy$$ that satisfy the requirements to be a ring.
2. $\varnothing \neq H \subseteq G$ is a subgroup of G if and only if $xy^{-1} \in H$ whenever $x, y \in H$.	2. $\varnothing \neq S \subseteq R$ is a subring of R if and only if $x - y, xy \in S$ whenever $x, y \in S$.

3. Normal subgroup H of G:
A subgroup H of G such that
$xH = Hx$ for all $x \in G$.

3. Ideal I of R: A subring I of R such
that $ax, xa \in I$ for all $x \in I$ and
every $a \in R$.

4. Factor groups G/N, where N is a
normal subgroup of G.

4. Factor rings R/I, where I is an
ideal of R.

5. Group homomorphism:
A mapping $f : G \to H$ such that
$f(xy) = f(x)f(y)$ for all $x, y \in G$.

5. Ring homomorphism:
A mapping $f : R \to S$ such that
$f(x + y) = f(x) + f(y)$ and
$f(xy) = f(x)f(y)$ for all $x, y \in R$.

6. N is a normal subgroup of G if and
only if N is the kernel of a group
homomorphism with domain G.

6. I is an ideal of R if and only if I is
the kernel of a ring
homomorphism with domain R.

7. If $f : G \to H$ is a group
epimorphism with kernel K, then
G/K and H are isomorphic
groups.

7. If $f : R \to S$ is a ring epimorphism
with kernel K, then R/K and S are
isomorphic rings.

8. Second and Third Isomorphism
Theorems for Groups.

8. Second and Third Isomorphism
Theorems for Rings.

9. Correspondence Theorem for
Groups.

9. Correspondence Theorem for
Rings.

Problem Set 3.3

1. Prove that $f : \mathbf{C} \to \mathbf{C}$, where $f(x + yi) = x - yi$, is a ring isomorphism.

2. Show that the rings $\mathbf{Q}[\sqrt{2}] = \{x + y\sqrt{2} \mid x,y \in \mathbf{Q}\}$ and $\mathbf{Q}[\sqrt{5}] = \{x + y\sqrt{5} \mid x,y \in \mathbf{Q}\}$ cannot be isomorphic.

3. Consider the ring $R = \left\{ \begin{pmatrix} x & y \\ 0 & z \end{pmatrix} \mid x, y, z \in \mathbf{Z} \right\}$ under matrix addition and multiplication and the ring $S = \mathbf{Z} \times \mathbf{Z}$ of Exercise 22 of Problem Set 3.1, where $R_1 = R_2 = \mathbf{Z}$. Determine whether $f : R \to S$, where $f\left(\begin{pmatrix} x & y \\ 0 & z \end{pmatrix} \right) = (x, z)$ is a ring homomorphism, and if it is, describe the kernel of f.

4. Assume that the set $\mathbf{Z}$ of integers is a ring under the binary operations $x \oplus y = x + y - 1$ and $x \otimes y = x + y - xy$. Prove that this ring is isomorphic to the ring $\mathbf{Z}$ of integers under the usual operations of addition and multiplication of integers.

5.   If $f : R \rightarrow S$ is a ring isomorphism and $f^{-1} : S \rightarrow R$ is the inverse function for f, show that f^{-1} is also a ring isomorphism.

6.   Show that $2\mathbf{Z}$ and $5\mathbf{Z}$ are isomorphic as additive groups, but not as rings.

7.   If $f : R \rightarrow S$ and $g : S \rightarrow T$ are ring homomorphisms, prove that $g \circ f : R \rightarrow T$ is a ring homomorphism.

8.   If R and R' are rings, each isomorphic to a ring S, prove that R and R' are isomorphic.

9.   If R is a ring with identity, Theorem 3.3.5 shows that if $f : R \rightarrow S$ is a ring epimorphism and $x \in R$ has a multiplicative inverse in R, then $f(x)$ has a multiplicative inverse in S. Show by example that the converse is false. That is, show that one can find rings R and S such that S is a homomorphic image of R and there is a $y \in S$ that has a multiplicative inverse in S, yet for any $x \in R$ such that $f(x) = y$, x does not have a multiplicative inverse in R.

10.   Prove that the mapping $f : \mathbf{Z} \rightarrow \mathbf{Z}$ given by $f(x) = 2x$ is a group homomorphism that is *not* a ring homomorphism.

11.   Prove that there are exactly two ring isomorphisms $f : \mathbf{C} \rightarrow \mathbf{C}$ that leave the elements of $\mathbf{R}$ fixed. A mapping $f : \mathbf{C} \rightarrow \mathbf{C}$ leaves the elements of $\mathbf{R}$ fixed if $f(x) = x$ for all $x \in \mathbf{R}$. Hint: Consider $f(i)^2$.

12.   Suppose that $S \neq \{0\}$ is a subring of $\mathbf{Z}$. If there exists a ring isomorphism $f : \mathbf{Z} \rightarrow S$, prove that $S = \mathbf{Z}$. Conclude that $\mathbf{Z}$ cannot be isomorphic to a proper subring of itself.

13.   Let R be a ring without identity, and consider the set $S = \mathbf{Z} \times R$. Define addition and multiplication on $\mathbf{Z} \times R$ by

$$(m, x) + (n, y) = (m + n, x + y) \qquad \text{and}$$
$$(m, x)(n, y) = (mn, my + nx + xy)$$

for all $(m, x), (n, y) \in \mathbf{Z} \times R$. We have seen in Exercise 15 of Problem Set 3.2 that S is a ring with identity $(1,0)$ and that $R' = \{(0, x) \mid x \in R\}$ is a subring of S. Show that $f : R \rightarrow R'$, where $f(x) = (0, x)$, is a ring isomorphism. Conclude that any ring can be embedded in a ring with identity.

14.   Let $S = \left\{ \begin{pmatrix} x & y \\ 2y & x \end{pmatrix} \middle| x, y \in \mathbf{Z} \right\}$. Prove that S is a subring of $M_2(\mathbf{Z})$ and that $f : \mathbf{Z}[\sqrt{2}] \rightarrow S$, where $f(x + y\sqrt{2}) = \begin{pmatrix} x & y \\ 2y & x \end{pmatrix}$, is a ring isomorphism.

15.   Suppose that $f : R \rightarrow S$ is a nonzero ring homomorphism, where R and S are rings with identity, and that both R and S have at least two elements. If

$f(e_R) \neq e_S$, prove that $f(e_R)$ is a zero-divisor in S. Conclude that if S is an integral domain, then every nonzero ring homomorphism from a ring R to S must send the identity of R to the identity of S.

16. If R is a commutative principal ideal ring and $f : R \rightarrow S$ is a ring epimorphism, prove that S must also be a principal ideal ring.

17. If $f : R \rightarrow S$ is a ring epimorphism and I is an ideal of S, prove each of the following:

   (a) If R is commutative and I is a prime ideal of S, then $f^{-1}(I)$ is a prime ideal of R.
   (b) If I is a maximal ideal of S, then $f^{-1}(I)$ is a maximal ideal of R.

18. If $a \in R$, prove that the mapping $\phi_a : R[x] \rightarrow R$ defined by $\phi_a(a_0 + a_1 x + a_2 x^2 + \cdots + a_n x^n) = a_0 + a_1 a + a_2 a^2 + \cdots + a_n a^n$ is a ring homomorphism. ϕ_a is the **evaluation homomorphism** at $a \in R$.

19. Prove that the mapping $f : R[x] \rightarrow R$ defined by $f(a_0 + a_1 x + a_2 x^2 + \cdots + a_n x^n) = a_0$ is a ring homomorphism. Is this a special case of Exercise 18?

20. Prove that the mapping $\phi : R[x] \rightarrow (R/I)[x]$ of Example 2 of 3.3.14 defined by $\phi(a_0 + a_1 x + \cdots + a_n x^n) = (a_0 + I) + (a_1 + I)x + \cdots + (a_n + I)x^n$ is a well-defined ring epimorphism.

21. Fill in the details for the proofs of Theorems 3.3.15 and 3.3.16.

22. Prove Theorem 3.3.17.

23. Consider the polynomial ring $R[x]$, and let $f(x) = x^2 + 1 \in R[x]$.

   (a) Prove that $I = \{f(x)g(x) \mid g(x) \in R[x]\}$ is an ideal of $R[x]$.
   (b) Show that if $g(x) \in R[x]$, then $g(x) + I = a + bx + I$, where $a, b \in R$.
   (c) Prove that $\phi : R[x]/I \rightarrow C$, where $\phi(a + bx + I) = a + bi$, is a ring isomorphism.
   (d) Explain why I is a maximal ideal of $R[x]$.

   Remark: We have also seen that the field C of complex numbers is isomorphic to the field $\mathcal{C}$ of Example 2 of 3.3.3. Hence, the fields $\mathcal{C}$, C, and $R[x]/I$ are all isomorphic.

24. Consider the polynomial ring $Q[x]$, and let $f(x) = x^2 - 2$. We have seen in Example 5 of 3.2.2 that $Q[\sqrt{2}] = \{a + b\sqrt{2} \mid a, b \in Q\}$ is a field.

   (a) Show that $I = \{f(x)g(x) \mid g(x) \in Q[x]\}$ is an ideal of $Q[x]$.

(b) Show that if $g(x) \in \mathbf{Q}[x]$, then $g(x) + I = a + bx + I$ for some $a, b \in \mathbf{Q}$.

(c) Prove that $\phi : \mathbf{Q}[x] / I \rightarrow \mathbf{Q}[\sqrt{2}\,]$ defined by $\phi(a + bx + I) = a + b\sqrt{2}$ is a ring isomorphism.

(d) Explain why I is a maximal ideal of $\mathbf{Q}[x]$.

TECHNOLOGY PROBLEMS

Use a computer algebra system to solve each of the following problems.

25. (a) Show that the mapping $f : \mathbf{Z}_{48} \rightarrow \mathbf{Z}_8$ defined by $f([x]_{48}) = [x]_8$ is a ring epimorphism.

(b) Determine the kernel K of f.

(c) Find all the subrings of $\mathbf{Z}_{48}$ that contain K and all the subrings of $\mathbf{Z}_8$.

26. Consider the ring $M_4(\mathbf{R})$ of 4×4 matrices with entries from $\mathbf{R}$ under the usual operations of addition and multiplication of matrices.

(a) Determine whether the set of all matrices of the form

$$\begin{pmatrix} 0 & 0 & 0 & 0 \\ a_{21} & a_{22} & a_{23} & a_{24} \\ 0 & 0 & 0 & 0 \\ 0 & 0 & 0 & 0 \end{pmatrix}$$

is a left ideal of $M_4(\mathbf{R})$, a right ideal of $M_4(\mathbf{R})$, or neither.

(b) Determine whether the set of all matrices of the form

$$\begin{pmatrix} 0 & 0 & a_{13} & 0 \\ 0 & 0 & a_{23} & 0 \\ 0 & 0 & a_{33} & 0 \\ 0 & 0 & a_{43} & 0 \end{pmatrix}$$

is a left ideal of $M_4(\mathbf{R})$, a right ideal of $M_4(\mathbf{R})$, or neither.

(c) Determine whether the set of all matrices of the form

$$\begin{pmatrix} a_{11} & a_{12} & a_{13} & a_{14} \\ 0 & a_{22} & a_{23} & a_{24} \\ 0 & 0 & a_{33} & a_{34} \\ 0 & 0 & 0 & a_{44} \end{pmatrix}$$

is a left ideal, a right ideal, or a subring of $M_4(\mathbf{R})$.

(d) Determine whether the set of all matrices of the form

$$\begin{pmatrix} 0 & a_{12} & a_{13} & 0 \\ 0 & 0 & a_{23} & a_{24} \\ 0 & 0 & 0 & a_{34} \\ 0 & 0 & 0 & 0 \end{pmatrix}$$

is a left ideal, a right ideal, or a subring of $M_4(\mathbf{R})$.

■4

The Rational, Real, and Complex Number Systems

*In the preceding chapters, properties of **Z**, **Q**, **R**, and **C** were used as examples of groups, rings, integral domains, and fields or to construct examples of these algebraic structures. This chapter presents important properties of our number systems. It is devoted to showing that **Z**, **Q**, **R**, and **C** are each unique up to isomorphism and to showing how the field **Q** can be constructed from **Z**, how **R** can be constructed from **Q**, and how **C** can be constructed from **R**. Sections 4.2 and 4.3 can be omitted without loss of continuity with the remainder of the text. We begin with the field of fractions of an integral domain.*

4.1   FIELDS OF FRACTIONS AND THE RATIONAL NUMBERS

Section Overview. In this section, the field of fractions of an integral domain is constructed, and it is shown that as such this field of fractions is unique up to isomorphism. This is a classical construction in abstract algebra, so study it carefully. The field of rational numbers **Q** is obtained from this construction by selecting the integral domain to be **Z**.

If a and b are elements of an additive group G, then an equation of the form $x + a = b$ can always be solved for x in G. However, if R is a ring and $a, b, c \in R$, $a \neq 0$, then $ax + b = c$ may not have a solution for x that lies in R. For example, the equation $3x + 2 = 7$ cannot be solved for x in $\mathbf{Z}$. If we can find a field F such that $\mathbf{Z} \subseteq F$, then an equation of the form $ax + b = c$, where $a, b, c \in \mathbf{Z}$, $a \neq 0$, always has $a^{-1}(c - b) \in F$ as a solution. This points out the need to find a field that contains a copy of $\mathbf{Z}$. Fortunately, such a field can be constructed. Since the method for "building" the field F from $\mathbf{Z}$ can be applied to any integral domain, the construction will be carried out for an arbitrary integral domain. The field we seek can then be obtained as a special case of this general construction by taking the integral domain to be $\mathbf{Z}$.

Before beginning the construction of the field of fractions of an integral domain, let's look at the concept of a rational number from a different point of view. This will require that you adjust your thinking somewhat, but this forms the skeleton of what we are going to do in our construction. We regard a rational number $\frac{a}{b}$, where a and b are integers with $b \neq 0$, to be a set of ordered pairs (c, d) of integers c and d with $d \neq 0$. Membership in $\frac{a}{b}$ is determined by $(c, d) \in \frac{a}{b}$ if and only if $ad = bc$. For example, the rational number $\frac{3}{4}$ is the following set:

$$\frac{3}{4} = \{\ldots, (-9, -12), (-6, -8), (-3, -4), (3, 4), (6, 8), (9, 12), \ldots\}.$$

(You are used to writing each ordered pair in $\frac{3}{4}$ as a fraction and setting them all equal, but remember, you are being asked to adjust your thinking.) Now let's look at this more generally. Let $\mathbf{Z}^*$ be the set of nonzero integers, and define the relation E on $\mathbf{Z} \times \mathbf{Z}^*$ by $(a, b)\mathrm{E}(c, d)$ if and only if $ad = bc$, where $(a, b), (c, d) \in \mathbf{Z} \times \mathbf{Z}^*$. This just says that the product of the two outside members of the ordered pairs in $(a, b)\mathrm{E}(c, d)$ is equal to the product of the inside members as indicated in the figure below.

$$\overset{\textit{outside}}{\overbrace{(a, b)\mathrm{E}(c, d)}}$$
$$\underset{\textit{inside}}{\underbrace{\phantom{(a, b)\mathrm{E}(c, d)}}}$$

Remember that order is important in a relation. In this instance, $(a, b)\mathrm{E}(c, d)$ means that $ad = bc$ and not $da = bc$, $ad = cb$, or $da = cb$. Of course, if the ring is commutative, this is not a concern. As we shall see, E is an equivalence relation on $\mathbf{Z} \times \mathbf{Z}^*$. Under this relation, all the ordered pairs in $\frac{3}{4}$ are related. For example, $(-6, -8)\mathrm{E}(9, 12)$ since $-6 \cdot 12 = -8 \cdot 9$. We have indicated that E is an equivalence relation on $\mathbf{Z} \times \mathbf{Z}^*$. The following lemma shows that this is actually true for an arbitrary integral domain D. In the following, D is an integral domain, and D^* denotes the set of nonzero elements of D.

4.1.1   Lemma

The relation E defined on $D \times D^*$ by

$$(a, b)E(c, d) \qquad \text{if and only if } ad = bc$$

is an equivalence relation on $D \times D^*$.

Proof. If $(a, b) \in D \times D^*$, then $ab = ba$ since D is commutative. Hence, $(a, b)E$ (a, b), so E is reflexive. Likewise, if $(a, b)E(c, d)$ with $(a, b), (c, d) \in D \times D^*$, then $ad = bc$. Hence, $cb = da$, so $(c, d)E(a, b)$. Thus, E is symmetric. Finally, suppose that $(a, b), (c, d), (f, g) \in D \times D^*$ are such that $(a, b)E(c, d)$ and $(c, d) E(f, g)$. Then (1) $ad = bc$ and (2) $cg = df$. Multiplying both sides of equation (1) by g and both sides of (2) by b produces $adg = bcg$ and $bcg = bdf$, respectively, so $adg = bdf$. Hence, $(ag - bf)d = 0$. But $d \in D^*$ indicates that $d \neq 0$, so because D has no zero-divisors, it must be the case that $ag = bf$. Therefore, $(a, b)E(f, g)$, so E is transitive. Hence, E is reflexive, symmetric, and transitive, and so E is an equivalence relation on $D \times D^*$. ∎

Let's return to the integers. Early in your mathematical training, you were taught that the definitions of addition and multiplication of fractions are given by $\frac{a}{b} + \frac{c}{d} = \frac{ad + bc}{bd}$ and $\frac{a}{b} \cdot \frac{c}{d} = \frac{ac}{bd}$, respectively. This provides the model for how we will define addition and multiplication of ordered pairs in $D \times D^*$. These definitions of addition and multiplication of fractions appear again when addition and multiplication of congruence classes are considered. Compare the definitions of addition and multiplication of ordered pairs in the following lemma with the addition and multiplication of fractions just given.

4.1.2   Lemma

If addition and multiplication are defined on $D \times D^*$ by $(a, b) + (c, d) = (ad + bc, bd)$ and $(a, b)(c, d) = (ac, bd)$ for all (a, b) and (c, d) in $D \times D^*$, then these operations are well-defined, and the equivalence relation E of Lemma 4.1.1 is a congruence relation on $D \times D^*$ with respect to both of these operations.

Proof. The proof has two parts: first it must be shown that the binary operations we have defined on the ordered pairs in $D \times D^*$ are well-defined. After this has been accomplished, the second step is to show that the equivalence relation of Lemma 4.1.1 is a congruence relation with respect to these operations. To show that the operations are well-defined, suppose $(a, b) = (a', b')$ and $(c, d) = (c', d')$. It follows from the definition of equality of ordered pairs that $a = a'$, $b = b'$, $c = c'$, and $d = d'$. Now addition and multiplication are well-defined on D, so we have $ad + bc = a'd' + b'c'$ and $bd = b'd'$. Hence, $(ad + bc, bd) = (a'd' + b'c', b'd')$ and thus $(a, b) + (c, d) = (a', b') + (c', d')$, which shows that addition is well-defined. A similar argument shows that $(a, b)(c, d) = (a', b')(c', d')$, so multiplication is well-defined.

Finally, we need to show that E is a congruence relation on $D \times D^*$ with respect to both of these operations. If $(a, b)E(a', b')$ and $(c, d)E(c', d')$, it must be shown that

$$[(a, b) + (c, d)]E[(a', b') + (c', d')] \qquad \text{and}$$
$$[(a, b)(c, d)]E[(a', b')(c', d')],$$

which is equivalent to showing that

$$(ad + bc, db)E(a'd' + b'c'b'd') \qquad \text{and}$$
$$(ac, bd)E(a'c'b'd').$$

If $(a, b)E(a', b')$ and $(c, d)E(c', d')$, then (1) $ab' = ba'$ and (2) $cd' = dc'$. Multiplying both sides of equation (1) by dd' and both sides of (2) by bb' produces $ab'dd' = ba'dd'$ and $cd'bb' = dc'bb'$, respectively. Adding the last two equations gives (3) $ab'dd' + cd'bb' = ba'dd' + dc'bb'$. Factoring $b'd'$ from the left side of equation (3) and bd from the right side gives $(ad + bc)b'd' = bd(a'd' + b'c')$. Hence, we see that $(ad + bc, bd)E(a'd' + b'c', b'd')$. Finally, if both sides of equation (1) are multiplied by the left side of equation (2) and both sides of (2) are multiplied by the right side of (1), then $ab'cd' = ba'dc'$, so $(ac)(b'd') = (bd)(a'c')$. Thus, $(ac, bd)E(a'c', b'd')$. Hence, E is a congruence relation on $D \times D^*$. ∎

4.1.3   Definition

Suppose that D is an integral domain, and let E be the congruence relation on $D \times D^*$ of Lemmas 4.1.1 and 4.1.2. The congruence class determined by $(a, b) \in D \times D^*$ is denoted by $\frac{a}{b}$. The ratio $\frac{a}{b}$ is referred to as a **fraction** or a **quotient** of elements of D. The set of all such fractions is denoted by $Q(D)$.

It will now be shown that the set of all such quotients of elements of an integral domain D can be made into a field that contains a subring that is ring isomorphic to D. The proof of the following lemma is left as an exercise. It provides additional information about elementary properties of congruence classes in $Q(D)$.

4.1.4 Lemma

If e is the multiplicative identity of D and $\frac{a}{b} \in Q(D)$, then:

1. For any nonzero $x \in D$, $\frac{a}{b} = \frac{ax}{bx}$.

2. The equality $\frac{a}{b} = \frac{0}{e}$ holds if and only if $a = 0$.

3. The equality $\frac{b}{b} = \frac{e}{e}$ holds for any $b \in D^*$.

4.1.5 Theorem

If D is an integral domain with identity e, then the set $Q(D)$ together with the binary operations of addition and multiplication of congruence classes defined by $\frac{a}{b} + \frac{c}{d} = \frac{ad+bc}{bd}$ and $\frac{a}{b} \cdot \frac{c}{d} = \frac{ac}{bd}$ is a field. Moreover, $Q(D)$ contains a subring D_e that is isomorphic to D.

Proof. Because of Lemmas 4.1.1 and 4.1.2 and Theorem 0.4.8, the binary operations are well-defined. Addition is associative since

$$\left(\frac{a}{b} + \frac{c}{d}\right) + \frac{g}{h} = \frac{ad+bc}{bd} + \frac{g}{h} = \frac{(ad+bc)h + (bd)g}{(bd)h}$$

$$= \frac{adh + b(ch+dg)}{b(dh)} = \frac{a}{b} + \frac{ch+dg}{dh} = \frac{a}{b} + \left(\frac{c}{d} + \frac{g}{h}\right).$$

The additive identity is $\frac{0}{e}$, and the additive inverse $-\frac{a}{b}$ of $\frac{a}{b}$ is $\frac{-a}{b} = \frac{a}{-b}$. Addition is easily shown to be commutative, so $Q(D)$ is an additive abelian group. It is also not difficult to show that multiplication is associative and multiplication is distributive over addition from the left since

$$\frac{a}{b}\left(\frac{c}{d} + \frac{g}{h}\right) = \frac{a}{b}\left(\frac{ch+dg}{dh}\right) = \frac{a(ch)+a(dg)}{b(dh)}$$

$$= \frac{a(ch)+a(dg)}{b(dh)} \cdot \frac{b}{b} = \frac{(ac)(bh)+(bd)(ag)}{(bd)(bh)} = \frac{ac}{bd} + \frac{ag}{bh} = \frac{a}{b} \cdot \frac{c}{d} + \frac{a}{b} \cdot \frac{g}{h}.$$

$\left(\text{Notice that we have used the fact from Lemma 4.1.4 that } \frac{b}{b} = \frac{e}{e}.\right)$ It is straightforward to show that multiplication is commutative, so it follows that multiplication is also distributive over addition from the right. Hence, $Q(D)$ is a commutative ring with identity $\frac{e}{e}$. If $\frac{a}{b} \neq \frac{0}{e}$, then, by Lemma 4.1.4, $a \neq 0$, so $\frac{b}{a} \in Q(D)$. Since $\frac{a}{b} \cdot \frac{b}{a} = \frac{ab}{ba} = \frac{e}{e}$, we see that $\left(\frac{a}{b}\right)^{-1} = \frac{b}{a}$. Hence, $Q(D)$ is a field. Finally, the set $D_e = \{\frac{a}{e} \mid a \in D\}$ is a subring of $Q(D)$, and the mapping $f : D \rightarrow D_e$ defined by $f(a) = \frac{a}{e}$ is a ring isomorphism. Thus, $Q(D)$ contains a subring isomorphic to D. ∎

If the proceeding theorem is applied to the integral domain $\mathbf{Z}$, we obtain the **field $Q(\mathbf{Z}) = \mathbf{Q}$ of rational numbers.** If $f : D \rightarrow Q(D)$ is the ring monomorphism of the theorem, then for any $b \neq 0$ in D, $f(b)^{-1} = \frac{e}{b}$. Hence, we can write $f(a)f(b)^{-1}$ for $\frac{a}{b}$. Theorem 4.1.5 also shows that any integral domain can be embedded in a field.

4.1.6   Definition

> If D is an integral domain, then a field F is a **field of fractions** of D if there is a ring monomorphism $f : D \rightarrow F$ and every element of F can be written as $f(a)f(b)^{-1}$ for some $a, b \in D$ with $b \neq 0$.

4.1.7   Theorem

If D is an integral domain, then D has a field of fractions, and any two fields of fractions of D are isomorphic.

Proof. It follows from Theorem 4.1.5 that $Q(D)$ is a field of fractions of D. Suppose that F is also a field of fractions of D. Then there is a ring monomorphism $f : D \rightarrow F$, and every element of F can be written as $f(a)f(b)^{-1}$, where $a, b \in D$ with $b \neq 0$. Let $\phi : Q(D) \rightarrow F$ be defined by $\phi\left(\frac{a}{b}\right) = f(a)f(b)^{-1}$. We claim that ϕ is a well-defined ring isomorphism. If $\frac{a}{b} = \frac{c}{d}$ in $Q(D)$, then $ad = bc$ in D. Hence, $f(ad) = f(bc)$, and from this it follows that $f(a)f(b)^{-1} = f(c)f(d)^{-1}$. Thus, $\phi\left(\frac{a}{b}\right) = \phi\left(\frac{c}{d}\right)$, so ϕ is well-defined. Next suppose that $\phi\left(\frac{a}{b}\right) = \phi\left(\frac{c}{d}\right)$. Then $f(a)f(b)^{-1} = f(c)f(d)^{-1}$ implies that $f(ad) = f(bc)$. But f is an injective mapping, so $ad = bc$ in D. By the definition of the equivalence relation on $D \times D^*$, $\frac{a}{b} = \frac{c}{d}$ in $Q(D)$. Thus, ϕ is injective. ϕ is clearly a surjective mapping, since if $x \in F$, then x can be written as $f(a)f(b)^{-1}$ for some a, $b \in D$ with $b \neq 0$. The quotient $\frac{a}{b}$ is therefore an element of $Q(D)$, and $\phi\left(\frac{a}{b}\right) = f(a)f(b)^{-1} = x$, so we have shown that ϕ is a bijective mapping. Finally, ϕ is a ring homomorphism since

$$\phi\left(\frac{a}{b} + \frac{c}{d}\right) = \phi\left(\frac{ad + bc}{bd}\right)$$
$$= f(ad + bc)f(bd)^{-1}$$
$$= [f(a)f(d) + f(b)f(c)]f(b)^{-1}f(d)^{-1}$$
$$= f(a)f(b)^{-1} + f(c)f(d)^{-1}$$
$$= \phi\left(\frac{a}{b}\right) + \phi\left(\frac{c}{d}\right) \qquad \text{and}$$
$$\phi\left(\frac{a}{b} \cdot \frac{c}{d}\right) = \phi\left(\frac{ac}{bd}\right)$$
$$= f(ac)f(bd)^{-1}$$
$$= [f(a)f(b)^{-1}][f(c)f(d)^{-1}]$$
$$= \phi\left(\frac{a}{b}\right)\phi\left(\frac{c}{d}\right). \blacksquare$$

4.1.8   Theorem

If D_1 and D_2 are isomorphic integral domains, then $Q(D_1)$ and $Q(D_2)$ are isomorphic fields.

Proof. If $f : D_1 \to D_2$ is a ring isomorphism, then $\phi : Q(D_1) \to Q(D_2)$ defined by $\phi\left(\frac{a}{b}\right) = \frac{f(a)}{f(b)}$ is a ring isomorphism. (You should show this.) $\blacksquare$

Theorem 4.1.7 shows that the field of fractions of an integral domain is unique up to isomorphism. Consequently, we can speak of *the* field of fractions of an integral domain. Applying these observations to **Z**, we obtain an important property of the field of rational numbers: *as the field of fractions of* **Z**, **Q** *is unique up to isomorphism.*

The converse of Theorem 4.1.8 is false. It will be shown in the exercises that if the construction of $Q(D)$ is relaxed to include domains that do not have an identity, then $Q(2\mathbf{Z})$ and $Q(\mathbf{Z})$ are isomorphic fields. Of course, we know that $2\mathbf{Z}$ and $\mathbf{Z}$ cannot be isomorphic since $\mathbf{Z}$ contains an identity while $2\mathbf{Z}$ does not. See Exercises 10 and 11 for the details.

4.1.9   Definition

A field E is said to be a **field extension** of a field F when F is a subfield of E. If R is a ring with identity e, then the smallest positive integer n such that $ne = 0$ is said to be the **characteristic** of R. If no such positive integer n exists, then R is said to have characteristic 0. The characteristic of R is denoted by char R.

4.1.10   Lemma

If D is an integral domain, then char $D = 0$ or char $D = p$ for some prime integer p.

Proof. If char $D = n \neq 0$ and n is not prime, there are positive integers $1 < n_1 < n$ and $1 < n_2 < n$ such that $n = n_1 n_2$. Hence, $0 = (n_1 n_2)e = (n_1 e) \cdot (n_2 e)$. But $n_1 e$, $n_2 e \in D$ and D is an integral domain, so it must be that either $n_1 e = 0$ or $n_2 e = 0$. This is a contradiction since n is the smallest positive integer such that $ne = 0$. Thus, if char $D \neq 0$, then char D must be prime. ■

In Problem Set 3.2, Exercise 3, we saw that it is possible for R to be a subring of a ring S, yet R and S can have different identities. This is not possible for fields.

4.1.11   Lemma

If E is a field extension of F, then the identities of E and F are equal.

Proof. Suppose e and e' are the identities of E and F, respectively. If $e' \neq e$, then $e' - e \neq 0$. But $e' \in F \subseteq E$, so $e'e = e'$. Now $e'(e' - e) = e'^2 - e'e = e' - e' = 0$, which implies that $e' = 0$ since E is a field and $e' - e \neq 0$. But this cannot be the case since $e' = 0$ implies that F contains only one element. This contradicts the definition of a field, which asserts that a field must contain at least two elements. Hence, $e = e'$. ■

4.1.12   Corollary

The identity of a field F is the identity of all the subfields of F.

If char $F = 0$, then $ne \neq 0$ for every positive integer n. This implies that $ne \neq 0$ for every nonzero integer n because if n is a negative integer such that $ne = 0$, then $-n$ is a positive integer and $(-n)e = -(ne) = -0 = 0$, a clear contradiction. We close this section on fields of fractions with the following observation.

4.1.13   Theorem

If F is a field and char $F = 0$, then F contains a copy of $\mathbf{Q}$, and if char $F = p$, then F contains a copy of the field $\mathbf{Z}_p$.

Proof. (Sketched) Suppose that char $F = 0$. Then $\mathbf{Z}e$ is a subring of F that is isomorphic to $\mathbf{Z}$ via the ring homomorphism $f : \mathbf{Z} \to \mathbf{Z}e$ given by $f(n) = ne$. Since $ne \neq 0$ for any nonzero integer n, ne has a multiplicative inverse $(ne)^{-1}$ in F. If $F' = \{(me)(ne)^{-1} \mid m, n \in \mathbf{Z}, n \neq 0\}$, then F' is a subfield of F and the mapping $\phi : \mathbf{Q} \to F'$, where $\phi\left(\dfrac{m}{n}\right) = (me)(ne)^{-1}$, is a ring isomorphism. Hence, F contains a subfield isomorphic to $\mathbf{Q}$.

If char $F = p$, then p is a prime and $pe = 0$. In this case, the ring homomorphism $f : \mathbf{Z} \to \mathbf{Z}e$ defined by $f(n) = ne$ is an epimorphism with kernel $p\mathbf{Z}$. Thus, by Theorem 3.3.12, $\mathbf{Z}/p\mathbf{Z}$ and $\mathbf{Z}e$ are isomorphic rings. But, by Example 1 of 3.3.14, $\mathbf{Z}_p$ and $\mathbf{Z}/p\mathbf{Z}$ are also isomorphic rings, so $\mathbf{Z}_p$ and $\mathbf{Z}e$ are isomorphic. Thus, F contains a copy of $\mathbf{Z}_p$ when char $F = p$. ∎

Problem Set 4.1

1. Prove Lemma 4.1.4.

2. (a) Let R be a ring. If there is a positive integer n such that $nx = 0$ for all $x \in R$, show that there is an infinite number of such integers. In this case, show there is a smallest positive integer with this property.

    (b) If R is a ring, define the characteristic of R to be the smallest positive integer n such that $nx = 0$ for all $x \in R$ whenever such a positive integer exists, and 0 otherwise. Prove that this definition of the characteristic of R coincides with the definition of the characteristic of R given in Definition 4.1.9 when the ring has an identity.

(c) Must the characteristic of a ring be unique? Explain your answer.

3. Fill in the details of the proof of Theorem 4.1.13.

4. Prove that the function ϕ of Theorem 4.1.8 is a ring isomorphism.

5. Prove that any field F that contains a subring that is isomorphic to an integral domain D must contain a subfield that is isomorphic to $Q(D)$.

6. Prove that $\mathbf{Z}$ and $\mathbf{Q}$ cannot be isomorphic rings.

7. Prove that neither of the fields $\mathbf{Q}$ and $\mathbf{Z}_p$, where p is a prime integer, can contain a proper subfield.

8. We have seen in Problem Set 3.1, Exercise 2, that $\mathbf{Z}[\sqrt{3}]$ is an integral domain. Prove that the field of fractions $Q(\mathbf{Z}[\sqrt{3}])$ of $\mathbf{Z}[\sqrt{3}]$ is isomorphic to the field $\mathbf{Q}[\sqrt{3}] = \{x + y\sqrt{3} \mid x, y \in \mathbf{Q}\}$, where the binary operations on $\mathbf{Q}[\sqrt{3}]$ are the usual ones.

9. (a) Suppose that $D = \{\frac{a}{b} \in \mathbf{Q} \mid 5 \nmid b\}$. Show that D is a subring of $\mathbf{Q}$ with identity 1 such that $\mathbf{Z} \subseteq D \subseteq \mathbf{Q}$. Prove that $Q(D) = \mathbf{Q}$.

   (b) Prove that any subring D of $\mathbf{Q}$ with identity $1 \neq 0$ must be an integral domain and that if $\mathbf{Z} \subseteq D \subseteq \mathbf{Q}$, then $Q(D) = \mathbf{Q}$.

10. Let R be a commutative ring. A **multiplicatively closed subset** of R is a subset S of R such that $0 \notin S$ and $st \in S$ whenever $s, t \in S$. (If R has an identity e, then it is also assumed that $e \in S$.)

    (a) If S is a multiplicatively closed subset of R, define the relation E on $R \times S$ by $(a, s)\mathrm{E}(b, t)$ if and only if there is a $u \in S$ such that $u(at - bs) = 0$. Prove that E is an equivalence relation on $R \times S$. If addition and multiplication are defined on $R \times S$ by $(a, s) + (b, t) = (at + bs, st)$ and $(a, s)(b, t) = (ab, st)$, prove that E is a congruence relation on $R \times S$.

    (b) Let $\frac{a}{s}$ denote the congruence class determined by the pair (a, s) in $R \times S$, and suppose that $S^{-1}R$ is the set of all such congruence classes. Define addition and multiplication on $S^{-1}R$ by $\frac{a}{s} + \frac{b}{t} = \frac{at + bs}{st}$ and $\frac{a}{s} \cdot \frac{b}{t} = \frac{ab}{st}$ for all $\frac{a}{s}, \frac{b}{t} \in S^{-1}R$. Explain why these operations on $S^{-1}R$ are well-defined.

    (c) Prove that $S^{-1}R$ together with the binary operations of addition and multiplication defined in (b) is a commutative ring with identity. Which elements of $S^{-1}R$ have multiplicative inverses in $S^{-1}R$? Must $S^{-1}R$ be a field?

(d) Show that $\phi : R \to S^{-1}R$, where $\phi(a) = \frac{as}{s}$ and s is a fixed element of S, is a ring homomorphism with kernel $K = \{a \in R \mid at = 0 \text{ for some } t \in S\}$. Show, furthermore, that R/K embeds in $S^{-1}R$. Hint: Consider $f : R/K \to S^{-1}R$ defined by $f(a + K) = \phi(a)$.

(e) Show that if D is a domain and $S = D^*$, the set of nonzero elements of D, then $S^{-1}D = Q(D)$ is a field of fractions of D. Conclude that the construction of a ring of fractions of a commutative ring with respect to a multiplicatively closed set S in R is a generalization of the construction of the field of fractions of an integral domain.

(f) If the requirement that $0 \notin S$ is dropped from the definition of a multiplicatively closed set S and S is a multiplicatively closed subset of R such that $0 \in S$, show that $S^{-1}R$ is the zero ring.

11. Apply the construction of Exercise 10 to $2\mathbf{Z}$ to obtain $Q(2\mathbf{Z})$. Show that $Q(2\mathbf{Z})$ and $Q(\mathbf{Z}) = \mathbf{Q}$ are isomorphic fields even though $2\mathbf{Z}$ and $\mathbf{Z}$ are not isomorphic.

4.2   ORDERED INTEGRAL DOMAINS

Section Overview. In Chapter 1, we studied properties of the integers. This section provides additional information on the ring of integers and shows that, up to isomorphism, there is only one ordered integral domain in which the set of positive elements is well-ordered. Theorem 4.2.6 shows that such a domain must be isomorphic to $\mathbf{Z}$.

A fundamental property of the integral domain $\mathbf{Z}$ is that it contains a set of elements $\mathbf{N}$ with the following properties:

1. The set of integers $\mathbf{N}$ is closed under addition and multiplication: if $x, y \in \mathbf{N}$, then $x + y$ and xy are in $\mathbf{N}$.
2. **Trichotomy Property** for $\mathbf{Z}$: If $x \in \mathbf{Z}$, then one and only one of the following holds: $x \in \mathbf{N}$, $x = 0$, $-x \in \mathbf{N}$.
3. The set $\mathbf{N}$ is well-ordered by the order relation $<$. (Recall that this is an axiom. See Section 1.1.)

If $x \in \mathbf{Z}$, then we say that x is a **positive integer** if $x \in \mathbf{N}$ and a **negative integer** if $-x \in \mathbf{N}$. We refer to $\mathbf{N}$ as the set of **positive elements** of $\mathbf{Z}$. In Section 1.1, the set $\mathbf{N}$ was used to define the relation **less than** on $\mathbf{Z}$ by setting $x < y$ if and only if $y - x \in \mathbf{N}$.

We have previously used properties of this order relation on **Z** quite freely without concern for a proof of these properties. To give a complete description of our number systems, it is necessary to give a formal development of this order relation on **Z**.

Before beginning the discussion of ordered integral domains, consider the field

$$\mathbf{Z}_3 = \{[0], [1], [2]\},$$

and suppose that we order the elements of $\mathbf{Z}_3$ as

$$[0] < [1] < [2].$$

One property that we would like an order relation on an integral domain to have is the ability to add an element from the domain to both sides of an inequality and have the result remain unequal in the same order. This property holds for the order relation defined earlier on **Z** (see Theorem 1.1.1), but not for our ordering of $\mathbf{Z}_3$. If we add [1] to both sides of [0] < [2], we see that

$$[0] + [1] < [2] + [1] \qquad \text{or}$$
$$[1] < [0].$$

This obviously contradicts the ordering [0] < [1] < [2]. It can be shown that the elements of $\mathbf{Z}_3$ cannot be ordered in a way that the familiar properties of order hold. In fact, it can be shown that it is not possible to define such an order relation on any finite integral domain because of similar difficulties. What is needed to order an integral domain is a set of "positive elements" of the domain that mimic the behavior of the elements of **N** under addition and multiplication and a trichotomy property. When such a set of positive elements exists in an integral domain D, it is possible to define an order relation $<$ on D in a way that properties that are similar to the familiar properties of $<$ on **Z** hold.

4.2.1   Definition

If D is an integral domain, then a nonempty set D^+ of elements of D is said to be a set of **positive elements** if

1. the set D^+ is closed under addition and multiplication and
2. a trichotomy property holds for elements of D with respect to D^+. That is, one and only one of $x \in D^+$, $x = 0$, and $-x \in D^+$ holds for each $x \in D$.

If $x \in D^+$, then x is a **positive element** of D, and if $-x \in D^+$, then x is a **negative element** of D. An integral domain D with a set of positive elements D^+ is said to be an **ordered integral domain**. A field with a set of positive elements is an **ordered field**. For such a domain D, define the relation **less than** on D by setting $x < y$ if $y - x \in D^+$. The notation $x \le y$ means that $x < y$ or $x = y$. We read $x < y$ as "x is **less than** y," and $x \le y$ is read as "x is **less than or equal to** y." Finally, $y > x$ means $x < y$, and $y \ge x$ indicates that $x \le y$. Likewise, $y > x$ is read "y is **greater than** x," and $y \ge x$ is read "y is **greater than or equal to** x." Let S be a nonempty subset of an ordered integral domain D. If there is an element $s \in S$ such that $s \le x$ for all $x \in S$, then s is said to be a **smallest element** of S. We say that D^+ is **well-ordered** if every nonempty subset of D^+ has a smallest element.

Because of the trichotomy property for an ordered integral domain D, each $x \in D$ satisfies one and only one of $x > 0$, $x = 0$, and $-x > 0$. Moreover, a smallest element of a set S (if such exists) in an ordered integral domain is unique. This follows from the fact that if s_1 and s_2 are smallest elements of S, then $s_1 \le s_2$ and $s_2 \le s_1$, so, because of the trichotomy property, it must be the case that $s_1 = s_2$. Hence, we can refer to a smallest element as *the* smallest element of a set S whenever such an element exists.

The following theorem lists several basic order properties that hold in every ordered integral domain. The proof of each of these properties is straightforward, requiring only the use of the definition of $<$ and the properties of D^+. We prove 1 through 3 to demonstrate the techniques involved and leave the proof of 4 through 8 as exercises. Additional order properties for ordered integral domains are given in the exercises.

4.2.2   Theorem

Let x, y, and z be arbitrary elements of an ordered integral domain D with identity e.

1. If $x > 0$ and $y > 0$, then $x + y > 0$ and $xy > 0$.
2. If $x \ne 0$, then $x^2 > 0$. In particular, $e > 0$.
3. If $x < y$, then $x + z < y + z$.
4. If $x < y$ and $y < z$, then $x < z$.
5. If $x < y$ and $z < w$, then $x + z < y + w$.
6. For every positive integer n, $ne > 0$.
7. If $x < y$ and $z > 0$, then $xz < yz$.
8. If $x < y$ and $z < 0$, then $xz > yz$.

Proof

1. This is the direct result of the fact that a set of positive elements D^+ of D is closed under addition and multiplication. The fact that $x > 0$ gives $x \in D^+$, and $y > 0$ means that $y \in D^+$. Hence, $x + y$ and xy are in D^+. Therefore, $x + y > 0$ and $xy > 0$.

2. If $x \in D$, $x \neq 0$, then $x \in D^+$ or $-x \in D^+$. Hence, either $x^2 \in D^+$ or $(-x)^2 \in D^+$ since D^+ is closed under multiplication. But $(-x)^2 = x^2$, so in either case $x^2 \in D^+$ Thus, $x^2 > 0$ when $x \neq 0$. Moreover, $e = e^2 > 0$.

3. If $x < y$, then $y - x \in D^+$. But then for any $z \in D$, $y - x = (y + z) - (x + z)$, so $(y + z) - (x + z) \in D^+$. Hence, $x + z < y + z$. ■

| 4.2.3 | **Example** |

An Ordered Integral Domain in Which the Set of Positive Elements Is Well-Ordered

Let $D = \{2^n \mid n \in \mathbf{Z}\}$, and define addition and multiplication on D by $2^m \oplus 2^n = 2^{m+n}$ and $2^m \otimes 2^n = 2^{mn}$ for all 2^m, $2^n \in D$. It is not difficult to show that the binary operations are well-defined and that addition and multiplication are both associative and commutative. Moreover, D is a group under the binary operation $\oplus$. The additive identity is $2^0 = 1$ since

$$1 \oplus 2^n = 2^n \oplus 1 = 2^n \oplus 2^0 = 2^{n+0} = 2^n.$$

It follows from

$$2^{-n} \oplus 2^n = 2^n \oplus 2^{-n} = 2^{n-n} = 2^0 = 1$$

that the additive inverse of 2^n is 2^{-n}. Notice also that

$$2^n \otimes (2^s \oplus 2^t) = 2^n \otimes 2^{s+t}$$
$$= 2^{n(s+t)}$$
$$= 2^{ns+nt}$$
$$= 2^{ns} \oplus 2^{nt}$$
$$= (2^n \otimes 2^s) \oplus (2^n \otimes 2^t),$$

so multiplication is distributive over addition. It is also the case that D has a multiplicative identity $e = 2$ since

$$2 \otimes 2^n = 2^n \otimes 2 = 2^n \otimes 2^1 = 2^{n1} = 2^n.$$

Hence, D is a commutative ring with identity. If

$$2^m \otimes 2^n = 2^{mn} = 1 = 2^0,$$

then $mn = 0$ implies that $m = 0$ or $n = 0$ since **Z** is an integral domain. Thus, 2^m or 2^n is 1, the additive identity of D. Therefore, D is an integral domain. Now let

$$D^+ = \{2^n \mid n \in \mathbf{N}\}.$$

If $2^m, 2^n \in D^+$, then $2^m \oplus 2^n = 2^{m+n}$ and $2^m \otimes 2^n = 2^{mn}$ indicate that D^+ is closed under addition and multiplication since $m + n$ and mn are both in **N**. If $2^n \in D$, then only one of $2^n \in D^+$, $2^n = 1$, or $2^{-n} \in D^+$ holds because, for the integer n, one and only one of $n > 0$, $n = 0$, or $n < 0$ holds. Therefore, D^+ is a set of positive elements of D, so D is an ordered integral domain. The order relation $<$ defined on D by using D^+ is given by $2^m < 2^n$ if and only if $m < n$ in **Z**. If S is a nonempty subset of D^+, then $S_\mathbf{N} = \{n \in \mathbf{N} \mid 2^n \in S\}$ is a nonempty subset of **N**. If n_0 is the smallest element of $S_\mathbf{N}$ (which exists, since **N** is well-ordered), then 2^{n_0} is the smallest element of S. Hence, D is an ordered integral domain in which the set of positive elements is well-ordered.

You may have noticed that we have always referred to D^+ as *a* set of positive elements of an ordered integral domain rather than *the* set of positive elements. The following theorem shows that an integral domain D can contain only one set of positive elements when D^+ is well-ordered.

4.2.4   Theorem

Let D be an ordered integral domain. If D^+ is a set of positive elements of D that is well-ordered, then

1. the smallest element of D^+ is e,
2. the set of positive elements D^+ is unique and $D^+ = \{ne \mid n \in \mathbf{N}\}$, and
3. every element of D is of the form me for some $m \in \mathbf{Z}$.

Proof

1. First, note that part 2 of Theorem 4.2.2 shows that $e \in D^+$. Since D^+ is well-ordered, D^+ has a smallest element, say, s. If $s = e$, there is nothing to prove, so suppose $0 < s < e$. Then, by parts 2 and 7 of Theorem 4.2.2, $0 < s^2 < se = s$, which contradicts the choice of s as the smallest element

of D^+. Therefore, such an $s \in D^+$ cannot exist, so e is the smallest element of D^+.

2. Let S be the set of positive integers n for which $ne \in D^+$. Since $1e = e$ is in D^+, $1 \in S$. Now suppose that $k \in S$; then $ke \in D^+$. But $e \in D^+$, so $ke + e \in D^+$ since D^+ is closed under addition. Therefore, $(k + 1)e \in D^+$, and so $k + 1 \in S$. Hence, $S = \mathbf{N}$ by the Principle of Induction, so $\{ne \mid n \in \mathbf{N}\} \subseteq D^+$. To show that $D^+ \subseteq \{ne \mid n \in \mathbf{N}\}$, let P be the set of all $x \in D^+$ such that x is not of the form ne for some $n \in \mathbf{N}$. If P is nonempty, then, since D^+ is well-ordered, P has a smallest element, say, x^*. Now, by part 1, e is the smallest element of D^+, so it must be the case that $x^* > e$ and thus $x^* - e > 0$. Hence,

$$e > 0 \Rightarrow -e + e > -e + 0 \qquad \text{by part 3 of Theorem 4.2.2}$$

$$\Rightarrow 0 > -e$$

$$\Rightarrow x^* = x^* + 0 > x^* - e > 0 \qquad \text{by part 3 of Theorem 4.2.2}$$

Therefore, $x^* > x^* - e \in D^+$. But x^* is the smallest element of D^+ that is not of the form ne for some $n \in \mathbf{N}$. Thus, there is a $k \in \mathbf{N}$ such that $x^* - e = ke$. But then $x^* = ke + e = (k + 1)e$, which is clearly a contradiction. Therefore $P = \varnothing$, so every element of D^+ is of the form ne for some $n \in \mathbf{N}$ and $D^+ = \{ne \mid n \in \mathbf{N}\}$. If D^* is another set of positive elements of D that is well-ordered by the order relation induced on D by D^*, then it can be shown in exactly the same manner that $D^* = \{ne \mid n \in \mathbf{N}\}$. Thus, $D^* = D^+$, so D^+ is unique.

3. Let $x \in D$. Then by the trichotomy property for D, $x \in D^+$, $x = 0$, or $-x \in D^+$. If $x \in D^+$, then by part 2, $x = me$ for some $m \in \mathbf{N}$. If $x = 0$, then $x = me$ with $m = 0$. If $-x \in D^+$, then $-x = ne$ for some $n \in \mathbf{N}$, so $x = -(ne) = (-n)e = me$ with $m = -n$. ∎

4.2.5	**Examples**

1. An Ordered Integral Domain in Which the Set of Positive Elements Is Not Well-Ordered

There exist ordered integral domains in which the set of positive elements is not well-ordered. For example, consider the field $\mathbf{Q}$ of rational numbers. We can use the ordering on $\mathbf{Z}$ to induce an ordering on $\mathbf{Q}$. If $\mathbf{Q}^+ = \{\frac{p}{q} \in \mathbf{Q} \mid pq \in \mathbf{N}\}$, then it is straightforward to show that $\mathbf{Q}^+$ is a set of positive elements in $\mathbf{Q}$. If $\frac{p}{q}, \frac{s}{t} \in \mathbf{Q}$ and we define $<$ on $\mathbf{Q}$ by $\frac{p}{q} < \frac{s}{t}$ if and only if $\frac{s}{t} - \frac{p}{q} \in \mathbf{Q}^+$, then $\mathbf{Q}$ is an ordered integral domain. The set $\mathbf{Q}^+$ is not well-ordered, however, because if $\mathbf{Q}^+$ were well-ordered under this ordering, then, by Theorem 4.2.4, $1 \in \mathbf{Q}^+$ would have to be the smallest element of $\mathbf{Q}^+$. But this clearly is not the case since $\frac{1}{2} < 1$ and $\frac{1}{2} \in \mathbf{Q}^+$. Consequently, $\mathbf{Q}^+$ is not well-ordered by this ordering.

2. An Infinite Integral Domain That Is Not an Ordered Integral Domain

It has already been mentioned that finite integral domains cannot be ordered. The field $\mathbf{C}$ of complex numbers is an example of an infinite integral domain that cannot be ordered. To see why, suppose that $\mathbf{C}$ has a set of positive elements $\mathbf{C}^+$, and define the relation $<$ on $\mathbf{C}$ by $z < w$ if and only if $w - z \in \mathbf{C}^+$. Since $i \neq 0$, $-1 = i^2 > 0$ by part 2 of Theorem 4.2.2. But $1 = 1^2 > 0$, also by part 2 of Theorem 4.2.2. Hence, both -1 and 1 are in $\mathbf{C}^+$, which clearly cannot be the case by the trichotomy property for ordered integral domains. Thus, $\mathbf{C}$ is not an ordered integral domain.

We now come to the main theorem of this section. It shows that, up to isomorphism, there is only one ordered integral domain that has a set of positive elements that is well-ordered.

4.2.6   Theorem

If D is an ordered integral domain in which the set of positive elements is well-ordered, then D and $\mathbf{Z}$ are isomorphic rings.

Proof. By Theorem 4.2.4, every element of D is of the form me for some $m \in \mathbf{Z}$. Let $f : \mathbf{Z} \to D$ be defined by $f(m) = me$. We claim that f is a well-defined ring isomorphism. If $m = n$, then $me = ne$, so $f(m) = f(n)$. Hence, f is well-defined. Note next that if $me \in D$, where $m \in \mathbf{Z}$, then $f(m) = me$. Consequently, f is a surjective mapping. To show that f is injective, suppose that $f(m) = f(n)$. Then $me = ne$, which gives $(m - n)e = 0$. If $m > n$, then $m - n$ is in $\mathbf{N}$, and thus, by Theorem 4.2.4, $(m - n)e \in D^+$, the set of positive elements of D. Hence, we have $(m - n)e = 0$ and $(m - n)e > 0$, which clearly cannot be the case because of the trichotomy property. Since a similar argument holds if $m < n$, it must be that $m = n$. Therefore, $f(m) = f(n)$ implies that $m = n$, so f is injective. Finally, let's show that f is a ring homomorphism. If $m, n \in \mathbf{Z}$, then

$$f(m + n) = (m + n)e$$
$$= me + ne$$
$$= f(m) + f(n) \qquad \text{and}$$
$$f(mn) = (mn)e$$
$$= (me)(ne)$$
$$= f(m)f(n). \ \blacksquare$$

Hence, we see that every ordered integral domain in which the set of positive elements is well-ordered is isomorphic to $\mathbf{Z}$. If D_1 and D_2 are both ordered integral domains in which the sets of positive elements are well-ordered, then D_1 and D_2 are both isomorphic to $\mathbf{Z}$. It follows that D_1 and D_2 are isomorphic, so all such domains are isomorphic to each other. This is often expressed by saying that *there is only one isomorphism class of ordered integral domains in which the set of positive elements is well-ordered.* This can also be expressed by saying that $\mathbf{Z}$ *is unique up to isomorphism.*

If D is an ordered integral domain, the order relation on D can be used to define an order relation on $Q(D)$, the field of fractions of D. Let $Q(D)^+ = \{\frac{a}{b} \mid ab \in D^+\}$ and define $<$ on $Q(D)$ by $\frac{a}{b} < \frac{c}{d}$ if and only if $\frac{c}{d} - \frac{a}{b} \in Q(D)^+$. Hence, $a < b$ in D if and only if $\frac{a}{e} < \frac{b}{e}$ in $Q(D)$. Therefore, when D is an ordered integral domain, $Q(D)$ can be made into an ordered field. If D is considered to be a subring of $Q(D)$ by identifying D with its image $D_e = \{\frac{a}{e} \mid a \in D\}$ in $Q(D)$, then the order on $Q(D)$ agrees with the order on D. In this setting, we say that **the order on D has been extended to an order on $Q(D)$.** The order relation on $\mathbf{Z}$ can be extended to an order relation on $\mathbf{Q}$ in exactly this manner so that $\mathbf{Q}$ is an ordered field. Since $\mathbf{Q}$ is an integral domain, $\mathbf{Q}^+$ cannot be well-ordered, for if it were, Theorem 4.2.6 would imply that $\mathbf{Q}$ and $\mathbf{Z}$ are isomorphic. This in turn would imply that $\mathbf{Z}$ is a field, and we know this is not the case.

Problem Set 4.2

1. Suppose that R is a commutative ring with identity. Prove that R is an integral domain if and only if $xy = xz$ implies $y = z$ for all $x, y, z \in R$ with $x \neq 0$.

2. Suppose that a ring R contains a set S of elements that are *not* zero-divisors. Prove that S is closed under multiplication but that S need not be closed under addition. Hint: Consider a subset of $\mathbf{Z}_6$.

3. Complete the proof of Theorem 4.2.2.

4. Suppose that D is an ordered integral domain. Prove that each of the following holds for any $x, y, z, w \in D$.
   (a) If $x < y$, then $-y < -x$.
   (b) If $0 < x < y$ and $0 < z < w$, then $xz < yw$.
   (c) If $x > 0$ and $xy < xz$, then $y < z$.
   (d) If $x > e$, then $x^2 > x$.
   (e) If $0 < x < y$ and x and y both have multiplicative inverses, then $y^{-1} < x^{-1}$.

5. If D is an ordered integral domain and $x \in D$, then the absolute value of x, denoted by $|x|$, is defined by
$$|x| = \begin{cases} x & \text{if } x \geq 0 \\ -x & \text{if } x < 0 \end{cases}.$$
   Prove each of the following for any $x, y \in D$.
   (a) $|x| = 0$ if and only if $x = 0$
   (b) $|x| \geq 0$
   (c) $|xy| = |x||y|$
   (d) $-|x| \leq x \leq |x|$
   (e) $|x + y| \leq |x| + |y|$

6. Suppose that D is an ordered integral domain. Show that if $x \in D$, then there is always a $y \in D$ such that $y > x$. That is, an ordered integral domain cannot contain a largest element. Show that a finite integral domain cannot contain a set of positive elements D^+. Conclude that $\mathbf{Z}_p$, where p is a prime, cannot be an ordered integral domain.

7. If D is an ordered integral domain, explain why the equation $x^2 + e = 0$ has no solution with x in D.

8. Let D^+ be the set of positive elements of an ordered integral domain D. Suppose that $f : D \to R$ is a ring isomorphism.
   (a) Prove that the ring R has $f(D^+)$ as a set of positive elements.
   (b) Is R an ordered integral domain?
   (c) Prove that if the set of positive elements of D is well-ordered, then the same is true of the set of positive elements of R.

9. Let $D = \left\{ \begin{pmatrix} x & 0 \\ 0 & x \end{pmatrix} \mid x \in \mathbf{Z} \right\}$. Show that D is an ordered integral domain in which the set of positive elements is well-ordered.

10. If D_1 and D_2 are the integral domains of Example 4.2.3 and Exercise 9, respectively, find a ring isomorphism $f : D_1 \to D_2$.

11. Suppose that R is a commutative ring, and let R^+ be a nonempty subset of R that has exactly the same properties as those specified for the set D^+ in Definition 4.2.1. Call a commutative ring R that has such a set an **ordered ring**. Show that if a commutative ring R has a zero-divisor, then R cannot be an ordered ring. Conclude that an ordered ring must be a domain.

12. In Example 4.2.3 it was shown that $D = \{2^n \mid n \in \mathbf{Z}\}$ is an ordered integral domain in which the set of positive elements is well-ordered. Hence, by Theorem 4.2.6, $\mathbf{Z}$ and D must be isomorphic rings. Find a ring isomorphism $f : \mathbf{Z} \to D$.

13. (a) Let $<$ be the order on $\mathbf{Q}$ that extends the order on $\mathbf{Z}$. If $x, y \in \mathbf{Q}$, with $x < y$, prove that there is an $a \in \mathbf{Q}$ such that $x < a < y$. Hint: Think average. Conclude that between any two distinct rational numbers, there is always another rational number.
     (b) If x and y are distinct rational numbers, prove that there are rational numbers $a_1, a_2, \ldots, a_n$ such that $x < a_1 < a_2 < \cdots < a_n < y$.
     (c) Prove that there can be no smallest positive rational number.

14. Let $<$ be the order on $\mathbf{Q}$ that extends the order on $\mathbf{Z}$.

     (a) If x and y are positive rational numbers such that $x < y$, show that $x^{-1} > y^{-1}$.
     (b) If x and y are negative rational numbers such that $x < y$, decide which of the following is true and then prove your conjecture:

     (i) $x^{-1} < y^{-1}$
     (ii) $y^{-1} < x^{-1}$.

4.3   THE FIELD OF REAL NUMBERS

Section Overview. We have seen that the integral domain $\mathbf{Z}$ is, up to isomorphism, the only integral domain with a set of positive elements that is well-ordered. We have also seen that the field of rational numbers $\mathbf{Q}$ is the field of fractions of $\mathbf{Z}$ and, as such, is unique up to isomorphism. These properties of $\mathbf{Z}$ and $\mathbf{Q}$ establish the first two important characterizations of our number system. In this section we show that the field $\mathbf{R}$ of real numbers can be developed from the field $\mathbf{Q}$ of rational numbers using Dedekind cuts. It is shown that $\mathbf{R}$ is complete in the sense that every nonempty set of real numbers that has an upper bound in $\mathbf{R}$ has a least upper bound in $\mathbf{R}$. It is also shown that there is only one isomorphism class of complete ordered fields and that this isomorphism class is determined by $\mathbf{R}$.

One reason for constructing the field of rational numbers from the ring of integers is so that equations of the form $ax + b = c$, where $a, b, c \in \mathbf{Z}$ with $a \neq 0$, will always have a solution. Such equations cannot always be solved for x in $\mathbf{Z}$, but a solution $a^{-1}(c - b)$ always exists in $\mathbf{Q}$. However, $\mathbf{Q}$ is not completely satisfactory as a field for solving equations. One defect of this field is that very simple equations such as $x^2 - 2 = 0$ cannot be solved for x in $\mathbf{Q}$. (Recall that $\sqrt{2}$ is not a rational number.) Hence, we find ourselves looking for a field that contains $\mathbf{Q}$ in which such equations have solutions. Another defect is that $\mathbf{Q}$ is incomplete in the sense that there are nonempty subsets of $\mathbf{Q}$ that are bounded above but that do not have a least upper bound in $\mathbf{Q}$. This is really a problem of continuity. If the rational numbers are plotted on a number line, there are points on the line that do not correspond to any rational number, and in this sense the rational line is incomplete. For example, there is no rational number that corresponds to the point on the line whose distance from the origin to the point in the positive direction is $\sqrt{2}$.

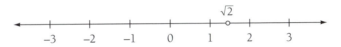

This underlying discontinuity of the rational number system causes difficulties at a very elementary level if one tries to develop mathematics using only the field of rational numbers. (For example, the Intermediate Value Theorem from calculus does not hold.) It is desirable, therefore, to seek a field that contains $\mathbf{Q}$ as a subfield that remedies these defects. In addition, we would like to construct this field using only elements of $\mathbf{Q}$. We will use hindsight and our knowledge of the real number system and of the irrational numbers to describe steps in the construction. However, only rational numbers can be used in the lemmas and theorems we prove if we are to be able to truly say that the field $\mathbf{R}$ of real numbers has been constructed from $\mathbf{Q}$.

4.3.1 Definition

Let S be a nonempty subset of an ordered field F. An element $b \in F$ such that $x \leq b$ for each $x \in S$ is said to be an **upper bound** for S. An element $b \in F$ such that $b \leq x$ for each $x \in S$ is said to be a **lower bound** for S. Neither a lower bound nor an upper bound need exist for a given set S. A nonempty subset of F that has at least one upper bound in F is said to be **bounded from above**. Similarly, a nonempty subset of F that has at least one lower bound in F is **bounded from below**.

The set $S = \left\{\frac{1}{2}, 1, \frac{3}{2}\right\}$ in $\mathbf{Q}$ has an infinite number of upper bounds and an infinite number of lower bounds. In fact, any number $b \in \mathbf{Q}$ such that $\frac{3}{2} \leq b$ is an upper bound for S, and any $b \in \mathbf{Q}$ such that $b \leq \frac{1}{2}$ is a lower bound for S. The set $S = \left\{\frac{1}{2}, 1, \frac{3}{2}, 2, \frac{5}{2}, \ldots\right\}$, however, has lower bounds but no upper bounds in $\mathbf{Q}$. Consequently, we see that a nonempty subset in an ordered field need not be bounded from above. Similar examples show that a nonempty set in an ordered field need not be bounded from below.

4.3.2   Definition

Let S be a nonempty subset of an ordered field F that is bounded from above. If $b \in F$ is an upper bound for S with the property that there is no upper bound for S that is less than b, then b is said to be a **least upper bound** of S. If S is bounded from below and $b \in F$ is a lower bound for S with the property that there is no lower bound for S that is greater than b, then b is said to be a **greatest lower bound** of S. The notation lub(S) and glb(S) is used for the least upper bound and the greatest lower bound of S, respectively, when such exist. An ordered field F in which every nonempty subset that is bounded from above has a least upper bound in F is said to be a **complete ordered field**. A complete ordered field is sometimes said to have the **least-upper-bound property**.

It is not difficult to argue that the greatest lower bound (least upper bound) of a nonempty subset S of an ordered field is unique whenever it exists. The field $\mathbf{Q}$ is an ordered field that is *not* complete. The set $S = \{x \in \mathbf{Q} \mid x^2 < 2\}$ is bounded from above since, for example, 3 is an upper bound for S. However, S does not have a least upper bound in $\mathbf{Q}$. The least upper bound of S is $\sqrt{2}$, but $\sqrt{2}$ is not a rational number. The irrational number $\sqrt{2}$ is an element of the yet-to-be described field $\mathbf{R}$ of real numbers. To extend $\mathbf{Q}$ to a complete ordered field, we must somehow "add" elements to $\mathbf{Q}$ that will serve as least upper bounds for the nonempty subsets of $\mathbf{Q}$ that are bounded from above but do not have a least upper bound in $\mathbf{Q}$. This "new" set of numbers must also be a field that, at the very least, contains a subfield isomorphic to $\mathbf{Q}$, and the order on $\mathbf{Q}$ must be extendable to this new field.

There are two methods that mathematicians use to construct the field of real numbers from the rational numbers: one method is to use equivalence classes of Cauchy sequences of rational numbers, and the other is to use **Dedekind cuts** of rational numbers. Although the method using Cauchy sequences is more elegant, the method using Dedekind cuts is more geometric and consequently more visual. For this reason, we use the method of Dedekind cuts. It was **Richard Dedekind** (1831–1916) who first realized that each real number r divides the set of rational numbers into distinct sets: those rational numbers x such that $x < r$ and those such that $x > r$. This can be viewed as cutting the line at the real number r. Make it a point to remember that we are considering only the rational numbers that are less than r and only the rational numbers that are greater than r. Irrational numbers less than r are not considered, nor are irrational numbers greater than r.

$$x \in \mathbf{Q} \ \& \ x < r \qquad\qquad r \qquad\qquad x \in \mathbf{Q} \ \& \ x > r$$

Dedekind went on to use these sets of rational numbers to develop the field of real numbers **R**. (Information on the life of Dedekind can be found in the Historical Notes at the end of Section 3.1.)

The idea behind Dedekind cuts is to specify each real number r by using an infinite set of rational numbers. The set of rational numbers x such that $x < r$ is defined to be a Dedekind cut.

4.3.3   **Definition**

A **Dedekind cut of rational numbers**, or simply a **cut**, is a nonempty proper subset r of $\mathbf{Q}$ that satisfies the following two conditions:

1.   If $x \in r$ and $y \in \mathbf{Q}$ is such that $y < x$, then $y \in r$. A subset of $\mathbf{Q}$ that satisfies this property is said to be **closed downward**.
2.   If $x \in r$, then there is always a $y \in r$ such that $y > x$; that is, r has no largest element.

The set of rational numbers $r = \{x \mid x \in \mathbf{Q} \text{ and } x < 1\}$ is a Dedekind cut. This follows because if $x \in r$ and $y \in \mathbf{Q}$ is such that $y < x$, then $y < 1$, so r is closed downward. If $x \in r$, then $x < \frac{x+1}{2} < 1$ shows that r has no largest element. If r_1 and r_2 are cuts, we write $r_1 < r_2$ if $r_1 \subset r_2$ and $r_1 = r_2$ if r_1 and r_2 are equal sets. The notations $r_1 \leq r_2$ and $r_1 \geq r_2$ have obvious meanings. For the time being, $\mathbf{Q}^c$ denotes the set of all Dedekind cuts of rational numbers. Care must be taken when describing a cut at a real number. For example, suppose that the line is cut at $\sqrt{2}$. Then we cannot describe the cut of all rational numbers that are less than $\sqrt{2}$ as the set $\{x \in \mathbf{Q} \mid x < \sqrt{2}\}$. Why? Because we have used the irrational number $\sqrt{2}$, which is an element of the field we seek to develop, to define the cut. Cuts must be described using only rational numbers. The cut at $\sqrt{2}$ can indeed be described by using only rational numbers since the sets $\{x \in \mathbf{Q} \mid x < \sqrt{2}\}$ and $\{x \in \mathbf{Q} \mid x^2 < 2 \text{ or } x < 0\}$ are equal. This latter set is a correct description of the cut at $\sqrt{2}$. However, it isn't obvious that the set described by $\{x \in \mathbf{Q} \mid x^2 < 2 \text{ or } x < 0\}$ is a cut. For example, we can't show that $\{x \in \mathbf{Q} \mid x^2 < 2 \text{ or } x < 0\}$ has no largest element in exactly the same way as we did for $\{x \mid x \in \mathbf{Q} \text{ and } x < 1\}$. Since $\sqrt{2}$ is not available, we cannot use $x < \frac{x + \sqrt{2}}{2} < \sqrt{2}$. You will be asked as an exercise to show that $\{x \in \mathbf{Q} \mid x^2 < 2 \text{ or } x < 0\}$ is indeed a cut. A final point is that since a cut must be a nonempty proper subset of $\mathbf{Q}$, $\mathbf{Q} \notin \mathbf{Q}^c$.

4.3.4   Lemma

For any cuts r_1 and r_2, exactly one of the three conditions $r_1 < r_2$, $r_1 = r_2$, $r_2 < r_1$ holds. That is, a trichotomy property holds for elements of $\mathbf{Q}^c$.

Proof. Clearly, at most one of the three conditions can hold. We need to show that when two of the three conditions fail, the other condition must hold.

Case 1. Suppose $r_1 \not< r_2$ and $r_1 \neq r_2$, and let's show that $r_2 < r_1$. Since $r_1 \not< r_2$ and $r_1 \neq r_2$, $r_1 \setminus r_2 \neq \varnothing$. Let $x \in r_1 \setminus r_2$. If y is an arbitrary element of r_2, then $x < y$, $x = y$, or $y < x$ since the trichotomy property holds for elements of $\mathbf{Q}$. Clearly, $x \neq y$ because if $x = y$, then $x \in r_2$, which is a contradiction of the choice of x. If $x < y$, then $x \in r_2$ because r_2 is closed downward. But this again contradicts the choice of x, so it must be the case that $y < x$. Hence, $y \in r_1$ since r_1 is closed downward. Therefore, $y \in r_2$ implies that $y \in r_1$, so $r_2 \subset r_1$. Hence, $r_2 < r_1$.

Case 2. If $r_2 \not< r_1$ and $r_1 \neq r_2$, then the proof that $r_1 < r_2$ is analogous to the proof for Case 1; simply interchange the subscripts.

Case 3. Finally, if $r_1 \not< r_2$ and $r_2 \not< r_1$, then either $r_1 = r_2$ or $r_1 \neq r_2$. Suppose $r_1 \neq r_2$. Then we have $r_1 \not< r_2$ and $r_1 \neq r_2$, so by Case 1, $r_2 < r_1$, which contradicts the assumption that $r_2 \not< r_1$. Hence, if $r_1 \not< r_2$ and $r_2 \not< r_1$, we must have $r_1 = r_2$. ∎

Any cut r such that $\text{lub}(r)$ exists in $\mathbf{Q}$ is said to be a **rational cut**. If r is a rational cut and $x \in \mathbf{Q}$ such that $\text{lub}(r) = x$, then $r = \{y \in \mathbf{Q} \mid y < x\}$. Such a cut is denoted by $\overline{x}$. In this notation, $\text{lub}(\overline{x}) = x$, $x \notin \overline{x}$, and $\overline{x}$ has no largest element because if $y \in \overline{x}$, then $y < \frac{y + x}{2} < x$. The notation $\overline{\mathbf{Q}} = \{\overline{x} \mid x \in \mathbf{Q}\}$ now denotes the set of all rational cuts. If each $x \in \mathbf{Q}$ is identified with $\bar{x} \in \overline{\mathbf{Q}}$, then $\mathbf{Q}$ can be considered to be a subset of $\mathbf{Q}^c$. For the moment, however, we continue to make a distinction between elements of $\mathbf{Q}$ and those of $\overline{\mathbf{Q}}$. If r is a cut in $\mathbf{Q}^c$ such that $\mathrm{lub}(r)$ does not exist in $\mathbf{Q}$, then r is said to be an **irrational cut**.

4.3.5   Examples

1. The set $r = \left\{ x \in \mathbf{Q} \, | \, x < \frac{15}{8} \right\}$ is the rational cut $\overline{\frac{15}{8}}$. The least upper bound

    of r is $\frac{15}{8}$ and $\frac{15}{8} \notin r$.

2. The set $r = \left\{ x \in \mathbf{Q} \, | \, x \leq \frac{3}{4} \right\}$ is not a cut since r has $\frac{3}{4}$ as a least upper

    bound and $\frac{3}{4} \in r$. It cannot be the case that a cut contains its least

    upper bound if the cut has a least upper bound in $\mathbf{Q}$.
3. The set $r = (x \in \mathbf{Q} \, | \, 0 < x < 3\}$ is not a cut since it is not closed downward. For example, $-1 < 3$, but $-1 \notin r$.
4. The set $r = \{ x \in \mathbf{Q} \, | \, x^2 < 2 \text{ or } x < 0 \}$ is an irrational cut. We know from our knowledge of the real number system that the least upper bound of r is $\sqrt{2}$ and that $\sqrt{2}$ does not exist in $\mathbf{Q}$.

It will now be shown that $\mathbf{Q}^c$ can be made into an ordered field if addition and multiplication are defined on $\mathbf{Q}^c$ in an appropriate manner. The first step is to define addition on $\mathbf{Q}^c$ and then to show that $\mathbf{Q}^c$ is an additive abelian group.

4.3.6   Definition

If $r_1, r_2 \in \mathbf{Q}^c$, then the **sum** of r_1 and r_2 is defined by $r_1 + r_2 = \{x_1 + x_2 \, | \, x_1 \in r_1 \text{ and } x_2 \in r_2\}$.

4.3.7   Example

Consider the cuts $r_1 = \{ x \in \mathbf{Q} \, | \, x^2 < 2 \text{ or } x < 0 \}$ and $r_2 = \{ x \in \mathbf{Q} \, | \, x < 4 \}$. According to Definition 4.3.6, $r_1 + r_2$ is the result of forming all possible sums of an element from r_1 with an element from r_2. If $x \in r_1$ is such that $x < 2$ and $y \in r_2$, then $x + y < 2 + 4 = 6$, so it follows that $r_1 + r_2 = \{x + y \, | \, x \in r_1, y \in r_2 \text{ and } x^2 + y < 6\}$.

4.3.8   Lemma

If r_1, r_2 are cuts, then $r_1 + r_2$ is a cut. That is, the operation of addition of cuts is a binary operation on $\mathbf{Q}^c$.

Proof. We first show that the addition of two cuts is a cut. If $x_1 \in r_1$ and $x_2 \in r_2$, then $x_1 + x_2 \in r_1 + r_2$, so $r_1 + r_2 \neq \emptyset$. Now $r_1 \neq \mathbf{Q}$ and $r_2 \neq \mathbf{Q}$, and so there exist y_1 and y_2 in $\mathbf{Q}$ such that $y_1 \notin r_1$ and $y_2 \notin r_2$. Hence, $y_1 > x_1$ and $y_2 > x_2$ for all $x_1 \in r_1$ and all $x_2 \in r_2$. This leads to $y_1 + y_2 > x_1 + x_2$ for all $x_1 + x_2$ in $r_1 + r_2$. Therefore, $y_1 + y_2 \notin r_1 + r_2$, so $r_1 + r_2$ is a proper subset of $\mathbf{Q}$. We now need to show that $r_1 + r_2$ is closed downward. Suppose that $x_1 + x_2 \in r_1 + r_2$, and let $y \in \mathbf{Q}$ be such that $y < x_1 + x_2$. If $x' \in \mathbf{Q}$ is such that $y = x_1 + x'$, then $x_1 + x' = y < x_1 + x_2$, so it follows that $x' < x_2$. Hence, $x' \in r_2$ since r_2 is closed downward. Therefore, $y \in r_1 + r_2$. We also need to show that $r_1 + r_2$ has no largest element. If $x_1 + x_2 \in r_1 + r_2$, then $x_1 \in r_1$ and $x_2 \in r_2$. Now r_1 has no largest element, so there is an $x'' \in r_1$ such that $x'' > x_1$. But then $x'' + x_2 > x_1 + x_2$, so $r_1 + r_2$ has no largest element since $x'' + x_2 \in r_1 + r_2$. The sum of two cuts in $\mathbf{Q}^c$ is therefore a cut. Since it is easy to argue that addition of cuts is well-defined, we have shown that addition of cuts is a well-defined binary operation on $\mathbf{Q}^c$. ■

4.3.9   Lemma

The rational cut $\overline{0}$ is an additive identity for $\mathbf{Q}^c$.

Proof. Let r be a cut in $\mathbf{Q}^c$, and suppose that $x_1 + x_2 \in r + \overline{0}$ with $x_1 \in r$ and $x_2 \in \overline{0}$. Since, x_2 is in $\overline{0}$, $x_2 < 0$, so $x_1 + x_2 < x_1$. Hence, $x_1 + x_2 \in r$ since r is closed downward. Therefore, $r + \overline{0} \subseteq r$. Conversely, let $x \in r$. Since r has no largest element, let $y \in r$ be such that $x < y$. If $z = x - y$, then $z < 0$ and thus $z \in \overline{0}$. Hence, $x = y + z \in r + \overline{0}$, so $r \subseteq r + \overline{0}$. It follows that $r + \overline{0} = r$. Similarly $\overline{0} + r = r$, so $\overline{0}$ is an additive identity for $\mathbf{Q}^c$. ■

If r is a cut, then we set $-r = \{x \in \mathbf{Q} \mid -x$ is an upper bound for r but not the least upper bound of r in case r is a rational cut$\}$. We refer to $-r$ as the **negative of the cut** r. It is not difficult to show that if $r > \overline{0}$, then $-r < \overline{0}$, and if $r < \overline{0}$, then $-r > \overline{0}$. The following lemma shows that $-r$ is actually the additive inverse of r. Since we have demonstrated the techniques involved in proving details about cuts in the preceding two lemmas, the proof is omitted.

4.3.10   Lemma

If $r \in \mathbf{Q}^c$, then $-r$ is a cut, and $r + (-r) = -r + r = \overline{0}$.

It can now be shown that $\mathbf{Q}^c$ is an additive abelian group.

4.3.11   Theorem

$\mathbf{Q}^c$ together with the binary operation of addition of cuts is an additive abelian group.

Proof. Lemma 4.3.8 shows that addition of cuts is a binary operation on $\mathbf{Q}^c$. Since addition of rational numbers is associative and commutative, addition of cuts is also associative and commutative. Lemmas 4.3.9 and 4.3.10 show that $\overline{0}$ is the additive identity of $\mathbf{Q}^c$ and that each element of $\mathbf{Q}^c$ has an additive inverse. Hence, $\mathbf{Q}^c$ is an additive abelian group under the operation of addition of cuts. ∎

The next step in showing that $\mathbf{Q}^c$ is a field is to define multiplication of cuts. The definition of this operation is not quite as straightforward as addition of cuts. Recall that the first thing we did after defining a Dedekind cut was to define the relation $\leq$ on $\mathbf{Q}^c$. It was then shown that a trichotomy property holds for cuts. Hence, if r is a cut, then exactly one of $r > \overline{0}, r = \overline{0}$, or $r < \overline{0}$ holds.

4.3.12   Definition

If $r_1, r_2 \in \mathbf{Q}^c$, then the product of r_1 and r_2 is defined as follows:

1. If $r_1 = \overline{0}$ or $r_2 = \overline{0}$, then $r_1 r_2 = \overline{0}$.
2. If $r_1 > \overline{0}$ and $r_2 > \overline{0}$, then

$$r_1 r_2 = \{xy \mid x \in r_1, x > 0, y \in r_2, y > 0\} \cup \{x \in \mathbf{Q} \mid x \leq 0\}.$$

3. If $r_1 < \overline{0}$ and $r_2 < \overline{0}$, then $r_1 r_2 = (-r_1)(-r_2)$, where $(-r_1)(-r_2)$ is as defined in item 2.
4. If $r_1 < \overline{0}$ and $r_2 > \overline{0}$, then $r_1 r_2 = -((-r_1)r_2)$, where $(-r_1)r_2$ is as defined in item 2.
5. If $r_1 > \overline{0}$ and $r_2 < \overline{0}$, then $r_1 r_2 = -(r_1(-r_2))$, where $r_1(-r_2)$ is as defined in item 2.

4.3.13   Example

For the sake of this example, if a and b are real numbers with $a < b$, $\mathbf{Q}(a, b)$ denotes the set of all rational numbers in the interval (a, b). Consider the rational cuts $\overline{\frac{1}{2}}$ and $\overline{\frac{1}{4}}$. Both are cuts that are greater than $\overline{0}$. According to Definition 4.3.12, to compute the product of these two cuts, we form the set of all possible products of elements from the sets $\mathbf{Q}\left(0, \frac{1}{2}\right)$ and $\mathbf{Q}\left(0, \frac{1}{4}\right)$ and then take the union of this set with the set $\{x \in \mathbf{Q} \mid x \leq 0\}$. If we form the set of all possible products of elements of $\mathbf{Q}\left(0, \frac{1}{2}\right)$ with elements of $\mathbf{Q}\left(0, \frac{1}{4}\right)$, the result is the set $\mathbf{Q}\left(0, \frac{1}{8}\right)$. Hence, the product of the cuts $\frac{\bar{1}}{2}$ and $\frac{\bar{1}}{4}$ is $\frac{\bar{1}}{2} \cdot \frac{\bar{1}}{4} = \mathbf{Q}\left(0, \frac{1}{8}\right) \cup \{x \in \mathbf{Q} \mid x \leq 0\} = \frac{\bar{1}}{8}$, a result that you probably anticipated before these computations were performed.

The proof of the following theorem is long and quite technical. One who encounters this construction of the real number system for the first time can easily become lost in a sea of technical details in the proof. For this reason, the theorem is offered without proof. This does not mean that the theorem is not important. In fact, it is quite important. Theorems 4.3.14 and 4.3.15 are the very reason we undertook this construction.

4.3.14   Theorem

Consider $\mathbf{Q}^c$ together with the operations of addition and multiplication of cuts. Then $\mathbf{Q}^c$ is a field under these operations, and the following hold:

1. The rational cut $\bar{1} = \{x \in \mathbf{Q} \mid x < 1\}$ is the multiplicative identity for $\mathbf{Q}^c$.

2. The multiplicative inverse of a nonzero cut in $\mathbf{Q}^c$ is given by: if $r \in \mathbf{Q}^c$ and $r > \bar{0}$, then $r^{-1} = \{x \in \mathbf{Q} \mid \frac{1}{x}$ is an upper bound for r but not the least upper bound of r in case r is a rational cut$\} \cup \{x \in \mathbf{Q} \mid x \leq 0\}$. If $r < \bar{0}$, then $-r > \bar{0}$, so $-(-r)^{-1}$ is the multiplicative inverse of r.

3. The set $\overline{\mathbf{Q}} = \{\bar{x} \mid x \in \mathbf{Q}\}$ of all rational cuts is a subfield of $\mathbf{Q}^c$ that is isomorphic to $\mathbf{Q}$. If each element $x \in \mathbf{Q}$ is identified with the corresponding element $\bar{x} \in \mathbf{Q}^c$, $\mathbf{Q}$ can be considered to be a subfield of $\mathbf{Q}^c$.

4. The set $\mathbf{Q}^{c+} = \{r \in \mathbf{Q}^c \mid r > \bar{0}\}$ is a set of positive elements for $\mathbf{Q}^c$.

5. $\mathbf{Q}^c$ is an ordered field, and the order on $\mathbf{Q}^c$ agrees with the order on $\mathbf{Q}$.

From Theorem 4.3.14, we have that $\mathbf{Q}^c$ is an ordered field that contains a subfield that is isomorphic to $\mathbf{Q}$. We now identify each element x of $\mathbf{Q}$ with the rational cut $\overline{x}$ in $\mathbf{Q}$ and consider $\mathbf{Q}$ to be a subfield of $\mathbf{Q}^c$. Under this identification, the order on $\mathbf{Q}^c$ agrees with the order on $\mathbf{Q}$. The following theorem is important. It shows that every nonempty subset of $\mathbf{Q}^c$ that has an upper bound in $\mathbf{Q}^c$ has a least upper bound that is in $\mathbf{Q}^c$.

4.3.15   Theorem

The ordered field $\mathbf{Q}^c$ is complete.

Proof. Let S be a nonempty set of cuts in $\mathbf{Q}^c$, and suppose that S is bounded from above. Then there is a cut $b \in \mathbf{Q}^c$ such that $r \leq b$ for each cut $r \in S$. If $b^* = \cup_{r \in S} r$, then clearly $r \leq b^*$ for each $r \in S$ since each $r \in S$ is a subset of b^*. We claim that b^* is a cut in $\mathbf{Q}^c$ and that b^* is the least upper bound of S. First, let's show that b^* is a cut in $\mathbf{Q}^c$; Since each $r \in S$ is nonempty, b^* is a nonempty subset of $\mathbf{Q}^c$. The upper bound b is a nonempty proper subset of $\mathbf{Q}$, so there is an element $x \in \mathbf{Q}$ such that $x \notin b$ and $y < x$ for all $y \in b$. From this we see that $b < \overline{x}$, so $r < \overline{x}$ for all $r \in S$. Therefore, $y < x$ for all $y \in b^*$, which implies that $x \notin b^*$. Thus, $b^* \neq \mathbf{Q}$, and so b^* is a proper subset of $\mathbf{Q}$. The set b^* is closed downward, for if $y \in b^*$ and $y' < y$, then $y \in r$ for some $r \in S$. Hence, $y' \in r$ since r is closed downward. Thus, $y' \in \cup_{r \in S} r = b^*$, so b^* is closed downward. Finally, if $y \in b^*$ and $y \in r$, where $r \in S$, then since r has no largest element, there is a $y' \in r$ such that $y < y'$. But then $y' \in b^*$, so b^* has no largest element. Hence, b^* is indeed a cut in $\mathbf{Q}^c$. Finally, let's show that $b^* = \mathrm{lub}(S)$. Suppose that b is any upper bound for S. Then $r \leq b$ for each $r \in S$, and so $r \subseteq b$ for each $r \in S$. Hence, $b^* = \cup_{r \in S} r \subseteq b$, so $b^* \leq b$. Hence, $b^* = \mathrm{lub}(S)$. ∎

We now have shown that $\mathbf{Q}^c$ is a complete ordered field that contains $\mathbf{Q}$ as a subfield. What we show next is that $\mathbf{Q}^c$ is unique up to isomorphism. For this we need the following theorem.

4.3.16   Theorem

Any ordered field F has characteristic 0 and consequently contains a copy of $\mathbf{Q}$.

Proof. Let $p \neq 0$ be the characteristic of F. Since F is an integral domain, it follows from Lemma 4.1.10 that p is prime. If e is the identity of F, then we know from part 6 of Theorem 4.2.2 that $pe > 0$. But $pe = 0$, so $0 > 0$, which is an obvious contradiction. Therefore, the characteristic of F must be 0. The fact that F contains a copy of $\mathbf{Q}$ is now a consequence of Theorem 4.1.13. ∎

4.3.17   Theorem (Archimedean Order Property)

If F is a complete ordered field and $r_1, r_2 \in F$ are such that $r_1 > 0$ and $r_2 > 0$, then a positive integer n exists such that $nr_1 > r_2$.

Proof. If $mr_1 = r_2$ for some positive integer m, then $(m + 1)r_1 > mr_1 = r_2$, so let $n = m + 1$. If such an integer m does not exist, then either $nr_1 > r_2$ or $nr_1 < r_2$ for every positive integer n. We show that it is not possible for $nr_1 < r_2$ for every positive integer n. Assume that $nr_1 < r_2$ for every positive integer n. Then r_2 is an upper bound for the set $S = \{nr_1 \mid n \in \mathbf{N}\}$. Since F is a complete ordered field and S is bounded from above, S has a least upper bound, say, r^*. Now $r^* - r_1 < r^*$, so $r^* - r_1$ is not an upper bound for the set S. Thus, there must be an element $mr_1 \in S$ such that $r^* - r_1 < mr_1$. Hence, we see that $r^* < (m + 1)r_1 \in S$, which clearly contradicts the fact that r^* is an upper bound for S. Hence, the assumption that $nr_1 < r_2$ for every positive integer n leads to a contradiction, so there must be a positive integer n such that $nr_1 > r_2$. ∎

An ordered field F is said to be **Archimedean ordered** if it satisfies the Archimedean Order Property. We see from Theorem 4.3.17 that every complete ordered field is Archimedean ordered. Since an ordered field contains a copy of $\mathbf{Q}$, to simplify notation, we now consider $\mathbf{Q}$ to be a subfield of every complete ordered field. Recall that the identity of any subfield of a field must coincide with the identity of the field (Lemma 4.1.11). Hence, if F is an ordered field with subfield $\mathbf{Q}$, 1 now denotes the identity of F.

4.3.18   Theorem

If F is a complete ordered field and $r_1, r_2 \in F$ are such that $r_1 < r_2$, then a rational number x exists such that $r_1 < x < r_2$.

Proof.   There are several cases to be considered:

Case 1. $r_1 > 0$: Since $r_2 - r_1 > 0$, Theorem 4.3.17 indicates that there is a positive integer n such that $n(r_2 - r_1) > 1$. Let n be such an integer, and apply Theorem 4.3.17 to nr_1 and 1. Let m be the smallest positive integer such that $m = m1 > nr_1$. Then $r_1 < \frac{m}{n}$. If we can show that $\frac{m}{n} < r_2$, that is, if we can show that $m < nr_2$, the proof will be complete for this case. Suppose $m \geq nr_2$. Since $n(r_2 - r_1) > 1$, $nr_2 > nr_1 + 1$, so $m > nr_1 + 1 > 1$ since $m \geq nr_2$. Hence, $m > 1$

and thus $m - 1 \in \mathbf{N}$ is such that $(m - 1) > nr_1$. But this contradicts our choice of m as the smallest positive integer such that $m \geq nr_1$. Thus, the assumption that $m \geq nr_2$ has led to a contradiction. Hence, $\frac{m}{n} < r_2$, so we have $r_1 < \frac{m}{n} < r_2$.

Case 2. $r_1 = 0$: By Theorem 4.3.17, there is a positive integer n such that $nr_2 > 1$, so $0 < \frac{1}{n} < r_2$.

Case 3. $r_1 < 0$ and $r_2 > 0$: In this case, $r_1 < 0 < r_2$.

Case 4. $r_1 < r_2 < 0$: If $r_1 < r_2 < 0$, then $-r_1 > -r_2 > 0$, and thus,by the first case, there is a rational number x such that $-r_1 > x > -r_2$. But then $r_1 < -x < r_2$, and this completes the proof. ■

4.3.19   Corollary

If F is a complete ordered field, then every element of F is the least upper bound of a nonempty set of rational numbers.

Proof. If $r \in F$, let $c_r = \{x \in \mathbf{Q} \mid x < r\}$. Since $r - 1$, $r \in F$, Theorem 4.3.18 indicates that there is a rational number y such that $r - 1 < y < r$, so $y \in c_r$. Hence, $c_r \neq \emptyset$. Now let's show that $r = \mathrm{lub}(c_r)$. If $r \neq \mathrm{lub}(c_r)$, there is an upper bound $b \in F$ of c_r such that $b < r$. By Theorem 4.3.18, there is an $x' \in \mathbf{Q}$ such that $b < x' < r$. But $x' < r$ implies that $x' \in c_r$, so b is not an upper bound for c_r, a contradiction. Therefore, it must be the case that $r = \mathrm{lub}(c_r)$. ■

The set c_r constructed in the proof of Corollary 4.3.19 is just a Dedekind cut of rational numbers in the field F. Hence, every element of a complete ordered field is the least upper bound of a Dedekind cut of rational numbers. This fact is central to the proof of the following theorem.

4.3.20   Theorem

Every complete ordered field is isomorphic to $\mathbf{Q}^c$.

Proof (Sketched). Let F be a complete ordered field. Then $\mathbf{Q}$ is a subfield of F, and by Corollary 4.3.19, every $r \in F$ is the least upper bound of a cut c_r of rational numbers. Since $\mathbf{Q}$ is also a subfield of the complete ordered field $\mathbf{Q}^c$, there is also an element r' of $\mathbf{Q}^c$ such that $r' = \mathrm{lub}(c_r)$. If we let $\phi: \mathbf{Q}^c \to F$ be defined by $f(r') = r$, then ϕ is a well-defined bijective mapping that is a ring homomorphism. Consequently, F and $\mathbf{Q}^c$ are isomorphic fields. ■

Because of Theorem 4.3.20, there is only one isomorphism class of complete ordered fields, and this isomorphism class is determined by $\mathbf{Q}^c$. We now call each element of $\mathbf{Q}^c$ a **real number** and set $\mathbf{Q}^c = \mathbf{R}$. Each rational number x is identified with the rational cut $\overline{x}$, and an irrational cut in $\mathbf{R}$ is referred to as an **irrational number**. The irrational number $\sqrt{2} = \{x \in \mathbf{Q} \mid x^2 < 2 \text{ or } x < 0\}$ contains the set of all rational numbers $\{1.4, 1.41, 1.414, 1.4142, 1.41421, \dots\}$. These rational numbers are often the ones used for approximations to $\sqrt{2}$.

This completes our development of the real number system using Dedekind cuts. One should remember that there is only one isomorphism class of complete ordered fields and that this class is determined by $\mathbf{R}$. Hence, *as a complete ordered field, $\mathbf{R}$ is unique up to isomorphism.*

As you know, every real number has a decimal expansion. There are (1) terminating decimals, (2) nonterminating but repeating decimals, and (3) nonterminating, nonrepeating decimals. It is interesting to note that decimals of the first two types are representations of rational numbers, while decimals of the third type are representations of irrational numbers. To see this, consider the following examples:

4.3.21 | **Examples**

1. The number 34.243 can easily be expressed as a rational number since

$$34.243 = \frac{34{,}243}{1000}$$

2. Consider the nonterminating but repeating decimal $x = 25.424242\dots$.
    Since this decimal has a repeating block with two places, multiply both sides of the equation by 100 to move the decimal one repeating block to the right. Now subtract $x = 25.424242\dots$ from this result.

$$100x = 2542.424242\dots$$
$$x = 25.424242\dots$$
$$99x = 2517$$

    Solving for x produces $x = \frac{2517}{99} = 25.424242\dots = 25.\overline{42}$. The bar over the 42 in $25.\overline{42}$ indicates that the number has the repeating block 42 after the decimal.

3. The nonrepeating, nonterminating decimals are decimal representations of irrational numbers. For example,

$$\sqrt{2} = 1.41421356237309504880168872420969807856967187537696\dots$$

and
$$\pi = 3.14159265358979323846264338327950288419716939937511\ldots$$
are both irrational numbers.

Finally, we state without proof the following important theorem regarding real numbers. One important result of the theorem is that equations of the form $x^n - r = 0$ always have a solution in **R** for every positive integer n and each positive real number r.

4.3.22   **Theorem**

If r is a positive real number and n is any positive integer, then there is exactly one positive real number a such that $a^n = r$.

The positive real number a of Theorem 4.3.22, written as $\sqrt[n]{r}$ or as $r^{1/n}$, is the **principal nth root** of r. For example, 4 has 2 and -2 as square roots, but 2 is the principal square root of 4. Theorem 4.3.18 shows that between any two real numbers there is a rational number. Thus, one can find a rational number between any two irrational numbers. It is also the case that there is an irrational number between any two rational numbers. (See Exercise 7.) Hence, we can conclude that the rational and the irrational numbers are "pretty well-shuffled" in the real number system.

Before concluding our remarks concerning the real number system, we would be remiss if we did not point out some of the reasons for the fundamental importance of this system in mathematics. As we indicated earlier, the completion of the field of rational numbers really deals with an underlying concept of continuity. When the rational numbers are plotted on a line, there are "holes" (which correspond to the irrational numbers) that are filled when the field of real numbers is developed. Since the field of real numbers is complete, this system has the least-upper-bound property. The least-upper-bound property allows one to prove that any closed and bounded set of real numbers that is contained in a union of a family of open sets is contained in a union of a finite subfamily of these open sets. This result, known as the Heine-Borel Theorem, states, in the language of topology, that any closed and bounded set in the real number system is compact. Compactness of closed and bounded intervals of real numbers is sufficient to show that any function that is continuous on such an interval is uniformly continuous there, and this in turn sets up machinery that is adequate to develop a theory of integration of continuous functions.

$$\begin{Bmatrix} \text{Completeness} \\ \text{of the Real} \\ \text{Number System} \end{Bmatrix} \Rightarrow \begin{Bmatrix} \text{Least Upper} \\ \text{Bound Property} \end{Bmatrix} \Rightarrow \begin{Bmatrix} \text{Heine-Borel} \\ \text{Theorem} \end{Bmatrix} \Rightarrow$$

$$\begin{Bmatrix} \text{Uniform} \\ \text{Continuity} \end{Bmatrix} \Rightarrow \begin{Bmatrix} \text{Integration of} \\ \text{Continuous Functions} \end{Bmatrix}$$

Problem Set 4.3

1. Find the decimal representation of each of the following rational numbers. For nonterminating decimals, find the repeating block.

   (a) $\frac{35}{8}$ (c) $\frac{57}{7}$

   (b) $\frac{23}{6}$ (d) $\frac{12}{123}$

2. Convert each of the following repeating, nonterminating decimals to a rational number. The bar over the set of integers designates the repeating block.

   (a) $45.\overline{341}$ (c) $0.002\overline{34}$
   (b) $34.5\overline{13}$ (d) $0.0000\overline{47}$

3. (a) Prove that $\sqrt{3}$ is an irrational number.

   (b) Prove that the set $r = \{x \in \mathbf{Q} \mid x^2 < 2 \text{ or } x < 0\}$ is a cut. Hint: Let

   $x > 0$ be a rational number such that $x^2 < 2$. If $y = \frac{x(x^2 + 6)}{3x^2 + 2}$, show

   that $x < y$ by showing that $y - x > 0$. Show that $y \in r$ by showing that $y^2 - 2 < 0$.

4. Prove or disprove each of the following statements. Give a counterexample for each statement that you decide is false.

   (a) If x and y are irrational numbers, then $x + y$ must be an irrational number.
   (b) If x and y are irrational numbers, then xy must be an irrational number.
   (c) If x is a rational number and y is an irrational number, then $x + y$ is an irrational number.
   (d) If x is a nonzero rational number and y is an irrational number, then xy must be an irrational number.
   (e) If x is a nonzero irrational number, then x^{-1} is an irrational number.

5. Prove that a least upper (greatest lower) bound of a nonempty subset of an ordered field is unique whenever it exists. For this reason we can refer to a least upper (lower) bound of S as *the* least upper (greatest lower) bound of S.

6. Prove that if an ordered field F has the least-upper-bound property, then every nonempty subset of F that has a lower bound has a greatest lower bound.

7. If $x, y \in \mathbf{R}$ and $x < y$, prove that $x < x + \dfrac{y-x}{\sqrt{2}} < y$. Does this show that between any two distinct real numbers (rational numbers) one can find an irrational number?

8. Show that 37.43 and $37.42\overline{9}$ represent the same rational number.

9. Let $r \in \mathbf{Q}^c$, and set $-r = \{x \in \mathbf{Q} \mid -x$ is an upper bound for r but not the least upper bound of r in case r is a rational cut$\}$. Show that $-r$ is a cut in $\mathbf{Q}^c$.

10. Show that addition of real numbers is commutative and associative by showing that if r_1, r_2, and r_3 are Dedekind cuts of rational numbers, then

    (a) $r_1 + r_2 = r_2 + r_1$.
    (b) $(r_1 + r_2) + r_3 = r_1 + (r_2 + r_3)$.

11. Prove that the rational cut $\overline{1}$ is such that $\overline{1}r = r\overline{1} = r$ for each cut $r \in \mathbf{Q}^c$ with $r > \overline{0}$.

12. Prove that $\mathbf{Q}^{c+} = \{r \in \mathbf{Q}^c \mid r > \overline{0}\}$ is a set of positive elements for $\mathbf{Q}^c$.

4.4   THE FIELD OF COMPLEX NUMBERS

Section Overview. In this section the field **C** of complex numbers is developed from the field **R** of real numbers. It is shown that **C** contains a subfield that is isomorphic to **R** and that **C** cannot be an ordered field. Consequently, the order on **R** cannot be extended to **C**. The field **C** is *unique up to isomorphism* in the sense that **C** is the *algebraic closure* of **R**, a fact not discussed until a later chapter. Be sure to take note of the summary given at the end of this section. It gives a list of the basic attributes of our number systems and the properties that characterize each system up to isomorphism.

It was pointed out in Section 4.3 that one reason for developing the field **Q** of rational numbers from the ring of integers **Z** is to find a field in which equations of the form $ax + b = c$ can always be solved, where a, b, $c \in \mathbf{Z}$, $a \neq 0$. This field is satisfactory for solving linear equations, but not for solving equations such as $x^2 - 2 = 0$. Since solutions of such equations do not always lie in $\mathbf{Q}$, the field of rational numbers was extended to the field of real numbers, which contains all the irrational numbers. We now find ourselves in a similar situation when trying to find solutions for equations of the form $ax^2 + bx + c = 0$, where a, b, $c \in \mathbf{R}$, $a \neq 0$. The development of the quadratic formula,

$$ x = \frac{-b \pm \sqrt{b^2 - 4ac}}{2a}, $$

provided for solutions of quadratic equations, but there are instances where the solutions are unsatisfactory in the sense that they do not lie in $\mathbf{R}$. For example, the very simple equation $x^2 + 1 = 0$ has solutions $x = \pm\sqrt{-1}$. But $\sqrt{-1}$ cannot be a real number since $\mathbf{R}$ is an ordered field, and this implies that $a^2 \geq 0$ for each $a \in \mathbf{R}$. Hence, one cannot find a real number whose square is -1. So we find ourselves looking for a field extending $\mathbf{R}$ that contains the solutions of all quadratic equations $ax^2 + bx + c = 0$ whose coefficients are in $\mathbf{R}$.

We have previously referred to expressions such as $x + yi$ with x, $y \in \mathbf{R}$ and $i = \sqrt{-1}$ as **complex numbers**. These numbers can be added and multiplied using the usual rules of algebra if i^2 is replaced with -1 wherever it occurs. One can show that if these numbers are added and multiplied in this fashion, they form a field that contains a subfield isomorphic to $\mathbf{R}$. This, however, leaves one fundamental question unanswered. Simply put, what is i? Since there is no real number whose square is -1, $i \notin \mathbf{R}$. Consequently, if we don't know the nature of i, it cannot be said that we understand the nature of complex numbers themselves. Because of this question, it is necessary to find a method of constructing the field of complex numbers from a known algebraic structure that will enable us to give a definitive rationale for i. Since the field of real numbers was developed in Section 4.3 and so is now "at hand," the method to be employed for constructing the field of complex numbers is to use ordered pairs of real numbers. One thing you should notice about this construction is that an equivalence relation is not involved.

4.4.1   Definition

A **complex number** z is an ordered pair (x, y) of real numbers. We call x the **real part** of z and y the **imaginary part**. Two complex numbers (x_1, y_1) and (x_2, y_2) are equal if and only if $x_1 = x_2$ and $y_1 = y_2$. The **conjugate** of the complex number $z = (x, y)$ is the complex number $\bar{z} = (x, -y)$, and the **norm** or **absolute value** of z is the real number $|z| = \sqrt{x^2 + y^2}$. If **C** is the set of all complex numbers, then addition and multiplication are defined on **C** by

$$(x_1, y_1) + (x_2, y_2) = (x_1 + x_2, y_1 + y_2)$$

$$(x_1, y_1)(x_2, y_2) = (x_1 x_2 - y_1 y_2, x_1 y_2 + x_2 y_1)$$

Since the real and imaginary parts of complex numbers are real numbers, it is easy to argue that the operations of addition and multiplication of complex numbers are well-defined since addition and multiplication of real numbers are well-defined. The following theorem tells the complete story of the development of complex numbers as ordered pairs of real numbers.

4.4.2   Theorem

The set **C** of complex numbers together with the binary operations of addition and multiplication defined in Definition 4.4.1 is a field that contains a subfield **R′** isomorphic to **R**. If every element of **R** is identified with the corresponding element in **R′**, then every complex number $z = (x, y)$ of **C** can be written as $z = x + yi$, where i is the complex number $(0, 1)$. Moreover, under this identification $i^2 = -1$.

Proof. The additive identity of **C** is clearly $(0, 0)$, and the additive inverse of the complex number (x, y) is the complex number $(-x, -y)$. The fact that addition of complex numbers is associative and commutative follows directly from the fact that addition is associative and commutative in **R**. Hence, **C** is an additive abelian group.

Multiplication of complex numbers is easily shown to be associative and commutative, and the multiplicative identity of **C** is the complex number $e = (1, 0)$. Indeed, if $z = (x, y)$ is any complex number, then

$$ze = (x, y)(1, 0) = (x1 - y0, x0 + y1) = (x, y) = z \qquad \text{and}$$

$$ez = (1, 0)(x, y) = (1x - 0y, 0x + 1y) = (x, y) = z.$$

It is also not difficult to show that multiplication is distributive over addition from the left and the right. (You will be asked to show in the exercises that multiplication is distributive over addition from the left. Why is this sufficient to show that multiplication is also distributive over addition from the right?) If $z = (x, y)$ is a nonzero complex number, then either x or y is a nonzero real number, so $x^2 + y^2 \neq 0$. Since

$$(x, y) \left(\frac{x}{x^2 + y^2}, \frac{-y}{x^2 + y^2} \right) = \left(\frac{x^2 + y^2}{x^2 + y^2}, \frac{xy - xy}{x^2 + y^2} \right) = (1, 0) = e,$$

$z^{-1} = \left(\dfrac{x}{x^2 + y^2}, \dfrac{-y}{x^2 + y^2} \right)$ is the multiplicative inverse of z, so **C** is a field.

The set $\mathbf{R'} = \{(x, 0) \mid x \in \mathbf{R}\}$ is a subfield of **C**, and $f : \mathbf{R} \to \mathbf{R'}$ defined by $f(x) = (x, 0)$ is a ring isomorphism. Therefore, **C** contains a subfield that is isomorphic to **R**. Now suppose that each real number x is identified with its image $(x, 0)$ in $\mathbf{R'}$ and that we agree to write x for $(x, 0)$. If we let $i = (0, 1)$, then for any complex number $z = (x, y)$

$$z = (x, y) = (x, 0) + (0, y) = (x, 0) + (y, 0)(0, 1) = x + yi \qquad \text{and}$$

$$i^2 = (0, 1)(0, 1) = (0 \cdot 0 - 1 \cdot 1, 1 \cdot 0 + 0 \cdot 1) = (-1, 0) = -1. \blacksquare$$

From the construction of **C**, the question regarding the nature of i has now been answered; i is the complex number $(0, 1)$. Furthermore, this construction provides for writing complex numbers in the form $x + yi$, where $x, y \in \mathbf{R}$. Under this construction when adding or multiplying complex numbers written in the form $x + yi$, one can still apply the usual rules of algebra and then replace i^2 by -1. For example,

$$(3 + 2i) + \left(-4 + \frac{1}{2} i \right) = (3 - 4) + \left(2i + \frac{1}{2} i \right)$$

$$= -1 + \frac{5}{2} i \qquad \text{and}$$

$$(3 + 2i) \left(-4 + \frac{1}{2} i \right) = 3 \left(-4 + \frac{1}{2} i \right) + 2i \left(-4 + \frac{1}{2} i \right)$$

$$= -12 + \frac{3}{2} i - 8i + i^2$$

$$= -12 + \frac{3}{2} i - \frac{16}{2} i + (-1)$$

$$= -13 + \frac{13}{2} i.$$

The multiplicative inverse of a complex number can be found formally using the usual rules of algebra and complex conjugates. If $z = x + yi$, then $\overline{z} = x - yi$ and $z\overline{z} = x^2 + y^2$. Hence,

$$\frac{1}{z} = \frac{\overline{z}}{z\overline{z}} = \frac{x - yi}{x^2 + y^2} = \frac{x}{x^2 + y^2} - \frac{y}{x^2 + y^2}i \qquad \text{and so} \qquad z^{-1} = \frac{1}{z}.$$

Division of complex numbers is actually defined by $\dfrac{z_1}{z_2} = z_1 z_2^{-1}$, but this can also be performed quickly using the usual rules of algebra and complex conjugates. As an illustration,

$$\frac{2 + 3i}{4 - 5i} = \frac{(2 + 3i)(4 + 5i)}{(4 - 5i)(4 + 5i)} = \frac{(8 - 15) + (10 + 12)i}{16 + 25}$$

$$= \frac{-7 + 22i}{41} = -\frac{7}{41} + \frac{22}{41}i.$$

Complex numbers can also be expressed in **trigonometric form.** Suppose that we plot a nonzero complex number $z = x + yi$ by locating the point with coordinates (x, y) in the x, y plane. If this point is joined to the origin by a line segment, an angle is formed with the positive x axis. The least positive measure of this angle is the (principal) **argument** of z. No argument is assigned to the complex number $z = 0$. The distance from the point (x, y) to the origin is the **norm** $r = |z| = \sqrt{x^2 + y^2}$ of z. It follows that $x = r\cos\theta$ and $y = r\sin\theta$. Hence, $z = r(\cos\theta + i\sin\theta)$. This is the trigonometric form of the complex

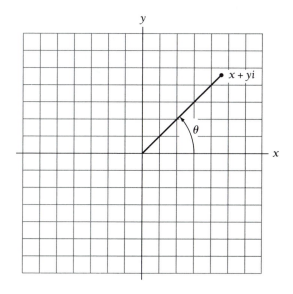

number z. The trigonometric form of a complex number is useful in finding the roots of complex numbers. To derive a formula for the roots of a complex number, we first need to be able to compute the nth power of a complex number in trigonometric form. Recall that

$$\cos(\alpha + \beta) = \cos \alpha \cos \beta - \sin \alpha \sin \beta \qquad \text{and}$$

$$\sin(\alpha + \beta) = \sin \alpha \cos \beta + \sin \beta \cos \alpha.$$

Hence, if $z_1 = r_1(\cos \theta_1 + i \sin \theta_1)$ and $z_2 = r_2(\cos \theta_2 + i \sin \theta_2)$, then

$$z_1 z_2 = r_1 r_2(\cos \theta_1 + i \sin \theta_1)(\cos \theta_2 + i \sin \theta_2)$$

$$= r_1 r_2[(\cos \theta_1 \cos \theta_2 - \sin \theta_1 \sin \theta_2) + i(\sin \theta_1 \cos \theta_2 + \sin \theta_2 \cos \theta_1)]$$

$$= r_1 r_2 [\cos (\theta_1 + \theta_2) + i \sin(\theta_1 + \theta_2)].$$

Similarly, it can be shown that $\dfrac{z_1}{z_2} = \dfrac{r_1}{r_2}[\cos(\theta_1 - \theta_2) + i \sin(\theta_1 - \theta_2)]$. Thus, we see that the norm of the product of two complex numbers is the product of the norms and that the argument of the product is the sum of the arguments. However, the sum of the arguments may have to be "reduced" to obtain the (principal) argument. For example, if the argument of z_1 is $\dfrac{7\pi}{4}$ and the argument of z_2 is $\dfrac{\pi}{2}$, then $\dfrac{7\pi}{4} + \dfrac{\pi}{2}$ must be replaced with $\dfrac{\pi}{4}$ to obtain the angle of smallest positive measure for the argument of $z_1 z_2$.

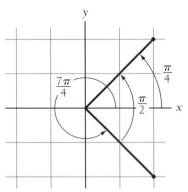

If z_1 and z_2 are the same complex number $z = r(\cos \theta + i \sin \theta)$, then

$$z^2 = r^2(\cos 2\theta + i \sin 2\theta).$$

The following more general result is due to **Abraham De Moivre** (1667–1754).

4.4.3   De Moivre's Theorem

If $z = r(\cos\theta + i\sin\theta)$, then for any positive integer n, $z^n = r^n(\cos n\theta + i\sin n\theta)$.

You will be asked to give a proof of the De Moivre's Theorem in the exercises. Use induction. The following examples demonstrate how De Moivre's Theorem can be used.

4.4.4 | Examples

1. Compute $(1 + i)^8$. The first step is to express $1 + i$ in trigonometric form. Consider the following figure, where we have located the complex number $1 + i$ in the (x, y)-plane. The norm of $1 + i$ is $\sqrt{1^2 + 1^2} = \sqrt{2}$, and the figure shows that the argument of $1 + i$ is $\frac{\pi}{4}$. Hence,

    $$(1+i)^8 = \left[\sqrt{2}\left(\cos\frac{\pi}{4} + i\sin\frac{\pi}{4}\right)\right]^8$$
    $$= (\sqrt{2})^8(\cos 2\pi + i\sin 2\pi)$$
    $$= 2^4(1 + 0) = 16.$$

    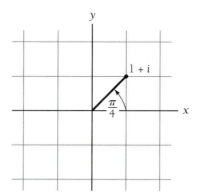

2. In this example we compute $\left(-\frac{1}{2} + \frac{\sqrt{3}}{2}i\right)^{25}$. The argument of $-\frac{1}{2} + \frac{\sqrt{3}}{2}i$ is $\frac{2\pi}{3}$ and the norm is 1. Hence,

$$\left(-\frac{1}{2}+\frac{\sqrt{3}}{2}i\right)^{25}=\left[1\left(\cos\frac{2\pi}{3}+i\sin\frac{2\pi}{3}\right)\right]^{25}$$

$$=1^{25}\left(\cos\frac{50\pi}{3}+i\sin\frac{50\pi}{3}\right)=\cos\frac{2\pi}{3}+i\sin\frac{2\pi}{3}=-\frac{1}{2}+\frac{\sqrt{3}}{2}i$$

4.4.5   Definition

If $n \in \mathbf{N}$ and z is a complex number, then a complex number u such that $u^n = z$ is said to be an **nth root** of z.

The preceding Example 1 shows that $1 + i$ is an 8th root of 16, while in the second of these examples we have the seemingly odd situation in which $-\frac{1}{2}+\frac{\sqrt{3}}{2}i$ is a 25th root of itself. But if we were asked to find the 8th roots of 16 in $\mathbf{C}$, how would we go about finding these roots? Fortunately, De Moivre's Theorem also allows us to compute the complex roots of a complex number. As we shall see, for each positive integer n, each nonzero complex number has n distinct roots in $\mathbf{C}$.

4.4.6   Theorem

Any complex number $z = r(\cos\theta + i\sin\theta)$ has n distinct nth roots that are given by

$$u_k = \sqrt[n]{r}\left(\cos\frac{\theta+2k\pi}{n}+i\sin\frac{\theta+2k\pi}{n}\right)$$

for $k = 0, 1, 2 \ldots, n-1$, where $\sqrt[n]{r}$ denotes the principal nth real root of the positive real number r.

Proof. If k is an integer such that $0 \le k \le n-1$, then, by De Moivre's Theorem,

$$u_k^n = \left(\sqrt[n]{r}\right)^n\left(\cos n\left[\frac{\theta+2k\pi}{n}\right]+i\sin n\left[\frac{\theta+2k\pi}{n}\right]\right)$$

$$= r(\cos[\theta+2k\pi]+i\sin[\theta+2k\pi])$$

$$= r(\cos\theta+i\sin\theta)$$

$$= z.$$

Hence, u_k is an nth root of z for $k = 0, 1, 2, \ldots, n - 1$. Substituting these values of k into $\frac{\theta + 2k\pi}{n}$ produces n distinct angles with radian measure

$$\frac{\theta}{n} = \frac{\theta}{n} + 0\frac{2\pi}{n}$$

$$\frac{\theta + 2\pi}{n} = \frac{\theta}{n} + 1\frac{2\pi}{n}$$

$$\frac{\theta + 4\pi}{n} = \frac{\theta}{n} + 2\frac{2\pi}{n}$$

$$\vdots$$

$$\frac{\theta + 2(n-1)\pi}{n} = \frac{\theta}{n} + (n-1)\frac{2\pi}{n}.$$

It follows that all of $u_0, u_1, \ldots, u_{n-1}$ are distinct. The proof will be complete if it can be shown that these are the only roots of z. If $u = s(\cos\beta + i\,\sin\beta)$ is another nth root of z, then $u^n = z$ and thus

$$s^n(\cos n\beta + i\sin n\beta) = r(\cos\theta + i\sin\theta).$$

It follows that $s^n = r$, $\cos n\beta = \cos\theta$, and $\sin n\beta = \sin\theta$. Therefore, $s = \sqrt[n]{r}$. Now $\cos n\beta = \cos\theta$ and $\sin n\beta = \sin\theta$ imply that $n\beta$ and θ differ by a multiple of 2π. Hence, $n\beta = \theta + 2k\pi$, which gives $\beta = \frac{\theta + 2k\pi}{n}$, and this completes the proof. ■

4.4.7 | **Examples**

1. Find the 4th roots of -1.

Solution. If we write -1 as $z = -1 + 0i$, then the norm of z is $|z| = r = 1$. The radian measure of the angle formed by the positive x axis and the line segment joining the point with coordinates $(-1, 0)$ to the origin is π. Using Theorem 4.4.6, we see that

$$u_k = \sqrt[4]{1}\left(\cos\frac{\pi + 2k\pi}{4} + i\sin\frac{\pi + 2k\pi}{4}\right) \qquad \text{for } k = 0, 1, 2, 3.$$

Hence,

$$u_0 = \sqrt[4]{1}\left(\cos\frac{\pi}{4} + i\sin\frac{\pi}{4}\right)$$

$$= \cos\frac{\pi}{4} + i\,\sin\frac{\pi}{4}$$

$$= \frac{\sqrt{2}}{2} + \frac{\sqrt{2}}{2}i,$$

$$u_1 = \sqrt[4]{1}\left(\cos\frac{\pi+2\pi}{4} + i\sin\frac{\pi+2\pi}{4}\right)$$

$$= \cos\frac{3\pi}{4} + i\sin\frac{3\pi}{4}$$

$$= -\frac{\sqrt{2}}{2} + \frac{\sqrt{2}}{2}i,$$

$$u_2 = \sqrt[4]{1}\left(\cos\frac{\pi+4\pi}{4} + i\sin\frac{\pi+4\pi}{4}\right)$$

$$= \cos\frac{5\pi}{4} + i\sin\frac{5\pi}{4}$$

$$= -\frac{\sqrt{2}}{2} - \frac{\sqrt{2}}{2}i, \qquad \text{and}$$

$$u_3 = \sqrt[4]{1}\left(\cos\frac{\pi+6\pi}{4} + i\sin\frac{\pi+6\pi}{4}\right)$$

$$= \cos\frac{7\pi}{4} + i\sin\frac{7\pi}{4}$$

$$= \frac{\sqrt{2}}{2} - \frac{\sqrt{2}}{2}i.$$

This is a nice place to demonstrate the symmetry involved with the roots of a complex number. The arguments of the 4th roots of 1 are $\frac{\pi}{4}$, $\frac{3\pi}{4}$, $\frac{5\pi}{4}$, and $\frac{7\pi}{4}$. The roots u_0, u_1, u_2, and u_3 lie on the unit circle since they each have norm 1. Moreover, since each angle differs from the one that follows it by $\frac{\pi}{2}$ they are equally spaced around the circle.

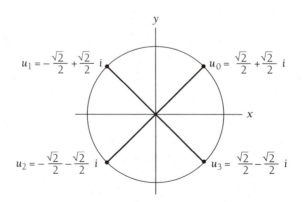

2. Find the cube roots of $z = 2\sqrt{3} + 2i$.

Solution. The norm of z is $|z| = r = \sqrt{2(\sqrt{3})^2 + 2^2} = \sqrt{12 + 4} = \sqrt{16} = 4$. Hence, the angle of radian measure θ determined by the positive x axis and the line segment joining the origin to the point with coordinates $(2\sqrt{3}, 2)$ has $\cos\theta = \frac{2\sqrt{3}}{4} = \frac{\sqrt{3}}{2}$ and $\sin\theta = \frac{2}{4} = \frac{1}{2}$. Hence, $\theta = \frac{\pi}{6}$. By Theorem 4.4.6, we see that

$$u_k = \sqrt[3]{4}\left(\cos\frac{\frac{\pi}{6} + 2k\pi}{3} + i\sin\frac{\frac{\pi}{6} + 2k\pi}{3}\right) \text{ for } k = 0, 1, 2.$$

Therefore,

$$u_0 = \sqrt[3]{4}\left(\cos\frac{\frac{\pi}{6} + 2\cdot 0\cdot\pi}{3} + i\sin\frac{\frac{\pi}{6} + 2\cdot 0\cdot\pi}{3}\right)$$

$$= \sqrt[3]{4}\left(\cos\frac{\pi}{18} + i\sin\frac{\pi}{18}\right)$$

$$\cong 1.56328 + 0.275649i,$$

$$u_1 = \sqrt[3]{4}\left(\cos\frac{\frac{\pi}{6} + 2\pi}{3} + i\sin\frac{\frac{\pi}{6} + 2\pi}{3}\right)$$

$$= \sqrt[3]{4}\left(\cos\frac{13\pi}{18} + i\sin\frac{13\pi}{18}\right)$$

$$\cong -1.02036 + 1.21602i, \qquad \text{and}$$

$$u_2 = \sqrt[3]{4}\left(\cos\frac{\frac{\pi}{6} + 4\pi}{3} + i\sin\frac{\frac{\pi}{6} + 4\pi}{3}\right)$$

$$= \sqrt[3]{4}\left(\cos\frac{25\pi}{18} + i\sin\frac{25\pi}{18}\right)$$

$$\cong -0.542923 - 1.49167i.$$

3.   If n is a positive integer, find the nth roots of 1.

Solution. If $z = 1 = 1 + 0i$, then the norm of z is 1, and the argument of z is 0 radians. Hence, we see that the nth roots of $z = 1$ are given by

$$u_k = \sqrt[n]{1}\left(\cos \frac{0 + 2k\pi}{n} + i \sin \frac{0 + 2k\pi}{n}\right)$$

$$= \cos \frac{2k\pi}{n} + i \sin \frac{2k\pi}{n} \qquad \text{for } k = 0, 1, 2, \ldots, n - 1.$$

These roots are known as the nth **roots of unity**.

Recall that the ring $q(\mathbf{R})$ of real quaternions is the set with elements of the form $a + bi + cj + dk$, where $a, b, c, d \in \mathbf{R}$. Elements of $q(\mathbf{R})$ are added and multiplied using the rules of algebra given in Example 1 of 3.1.9. The additive identity of $q(\mathbf{R})$ is $0 = 0 + 0i + 0j + 0k$, and the multiplicative identity is $1 = 1 + 0i + 0j + 0k$. If $z = a + bi + cj + dk \neq 0$ and $|z| = \sqrt{a^2 + b^2 + c^2 + d^2} \neq 0$, it follows that

$$z^{-1} = \frac{a}{|z|^2} - \frac{b}{|z|^2}i - \frac{c}{|z|^2}j - \frac{d}{|z|^2}k$$

is the multiplicative inverse of z. Hence, $q(\mathbf{R})$ is a division ring.

However, $q(\mathbf{R})$ is not a field since, for example, $ij \neq ji$. The field of complex numbers $\mathbf{C}$ embeds in $q(\mathbf{R})$ via the mapping $f : \mathbf{C} \rightarrow q(\mathbf{R})$ defined by $f(a + bi) = a + bi + 0j + 0k$. If each element of $\mathbf{C}$ is identified with the corresponding element of $q(\mathbf{R})$, $\mathbf{C}$ can be considered to be a subfield of the division ring $q(\mathbf{R})$.

Summary: Properties of Our Number Systems

1.   The ring of integers $\mathbf{Z}$ is an ordered integral domain.

   (a)   It was taken as an axiom that the set of positive elements $\mathbf{N}$ is well-ordered.

(b)   The integral domain **Z** is *unique up to isomorphism* in the sense that every ordered integral domain in which the set of positive elements is well-ordered is isomorphic to **Z**.

2.   The field of rational numbers **Q** is the field of fractions of **Z**.

(a)   The field **Q** can be constructed from **Z**, and **Q** contains a subring that is isomorphic to **Z**. If **Z** is identified with this subring, then **Z** can be considered a subring of **Q**.

(b)   The field **Q** can be ordered in such a way that the ordering on **Q** agrees with the ordering on **Z**. The set $\mathbf{Q}^{+} = \left\{ \frac{p}{q} \,\middle|\, pq \in \mathbf{N} \right\}$ is a set of positive elements of **Q**, but $\mathbf{Q}^{+}$ is not well-ordered.

(c)   The field **Q** is not a complete field. There are nonempty subsets of **Q** that have upper bounds in **Q** that do not have least upper bounds in **Q**.

(d)   As a *field of fractions* of **Z**, **Q** is *unique up to isomorphism*.

3.   The field of real numbers **R** is the completion of the field **Q** of rational numbers.

(a)   The field **R** can be constructed from **Q**, and **R** contains a subfield that is isomorphic to **Q**. If **Q** is identified with this subfield, then **Q** can be viewed as a subfield of **R**.

(b)   An order can be defined on **R** in a way that the ordering on **R** agrees with the ordering on **Q**.

(c)   As an ordered field, **R** is complete. That is, each nonempty subset of **R** that has an upper bound in **R** has a least upper bound in **R**.

(d)   The *complete ordered field* **R** is *unique up to isomorphism* since every complete ordered field is isomorphic to **R**.

4.   The field **C** of complex numbers can be constructed from the field of real numbers.

(a)   The field **C** contains a subfield isomorphic to **R**. If **R** is identified with this subfield, then **R** can be considered to be a subfield of **C**.

(b)   We have seen that it is not possible for **C** to contain a set of positive elements, so **C** is not an ordered field.

(c)   The *field* **C** is *unique up to isomorphism* in the sense that it is the **algebraic closure** of **R**. The algebraic closure of a field is discussed in Chapter 8.

(d)   The field **C** can be identified with a subfield of the division ring of real quaternions.

We conclude from the preceding properties that if the appropriate identifications are made, then $\mathbf{Z} \subset \mathbf{Q} \subset \mathbf{R} \subset \mathbf{C} \subset q(\mathbf{R})$. We assume these identifications when we work with our number systems on a day-to-day basis.

HISTORICAL NOTES

Historically, the discovery of the quadratic formula and the ensuing complex numbers came about quite slowly. Early on, complex numbers caused a great deal of consternation among mathematicians, and numbers of the form $a\sqrt{-1}$, where a is a real number, were often dismissed as not being "of this world." They were referred to as "imaginary numbers," a term still in use today. **Gerome Cardan** was one of the first mathematicians to use complex numbers in his calculations when solving cubic equations. Complex numbers slowly gained acceptance, and today we know that complex analysis is a valid area of mathematics that enjoys a wide variety of practical applications.

Gerome Cardan (1501–1576). Gerome Cardan is actually the English name of Girolamo Cardano, who was born in what is now Pavia, Italy. His father, a lawyer and mathematician, lectured on mathematics at the University of Pavia, and Cardan received his first instruction in mathematics from his father. Cardan went on to study medicine at the University of Pavia and at Padua University, and he received his doctorate in medicine in 1525. He was a brilliant student but outspoken and difficult. His application to join the College of Physicians in Milan was rejected because of his reputation. However, Cardan was eventually accepted to the college, and he began a distinguished career that included writings in medicine, theology, astronomy, philosophy, and mathematics. In 1545 he published his greatest mathematical work, *Ars Magna*, in which he gave methods for solving cubic and quartic equations, where he did some of the first calculations with complex numbers.

Abraham De Moivre (1667–1754). Abraham De Moivre was born in Vitry, a town in France about 100 miles east of Paris, in 1667. His early education was in the classics, and he later studied logic and probability theory. Life became very difficult for Protestants in France, and he was imprisoned for two years. After being released, he left France for England, never to return. In England, he was given a position by the Royal Society to review the rival claims of Newton and Leibniz as to who discovered the calculus. De Moivre was never able to achieve a university position, but he made significant contributions to mathematics. De Moivre's major mathematical work was the book *Doctrine of Chance*, first published in 1718, in which he set forth the rules for games of chance.

Problem Set 4.4

1. Compute each of the following:

    (a) $(3 + 2i) + (-\frac{1}{2} - 4i)$

    (c) $|3 - \sqrt{2}i|$

    (b) $(-2 + 3i)(5 - 4i)$

    (d) $\dfrac{5 - i}{2 + i} - \dfrac{1 - 5i}{6 + 7i}$

2. Express each of the following in trigonometric form.

    (a) $-1 + i$

    (c) 4

    (b) $-\dfrac{\sqrt{3}}{2} - \dfrac{1}{2}i$

    (d) $2 - 2i$

3. Compute each of the following. Convert your answer to the form $x + yi$.

    (a) $3\left(\cos\dfrac{5\pi}{6} + i\,\sin\dfrac{5\pi}{6}\right) \cdot 2\left(\cos\dfrac{7\pi}{6} + i\,\sin\dfrac{7\pi}{6}\right)$

    (b) $5\left(\cos\dfrac{3\pi}{4} - i\,\sin\dfrac{3\pi}{4}\right) \cdot 3\left(\cos\dfrac{-5\pi}{4} - i\,\sin\dfrac{-5\pi}{4}\right)$

    (c) $2\left(\cos\dfrac{2\pi}{3} - i\,\sin\dfrac{2\pi}{3}\right) \cdot 4\left(\cos\dfrac{5\pi}{4} + i\,\sin\dfrac{5\pi}{4}\right)$

4. Use De Moivre's Theorem to compute each of the following. Leave your answer in the form $x + yi$.

    (a) $\left(\sqrt{3} - \sqrt{3}i\right)^5$

    (c) $\left(1 - \sqrt{3}i\right)^{10}$

    (b) $\left(\dfrac{\sqrt{3}}{2} + \dfrac{1}{2}i\right)^{12}$

    (d) $\left(\sqrt{3} - i\right)^{23}$

5. Prove Theorem 4.4.3.

6. Find the indicated roots of each of the following:

    (a) The 4th roots of $1 + i$

    (b) The cube roots of $-\dfrac{1}{2} + \dfrac{\sqrt{3}}{2}i$

    (c) The 4th roots of 1

7. In the field of complex numbers, show that multiplication is distributive over addition from the left.

8. Prove that the set G of 5th roots of 1 form an abelian group under multiplication. Find a 5th root ζ of 1 that generates G. That is, find a 5th root ζ such that every element of G is of the form ζ^k, $k = 0, 1, 2, 3, 4$. Such a root ζ is called a **primitive 5th root of unity.** In general, show that the nth roots of unity form an abelian group under multiplication and that
    $$\omega = \cos\left(\frac{2\pi}{n}\right) + i\sin\left(\frac{2\pi}{n}\right) \text{ is a } \textbf{primitive } n\textbf{th root of unity.}$$

9. Find all the complex numbers z that satisfy the following equations. Leave your answer in the form $x + yi$.

    (a) $z^3 + i = 0$ (c) $z^4 - 1 + \sqrt{3}i = 0$

    (b) $z^4 + 16i = 0$

10. Let z be a complex number, and suppose that $\sqrt[c]{z}$ denotes the complex square roots of z. If r is a real number, show that $\sqrt[c]{r} = \pm\sqrt{r}$ if r is positive and that $\sqrt[c]{r} = \pm\sqrt{|r|}i$ when r is negative. If $ax^2 + bx + c = 0$ is a quadratic equation with its coefficients in $\mathbf{C}$, show that this equation has solutions given by
    $$x = \frac{-b + \sqrt[c]{b^2 - 4ac}}{2a}$$
    Does this agree with the usual quadratic formula if $a, b, c \in \mathbf{R}$? Justify your answer.

11. If $z = r(\cos\theta + i\sin\theta)$ is a nonzero complex number, prove each of the following:

(a) $z^{-1} = r^{-1}(\cos \theta - i \sin \theta)$

(b) $z^{-n} = r^{-n}(\cos n\theta - i \sin n\theta)$

12.   If $z_1 = r_1(\cos \theta_1 + i \sin \theta_1)$ and $z_2 = r_2(\cos \theta_2 + i \sin \theta_2)$ are complex numbers with $z_2 \neq 0$, show that

$$\frac{z_1}{z_2} = \frac{r_1}{r_2}[\cos(\theta_1 - \theta_2) + i \sin(\theta_1 - \theta_2)].$$

13.   If $z = x + yi$ is a complex number, show that

(a) $z = \bar{z}$ if and only if z is a real number.

(b) $z + \bar{z} = 2x$.

(c) $z - \bar{z} = 2yi$.

(d) $\bar{\bar{z}} = z$.

14.   Prove each of the following, where z, z_1 and z_2 are complex numbers.

(a) $|z| = |\bar{z}|$

(d) $\left|\dfrac{z_1}{z_2}\right| = \dfrac{|z_1|}{|z_2|}$, where $z_2 \neq 0$

(b) $|z|^2 = z\bar{z}$

(e) $z^{-1} = \dfrac{\bar{z}}{z\bar{z}}$, where $z \neq 0$

(c) $|z_1 z_2| = |z_1||z_2|$

15.   Let $z_1, z_2, \ldots, z_n$ be complex numbers, where n is any positive integer. Prove that

(a) $\overline{z_1 + z_2 + \cdots + z_n} = \bar{z}_1 + \bar{z}_2 + \cdots + \bar{z}_n.$

(b) $\overline{z_1 z_2 \cdots z_n} = \bar{z}_1 \bar{z}_2 \cdots \bar{z}_n.$

(c) $\overline{\left(\dfrac{z_1}{z_2}\right)} = \dfrac{\bar{z}_1}{\bar{z}_2}.$

(d) For any positive integer n and any complex number z, $\overline{n \cdot z} = n\bar{z}$ and $\overline{z^n} = \bar{z}^{n}.$

16. Let $f(x) = a_0 + a_1x + a_2x^2 + \cdots + a_nx^n$ be a polynomial with its coefficients in $\mathbf{R}$, and suppose that $z \in \mathbf{C}$ is root of $f(x)$. Show that $\bar{z}$ is also a root of $f(x)$. Hint: Use Exercises 13 and 15 to show that $f(\bar{z}) = \overline{f(z)} = \bar{0} = 0$.

17. Prove that the set $\mathbf{C}^* = \{a + bi + 0j + 0k \mid a, b \in \mathbf{R}\}$ is a subfield of the division ring $q(\mathbf{R})$ and that $\mathbf{C}$ and $\mathbf{C}^*$ are isomorphic.

5

Groups Again

A brief introduction to groups was presented in Chapter 2. Group theory is an active area of study by mathematicians, and much more can be said about groups than was presented previously. Even though the concept of a group dates back over 100 years, group theory continues to be an active area of research. To add to our knowledge of group theory, we now investigate three types of groups: permutation groups, cyclic groups, and finite abelian groups, as well as groups that can be formed from other groups.

5.1 PERMUTATIONS

Section Overview. In this section, permutation groups are studied. An interesting theorem from this section is Cayley's Theorem, which establishes an isomorphism between an arbitrary group and a group of permutations. The important concepts presented are the product of permutations, even and odd permutations, cycles and transpositions, and the group of even permutations.

The theory of permutations began with the study of combinatorics and dates back to around A.D. 1300. But it was **Augustin-Louis Cauchy** (1789–1857) who recognized the importance of rearranging the objects in a set. He defined the order of a permutation, the product of two permutations, and the inverse of a permutation. He also defined a circular (cyclic) permutation and introduced the notation $(x_1, x_2, \ldots, x_n)$ for such a permutation.

5.1.1   Definition

If S is a set, then a **permutation** on S is a bijective function $\alpha : S \to S$. The set of all permutations on a set S is denoted by $\Gamma(S)$ and by $\mathbf{S}_n$ when $S = \{1, 2, \ldots, n\}$. If $\alpha, \beta \in \Gamma(S)$, we simplify notation by writing $\alpha\beta$ for $\alpha \circ \beta$, and $\alpha\beta$ is referred to as the product of α and β rather than α composed with β.

Since the composition of two bijective functions is a bijective function (Problem Set 0.3, Exercise 11), the product of permutations is a binary operation on $\Gamma(S)$. Furthermore, by Theorem 0.3.7, each permutation $\alpha \in \Gamma(S)$ has an inverse $\alpha^{-1} : S \to S$. Since α is the inverse function for α^{-1}, α^{-1} must also be a bijection, again by Theorem 0.3.7, so $\alpha^{-1} \in \Gamma(S)$. If 1_S denotes the identity function on S, then the following hold:

(a) If $\alpha, \beta \in \Gamma(S)$, then $\alpha\beta \in \Gamma(S)$.
(b) If $\alpha, \beta, \gamma \in \Gamma(S)$, then $\alpha(\beta\gamma) = (\alpha\beta)\gamma$.
(c) The identity mapping 1_S is in $\Gamma(S)$.
(d) If $\alpha \in \Gamma(S)$, then $\alpha^{-1} \in \Gamma(S)$.

This shows that $\Gamma(S)$ is a group under the binary operation of function composition. The group $\Gamma(S)$ is known as the **permutation group** on S, and when S is finite and contains n elements, S is the **symmetric group on n letters**.

If $S = \{x_1, x_2, x_3, x_4, x_5\}$, then the permutation $\alpha \in \mathbf{S}_n$ defined by

$$x_1 \longmapsto x_4$$
$$x_2 \longmapsto x_5$$
$$x_3 \longmapsto x_1$$
$$x_4 \longmapsto x_3$$
$$x_5 \longmapsto x_2$$

can be simplified by using only the subscripts of the elements that define α:

$$1 \longmapsto 4$$
$$2 \longmapsto 5$$
$$3 \longmapsto 1$$
$$4 \longmapsto 3$$
$$5 \longmapsto 2$$

Since this can obviously be done for any permutation in $\mathbf{S}_n$, S can be replaced by $\{1, 2, 3, 4, 5\}$. In general, if $S = \{x_1, x_2, \ldots, x_n\}$, then S can be replaced by $\{1, 2, \ldots, n\}$. No generality is lost when making this simplification. The permutation

$\alpha \in \mathbf{S}_5$ defined earlier can be written as $\alpha = \begin{pmatrix} 1\,2\,3\,4\,5 \\ 4\,5\,1\,3\,2 \end{pmatrix}$, which we will call the **two-row notation** of α. This notation indicates that each element of the first row of α is mapped to the element that lies directly beneath it. If $\beta = \begin{pmatrix} 1\,2\,3\,4\,5 \\ 4\,3\,1\,2\,5 \end{pmatrix}$, then the product $\alpha\beta$ is the composition of α with β, using the usual convention that composition is read from right to left. To find the permutation $\alpha\beta$ determined by the product of α and β, start by looking at 1 in the top row of β and tracing along the path shown in the following figure. From this we see that β maps 1 to 4 and then α maps 4 to 3. Hence, the product $\alpha\beta$ maps 1 to 3, so $\alpha\beta(1) = 3$.

$$\begin{pmatrix} 1\,2\,3\,4\,5 \\ 4\,5\,1\,3\,2 \end{pmatrix}\begin{pmatrix} 1\,2\,3\,4\,5 \\ 4\,3\,1\,2\,5 \end{pmatrix} = \begin{pmatrix} 1\,2\,3\,4\,5 \\ 3\,-\,-\,-\,- \end{pmatrix}$$

Similarly, $\beta(2) = 3$ and $\alpha(3) = 1$, so $\alpha\beta(2) = 1$; $\beta(3) = 1$ and $\alpha(1) = 4$, so $\alpha\beta(3) = 4$; $\beta(4) = 2$ and $\alpha(2) = 5$, so $\alpha\beta(4) = 5$. Finally, $\beta(5) = 5$ and $\alpha(5) = 2$, so $\alpha\beta(5) = 2$. This process of **tracing through the elements of the permutations**, as demonstrated in the figure, can be used to fill in the blanks shown in the figure. This yields

$$\alpha\beta = \begin{pmatrix} 1\ 2\ 3\ 4\ 5 \\ 4\ 5\ 1\ 3\ 2 \end{pmatrix}\begin{pmatrix} 1\ 2\ 3\ 4\ 5 \\ 4\ 3\ 1\ 2\ 5 \end{pmatrix} = \begin{pmatrix} 1\ 2\ 3\ 4\ 5 \\ 3\ 1\ 4\ 5\ 2 \end{pmatrix}.$$

Note that

$$\beta\alpha = \begin{pmatrix} 1\ 2\ 3\ 4\ 5 \\ 4\ 3\ 1\ 2\ 5 \end{pmatrix}\begin{pmatrix} 1\ 2\ 3\ 4\ 5 \\ 4\ 5\ 1\ 3\ 2 \end{pmatrix} = \begin{pmatrix} 1\ 2\ 3\ 4\ 5 \\ 2\ 5\ 4\ 1\ 3 \end{pmatrix},$$

so $\alpha\beta \neq \beta\alpha$. Therefore, the product of permutations is not a commutative operation. The **identity permutation** is denoted simply by (1). For example, in $\mathbf{S}_5$

$$(1) = \begin{pmatrix} 1\ 2\ 3\ 4\ 5 \\ 1\ 2\ 3\ 4\ 5 \end{pmatrix},$$

and in general, in $\mathbf{S}_n$,

$$(1) = \begin{pmatrix} 1\ 2\ 3\ \cdots\ n \\ 1\ 2\ 3\ \cdots\ n \end{pmatrix}.$$

The order of the elements in the first row of a permutation is not important. For example, inspection of $\alpha = \begin{pmatrix} 1\,2\,3\,4\,5 \\ 4\,5\,1\,3\,2 \end{pmatrix}$ and $\alpha' = \begin{pmatrix} 4\,1\,3\,2\,5 \\ 3\,4\,1\,5\,2 \end{pmatrix}$ shows that α can be obtained from α' by a rearrangement of the columns of α'.

Hence, $\alpha = \alpha'$. To find the inverse of a permutation, simply interchange the two rows of the permutation. As an illustration, the inverse of the permutation $\alpha = \begin{pmatrix} 1\ 2\ 3\ 4\ 5 \\ 4\ 5\ 1\ 3\ 2 \end{pmatrix}$ is the permutation $\alpha^{-1} = \begin{pmatrix} 4\ 5\ 1\ 3\ 2 \\ 1\ 2\ 3\ 4\ 5 \end{pmatrix} = \begin{pmatrix} 1\ 2\ 3\ 4\ 5 \\ 3\ 5\ 4\ 1\ 2 \end{pmatrix}$. A quick calculation shows that $\alpha\alpha^{-1} = \alpha^{-1}\alpha = (1) = \begin{pmatrix} 1\ 2\ 3\ 4\ 5 \\ 1\ 2\ 3\ 4\ 5 \end{pmatrix}$.

5.1.2   **Definition**

A permutation $\alpha \in \mathbf{S}_n$ is said to be a **cycle** if there is a set of distinct integers $\{n_1, n_2, \ldots, n_k\} \subseteq S = \{1, 2, \ldots, n\}$ such that $\alpha(n_1) = n_2$, $\alpha(n_2) = n_3$, $\ldots$, $\alpha(n_{k-1}) = n_k$, $\alpha(n_k) = n_1$, and α leaves all other integers in S fixed. We say that k is the **length** of the cycle, and a cycle of length k is referred to as a k-**cycle**. A 2-cycle is called a **transposition**. If α is a k-cycle in $\mathbf{S}_n$ and $\{n_1, n_2, \ldots, n_k\}$ is the set of integers in S such that $\alpha(n_1) = n_2$, $\alpha(n_2) = n_3$, $\ldots$, $\alpha(n_{k-1}) = n_k$, $\alpha(n_k) = n_1$, then α is denoted by $(n_1, n_2, \ldots, n_k)$.

The 4-cycle $(2, 5, 3, 1)$ in $\mathbf{S}_6$ can be written in two-row notation as $(2, 5, 3, 1) = \begin{pmatrix} 1\ 2\ 3\ 4\ 5\ 6 \\ 2\ 5\ 1\ 4\ 3\ 6 \end{pmatrix}$. From the definition of a cycle, we see that

$$2 \longmapsto 5 \longmapsto 3 \longmapsto 1 \longmapsto 2 \quad \text{or} \quad 2 \longmapsto 5 \longmapsto 3 \longmapsto 1$$

A permutation often determines one or more cycles. In the permutation $\alpha = \begin{pmatrix} 1\ 2\ 3\ 4\ 5\ 6 \\ 4\ 5\ 1\ 3\ 2\ 6 \end{pmatrix}$, begin with 1 and trace through the permutation as shown in the following figure.

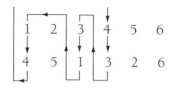

By tracing through the permutation, we see that $1 \longmapsto 4 \longmapsto 3 \longmapsto 1$, so we have the 3-cycle $(1, 4, 3)$. The permutation α also determines a 2-cycle $(2, 5)$ and leaves 6 fixed. In a cycle, each element in the list maps to the next element, with the last element mapping back to the first. The order of the components of a cycle is not unique. For example, $(1, 4, 3) = (3, 1, 4) = (4, 3, 1)$.

Moreover, two cycles are said to be **disjoint** if they have no integer in common. The cycles $(2, 5)$ and $(1, 4, 3)$ are disjoint.

Care must be taken when working with cycles. Given a cycle such as $(2, 4, 1)$, where does this cycle belong? It can represent an element of $\mathbf{S}_n$ for any integer $n \geq 4$.

$$\text{For } n = 4: \quad (2, 4, 1) = \begin{pmatrix} 1\ 2\ 3\ 4 \\ 2\ 4\ 3\ 1 \end{pmatrix}$$

$$\text{For } n = 5: \quad (2, 4, 1) = \begin{pmatrix} 1\ 2\ 3\ 4\ 5 \\ 2\ 4\ 3\ 1\ 5 \end{pmatrix}$$

$$\text{For } n = 6: \quad (2, 4, 1) = \begin{pmatrix} 1\ 2\ 3\ 4\ 5\ 6 \\ 2\ 4\ 3\ 1\ 5\ 6 \end{pmatrix}$$

$$\vdots \qquad\qquad \vdots$$

Since $(2, 4, 1)$ can represent an element of $\mathbf{S}_n$ for any integer $n \geq 4$, the context of the discussion must be used to determine the permutation represented by $(2, 4, 1)$.

We state the following theorem without proof. It indicates that any permutation can be decomposed into a product of cycles.

5.1.3 Theorem

For any positive integer n, each permutation in $\mathbf{S}_n$ can be decomposed into a product of pairwise disjoint cycles. This decomposition is unique except for the order in which the cycles appear in the decomposition.

5.1.4 Example

Decompose $\delta = \begin{pmatrix} 1\ 2\ 3\ 4\ 5\ 6\ 7 \\ 4\ 5\ 2\ 1\ 3\ 7\ 6 \end{pmatrix}$ into a product of disjoint cycles in $\mathbf{S}_7$.

Solution. Start with the integer 1. Inspection of δ shows that $\delta(1) = 4$ and $\delta(4) = 1$. Hence, $1 \longmapsto 4 \longmapsto 1$, which we can write as $(1, 4)$. Next, pick the smallest integer from 1 to 7 that does not appear in the cycle $(1, 4)$. This is 2, so trace through the permutation beginning with 2. This shows that $2 \longmapsto 5 \longmapsto 3 \longmapsto 2$, and we can write this as the cycle $(2, 5, 3)$. The smallest integer from 1 to 7 that does not appear in either of these cycles is 6, and 6 determines the cycle $(6, 7)$. Every integer from 1 to 7 appears in one of the cycles, so all the cycles in δ have been determined. We claim that $\delta = (1, 4)(2, 5, 3)(6, 7)$. If

$$\alpha = (1, 4) = \begin{pmatrix} 1\ 2\ 3\ 4\ 5\ 6\ 7 \\ 4\ 2\ 3\ 1\ 5\ 6\ 7 \end{pmatrix},$$

$$\beta = (2, 5, 3) = \begin{pmatrix} 1\ 2\ 3\ 4\ 5\ 6\ 7 \\ 1\ 5\ 2\ 4\ 3\ 6\ 7 \end{pmatrix}, \quad \text{and}$$

$$\gamma = (6, 7) = \begin{pmatrix} 1\ 2\ 3\ 4\ 5\ 6\ 7 \\ 1\ 2\ 3\ 4\ 5\ 7\ 6 \end{pmatrix}, \quad \text{then}$$

$\alpha\beta\gamma = (1, 4)\ (2, 5, 3)\ (6, 7)$

$$= \begin{pmatrix} 1\ 2\ 3\ 4\ 5\ 6\ 7 \\ 4\ 2\ 3\ 1\ 5\ 6\ 7 \end{pmatrix}\begin{pmatrix} 1\ 2\ 3\ 4\ 5\ 6\ 7 \\ 1\ 5\ 2\ 4\ 3\ 6\ 7 \end{pmatrix}\begin{pmatrix} 1\ 2\ 3\ 4\ 5\ 6\ 7 \\ 1\ 2\ 3\ 4\ 5\ 7\ 6 \end{pmatrix}$$

$$= \begin{pmatrix} 1\ 2\ 3\ 4\ 5\ 6\ 7 \\ 4\ 5\ 2\ 1\ 3\ 7\ 6 \end{pmatrix}$$

$$= \delta.$$

A permutation may itself be a cycle. For example, the permutation $\zeta = \begin{pmatrix} 1\ 2\ 3\ 4\ 5 \\ 3\ 1\ 5\ 2\ 4 \end{pmatrix} \in \mathbf{S}_5$ is a cycle that can be written as $(1, 3, 5, 4, 2)$. In this case, ζ is a product of cycles, but there is only one factor in the product, namely, ζ itself.

Next, suppose that $\alpha = (2, 4, 6)$ and $\beta = (1, 3, 5, 2)$ are cycles in $\mathbf{S}_6$. Then $\alpha\beta = (2, 4, 6)(1, 3, 5, 2) = (1, 3, 5, 4, 6, 2) = \gamma$. This product can be computed mentally, without writing α and β in two-row notation, by tracing through the elements of the permutations. We write out the beginning of what can actually be done mentally: in $\beta = (1, 3, 5, 2)$, $1 \longmapsto 3$, and in $\alpha = (2, 4, 6)$, 3 is not moved. Hence, $1 \longmapsto 3$ in the product. Thus, we write the first components of the product $\alpha\beta$ as $(1, 3, \ldots)$. Now we need to see where 3 is mapped in order to obtain the next element in the product. In $\beta = (1, 3, 5, 2)$, $3 \longmapsto 5$, and in $\alpha = (2, 4, 6)$, 5 is not moved, so $\alpha\beta = (1, 3, 5, \ldots)$. Next, we need to compute the image of 5 to obtain the next element of the product. In $\beta = (1, 3, 5, 2)$, $5 \longmapsto 2$, and in $\alpha = (2, 4, 6)$, $2 \longmapsto 4$, so $5 \longmapsto 4$ in the product. Hence, $\alpha\beta = (1, 3, 5, 4, \ldots)$ and so on to complete the product $\alpha\beta$. In general, if this produces a cycle that does not contain all the integers in S, pick the smallest integer from S that is not in the cycle, and repeat the process. Continue this procedure whenever a cycle is encountered until all the integers in S are exhausted.

It can easily be shown that if $\zeta = (1, 3, 5, 4, 6, 2) \in \mathbf{S}_6$, then $\zeta = (1, 3, 5, 4, 6, 2) = (2, 4, 6)(1, 3, 5, 2)$, so ζ is the product of two cycles that are not disjoint. This does not violate Theorem 5.1.3, which indicates that each permutation can be written as a product of disjoint cycles. *What Theorem 5.1.3 says is that among all the decompositions of a permutation as a product of cycles, there is exactly one (except for the order of the cycles involved) in which the cycles are pairwise disjoint.*

5.1.5 | **Example**

Show that the number of permutations in $\mathbf{S}_n$ is $n!$. In particular, determine the number of permutations in $\mathbf{S}_5$.

Solution. This is what is known as a counting problem, and the solution invokes the **First Principle of Counting**: if event E_1 can be done in n_1 ways, event E_2 can be done in n_2 ways, ... and event E_k can be done in n_k ways, then all the events can be accomplished together in $n_1 n_2 \cdots n_k$ ways. To form a permutation

$$\alpha = \left(\begin{array}{cccccc} 1 & 2 & 3 & \cdots & n-1 & n \\ \underline{} & \underline{} & \underline{} & \cdots & \underline{} & \underline{} \end{array} \right)$$

in $\mathbf{S}_n$, the element 1 can be mapped to an element of $\{1, 2, \ldots, n\}$ in n different ways. After an element has been chosen to be the image of 1, this leaves $n - 1$ choices for the image of 2. After the image of 2 has been chosen, there are $n - 2$ choices for the image of 3, and so on. Hence, by the First Principle of Counting, the number of ways a permutation in $\mathbf{S}_n$ can be constructed is $n(n - 1)(n - 2) \cdots 2 \cdot 1 = n!$ Thus, $n!$ is the number of permutations in $\mathbf{S}_n$. Since $5! = 5 \cdot 4 \cdot 3 \cdot 2 \cdot 1 = 120$, there are 120 permutations in $\mathbf{S}_5$.

If α is a permutation, then the smallest positive integer k such that $\alpha^k = (1)$ is the **order of the permutation**, where $\alpha^1 = \alpha$, $\alpha^2 = \alpha\alpha$, $\alpha^3 = \alpha\alpha\alpha$, and so on. *One can show that if α is a cycle of length k, then k is the order of α. Furthermore, if a permutation α is decomposed into a product of disjoint cycles, then the order of α is the least common multiple of the orders of the disjoint cycles of α.*

5.1.6 | **Example**

Find the order of the permutation

$$\alpha = \left(\begin{array}{ccccccccc} 1 & 2 & 3 & 4 & 5 & 6 & 7 & 8 & 9 \\ 3 & 4 & 5 & 6 & 1 & 8 & 9 & 2 & 7 \end{array} \right).$$

Solution. The permutation α can be decomposed into a product of disjoint cycles as $\alpha = (2, 4, 6, 8)(1, 3, 5)(7, 9)$. The orders of the cycles of α are 4, 3, and 2. The least common multiple of 4, 3, and 2 is 12. Hence, the smallest positive integer k for which $\alpha^k = (1)$ is 12.

Recall that a 2-cycle is a transposition. Every permutation can be written as a product of transpositions. This follows from the fact that every permutation can be factored into a product of cycles and any k-cycle can be written as

$$(n_1, n_2, \ldots, n_k) = (n_1, n_k)(n_1, n_{k-1}) \cdots (n_1, n_2).$$

In particular, a 4-cycle such as $(5, 2, 6, 3)$ can be written as $(5, 2, 6, 3) = (5, 3)(5, 6)(5, 2)$. The following example shows, however, that a factorization of a permutation into a product of transpositions is *not* unique.

5.1.7 | **Example**

Show that a permutation can be written as a product of transpositions in more than one way.

Solution. Consider the following product of transpositions in $\mathbf{S}_6$.

$$\alpha = (1, 3)(1, 4)(1, 5)(1, 4) = \begin{pmatrix} 1\ 2\ 3\ 4\ 5\ 6 \\ 3\ 2\ 1\ 5\ 4\ 6 \end{pmatrix}$$

$$= (1, 3)(4, 5)$$

As a second example, in $\mathbf{S}_7$, we see that

$$\beta = (1, 4)(1, 2)(1, 3)(1, 2)(1, 6)(1, 7)(1, 6) = \begin{pmatrix} 1\ 2\ 3\ 4\ 5\ 6\ 7 \\ 4\ 3\ 2\ 1\ 5\ 7\ 6 \end{pmatrix}$$

$$= (1, 4)(2, 3)(6, 7).$$

In the preceding example, even though α was expressed in two different ways as a product of transpositions, the number of transpositions in each expression for α was even. Likewise, the number of transpositions in each expression for β is odd. We now show that it is always the case that the number of transpositions in an expression for a given permutation is always even or always odd. Before a proof can be given, new notation is needed. The proof involves consideration of a polynomial P in n variables $x_1, x_2, \ldots, x_n$ with factors of the form $(x_i - x_j)$ with $1 \leq i < j \leq n$. If Π denotes a product, define P by

$$P = \Pi_{1 \le i < j \le n}(x_i - x_j)$$

If α is a permutation in $\mathbf{S}_n$, then we define $\alpha(P)$ to be the new polynomial of this form derived from applying α to the subscripts of P:

$$\alpha(P) = \Pi_{1 \le i < j \le n}(x_{\alpha(i)} - x_{\alpha(j)}).$$

For example when $n = 4$,

$$P = (x_1 - x_2)(x_1 - x_3)(x_1 - x_4)(x_2 - x_3)(x_2 - x_4)(x_3 - x_4),$$

and if α is the transposition $\alpha = (1, 3) \in \mathbf{S}_4$, then

$$\alpha(P) = (x_3 - x_2)(x_3 - x_1)(x_3 - x_4)(x_2 - x_1)(x_2 - x_4)(x_1 - x_4).$$

The product of the factors of $\alpha(P)$ that are unchanged from those of P is

$$(x_3 - x_4)(x_2 - x_4)(x_1 - x_4),$$

while the product of the factors of $\alpha(P)$ that are changed from those of P is

$$(x_3 - x_2)(x_3 - x_1)(x_2 - x_1).$$

But

$$(x_3 - x_2)(x_3 - x_1)(x_2 - x_1) = -(x_2 - x_3)(x_1 - x_3)(x_1 - x_2),$$

so it follows that

$$\alpha(P) = -(x_2 - x_3)(x_1 - x_3)(x_1 - x_2)(x_3 - x_4)(x_2 - x_4)(x_1 - x_4)$$
$$= -(x_1 - x_2)(x_1 - x_3)(x_1 - x_4)(x_2 - x_3)(x_2 - x_4)(x_3 - x_4)$$
$$= -P.$$

Consequently, applying the transposition $\alpha = (1, 3)$ to P has the effect of changing the sign of P. If β and γ are also transpositions in $\mathbf{S}_4$, then in a similar fashion it can be shown that $\beta\alpha(P) = \beta(-P) = -\beta(P) = -(-P) = P$ and that $\gamma\beta\alpha(P) = \gamma(P) = -P$. Thus, it follows that applying an even number of transpositions in $\mathbf{S}_4$ to P leaves the sign of P unchanged, while applying an odd number of transpositions in $\mathbf{S}_4$ to P has the effect of changing the sign of P. An analysis of this procedure shows that this is true for any polynomial $P = \Pi_{1 \le i < j \le n}(x_i - x_j)$ and any finite number of transpositions in $\mathbf{S}_n$. In fact, if $\alpha_1, \alpha_2, ..., \alpha_k$ are transpositions in $\mathbf{S}_n$, then

$$(\alpha_1\alpha_2 \cdots \alpha_k)(P) = (-1)^k P.$$

5.1.8   Theorem

Let α be a permutation in $\mathbf{S}_n$. If α can be decomposed into a product of transpositions with an even (odd) number of factors, then every decomposition of α into a product of transpositions must contain an even (odd) number of factors.

Proof. Let $P = \Pi_{1 \leq i < j \leq n}(x_i - x_j)$ be a polynomial of the form described earlier, and suppose that α is decomposed as a product of transpositions in $\mathbf{S}_n$ as $\alpha = \alpha_1\alpha_2 \cdots \alpha_j$ and as $\alpha = \beta_1\beta_2 \cdots \beta_k$. Then we see that $\alpha(P) = \alpha_1\alpha_2 \cdots \alpha_j(P) = (-1)^j P$ and $\alpha(P) = \beta_1\beta_2 \cdots \beta_k(P) = (-1)^k P$, so $(-1)^j P = (-1)^k P$. But this clearly implies that j and k both must be even integers or they both must be odd. ∎

5.1.9   Definition

A permutation that can be decomposed into a product of an even number of transpositions is an **even permutation**, and one that can be decomposed into a product of an odd number of transpositions is an **odd permutation**. The set of all even permutations on n letters is denoted by $\mathbf{A}_n$.

We have seen that a k-cycle $(n_1, n_2, \ldots, n_k)$ can be expressed as $(n_1, n_2, \ldots, n_k) = (n_1, n_k)(n_1, n_{k-1}) \cdots (n_1, n_2)$. Since there are $k - 1$ transpositions on the right side of this equation, it follows that a k-cycle is an even permutation if k is an odd integer and an odd permutation if k is an even integer. For example, the 5-cycle $(3, 5, 1, 7, 4)$ is an even permutation since $(3, 5, 1, 7, 4) = (3, 4)(3, 7)(3, 1)(3, 5)$ and the 4-cycle $(1, 5, 3, 6)$ is an odd permutation since $(1, 5, 3, 6) = (1, 6)(1, 3)(1, 5)$.

5.1.10   Theorem

The set $\mathbf{A}_n$ of even permutations on n letters is a group under composition of permutations.

Proof. The operation of composition of even permutations is closed since the product of two even permutations is clearly an even permutation. Thus, composition of even permutations is a binary operation on $\mathbf{A}_n$. The identity permutation $e = (1)$ is in $\mathbf{A}_n$ since $(1) = (1, 2)(2, 1)$. Suppose that $\alpha \in \mathbf{A}_n$ and that α can be expressed as $\alpha = (n_1, m_1)(n_2, m_2) \cdots (n_k, m_k)$, where k is an even integer less than n. Then $\alpha^{-1} = (m_k, n_k)(m_{k-1}, n_{k-1}) \cdots (m_1, n_1)$, so α^{-1} is an even permutation. Thus, $\alpha^{-1} \in \mathbf{A}_n$. This shows that $\mathbf{A}_n$ is a group. ∎

The group of even permutations A_n is obviously a subgroup of S_n. In fact, A_n is a normal subgroup of S_n, but the proof of this fact is left to the exercises. The group A_n is known as the **alternating group** of permutations on n letters.

Permutation groups are very closely related to general groups, a fact demonstrated by the following theorem due to **Arthur Cayley** (1821–1895). (Cayley was the first mathematician to define an abstract group, even though other mathematicians had worked with properties of groups without using the term *group* and without forming the concept of a group. See the Historical Notes in Section 1.3 for a brief sketch of the life of Cayley.)

5.1.11 | Cayley's Theorem

Every group is isomorphic to a group of permutations.

Proof. Let G be a group. If $x \in G$, define $f_x : G \to G$ by $f(y) = xy$. We claim that f_x is a bijective function. First, note that f_x is well-defined, for if $y = z$, then $xy = xz$ since the binary operation on G is well-defined. Thus, $f_x(y) = f_x(z)$, so f_x is well-defined. The mapping f_x is injective because if $f_x(y) = f_x(z)$, then $xy = xz$. Hence, $x^{-1}xy = x^{-1}xz$, so $y = z$. Finally, if $y \in G$, then $x^{-1}y \in G$ and $f_x(x^{-1}y) = xx^{-1}y = y$. Consequently, f_x is surjective. Therefore, f_x is a bijective function from G to G and so is a permutation on G. Hence, every element of G determines a permutation on G. Let $\mathbf{P}(G)$ be this set of permutations, and define multiplication on $\mathbf{P}(G)$ by $f_x \circ f_y$, the composition of f_x with f_y. We claim that $\mathbf{P}(G)$ is a group under this operation and that G is isomorphic to $\mathbf{P}(G)$. To add clarity, the proof is presented as a sequence of steps. The first four steps show that $\mathbf{P}(G)$ is a group, and the last four steps show that $\mathbf{P}(G)$ and G are isomorphic.

1. The operation $\circ$ on $\mathbf{P}(G)$ is closed: if $f_x, f_y \in \mathbf{P}(G)$ and $z \in G$, then $(f_x \circ f_y)(z) = f_x(yz) = xyz = f_{xy}(z)$. Hence, $f_x \circ f_y = f_{xy} \in \mathbf{P}(G)$, so the operation is closed. The operation is clearly well-defined since composition of functions is a well-defined operation.
2. The operation $\circ$ is associative: if $f_x, f_y, f_z \in \mathbf{P}(G)$, then $f_x \circ (f_y \circ f_z) = f_x \circ f_{yz} = f_{x(yz)} = f_{(xy)z} = f_{xy} \circ f_z = (f_x \circ f_y) \circ f_z$. Notice that the proof relies on the fact that $x(yz) = (xy)z$ in G.
3. $\mathbf{P}(G)$ has a multiplicative identity: if e is the identity of G and f_x is any element of $\mathbf{P}(G)$, then $(f_e \circ f_x)(z) = f_e(xz) = exz = xz = f_x(z)$ for any $z \in G$. Hence, $f_e \circ f_x = f_x$. Similarly, $f_x \circ f_e = f_x$, so f_e is the multiplicative identity of $\mathbf{P}(G)$.
4. Every element of $\mathbf{P}(G)$ has a multiplicative inverse: if $f_x \in \mathbf{P}(G)$, then $f_{x^{-1}} \in \mathbf{P}(G)$ and $(f_x \circ f_{x^{-1}})(z) = f_x(x^{-1}z) = xx^{-1}z = ez = f_e(z)$ for any

$z \in G$. Hence, $f_x \circ f_{x^{-1}} = f_e$. In a similar manner, one can show that $f_{x^{-1}} \circ f_x = f_e$. Thus, every $f_x \in \mathbf{P}(G)$ has an inverse in $\mathbf{P}(G)$, so $\mathbf{P}(G)$ is a group under the operation $\circ$.

Finally, define $\phi : G \to \mathbf{P}(G)$ by $\phi(x) = f_x$. The proof will be complete if we can show that ϕ is a well-defined group isomorphism.

5. The mapping ϕ is well-defined: suppose that $x = y$; then for any $z \in G$, $xz = yz$, so $f_x(z) = f_y(z)$. Since z was chosen arbitrarily in G, $f_x = f_y$, so $\phi(x) = \phi(y)$.

6. The binary operation is preserved by ϕ: we have already seen in step 1 that $f_{xy} = f_x \circ f_y$. Hence, $\phi(xy) = f_{xy} = f_x \circ f_y = \phi(x) \circ \phi(y)$.

7. The mapping ϕ is injective: suppose that $\phi(x) = \phi(y)$; then $f_x = f_y$. Thus, $f_x(e) = f_y(e)$, so $xe = ye$ or $x = y$.

8. The mapping ϕ is surjective: suppose that $f_x \in \mathbf{P}(G)$; then $\phi(x) = f_x$, so ϕ is surjective.

Therefore, ϕ is a group isomorphism, so G and $\mathbf{P}(G)$ are isomorphic groups. ∎

We have just seen that an arbitrary group G is really a group of permutations $\mathbf{P}(G)$. However, $\mathbf{P}(G)$ need not coincide with the group $\Gamma(G)$ of all permutations on G since bijective functions $f : G \to G$ may exist that are not of the form f_x for some $x \in G$. Nevertheless, $\mathbf{P}(G)$ is a subgroup of $\Gamma(G)$ since function composition is the binary operation on both groups. Since there is a one-to-one correspondence among the permutations in $\mathbf{P}(G)$ and the elements of G, the number of elements of $\mathbf{P}(G)$ is equal to the number of elements in G, that is, $|\mathbf{P}(G)| = |G|$. When G is a finite group, Example 5.1.5 shows that $|\Gamma(G)| = |G|!$, so for groups of order 2, $\mathbf{P}(G) = \Gamma(G)$, but $\mathbf{P}(G)$ is a proper subgroup of $\Gamma(G)$ when G has finite order ≥ 3.

Permutations can also be used to represent the reflections and rotations of symmetric figures, such as an equilateral triangle, a square, or a regular pentagon. The reflections and rotations of a square were used in Chapter 2 to introduce the concept of a group. When one of these actions is performed on such a figure, the figure that results must be in exactly the same position as the original with the exception that the numbering on the vertices may be different. Such reflections and rotations of asymmetric figure are referred to as **rigid motions** of the figure. This is demonstrated with an equilateral triangle. In the exercises, you will be asked to use permutations to describe the rigid motions of two symmetric figures.

The following figures represent either the reflection of an equilateral triangle about a line or a rotation through an angle about the center of the triangle. The permutation that describes the reflection or rotation is located below or beside the figure.

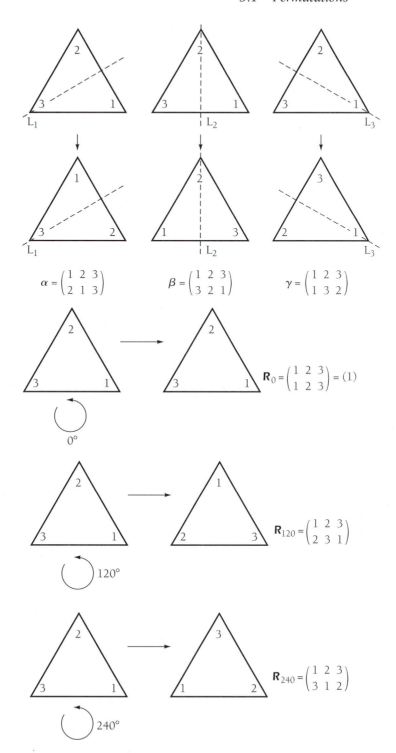

Of course, rotation of the triangle through 0° or 360° is the identity permutation (1) of S_3. This is precisely the group S_3 whose multiplication table is given below.

•	(1)	α	β	γ	R_{120}	R_{240}
(1)	(1)	α	β	γ	R_{120}	R_{240}
α	α	(1)	R_{240}	R_{120}	γ	β
β	β	R_{120}	(1)	R_{240}	α	γ
γ	γ	R_{240}	R_{120}	(1)	β	α
R_{120}	R_{120}	β	γ	α	R_{240}	(1)
R_{240}	R_{240}	γ	α	β	(1)	R_{120}

HISTORICAL NOTES

Augustin-Louis Cauchy (1789–1857). Augustin-Louis Cauchy was born in Paris, France, during the year the French Revolution began. His father, fearing the terrors of the Revolution, took the family and fled to the French countryside, where they were to live for the next 11 years. The elder Cauchy assumed responsibility for the education of his son and ensured that the boy received a strong literary grounding before allowing him to study mathematics. Young Cauchy's classical schooling continued when, at the age of 13, he entered the Ecole Centrale du Pantheon. He then spent two years at the Ecole Polytechnique before entering a civil engineering school in 1807. An outstanding student and a successful engineer, Cauchy found that his interest lay in pure mathematics. Eventually, he was able to obtain teaching positions at the Ecole Polytechnique and later at the Collège de France. By 1816, Cauchy had advanced to the top rank of the mathematicians of his era. His mathematical production was prodigious, and he was elected to the French Academy of Science at the age of 27. Cauchy also acquired the reputation of being a difficult, contentious man whose excessive piety irritated many of his colleagues. During his lifetime, Cauchy produced 789 papers in mathematics, an incredible achievement. His name is attached to many mathematical concepts and theorems, for example, Cauchy sequences, the Cauchy integral theorem, and the Cauchy-Riemann equations. (It was mentioned early in Section 4.3 that equivalence classes of Cauchy sequences of rational numbers can be used to develop the field of real numbers.)

Problem Set 5.1

1. If $\alpha = \begin{pmatrix} 1\ 2\ 3\ 4\ 5 \\ 4\ 2\ 5\ 3\ 1 \end{pmatrix}$, $\beta = \begin{pmatrix} 1\ 2\ 3\ 4\ 5 \\ 3\ 4\ 1\ 5\ 2 \end{pmatrix}$, and $\gamma = \begin{pmatrix} 1\ 2\ 3\ 4\ 5 \\ 5\ 2\ 4\ 1\ 3 \end{pmatrix}$,

    compute each of the following:

    (a) $\alpha\beta$ (b) $\beta\gamma$ (c) α^2 (d) $\alpha\gamma$ (e) $\gamma\alpha$

2. Compute each of the following in $\mathbf{S}_6$.

    (a) $(2, 5, 3)(1, 4, 2)$ (c) $(2, 5, 4, 6, 3)(3, 6, 4, 5, 2)$

    (b) $(1, 6, 2)(1, 5, 6, 4)$ (d) $(1, 4, 5, 6)(3, 2)$

3. Decompose each of the following into a product of disjoint cycles in $\mathbf{S}_7$.

    (a) $\begin{pmatrix} 1\ 2\ 3\ 4\ 5\ 6\ 7 \\ 3\ 4\ 5\ 6\ 1\ 2\ 7 \end{pmatrix}$ (c) $\begin{pmatrix} 1\ 2\ 3\ 4\ 5\ 6\ 7 \\ 4\ 3\ 2\ 1\ 5\ 6\ 7 \end{pmatrix}\begin{pmatrix} 1\ 2\ 3\ 4\ 5\ 6\ 7 \\ 1\ 3\ 4\ 2\ 5\ 6\ 7 \end{pmatrix}$

    (b) $(1, 4, 5, 6)(4, 1, 3, 5, 7)$ (d) $(2, 4)(2, 5)(2, 5)$

4. Show that if $\alpha = (3, 2, 4, 1, 5)$, then $\alpha^{-1} = (5, 1, 4, 2, 3)$. Generalize this, and show that if $\alpha = (n_1, n_2, \ldots, n_k)$, then $\alpha^{-1} = (n_k, n_{k-1}, \ldots, n_1)$.

5. Write each of the following as a product of transpositions, and classify each as an even or an odd permutation.

    (a) $\begin{pmatrix} 1\ 2\ 3\ 4\ 5 \\ 5\ 2\ 4\ 1\ 3 \end{pmatrix}$ (b) $\begin{pmatrix} 1\ 2\ 3\ 4\ 5\ 6\ 7 \\ 3\ 4\ 5\ 6\ 1\ 2\ 7 \end{pmatrix}$ (c) $\begin{pmatrix} 1\ 2\ 3\ 4\ 5\ 6\ 7 \\ 3\ 6\ 5\ 1\ 4\ 2\ 7 \end{pmatrix}$

6. Classify each of the following as an even or an odd permutation without expressing it as a product of transpositions.

    (a) $(1, 6, 2)(1, 5, 6, 4)$

    (b) $(2, 5, 4, 6, 3)(3, 6, 4, 5, 2)$

7. Prove that $\mathbf{A}_n$ is a normal subgroup of $\mathbf{S}_n$. Hint: Consider $\mathbf{Z}_2$ as a group under addition modulo 2 and show that the function

    $$f: \mathbf{S}_n \to \mathbf{Z}_2, \text{ where } f(\alpha) = \begin{cases} [0] \text{ if } \alpha \text{ is even} \\ [1] \text{ is } \alpha \text{ is odd} \end{cases}$$

    is a group homomorphism. What is the kernel of f?

8. Prove that $\left| \mathbf{A}_n \right| = \dfrac{n!}{2}$.

9. Compute the order of each of the following permutations.

    (a) $\alpha = \begin{pmatrix} 1\ 2\ 3\ 4\ 5 \\ 3\ 4\ 2\ 5\ 1 \end{pmatrix}$

(b) $\beta = (2, 4, 3)(1, 5)(6, 7, 8, 9)$

(c) $\gamma = \begin{pmatrix} 1\ 2\ 3\ 4\ 5\ 6\ 7 \\ 3\ 4\ 5\ 6\ 1\ 2\ 7 \end{pmatrix}$

10. Express the rigid motions of each of the following figures as permutations, and in each case, construct a multiplication table for each set of permutations associated with the figure.

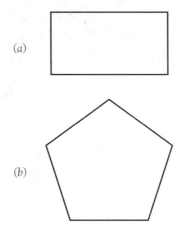

(a)

(b)

11. The multiplication table for the group of rigid motions of an equilateral triangle is given on page 234. Use the table to identity the subgroup $\mathbf{A}_3$ of $\mathbf{S}_3$. Hint: Show that each rotation is an even permutation and that these are the only even permutations.

TECHNOLOGY PROBLEMS

Use a computer algebra system to solve each of the following problems.

12. Find the cycle decomposition of each of the following permutations.

(a) $\begin{pmatrix} 1\ \ 2\ 3\ 4\ 5\ 6\ 7\ 8\ 9\ 10 \\ 10\ 9\ 8\ 5\ 6\ 7\ 4\ 3\ 2\ \ 1 \end{pmatrix}$

(b) $\begin{pmatrix} 1\ \ 2\ \ 3\ \ 4\ 5\ 6\ 7\ 8\ \ 9\ \ 10\ 11\ 12\ 13\ 14\ 15 \\ 14\ 2\ 15\ 4\ 6\ 7\ 5\ 10\ 13\ 11\ 12\ 10\ \ 9\ \ 3\ \ \ 1 \end{pmatrix}$

(c) Find the inverse of the permutations of parts (a) and (b).

13. Find the permutation given by the following products in $\mathbf{S}_{12}$.

(a) $(2, 4, 6, 8, 10, 12)(1, 3, 5, 7, 9, 11)$

(b) $(2, 4, 6)(8, 10, 12)(1, 3, 5)(7, 9, 11)$

5.2   CYCLIC GROUPS

Section Overview. The second type of group we study is known as a cyclic group. A cyclic group is simply a group that is generated by a single element. Before beginning this section, make sure that you understand the concept of the order of an element in a group. Theorem 5.2.4 shows that every subgroup of a cyclic group is cyclic, and Theorem 5.2.8 provides a quick way to determine the order of an element in a cyclic group. Theorem 5.2.10 is the converse of Lagrange's Theorem for cyclic groups.

Cyclic groups were introduced and defined in Examples 2 through 5 of 2.2.2. For your convenience, the following definition includes the definitions presented there.

5.2.1   Definition

If G is a group and $x \in G$, then $\langle x \rangle = \{x^n \mid n \in \mathbf{Z}\} = \{ \ldots, x^{-2}, x^{-1}, x^0 = e, x^1, x^2, \ldots \}$. The group G is said to be **cyclic** if there is an $x \in G$ such that $G = \langle x \rangle$. When $G = \langle x \rangle$, we say that G is a **cyclic group generated by** x and that x is a **generator of** G. If H is a subgroup of G and there is an $x \in H$ such that $\langle x \rangle = H$, then H is a **cyclic subgroup** of G. If $x \in G$ and there is an integer n such that $x^n = e$, then the smallest positive integer n such that $x^n = e$ is said to be the **order** of x. If an integer n exists such that $x^n = e$, then x is said to be of **finite order**, and if no such integer exists, then x is said to have **infinite order**. The order of $x \in G$ is denoted by $\mathbf{o}(x)$.

Now let's investigate cyclic groups in more detail through the following examples.

5.2.2   Examples

1. The additive group $\mathbf{Z}_6$ is a cyclic group generated by [1]. By saying that $\mathbf{Z}_6$ is generated by [1], we mean that every element of $\mathbf{Z}_6$ can be obtained by repeatedly adding [1] to itself:

$$0[1] = [0]$$
$$1[1] = [1]$$
$$2[1] = [1] + [1] = [2]$$
$$3[1] = [1] + [1] + [1] = [3]$$
$$\vdots$$
$$6[1] = [6] = [0].$$

By the process of repeatedly adding $[1]$ to itself, we have generated all the elements of $\mathbf{Z}_6$. Since 6 is the smallest positive integer such that $6[1] = [0]$, $\mathbf{o}([1]) = 6$. If this procedure is continued, nothing new is gained. The elements begin to repeat, and $\mathbf{Z}_6$ is generated again:

$$7[1] = [7] = [1]$$
$$8[1] = [8] = [2]$$
$$9[1] = [9] = [3]$$
$$\vdots$$
$$12[1] = [0]$$
$$\vdots$$

The same is true if we use negative integers:

$$-1[1] = -[1] = [5]$$
$$-2[1] = -[2] = [4]$$
$$-3[1] = -[3] = [3]$$
$$\vdots$$
$$-6[1] = -[6] = [0]$$
$$\vdots$$

This shows that $\mathbf{Z}_6 = \{\ldots, -4[1], -3[1], -2[1], -1[1], 0[1], 1[1], 2[1], 3[1], 4[1],\ldots\} = \{n[1] \mid n \in \mathbf{Z}\}$. Not every element of $\mathbf{Z}_6$ generates $\mathbf{Z}_6$. For example, $\langle[2]\rangle = \{[0], [2], [4]\}$ and $\mathbf{o}([2]) = 3$. It is not difficult to calculate all of the subgroups of $\mathbf{Z}_6$. Examination of all the subgroups of $\mathbf{Z}_6$ shows that each is cyclic.

2. For this example, consider the multiplicative group $U(\mathbf{Z}_{12})$ of units of $\mathbf{Z}_{12}$. Recall that $U(\mathbf{Z}_{12})$ is the group of all elements of $\mathbf{Z}_{12}$ that have multiplicative inverses in $\mathbf{Z}_{12}$ and that an element $[x] \in U(\mathbf{Z}_{12})$ if and only if $\gcd(x, 12) = 1$. Hence, $U(\mathbf{Z}_{12}) = \{[1], [5], [7], [11]\}$. $U(\mathbf{Z}_{12})$ is *not* generated by a single element

under multiplication. Each element of $U(\mathbf{Z}_{12})$ generates a proper subgroup of $U(\mathbf{Z}_{12})$.

$[1]^1 = [1]$, so $\langle[1]\rangle = \{[1]\}$ is the subgroup generated by $[1]$.

$[5]^1 = [5]$

$[5]^2 = [25] = [1]$, so $\langle[5]\rangle = \{[1],[5]\}$ is the subgroup generated by $[5]$.

$[7]^1 = [7]$

$[7]^2 = [49] = [1]$, so $\langle[7]\rangle = \{[1],[7]\}$ is the subgroup generated by $[7]$.

$[11]^1 = [11]$

$[11]^2 = [121] = [1]$, so $\langle[11]\rangle = \{[1],[11]\}$.

Hence, $U(\mathbf{Z}_{12})$ is not a cyclic group and thus groups exists that are not cyclic. Note that $\mathbf{o}([1]) = 1$ and that $[5]$, $[7]$, and $[11]$ each have order 2.

3. The additive group $\mathbf{Z}$ is a cyclic group. It is generated by 1 and by -1 since $\mathbf{Z} = \{n1 \mid n \in \mathbf{Z}\} = \{n(-1) \mid n \in \mathbf{Z}\}$. Hence, $\mathbf{Z}$ is a cyclic group. Every subgroup of $\mathbf{Z}$ looks like $k\mathbf{Z}$ for some integer k. Therefore, every subgroup of $\mathbf{Z}$ is cyclic.

4. For this example, review the material on the group of rotations and reflections of a square given at the beginning of Chapter 2, and let $\mathcal{R} = \{R_0, R_{90}, R_{180}, R_{270}\}$ be the set of all rotations of a square. The group of rotations $\mathcal{R}$ is a cyclic group generated by R_{90} since

$$R_{90}{}^1 = R_{90},$$

$$R_{90}{}^2 = R_{90}\, R_{90} = R_{180},$$

$$R_{90}{}^3 = R_{90}\, R_{90}\, R_{90} = R_{270}, \qquad \text{and}$$

$$R_{90}{}^4 = R_{90}\, R_{90}\, R_{90}\, R_{90} = R_{360} = R_0.$$

Hence, $\langle R_{90}\rangle = \mathcal{R}$. $\mathcal{R}$ is also generated by R_{270}, but R_{180} generates a proper subgroup of $\mathcal{R}$ since $\langle R_{180}\rangle = \{R_0, R_{180}\}$. Thus, $\mathbf{o}(R_{90}) = \mathbf{o}(R_{270}) = 4$ and $\mathbf{o}(R_{180}) = 2$.

Since $\mathcal{R}$ is a finite group, the order of a subgroup of $\mathcal{R}$ must divide the order of $\mathcal{R}$. Because of this, a little thought should convince you that $\langle R_0\rangle, \langle R_{180}\rangle$, and $\mathcal{R}$ are the only subgroups of $\mathcal{R}$. Hence, $\mathcal{R}$ is a cyclic group, and every subgroup of $\mathcal{R}$ is cyclic.

If G is a group and $x \in G$, then Examples 1 and 2 show that it may be the case that $\langle x\rangle \neq G$. However, the following theorem shows that $\langle x\rangle$ is always a subgroup of G, as you may have already guessed.

5.2.3   Theorem

If G is a group, then for any $x \in G$, $\langle x \rangle$ is a subgroup of G.

Proof. Let x be any element of a group G. Then $\langle x \rangle \neq \varnothing$ since x^n is in $\langle x \rangle$ for every integer n. Next, if $x^n, x^m \in \langle x \rangle$, then $x^{-m} \in \langle x \rangle$ since $-m \in \mathbf{Z}$. Finally, $x^n x^{-m} = x^{n-m} \in \langle x \rangle$ since $n - m \in \mathbf{Z}$. Thus, by Theorem 2.2.3, $\langle x \rangle$ is a subgroup of G. ∎

The groups in the Examples 1, 3, and 4 are examples of cyclic groups that have the property that their subgroups are cyclic. This did not happen by chance. It is always true that a subgroup of a cyclic group is cyclic.

5.2.4   Theorem

Every subgroup of a cyclic group is cyclic.

Proof. Suppose that G is a cyclic group G with generator x. If H is a subgroup of G and $H = \{e\}$, then $H = \langle e \rangle$, so H is cyclic. If $H \neq \{e\}$, let k be the smallest positive integer such that $x^k \in H$. Then $\langle x^k \rangle \subseteq H$. If $y \in H$, then $y = x^n$ for some $n \in \mathbf{Z}$. Now H is a subgroup of G, so the multiplicative inverse x^{-k} of x^k is also in H. By the Division Algorithm 1.1.3, there are integers q and r such that $n = kq + r$, where $0 \leq r < k$. Hence, we see that $x^r = x^{n-kq} = x^n(x^{-k})^q \in H$. Consequently, $r = 0$ since k is the smallest positive integer such that $x^k \in H$. Therefore, $n = kq$, which implies that $y = x^n = x^{kq} = (x^k)^q \in \langle x^k \rangle$. Hence, $H = \langle x^k \rangle$. ∎

The following interesting theorem tells us exactly what the structure of a finite cyclic group looks like.

5.2.5   Theorem

Let $G = \langle x \rangle$ be a finite cyclic group. If $\mathbf{o}(x) = n$, then the following hold:

1. If j and k are integers, then $x^j = x^k$ if and only if n divides $j - k$.
2. The elements of $\{e, x, x^2, \ldots, x^{n-1}\}$ are distinct.
3. $G = \{e, x, x^2, \ldots, x^{n-1}\}$ and $|G| = \mathbf{o}(x)$.

Proof

1. Let j and k be integers such that $x^j = x^k$. By the Division Algorithm 1.1.3, there are integers q and r such that $j - k = nq + r$, where $0 \leq r < n$. From this we see that $e = x^{j-k} = x^{nq+r} = (x^n)^q(x^r) = x^r$. Hence, $r = 0$ since n is

the smallest positive integer such that $x^n = e$. Therefore, $j - k = nq$, so $n \mid (j - k)$. Conversely, if $n \mid (j - k)$, then $j - k = nq$ for some integer q, and we have $x^{j-k} = x^{nq} = (x^n)^q = e$. Thus, $x^j = x^k$.

2. If $x^j = x^k$ with $0 \leq j \leq k < n$, then $x^{k-j} = e$. But this implies that $n \mid (k - j)$, which is possible only if $k - j = 0$. Thus, $j = k$ and so the elements of $\{e, x, x^2, \ldots, x^{n-1}\}$ are distinct.

3. Let $x^m \in G$, and suppose that $m > n$. Let q and r be integers such that $m = nq + r$, where $0 \leq r < n$. Then $x^m = x^{nq+r} = x^{nq} x^r = (x^n)^q x^r = e^q x^r = e x^r = x^r$. But since $0 \leq r < n$, $x^r \in \{e, x, x^2, \ldots, x^{n-1}\}$, so $x^m \in \{e, x, x^2, \ldots, x^{n-1}\}$. Hence, $G \subseteq \{e, x, x^2, \ldots, x^{n-1}\}$. Since $\{e, x, x^2, \ldots, x^{n-1}\} \subseteq G$, we have equality. Clearly, $|G| = \mathbf{o}(x)$. ∎

Let $G = \langle x \rangle$ be a cyclic group. If $\mathbf{o}(x) = n$, then $|G| = n$, and we say that G is a **cyclic group of finite order** n. If x has infinite order, then we say that G has **infinite order.**

5.2.6   Corollary

If G is a finite group and $x \in G$, then $\mathbf{o}(x)$ divides $|G|$.

Proof. Since G contains a finite number of elements, $\mathbf{o}(x)$ must be finite, and by Theorem 5.2.5, $|\langle x \rangle| = \mathbf{o}(x)$. Now invoke Lagrange's Theorem 2.2.9. ∎

5.2.7   Example

Consider the additive group $\mathbf{Z}_8$. Any element of $\mathbf{Z}_8$ generates a cyclic subgroup of $\mathbf{Z}_8$. The following is a list of all the subgroups of $\mathbf{Z}_8$ that are generated by its elements and the order of each generator.

$\langle [0] \rangle = \{[0]\}$	$\mathbf{o}([0]) = 1$
$\langle [1] \rangle = \{[0], [1], [2], [3], [4], [5], [6], [7]\}$	$\mathbf{o}([1]) = 8$
$\langle [2] \rangle = \{[0], [2], [4], [6]\}$	$\mathbf{o}([2]) = 4$
$\langle [3] \rangle = \{[0], [1], [2], [3], [4], [5], [6], [7]\}$	$\mathbf{o}([3]) = 8$
$\langle [4] \rangle = \{[0], [4\}$	$\mathbf{o}([4]) = 2$
$\langle [5] \rangle = \{[0], [1], [2], [3], [4], [5], [6], [7]\}$	$\mathbf{o}([5]) = 8$
$\langle [6] \rangle = \{[0], [2], [4], [6]\}$	$\mathbf{o}([6]) = 4$
$\langle [7] \rangle = \{[0], [1], [2], [3], [4], [5], [6], [7]\}$	$\mathbf{o}([7]) = 8$

Not all of these subgroups are distinct. Indeed, $\langle[1]\rangle = \langle[3]\rangle = \langle[5]\rangle = \langle[7]\rangle$ and $\langle[2]\rangle = \langle[6]\rangle$. Notice that the order of each element of $\mathbf{Z}_8$ divides the order of $\mathbf{Z}_8$, as asserted by Corollary 5.2.6.

Example 5.2.7 shows that certain elements of the finite cyclic group $\mathbf{Z}_8$ generate $\mathbf{Z}_8$, while others do not. Furthermore, not all of the subgroups of $\mathbf{Z}_8$ generated by the elements of $\mathbf{Z}_8$ are distinct. This raises the question of how one determines the elements of a finite cyclic group G that generate distinct subgroups of G. This question can be addressed by proving the converse of Lagrange's Theorem for cyclic groups. To prove this converse, we need the following theorem that provides information about the order of the elements of a cyclic group.

5.2.8   Theorem

Let G be a group, and suppose that x is an element of G such that $\mathbf{o}(x) = n$.

1. For any integer m, $x^m = e$ if and only if n divides m.
2. For every positive integer m, $\mathbf{o}(x^m) = \dfrac{n}{\gcd(m, n)}$. (Recall that $\gcd(m, n)$ denotes the greatest common divisor of m and n.)

Proof

1. Suppose that m is an integer such that $x^m = e$. By the Division Algorithm 1.1.3, there are integers q and r such that $m = nq + r$, where $0 \le r < n$. Thus, $e = x^m = x^{nq+r} = (x^n)^q x^r = e^q x^r = x^r$. But n is the smallest positive integer such that $x^n = e$, so $r = 0$. Hence, $m = nq$ and thus $n \mid m$. Conversely, if $n \mid m$, then there is an integer k such that $m = nk$. But this yields $x^m = x^{nk} = (x^n)^k = e^k = e$.
2. If $\mathbf{o}(x^m) = k$, then $x^{mk} = e$ and so, by part 1, $n \mid mk$. Hence, there is a $q \in \mathbf{Z}$ such that $mk = nq$. Let $d = \gcd(m, n)$. Then there exist integers a and b such that $m = da$ and $n = db$ with $\gcd(a, b) = 1$. Now $mk = nq$ implies that $dak = dbq$, so $ak = qb$. Hence, $b \mid ka$. But $\gcd(a, b) = 1$, and so $b \mid k$. Thus, $\frac{n}{d}$ divides k. Note also that $(x^m)^{n/d} = x^{mn/d} = x^{nda/d} = x^{na} = (x^n)^a = e^a = e$, so by part 1, k divides $\frac{n}{d}$. Thus, $k \mid \frac{n}{d}$ and $\frac{n}{d} \mid k$, so because k and $\frac{n}{d}$ are positive integers, it must be the case that $k = \frac{n}{d}$. Hence, $\mathbf{o}(x^m) = \dfrac{n}{\gcd(m, n)}$. ∎

5.2.9   Example

1. Consider the group $\mathcal{R} = \{R_0, R_{90}, R_{180}, R_{270}\}$ of the rotations of a square. (See Example 4 of 5.2.2.) We know that R_{180} has order 2, so what is the

order of R_{180}^{12}? Theorem 5.2.8 indicates that the order of R_{180}^{12} is given by
$\mathbf{o}(R_{180}^{12}) = \frac{2}{\gcd(2,\,12)} = \frac{2}{2} = 1$, so $(R_{180}^{12})^1 = R_0$. This should be
obvious without doing the mathematics. Since $R_{180}^2 = R_{360} = R_0$,
$R_{180}^{12} = (R_{180}^2)^6 = R_0^6 = R_0$.

2. Consider $[20] \in \mathbf{Z}_{32}$, which has order 8 since $8[20] = [8 \cdot 4 \cdot 5] = [32][5] = [0][5] = [0]$ and 8 is the smallest such positive integer that produces $[0]$ when multiplied by $[20]$. Now pick an arbitrary integer, say, 12. What is the order of $12[20]$? Since $\gcd(8, 12) = 4$, $\mathbf{o}(12[20]) = \frac{8}{4} = 2$. Note that $2(12[20]) = [32][15] = [0]$.

Lagrange's Theorem indicates that if G is a finite group and H is a subgroup of G, then the order of H divides the order of G. The converse of Lagrange's Theorem states that if d is a divisor of the order of G, then G has a subgroup of order d. *The converse of Lagrange's Theorem does not hold for arbitrary groups.* However, the following theorem shows that the converse of Lagrange's Theorem does hold for finite cyclic groups. The theorem also characterizes the elements of a finite cyclic group that generate its cyclic subgroups.

5.2.10 Theorem

Let $G = \langle x \rangle$ be a finite cyclic group of order n. Then for every positive divisor d of n, there is a unique cyclic subgroup H of G of order d. Moreover if $n = kd$, then $H = \langle x^k \rangle$.

Proof. Suppose that d is a positive divisor of n. Then $n = dk$ for some positive integer k and, by Theorem 5.2.8, $\mathbf{o}(x^k) = \frac{n}{\gcd(k,\,n)} = \frac{n}{k} = d$. Let $H = \langle x^k \rangle$; then $|H| = \mathbf{o}(x^k) = d$, so G has a subgroup of order d. Now let's show that H is unique. Let K be any subgroup of G of order d, and suppose that m is the smallest positive integer such that $x^m \in K$. Then it follows exactly as in the proof of Theorem 5.2.4 that $K = \langle x^m \rangle$, and so $|K| = \mathbf{o}(x^m) = d$. But by Theorem 5.2.8, $\mathbf{o}(x^m) = \frac{n}{\gcd(m,\,n)} = d$, so $\gcd(m, n) = \frac{n}{d} = k$. Thus, $k \mid m$. If $m = kq$, then $x^m = (x^k)^q \in H$, and so $K \subseteq H$. But since $|K| = |H|$, it must be the case that $K = H$, so H is indeed unique. ∎

5.2.11 Corollary

If $\langle x \rangle = G$ is a cyclic group of order n, then for any integer m, x^m generates G if and only if m and n are relatively prime.

5.2.12   Examples

1.   Find all the cyclic subgroups of the finite additive cyclic group $\mathbf{Z}_{12}$.

Solution. Since $\mathbf{Z}_{12} = \langle [1] \rangle$, $\mathbf{Z}_{12}$ is cyclic, and Theorem 5.2.10 shows that if d is a positive divisor of 12 and $12 = dk$, then $\mathbf{Z}_{12}$ has a subgroup of order d that is generated by $k[1] = [k]$. Switching the roles of d and k and considering k to be the positive divisor, then $\mathbf{Z}_{12}$ has a subgroup of order k with generator $d[1] = [d]$. We see that the following are true:

$12 = 1(12)$, so $[1]$ generates a subgroup of order 12:

$$\langle [1] \rangle = \{[1], [2], [3], [4], [5], [6], [7], [8], [9], [10], [11], [0]\} = \mathbf{Z}_{12}$$

$12 = 2(6)$, so $[2]$ generates a subgroup of order 6:

$$\langle [2] \rangle = \{[2], [4], [6], [8], [10], [0]\}.$$

$12 = 3(4)$, so $[3]$ generates a subgroup of order 4:

$$\langle [3] \rangle = \{[3], [6], [9], [0]\}.$$

$12 = 4(3)$, so $[4]$ generates a subgroup of order 3:

$$\langle [4] \rangle = \{[4], [8], [0]\}.$$

$12 = 6(2)$, so $[6]$ generates a subgroup of order 2:

$$\langle [6] \rangle = \{[6], [0]\}.$$

$12 = 12(1)$, so $[12]$ generates a subgroup of order 1:

$$\langle [12] \rangle = \{[12], [0]\}.$$

2.   Find all the subgroups of the cyclic group $G = \langle x \rangle$ if $\mathbf{o}(x) = 8$.

Solution. Since $8 = 1(8) = 2(4)$, it follows that:

x^1 generates a subgroup of order 8:

$$\langle x \rangle = \{x, x^2, x^3, x^4, x^5, x^6, x^7, x^8 = e\} = G.$$

x^2 generates a subgroup of order 4:

$$\langle x^2 \rangle = \{x^2, x^4, x^6, x^8 = e\}.$$

x^4 generates a subgroup of order 2:

$$\langle x^4 \rangle = \{x^4, x^8 = e\}.$$

x^8 generates a subgroup of order 1:

$$\langle x^8 \rangle = \{x^8 = e\}.$$

The following theorem characterizes cyclic groups. You will see it again when we study finite abelian groups.

5.2.13   **Theorem**

Let $G = \langle x \rangle$ be a cyclic group.

1. If G has finite order n, then G is isomorphic to the additive group $\mathbf{Z}_n$.
2. If G has infinite order, then G is isomorphic to the additive group $\mathbf{Z}$.

Proof

1. We know from Theorem 5.2.5 that $G = \{e = x^0, x, x^2, \ldots, x^{n-1}\}$ and that the elements of G are distinct. Let $f : G \to \mathbf{Z}_n$ be such that $f(x^k) = [k]$. The assertion is that f is a well-defined group isomorphism. Suppose that $x^j = x^k$, where $0 \le j \le k < n$. Then, by Theorem 5.2.5, n divides $k - j$, which is possible only if $j = k$. Hence, $[j] = [k]$, so f is well-defined. Next, note that $f(x^j x^k) = f(x^{j+k}) = [j + k] = [j] + [k] = f(x^j) + f(x^k)$ and so f is indeed a group homomorphism. If $x^k \in \ker f$, where $0 \le k < n$, then $f(x^k) = [k] = [0]$, so $n \mid k$. But this implies that $k = 0$ and so $x^k = e$. Therefore, $\ker f = \{e\}$ and thus f is a group monomorphism by Theorem 2.3.5. Finally, if $[k] \in \mathbf{Z}_n$, then k can be chosen so that $0 \le k < n$. Thus, $x^k \in G$ and $f(x^k) = [k]$, so f is a group epimorphism.
2. If $f : G \to \mathbf{Z}$ is defined by $f(x^k) = k$, then as in part 1, we can show that f is a group isomorphism. The details are left as an exercise. ∎

The cyclic groups have now been completely determined. An infinite cyclic group is isomorphic to the additive group $\mathbf{Z}$, so there is only one isomorphism class of these groups. Any finite cyclic group is isomorphic to $\mathbf{Z}_n$ for some integer n. Consequently, for each positive integer n, there is only one isomorphism class of cyclic groups of order n.

Problem Set 5.2

1. Show that every cyclic group is abelian.
2. Let x and y be elements of a group G. If $x \in \langle y \rangle$, show that $\langle x \rangle \subseteq \langle y \rangle$.
3. The set $\mathbf{Z}$ is a cyclic group under addition with generators 1 and -1. Describe all the cyclic subgroups of $\mathbf{Z}$.
4. Complete the proof of part 2 of Theorem 5.2.13.
5. Prove that each of the following is *not* a cyclic group.
    (a) $\mathbf{R}$ under addition
    (b) $\mathbf{C}$ under addition
    (c) $\mathbf{Q}^*$, the set of nonzero rational numbers under multiplication

6. (a) Let G be a cyclic group of order 24. If $G = \langle x \rangle$, determine the order of each of the following subgroups of G, without computing the subgroup.

   (i) $\langle x^3 \rangle$ (ii) $\langle x^6 \rangle$ (iii) $\langle x^5 \rangle$

   (b) Compute each of the following subgroups of G.

   (i) $\langle x^2 \rangle$ (ii) $\langle x^3 \rangle$ (iii) $\langle x^5 \rangle$

7. The integers 1 and -1 both generate the additive cyclic group $\mathbf{Z}$. Prove that these are the only generators of $\mathbf{Z}$.

8. Suppose that G is a cyclic group of order 42. Find all the elements of G of order 6.

9. (a) Prove Corollary 5.2.11.
   (b) Find all the generators of $\mathbf{Z}_{10}$.

10. If G is a cyclic group of order p, where p is a prime, show that every element of G generates G. Conversely, if G is a cyclic group of order n and every element of G generates G, must n be a prime?

11. Show that for every positive integer n, there is a cyclic group of order n.

12. Let G be a finite group that is not necessarily cyclic. Show that if G has exactly one nontrivial cyclic subgroup, then $|G| = p^2$, where p is a prime.

13. Suppose that x and y are elements of an abelian group G. If $\langle x \rangle \cap \langle y \rangle = \{e\}$, prove that $\mathbf{o}(xy) = \operatorname{lcm}(\mathbf{o}(x), \mathbf{o}(y))$, where $\operatorname{lcm}(\mathbf{o}(x), \mathbf{o}(y))$ is the **least common multiple** of $\mathbf{o}(x)$ and $\mathbf{o}(y)$. (See Exercise 14 of Problem Set 1.1.)

14. Let G be an abelian group. Prove that the set of all elements of G that have finite order is a subgroup of G.

15. If G and H are cyclic groups of the same order, $G = \langle x \rangle$, and $H = \langle y \rangle$, show there is a group isomorphism $f : G \to H$ such that $f(x) = y$.

16. Suppose that $G = \langle x \rangle$ is a cyclic group, and let $f : G \to H$ be a group epimorphism.

    (a) Prove that H is a cyclic group and $H = \langle f(x) \rangle$. This shows that *every homomorphic image of a cyclic group is cyclic.*
    (b) If G is of finite order, prove that $\mathbf{o}(f(x))$ divides the order of G.

TECHNOLOGY PROBLEMS

Use a computer algebra system to solve each of the following problems.

17. Find the order of each of the following elements in the additive group $\mathbf{Z}_n$ for the given value of n.

    (a) [4] for $n = 108$
    (b) [8] for $n = 308$
    (c) [124] for $n = 10{,}232$

18. Find a generator $[x]$ other than $[1]$ that generates the additive group $\mathbf{Z}_{36}$, and print out $\langle [x] \rangle$.

19. In $\mathbf{Z}_{128}$ print out the subgroups generated by $[2]$ and $[8]$.

5.3   DIRECT PRODUCTS OF GROUPS

Section Overview. A new group can be constructed from two groups through a procedure known as forming the external direct product. One purpose of this section is to show how this can be accomplished. Sometimes each element of a group G can be expressed uniquely as a product of elements from two subgroups. When this is the case, the group is said to be the internal direct product of the two subgroups. As we shall see, the concept of the external direct product of two groups is very closely connected to the internal direct product of two subgroups. After these concepts have been developed for two groups, the procedure will be generalized to a finite number of groups. This mathematical machinery is being developed for a reason; it will be used when finite abelian groups are studied, so make sure you understand these concepts completely.

If G_1 and G_2 are groups, then $G_1 \times G_2 = \{(x_1, x_2) \mid x_1 \in G_1, x_2 \in G_2\}$ can be made into a group by defining a binary operation componentwise on the ordered pairs of $G_1 \times G_2$ as

$$(x_1, x_2)(y_1, y_2) = (x_1 y_1, x_2 y_2).$$

This operation is well-defined since the binary operations on G_1 and G_2 are well-defined, and the associativity of this operation follows directly from the fact that the operations on G_1 and G_2 are associative. Indeed, if $x = (x_1, x_2)$, $y = (y_1, y_2)$, and $z = (z_1, z_2)$, then

$$(xy)z = [(x_1, x_2)(y_1, y_2)](z_1, z_2) = (x_1y_1, x_2y_2)(z_1, z_2)$$
$$= ([x_1y_1]z_1, [x_2y_2]z_2)$$
$$= (x_1[y_1z_1], x_2[y_2z_2])$$
$$= (x_1, x_2)(y_1z_1, y_2z_2)$$
$$= (x_1, x_2)[(y_1, y_2)(z_1, z_2)]$$
$$= x(yz).$$

The identity of $G_1 \times G_2$ is (e_1, e_2), where e_1 and e_2 are the identities of G_1 and G_2, respectively, and the multiplicative inverse of an element (x_1, x_2) in $G_1 \times G_2$ is $(x_1, x_2)^{-1} = (x_1^{-1}, x_2^{-1})$. The group $G_1 \times G_2$ has two fundamental properties that always hold:

1. If $H_1 = G_1 \times \{e_2\}$ and $H_2 = \{e_1\} \times G_2$, then H_1 and H_2 are normal subgroups of $G_1 \times G_2$.

Proof. The first step in the proof is to show that H_1 is a subgroup of $G_1 \times G_2$. Since $(e_1, e_2) \in H_1$, H_1 is nonempty. If $x = (x_1, e_2)$ and $y = (y_1, e_2)$ are in H_1, then $xy^{-1} = (x_1, e_2)(y_1^{-1}, e_2) = (x_1 y_1^{-1}, e_2)$ is in H_1 since $x_1y_1^{-1} \in G_1$. Hence, by Theorem 2.2.3, H_1 is a subgroup of $G_1 \times G_2$. Next let's show that H_1 is normal in $G_1 \times G_2$. If $x = (x_1, x_2)$ is in $G_1 \times G_2$ and $h = (y_1, e_2)$, then $xhx^{-1} = (x_1, x_2)(y_1, e_2)(x_1^{-1}, x_2^{-1}) = (x_1y_1x_1^{-1}, x_2e_2x_2^{-1}) = (x_1y_1x_1^{-1}, e_2) \in H_1$ since $x_1y_1x_1^{-1} \in G_1$. Therefore, $xhx^{-1} \in H_1$ and thus, by Theorem 2.2.15, H_1 is normal in $G_1 \times G_2$. A similar proof shows that H_2 is a normal subgroup of G. ∎

The second property deals with the uniqueness of an expression for an element $x \in G_1 \times G_2$ as a product of an element of H_1 and an element of H_2. Before this property can be stated, the following definition is required to make the concept of uniqueness more precise.

5.3.1   Definition

Let H_1 and H_2 be subgroups of a group G, and suppose that $x \in G$. Then we say that x can be **expressed as a product of an element of H_1 and an element of H_2** if there exists $h_1 \in H_1$ and $h_2 \in H_2$ such that $x = h_1h_2$. We say that the **expression** $x = h_1h_2$ for x is **unique** if, whenever $x = h_1'h_2'$ is another expression for x with $h_1' \in H_1$, $h_2' \in H_2$, then $h_1 = h_1'$ and $h_2 = h_2'$.

Now for the second property.

2. Each element $G_1 \times G_2$ has a unique expression as a product of an element of H_1 and an element of H_2.

Proof. Suppose that $x \in G_1 \times G_2$ and $x = (x_1, x_2)$. Then

$$x = (x_1, e_2)(e_1, x_2),$$

where $(x_1, e_2) \in H_1$ and $(e_1, x_2) \in H_2$, so x can be expressed as an element of H_1 times an element of H_2. We claim that this expression for x is unique. If

$$x = (x_1', e_2)(e_1, x_2')$$

is another expression for x with$(x_1', e_2) \in H_1$ and$(e_1, x_2') \in H_2$, then

$$(x_1, x_2) = (x_1, e_2)(e_1, x_2) = x = (x_1', e_2)(e_1, x_2') = (x_1', x_2').$$

Therefore, $x_1 = x_1'$ and $x_2 = x_2'$, and from this it follows immediately that $(x_1, e_2) = (x_1', e_2)$ and $(e_1, x_2') = (e_1, x_2)$, so the expression for x as a product of an element from H_1 and an element from H_2 is indeed unique. ■

5.3.2   Definition

The group $G_1 \times G_2$ is the **external direct product** of the groups G_1 and G_2.

5.3.3   Example

Consider the additive group $\mathbf{Z}_2 = \{[0], [1]\}$ and the multiplicative group $G = \{-i, -1, 1, i\}$. The eight elements of the group $G \times \mathbf{Z}_2$ are $(-i, [0])$, $(-i, [1])$, $(-1, [0])$, $(-1, [1])$, $(1, [0])$, $(1, [1])$, $(i, [0])$ and $(i, [1])$. The following are illustrations of multiplication in the group $G \times \mathbf{Z}_2$.

$$(-1, [1])(i,[0]) = ((-1)i, [1] + [0]) = (-i, [1])$$
$$(-i, [1])(i,[1]) = ((-i)i, [1] + [1]) = (1, [0])$$

The identity of the group is $(1, [0])$. Notice that the operation with the first components of the ordered pairs is multiplication of elements of G, while the operation with the second components of the ordered pairs is addition of elements in $\mathbf{Z}_2$.

The two equivalent conditions given in the following lemma are often quite useful for showing that an expression for x in a group G is unique.

5.3.4   Lemma

Let N_1 and N_2 be subgroups of a group G. If $G = N_1N_2$, then the following two conditions are equivalent:

1. If $x \in G$ and $x = n_1n_2$, where $n_1 \in N_1$ and $n_2 \in N_2$, then this expression for x is unique.
2. $N_1 \cap N_2 = \{e\}$.

Proof

$1 \Rightarrow 2$. Suppose that $x \in N_1 \cap N_2$. Then $x \in N_1$ and $x \in N_2$. Now N_2 is a subgroup of G, so x^{-1} is in N_2. Thus, $xx^{-1} = e = ee$ with $x \in N_1$ and $x^{-1} \in N_2$. But by part 1, e can be written in only one way as a product of an element of N_1 and an element of N_2. Hence, $x = x^{-1} = e$, so $N_1 \cap N_2 = \{e\}$.

$2 \Rightarrow 1$. Let $x = n_1n_2 = n_1'n_2'$, where $n_1, n_1' \in N_1$ and $n_2, n_2' \in N_2$. Then $n_1'^{-1}n_1 = n_2'n_2^{-1}$. But $n_1'^{-1}n_1 \in N_1$ and $n_2'n_2^{-1} \in N_2$ since N_1 and N_2 are subgroups of G. Hence, $n_1'^{-1}n_1 = n_2'n_2^{-1} \in N_1 \cap N_2 = \{e\}$, and from this it follows that $n_1 = n_1'$ and $n_2 = n_2'$. Thus, the expression for x is unique. Therefore, $1 \Rightarrow 2$ and $2 \Rightarrow 1$, so the two conditions are equivalent. ∎

We have just seen that if G_1 and G_2 are groups, then we can form the direct product $G_1 \times G_2$. If G is a group and G is isomorphic to $G_1 \times G_2$, our goal now is to see what effect, if any, this has on the internal structure of G.

5.3.5   Theorem

If G is isomorphic to $G_1 \times G_2$, then there are normal subgroups N_1 and N_2 of G such that $G = N_1N_2$, and each element of G can be expressed uniquely as a product of an element from N_1 and an element from N_2.

Proof. Let $f : G_1 \times G_2 \to G$ be a group isomorphism, and suppose that $H_1 = G_1 \times \{e_2\}$ and $H_2 = \{e_1\} \times G_2$. We have seen that H_1 and H_2 are normal subgroups of $G_1 \times G_2$, so Theorem 2.3.8 and the Correspondence Theorem 2.3.16 show that $N_1 = f(H_1)$ and $N_2 = f(H_2)$ are normal subgroups of G.

It remains only to show that each $y \in G$ can be expressed uniquely as a product of an element from N_1 and an element from N_2. If $y \in G$, then there is an

$x \in G_1 \times G_2$ such that $f(x) = y$. If $x = (x_1, x_2) = (x_1, e_2)(e_1, x_2)$, $n_1 = f((x_1, e_1))$, and $n_2 = f((e_1, x_2))$, then $y = f(x) = f((x_1, e_2)(e_1, x_2)) = f((x_1 e_2))f((e_1, x_2)) = n_1 n_2 \in N_1 N_2$. This shows that each $y \in G$ can be expressed as a product of an element from N_1 and an element from N_2 and, consequently, that $G = N_1 N_2$.

The last step is to show that this expression for y is unique. Because of Lemma 5.3.4, it suffices to show that $N_1 \cap N_2 = \{e\}$. Since f is a group isomorphism, the inverse mapping f^{-1} is also a group isomorphism. The inverse image of N_1 is H_1, and the inverse image of N_2 is H_2. If $z \in N_1 \cap N_2$, then $f^{-1}(z) \in H_1 \cap H_2 = \{(e_1, e_2)\}$. Hence, $z = f((e_1, e_2)) = e$, so $N_1 \cap N_2 = \{e\}$. ∎

When G and $G_1 \times G_2$ are isomorphic, they have internal structures that are quite similar. Because of the Correspondence Theorem for Groups, the subgroup structure of $G_1 \times G_2$ is transferred to G through the isomorphism. This can be summarized as follows:

$$G_1 \times G_2 \qquad\qquad\qquad G$$

1. $G_1 \times G_2$ has two normal subgroups $H_1 = G_1 \times \{e\}$ and $H_2 = \{e_1\} \times G_2$ such that $H_1 \cap H_2 = \{(e_1, e_2)\}$.

1. G has two normal subgroups N_1 and N_2 such that $N_1 \cap N_2 = \{e\}$.

2. Each element of $G_1 \times G_2$ can be written uniquely as a product of an element from H_1 and an element from H_2.

2. Each element of G can be written uniquely as a product of an element from N_1 and an element from N_2.

3. The subgroup structure of $G_1 \times G_2$ is transferred to G through the isomorphism. In particular, if $f : G_1 \times G_2 \to G$ is a group isomorphism, then we can let $N_1 = f(H_1)$ and $N_2 = f(H_2)$.

5.3.6   Example

The additive groups $\mathbf{Z}_{12}$ and $\mathbf{Z}_3 \times \mathbf{Z}_4$ are isomorphic via the mapping $f : \mathbf{Z}_3 \times \mathbf{Z}_4 \to \mathbf{Z}_{12}$ defined by $f(([x]_3, [y]_4)) = [4x]_{12} + [3y]_{12}$. You should show this! Now $H_1 = \mathbf{Z}_3 \times \{[0]_4\}$ and $H_2 = \{[0]_3\} \times \mathbf{Z}_4$ are normal subgroups of $\mathbf{Z}_3 \times \mathbf{Z}_4$, so $N_1 = f(H_1)$ and $N_2 = f(H_2)$ are normal subgroups of $\mathbf{Z}_{12}$. Let's compute N_1 and N_2. To simplify notation, we drop the subscripts on the congruence classes. You should be able to tell where each congruence class belongs. The elements of H_1 and H_2 are $([0], [0])$, $([1], [0])$, $([2], [0])$ and $([0], [0])$, $([0], [1])$, $([0], [2])$, $([0], [3])$, respectively. Hence,

$$f(([0], [0])) = [4 \cdot 0] + [3 \cdot 0] = [0] + [0] = [0],$$
$$f(([1], [0])) = [4 \cdot 1] + [3 \cdot 0] = [4] + [0] = [4],$$
$$f(([2], [0])) = [4 \cdot 2] + [3 \cdot 0] = [8] + [0] = [8]$$

and

$$f(([0], [0])) = [4 \cdot 0] + [3 \cdot 0] = [0] + [0] = [0],$$
$$f(([0], [1])) = [4 \cdot 0] + [3 \cdot 1] = [0] + [3] = [3],$$
$$f(([0], [2])) = [4 \cdot 0] + [3 \cdot 2] = [0] + [6] = [6],$$
$$f(([0], [3])) = [4 \cdot 0] + [3 \cdot 3] = [0] + [0] = [9].$$

Therefore, $N_1 = f(H_1) = \{[0], [4], [8]\}$ and $N_2 = f(H_2) = \{[0], [3], [6], [9]\}$. We claim that every element of $\mathbf{Z}_{12}$ can be written uniquely as the sum of an element from N_1 and an element from N_2. This follows since $N_1 \cap N_2 = \{[0]\}$ and

$[0] = [0] + [0]$	$[3] = [0] + [3]$	$[6] = [0] + [6]$	$[9] = [0] + [9]$
$[1] = [4] + [9]$	$[4] = [4] + [0]$	$[7] = [4] + [3]$	$[10] = [4] + [6]$
$[2] = [8] + [6]$	$[5] = [8] + [9]$	$[8] = [8] + [0]$	$[11] = [8] + [3]$

Notice that $\mathbf{Z}_{12}$ is an additive group, and we have shown that each element of $\mathbf{Z}_{12}$ can be expressed uniquely as the sum of an element of N_1 and an element of N_2 rather than the product of an element of N_1 and an element of N_2.

5.3.7   Definition

If a multiplicative group G has two normal subgroups N_1 and N_2 such that $G = N_1 N_2$ and $N_1 \cap N_2 = \{e\}$, then we say that G is the **internal direct product** of N_1 and N_2. This is denoted by $G = N_1 \times N_2$. If G is an additive group and N_1 and N_2 are normal subgroups of G such that $G = N_1 + N_2$ and $N_1 \cap N_2 = \{0\}$, then G is the **internal direct sum** of N_1 and N_2. This is denoted by $G = N_1 \oplus N_2$.

The only difference between the internal direct sum and the internal direct product is in the notation selected for the binary operation for the group. Internal direct sum deals with additive groups, and internal direct product deals with multiplicative groups. We make this distinction since it seems somewhat awkward to refer to $\mathbf{Z}_{12}$ in Example 5.3.6 as being the internal direct product of N_1 and N_2 since $\mathbf{Z}_{12} = N_1 + N_2$ and the operation on $\mathbf{Z}_{12}$ is addition. *Be warned that terminology and notation vary widely in undergraduate abstract algebra texts with respect to internal direct products and internal direct sums. Some authors refer to what we call an internal direct sum as an internal direct product, making no distinction with regard to the operation on the group.*

Now let's begin with a group G, and ask about the converse of Theorem 5.3.5. That is, if G is the internal direct product of two normal subgroups N_1 and N_2, must it be the case that G and $N_1 \times N_2$ are isomorphic? This question is answered by the following theorem.

5.3.8   Theorem

If a group G is the internal direct product of two normal subgroups N_1 and N_2, then G and $N_1 \times N_2$ are isomorphic groups.

Proof. Since $G = N_1 N_2$, each $x \in G$ can be written as $x = n_1 n_2$, where $n_1 \in N_1$ and $n_2 \in N_2$. Furthermore, by Lemma 5.3.4, this expression for x is unique since $N_1 \cap N_2 = \{e\}$. Hence, define $f : N_1 N_2 \to N_1 \times N_2$ by $f(n_1 n_2) = (n_1, n_2)$. The mapping f is well-defined since each expression $n_1 n_2$ is unique. Now let's show that f is a group homomorphism. To prove this, we must recall a previous theorem. Theorem 2.2.19 indicates that since N_1 and N_2 are normal subgroups of G such that $N_1 \cap N_2 = \{e\}$, elements of N_1 and N_2 commute. With this in mind, if $n_1 n_2, n_1' n_2' \in N_1 N_2$, then $f((n_1 n_2)(n_1' n_2')) = f(n_1(n_2 n_1')n_2') = f((n_1(n_1' n_2)n_2')) = f((n_1 n_1')(n_2 n_2')) = (n_1 n_1', n_2 n_2') = (n_1, n_2)(n_1', n_2') = f(n_1 n_2)f(n_1' n_2')$, so f is indeed a group homomorphism. The fact that f is a bijective function follows easily, and thus G and $N_1 \times N_2$ are isomorphic groups. ∎

5.3.9   Example

Consider the additive group $\mathbf{Z}_6 = \{[0], [1], [2], [3], [4], [5]\}$. $N_1 = \{[0], [3]\}$ and $N_2 = \{[0], [2], [4]\}$ are normal subgroups of $\mathbf{Z}_6$, and every element of $\mathbf{Z}_6$ can be written uniquely as the sum of an element from N_1 and N_2. This follows since

$$[0] = [0] + [0] \qquad [3] = [3] + [0]$$
$$[1] = [3] + [4] \qquad [4] = [0] + [4]$$
$$[2] = [0] + [2] \qquad [5] = [3] + [2],$$

and each expression is unique since $N_1 \cap N_2 = \{[0]\}$. Hence, $\mathbf{Z}_6 = N_1 \oplus N_2$. We claim that $\mathbf{Z}_6$ is isomorphic to the external direct product of N_1 and N_2.

$$N_1 \times N_2 = \{([0], [0]), ([0], [2]), ([0], [4]), ([3], [0]), ([3], [2]), ([3], [4])\}$$

The mapping $f : N_1 \times N_2 \rightarrow \mathbf{Z}_6$, where $f(([x], [y])) = [x] + [y]$, is an isomorphism. Note that

$$([0], [0]) \longmapsto [0] + [0] = [0] \quad ([1], [0]) \longmapsto [1] + [0] = [1]$$
$$([0], [2]) \longmapsto [0] + [2] = [2] \quad ([1], [2]) \longmapsto [1] + [2] = [3]$$
$$([0], [4]) \longmapsto [0] + [4] = [4] \quad ([1], [4]) \longmapsto [1] + [4] = [5].$$

Make note of the difference between the external direct product, the internal direct product, and the internal direct sum. They are closely connected but different.

Summary of the External Direct Product and the Internal Direct Product and Sum

1. If an additive group G is isomorphic to the external direct product $G_1 \times G_2$ of two groups G_1 and G_2, then G has two normal subgroups N_1 and N_2 such that $G = N_1 \oplus N_2$. If G is a multiplicative group, then $G = N_1 \dot\times N_2$.
2. If G is an additive group with normal subgroups N_1 and N_2 such that $G = N_1 \oplus N_2$, then G is isomorphic to $N_1 \times N_2$. If G is a multiplicative group and $G = N_1 \dot\times N_2$, then G is isomorphic to $N_1 \times N_2$.

The method of constructing a new group from a pair of groups can be extended to three groups, to four groups, or, for that matter, to any finite number of groups.

5.3.10   Example

Consider the additive group $\mathbf{Z}_2 = \{[0]_2, [1]_2\}$, the multiplicative group of units $U(\mathbf{Z}_8) = \{[1]_8, [3]_8, [5]_8, [7]_8\}$ of $\mathbf{Z}_8$, and the multiplicative group $G = \{-i, -1, 1, i\}$. The direct product of these three groups is the group

$$\mathbf{Z}_2 \times U(\mathbf{Z}_8) \times G = \{([x]_2, [y]_8, z) \mid [x]_2 \in \mathbf{Z}_2, [y]_8 \in U(\mathbf{Z}_8), z \in G\}.$$

The binary operation is componentwise: addition in $\mathbf{Z}_2$ of the first components of the ordered triples, multiplication in $U(\mathbf{Z}_8)$ of the second components, and multiplication in G of the third components of the ordered triples. For example, $([1]_2, [5]_8, -i)([1]_2, [7]_8, 1) = ([1]_2 + [1]_2, [5]_8 [7]_8, (-i)1) = ([0]_2, [3]_8, -i)$. The identity of

$\mathbf{Z}_2 \times U(\mathbf{Z}_8) \times G$ is $([0]_2, [1]_8, 1)$, and the inverse of

$([x]_2, [y]_8, z)$ is $([x]_2, [y]_8, z)^{-1} = \{-[x]_2, [y]_8^{-1}, z^{-1}\}.$

In general, if $\{G_i\}_{i=1}^n$ is a finite family of groups, then $\times_{i=1}^n G_i = G_1 \times G_2 \times \cdots \times G_n$ is a group under the componentwise binary operation

$$(x_1, x_2, \ldots, x_n)(y_1, y_2, \ldots, y_n) = (x_1 y_1, x_2 y_2, \ldots, x_n y_n).$$

The group $\times_{i=1}^n G_i$ is the **external direct product** of the finite family $\{G_i\}_{i=1}^n$ of groups. Conversely, if $\{N_i\}_{i=1}^n$ is a finite family of normal subgroups of an additive group G such that $G = N_1 + N_2 + \cdots + N_n$ and each $x \in G$ can be expressed uniquely as $x = n_1 + n_2 + \cdots + n_n$, where $n_i \in N_i$ for $i = 1, 2, \ldots, n$, then G is said to be the **internal direct sum** of the finite family $\{N_i\}_i^n$ of normal subgroups of G. This is denoted by $G = \oplus_{i=1}^n N_i = N_1 \oplus N_2 \oplus \cdots \oplus N_n$. If G is a multiplicative group and $G = N_1 N_2 \cdots N_n$ and each element of G can be expressed uniquely as $x = n_1 n_2 \cdots n_n$, then G is the **internal direct product** of the finite family of normal subgroups $\{N_i\}_{i=1}^n$. This is expressed by $G = \dot{\times}_{i=1}^n N_i = N_1 \dot{\times} N_2 \dot{\times} \cdots \dot{\times} N_n$.

The following lemma and theorem are a generalization of Lemma 5.3.4 and Theorems 5.3.5 and 5.3.8. A proof of each can be modeled along the same lines as the proofs given earlier. Both are stated for multiplicative groups. You should translate each to a valid statement about additive groups.

5.3.11   Lemma

Suppose that $\{N_i\}_{i=1}^n$ is a finite family of normal subgroups of a group G such that $G = N_1 N_2 \cdots N_n$. Then the following are equivalent:

1. Each element x of G has a unique expression as $x = n_1 n_2 \cdots n_n$, where $n_i \in N_i$ for $i = 1, 2, \ldots, n$.
2. $N_i \cap (N_1 \cdots N_{i-1} N_{i+1} \cdots N_n) = \{e\}$ for $i = 1, 2, \ldots, n$.

5.3.12   Theorem

If a group G is isomorphic to the external direct product $\times_{i=1}^n G_i$ of the finite family of groups $\{G_i\}_{i=1}^n$, then there is a finite family $\{N_i\}_{i=1}^n$ of normal subgroups of G such that G is the internal direct product of $\{N_i\}_{i=1}^n$. Conversely, if G is the internal direct product of a finite family $\{N_i\}_{i=1}^n$ of normal subgroups, then G is isomorphic to the external direct product $\times_{i=1}^n N_i$.

Problem Set 5.3

1. Prove that the external direct product of two groups G_1 and G_2 is abelian if and only if G_1 and G_2 are abelian.

2. (a) Prove that the additive cyclic group $\mathbf{Z}_6$ is isomorphic to $\mathbf{Z}_2 \times \mathbf{Z}_3$.
   (b) Show that there are subgroups G_1 and G_2 of $\mathbf{Z}_6$ such that $|G_1| = 2$, $|G_2| = 3$, and $\mathbf{Z}_6 = G_1 \oplus G_2$. Are G_1 and $\mathbf{Z}_2$ isomorphic? G_2 and $\mathbf{Z}_3$?

3. (a) Generalize your proof given in Exercise 2 to show that if m and n are relatively prime positive integers and G is a cyclic group of order mn, then G is isomorphic to $\mathbf{Z}_m \times \mathbf{Z}_n$.
   (b) Let m and n be positive integers that are relatively prime. If G is a cyclic group of order mn, prove that there are subgroups G_1 and G_2 of G such that $|G_1| = m$, $|G_2| = n$, and $G = G_1 \dot\times G_2$. Are G_1 and $\mathbf{Z}_m$ isomorphic? G_2 and $\mathbf{Z}_n$?

4. Show that the additive group $\mathbf{Z}$ cannot be expressed as an internal direct sum of two nontrivial subgroups.

5. Let G_1, G_2, H_1, and H_2 be groups, and suppose that G_1 and H_1 are isomorphic and G_2 and H_2 are isomorphic. Prove that $G_1 \times G_2$ and $H_1 \times H_2$ are isomorphic.

6. Suppose that $G = G_1 \dot\times G_2 \dot\times \cdots \dot\times G_n$. If $H_k \subseteq G_k$ is a subgroup of G for $k = 1, 2, \ldots, n$, prove that the product $H = H_1 H_2 \cdots H_n$ is direct.

7. Let G be a group, and suppose that $G = G_1 \dot\times G_2$, where G_1 and G_2 are normal subgroups of G. If $H_1 \subseteq G_1$ and $H_2 \subseteq G_2$ are normal subgroups of G, prove that $(G_1/H_1) \times (G_2/H_2)$ is isomorphic to $G/(H_1 H_2)$.

8. Let G_1 and G_2 be normal subgroups of G such that $G = G_1 \dot\times G_2$. Prove that G/G_2 and G_1 are isomorphic groups. Likewise, show that G/G_1 and G_2 are isomorphic.

9. Let G be a finite group. Prove that G is the internal direct product of two normal subgroups G_1 and G_2 if and only if $G_1 \cap G_2 = \{e\}$ and $|G| = |G_1| |G_2|$.

10. Prove Lemma 5.3.11.

11. Prove Theorem 5.3.12.

TECHNOLOGY PROBLEMS

Use a computer algebra system to solve each of the following problems.

12.   Find the order of each of the following elements of the additive group.

(a) $([8], [22]) \in \mathbf{Z}_{12} \times \mathbf{Z}_{26}$
(b) $([45], [12]) \in \mathbf{Z}_{128} \times \mathbf{Z}_{1242}$

13.   Print out the subgroup of $\mathbf{Z}_6 \times \mathbf{Z}_{12}$ generated by the element $([4]_6, [9]_{12})$.

14.   Show that $([3]_5, [11]_{12})$ generates $\mathbf{Z}_5 \times \mathbf{Z}_{12}$ by printing out the elements generated by $([3]_5, [11]_{12})$. Check your result by computing the order of $([3]_5, [11]_{12})$.

5.4   FINITE ABELIAN GROUPS

Section Overview. If G is a cyclic group of infinite order, then G is isomorphic to the additive group $\mathbf{Z}$, and if G is a finite cyclic group of order n, then G is isomorphic to $\mathbf{Z}_n$. Hence, all cyclic groups have been determined up to isomorphism. Our goal now is to classify all finite abelian groups. We prove the **Fundamental Theorem of Finite Abelian Groups**, which asserts that if G is a finite abelian group, then G is the internal direct sum of cyclic subgroups, each of which has order the power of a prime. This characterization of finite abelian groups, together with the classification of finite cyclic groups, permits us to classify all finite abelian groups of a given order.

Before proving the Fundamental Theorem of Finite Abelian Groups, we state the theorem and show how it can be used to classify finite abelian groups. *The binary operation on each finite abelian group is written additively, except in those cases when permutation groups are considered. In the case of permutation groups, it is standard practice to write the operation multiplicatively.*

5.4.1   The Fundamental Theorem of Finite Abelian Groups

If G is any finite abelian group, then G can be expressed as an internal direct sum of cyclic subgroups, each of which has order the power of a prime. Furthermore, this representation of G is unique up to rearrangement of the summands.

To simplify terminology, we refer to the preceding theorem as the Fundamental Theorem. Let's investigate this theorem and see what it means in light of what we know about finite cyclic groups, internal direct sums, and external direct products.

An Analysis of the Fundamental Theorem

1. The Fundamental Theorem indicates that if G is a finite abelian group, then a family of cyclic subgroups $\{C_k\}_{k=1}^{m}$ of G exists such that $G = \oplus_{k=1}^{m} C_k$. The order of C_k is $p_k^{n_k}$, where p_k is prime and n_k is a positive integer for $k = 1, 2, \ldots, m$. In addition, this expression for G is unique up to rearrangement of the order of the summands in $\oplus_{k=1}^{m} C_k$. There is nothing in the statement of the theorem to prevent one or more of the C_k's from being isomorphic. However, no two of the C_k's can be equal, for then the sum would not be direct.

2. We know from the previous section that since G is the internal direct sum of the family of subgroups $\{C_k\}_{k=1}^{m}$, G is isomorphic to the external direct product $\times_{k=1}^{m} C_k$.

3. Since C_k is a cyclic group of order $p_k^{n_k}$, Theorem 5.2.13 shows that C_k is isomorphic to the additive group $\mathbf{Z}_{p_k}^{n_k}$ for $k = 1, 2, \ldots, m$.

4. Therefore, any finite abelian group G is isomorphic to a group of the form

$$\mathbf{Z}_{p_1}^{n_1} \times \mathbf{Z}_{p_1}^{n_1} \times \cdots \times \mathbf{Z}_{p_m}^{n_m} \qquad (*)$$

If n is the order of G, then $n = p_1^{n_1} p_2^{n_2} \cdots p_m^{n_m}$, so each p_k is a prime divisor of n and not all of the primes need be distinct since it is possible for one or more of the C_k's to be isomorphic. This expression for G is unique up to rearrangement of the factors in equation $(*)$.

These observations allow us to classify, up to isomorphism, all finite abelian groups. However, before we can demonstrate how these results can be used to classify finite abelian groups, we need the following theorem. The proof uses the fact that if m and n are positive integers, then $mn = \gcd(m, n)\operatorname{lcm}(m, n)$. (See Exercise 14 of Problem Set 1.1.)

5.4.2   Theorem

For any positive integers m and n, $\gcd(m, n) = 1$ if and only if $\mathbf{Z}_{mn}$ is isomorphic to $\mathbf{Z}_m \times \mathbf{Z}_n$.

Proof. We know that $[1]_m$ and $[1]_n$ are generators of $\mathbf{Z}_m$ and $\mathbf{Z}_n$, respectively, so every element of $\mathbf{Z}_m$ looks like $a[1]_m$ for some integer a, $0 \leq a \leq m - 1$, and every element of $\mathbf{Z}_n$ looks like $b[1]_n$, where b is an integer such that $0 \leq b \leq n - 1$. Thus, every element of $\mathbf{Z}_m \times \mathbf{Z}_n$ is of the form $(a[1]_m, b[1]_n)$, where a and b satisfy the stated conditions. If $k = \operatorname{lcm}(m, n)$, then $k(a[1]_m, b[1]_n) = (ka[1]_m, kb[1]_n) = ([0]_m, [0]_n)$ since $m \mid k$ and $n \mid k$. Thus, it follows from Theorem 5.2.8 that every element of $\mathbf{Z}_m \times \mathbf{Z}_n$ has order at most k.

If $\gcd(m, n) > 1$, then $k < mn = \left|\mathbf{Z}_{mn}\right|$, so the order of every element of $\mathbf{Z}_m \times \mathbf{Z}_n$ is less than mn. Thus, $\mathbf{Z}_m \times \mathbf{Z}_n$ and $\mathbf{Z}_{mn}$ cannot be isomorphic. Why? Because $\mathbf{Z}_{mn}$ is cyclic and is generated by an element of order mn. If $\mathbf{Z}_m \times \mathbf{Z}_n$ and $\mathbf{Z}_{mn}$ are isomorphic, then $\mathbf{Z}_m \times \mathbf{Z}_n$ would also be cyclic and generated by an element of order mn, which is impossible since we have just shown that the order of every element of $\mathbf{Z}_m \times \mathbf{Z}_n$ is less than mn. Hence, $\gcd(m, n) \neq 1$ implies that $\mathbf{Z}_m \times \mathbf{Z}_n$ and $\mathbf{Z}_{mn}$ are not isomorphic. Therefore by the contrapositive, if $\mathbf{Z}_m \times \mathbf{Z}_n$ and $\mathbf{Z}_{mn}$ are isomorphic, then $\gcd(m, n) = 1$.

To complete the proof we need to show that if $\gcd(m, n) = 1$, then $\mathbf{Z}_m \times \mathbf{Z}_n$ and $\mathbf{Z}_{mn}$ are isomorphic. If $\gcd(m, n) = 1$, then $\mathrm{lcm}(m, n) = mn$ and so $\mathbf{o}(([1]_m, [1]_n)) = mn$. Hence, $\langle ([1]_m, [1]_n) \rangle = \mathbf{Z}_m \times \mathbf{Z}_n$, so $\mathbf{Z}_m \times \mathbf{Z}_n$ is a cyclic group of order mn. Thus, $\mathbf{Z}_{mn}$ and $\mathbf{Z}_m \times \mathbf{Z}_n$ are isomorphic since any two cyclic groups of the same order are isomorphic.   ∎

5.4.3   Example

The group $\mathbf{Z}_{72}$ is isomorphic to $\mathbf{Z}_8 \times \mathbf{Z}_9$ since $\gcd(8, 9) = 1$.

The classification procedure for finite abelian groups is illustrated in the following examples.

5.4.4   Examples

1.   Find, up to isomorphism, all the finite abelian groups G of order 12.

Solution. As observed earlier, G is isomorphic to a group of the form $\mathbf{Z}_{p_1}^{n_1} \times \mathbf{Z}_{p_2}^{n_2} \times \cdots \times \mathbf{Z}_{p_m}^{n_m}$ with repetitions allowed among the prime divisors p_i of 12. Since 12 can be factored as $2 \cdot 2 \cdot 3$ and as $2^2 \cdot 3$, where each factor in each factorization is the power of a prime divisor of 12, every finite abelian group of order 12 is isomorphic to one of the two abelian groups given in the following table.

Factorization	Abelian Group
$2 \cdot 2 \cdot 3$	$\rightarrow \mathbf{Z}_2 \times \mathbf{Z}_2 \times \mathbf{Z}_3$
$2^2 \cdot 3$	$\rightarrow \mathbf{Z}_4 \times \mathbf{Z}_3$

But why can't $\mathbf{Z}_2 \times \mathbf{Z}_2 \times \mathbf{Z}_3$ and $\mathbf{Z}_4 \times \mathbf{Z}_3$ be isomorphic? If $\mathbf{Z}_2 \times \mathbf{Z}_2 \times \mathbf{Z}_3$ and $\mathbf{Z}_4 \times \mathbf{Z}_3$ are isomorphic, then $\mathbf{Z}_2 \times \mathbf{Z}_2$ and $\mathbf{Z}_4$ are isomorphic. But Theorem 5.4.2 indicates that this is impossible. You might also wonder why $\mathbf{Z}_{12}$ does not seem to appear in the list. Well, it does, through isomorphism. Since $\gcd(3, 4) = 1$, $\mathbf{Z}_{12}$ and $\mathbf{Z}_4 \times \mathbf{Z}_3$ are isomorphic. Remember that 12 must be expressed as a product of powers of its prime divisors, with repetitions allowed, when forming the groups $\mathbf{Z}_{p_1}^{n_1} \times \mathbf{Z}_{p_2}^{n_2} \times \cdots \times \mathbf{Z}_{p_m}^{n_m}$. For this reason the factorization $2 \cdot 6$ of 12 is not considered.

The prime powers are called the **elementary divisors** of G. For example, 2, 2, and 3 are the elementary divisors of every group G isomorphic to $\mathbf{Z}_2 \times \mathbf{Z}_2 \times \mathbf{Z}_3$. Likewise, 2^2 and 3 are the elementary divisors of any group that is isomorphic to $\mathbf{Z}_4 \times \mathbf{Z}_3$.

2. Find, up to isomorphism, all the finite abelian groups G of order 36.

Solution. The factorizations of 36 are $2 \cdot 2 \cdot 3 \cdot 3$, $2^2 \cdot 3 \cdot 3$, $2 \cdot 2 \cdot 3^2$ and $2^2 \cdot 3^2$, where each factor in each factorization is the power of a prime divisor of 36. Every finite abelian group of order 36 is isomorphic to one of the following groups.

FACTORIZATION	ELEMENTARY DIVISORS	ABELIAN GROUP
$2 \cdot 2 \cdot 3 \cdot 3$	$\rightarrow 2, 2, 3, 3$	$\rightarrow \mathbf{Z}_2 \times \mathbf{Z}_2 \times \mathbf{Z}_3 \times \mathbf{Z}_3$
$2^2 \cdot 3 \cdot 3$	$\rightarrow 2^2, 3, 3$	$\rightarrow \mathbf{Z}_4 \times \mathbf{Z}_3 \times \mathbf{Z}_3$
$2 \cdot 2 \cdot 3^2$	$\rightarrow 2, 2, 3^2$	$\rightarrow \mathbf{Z}_2 \times \mathbf{Z}_2 \times \mathbf{Z}_9$
$2^2 \cdot 3^2$	$\rightarrow 2^2, 3^2$	$\rightarrow \mathbf{Z}_4 \times \mathbf{Z}_9$

$\mathbf{Z}_{36}$ is isomorphic to the last group $\mathbf{Z}_4 \times \mathbf{Z}_9$ in the list, since $\gcd(4, 9) = 1$.

3. Determine the elementary divisors of any finite abelian group that is isomorphic to $\mathbf{Z}_6 \times \mathbf{Z}_{12} \times \mathbf{Z}_{15}$.

Solution. Since $6 = 2 \cdot 3$, $12 = 2^2 \cdot 3$, and $15 = 3 \cdot 5$, Theorem 5.4.2 indicates that $\mathbf{Z}_6$ is isomorphic to $\mathbf{Z}_2 \times \mathbf{Z}_3$, $\mathbf{Z}_{12}$ is isomorphic to $\mathbf{Z}_{2^2} \times \mathbf{Z}_3$, and $\mathbf{Z}_{15}$ is isomorphic to $\mathbf{Z}_3 \times \mathbf{Z}_5$, as in the following illustration.

$$\overbrace{\mathbf{Z}_6}^{} \times \overbrace{\mathbf{Z}_{12}}^{} \times \overbrace{\mathbf{Z}_{15}}^{}$$
$$\underbrace{\mathbf{Z}_2 \times \mathbf{Z}_3}_{} \times \underbrace{\mathbf{Z}_{2^2} \times \mathbf{Z}_3}_{} \times \underbrace{\mathbf{Z}_3 \times \mathbf{Z}_5}_{}$$

Hence, $\mathbf{Z}_6 \times \mathbf{Z}_{12} \times \mathbf{Z}_{15}$ is isomorphic to $\mathbf{Z}_2 \times \mathbf{Z}_3 \times \mathbf{Z}_{2^2} \times \mathbf{Z}_3 \times \mathbf{Z}_3 \times \mathbf{Z}_5$, and consequently the elementary divisors of $\mathbf{Z}_6 \times \mathbf{Z}_{12} \times \mathbf{Z}_{15}$ are 2, 2^2, 3, 3, 3, and 5. There is another way of writing $\mathbf{Z}_6 \times \mathbf{Z}_{12} \times \mathbf{Z}_{15}$, using what are known as the **invariant factors** of $\mathbf{Z}_6 \times \mathbf{Z}_{12} \times \mathbf{Z}_{15}$. The invariant factors can be determined from the elementary divisors of $\mathbf{Z}_6 \times \mathbf{Z}_{12} \times \mathbf{Z}_{15}$, and there is a systematic way of doing this that really is quite simple. First, arrange the elementary divisors in ascending order according to the primes and the power of the primes. For example, 2, 2^2, 3, 3, 3, 5 is such an arrangement of the elementary divisors of $\mathbf{Z}_6 \times \mathbf{Z}_{12} \times \mathbf{Z}_{15}$. Next, write the primes in a figure such as the following one, with each prime on a separate line in ascending order of its power. Make sure that the last column contains no blank spaces. That is, the figure should be flush on the right.

$$
\begin{array}{rrr}
 & 2 & 2^2 \\
3 & 3 & 3 \\
 & & 5 \\
\hline
3 & 6 & 60
\end{array}
$$

Finally, form the product of numbers in each column. Since the numbers in each column are pairwise relatively prime, it follows that the group $\mathbf{Z}_6 \times \mathbf{Z}_{12} \times \mathbf{Z}_{15}$ is isomorphic to the group $\mathbf{Z}_3 \times \mathbf{Z}_6 \times \mathbf{Z}_{60}$. The integers 3, 6, and 60 are the invariant factors of $\mathbf{Z}_6 \times \mathbf{Z}_{12} \times \mathbf{Z}_{15}$. Notice that $3 \,|\, 6$ and $6 \,|\, 60$, that is, each invariant factor divides the one that follows it. This is always the case if the invariant factors have been computed correctly using the procedure just described.

4. Find, up to isomorphism, all finite abelian groups of order 72. Compute the elementary divisors and the invariant factors, and express each group that represents each isomorphism class in terms of its invariant factors.

Solution. The solution is displayed in the following table. You should compute each set of invariant factors for the corresponding set of elementary divisors. Notice that the expression for the abelian group in terms of invariant factors is more compact than when expressed using elementary divisors.

FACTORIZATION OF 72	ELEMENTARY DIVISORS	INVARIANT FACTORS	ABELIAN GROUP
$2 \cdot 2 \cdot 2 \cdot 3 \cdot 3$	$\rightarrow 2, 2, 2, 3, 3$	$\rightarrow 2, 6, 6$	$\rightarrow \mathbf{Z}_2 \times \mathbf{Z}_6 \times \mathbf{Z}_6$
$2 \cdot 2^2 \cdot 3 \cdot 3$	$\rightarrow 2, 2^2, 3, 3$	$\rightarrow 6, 12$	$\rightarrow \mathbf{Z}_6 \times \mathbf{Z}_{12}$
$2^3 \cdot 3 \cdot 3$	$\rightarrow 2^3, 3, 3$	$\rightarrow 3, 24$	$\rightarrow \mathbf{Z}_3 \times \mathbf{Z}_{24}$
$2 \cdot 2 \cdot 2 \cdot 3^2$	$\rightarrow 2, 2, 2, 3^2$	$\rightarrow 2, 2, 18$	$\rightarrow \mathbf{Z}_2 \times \mathbf{Z}_2 \times \mathbf{Z}_{18}$
$2 \cdot 2^2 \cdot 3^2$	$\rightarrow 2, 2^2, 3^2$	$\rightarrow 2, 36$	$\rightarrow \mathbf{Z}_2 \times \mathbf{Z}_{36}$
$2^3 \cdot 3^2$	$\rightarrow 2^3, 3^2$	$\rightarrow 72$	$\rightarrow \mathbf{Z}_{72}$

The development of the mathematical machinery necessary to prove the Fundamental Theorem can often be quite tedious. For this reason, the following material can be omitted on first reading without consequence with regard to the material in later chapters. However, it is worthwhile for you to work your way through as much of this material as time permits.

The first step in proving the Fundamental Theorem is to show that a finite abelian group can be expressed as an internal direct sum of a certain type of subgroup known as a p-subgroup.

5.4.5   Definition

Let G be a group and p a prime integer. If every element of G has order that is a power of p, then G is said to be a p-**group**. A subgroup of G that is itself a p-group is referred to as a p-**subgroup** of G.

5.4.6   Theorem

If G is an abelian group, p is a prime integer, and $G(p)$ is the set of all elements of G whose orders are a power of p, then $G(p)$ is a subgroup of G.

Proof. Since $0 = 1 \cdot 0$, 0 has order $1 = p^0$. Thus, $0 \in G(p)$, so $G(p)$ is non-empty. If $x, y \in G(p)$ have order p^m and p^n, respectively, and $m \geq n$, then $p^m(x - y) = p^m x - p^m y = p^m x - p^{m-n}p^n y = 0 - 0 = 0$. Hence, by Theorem 5.2.8, the order of $x - y$ divides p^m and, consequently, must be a power of p since p is a prime. Thus, $x - y \in G(p)$, so by the additive version of Theorem 2.2.3, $G(p)$ is a subgroup of G. The proof is similar if $n \geq m$. (In the theorem,

G is assumed to be abelian. Read the proof carefully, and see if you can determine where this assumption was indirectly used.) ∎

5.4.7   **Examples**

1. The group $\{0\}$ is a p-subgroup of G for every prime p since $0 = 1 \cdot 0 = p^0 \cdot 0$. Since 1 is obviously the smallest positive integer such that $1 \cdot 0 = 0$, 0 has order $1 = p^0$. Hence, $\{0\}$ is a p-subgroup of G. This also indicates that $0 \in G(p)$ for every prime p.

2. Consider the permutation $\alpha = (1, 2, 3, 4, 5, 6, 7, 8)$. Recall that the order of a cycle is the length of the cycle. Hence, $\alpha^8 = (1)$, so α generates a cyclic group $G = \{e = (1),\ \alpha,\ \alpha^2,\ \alpha^3,\ \alpha^4,\ \alpha^5,\ \alpha^6,\ \alpha^7\}$ of order 8. Using the formula $\mathbf{o}(\alpha^m) = \dfrac{8}{\gcd(m, 8)}$ of Theorem 5.2.8, we see that

    > $\alpha,\ \alpha^3,\ \alpha^5,$ and α^7 each have order $8 = 2^3$,
    >
    > α^2 and α^6 each have order $4 = 2^2$,
    >
    > α^4 has order $2 = 2^1$, and
    >
    > e has order $1 = 2^0$.

    Hence, G is a 2-group. Note that G is isomorphic to the additive group $\mathbf{Z}_8 = \mathbf{Z}_{2^3}$. The isomorphism is $f : G \rightarrow \mathbf{Z}_8$, where $f(\alpha^k) = [k]$.

3. Consider the permutation $\alpha = (1, 2, 3, 4, 5, 6, 7, 8, 9, 10, 11, 12)$. Since $\alpha^{12} = (1)$, α generates a cyclic group $G = \{e = (1),\ \alpha,\ \alpha^2,\ \alpha^3,\ \alpha^4,\ \alpha^5,\ \alpha^6,\ \alpha^7,\ \alpha^8,\ \alpha^9,\ \alpha^{10},\ \alpha^{11}\}$ of order 12.

    > $\alpha,\ \alpha^5,\ \alpha^7,$ and α^{11} each has order 12.
    >
    > α^2 and α^{10} each has order 6.
    >
    > α^3 and α^9 each has order $4 = 2^2$.
    >
    > α^4 and α^8 each has order $3 = 3^1$.
    >
    > α^6 has order $2 = 2^1$.
    >
    > e has order 1.

    The group G is *not* a p-group for any prime p since a fixed prime p cannot be found such that the order of each element of G is a power of p. However, G does have a 2-subgroup and a 3-subgroup. These are

    $$G(2) = \{e, \alpha^3, \alpha^6, \alpha^9\} \quad \text{and} \quad G(3) = \{e, \alpha^4, \alpha^8\},$$

respectively. Observe that $G = G(2) \dot\times G(3)$. This follows since $G(2) \cap G(3) = \{e\}$ and since the following list shows that each element of G can be expressed as a product of an element of $G(2)$ and an element of $G(3)$.

$$e = ee \qquad\qquad \alpha^6 = \alpha^6 e$$
$$\alpha = \alpha^9 \alpha^4 \qquad\qquad \alpha^7 = \alpha^3 \alpha^4$$
$$\alpha^2 = \alpha^6 \alpha^8 \qquad\qquad \alpha^8 = e\alpha^8$$
$$\alpha^3 = \alpha^3 e \qquad\qquad \alpha^9 = \alpha^9 e$$
$$\alpha^4 = e\,\alpha^4 \qquad\qquad \alpha^{10} = \alpha^6 \alpha^4$$
$$\alpha^5 = \alpha^9 \alpha^8 \qquad\qquad \alpha^{11} = \alpha^3 \alpha^8$$

Notice that $G(2)$ is isomorphic to $\mathbf{Z}_4$ and $G(3)$ is isomorphic to $\mathbf{Z}_3$ via the isomorphisms

$$g : G(2) \to \mathbf{Z}_4, \text{ where } g(\alpha^k) = [k]_4, \quad \text{and}$$
$$h : G(3) \to \mathbf{Z}_3, \text{ where } h(\alpha^k) = [k]_3.$$

Therefore since $G = G(2) \dot\times G(3)$, we see that G is isomorphic to $\mathbf{Z}_4 \times \mathbf{Z}_3$.

The preceding Example 3 shows that the finite abelian group

$$G = \{e, \alpha, \alpha^2, \alpha^3, \alpha^4, \alpha^5, \alpha^6, \alpha^7, \alpha^8, \alpha^9, \alpha^{10}, \alpha^{11}\}$$

is the internal direct product of its p-subgroups $G(2)$ and $G(3)$. Notice that 2 and 3 are the prime divisors of the order 12 of G. It will now be shown that this is true for every finite abelian group.

5.4.8   Theorem

The following hold for any abelian group G.

1. If $x \in G$ has finite order, m is an integer such that $\gcd(m, \mathbf{o}(x)) = 1$, and $mx = 0$, then $x = 0$.
2. If G is finite and $|G| = n$, then $nx = 0$ for every $x \in G$.
3. Let x be an element of G with order n. If $n = n_1 n_2$ and $\gcd(n_1, n_2) = 1$, then x can be expressed as $x = x_1 + x_2$, where $\mathbf{o}(x_1) = n_1$ and $\mathbf{o}(x_2) = n_2$.

Proof

1. Since $\gcd(m, \mathbf{o}(x)) = 1$, Theorem 1.1.7 shows that there are integers a and b such that $1 = ma + \mathbf{o}(x)b$. But then $x = max + \mathbf{o}(x)bx = 0$ since $mx = \mathbf{o}(x)x = 0$.

2. Since G is finite, if $x \in G$, then x generates a cyclic subgroup of G of order $\mathbf{o}(x) = k$. By Lagrange's Theorem 2.2.9, $k \mid n$. If $n = mk$, then $nx = m(kx) = 0$.

3. If either $n_1 = 1$ or $n_2 = 1$, there is nothing to prove. For example, if $n_1 = 1$ and $n_2 = n$, then $x = 0 + x$ with $\mathbf{o}(0) = 1$ and $\mathbf{o}(x) = n$. Suppose that $n_1 \neq 1$ and $n_2 \neq 1$. Since $\gcd(n_1, n_2) = 1$, there are integers a and b such that $1 = n_2 a + n_1 b$. Hence, $x = (n_2 a)x + (n_1 b)x$. We claim that $(n_2 a)x$ has order n_1. The additive version of Theorem 5.2.8 indicates that $\mathbf{o}((n_2 a)x) = \frac{n}{\gcd(n_2 a, n)}$. Since $1 = n_2 a + n_1 b$, a and n_1 must be relatively prime. If not, there is a positive integer $d > 1$ such that $d \mid a$ and $d \mid n_1$, in which case $d \mid 1$, which obviously cannot be the case. From this it follows that $\gcd(n_2 a, n) = n_2$, so $\mathbf{o}((n_2 a)x) = n_1$. Similarly, $\mathbf{o}((n_1 b)x) = n_2$, so if we let $x_1 = (n_2 a)x$ and $x_2 = (n_1 b)x$, then $x = x_1 + x_2$ with $\mathbf{o}(x_1) = n_1$ and $\mathbf{o}(x_2) = n_2$. ■

5.4.9   Lemma

Let G be an abelian group, and suppose that $x \in G$ has order $n = p_1^{n_1} p_2^{n_2} \cdots p_m^{n_m}$, where $p_1, p_2, \ldots, p_m$ are the prime divisors of n, and $n_1, n_2, \ldots, n_m$ are positive integers. Then $x = x_1 + x_2 + \cdots + x_m$, where $x_k \in G(p_k)$ for $k = 1, 2, \ldots, m$.

Proof. The proof is by induction on the number k of prime divisors of n. If $k = 1$, then n is divisible by a single prime p_1, so $n = p_1^{n_1}$ for some positive integer n_1. Hence, $x \in G(p_1)$. Now assume that the theorem is true for all $x \in G$ whose order n is divisible by k primes, and let $x \in G$ be such that n is divisible by $k + 1$ primes. It follows that $n = p_1^{n_1} p_2^{n_2} \cdots p_{k+1}^{n_{k+1}}$, where $n_1, n_2, \ldots, n_{k+1}$ are positive integers. Now $p_2^{n_2} \cdots p_{k+1}^{n_{k+1}} x$ has order $p_1^{n_1}$, and $p_1^{n_1}$ and $p_2^{n_2} \cdots p_{k+1}^{n_{k+1}}$ are relatively prime. Hence, Theorem 5.4.8 shows that x can be expressed as $x = x_1 + y$, where $\mathbf{o}(x_1) = p_1^{n_1}$ and $\mathbf{o}(y) = p_2^{n_2} \cdots p_{k+1}^{n_{k+1}}$. Since there are k prime divisors of $\mathbf{o}(y)$, by the induction hypothesis, there exist x_k in $G(p_k)$ for $k = 2, 3, \ldots, k + 1$, such that $y = x_2 + x_3 + \cdots + x_{k+1}$. Since $\mathbf{o}(x_1) = p_1^{n_1}$ implies that $x_1 \in G(p_1)$, we have shown that $x = x_1 + x_2 + \cdots + x_{k+1}$ with $x_k \in G(p_k)$ for $k = 1, 2, \ldots, k + 1$. This proves the lemma. ■

5.4.10   Theorem

Let G be a finite abelian group of order n. Then $G = G(p_1) \oplus G(p_2) \oplus \cdots \oplus G(p_m)$, where $p_1, p_2, \ldots, p_m$ are the prime divisors of n.

Proof. Lemma 5.4.9 shows that $G = G(p_1) + G(p_2) + \cdots + G(p_m)$, so it remains only to show that the sum is direct. Suppose that $x_1 + x_2 + \cdots + x_m = 0$, where $x_k \in G(p_k)$ for $k = 1, 2, \ldots, m$. The proof will be complete if it can be shown that $x_1 = x_2 = \cdots = x_m = 0$. (Do you know why?) Since $x_k \in G(p_k)$, each x_k has an order that is a power of p_k. Suppose that $\mathbf{o}(x_k) = p_k^{n_k}$ for $k = 1, 2, \ldots, m$, and let's show that $x_1 = 0$. The proofs that each of the other summands in x is zero are similar. Since $x_1 + x_2 + \cdots + x_m = 0$,

$$0 = p_2^{n_2} p_3^{n_3} \cdots p_m^{n_m}(x_1 + x_2 + \cdots + x_m)$$

$$= p_2^{n_2} p_3^{n_3} \cdots p_m^{n_m} x_1 + p_2^{n_2} p_3^{n_3} \cdots p_m^{n_m} x_2 + \cdots + p_2^{n_2} p_3^{n_3} \cdots p_m^{n_m} x_m$$

$$= p_2^{n_2} p_3^{n_3} \cdots p_m^{n_m} x_1 + p_3^{n_3} \cdots p_m^{n_m}(p_2^{n_2} x_2) + \cdots + p_2^{n_2} p_3^{n_3} \cdots (p_m^{n_m} x_m)$$

$$= p_2^{n_2} p_3^{n_3} \cdots p_m^{n_m} x_1.$$

But $\mathbf{o}(x_1) = p_1^{n_1}$ and $p_2^{n_2} p_3^{n_3} \cdots p_m^{n_m}$ are relatively prime, so Theorem 5.4.8 indicates that $x_1 = 0$. Hence, $G = G(p_1) \oplus G(p_2) \oplus \cdots \oplus G(p_m)$. ∎

Thus, we see that any finite abelian group can be decomposed as an internal direct sum of its p-subgroups. More can be said about the individual p-subgroups. In fact, each p-subgroup of G can be further decomposed as an internal direct sum of cyclic p-subgroups. We now show how this decomposition can be carried out.

If G is a finite abelian p-group, then the order of every element of G is a power of p. Hence, among the elements of G, there is an element x whose order is maximal, so $\mathbf{o}(y) \leq \mathbf{o}(x) = p^m$ for all $y \in G$. If $y \in G$ and $\mathbf{o}(y) = p^k$, then $k \leq m$ and $p^m y = p^{m-k}(p^k y) = 0$. The proof of the following lemma is one of the most difficult proofs encountered up to this point in the text. So sit down with a pencil and paper and work your way through it. The lemma allows us to take a very basic step in the proof of the Fundamental Theorem.

5.4.11 Lemma

Let G be a finite abelian p-group, and suppose that x is an element of G whose order is maximal among the orders of the elements of G. Then there is a subgroup H of G such that $G = \langle x \rangle \oplus H$.

Proof. Suppose that $\mathcal{C} = \{G' \mid G' \text{ is a subgroup of } G \text{ and } \langle x \rangle \cap G' = \{0\}\}$. Note that $\mathcal{C} \neq \varnothing$ since $G' = \{0\}$ is in $\mathcal{C}$. Furthermore, since G is a finite group, $\mathcal{C}$ can contain only a finite number of subgroups of G, and there is a subgroup

H in $\mathcal{C}$ that is maximal. That is, if H' is a subgroup of G in $\mathcal{C}$ such that $H \subseteq H'$, then $H = H'$. If $\mathbf{o}(x) = p^m$, then by the observation immediately preceding the lemma, $p^m y = 0$ for each $y \in G$. If $y \in G$ and $y \notin \langle x \rangle \oplus H$, then $p^m y = 0 = 0 + 0 \in \langle x \rangle \oplus H$. Let k be the smallest positive integer such that $p^k y \in \langle x \rangle \oplus H$. From this it follows that

$$y' = p^{k-1}y \notin \langle x \rangle \oplus H \tag{1}$$

and that $py' \in \langle x \rangle \oplus H$. The assumption that there is a $y \in G$ such that $y \notin \langle x \rangle \oplus H$ has led us to the conclusion that there is a $y' \in G$ such that $py' \in \langle x \rangle \oplus H$, but $y' \notin \langle x \rangle \oplus H$. If it can be shown that $y' \in \langle x \rangle \oplus H$, then we will have a contradiction. Hence, it cannot be the case that there is a $y \in G$ such that $y \notin \langle x \rangle \oplus H$. Consequently, $G = \langle x \rangle \oplus H$, and the proof of the theorem will be finished. Now let's show that $y' \in \langle x \rangle \oplus H$. Suppose that

$$py' = nx + h, \text{ where } n \in \mathbf{Z} \text{ and } h \in H. \tag{2}$$

Since x has maximal order, by the observation immediately preceding the lemma, $p^m y = 0$ for all $y \in G$. Thus, we see from equation (2) that

$$0 = p^m y' = p^{m-1}(py') = p^{m-1}nx + p^{m-1}h,$$

so $p^{m-1}nx = -p^{m-1}h \in \langle x \rangle \cap H = \{0\}$. Therefore, by Theorem 5.2.8, $p^m | p^{m-1}n$ and so it follows that $p | n$. If $n = ps$, then $py' = psx + h$, so $h = p(y' - sx)$. If

$$d = y' - sx, \tag{3}$$

then $pd = h \in H$. But $d \notin H$ because if $d \in H$, then $y' = sx + d \in \langle x \rangle \oplus H$, which would contradict equation (1). It follows that $K = \{h' + qd | h' \in H \text{ and } q \in \mathbf{Z}\}$ is a subgroup of G that properly contains H. Thus, it must be the case that $\langle x \rangle \cap K \neq \{0\}$ since H is maximal with respect to this property. Suppose that $0 \neq z \in \langle x \rangle \cap K$. Then

$$z = tx = ud + h', \text{ where } t, u \in \mathbf{Z} \text{ and } h' \in H. \tag{4}$$

If $p | u$ and $u = pv$, then since $pd \in H$, $0 \neq z = tx = pvd + h' \in \langle x \rangle \cap H = \{0\}$, which is a contradiction. Thus, $p \nmid u$, so $\gcd(p, u) = 1$. Hence, by Theorem 1.1.7, there are integers a and b such that $1 = pa + ub$. Consequently,

$$\begin{aligned} y' = 1y' = (pa + ub)y' &= a(py') + b(uy') \\ &= a(nx + h) + b(u(d + sx)) && \text{by (2) and (3)} \\ &= a(nx + h) + b(ud + usx) \\ &= a(nx + h) + b(tx - h' + usx) && \text{by (4)} \\ &= (an + bt + bus)x + (uh - bh') \in \langle x \rangle \oplus H \end{aligned}$$

Hence, we have shown that there cannot exist an element $y \in G$ such that $y \notin \langle x \rangle \oplus H$, so $G = \langle x \rangle \oplus H$. ∎

Before we can prove the Fundamental Theorem, a couple of additional observations will be helpful.

5.4.12   Lemma

If G is a nontrivial finite abelian group and p is a prime that divides the order of G, then G has a nonzero element of order p.

Proof. The proof is by induction on the order of G. If $|G| = 2$, then $G = \{0, x\}$, and the only prime dividing 2 is 2. Since $x + x = 0$ or $2x = 0$, x is an element of order 2. Hence, the theorem is true for $n = 2$. Let $|G| = n$, and make the induction hypothesis that the theorem is true for all finite abelian groups H such that $2 \leq |H| < n$. Let p be a prime that divides n. Pick any $x \in G$, $x \neq 0$, and suppose that $\mathbf{o}(x) = m$. Then either $p \mid m$ or $p \nmid m$. If $p \mid m$, then $m = pk$ for some integer k, so $p(kx) = mx = 0$. Therefore, $kx \neq 0$ has order p. If $p \nmid m$, then $\langle x \rangle$ is a normal subgroup of G that has m elements, and $n = m[G : \langle x \rangle]$, where $[G : \langle x \rangle]$ is the index of $\langle x \rangle$ in G. Since p divides n, we see that p divides $m[G : \langle x \rangle]$. But $p \nmid m$, so since p is prime, $p \mid [G : \langle x \rangle]$. Therefore, p divides $|G/\langle x \rangle|$, and since $|G/\langle x \rangle| < n$, the induction hypothesis implies that there is a nonzero element $y + \langle x \rangle \in G/\langle x \rangle$ that has order p. But $py + \langle x \rangle = p(y + \langle x \rangle) = 0$, so $py \in \langle x \rangle$. Therefore $p(my) = m(py) = 0$, so either $my = 0$ or $my \neq 0$ has order p. The proof will be complete if we can show that $my \neq 0$. If $my = 0$, then $m(y + \langle x \rangle) = my + \langle x \rangle = 0$, so because $y + \langle x \rangle$ has order p, it follows from Theorem 5.2.8 that $p \mid m$. But we have assumed that $p \nmid m$, so we have a contradiction. Therefore $my \neq 0$. ∎

5.4.13   Lemma

If p is a prime and G is a finite nontrivial abelian group, then the following are true:

1. The group G is a p-group if and only if $|G| = p^k$ for some positive integer k.
2. If p is a prime that divides the order of G and $pG = \{px \mid x \in G\}$, then $|pG| < |G|$.

Proof

1. Suppose that G is a nontrivial finite p-group. If q is a prime different from p and if q divides $|G|$, then by Lemma 5.4.12, G has at least one

nonzero element of order q. But since G is a p-group, this contradicts the fact that every element of G has order that is a power of p. Therefore, p is the only prime divisor of $|G|$, so $|G| = p^k$ for some positive integer k.

Conversely, suppose that $|G| = p^k$ for some positive integer k. If $x \in G$ and $x = 0$, then $0 = p^0 0 = 1 \cdot 0 = 0$ and $\mathbf{o}(x)$ is a power of p. If $x \neq 0$, then $\langle x \rangle$ is a subgroup of G, so by Lagrange's Theorem 2.2.9, $\mathbf{o}(x)$ divides the order of G. Hence, $\mathbf{o}(x) \mid p^k$, and so $\mathbf{o}(x)$ must be a power of p, which tells us that G is a p-group.

2. Let $|G| = n$. If p divides the order of G, then by Lemma 5.4.12, G has at least one nonzero element x of order p. But then $px = 0$, and from this we see that pG can have at most $n - 1$ elements. Hence, $|pG| < |G|$. ∎

The proof of the Fundamental Theorem is long, so we break the theorem into two parts. The first part states that any finite abelian group G can be expressed as an internal direct sum of cyclic subgroups. The second part of the theorem indicates that such an expression for G is unique.

5.4.14   The Fundamental Theorem of Finite Abelian Groups

Part 1. If G is any finite abelian group, then G can be expressed as an internal direct sum of cyclic subgroups, each of which has order the power of a prime.

Proof of Part 1. Let G be a finite abelian group G of order n. Then, by Theorem 5.4.10, $G = G(p_1) \oplus G(p_2) \oplus \cdots \oplus G(p_m)$, where $n = p_1^{n_1} p_2^{n_2} \cdots p_m^{n_m}$, $p_1, p_2, \ldots, p_m$ are the prime divisors of n, and $n_1, n_2, \ldots, n_m$ are positive integers. It will be shown that G can be written as an internal direct sum of cyclic subgroups by showing that each $G(p_i)$ can be expressed as an internal direct sum of cyclic subgroups, each of which has order the power of p_i. Let p be one of the primes p_1, $p_2, \ldots, p_m$. The proof is by induction using the Second Principle of Induction (see Problem Set 1.2, Exercise 18). Lemma 5.4.13 shows that $|G(p)| = p^k$ for some positive integer k. We induct on k. If $k = 1$, then $|G(p)| = p$, and $G(p)$ is a cyclic group of order p. In this case, $G(p)$ is isomorphic to $\mathbf{Z}_p$, and it is obvious that $G(p)$ can be expressed as an internal direct sum of cyclic subgroups of G, each having order that is a power of p. Now make the induction hypothesis that any p-group of G of order of p^m with $m < k$ can be expressed as an internal direct sum of cyclic subgroups of G, each having an order that is a power of p. Suppose that $|G(p)| = p^k$, and let $x \in G(p)$ be an element of $G(p)$ whose order is maximal among the orders of the elements of $G(p)$. Then $\mathbf{o}(x) = p^t$, where t is a positive integer and $\langle x \rangle$ is a cyclic subgroup of $G(p)$ of order p^t. Lemma 5.4.11 shows that $G(p) = \langle x \rangle \oplus H$ for some subgroup H of $G(p)$. Moreover, each element of H is also an element of $G(p)$ and as such has order that is a power of p. Thus, H is also a p-subgroup of G. Obviously, $|H| < |G(p)|$, by Lemma 5.4.13, $|H| = p^m$, and it must be the case that $m < k$. Hence, by the induction hypothesis, H can be written as an internal direct sum of cyclic subgroups, each having an order that is a power of p. If $H = \langle x_1 \rangle \oplus \langle x_2 \rangle \oplus \cdots \oplus \langle x_s \rangle$, then $G(p) = \langle x \rangle \oplus \langle x_1 \rangle \oplus \langle x_2 \rangle \oplus \cdots \oplus \langle x_s \rangle$, and the proof is complete. ∎

Finally, we come to the proof of the second part of the Fundamental Theorem, the proof of uniqueness.

Part 2. If G is any finite abelian group, then the expression for G as an internal direct sum of cyclic subgroups, each of which has order the power of a prime, is unique up to the order of the summands in the expression.

Proof of Part 2. Observe that each $G(p_i)$ in the decomposition

$$G = G(p_1) \oplus G(p_2) \oplus \cdots \oplus G(p_m)$$

of part 1 is uniquely determined by G. This follows since each $G(p_i)$ is composed of all of the elements of G whose order is a power of p_i. Hence, we can show uniqueness of the decomposition of G into an internal direct sum of cyclic subgroups by showing uniqueness of the decomposition of each $G(p_i)$. Let p be any one of the primes $p_1, p_2, \ldots, p_m$. Then we need to show that if

$$G(p) = \langle x_1 \rangle \oplus \langle x_2 \rangle \oplus \cdots \oplus \langle x_s \rangle \qquad \text{and}$$
$$G(p) = \langle y_1 \rangle \oplus \langle y_2 \rangle \oplus \cdots \oplus \langle y_t \rangle,$$

then $s = t$ and that after a suitable reindexing of the summands in the second (or the first, if you like) expression for $G(p)$, $\langle x_i \rangle$ and $\langle y_i \rangle$ are isomorphic for $i = 1, 2, \ldots, s$. As a first step, reindex the summands in each expression so that

$$|\langle x_1 \rangle| \geq |\langle x_2 \rangle| \geq \cdots \geq |\langle x_s \rangle| \qquad \text{and}$$
$$|\langle y_1 \rangle| \geq |\langle y_2 \rangle| \geq \cdots \geq |\langle y_s \rangle|.$$

By Lemma 5.4.13, $|G(p)| = p^k$, so the proof is by induction on k. If $k = 1$, then $|G(p)| = p$, and $G(p) = \langle x_1 \rangle$ since $\langle x_1 \rangle$ can have order no less than p. Similarly, $G(p) = \langle y_1 \rangle$, so $\langle x_1 \rangle$ and $\langle y_1 \rangle$ are both cyclic groups of order p. They are thus both isomorphic to $\mathbf{Z}_p$ and so areorphic to each other. Hence, we have the desired result when $k = 1$. Make the induction hypothesis that a unique decomposition exists for any abelian p-group G of order p^m, where $m < k$ is unique. Suppose that $|G(p)| = p^k$ and that

$$G(p) = \langle x_1 \rangle \oplus \langle x_2 \rangle \oplus \cdots \oplus \langle x_s \rangle \quad \text{and}$$

$$G(p) = \langle y_1 \rangle \oplus \langle y_2 \rangle \oplus \cdots \oplus \langle y_t \rangle.$$

In the representation $G(p) = \langle x_1 \rangle \oplus \langle x_2 \rangle \oplus \cdots \oplus \langle x_s \rangle$, there may be cyclic subgroups $\langle x_i \rangle$ such that $|\langle x_i \rangle| = p$, in which case $p(x_i) = 0$. Let s' be the largest integer such that $|\langle x_i \rangle| > p$ if $i \leq s'$, and suppose that t' is the largest integer such that $|\langle y_i \rangle| > p$ if $i \leq t'$. By Exercise 9, $pG(p) = \{px \mid x \in G(p)\}$ is a p-subgroup of $G(p)$, and by Exercise 10,

$$pG(p) = p\langle x_1 \rangle \oplus p\langle x_2 \rangle + \cdots \oplus p\langle x_{s'} \rangle \quad \text{and}$$

$$pG(p) = p\langle y_1 \rangle \oplus p\langle y_2 \rangle \oplus \cdots \oplus p\langle y_{t'} \rangle.$$

What we have done is to remove the trivial summands $\{0\}$ from the expressions for $pG(p)$ so that the sums will continue to be direct. Lemma 5.4.13 shows that $|pG(p)| < |G(p)|$, so $|pG(p)| = p^m$, where $m < k$. Hence, it follows from the induction hypothesis that $s' = t'$ and that we can reindex the summands so that $p\langle x_i \rangle$ and $p\langle y_i \rangle$ are isomorphic for $i = 1, 2, \ldots, s'$. Thus, $p|\langle x_i \rangle| = |p\langle x_i \rangle| = |p\langle y_i \rangle| = p|\langle y_i \rangle|$ for each i, and from this we see that $|\langle x_i \rangle| = |\langle y_i \rangle|$. Therefore, $\langle x_i \rangle$ and $\langle y_i \rangle$ are cyclic groups of the same order, so they must be isomorphic. Hence, $\langle x_i \rangle$ and $\langle y_i \rangle$ are isomorphic for $i = 1, 2, \ldots, s'$. All that remains to be shown is that the number $s - s'$ of $\langle x_i \rangle$ with $|\langle x_i \rangle| = p$ is equal to the number $t - t'$ of $\langle y_i \rangle$ with $|\langle y_i \rangle| = p$. Clearly, these groups are isomorphic since they are all cyclic groups of order p. If we can show that $s - s' = t - t'$, then the fact that $s' = t'$ will imply that $s = t$, and the proof will be complete. Since

$$|\langle x_1 \rangle||\langle x_2 \rangle| \cdots |\langle x_{s'} \rangle| p^{s-s'} = |G| = |\langle y_1 \rangle||\langle y_2 \rangle| \cdots |\langle y_{t'} \rangle| p^{t-t'} \quad \text{and}$$

$$|\langle x_1 \rangle||\langle x_2 \rangle| \cdots |\langle x_{s'} \rangle| = |\langle y_1 \rangle||\langle y_2 \rangle| \cdots |\langle y_{t'} \rangle|,$$

we see that

$$p^{s-s'} = p^{t-t'}, \text{ so } s - s' = t - t'. \quad \blacksquare$$

Problem Set 5.4

1. Give an example of an additive p-group of order 8.

2. Show that each of the following additive groups can be written as an external direct product of two cyclic groups.

(a) $\mathbf{Z}_{12}$ (c) $\mathbf{Z}_{35}$
(b) $\mathbf{Z}_{15}$ (d) $\mathbf{Z}_{45}$

3. For each of the following values of n, classify, up to isomorphism, all finite additive abelian groups of order n. Express each representative of each isomorphism class as an external product of cyclic groups by using elementary divisors and then by using invariant factors.

    (a) $n = 6$ (c) $n = 90$
    (b) $n = 24$ (d) $n = 600$

4. List the elementary divisors of the given additive group.

    (a) $\mathbf{Z}_{350}$ (c) $\mathbf{Z}_{12} \times \mathbf{Z}_{18} \times \mathbf{Z}_{24}$
    (b) $\mathbf{Z}_{20} \times \mathbf{Z}_{30} \times \mathbf{Z}_{40}$ (d) $\mathbf{Z}_{12} \times \mathbf{Z}_{30} \times \mathbf{Z}_{150} \times \mathbf{Z}_{250}$

5. Write each of the groups of Exercise 4 as an external product of cyclic groups using its elementary divisors and then by using its invariant factors.

6. If G is an additive abelian p-group, prove that $f : G \rightarrow G$ defined by $f(x) = nx$ is a group isomorphism for each integer n such that $\gcd(n, p) = 1$.

7. If G is the additive group $\mathbf{Q}/\mathbf{Z}$, describe the elements of $G(2)$. What are the elements of $G(p)$ for any prime p?

8. Characterize the positive integers n for which there is exactly one additive abelian group (up to isomorphism) of order n.

9. If G is an additive abelian group and p is a prime, prove that $pG = \{px \mid x \in G\}$ is a subgroup of G. If G is a p-group, show that pG is a p-subgroup of G.

10. If p is a prime and G is an additive abelian group for which there are subgroups G_1 and G_2 such that $G = G_1 \oplus G_2$, prove that $pG = pG_1 \oplus pG_2$. (See Exercise 9.) Generalize your proof to show that if $G = G_1 \oplus G_2 \oplus \cdots \oplus G_n$, then $pG = pG_1 \oplus pG_2 \oplus \cdots \oplus pG_k$, where $k \leq n$ and $pG = pG_1 \oplus pG_2 \oplus \cdots \oplus pG_k$ is a reindexed $pG = pG_1 \oplus pG_2 \oplus \cdots \oplus pG_n$ with the summands for which $pG_i = 0$ removed.

11. Determine, up to isomorphism, all additive abelian groups of order p^3, where p is a prime.

12. Determine, up to isomorphism, all additive abelian groups of order p^2q^3, where p and q are prime integers.

13. Prove that if G is an additive cyclic group of order p^k, where p is a prime, then G has a subgroup of order p^t for each integer t such that $0 \leq t \leq k$.

6

Polynomial Rings

In many areas of mathematics, polynomials play an important role. The purpose of this chapter is to study polynomials in a general setting where the coefficients belong to an arbitrary commutative ring R with identity. It is not always necessary to assume that R is commutative or that R has an identity. However, much of the material presented in this chapter is simplified somewhat when R is commutative and when R has an identity. Hence, we assume throughout this chapter, unless stated otherwise, that R is a commutative ring with identity.

6.1   POLYNOMIAL RINGS

Section Overview. In Section 3.1 it was shown that the set $R[x]$ of all polynomials in x with their coefficients in a ring R is itself a ring. No mention was made there of some of the fundamental questions that arise regarding $R[x]$. For example, what is the exact nature of the indeterminate x, and why should we assume that $(a + b)x = ax + bx$ for all $a, b \in R$? Is it always the case that $x \in R[x]$? To answer these and similar questions regarding $R[x]$, a formal development of polynomial rings is required.

If $f(x) = a_0 + a_1x + a_2x^2 + \cdots + a_nx^n$ is a polynomial with its coefficients taken from a ring R, it is legitimate to inquire about the nature of the indeterminate x. Exactly what is an indeterminate x, and what does it mean to add x to an element of R or to multiply an element of the ring R by x? Should x^k be interpreted as representing $x \cdot x \cdot x \cdots x$ with k factors? It is common practice to write $(a + b)x = ax + bx$, but if we don't know the exact nature of x, why should we assume that x is distributive over addition? One can also ask, "$a_0 + a_1x + \cdots + a_nx^n$ is a sum of elements of what? A ring? If so, what ring?" One approach to answering these questions is to consider x to be an arbitrary element of the ring R so that the properties of the ring can be applied to x. This approach is not entirely satisfactory since, at times, we wish to consider a polynomial as a formal object with x representing no element of R whatsoever.

We begin the development of polynomial rings with the following definition.

6.1.1   Definition

> A sequence of elements of R of the form $(a_0, a_1, a_2, \ldots)$ is said to be a **polynomial** over R if there is a nonnegative integer n such that $a_k = 0$ for all $k > n$. The integer n is not fixed and depends on the particular polynomial. The notation $R[x]$ denotes the set of all such polynomials over R. For each integer $k \geq 0$, we refer to a_k as the **kth coefficient** of the polynomial $(a_0\, a_1, a_2, \ldots)$. Two polynomials $(a_0\, a_1, a_2, \ldots)$ and $(b_0\, b_1, b_2, \ldots)$ in $R[x]$ are **equal** if $a_k = b_k$ for $k = 0,1,2, \ldots$.

6.1.2   Example

> The condition on a polynomial that there is a nonnegative integer n such that $a_k = 0$ for all $k > n$ does *not* mean that $a_k \neq 0$ for all $k \leq n$. For example, $(1, 0, 3, 0, 0, \ldots)$ is a polynomial in $\mathbf{Z}[x]$ since $a_k = 0$ for all $k > 2$. Likewise, $(0, 0, 0, \ldots)$ is a polynomial in $\mathbf{Z}[x]$ since $a_k = 0$ for all $k > 0$. $\left(1, -5, \frac{1}{2}, 0, 0, 0, \ldots\right)$ and $\left(\sqrt{2}, 0, -5, 7, \frac{7}{8}, 0, 0, 0, \ldots\right)$ are polynomials in $\mathbf{R}[x]$. In the first case, $a_k = 0$ when $k > 2$, and in the second, $a_k = 0$ if $k > 4$. The polynomial $\left(1, -5, \frac{1}{2}, 0, 0, 0, \ldots\right)$ is also in $\mathbf{Q}[x]$, but $\left(\sqrt{2}, 0, -5, 7, \frac{7}{8}, 0, 0, 0, \ldots\right) \notin \mathbf{Q}[x]$ since $\sqrt{2} \notin \mathbf{Q}$.

Addition and multiplication of polynomials in $R[x]$ are defined as follows:

$$(a_0, a_1, a_2, \ldots) + (b_0, b_1, b_2, \ldots) = (a_0 + b_0, a_1 + b_1, a_2 + b_2, \ldots)$$

and

$$(a_0, a_1, a_2, \ldots)(b_0, b_1, b_2, \ldots) = (c_0, c_1, c_2, \ldots),$$

where

$$c_k = a_0 b_k + a_1 b_{k-1} + a_2 b_{k-2} + \cdots + a_k b_0$$

for $k = 0, 1, 2, \ldots$. To form c_k we look for all possible products $a_i b_j$, where the subscripts i and j are such that $i + j = k$, and then we add these products together. Thus,

$$c_0 = a_0 b_0 \qquad\qquad c_3 = a_0 b_3 + a_1 b_2 + a_2 b_1 + a_3 b_0$$

$$c_1 = a_0 b_1 + a_1 b_0 \qquad\qquad c_4 = a_0 b_4 + a_1 b_3 + a_2 b_2 + a_3 b_1 + a_4 b_0$$

$$c_2 = a_0 b_2 + a_1 b_1 + a_2 b_0 \qquad\qquad \vdots$$

The notation $(a_0, a_1, a_2, \ldots)$ can often be cumbersome when writing the general expression for a polynomial. A space-saving device for writing such polynomials is to simply display the kth coefficient of the polynomial and write the polynomial as (a_k), where it is understood that the subscript k starts at the integer 0 and then proceeds through the nonnegative integers.

6.1.3   Theorem

The set $R[x]$ of all polynomials in x with their coefficients in R is a commutative ring with identity that contains a subring isomorphic to R.

Proof. Let (a_k), (b_k), and (c_k) be polynomials in $R[x]$. Then the operation of addition is closed since $(a_k) + (b_k) = (a_k + b_k)$ is clearly a polynomial in $R[x]$. Addition is a well-defined binary operation on $R[x]$ since the operation of addition is well-defined on R. Addition is associative since

$$(a_k) + [(b_k) + (c_k)] = (a_k) + (b_k + c_k)$$

$$= (a_k + [b_k + c_k])$$

$$= ([a_k + b_k] + c_k)$$

$$= (a_k + b_k) + (c_k)$$

$$= [(a_k) + (b_k)] + (c_k).$$

The additive identity of $R[x]$ is the **zero polynomial** $(0) = (0, 0, 0, \ldots)$, and the additive inverse of (a_k) is the polynomial $(-a_k)$. Since addition is clearly commutative, $R[x]$ is an additive abelian group. One can easily show that multiplication of polynomials is a well-defined binary operation on $R[x]$, so to complete the proof that $R[x]$ is a ring, we need to show that multiplication of polynomials is associative and that multiplication of polynomials is distributive

over addition of polynomials. To see that multiplication of polynomials is associative, first consider $(a_k)[(b_k)(c_k)]$. To compute the mth coefficient in the
product $(b_k)(c_k)$, we look for all possible products $b_j c_k$, where $j + k = m$, and
then form the sum of these products. Thus, the mth coefficient of the product
$(b_k)(c_k)$ looks like

$$\sum_{j+k=m} b_j \, c_k.$$

After we have computed the product $(b_k)(c_k)$, to compute the nth coefficient of the product $(a_k)[(b_k)(c_k)]$, we look for all products of the form
$a_i\left(\sum_{j+k=m} b_j c_k\right)$, where $i + m = n$, and then add these together to obtain

$$\sum_{i+m=n} \left[a_i \left(\sum_{j+k=m} b_j c_k \right) \right].$$

But $j + k = m$, so $i + j + k = n$, and by using the distributive property in R,
$\sum_{i+m=n} \left[a_i \left(\sum_{j+k=m} b_j c_k \right) \right]$ becomes

$$\sum_{i+j+k=n} a_i(b_j \, c_k).$$

Hence, the nth coefficient of $(a_k)[(b_k)(c_k)]$ looks like the sum of all possible
products of the form $a_i(b_j c_k)$, where $i + j + k = n$. A similar argument leads to
the conclusion that the nth coefficient of $[(a_k)(b_k)](c_k)$ is the sum of all possible products of the form $(a_i b_j)c_k$, where $i + j + k = n$. Thus, the nth coefficient
is

$$\sum_{i+j+k=n} (a_i b_j)c_k.$$

But multiplication is associative in R, so $a_i(b_j c_k) = (a_i b_j)c_k$. Hence, we have
$(a_k)[(b_k)(c_k)] = [(a_k)(b_k)](c_k)$. Similar arguments can be used to show that
multiplication of polynomials is distributive over addition of polynomials
and that multiplication of polynomials is a commutative operation. The
proofs of these two properties of $R[x]$ are left as exercises. Hence, $R[x]$ is a
commutative ring. Next, let $(a_0, a_1, \ldots) = (e, 0, 0, \ldots)$. The kth coefficient
of the product $(a_k)(b_k)$ is $c_k = a_0 b_k + a_1 b_{k-1} + a_2 b_{k-2} + \cdots + a_k b_0$. Since
$a_i = 0$ for $i = 1, 2, \ldots, k$, the only nonzero summand in c_k is
$a_0 b_k = e b_k = b_k$. Therefore, $c_k = b_k$ for each $k \geq 0$, so $(a_k)(b_k) = (b_k)$. Thus,
$R[x]$ has $(e, 0, 0, \ldots)$ as a multiplicative identity.

Finally, the subset $R' = \{(a, 0, 0, \ldots) \mid a \in R\}$ is easily shown to be a subring of $R[x]$ and the mapping $\phi : R \longrightarrow R'$, where $\phi(a) = (a, 0, 0, \ldots)$ is a ring isomorphism. Hence, $R[x]$ contains a copy of R. ∎

The polynomials in the ring R' are referred to as the **constant polynomials** of $R[x]$.

In Definition 6.1.1, we indicated that x would be specified later. Our goal now is to define x and show that every polynomial $(a_0, a_1, \ldots, a_n, 0, 0, \ldots)$ in $R[x]$ can be written as $a_0 + a_1 x + \cdots + a_n x^n$.

Let

$$x^0 = (e, 0, 0, 0, 0, 0, \ldots) \qquad \text{and}$$

$$x^1 = x = (0, e, 0, 0, 0, 0, \ldots).$$

Then by using the definition of multiplication in $R[x]$, it follows easily that

$$x^2 = x \cdot x = (0, 0, e, 0, 0, 0, \ldots)$$

$$x^3 = x \cdot x \cdot x = (0, 0, 0, e, 0, 0, \ldots)$$

$$\vdots$$

Note that the elements $x^0, x, x^2, x^3, \ldots$ are all elements of $R[x]$. The polynomial x is referred to as the **indeterminate** of $R[x]$. Since R and R' are isomorphic rings, there is no longer a need to distinguish between a and $(a, 0, 0, 0, \ldots)$. If these elements of R and R' are identified, then

$$ax^0 = (a, 0, 0, 0, 0, 0, \ldots) = x^0 a = a$$

$$ax = (0, a, 0, 0, 0, 0, \ldots) = xa$$

$$ax^2 = (0, 0, a, 0, 0, 0, \ldots) = x^2 a$$

$$ax^3 = (0, 0, 0, a, 0, 0, \ldots) = x^3 a$$

$$\vdots$$

If (a_k) is a polynomial in $R[x]$, let n be the smallest nonnegative integer such that $a_k = 0$ for all $k > n$. Then $a_n \neq 0$, and the polynomial can be written as $(a_0, a_1, a_2, \ldots, a_n, 0, 0, \ldots)$. The integer n is called the **degree of the polynomial** and a_n is the **leading coefficient**. No degree is assigned to the zero polynomial (0) since there is no smallest nonnegative integer n such that $a_k = 0$ for all $k > n$ and $a_n \neq 0$. Since

$$(a_0, a_1, a_2, \ldots, a_n, 0, 0, \ldots) = (a_0, 0, 0, 0, \ldots) + (0, a_1, 0, 0, \ldots)$$

$$+ (0, 0, a_2, 0, \ldots) + \cdots + (0, 0, 0, \ldots, a_n, 0, 0, \ldots)$$

$$= a_0 x^0 + a_1 x + a_2 x^2 + \cdots + a_n x^n,$$

we see that any nonzero polynomial $(a_k) \in R[x]$ can be written in the form

$$a_0 + a_1x + a_2x^2 + \cdots + a_nx^n \qquad\qquad (*)$$

The polynomial $(*)$ is not unique since we could add terms such as $0x^{n+1}$. But this yields no additional information about the polynomial, so we omit such terms. With this in mind, we denote the zero polynomial (0) simply by 0. If R is identified with the ring R' of Theorem 6.1.3, R can be considered a subring of $R[x]$. The questions raised previously about the nature of the indeterminate x and its relation to the elements of R can now be easily answered by considering R to be a subring of $R[x]$ and applying properties of $R[x]$ to x.

If the polynomial $(a_0, a_1, \ldots, a_n, 0, 0, \ldots)$ is written in the form $a_0 + a_1x + a_2x^2 + \cdots + a_nx^n$, we say that $a_0, a_1x, a_2x^2, \ldots, a_nx^n$ are the **terms** of the polynomial, with a_0 referred to as the **constant term**. When the polynomial is written in this form, a_k is the **coefficient** of x^k for $k = 1, 2, \ldots, n$, a_n is the leading coefficient, and the degree of the polynomial is n. If the coefficient a_k of x^k is e, we write x^k for ex^k with $k \geq 1$. Finally, two polynomials

$$a_0 + a_1x + a_2x^2 + \cdots + a_nx^n \text{ and } b_0 + b_1x + b_2x^2 + \cdots + b_mx^m$$

are equal if $m = n$ and $a_k = b_k$ for $k = 0, 1, \ldots, n$.

When two polynomials of $R[x]$ are written in the form given in $(*)$, they can be added and multiplied in the usual fashion. In the case of addition, add the coefficients of the like powers of x, and for multiplication, multiply in the usual fashion and then simplify the result. Polynomials in $R[x]$ are denoted by $f(x), g(x), h(x)$, and so on, and the **degree of a polynomial** $f(x)$ **is denoted by** $\deg(f(x))$. A polynomial of degree 1 or 2 is referred to as a **linear** or **quadratic polynomial**, respectively.

6.1.4	**Examples**

1. If $f(x) = 2 + 3x - 4x^3 + x^5$, then $f(x) \in \mathbf{Z}[x], f(x) \in \mathbf{Q}[x], f(x) \in \mathbf{R}[x]$, and $f(x) \in \mathbf{C}[x]$. In fact, if R is a subring of a ring S such that $e_R = e_S$ and $g(x) \in R[x]$, then $g(x) \in S[x]$.
2. If $f(x) = \frac{1}{2} - 2x$, then $f(x) \in \mathbf{Q}[x]$, but $f(x) \notin \mathbf{Z}[x]$.
3. The polynomial $f(x) = [4] + [2]x + [5]x^4$ is an element of $\mathbf{Z}_6[x]$, where $[4], [2]$, and $[5]$ are congruence classes in $\mathbf{Z}_6$.
4. If $M_2(\mathbf{R})$ is the ring of 2×2 matrices over $\mathbf{R}$, then the polynomial

$$f(x) = \begin{pmatrix} \sqrt{2} & \frac{3}{7} \\ 2 & 0 \end{pmatrix} x^2 - \begin{pmatrix} 2 & \sqrt[3]{4} \\ 1 & 2 \end{pmatrix} x^3$$

is a polynomial in $M_2(\mathbf{R})[x]$. This is an example of a polynomial over a noncommutative ring.

It is customary to omit terms from a polynomial whose coefficients are the zero element of the ring. In Example 1 above, we have written $f(x) = 2 + 3x - 4x^3 + x^5$ for $f(x) = 2 + 3x + 0x^2 - 4x^3 + 0x^4 + x^5$, and in Example 4,

$$f(x) = \begin{pmatrix} \sqrt{2} & \frac{3}{7} \\ 2 & 0 \end{pmatrix} x^2 - \begin{pmatrix} 2 & \sqrt[3]{4} \\ 1 & 2 \end{pmatrix} x^3$$

denotes the polynomial

$$f(x) = \begin{pmatrix} 0 & 0 \\ 0 & 0 \end{pmatrix} + \begin{pmatrix} 0 & 0 \\ 0 & 0 \end{pmatrix} x + \begin{pmatrix} \sqrt{2} & \frac{3}{7} \\ 2 & 0 \end{pmatrix} x^2 - \begin{pmatrix} 2 & \sqrt[3]{4} \\ 1 & 2 \end{pmatrix} x^3.$$

6.1.5   Example

As an illustration of adding and multiplying polynomials, suppose that $f(x) = [2]x + [5]x^2$ and $g(x) = [3]x^2 + [4]x^3$ are polynomials in $\mathbf{Z}_6[x]$. Then

$$\begin{aligned}
f(x) + g(x) &= ([2]x + [5]x^2) + ([3]x^2 + [4]x^3) \\
&= [2]x + ([5] + [3])x^2 + [4]x^3 \\
&= [2]x + [8]x^2 + [4]x^3 \\
&= [2]x + [2]x^2 + [4]x^3
\end{aligned}$$

and

$$\begin{aligned}
f(x)g(x) &= ([2]x + [5]x^2)([3]x^2 + [4]x^3) \\
&= [2]x([3]x^2 + [4]x^3) + [5]x^2([3]x^2 + [4]x^3) \\
&= [2][3]x^3 + [2][4]x^4 + [5][3]x^4 + [5][4]x^5 \\
&= [6]x^3 + [8]x^4 + [15]x^4 + [20]x^5 \\
&= [0]x^3 + [23]x^4 + [2]x^5 \\
&= [5]x^4 + [2]x^5.
\end{aligned}$$

| 6.1.6 | **Examples** |

The notation $R[c]$ is also used to denote the set of all polynomials of the form $f(c) = a_0 + a_1c + a_2c^2 + \cdots + a_nc^n$, where c is an element of a larger ring that contains R as a subring. It is not difficult to show that $R[c]$ is also a ring. Sometimes these rings reduce to a particularly nice form.

1. Let $\mathbf{Z}[i]$ denote the ring of all polynomials in the complex number i with their coefficients in $\mathbf{Z}$. The larger ring is $\mathbf{C}$ since $\mathbf{Z} \subseteq \mathbf{C}$ and $i \in \mathbf{C}$. Since $i^n = \pm 1$ or $i^n = \pm i$ for every positive integer n, it follows that $\mathbf{Z}[i] = \{a + bi \mid a, b \in \mathbf{Z}\}$. We refer to $\mathbf{Z}[i]$ as the ring of Gaussian integers.

2. Let $\mathbf{Z}[\sqrt{2}]$ denote the ring of all polynomials in $\sqrt{2}$ with their coefficients in $\mathbf{Z}$. In this case, the larger ring can be taken to be either $\mathbf{R}$ or $\mathbf{C}$ since $\mathbf{Z} \subseteq \mathbf{R} \subseteq \mathbf{C}$ and $\sqrt{2} \in \mathbf{R} \subseteq \mathbf{C}$. If $p = a_0 + a_1\sqrt{2} + a_2(\sqrt{2})^2 + \cdots + a_n(\sqrt{2})^n$ is a polynomial in $\mathbf{Z}[\sqrt{2}]$, then p reduces to a polynomial in $\sqrt{2}$ of degree 1. Indeed, if $n = 2k$ is a positive even integer, $(\sqrt{2}^n) = (\sqrt{2})^{2k} = 2^k$, and if $n = 2k + 1$ is a positive odd integer, then $(\sqrt{2})^n = (\sqrt{2})^{2k+1} = 2^k\sqrt{2}$. Hence, p reduces to $a + b\sqrt{2}$, where $a, b \in \mathbf{Z}$. Therefore, the ring $\mathbf{Z}[\sqrt{2}]$ can be expressed as $\mathbf{Z}[\sqrt{2}] = \{a + b\sqrt{2} \mid a, b \in \mathbf{Z}\}$. In general, if n is a nonzero, square free integer, the ring $\mathbf{Z}[\sqrt{n}] = \{a + b\sqrt{n} \mid a, b \in \mathbf{Z}\}$ is often referred to as a ring of **quadratic integers**. (Recall that a square free integer is an integer n different from 0 and 1 that is not divisible by the square of any integer $k \neq \pm 1$.)

There is one final question that must be answered before we are finished with the formal development of $R[x]$. How do we think of $R[x]$, if R is a commutative ring that does not have an identity? The development of the indeterminate $x = (0, e, 0, 0, \ldots)$ in $R[x]$ depends on R having an identity e. So does this preclude us from considering polynomials whose coefficients come from a ring that does not have an identity, such as the ring $2\mathbf{Z}$ of even integers? Since $2\mathbf{Z}$ is a subring of the ring of integers $\mathbf{Z}$ that does have an identity, we can consider any polynomial whose coefficients are even integers to be a polynomial in $\mathbf{Z}[x]$, where $x = (0, 1, 0, 0, \ldots)$. In the more general case, where R is a commutative ring without an identity, we have seen in Exercise 13 of Problem Set 3.3 that R can be embedded in a ring S with an identity. Hence, when R does not have an identity, there is a "larger" ring S with an identity containing a subring R' that is isomorphic to R. If the elements of R are identified with those of R', then any polynomial in x with coefficients in R

can be considered to be a polynomial in $S[x]$. Under these conditions, $R[x]$ is a subring of $S[x]$, $x \in S[x]$, but $x \notin R[x]$.

The following lemma gives us information about the behavior of the degree function under addition and multiplication of polynomials.

6.1.7   **Lemma**

Let $f(x), g(x) \in R[x]$.

1. If $f(x)g(x) \neq 0$, then $\deg(f(x)g(x)) \leq \deg(f(x)) + \deg(g(x))$. Equality holds if the product of the leading coefficients of $f(x)$ and $g(x)$ is not zero.
2. If $f(x) \pm g(x) \neq 0$, then $\deg(f(x) \pm g(x)) \leq \max\{\deg(f(x)), \deg(g(x))\}$.

Proof. Suppose that $\deg(f(x)) = n$ and $\deg(g(x)) = m$.

1. If $f(x) = a_0 + a_1 x + \cdots + a_n x^n$ and $g(x) = b_0 + b_1 x + \cdots + b_m x^m$, then $f(x)g(x) = a_0 b_0 + (a_0 b_1 + a_1 b_0)x + \cdots + a_n b_m x^{m+n}$. Since $f(x)g(x) \neq 0$, at least one of the coefficients of $f(x)g(x)$ must be nonzero. If $a_n b_m \neq 0$, then

$$\deg(f(x)g(x)) = n + m = \deg(f(x)) + \deg(g(x)).$$

If $a_n b_m = 0$, the degree of $f(x)g(x)$ is determined by the nonzero term of $f(x)g(x)$ in which x^k has the largest exponent. In this case

$$\deg(f(x)g(x)) < \deg(f(x)) + \deg(g(x)).$$

2. Because of the trichotomy property of the integers, only one of $m > n, m = n$, or $m < n$ can hold. If $m < n$, then $f(x) + g(x) = (a_0 + b_0) + (a_1 + b_1)x + \cdots + (a_m + b_m)x^m + a_{m+1}x^{m+1} + \cdots + a_n x^n$ and

$$\deg(f(x) + g(x)) = n = \max\{\deg(f(x)), \deg(g(x))\}.$$

The proof is similar if $n < m$. If $m = n$, then $f(x) + g(x) = (a_0 + b_0) + (a_1 + b_1)x + \cdots + (a_n + b_n)x^n$. Since $f(x) + g(x) \neq 0$, at least one of the coefficients of the terms of $f(x) + g(x)$ is nonzero. If $a_n + b_n \neq 0$, then

$$\deg(f(x) + g(x)) = n = \max\{\deg(f(x)), \deg(g(x))\}.$$

If $a_n + b_n = 0$, then the degree of $f(x) + g(x)$ is determined by the nonzero term of $f(x) + g(x)$ for which x^k has the largest exponent. Thus,

$$\deg(f(x) + g(x)) < \max\{\deg(f(x)), \deg(g(x))\}.$$

The proof that $\deg(f(x) - g(x)) \leq \max\{\deg(f(x)), \deg(g(x))\}$ follows from the definition of subtraction in $R[x]$. Since $f(x) - g(x) = f(x) + [-g(x)]$ and $\deg(-g(x)) = \deg(g(x))$, from what we have just shown for addition of polynomials, we see that $\deg(f(x) - g(x)) = \deg(f(x) + [-g(x)]) \leq \max\{\deg(f(x)), \deg(-g(x))\} = \max\{\deg(f(x)), \deg(g(x))\}$. ∎

6.1.8   Examples

1. Consider $f(x) = [2] + [3]x^2$ and $g(x) = [4]x + [2]x^3$ in $\mathbf{Z}_6[x]$. Then $\deg(f(x)) = 2$ and $\deg(g(x)) = 3$. Since

$$f(x) + g(x) = ([2] + [3]x^2) + ([4]x + [2]x^3)$$
$$= [2] + [4]x + [3]x^2 + [2]x^3,$$

    $\deg(f(x) + g(x)) = 3$, so $\deg(f(x) + g(x)) = \max\{\deg(f(x)), \deg(g(x))\}$. We also see that

$$f(x)g(x) = ([2] + [3]x^2)([4]x + [2]x^3)$$
$$= [2][4]x + [2][2]x^3 + [3][4]x^3 + [3][2]x^5$$
$$= [2]x + [4]x^3,$$

    so $\deg(f(x)g(x)) = 3 < 2 + 3 = \deg(f(x)) + \deg(g(x))$. The ring $\mathbf{Z}_6$ has zero-divisors. Inspection of the computation of $f(x)g(x)$ shows that the reason $\deg(f(x)g(x)) < \deg(f(x)) + \deg(g(x))$ is that $[2][3] = [0]$ in $\mathbf{Z}_6$.

2. Let $f(x) = 2 + 3x^2$ and $g(x) = x - 3x^2$ be polynomials in $\mathbf{Z}[x]$. Then

$$f(x) + g(x) = 2 + x \qquad \text{and}$$
$$f(x)g(x) = 2x - 6x^2 + 3x^3 - 9x^4.$$

    Hence, $\deg(f(x) + g(x)) = 1 < 2 = \max\{\deg(f(x)), \deg(g(x))\}$ and $\deg(f(x)g(x)) = 4 = 2 + 2 = \deg(f(x)) + \deg(g(x))$.

If R is free of zero-divisors, the product of two nonzero polynomials in $R[x]$ is nonzero. This observation leads to the following theorem.

6.1.9   Theorem

A ring R is an integral domain if and only if $R[x]$ is an integral domain. Furthermore, when R is an integral domain, $\deg(f(x)g(x)) = \deg(f(x)) + \deg(g(x))$ for all nonzero $f(x), g(x) \in R[x]$.

Proof.   Suppose that R is an integral domain and that $f(x)$, $g(x) \in R\{x]$ are nonzero polynomials. If $\deg(f(x)) = n$ and $\deg(g(x)) = m$, then the leading coefficients a_n and b_m of $f(x)$ and $g(x)$, respectively, are nonzero elements of R. Hence, $a_n b_m \neq 0$ since R is an integral domain. But $a_n b_m$ is the coefficient of x^{m+n} in $f(x)g(x)$, so $f(x)g(x) \neq 0$. Thus, the product of nonzero polynomials in $R[x]$ is nonzero, and so $R[x]$ is an integral domain. Lemma 6.1.7 now shows that

$$\deg(f(x)g(x)) = n + m = \deg(f(x)) + \deg(g(x)).$$

Conversely, if $R[x]$ is an integral domain and a, $b \in R$ are both nonzero elements of R, then a and b can be viewed as constant polynomials in $R[x]$. As such, $ab \neq 0$ in $R[x]$, so $ab \neq 0$ in R. Hence, R is an integral domain. ∎

There is one property that holds in $R[x]$ that is reminiscent of the Division Algorithm 1.1.3 for the integers. If $f(x)$, $g(x) \in R[x]$, it is sometimes possible to find polynomials $q(x)$, $r(x) \in R[x]$ such that $f(x) = g(x)q(x) + r(x)$, where $r(x) = 0$ or $\deg(r(x)) < \deg(g(x))$. *If the leading coefficient of $g(x)$ has a multiplicative inverse in R, then we are assured of being able to find such a $q(x)$ and $r(x)$ in $R[x]$.* Before proving this in general, let's look at an example to see how this can actually be accomplished.

6.1.10   Example

If $f(x) = [2] + [4]x^2 + [3]x^4$ and $g(x) = [3] + [4]x + [5]x^2$ are polynomials in $\mathbf{Z}_9$, find polynomials $q(x)$ and $r(x)$ in $\mathbf{Z}_9$ such that $f(x) = g(x)q(x) + r(x)$, where either $r(x) = 0$ or $\deg(r(x)) < \deg(g(x))$.

Solution. Note that the leading coefficient $[5]$ of $g(x)$ has a multiplicative inverse in $\mathbf{Z}_9$, so we can find the polynomials $q(x)$ and $r(x)$ by the usual process of long division. Before proceeding, consider the following:

1.   To subtract an element $[b]$ from an element $[a]$ in $\mathbf{Z}_9$, simply add the additive inverse of $[b]$ to $[a]$. For example, $[2] - [3] = [2] + [6] = [8]$.
2.   If an element $[b] \in \mathbf{Z}_9$ has a multiplicative inverse, then it is possible to divide $[a] \in \mathbf{Z}_9$ by $[b]$. This can be accomplished by multiplying $[a]$ by $[b]^{-1}$. For instance, $[5]^{-1} = [2]$ in $\mathbf{Z}_9$, so $[2] \div [5] = [2][5]^{-1} = [2][2] = [4]$.

The result of dividing $f(x) = [2] + [4]x^2 + [3]x^4$ by $g(x) = [3] + [4]x + [5]x^2$ using long division is given below. The column on the right is where we have computed the coefficients of the quotient at each step in the process of finding $q(x)$.

$$[6]x^2 + [6]x + [5] = q(x)$$

$[5]x^2 + [4]x + [3] \,\big|\, [3]x^4 + [0]x^3 + [4]x^2 + [0]x + [2]$ $\qquad [3][5]^{-1} = [3][2] = [6]$

$\qquad\quad \underline{[3]x^4 + [6]x^3 + [0]x^2}$

$\qquad\qquad\quad [3]x^3 + [4]x^2 + [0]x + [2]$ $\qquad [3][5]^{-1} = [3][2] = [6]$

$\qquad\qquad\quad \underline{[3]x^3 + [6]x^2 + [0]x}$

$\qquad\qquad\qquad\quad [7]x^2 + [0]x + [2]$ $\qquad [7][5]^{-1} = [7][2] = [5]$

$\qquad\qquad\qquad\quad \underline{[7]x^2 + [2]x + [6]}$

$\qquad\qquad\qquad\qquad r(x) = [7]x + [5]298$

From the preceding work, we see that $q(x) = [6]x^2 + [6]x + [5]$ and $r(x) = [7]x + [5]$. You should actually perform each of the computations given to make sure that you understand each step in the procedure. Also compute $g(x)q(x) + r(x)$ to show that this does indeed yield $f(x)$. In the process of long division, $f(x)$ is the **dividend**, $g(x)$ the **divisor**, $q(x)$ the **quotient**, and $r(x)$ the **remainder**.

The condition that the leading coefficient of $g(x)$ has a multiplicative inverse in the ring is sufficient to ensure that we can find $q(x)$ and $r(x)$. This condition is not necessary, however, for such a $q(x)$ and $r(x)$ to exist. For example, if $f(x) = 5 + 3x + 5x^2 + 2x^3$ and $g(x) = 3 + 2x$ are polynomials in $\mathbf{Z}[x]$, then $q(x) = x^2 + x$ and $r(x) = 5$ are polynomials in $\mathbf{Z}[x]$ such that $f(x) = g(x)q(x) + r(x)$, and the leading coefficient of $g(x)$, 2, does not have a multiplicative inverse in $\mathbf{Z}$.

We now prove that such a $q(x)$ and $r(x)$ always exist when the leading coefficient of $g(x)$ has a multiplicative inverse in R. The proof of the Division Algorithm 1.1.3 for the integers was effected by picking the smallest element r from the set $S = \{x - yn \mid y \in \mathbf{Z} \text{ and } x - yn \geq 0\}$ and using this element to find q and r. We could use the same argument here by choosing the polynomial of smallest degree from the set $S = \{f(x) - h(x)g(x) \mid h(x) \in R[x]\}$ as $r(x)$. However, we use a different argument since we want to provide an algorithm for finding $q(x)$ and $r(x)$. The procedure used in the proof is just polynomial long division illustrated in Example 6.1.10.

6.1.11 Division Algorithm for Polynomials

If $f(x)$, $g(x) \in R[x]$ and the leading coefficient of $g(x)$ has a multiplicative inverse in R, then there exist unique polynomials $q(x)$ and $r(x)$ in $R[x]$ such that $f(x) = g(x)q(x) + r(x)$, where $r(x) = 0$ or $\deg(r(x)) < \deg(g(x))$.

Proof. Note that since the leading coefficient of $g(x)$ has a multiplicative inverse in R, $g(x) \neq 0$. First, let's show that $q(x)$ and $r(x)$ exist. There are three cases to be considered: case 1, $f(x) = 0$; case 2, $f(x) \neq 0$ and $\deg(g(x)) > \deg(f(x))$; and case 3, $f(x) \neq 0$ and $\deg(g(x)) \leq \deg(f(x))$. If $f(x) = 0$ or $\deg(g(x)) > \deg(f(x))$, we can choose $q(x) = 0$ and $r(x) = f(x)$. In either of these two cases, $f(x) = 0 \cdot g(x) + f(x)$, so the interesting case is case 3, when $\deg(g(x)) \leq \deg(f(x))$. The proof is by induction on the degree of $f(x)$. Assume that $f(x) \neq 0$ and $\deg(g(x)) \leq \deg(f(x))$. If $\deg(f(x)) = 0$, then $f(x)$ is a constant polynomial. Since $g(x) \neq 0$ and $\deg(g(x)) \leq \deg(f(x))$, $g(x)$ must also be a constant polynomial. Suppose that $f(x) = a$ and $g(x) = b$. Then b has a multiplicative inverse in R. Hence, $f(x) = b(ab^{-1}) + 0$, so we can let $q(x) = ab^{-1}$ and $r(x) = 0$. Therefore, we can find $q(x)$ and $r(x)$ when $\deg(f(x)) = 0$. Now make the induction hypothesis that we can find $q(x)$ and $r(x)$ for all nonzero polynomials of degree less than n, where $n \geq 0$. Let $f(x) = a_0 + a_1 x + a_2 x^2 + \cdots + a_n x^n$, $g(x) = b_0 + b_1 x + b_2 x^2 + \cdots + b_m x^m$, where $m \leq n$, and define the polynomial $f_1(x)$ by

$$f_1(x) = f(x) - g(x)(a_n b_m^{-1})x^{n-m}. \tag{*}$$

The last term in $f_1(x)$ is set up to cancel the leading term from $f(x)$. To see how this works, let's carry out the arithmetic.

$$f_1(x) = f(x) - g(x)(a_n b_m^{-1})x^{n-m}$$
$$= (a_0 + a_1 x + \cdots + a_n x^n) - (a_n b_m^{-1})x^{n-m}(b_0 + b_1 x + \cdots + b_m x^m)$$
$$= a_0 + a_1 x + \cdots + a_n x^n - (a_n b_m^{-1})x^{n-m}b_0 - \cdots - (a_n b_m^{-1})x^{n-m}b_m x^m$$
$$= a_0 + a_1 x + \cdots + a_n x^n - (a_n b_m^{-1}b_0)x^{n-m} - \cdots - (a_n b_m^{-1}b_{m-1})x^{n-1} - a_n x^n$$
$$= a_0 + a_1 x + \cdots + a_{n-1}x^{n-1} - (a_n b_m^{-1}b_0)x^{n-m} - \cdots - (a_n b_m^{-1}b_{m-1})x^{n-1}$$

has degree less than n, or $f_1(x) = 0$. $f_1(x)$ is just the result of the first step in the process you know as long division. Compare how $f_1(x)$ is computed using long division, as demonstrated below with what is given above for $f_1(x)$ in equation $(*)$.

$$a_n b_m^{-1} x^{n-m}$$

$$g(x) = b_m x^m + b_{m-1} x^{m-1} + \cdots + b_1 x + b_0 \overline{)a_n x^n + a_{n-1} x^{n-1} + \cdots + a_1 x + a_0 = f(x)}$$

$$a_n x^n + a_n b_m^{-1} b_{m-1} x^{n-1} + \cdots$$

$$\overline{\phantom{a_n x^n + a_n b_m^{-1} b_{m-1} x^{n-1} + \cdots}}$$

$$f_1(x) = f(x) - g(x)(a_n b_m^{-1}) x^{n-m}$$

If $f_1(x) = 0$, then $f(x) = g(x)q(x) + r(x)$, where $q(x) = (a_n b_m^{-1})x^{n-m}$ and $r(x) = 0$. If $f_1(x) \neq 0$, then $\deg(f_1(x)) < n$, and the induction hypothesis shows that there are polynomials $q_1(x)$ and $r(x)$ in $R[x]$ such that $f_1(x) = g(x)q_1(x) + r(x)$, where $r(x) = 0$ or $\deg(r(x)) < \deg(g(x))$. From this we see that

$$f(x) = f_1(x) + g(x)(a_n b_m^{-1}) x^{n-m}$$

$$= g(x)q_1(x) + r(x) + (a_n b_m^{-1}) x^{n-m} g(x)$$

$$= g(x)[q_1(x) + (a_n b_m^{-1}) x^{n-m}] + r(x)$$

$$= g(x)q(x) + r(x),$$

where $q(x) = q_1(x) + (a_n b_m^{-1}) x^{n-m}$. Thus, we have expressed $f(x)$ in the required form when $\deg(f(x)) = n$. Consequently, by the Principle of Mathematical Induction, we have shown that polynomials $q(x)$ and $r(x)$ can be found for all $f(x)$ and $g(x)$, provided that the leading coefficient of $g(x)$ has a multiplicative inverse in R.

Finally, we need to show that $q(x)$ and $r(x)$ are unique. As usual with uniqueness proofs, we assume that there is more than one choice for $q(x)$ and $r(x)$ and then show that our choices are actually equal. Suppose that $q(x), q'(x), r(x), r'(x) \in R[x]$ are such that

$$f(x) = g(x)q(x) + r(x) = g(x)q'(x) + r'(x),$$

where either $r(x) = 0$ or $\deg(r(x)) < \deg(g(x))$ and where either $r'(x) = 0$ or $\deg(r'(x)) < \deg(g(x))$. Then

$$r(x) - r'(x) = (q'(x) - q(x))g(x). \qquad (**)$$

If $q'(x) \neq q(x)$ and c is the leading coefficient of $q'(x) - q(x)$, then $c \neq 0$. Since the leading coefficient b_m of $g(x)$ has a multiplicative inverse in R, $b_m \neq 0$, so $cb_m \neq 0$. Hence, $(q'(x) - q(x))g(x) \neq 0$. Equation $(**)$ and Lemma 6.1.7 now show that

$$\deg(r(x) - r'(x)) = \deg((q'(x) - q(x))g(x))$$

$$= \deg(q'(x) - q(x)) + \deg(g(x))$$

$$\geq \deg(g(x)).$$

But this is impossible since $\deg(r(x)) < \deg(g(x))$, $\deg(r'(x)) < \deg(g(x))$, and Lemma 6.1.7 imply that

$$\deg(r(x) - r'(x)) \le \max\{\deg(r(x)), \deg(-r'(x))\} < \deg(g(x)).$$

Hence, the assumption that $q'(x) \ne q(x)$ leads to a contradiction, so it must be the case that $q'(x) = q(x)$. But if $q'(x) = q(x)$, then $r(x) = r'(x)$, and the proof is complete. ∎

6.1.12   Corollary

If F is a field and $f(x), g(x) \in F[x]$, and $g(x) \ne 0$, then there exist unique $q(x)$, $r(x) \in F[x]$ such that $f(x) = g(x)q(x) + r(x)$, where $r(x) = 0$ or $\deg(r(x)) < \deg(g(x))$.

Proof. If $g(x) \ne 0$, the leading coefficient of $g(x)$ is a nonzero element of F and so has a multiplicative inverse in F. ∎

6.1.13   Example

If p is a prime, then for any $f(x), g(x) \in \mathbf{Z}_p[x]$, $g(x) \ne 0$, we can find polynomials $q(x)$ and $r(x)$ in $\mathbf{Z}_p[x]$ such that $f(x) = g(x)q(x) + r(x)$, where $r(x) = 0$ or $\deg(r(x)) < \deg(g(x))$.

In general, for a commutative ring R with identity, if the leading coefficient of $g(x)$ does *not* have a multiplicative inverse in R, then as was shown in the last paragraph of Example 6.1.10, it is still possible for $q(x), r(x) \in R[x]$ to exist such that $f(x) = g(x)q(x) + r(x)$, where $r(x) = 0$ or $\deg(r(x)) < \deg(g(x))$. However, there is no algorithm for finding $q(x)$ and $r(x)$. When the leading coefficient of $g(x)$ is invertible, Theorem 6.1.11 provides an algorithm for finding $q(x)$ and $r(x)$. This is the familiar process you know as long division.

Problem Set 6.1

All rings are commutative rings with an identity.

1. Compute $f(x) + g(x)$, and determine whether $\deg(f(x) + g(x)) = \max\{\deg(f(x)), \deg(g(x))\}$ or $\deg(f(x) + g(x)) < \max\{\deg(f(x)), \deg(g(x))\}$.

(a) $f(x) = [4]x^2 + [3]x + [6]$, $g(x) = [8]x^2 + [9]x + [2]$ in $\mathbf{Z}_{12}[x]$
(b) $f(x) = [11]x^3 + [8]x^2 + [3]$, $g(x) = [4]x^3 + [4]x^2 + [2]x + [5]$ in $\mathbf{Z}_{10}[x]$
(c) $f(x) = [3]x^2 + [4]x + [1]$, $g(x) = [4]x^2 + [1]x + [4]$ in $\mathbf{Z}_5[x]$

2. Compute $f(x)g(x)$, and determine whether $\deg(f(x)g(x)) = \deg(f(x)) + \deg(g(x))$ or $\deg(f(x)g(x)) < \deg(f(x)) + \deg(g(x))$.

(a) $f(x) = [3]x^2 + [2]x + [4]$, $g(x) = [2]x + [1]$ in $\mathbf{Z}_5[x]$
(b) $f(x) = [2]x^4 + [5]x$, $g(x) = [3]x^2 + [2]x + [4]$ in $\mathbf{Z}_6[x]$
(c) $f(x) = [4]x^3 + [2]x^2 + [3]$, $g(x) = [3]x^2 + [4]$ in $\mathbf{Z}_{12}[x]$

3. Determine the form of the polynomials in $\mathbf{Z}[\sqrt[3]{2}]$.

4. Use long division to compute each of the following.

(a) $([4]x^5 + [2]x^3 + [1]) \div ([2]x^2 + [4])$ in $\mathbf{Z}_5[x]$
(b) $([5]x^6 + [3]x^3 + [2]x) \div ([3]x^2 + [6]x + [1])$ in $\mathbf{Z}_8[x]$
(c) $(x^4 + [1]) \div (x + [1])$ in $\mathbf{Z}_2[x]$

5. Answer the questions concerning the nature of the indeterminate x that were raised at the beginning of this section by using the fact that $x \in R[x]$ and $R \subseteq R[x]$.

6. If $x = (0, e, 0, \ldots)$, where e is the identity of the ring R, prove that for any positive integer n, $x^n = (0, 0, \ldots, 0, e, 0, \ldots)$, where e is in the nth position. (We begin counting the positions with zero.)

7. In $R[x]$, where R is a ring with identity e, prove that $(a, 0, 0, \ldots)x^n = x^n(a, 0, 0, \ldots)$ for any positive integer n, where $x = (0, e, 0, \ldots)$. Does R have to be a commutative ring for this to be true?

8. If $f(x), g(x), h(x) \in R[x]$, prove that $f(x)[g(x) + h(x)] = f(x)g(x) + f(x)h(x)$.

9. List all the polynomials of $\mathbf{Z}_3[x]$ that have degree 1. How many polynomials of degree m are there in $\mathbf{Z}_n[x]$?

10. If the ring R has zero-divisors, show that there are nonconstant polynomials $f(x), g(x) \in R[x]$ such that $f(x)g(x) = 0$.

11. If R is a commutative ring, show that multiplication is commutative in $R[x]$.

12. Decide which, if any, of the following subsets are subrings of $R[x]$.

(a) The set $R[x]^{\text{even}}$ of all polynomials in $R[x]$ that have exponents that are even integers
(b) The set $R[x]^{\text{odd}}$ of all polynomials in $R[x]$ that have exponents that are odd integers
(c) The set $R_0[x]$ of all polynomials in $R[x]$ that have zero constant term
(d) The set $\mathbf{Z}_{\text{even}}[x]$ of all polynomials in $\mathbf{Z}[x]$ whose coefficients are even integers (this includes the constant term of a polynomial since it is the coefficient of x^0)
(e) The set $\mathbf{Z}_{\text{odd}}[x]$ of all polynomials in $\mathbf{Z}[x]$ whose coefficients are odd integers

13. Complete the proof of 6.1.3 by showing that $R' = \{ (a, 0, 0, \ldots) \mid a \in R\}$ is a subring of $R[x]$ and that the mapping $\phi : R \longmapsto R' : a \longmapsto (a, 0, 0, \ldots)$ is a well-defined ring isomorphism.

14. Prove that $\phi : \mathbf{Z}[x] \longrightarrow \mathbf{Z}_n[x] : a_0 + a_1x + \cdots + a_nx^m \longmapsto [a_0] + [a_1]x + \cdots + [a_n]x^m$ is a well-defined ring homomorphism for each positive integer n.

15. Theorem 6.1.9 shows that $R[x]$ is an integral domain if and only if R is an integral domain. Hence, if F is a field, then F is an integral domain, so $F[x]$ is an integral domain. Prove, however, that $F[x]$ is not a field. Hint: Show that $x \in F[x]$ does not have a multiplicative inverse in $F[x]$. Prove that $f(x) \in F[x]$ has a multiplicative inverse in $F[x]$ if and only if $f(x)$ is a nonzero constant polynomial.

16. Suppose that R is a subring of a commutative ring S. If $s \in S$ and $R[s]$ denotes the ring of all polynomials in s, prove that $R[s]$ is the intersection of all the subrings of S that contains both R and s.

17. Suppose that $f\ R \longrightarrow S$ is a well-defined ring homomorphism. Prove that $\phi : R[x] \longrightarrow S[x] : a_0 + a_1x + \cdots + a_nx^n \longmapsto f(a_0) + f(a_1)x + \cdots + f(a_n)x^n$ is a well-defined ring homomorphism. Show that if f is an epimorphism, then so is ϕ. If f is an isomorphism, must ϕ be an isomorphism?

18. For a polynomial $f(x) = a_0 + a_1x + a_2x^2 + \cdots + a_{n-1}x^{n-1} + a_nx^n \in R[x]$ define the derivative of $f(x)$ formally by

$$D[f(x)] = a_1 + 2a_2x + \cdots + (n - 1)a_{n-1}x^{n-2} + na_nx^{n-1}.$$

Show that for any $f(x), g(x) \in R[x]$, $D[f(x) + g(x)] = D[f(x)] + D[g(x)]$ and $D[f(x)g(x)] = D[f(x)]g(x) + f(x)D[g(x)]$.

TECHNOLOGY PROBLEMS

Use a computer algebra system to solve each of the following problems.

19. Compute $f(x) + g(x)$, and determine whether $\deg(f(x) + g(x)) = \max\{\deg(f(x)), \deg(g(x))\}$ or $\deg(f(x) + g(x)) < \max\{\deg(f(x)), \deg(g(x))\}$.

    (a) $f(x) = [134]x^2 + [7]x + [132], g(x) = [123]x^2 + [127]x + [4]$ in $\mathbf{Z}_{135}[x]$
    (b) $f(x) = [75]x^3 + [23]x^2 + [3], g(x) = [48]x^3 + [72]x^2 + [22]x + [5]$ in $\mathbf{Z}_{80}[x]$
    (c) $f(x) = [15]x^2 + [42]x + [15], g(x) = [5]x^2 + [18]x + [24]$ in $\mathbf{Z}_{25}[x]$

20. Compute $f(x)g(x)$, and determine whether $\deg(f(x)g(x)) = \deg(f(x)) + \deg(g(x))$ or $\deg(f(x)g(x)) < \deg(f(x)) + \deg(g(x))$.

    (a) $f(x) = [15]x^2 + [24]x + [18],\ g(x) = [5]x + [12]$ in $\mathbf{Z}_{25}[x]$
    (b) $f(x) = [28]x^4 + [45]x,\qquad g(x) = [35]x^2 + [28]x + [6]$ in $\mathbf{Z}_{120}[x]$
    (c) $f(x) = [30]x^3 + [24]x^2 + [3],\ g(x) = [42]x^2 + [4]$ in $\mathbf{Z}_{60}[x]$

6.2   ROOTS, DIVISIBILITY, AND THE GREATEST COMMON DIVISOR

Section Overview. When F is a field, the integral domain $F[x]$ has many properties that are similar to those of the ring of integers. We have seen in Section 6.1 that there is a division algorithm for polynomials in $F[x]$ that bears a striking resemblance to the Division Algorithm for the integers. We shall also see that one can define and find the greatest common divisor for a pair of polynomials in $F[x]$. In many cases, the proofs of the results given in this section run parallel to the proofs of corresponding properties of the integers given in Section 1.1. You should compare the results of this section with those of Section 1.1 and note the similarities and the differences.

We continue to let R designate a commutative ring with identity, and F denotes a field. The letter e always denotes the identity of R and F except when **Z**, **Q**, **R**, or **C** is being considered. In each of these cases, 1 denotes the multiplicative identity.

6.2.1   Definition

If $f(x), g(x) \in R[x], g(x) \neq 0$, then $g(x)$ **divides** $f(x)$ or $g(x)$ is a **factor** of $f(x)$ if there is a polynomial $h(x) \in R[x]$ such that $f(x) = g(x)h(x)$. The notation $g(x) \mid f(x)$ indicates that $g(x)$ divides $f(x)$, and $g(x) \nmid f(x)$ indicates that $g(x)$ does *not* divide $f(x)$. If $f(x) \in R[x], c \in R$, and $f(x) = a_0 + a_1 x + \cdots + a_n x^n$, then $f(c)$ is the element $a_0 + a_1 c + \cdots + a_n c^n$ of R. If $f(c) = 0$, then c is said to be a **root** of $f(x)$.

The symbol 0 now has double duty. It is being used to denote the zero polynomial and the additive identity in the ring. You must determine how it is being used from the context of the discussion. For example, the expression $f(x) \neq 0$ means that $f(x)$ is not the zero polynomial. If $f(x) = a_0 + a_1 x + \cdots + a_n x^n$ and we say that $f(x) = 0$ is the zero polynomial, then, of course, $a_0 = a_1 = \cdots = a_n = 0$, where 0 denotes the additive identity of the ring. On the other hand, if we are asked to find the solutions in R of the equation $f(x) = 0$, this means that we are to find all $c \in R$ that are roots of $f(x)$.

The zero polynomial $f(x) = 0$ has every $c \in R$ as a root. Otherwise, if $f(x)$ has a root in R, then $f(x)$ must have degree at least 1. In previous mathematics courses, you have probably learned that a polynomial in **R**[x] of positive degree n can have at most n distinct roots in **R**. It will be shown that this is true for all polynomials in

$F[x]$ when the coefficient ring F is a field. The following two examples show that this need not be true if the coefficient ring is not a field.

6.2.2 | Examples

1. Suppose that S is a finite set with three elements. Then the number of elements of the power set $\mathcal{P}(S)$ of S is given by $\#\mathcal{P}(S) = 2^3 = 8$. (See Problem Set 1.2, Exercise 13.) The set $\mathcal{P}(S)$ can be made into a commutative ring R with identity by defining addition and multiplication on $R = \mathcal{P}(S)$ by

$$A + B = (A \cup B) \setminus (A \cap B)$$
$$A \bullet B = A \cap B$$

for all $A, B \in R$. The additive identity of R is $0 = \varnothing$, and the multiplicative identity is $e = S$. Every element $A \in R$ is its own additive inverse since $A + A = (A \cup A) \setminus (A \cap A) = A \setminus A = \varnothing$ and it is obvious that $A^2 = A$ for all $A \in R$. Now consider the polynomial $f(x) = x^2 + x$ in $R[x]$. For any $A \in R, f(A) = A^2 + A = A + A = \varnothing$, so every $A \in R$ is a root of $f(x)$. Hence, $f(x)$, which is a polynomial of degree 2, has 8 roots in R. In fact, if the number of elements of S is n, then $f(x)$ has 2^n roots in R.

2. Consider $f(x) = [4] + [3]x + [5]x^2$ in $\mathbf{Z}_6[x]$, and note that

$$f([0]) = [4] + [3][0] + [5][0]^2 = [4],$$
$$f([1]) = [4] + [3][1] + [5][1]^2 = [12] = [0],$$
$$f([2]) = [4] + [3][2] + [5][2]^2 = [30] = [0],$$
$$f([3]) = [4] + [3][3] + [5][3]^2 = [58] = [4],$$
$$f([4]) = [4] + [3][4] + [5][4]^2 = [96] = [0], \quad \text{and}$$
$$f([5]) = [4] + [3][5] + [5][5]^2 = [144] = [0].$$

$f(x)$ is a polynomial in $\mathbf{Z}_6[x]$ of degree 2 that has 4 roots in $\mathbf{Z}_6$.

The preceding examples show that for $f(x) \in R[x]$, it is possible for the number of roots of $f(x)$ to exceed the degree of $f(x)$. It will now be shown that this is not possible if the coefficient ring is a field. This fact depends on the following two theorems. Both are a direct consequence of the Division Algorithm for Polynomials 6.1.11. Before proving these theorems, let's look at two examples.

6.2.3 | Examples

In the following two examples, the polynomials are in the polynomial ring $\mathbf{Z}[x]$.

1. If $f(x) = 2x^3 - 2x^2 - 20x + 44$ is divided by $g(x) = x - 3$, the quotient is $q(x) = 2x^2 + 4x - 8$, and the remainder is $r(x) = 20$. Hence, we see that $f(x) = g(x)q(x) + r(x) = (x - 3)q(x) + 20$, so $f(3) = (3 - 3)\,q(3) + 20 = 20$. Thus, when $f(x)$ is divided by $x - 3$, the remainder is $f(3)$.
2. If $f(x) = 3x^3 - 11x^2 + 20x - 20$ is divided by $g(x) = x - 2$, the quotient is $q(x) = 3x^2 - 5x + 10$, and the remainder is $r(x) = 0$. From this it follows that $f(x) = g(x)q(x) + r(x) = (x - 2)q(x) + 0 = (x - 2)q(x)$. Hence, $x - 2$ is a factor of $f(x)$, and $f(2) = (2 - 2)\,q(2) = 0$. Thus, 2 is a root of $f(x)$. The converse also holds. If 2 is a root of $f(x)$, then $f(2) = 0$, so when $f(x)$ is divided by $x - 2$, the remainder is zero. Therefore, $x - 2$ is a factor of $f(x)$.

The two preceding examples are illustrations of the following two theorems. The first theorem shows that $f(c)$ is the remainder of the division of $f(x)$ by $x - c$, and the second shows that $x - c$ is a factor of $f(x)$ if and only if c is a root of $f(x)$.

6.2.4   Remainder Theorem

If $f(x) \in R[x]$ is a polynomial of positive degree and $c \in R$, then $f(c)$ is the remainder of the division of $f(x)$ by $x - c$.

Proof. Since the leading coefficient of $x - c$ is e and e has a multiplicative inverse in R, Theorem 6.1.11 indicates that there are $q(x), r(x) \in R[x]$ such that $f(x) = (x - c)q(x) + r(x)$, where $r(x) = 0$ or $\deg(r(x)) < \deg(x - c) = 1$. Hence, $r(x) = 0$ or $\deg(r(x)) = 0$, and this indicates that $r(x)$ is a constant polynomial that we will write as r. Therefore, $f(x) = (x - c)q(x) + r$, so we have $f(c) = (c - c)q(c) + r = r$. ∎

6.2.5   Factor Theorem

If $f(x) \in R[x]$ is a polynomial of positive degree, then $c \in R$ is a root of $f(x)$ if and only if $x - c$ is a factor of $f(x)$.

Proof. By Theorem 6.1.11, $f(x) = (x - c)q(x) + r$, where $r \in R$. But then $f(c) = 0$ if and only if $r = 0$ if and only if $f(x) = (x - c)q(x)$. Hence, c is a root of $f(x)$ if and only if $f(x) = (x - c)q(x)$. Thus, $c \in R$ is a root of $f(x)$ if and only if $x - c$ is a factor of $f(x)$. ∎

In the following theorem, the coefficient ring is a field. Inspection of the proof indicates the role that this condition plays in the proof.

6.2.6   Theorem

Let $f(x) \in F[x]$ be any polynomial of positive degree n. If $f(x)$ has n distinct roots $c_1, c_2, \ldots, c_n$ in F, then $f(x)$ can be written as $f(x) = a(x - c_1)(x - c_2) \cdots (x - c_n)$, where a is the leading coefficient of $f(x)$.

Proof.   The proof is by mathematical induction on the degree of the polynomial $f(x) \in F[x]$. If $\deg(f(x)) = 1$, then $f(x) = ax + b$, where $a, b \in F$. If $c_1 \in F$ is a root of $f(x)$, then $0 = f(c_1) = ac_1 + b$, which implies that $b = -ac_1$. Thus, $f(x) = ax - ac_1 = a(x - c_1)$, so the theorem is true for any polynomial in $F[x]$ of degree 1. Now make the induction hypothesis that any polynomial $g(x)$ in $F[x]$ of degree k with $c_1, c_2, \ldots, c_k$ distinct roots in F can be written as $a(x - c_1)(x - c_2) \cdots (x - c_k)$, where a is the leading coefficient of $g(x)$. Then if $f(x) \in F[x]$ is a polynomial of degree $k + 1$ with leading coefficient a and distinct roots $c_1, c_2, \ldots, c_{k+1} \in F$, the Factor Theorem indicates that $f(x) = (x - c_1)g(x)$ for some $g(x) \in F[x]$. Theorem 6.1.9 shows that $k + 1 = \deg(f(x)) = \deg(x - c_1) + \deg(g(x)) = 1 + \deg(g(x))$, so $\deg(g(x)) = k$. Furthermore, $0 = f(c_i) = (c_i - c_1)g(c_i) \in F$ for $i = 2, 3, \ldots, k + 1$, and since a field has no zero-divisors and the roots of $f(x)$ are distinct, $c_i - c_1 \neq 0$ implies that $g(c_i) = 0$ for $i = 2, 3, \ldots, k + 1$. Finally, it follows that the leading coefficient of $g(x)$ is a, so by the induction hypothesis, $g(x)$ can be written as $g(x) = a(x - c_2)(x - c_3) \cdots (x - c_{k+1})$. Consequently, $f(x) = a(x - c_1)(x - c_2)(x - c_3) \cdots (x - c_{k+1})$. Thus, by induction, the theorem holds for polynomials of any positive degree. ∎

This leads to the following corollary that fulfills our earlier promise to show that when F is a field, $f(x) \in F[x]$ can have at most n distinct roots in F.

6.2.7   Corollary

If $f(x) \in F[x]$ is of degree $n \geq 0$, then $f(x)$ can have at most n distinct roots in F.

Proof.   If $f(x)$ has degree $n \geq 0$, then $f(x) \neq 0$ since the zero polynomial has no degree. If $n = 0$, then $f(x)$ is a nonzero constant polynomial that has no roots. Hence, the theorem is true when $n = 0$. Assume that $n \geq 1$. If $f(x)$ has fewer than n distinct roots in F, there is nothing to prove. Suppose that $f(x)$ has exactly n distinct roots in F, and let's show that $f(x)$ can have no additional roots. If $c_1, c_2, \ldots, c_n \in F$ are the roots of $f(x)$, then Theorem 6.2.6 shows that $f(x) = a(x - c_1)(x - c_2)\cdots(x - c_n)$, where a is the leading coefficient of $f(x)$. If $c \in F$ is also a root of $f(x)$ distinct from the n roots already given, then $0 = f(c) = a(c - c_1)(c - c_2)\cdots(c - c_n)$. But all of the factors in this product are nonzero since $c \neq c_k$ for $k = 1, 2, \ldots, n$. Thus, we are led to the conclusion that F has zero-divisors, which is impossible since F is a field. Therefore, $f(x)$ cannot have an additional root. ∎

6.2.8   Corollary

Suppose that $f(x), g(x) \in F[x]$ are such that $f(c) = g(c)$ for every $c \in F$. If the number of elements of F exceeds both $\deg(f(x))$ and $\deg(g(x))$, then $f(x) = g(x)$.

Proof. Either $f(x) = g(x)$ or $f(x) \neq g(x)$. If $f(x) \neq g(x)$, let $h(x) = f(x) - g(x)$. Then $h(x)$ is a nonzero polynomial in $F[x]$, and $h(c) = 0$ for all $c \in F$. But by Lemma 6.1.7, $\deg(h(x)) = \deg(f(x) - g(x)) \leq \max\{\deg(f(x), \deg(g(x))\} <$ number of elements of F. Therefore, the number of roots of $h(x)$ exceeds the degree of $h(x)$, and this is impossible by Corollary 6.2.7. Hence, it cannot be the case that $f(x) \neq g(x)$, so $f(x) = g(x)$. ∎

The following example shows that Corollary 6.2.8 is not true if the number of elements in the field is not larger than the degree of both $f(x)$ and $g(x)$.

6.2.9   Example

Suppose that $f(x) = x^5$ and $g(x) = x$ are in $\mathbf{Z}_3[x]$. Then

$$f([0]) = [0]^5 = [0] \qquad g([0]) = [0]$$
$$f([1]) = [1]^5 = [1] \qquad g([1]) = [1]$$
$$f([2]) = [2]^5 = [2] \qquad g([2]) = [2],$$

but $f(x)$ and $g(x)$ are not the same polynomial in $\mathbf{Z}_3[x]$.

If x and y are integers, at least one of which is nonzero, then it was shown in Section 1.1 that a unique greatest common divisor d of x and y exists and that there are integers a and b such that $d = ax + by$. A similar concept can be developed for polynomials whose coefficient ring is a field. In the case of the integers, d is the largest integer that divides both x and y, so, for polynomials, we might suspect that the greatest common divisor of $f(x)$ and $g(x)$ would be a polynomial $d(x)$ of highest degree that divides both $f(x)$ and $g(x)$. But if $d(x)\,|\,f(x)$, $d(x)\,|\,g(x)$, and c is a nonzero element of F, then $cd(x)$ also divides $f(x)$ and $g(x)$. This is illustrated in the following simple example.

6.2.10 **Example**

Suppose that $f(x) = 3x^2 + 4x + 1$ and $g(x) = 6x^2 + 5x + 1$ are polynomials in $\mathbf{Q}[x]$. Since $f(x) = (3x + 1)(x + 1)$ and $g(x) = (3x + 1)(2x + 1)$, it is obvious that a polynomial of highest degree that divides both $f(x)$ and $g(x)$ is $3x + 1$. But it is also the case that

$$f(x) = (6x + 2)\left(\tfrac{1}{2}x + \tfrac{1}{2}\right) \quad \text{and}$$

$$g(x) = (6x + 2)\left(x + \tfrac{1}{2}\right).$$

Hence, $6x + 2$ divides both $f(x)$ and $g(x)$. Moreover,

$$f(x) = \left(\tfrac{3}{4}x + \tfrac{1}{4}\right)(4x + 4) \quad \text{and}$$

$$g(x) = \left(\tfrac{3}{4}x + \tfrac{1}{4}\right)(8x + 4).$$

Thus, $\tfrac{3}{4}x + \tfrac{1}{4}$ also divides both $f(x)$ and $g(x)$. In fact, if c is any nonzero constant in $\mathbf{Q}$, then $3cx + c$ will divide both $f(x)$ and $g(x)$.

The preceding example demonstrates that a polynomial of highest degree that divides both $f(x)$ and $g(x)$ is not unique. If the greatest common divisor of polynomials is to be unique, it is necessary to designate a particular polynomial of highest degree that divides $f(x)$ and $g(x)$ as *the* greatest common divisor of $f(x)$ and $g(x)$.

6.2.11 **Definition**

A polynomial $f(x) \in F[x]$ is said to be a **monic polynomial** if the leading coefficient of $f(x)$ is e. If $f(x)$ and $g(x)$ are polynomials in $F[x]$, at least one of which is nonzero, then a **monic polynomial** $d(x)$ in $F[x]$ is the **greatest common divisor** of $f(x)$ and $g(x)$, provided that the following conditions are satisfied:

(i) $d(x)\,|\,f(x)$ and $d(x)\,|\,g(x)$.
(ii) If $h(x) \in F[x]$, $h(x)\,|\,f(x)$, and $h(x)\,|\,g(x)$, then $h(x)\,|\,d(x)$.

The greatest common divisor of $f(x)$ and $g(x)$ is denoted by $\gcd(f(x), g(x))$. If $\gcd(f(x), g(x)) = e$, then $f(x)$ and $g(x)$ are said to be **relatively prime**.

The fact that the greatest common divisor of $f(x)$ and $g(x)$ (when it can be shown to exist) must be a monic polynomial is sufficient to show that it is unique. Note that a constant polynomial $f(x)$ is monic if and only if $f(x) = e$.

6.2.12   Theorem

Let $f(x)$ and $g(x)$ be polynomials in $F[x]$. Then the greatest common divisor of $f(x)$ and $g(x)$, when it exists, is unique.

Proof. If $d_1(x)$ and $d_2(x)$ are greatest common divisors of $f(x)$ and $g(x)$, it follows from the definition of the greatest common divisor that $d_1(x) \mid d_2(x)$ and $d_2(x) \mid d_1(x)$. Hence, $d_2(x) = k_1(x)d_1(x)$ and $d_1(x) = k_2(x)d_2(x)$ for polynomials $k_1(x)$ and $k_2(x)$ in $F[x]$. Thus, $d_1(x) = k_1(x)k_2(x)d_1(x)$, so $e = k_1(x)k_2(x)$ since $F[x]$ is an integral domain. Hence,

$$0 = \deg(e) = \deg(k_1(x)k_2(x))$$
$$= \deg(k_1(x)) + \deg(k_2(x)).$$

But the degree of a polynomial is a nonnegative integer, so $\deg(k_1(x)) = \deg(k_2(x)) = 0$. Therefore, $k_1(x)$ and $k_2(x)$ are nonzero constant polynomials. Hence, $k_1(x) = k_2(x) = e$ since $d_1(x)$ and $d_2(x)$ are monic polynomials. Consequently, $d_1(x) = d_2(x)$, so the greatest common divisor of $f(x)$ and $g(x)$ is unique. ■

Let $f(x), g(x) \in F[x]$ be such that $f(x) \neq 0$ and $g(x) = 0$. If the leading coefficient of $f(x)$ is a, then $a^{-1}f(x)$ is a monic polynomial, and it is obvious that $\gcd(f(x), 0) = a^{-1}f(x)$. Hence, to show the existence of $\gcd(f(x), g(x))$, it can be assumed that both $f(x)$ and $g(x)$ are nonzero.

In the following example, to find the greatest common divisor, we look for a monic polynomial of highest degree that divides both polynomials.

6.2.13   Example

Consider $f(x) = 9x^4 + 12x^3 - 5x^2 - 12x - 4$ and $g(x) = 3x^3 + 5x^2 - 4x - 4$ in $\mathbf{Q}[x]$. The polynomials $f(x)$ and $g(x)$ factor as

$$f(x) = (3x + 2)^2(x - 1)(x + 1) \qquad \text{and}$$
$$g(x) = (3x + 2)(x - 1)(x + 2).$$

Inspection of these factorizations shows that a common divisor of $f(x)$ and

$g(x)$ of highest degree is $(3x + 2)(x - 1) = 3x^2 - x - 2$. Hence, the greatest common divisor of $f(x)$ and $g(x)$ is

$$d(x) = \tfrac{1}{3}(3x^2 - x - 2) = x^2 - \tfrac{1}{3}x - \tfrac{2}{3}.$$

Of course, finding the greatest common divisor of two polynomials by factoring depends on being able to factor the polynomials easily. We shall subsequently show that the greatest common divisor can be found by a process that is similar to the procedure for finding the greatest common divisor of two integers.

In the case of integers x and y, both of which are nonzero, to show the existence of $\gcd(x, y)$, we picked the smallest positive integer in the set $S = \{ax + by \mid a, b \in \mathbf{Z}\}$ and showed that this was the greatest common divisor of x and y (see Theorem 1.1.7). Somewhat ironically, the *smallest* positive integer in S turns out to be the *largest* positive integer that divides both x and y. A similar technique works for polynomials, but the argument shifts to the degree of the polynomial since we do not have an order relation among the polynomials in $F[x]$. You should compare the proofs of Theorems 6.2.14 and 1.1.7 and note the similarities and the differences.

6.2.14   **Theorem**

Suppose $f(x)$ and $g(x)$ are nonzero polynomials in $F[x]$. Then $d(x) = \gcd(f(x), g(x))$ exists, is unique and there are not necessarily unique polynomials $a(x)$, $b(x) \in F[x]$ such that $d(x) = a(x)f(x) + b(x)g(x)$. Furthermore, $d(x)$ is the monic polynomial of smallest degree that can be written in the form $a(x)f(x) + b(x)g(x)$.

Proof.   Consider the set $S = \{a(x)f(x) + b(x)g(x) \mid a(x), b(x) \in F[x]\}$. First, note that $f(x) = ef(x) + 0g(x)$ and $g(x) = 0f(x) + eg(x)$, so both $f(x)$ and $g(x)$ are in S. Now let $d(x)$ be a monic polynomial that has the smallest degree among the polynomials in S. Clearly such a polynomial exists, for if we let D be the set of the degrees of the polynomials in S, then D is a nonempty subset of the well-ordered set $\mathbf{N}_0$, so D has a smallest element. We can assume that $d(x)$ is a monic polynomial. If $d(x)$ is not monic and a is the leading coefficient of $d(x)$, we can replace $d(x)$ by the monic polynomial $a^{-1}d(x)$. (Since $d(x) \in S$, it is also the case that $a^{-1}d(x)$ is in S.) The claim is that $d(x) = \gcd(f(x), g(x))$. By the Division Algorithm for Polynomials, there are polynomials $q(x)$, $r(x) \in F[x]$ such that $f(x) = d(x)q(x) + r(x)$. Since $d(x) \in S$, $d(x) = a(x)f(x) + b(x)g(x)$ for a pair of polynomials $a(x), b(x) \in F[x]$. Thus,

$$r(x) = f(x) - d(x)q(x)$$
$$= f(x) - [a(x)f(x) + b(x)g(x)]q(x)$$
$$= [e - a(x)q(x)]f(x) + [b(x)q(x)]g(x),$$

so $r(x) \in S$. But $\deg(r(x)) < \deg(d(x))$, so $r(x) = 0$ since $d(x)$ is a polynomial in S with smallest degree. Hence, $f(x) = d(x)q(x)$ and thus $d(x) \mid f(x)$. Similarly, $d(x) \mid g(x)$. Thus, $d(x)$ is a common divisor of $f(x)$ and $g(x)$. Next, suppose that $c(x) \in F[x]$, $c(x) \mid f(x)$, and $c(x) \mid g(x)$. Then $f(x) = a'(x)c(x)$ and $g(x) = b'(x)c(x)$ for some $a'(x), b'(x) \in F[x]$. Hence,

$$\begin{aligned}
d(x) &= a(x)f(x) + b(x)g(x) \\
&= a(x)a'(x)c(x) + b(x)b'(x)c(x) \\
&= [a(x)a'(x) + b(x)b'(x)]c(x).
\end{aligned}$$

Therefore, $c(x) \mid d(x)$, so $d(x)$ satisfies conditions (i) and (ii) of Definition 6.2.11. Hence, $d(x)$ is the greatest common divisor of $f(x)$ and $g(x)$. Uniqueness of $d(x)$ was established in Theorem 6.2.12, and

$$d(x) = [a(x) + g(x)]f(x) + [b(x) - f(x)]g(x)$$

shows that $a(x)$ and $b(x)$ are not unique. The fact that $d(x)$ is the polynomial of smallest degree that can be written in the form $a(x)f(x) + b(x)g(x)$ is also evident since any polynomial that can be written in this form is an element of S. ∎

If $d(x) = \gcd(f(x), g(x))$, we have just seen in the proof of Theorem 6.2.14 that every common divisor $c(x)$ of $f(x)$ and $g(x)$ divides $d(x)$. Hence, for any common divisor $c(x)$ of $f(x)$ and $g(x)$, $\deg(c(x)) \leq \deg(d(x))$. Thus, by selecting the monic polynomial with the smallest degree that can be written in the form $a(x)f(x) + b(x)g(x)$, we have found the unique monic polynomial with largest degree that divides both $f(x)$ and $g(x)$. In the sense that $d(x)$ is the polynomial of largest degree that divides both $f(x)$ and $g(x)$, $d(x)$ is called the greatest common divisor of $f(x)$ and $g(x)$.

Corollary 1.1.15 shows that if x, y, and z are integers such that x and y are relatively prime and $x \mid yz$, then $x \mid z$. A similar result holds for polynomials.

6.2.15   Theorem

Let $f(x), g(x), h(x) \in F[x]$ be such that $f(x) \mid g(x)h(x)$. If $f(x)$ and $g(x)$ are relatively prime, then $f(x) \mid h(x)$.

Proof.   If $f(x)$ and $g(x)$ are relatively prime, then there exist $a(x), b(x)$ in $F[x]$ such that $e = a(x)f(x) + b(x)g(x)$. Hence,

$$eh(x) = [a(x)f(x) + b(x)g(x)]h(x) \qquad \text{and thus}$$

$$h(x) = a(x)f(x)h(x) + b(x)g(x)h(x).$$

But since $f(x) \mid g(x)h(x)$, there is a polynomial $c(x) \in F[x]$ such that $g(x)h(x) = c(x)f(x)$. From this it follows that

$$h(x) = a(x)f(x)h(x) + b(x)c(x)f(x)$$
$$= [a(x)h(x) + b(x)c(x)]f(x),$$

so $f(x) \mid h(x)$. ∎

The Euclidean Algorithm for integers developed in Section 1.1 also carries over to polynomials. The procedure provides a method for not only finding the greatest common divisor $d(x)$ of two polynomials $f(x)$ and $g(x)$ in $F[x]$, but also for finding polynomials $a(x)$ and $b(x)$ in $F[x]$ such that $d(x) = a(x)f(x) + b(x)g(x)$. The **Euclidean Algorithm for Polynomials** that will now be developed may not actually produce the greatest common divisor of $f(x)$ and $g(x)$, but it will produce a polynomial that is a constant multiple of $d(x)$. The difficulty is that the polynomial $d'(x)$ produced by the Euclidean Algorithm for Polynomials may not be monic. However, this slight difficulty is easily overcome: if $d'(x)$ is monic, then $d(x) = d'(x)$ is the greatest common divisor of $f(x)$ and $g(x)$. If $d'(x)$ is not monic and a is the leading coefficient of $d'(x)$, then we need only let $d(x) = a^{-1}d'(x)$.

The validity of the Euclidean Algorithm for Polynomials depends on the following lemma. A proof can be fashioned along the same lines given in the proof of Lemma 1.1.9. You should provide a proof.

6.2.16 Lemma

If $f(x)$, $g(x)$, $q(x)$, and $r(x)$ are polynomials in $F[x]$ such that $f(x) = g(x)q(x) + r(x)$, then $\gcd(f(x), g(x)) = \gcd(g(x), r(x))$.

6.2.17 Euclidean Algorithm for Polynomials

Let $f(x)$ and $g(x)$ be nonzero polynomials in $F[x]$, and suppose that $\deg(g(x)) \leq \deg(f(x))$. If $g(x) \mid f(x)$, then $\gcd(f(x), g(x)) = a^{-1}g(x)$, where a is the leading coefficient of $g(x)$. If $g(x) \nmid f(x)$, then repeated application of the Division Algorithm for Polynomials produces the following:

$$f(x) = g(x)q_0(x) + r_0(x), \qquad \text{where} \quad \deg(r_0(x)) < \deg(g(x))$$

$$g(x) = r_0(x)q_1(x) + r_1(x), \qquad \text{where} \quad \deg(r_1(x)) < \deg(r_0(x))$$

$$r_0(x) = r_1(x)q_2(x) + r_2(x), \qquad \text{where} \quad \deg(r_2(x)) < \deg(r_1(x))$$

$$r_1(x) = r_2(x)q_3(x) + r_3(x), \qquad \text{where} \quad \deg(r_3(x)) < \deg(r_2(x))$$

$$r_2(x) = r_3(x)q_4(x) + r_4(x), \qquad \text{where} \quad \deg(r_4(x)) < \deg(r_3(x))$$

$$\vdots \qquad\qquad\qquad\qquad \vdots$$

Since $\deg(r_0(x)) > \deg(r_1(x)) > \deg(r_2(x)) > \cdots$ is a decreasing sequence of nonnegative integers, a positive integer k exists such that

$$\vdots \qquad\qquad\qquad\qquad \vdots$$

$$r_{k-2}(x) = r_{k-1}(x)q_k(x) + r_k(x), \quad \text{where} \quad \deg(r_k(x)) < \deg(r_{k-1}(x))$$

$$r_{k-1}(x) = r_k(x)q_{k+1}(x) + 0.$$

Furthermore, if a is the leading coefficient of $r_k(x)$, then $\gcd(f(x), g(x)) = a^{-1}r_k(x)$.

Proof. Lemma 6.2.16 shows that $\gcd(f(x), g(x)) = \gcd(g(x), r_0(x)) = \gcd(r_0(x), r_1(x)) = \gcd(r_1(x), r_2(x)) = \cdots = \gcd(r_{k-1}(x), r_k(x)) = \gcd(r_k(x), 0) = a^{-1}r_k(x)$ where a is the leading coefficient of $r_k(x)$. ∎

Hence, the greatest common divisor of $f(x)$ and $g(x)$ is the last nonzero remainder times the multiplicative inverse of its leading coefficient.

The following examples show how the **Euclidean Algorithm for Polynomials** can be used to find the greatest common divisor of two polynomials $f(x)$ and $g(x)$. You should review the Euclidean Algorithm 1.1.10 and Example 1.1.11 before considering these examples.

6.2.18 | Examples

1. Find the greatest common divisor of the polynomials $f(x) = 9x^4 + 12x^3 - 5x^2 - 12x - 4$ and $g(x) = 3x^3 + 5x^2 - 4x - 4$ in $\mathbf{Q}[x]$. Also find $a(x), b(x) \in \mathbf{Q}[x]$ such that $d(x) = a(x)f(x) + b(x)g(x)$.

Solution. These are the polynomials of Example 6.2.13, where we found that the greatest common divisor was $d(x) = x^2 - \frac{1}{3}x - \frac{2}{3}$ by factoring the polynomials. We now show how to find the greatest common divisor of $f(x)$ and $g(x)$ by using the Euclidean Algorithm for Polynomials. By long division,

$$
\begin{array}{ccccc}
f(x) & = & g(x) & q_0(x) \; + & r_0(x) \\
\downarrow & & \downarrow & \downarrow & \downarrow \\
9x^4 + 12x^3 - 5x^2 - 12x - 4 & = & (3x^3 + 5x^2 - 4x - 4)\,(3x - 1) & + & (12x^2 - 4x - 8)
\end{array}
$$

and

$$
\begin{array}{ccccc}
g(x) & = & r_0(x) & q_1(x) & + \ r_1(x) \\
\downarrow & & \downarrow & \downarrow & \downarrow \\
3x^3 + 5x^2 - 4x - 4 & = & (12x^2 - 4x - 8)\left(\tfrac{1}{4}x + \tfrac{1}{2}\right) + & & 0.
\end{array}
$$

The last nonzero remainder is $12x^2 - 4x - 8$, so the greatest common divisor of $f(x)$ and $g(x)$ is

$$
d(x) = \tfrac{1}{12}(12x^2 - 4x - 8) = x^2 - \tfrac{1}{3}x - \tfrac{2}{3}.
$$

In this case it is easy to find $a(x)$ and $b(x)$ in $\mathbf{Q}[x]$ such that $d(x) = a(x)f(x) + b(x)g(x)$. Solve the first equation above for $r_0(x)$, and then multiply both sides of the equation by $\tfrac{1}{12}$. This gives

$$
12x^2 - 4x - 8 = f(x) + (-3x + 1)g(x) \qquad \text{and thus}
$$

$$
\tfrac{1}{12}(12x^2 - 4x - 8) = \tfrac{1}{12}f(x) + \tfrac{1}{12}(-3x + 1)g(x) \qquad \text{or}
$$

$$
x^2 - \tfrac{1}{3}x - \tfrac{2}{3} = \tfrac{1}{12}f(x) + (-\tfrac{1}{4}x + \tfrac{1}{12})g(x).
$$

Hence,

$$
d(x) = \tfrac{1}{12}f(x) + (-\tfrac{1}{4}x + \tfrac{1}{12})g(x) \qquad \text{and so}
$$

$$
a(x) = \tfrac{1}{12} \quad \text{and} \quad b(x) = -\tfrac{1}{4}x + \tfrac{1}{12}
$$

2. Find the greatest common divisor of $f(x) = x^5 + x^4 + [2]x^2 + [2]x$ and $g(x) = x^3 + [4]$ in $\mathbf{Z}_5[x]$. Also find $a(x), b(x)$ in $\mathbf{Z}_5[x]$ such that $d(x) = a(x)f(x) + b(x)g(x)$.

Solution. Using long division in $\mathbf{Z}_5[x]$, we see that

$$
\begin{array}{ccccc}
f(x) & = & g(x) & q_0(x) & + & r_0(x) \\
\downarrow & & \downarrow & \downarrow & & \downarrow \\
x^5 + x^4 + [2]x^2 + [2]x & = & (x^3 + [4]) & (x^2 + x) & + & ([3]x^2 + [3]x), \quad (1)
\end{array}
$$

$$
\begin{array}{ccccc}
g(x) & = & r_0(x) & q_1(x) & + & r_1(x) \\
\downarrow & & \downarrow & \downarrow & & \downarrow \\
x^3 + [4] & = & ([3]x^2 + [3]x) & ([2]x + [3]) & + & (x + [4]), \quad (2)
\end{array}
$$

$$
\begin{array}{ccccc}
r_0(x) & = & r_1(x) & q_2(x) & + & r_2(x) \\
\downarrow & & \downarrow & \downarrow & & \downarrow \\
[3]x^2 + [3]x & = & (x + [4]) & ([3]x + [1]) & + & [1], \quad (3)
\end{array}
$$

$$r_1(x) = r_2(x) \ q_3(x) + r_3(x)$$
$$\downarrow \qquad \downarrow \qquad \downarrow \qquad \downarrow$$
$$x + [4] = [1] \ (x + [4]) + [0].$$

Hence, $d(x) = \gcd(f(x), g(x)) = [1]$, so $f(x)$ and $g(x)$ are relatively prime in $\mathbf{Z}_5[x]$. To find $a(x)$ and $b(x)$, we prepare for back-substitution by solving equations (1), (2) and (3) for the remainders.

$$[1] = ([3]x^2 + [3]x) - (x + [4])([3]x + [1]) \tag{4}$$
$$x + [4] = g(x) - ([3]x^2 + [3]x)([2]x + [3]) \tag{5}$$
$$[3]x^2 + [3]x = f(x) - g(x)(x^2 + x) \tag{6}$$

Now substitute the expression for the remainder from equation (5) into (4) and simplify.

$$[1] = ([3]x^2 + [3]x) - (g(x) - ([3]x^2 + [3]x)([2]x + [3]))([3]x + [1])$$
$$= ([1] + ([2]x + [3])([3]x + [1]))([3]x^2 + [3]x) - ([3]x + [1])g(x)$$
$$= (x^2 + x + [4])([3]x^2 + [3]x) - ([3]x + [1])g(x)$$
$$= (x^2 + x + [4])([3]x^2 + [3]x) + ([2]x + [4])g(x) \tag{7}$$

Finally, substitute from equation (6) into (7) and simplify.

$$[1] = (x^2 + x + [4])(f(x) - g(x)(x^2 + x)) + ([2]x + [4])g(x)$$
$$= (x^2 + x + [4])(f(x) + g(x)([4]x^2 + [4]x)) + ([2]x + [4])g(x)$$
$$= (x^2 + x + [4])f(x) + (x^2 + x + [4])([4]x^2 + [4]x)g(x) + ([2]x + [4])g(x)$$
$$= (x^2 + x + [4])f(x) + ((x^2 + x + [4])([4]x^2 + [4]x) + ([2]x + [4]))g(x)$$
$$= (x^2 + x + [4])f(x) + ([4]x^4 + [3]x^3 + [3]x + [4])g(x).$$

Hence, $a(x) = x^2 + x + [4]$ and $b(x) = [4]x^4 + [3]x^3 + [3]x + [4]$. You should compute $(x^2 + x + [4])f(x) + ([4]x^4 + [3]x^3 + [3]x + [4])g(x)$ to show that $[1]$ is the result.

In the opening remarks of this section, we indicated that when F is field, the integral domain $F[x]$ has many properties that are similar to those of the integral domain $\mathbf{Z}$. In Theorem 1.1.17 it was shown that any integer $a > 1$ can be factored as $a = p_1^{n_1} p_2^{n_2} \cdots p_k^{n_k}$, where $p_1, p_2, \ldots, p_k$ are the unique prime factors of a and $n_1, n_2, \ldots, n_k$ are the positive integers that represent the multiplicity of each unique prime factor occurring in the factorization of a. Now let's look at how a polynomial can be factored in $\mathbf{R}[x]$. The observations made here are used as a model for the definitions that follow. Consider $f(x) = x^2 + 4x + 3$ in $\mathbf{R}[x]$ that factors as

$$f(x) = (x + 1)(x + 3). \tag{1}$$

But this is not the only way that $f(x)$ can be factored. It also factors as

$$f(x) = \left(\tfrac{1}{2}x + \tfrac{1}{2}\right)(2x + 6). \tag{2}$$

If you look at the second factorization, (2), of $f(x)$, you will notice that this factorization of $f(x)$ can be obtained from the first by multiplying the first factor of equation (1) by $\tfrac{1}{2}$ and the second factor of equation (1) by 2, the multiplicative inverse of $\tfrac{1}{2}$. In fact, if c is any nonzero real number, then

$$f(x) = (cx + c)(c^{-1}x + 3c^{-1})$$

is a factorization of $f(x)$. The factorizations (1) and (2) of $f(x)$ are different, but obviously they are very closely associated. More specifically, the factors $x + 1$ and $\tfrac{1}{2}x + \tfrac{1}{2}$ and the factors $x + 3$ and $2x + 6$ are associated in the sense that one can be obtained from the other by multiplication by the appropriate constant. Clearly what makes this process work is that we are multiplying one factor by a nonzero real number and then multiplying the other factor by its multiplicative inverse. Since we are dealing with factorization of polynomials in $F[x]$, the first thing we need to do is to classify the polynomials in $F[x]$ that have a multiplicative inverse.

6.2.19   **Lemma**

If D is an integral domain, then the only polynomials $f(x) \in D[x]$ that have multiplicative inverses in $D[x]$ are the nonzero constant polynomials $f(x) = a$, where a has a multiplicative inverse in D.

Proof. If $g(x) \in D[x]$ is a multiplicative inverse of $f(x)$, then by Theorem 6.1.9, $0 = \deg(e) = \deg(f(x)g(x)) = \deg(f(x)) + \deg(g(x))$. Since $\deg(f(x)) \geq 0$ and $\deg(g(x)) \geq 0$, it must be the case that $\deg(f(x)) = \deg(g(x)) = 0$. Hence, both $f(x)$ and $g(x)$ are constant polynomials. If $f(x) = a$ and $g(x) = b$, then $ab = e$, so a has a multiplicative inverse in D. The converse is obvious. ∎

6.2.20   **Definition**

Let D be an integral domain. A constant polynomial $u \in D[x]$ is said to be a **unit** in $D[x]$ if u has a multiplicative inverse in $D[x]$. Two polynomials $f(x), g(x) \in D[x]$ are **associates** if there is a unit $u \in D[x]$ such that $f(x) = ug(x)$. A nonconstant polynomial $f(x) \in D[x]$ is **irreducible** in $D[x]$ if, whenever $f(x) = g(x)h(x)$, either $g(x)$ or $h(x)$ is a unit in $D[x]$. A polynomial that is not irreducible in $D[x]$ is referred to as a **reducible** polynomial in $D[x]$.

It is obvious from the preceding definition that when F is a field, polynomials in $F[x]$ of degree 1 are irreducible. Hence, a reducible polynomial must have degree of at least 2. Our goal is to show that every polynomial in $F[x]$ can be factored into a product of irreducible polynomials and that this factorization is in some sense unique. We already know from the Factor Theorem 6.2.5 that if c is a root of a polynomial $f(x)$, then $x - c$ is a factor of $f(x)$. This can be used to help with factoring polynomials.

6.2.21   **Examples**

1. The polynomial $2x^2 - 6$ is irreducible in $\mathbf{Q}[x]$. $2x^2 - 6$ can be factored as $2(x^2 - 3)$, but 2 is a unit in $\mathbf{Q}$. Since

    $$2x^2 - 6 = (2x - 2\sqrt{3})(x + \sqrt{3}), \tag{1}$$

    $2x^2 - 6$ is reducible in $\mathbf{R}[x]$. Note that $2x^2 - 6$ is expressed as a product of irreducible polynomials in $\mathbf{R}[x]$. The polynomial $2x^2 - 6$ can be factored in other ways, such as

    $$2x^2 - 6 = (4x - 4\sqrt{3})\left(-\tfrac{1}{2}x + \tfrac{1}{2}\sqrt{3}\right). \tag{2}$$

    Comparing the factors in equations (1) and (2), we see that

    $$2x - 2\sqrt{3} \qquad \text{and} \qquad 4x - 4\sqrt{3}$$

    are associates in $\mathbf{R}[x]$, as are

    $$x + \sqrt{3} \qquad \text{and} \qquad \tfrac{1}{2}x + \tfrac{1}{2}\sqrt{3}.$$

2. The polynomial $f(x) = x^2 + x + 1$ is irreducible in $\mathbf{Q}[x]$ and $\mathbf{R}[x]$ but reducible in $\mathbf{C}[x]$. The quadratic formula shows that the roots of $f(x)$ are

    $$x = \frac{-b \pm \sqrt{b^2 - 4ac}}{2a} = \frac{-1 \pm \sqrt{1 - 4}}{2} = -\frac{1}{2} \pm \frac{\sqrt{3}}{2}i.$$

    Hence,

    $$f(x) = \left(x + \frac{1}{2} - \frac{\sqrt{3}}{2}i\right)\left(x + \frac{1}{2} + \frac{\sqrt{3}}{2}i\right) \tag{3}$$

    in $\mathbf{C}[x]$, so $f(x)$ is expressible as a product of irreducible polynomials in $\mathbf{C}[x]$. If c is any nonzero complex number, then $f(x)$ can also be factored as

$$f(x) = \left(cx + \frac{c}{2} + -\frac{c\sqrt{3}}{2}i\right)\left(c^{-1}x + \frac{c^{-1}}{2} + \frac{c^{-1}\sqrt{3}}{2}i\right). \quad (4)$$

Notice that the factors

$$x + \frac{1}{2} - \frac{\sqrt{3}}{2}i \qquad \text{and} \qquad cx + \frac{c}{2} - \frac{c\sqrt{3}}{2}i$$

in equations (3) and (4) are associates and that

$$x + \frac{1}{2} + \frac{\sqrt{3}}{2}i \qquad \text{and} \qquad c^{-1}x + \frac{c^{-1}}{2} + \frac{c^{-1}\sqrt{3}}{2}i$$

are associates.

3. Consider the polynomial $f(x) = x^3 + x + [1]$ in $\mathbf{Z}_2[x]$. If $f(x)$ is reducible, then it is expressible as a polynomial of degree 1 times a polynomial of degree 2, or it is expressible as a product of three polynomials each having degree 1. In any case, because of the Factor Theorem 6.2.5, if $f(x)$ is reducible, it will have a root in $\mathbf{Z}_2$. But $f([0]) = [1]$ and $f([1]) = [1]$, so $f(x)$ has no roots in $\mathbf{Z}_2$. Therefore, $f(x)$ is irreducible in $\mathbf{Z}_2[x]$.

4. Consider $f(x) = x^3 + x + [1]$ as a polynomial in $\mathbf{Z}_3[x]$. Since $f([1]) = [1]^3 + [1] + [1] = [3] = [0]$, $x = [1]$ is a root of $f(x)$. Consequently, $f(x)$ is divisible by $x - [1] = x + [2]$. The quotient $q(x) = x^2 + x + [2]$ can be found by using the Division Algorithm for Polynomials. Hence,

$$f(x) = (x + [2])(x^2 + x + [2])$$

in $\mathbf{Z}_3[x]$, so $f(x)$ is reducible in $\mathbf{Z}_3[x]$. The question now arises as to whether or not $g(x) = x^2 + x + [2]$ is irreducible in $\mathbf{Z}_3[x]$. If $g(x)$ is reducible in $\mathbf{Z}_3[x]$, then $g(x)$ is expressible as a product of two polynomials each having degree 1. Consequently, $g(x)$ must have two roots or no roots in $\mathbf{Z}_3$. Observe that

$$g([0]) = [0]^2 + [0] + [2] = [2] \neq [0],$$
$$g([1]) = [1]^2 + [1] + [2] = [4] = [1] \neq [0],$$
$$g([2]) = [2]^2 + [2] + [2] = [8] = [2] \neq [0],$$

so $g(x)$ does not have a root in $\mathbf{Z}_3$. Hence, $g(x)$ is irreducible in $\mathbf{Z}_3[x]$. Therefore, we see that $f(x)$ can be factored into a product of irreducible polynomials in $\mathbf{Z}_3[x]$. The polynomial $f(x)$ can also be factored as

$$f(x) = ([2]x + [1])([2]x^2 + [2]x + [1]),$$

but the pairs $x + [2], [2]x + 1$ and $x^2 + x + [2], [2]x^2 + [2]x + [1]$ are associates in $\mathbf{Z}_3[x]$. Do you see why?

In each of the preceding examples, the polynomial under consideration can be factored into a product of irreducible polynomials, and when the polynomial is factored in a different way, the factors in the factorizations can be paired in such a way that the pairings are associates. We have also seen in these examples that polynomials of degree 2 or 3 are reducible in the polynomial ring, provided that they have a root in the coefficient field. The following example shows that this need not be true if the polynomial has degree 4 or greater.

5. The polynomial $f(x) = x^4 + [2]x^3 + [2]x^2 + x + [1]$ in $\mathbf{Z}_3[x]$ can be factored as $f(x) = (x^2 + x + [2])(x^2 + x + [2])$ in $\mathbf{Z}_3[x]$. We have seen in Example 4 that $x^2 + x + [2]$ is irreducible in $\mathbf{Z}_3[x]$ and so cannot be factored further. Thus, $f(x)$ is reducible and can be expressed as a product of two irreducible polynomials in $\mathbf{Z}_3[x]$, both of degree 2. There is another way that $f(x)$ can be factored in $\mathbf{Z}_3[x]$. Can you find this factorization and how is it related to the factorization $(x^2 + x + [2])(x^2 + x + [2])$ of $f(x)$?

We have observed that the factorization of a polynomial into a product of irreducible polynomials is not unique. However, these factorizations are unique in the sense that any one factorization of the polynomial has factors that are associates of the factors of any other factorization of the polynomial. These observations are made more specific by the following theorem. The proof, which is left as an exercise, is a modification of the proof of the Fundamental Theorem of Arithmetic 1.1.17.

6.2.22   Theorem

Let $f(x) \in F[x]$ be a polynomial with positive degree.

1. Then there exist irreducible polynomials, $f_1(x), f_2(x), \ldots, f_m(x)$ in $F[x]$ such that $f(x) = f_1(x) f_2(x) \cdots f_m(x)$. Furthermore, if $f(x) = g_1(x)g_2(x) \cdots g_n(x)$ is another factorization of $f(x)$, where $g_1(x), g_2(x), \ldots, g_n(x) \in F[x]$ are irreducible, then $m = n$, and after a suitable reindexing of the $g_i(x), f_i(x)$ and $g_i(x)$ are associates for $i = 1, 2, \ldots, n$.
2. If a is the leading coefficient of $f(x)$, then there exist monic irreducible polynomials $h_1(x), h_2(x), \ldots, h_n(x)$ in $F[x]$ such that $f(x) = ah_1(x)h_2(x) \cdots h_n(x)$. Moreover, this particular factorization of $f(x)$ is unique except for the order of the monic irreducible factors.

Theorem 6.2.22 indicates that if $f(x)$ is a polynomial in $F[x]$ of positive degree with leading coefficient a, then $f(x) = af_1(x)f_2(x)\cdots f_n(x)$, where $f_1(x), f_2(x)$, and $f_n(x)$ are monic irreducible polynomials in $F[x]$. There may be repetitions among the factors of $f(x)$. Consequently,

$$f(x) = ag_1(x)^{n_1}g_2(x)^{n_2}\cdots g_k(x)^{n_k},$$

where the $g_i(x)$ are the distinct monic irreducible factors among the factors $f_1(x)$, $f_2(x), \ldots, f_n(x)$ of $f(x)$ and the positive integer n_i represents the multiplicity of the factor $g_i(x)$ for $i = 1, 2, \ldots, k$.

Problem Set 6.2

1. If $f(x) = x^5 + [4]x$ is a polynomial in $\mathbf{Z}_5[x]$, show that every element of $\mathbf{Z}_5$ is a root of $g(x)$ so that

    $$f(x) = x(x - [1])(x - [2])(x - [3])(x - [4])$$
    $$= x(x + [4])(x + [3])(x + [2])(x + [1])$$

    Verify these expressions for f(x) by actually computing the product.

2. Find the greatest common divisor $d(x)$ of each of the following pairs of polynomials. In each case, find polynomials $a(x)$ and $b(x)$ such that $d(x) = a(x)f(x) + b(x)g(x)$.

    (a) $f(x) = x^3 + 6x^2 + 12x + 8$ and $g(x) = x^2 + 3x + 2$ in $\mathbf{Q}[x]$
    (b) $f(x) = x^2 + (1 - \sqrt{2})x - \sqrt{2}$ and $g(x) = x^2 - 2$ in $\mathbf{R}[x]$
    (c) $f(x) = x^4 + x + [1]$ and $g(x) = x^2 + x + [1]$ in $\mathbf{Z}_2[x]$

3. Example 2 of 6.2.2 shows that $x = [1], [2], [4], [5]$ are roots of the polynomial $f(x) = [5]x^2 + [3]x + [4]$ in $\mathbf{Z}_6[x]$ and that these are the only roots of $f(x)$ in $\mathbf{Z}_6$. Find all the pairs of roots $c_1, c_2 \in \mathbf{Z}_6$ of $f(x)$ such that $f(x) = [5](x - c_1)(x - c_2)$. Is it true for every pair of roots $c_1, c_2 \in \mathbf{Z}_6$ of $f(x)$ that $f(x) = [5](x - c_1)(x - c_2)$?

4. If $f(x) \in R[x]$ and $c \in R$, then c is said to be a **root of multiplicity** m if $(x - c)^m$ divides $f(x)$ but $(x - c)^{m+1}$ does not. A root of $f(x)$ of multiplicity 1 is a **simple root** of $f(x)$, and a root of multiplicity 2 is a **double root** of $f(x)$. For each of the following, find the multiplicity of the given root.

    (a) $12 - 8x - x^2 + x^3 \in \mathbf{Z}[x], c = 2$
    (b) $[2] + x^3 \in \mathbf{Z}_3[x], c = [1]$
    (c) $-81 + 108x - 48x^3 + 16x^4 \in \mathbf{Q}[x], c = -\dfrac{3}{2}$

(d) $-2\sqrt{2}x + 6x^2 - 3\sqrt{2}x^3 + x^4 \in \mathbf{R}[x]$, $c = \sqrt{2}$

(e) $-2 + 4x - 3x^2 + x^3 \in \mathbf{C}[x]$, $c = 1 + i$

5. Let $d(x) = \gcd[f(x), g(x)]$ in $F[x]$. If $f(x) = d(x)a(x)$ and $g(x) = d(x)b(x)$, where $a(x), b(x) \in F[x]$, must it be the case that $\gcd[a(x), b(x)] = e$? Prove or disprove your answer. Hint: See Corollary 1.1.8.

6. Prove Lemma 6.2.16.

7. Prove that if $f(x) \in F[x]$ has degree 2 or 3, then $f(x)$ is reducible in $F[x]$ if and only if $f(x)$ has a root in F.

8. (a) Let $f(x), g(x), h(x) \in F[x]$, and suppose that $f(x)$ is irreducible. If $f(x) \mid g(x)h(x)$, show that $f(x) \mid g(x)$ or $f(x) \mid h(x)$.

    (b) If $f(x) \in F[x]$ is irreducible and $f(x) \mid f_1(x)f_2(x), \ldots, f_n(x)$, prove that $f(x)$ divides $f_i(x)$ for some $i = 1, 2, \ldots, n$.

9. Suppose that F is a field with characteristic $\neq 2$. If $f(x) = ax^2 + bx + c$ is a quadratic polynomial in $F[x]$, show that $f(x)$ has two distinct roots in F if $b^2 - 4ac \neq 0$ and $b^2 - 4ac$ is a perfect square in F. An element $c \in F$ is said to be a **perfect square** in F if there is an element $s \in F$ such that $s^2 = c$. If $b^2 - 4ac$ is a perfect square in F, show that the roots of $f(x)$ are given by the quadratic formula.

10. If F is a field, prove that $f(x) \in F[x]$ is irreducible if and only if $M = (f(x)) = \{f(x)g(x) \mid g(x) \in F[x]\}$ is a maximal ideal of $F[x]$.

11. Determine whether each of the following polynomials has a root in the given field. If the polynomial has a root in the field, use the quadratic formula to find the roots of the polynomial.

    (a) $f(x) = [2] + [3]x + x^2$ in $\mathbf{Z}_5$

    (b) $f(x) = [1] + x + [2]x^2$ in $\mathbf{Z}_7$

    (c) $f(x) = [6] + [2]x + x^2$ in $\mathbf{Z}_7$

12. The element $c \in F$ is a **multiple root** of $f(x) \in F[x]$ if c is a root of $f(x)$ of multiplicity at least 2. Prove that if c is a multiple root of $f(x)$, then c is a root of both $f(x)$ and $D[f(x)]$, the formal derivative of $f(x)$. Show also that if $f(x)$ and $D[f(x)]$ are relatively prime, then $f(x)$ has no multiple roots in F. See Exercise 18 of Problem Set 6.1.

13. Let $f(x) \in F[x]$ be a polynomial of degree n. If F has more than n elements, prove that there is an element $c \in F$ such that $f(c) \neq 0$.

14. Suppose that R is a ring with identity e that is not necessarily commutative, and let

$$f(x) = a_nx^n + a_{n-1}x^{n-1} + \cdots + a_1x + a_0 \quad \text{and}$$

$$g(x) = b_mx^m + b_{m-1}x^{m-1} + \cdots + b_1x + b_0$$

be nonzero polynomials in $R[x]$. Suppose also that $ax^k = x^k a$ for all $a \in R$ and each positive integer k. If b_m has a multiplicative inverse in R and $m < n$, a "right-hand" division of $f(x)$ by $g(x)$ can be performed as follows:

Step 1. Divide the first term of the divisor $g(x)$ into the first term of $f(x)$ as shown below. Make note of the fact that b_m^{-1} is on the *right* of a_n in the product $a_n b_m^{-1}$. Now multiply $g(x)$ on the *left* by $a_n b_m^{-1} x^{n-m}$, placing each term of this product under the corresponding term in the dividend $f(x)$. Subtract to obtain a new dividend.

$$a_n b_m^{-1} x^{n-m}$$

$$b_m x^m + b_{m-1}x^{m-1} + \cdots + b_1 x + b_0 \overline{\big)\, a_n x^n + a_{n-1}x^{n-1} + \cdots + a_1 x + a_0}$$

$$\underline{a_n x^n + a_n b_m^{-1} b_{m-1} x^{n-1} + \cdots}$$

$$(a_{n-1} - a_n b_m^{-1} b_{m-1})x^{n-1} + \cdots$$

Step 2. Repeat step 1 using the new dividend, and then repeat the process until a remainder $r(x)$ is obtained such that either $r(x) = 0$ or $\deg(r(x)) < \deg(g(x))$. It follows that $f(x) = q(x)g(x) + r(x)$. Note that $q(x)$ is on the *left* of $g(x)$ in the expression $q(x)g(x) + r(x)$. When $g(x)$ is multiplied on the right by $q(x)$, it may be that $f(x) \neq g(x)q(x) + r(x)$. For this reason, the process just outlined is referred to as the **Left-Hand Division Algorithm for Polynomials** over a noncommutative ring with identity. A **Right-Hand Division Algorithm** can also be developed in a similar manner. When the Left- and Right-Hand Division Algorithms are applied to two polynomials, the quotient and the remainder in each case may be quite different, as shown by the following example and problem (b) .
Let

$$g(x) = \begin{pmatrix} 1 & 2 \\ -1 & -1 \end{pmatrix} x + \begin{pmatrix} 2 & -1 \\ 1 & 1 \end{pmatrix} \quad \text{and} \quad f(x) = \begin{pmatrix} 1 & 3 \\ 2 & 4 \end{pmatrix} x^2 + \begin{pmatrix} 1 & 4 \\ 0 & 2 \end{pmatrix} x + \begin{pmatrix} 2 & 1 \\ 4 & 0 \end{pmatrix}$$

be polynomials over the noncommutative ring $M_2(\mathbf{R})$. For a 2×2 matrix $\begin{pmatrix} a & b \\ c & d \end{pmatrix}$,

$$\begin{pmatrix} a & b \\ c & d \end{pmatrix}^{-1} = \begin{pmatrix} \dfrac{d}{ad-bc} & \dfrac{-b}{ad-bc} \\[2mm] \dfrac{-c}{ad-bc} & \dfrac{a}{ad-bc} \end{pmatrix},$$

so it follows that $\begin{pmatrix} 1 & 2 \\ -1 & -1 \end{pmatrix}^{-1}$ exists in $M_2(\mathbf{R})$ and that $\begin{pmatrix} 1 & 2 \\ -1 & -1 \end{pmatrix}^{-1} = \begin{pmatrix} -1 & -2 \\ 1 & 1 \end{pmatrix}$.

Since $\begin{pmatrix} 1 & 3 \\ 2 & 4 \end{pmatrix} \begin{pmatrix} -1 & -2 \\ 1 & 1 \end{pmatrix} = \begin{pmatrix} 2 & 1 \\ 2 & 0 \end{pmatrix}$, the first step of the Left-Hand Division Algorithm produces

$$
\begin{array}{r}
\begin{pmatrix} 2 & 1 \\ 2 & 0 \end{pmatrix}x \\ \hline
\left[\begin{pmatrix} 1 & 2 \\ -1 & -1 \end{pmatrix}x + \begin{pmatrix} 2 & -1 \\ 1 & 1 \end{pmatrix} \right] \overline{\left| \begin{pmatrix} 1 & 3 \\ 2 & 4 \end{pmatrix}x^2 + \begin{pmatrix} 1 & 4 \\ 0 & 2 \end{pmatrix}x + \begin{pmatrix} 2 & 1 \\ 4 & 0 \end{pmatrix} \right.} \\
\begin{pmatrix} 1 & 3 \\ 2 & 4 \end{pmatrix}x^2 + \begin{pmatrix} 5 & -1 \\ 4 & -2 \end{pmatrix}x \\ \hline
\begin{pmatrix} -4 & 5 \\ -4 & 4 \end{pmatrix}x + \begin{pmatrix} 2 & 1 \\ 4 & 0 \end{pmatrix}.
\end{array}
$$

Now $\begin{pmatrix} -4 & 5 \\ -4 & 4 \end{pmatrix} \begin{pmatrix} -1 & -2 \\ 1 & 1 \end{pmatrix} = \begin{pmatrix} 9 & 13 \\ 8 & 12 \end{pmatrix}$, so the second step gives

$$
\begin{array}{r}
\begin{pmatrix} 2 & 1 \\ 2 & 0 \end{pmatrix}x + \begin{pmatrix} 9 & 13 \\ 8 & 12 \end{pmatrix} \\ \hline
\left[\begin{pmatrix} 1 & 2 \\ -1 & -1 \end{pmatrix}x + \begin{pmatrix} 2 & -1 \\ 1 & 1 \end{pmatrix} \right] \overline{\left| \begin{pmatrix} 1 & 3 \\ 2 & 4 \end{pmatrix}x^2 + \begin{pmatrix} 1 & 4 \\ 0 & 2 \end{pmatrix}x + \begin{pmatrix} 2 & 1 \\ 4 & 0 \end{pmatrix} \right.} \\
\begin{pmatrix} 1 & 3 \\ 2 & 4 \end{pmatrix}x^2 + \begin{pmatrix} 5 & -1 \\ 4 & -2 \end{pmatrix}x \\ \hline
\begin{pmatrix} -4 & 5 \\ -4 & 4 \end{pmatrix}x + \begin{pmatrix} 2 & 1 \\ 4 & 0 \end{pmatrix} \\
\begin{pmatrix} -4 & 5 \\ -4 & 5 \end{pmatrix}x + \begin{pmatrix} 31 & 4 \\ 28 & 4 \end{pmatrix} \\ \hline
\begin{pmatrix} -29 & -3 \\ -24 & -4 \end{pmatrix}.
\end{array}
$$

Hence, $q(x) = \begin{pmatrix} 2 & 1 \\ 2 & 0 \end{pmatrix}x + \begin{pmatrix} 9 & 13 \\ 8 & 12 \end{pmatrix}$ and $r(x) = \begin{pmatrix} -29 & -3 \\ -24 & -4 \end{pmatrix}$.

(a) Show that $f(x) = q(x)g(x) + r(x)$, but that $f(x) \neq g(x)q(x) + r(x)$.

(b) Apply the Right-Hand Division Algorithm to find polynomials $q(x)$ and $r(x)$ in $M_2(\mathbf{R})[x]$ such that $f(x) = g(x)q(x) + r(x)$, where $r(x) = 0$ or $\deg(r(x)) < \deg(g(x))$.

(c) Let $f(x) = \begin{pmatrix} 1 & 1 \\ 0 & 1 \end{pmatrix} x^3 + \begin{pmatrix} 2 & 0 \\ -2 & 0 \end{pmatrix} x^2 + \begin{pmatrix} 2 & 1 \\ 4 & 0 \end{pmatrix} x + \begin{pmatrix} 0 & 1 \\ 2 & 0 \end{pmatrix}$ and

$g(x) = \begin{pmatrix} 2 & 2 \\ 1 & 2 \end{pmatrix} x^2 + \begin{pmatrix} 2 & 2 \\ 4 & 0 \end{pmatrix} x + \begin{pmatrix} 0 & 4 \\ -2 & 2 \end{pmatrix}$ be polynomials in $M_2(\mathbf{R})[x]$.

Apply the Left-Hand Division Algorithm to find polynomials $q(x)$, $r(x) \in M_2(\mathbf{R})[x]$ such that $f(x) = q(x)g(x) + r(x)$, where either $r(x) = 0$ or $\deg(r(x)) < \deg(g(x))$.

15. Determine which, if any, of Theorem 6.2.6 and Corollaries 6.2.7 and 6.2.8 remain valid if $F[x]$ is replaced by $R[x]$, where R is an integral domain.

TECHNOLOGY PROBLEMS

Use a computer algebra system to solve each of the following problems.

16. Find the roots of the given polynomial in the given coefficient ring.

(a) $f(x) = x^4 + [19]x^3 + [80]x^2 + [124]x + [32]$ in $\mathbf{Z}_{128}$
(b) $g(x) = [12]x^5 + [19]x^4 + [40]x^3 + [12]x^2$ in $\mathbf{Z}_{85}$
(c) The roots of $h(x) = [123]x^4 + [256]$ in $\mathbf{Z}_{512}$ such that their smallest positive representatives lie between 202 and 310

17. Find the remainder of each of the following divisions.

(a) $f(x) = [5]x^4 + [3]x^3 + [4]x \div g(x) = [3]x^2 + [4]x + [2]$ in $\mathbf{Z}_7[x]$
(b) $f(x) = [123]x^6 + [132]x^4 + [12]x^2 + [23] \div g(x) = [16]x^2 + [35]$ in $\mathbf{Z}_{127}[x]$
(c) $f(x) = x^4 + [1] \div g(x) = x^2 + [1]$ in $\mathbf{Z}_{31}[x]$

18. The definition of the greatest common divisor can be extended, in an obvious way, to a finite set of polynomials in $F[x]$. (You should rewrite Definition 6.2.11 for a finite set of polynomials in $F[x]$.) Find the greatest common division of each of the following sets of polynomials in $\mathbf{R}[x]$.

(a) $f(x) = x^4 - 42x^3 + 241x^2 + 4200x + 10{,}000$
   $g(x) = x^3 - 17x^2 - 184x - 400$

(b) $f(x) = x^4 + 3x^3 - 9x^2 - 23x - 12$

$g(x) = x^4 - x^3 - 17x^2 + 21x + 36$

$h(x) = x^4 + 6x^3 - 3x^2 - 56x - 48$

6.3   POLYNOMIALS IN Q[x], R[x], AND C[x]

Section Overview. In Section 6.2, polynomials with their coefficients in a commutative ring with identity were considered. We now restrict our attention to polynomials with their coefficients in **Q**, **R**, or **C**. It will be shown that every polynomial in **C**[x] can be factored into a product of linear factors in **C**[x]. This fact depends on the Fundamental Theorem of Algebra, an extremely important theorem. Make sure you understand what it says and its ramifications for polynomials in **R**[x]. Polynomials in **Z**[x] are also investigated, and an irreducibility criterion is developed for these polynomials.

We now turn our attention to polynomials with their coefficients in the familiar fields **Q**, **R**, or **C** and to the Fundamental Theorem of Algebra.

6.3.1   Fundamental Theorem of Algebra

If $f(x) \in$ **C**[x] is a polynomial of positive degree, then $f(x)$ has a root in **C**.

Unfortunately this theorem cannot be proven without introducing topics that are beyond the scope of this text. One method of proof, and probably the proof most often cited, uses advanced topics in complex analysis. At the time that the Fundamental Theorem was first proven, complex numbers were not completely understood and, over time, **Carl Friedrich Gauss** (1777–1855) gave four different proofs of this theorem. (Information on the life of Gauss can be found in the Historical Notes at the end of Section 1.3.) Each of his proofs used some form of an argument with complex numbers. Gauss first proved that any polynomial in **R**[x] has a real or a nonreal root. In his last proof he permitted the polynomial to be in **C**[x], thereby establishing the theorem we know today.

The Fundamental Theorem of Algebra actually indicates much more about the roots of a polynomial than might be expected when it is first considered. It actually implies that if $f(x) \in$ **C**$[x]$ is a polynomial of positive degree n, then $f(x)$ has n roots in **C**, provided that the roots of $f(x)$ are counted according to their multiplicity. **Counting roots according to their multiplicity** means that if c is a root of a polynomial $f(x)$ of multiplicity k, then c is counted k times when determining the total number of roots of $f(x)$. (See Problem Set 6.2, Exercise 4.)

6.3.2   Corollary

If $f(x) \in$ **C**$[x]$ is a polynomial of positive degree n with leading coefficient a, then $f(x)$ can be factored as

$$f(x) = a(x - c_1)(x - c_2) \cdots (x - c_n),$$

where $c_1, c_2, \ldots, c_n \in$ **C** are the n not necessarily distinct roots of $f(x)$.

Proof. The proof is by induction on the degree of the polynomial $f(x)$. If $f(x)$ has degree 1, then $f(x) = ax + b$, where $a, b \in$ **C**, $a \neq 0$. It is obvious that $f(x) = a(x - (-a^{-1}b))$ and that $x = -a^{-1}b$ is a root of $f(x)$ in **C**. Now make the induction hypothesis that the theorem holds for all polynomials in **C**$[x]$ of degree k, where k is a positive integer, and let $f(x) \in$ **C**$[x]$ be a polynomial of degree $k + 1$ with leading coefficient a. The Fundamental Theorem of Algebra indicates that $f(x)$ has a root in **C**. Suppose $c \in$ **C** is a root of $f(x)$. Then by the Factor Theorem 6.2.5, $x - c$ is a factor of $f(x)$. Hence, there is a $g(x) \in$ **C**$[x]$ such that $f(x) = (x - c)g(x)$. Clearly, a must also be the leading coefficient of $g(x)$. Since $k + 1 = \deg(f(x)) = \deg(x - c) + \deg(g(x)) = 1 + \deg(g(x))$, it follows that $\deg(g(x)) = k$. Thus, by the induction hypothesis, $g(x)$ can be factored as $g(x) = a(x - c_1)(x - c_2) \cdots (x - c_k)$, where $c_1, c_2, \ldots, c_k \in$ **C** are the k not necessarily distinct roots of $g(x)$. Thus, $f(x) = a(x - c)(x - c_1)(x - c_2) \cdots (x - c_k)$, where $c, c_1, \ldots, c_k$ are the $k + 1$ not necessarily distinct roots of $f(x)$. Hence, the theorem follows by induction. ∎

Since roots of $f(x)$ are not necessarily distinct, if $f(x) \in$ **C**$[x]$ is of positive degree n, then $f(x)$ can be written as $f(x) = a(x - c_1)^{n_1}(x - c_2)^{n_2} \cdots (x - c_k)^{n_k}$, where the c_i's are distinct and a is the leading coefficient of $f(x)$. The positive integer n_i represents the multiplicity of the root c_i for $i = 1, 2, \ldots, k$ and $n = n_1 + n_2 + \cdots + n_k$.

6.3.3   Example

The polynomial $2x^4 + 12x^3 + 44x^2 + 96x + 80$ factors as $f(x) = 2(x+1+3i)(x+1-3i)(x+2)^2$, and each factor is linear and consequently irreducible in $\mathbf{C}[x]$. If we multiply the first two factors of $f(x)$ together, then $f(x) = 2(x^2+2x+10)(x+2)^2$. Since $x^2+2x+10$ is irreducible in $\mathbf{R}[x]$, this is a factorization of $f(x)$ into a product of irreducible factors in $\mathbf{R}[x]$.

The preceding example illustrates that any polynomial in $\mathbf{R}[x]$ can be factored into a product of linear factors in $\mathbf{C}[x]$. One thing you should notice about the example is that the complex number $-1+3i$ and its conjugate $-1-3i$ are both roots of $f(x)$. It is shown in Theorem 6.3.4 that this is always the case. We show that if a complex number is the root of a polynomial with real coefficients, then so is its conjugate. From this it follows that the number of nonreal roots of a polynomial with real coefficients must be an even integer. We also show that any polynomial in $\mathbf{R}[x]$ can be factored into a product of irreducible linear and quadratic polynomials in $\mathbf{R}[x]$. Before these concepts are considered, let's review some elementary facts about complex numbers that were presented in Exercise 15 of Problem Set 4.4. Recall that if $z = a + bi$ is a complex number, the conjugate of z is the complex number $\bar{z} = a - bi$. If $z_1, z_2, \ldots, z_n$ are complex numbers, where n is any positive integer, then

$$\overline{z_1 + z_2 + \cdots + z_n} = \bar{z}_1 + \bar{z}_2 + \cdots + \bar{z}_n \qquad \text{and} \qquad (1)$$

$$\overline{z_1 z_2 \cdots z_n} = \bar{z}_1 \bar{z}_2 \cdots \bar{z}_n . \qquad (2)$$

Hence, if $z = z_1 = z_2 = \cdots = z_n$, then equations (1) and (2) yield

$$\overline{n \cdot z} = n \cdot \bar{z} \quad \text{and} \quad \overline{z^n} = \bar{z}^n \qquad (3)$$

6.3.4   Theorem

If $f(x)$ is a polynomial in $\mathbf{R}[x]$ and $z \in \mathbf{C}$ is a root of $f(x)$, then $\bar{z}$ is a root of $f(x)$.

Proof. Let $f(x) = a_0 + a_1 x + a_2 x^2 + \cdots + a_n x^n \in \mathbf{R}[x]$. If $z \in \mathbf{C}$ is a root of $f(x)$ and also a real number, then $\bar{z} = z$ and there is nothing to prove. So suppose that $z \in \mathbf{C}$ is a root of $f(x)$ and $z \notin \mathbf{R}$. Since, for each i, $\bar{a}_i = a_i$ and since $f(z) = 0$, $\overline{f(z)} = \bar{0} = 0$. Using the preceding equations (1), (2) and (3), it follows that

$$0 = \overline{a_0 + a_1 z + a_2 z^2 + \cdots + a_n z^n}$$

$$= \overline{a_0} + \overline{a_1 z} + \overline{a_2 z^2} + \cdots + \overline{a_n z^n}$$

$$= \overline{a_0} + \overline{a_1}\,\overline{z} + \overline{a_2}\,\overline{z}^2 + \cdots + \overline{a_n}\,\overline{z}^n$$

$$= a_0 + a_1 \overline{z} + a_2 \overline{z}^2 + \cdots + a_n \overline{z}^n$$

$$= a_0 + a_1 \overline{z} + a_2 \overline{z}^2 + \cdots + a_n \overline{z}^n$$

$$= f(\overline{z}) \quad \blacksquare$$

Because of Theorem 6.3.4, the nonreal roots of a polynomial in $\mathbf{R}[x]$ occur in pairs whenever they exist, so the number of nonreal roots of a polynomial with real coefficients must be an even integer. This tells us something about the real roots of a polynomial over $\mathbf{R}$ of odd degree:

Any polynomial in $\mathbf{R}[x]$ *of odd degree must have at least one real root.*

Furthermore, the number of real roots of such a polynomial must be odd. For example, if $f(x) \in \mathbf{R}[x]$ and $\deg(f(x)) = 5$, then $f(x)$ must have one real root and four nonreal roots, three real roots and two nonreal roots, or five real roots. Although at least one real root of a polynomial over $\mathbf{R}$ of odd degree always exists, there is no algebraic method that can always be used to find such a root of $f(x)$. The Fundamental Theorem of Algebra tells us that any polynomial in $\mathbf{R}[x]$ can be factored in $\mathbf{C}[x]$ into a product of linear factors. Now let's show that such a polynomial can be factored in $\mathbf{R}[x]$ into a product of linear and irreducible quadratic polynomials. To do this we need the following lemma.

6.3.5   **Lemma**

If the complex number $z = a + bi$, $b \neq 0$, is a root of $f(x) \in \mathbf{R}[x]$, then $x^2 - 2ax + (a^2 + b^2)$ is an irreducible quadratic factor of $f(x)$ in $\mathbf{R}[x]$.

Proof. If $z = a + bi$, $b \neq 0$, is a root of $f(x)$, then $\overline{z} = a - bi$ is also a root of $f(x)$. The corresponding factors of $f(x)$ are $x - a - bi$ and $x - a + bi$, so

$$(x - a - bi)(x - a + bi) = x^2 - 2ax + (a^2 + b^2)$$

is a factor of $f(x)$. Since $a, b \in \mathbf{R}$ and since both roots of $x^2 - 2ax + (a^2 + b^2)$ are not real numbers, $x^2 - 2ax + (a^2 + b^2)$ has no roots in $\mathbf{R}$. Thus, $x^2 - 2ax + (a^2 + b^2)$ is irreducible in $\mathbf{R}[x]$. $\blacksquare$

6.3.6   Theorem

Every polynomial in $\mathbf{R}[x]$ can be factored into a product of its leading coefficient and linear and/or irreducible quadratic polynomials in $\mathbf{R}[x]$.

Proof. Suppose that $f(x) \in \mathbf{R}[x]$ is such that $\deg(f(x)) = n$ is odd, and let a be the leading coefficient of $f(x)$. Since $\mathbf{R}[x] \subseteq \mathbf{C}[x]$, $f(x) \in \mathbf{C}[x]$. Corollary 6.3.2 indicates that $f(x)$ can be factored as

$$f(x) = a(x - c_1)(x - c_2) \cdots (x - c_n), \qquad (*)$$

where $c_1, c_2, \ldots, c_n \in \mathbf{C}$ are the n not necessarily distinct roots of $f(x)$. Since $\deg(f(x))$ is odd, $f(x)$ must have at least one real root. If all the roots of $f(x)$ are real numbers, then the factorization of $f(x)$ given in equation $(*)$ is a factorization of $f(x)$ into a product of linear polynomials in $\mathbf{R}[x]$. If one or more, but not all, of the roots are real numbers, then the roots $c_1, c_2, \ldots, c_n$ can be reindexed so that $c_1, c_2, \ldots, c_k$ are the real roots of $f(x)$ and $c_{k+1}, c_{k+2}, \ldots, c_n$ are the nonreal roots of $f(x)$. Because of Theorem 6.3.4, the nonreal roots of $f(x)$ occur in conjugate pairs, so suppose that when the roots of $f(x)$ are reindexed, the non-real roots are arranged so that $\overline{c_{k+1}} = c_{k+2}$, $\overline{c_{k+3}} = c_{k+4}, \ldots, \overline{c_{n-1}} = c_n$. Lemma 6.3.5 now shows that the result of each product indicated by the following groupings is an irreducible quadratic polynomial in $\mathbf{R}[x]$.

$$f(x) = a(x - c_1) \cdots (x - c_k)\underline{(x - c_{k+1})(x - c_{k+2})} \cdots \underline{(x - c_{n-1})(x - c_n)}$$

↑
This product produces an irreducible quadratic polynomial in $\mathbf{R}[x]$.

↑
This product produces an irreducible quadratic polynomial in $\mathbf{R}[x]$.

If $\deg(f(x))$ is even, then $f(x)$ may have a real root or it may not. In this case, if $f(x)$ has a real root, then the number of real roots must be even. An analysis similar to that just presented applies, and $f(x)$ can be factored into a product of its leading coefficient and linear factors in $\mathbf{R}[x]$ or into a product of its leading coefficient and linear and irreducible quadratic polynomials in $\mathbf{R}[x]$. If $f(x)$ has no real roots, then the linear factors in $\mathbf{C}[x]$ of $f(x)$ can be grouped in such a manner that the product of each grouping produces a quadratic polynomial that is irreducible in $\mathbf{R}[x]$. ∎

6.3.7   Examples

1. The degree of the polynomial $f(x) = 2x^3 + 11x^2 + 10x - 8 \in \mathbf{R}[x]$ is odd, so $f(x)$ must have at least one real root. In fact, $f(x)$ has three real roots, namely, $c_1 = -4$, $c_2 = -2$ and $c_3 = \frac{1}{2}$. Hence,

$$f(x) = 2(x+4)(x+2)\left(x-\tfrac{1}{2}\right)$$

is a factorization of $f(x)$ into a product of its leading coefficient 2 and 3 linear polynomials in **R**[*x*].

2. The degree of $f(x) = x^3 + \left(1 - \sqrt{2}\right)x^2 + \left(1 - \sqrt{2}\right)x - \sqrt{2} \in \mathbf{R}[x]$ is odd, so $f(x)$ must have at least one real root. The roots of $f(x)$ are $c_1 = \sqrt{2}$, $c_2 = -\tfrac{1}{2} + \tfrac{\sqrt{3}}{2}i$, and $c_3 = -\tfrac{1}{2} - \tfrac{\sqrt{3}}{2}i$. Hence,

$$f(x) = \left(x - \sqrt{2}\right)\left(x + \tfrac{1}{2} - \tfrac{\sqrt{3}}{2}i\right)\left(x + \tfrac{1}{2} + \tfrac{\sqrt{3}}{2}i\right)$$

$$= \left(x - \sqrt{2}\right)\left(x^2 + x + 1\right).$$

$f(x) = \left(x - \sqrt{2}\right)\left(x + \tfrac{1}{2} - \tfrac{\sqrt{3}}{2}i\right)\left(x + \tfrac{1}{2} + \tfrac{\sqrt{3}}{2}i\right)$ is the factorization of $f(x)$ into a product of linear polynomials in **C**[*x*], and $f(x) = \left(x - \sqrt{2}\right)(x^2 + x + 1)$ is the factorization of $f(x)$ into a product of a linear and an irreducible quadratic polynomial in **R**[*x*].

3. The polynomial $f(x) = x^4 - 2x^3 + 6x^2 - 8x + 8 \in \mathbf{R}[x]$ has no real roots. The roots of $f(x)$ are $c_1 = 2i$, $c_2 = -2i$, $c_3 = 1 + i$ and $c_4 = 1 - i$. The factorization of $f(x)$ in **C**[*x*] is

$$f(x) = (x - 2i)(x + 2i)(x - 1 - i)(x - 1 + i),$$

and the factorization of $f(x)$ in **R**[*x*] is

$$f(x) = (x^2 + 4)(x^2 - 2x + 2).$$

We have seen that a polynomial in **R**[*x*] can be factored into a product of linear and irreducible quadratic polynomials in **R**[*x*]. But this factorization is often not easily found. There is no simple algorithm that enables one to factor a polynomial in **R**[*x*]. However, there is one result concerning polynomials in **Q**[*x*] that is often helpful in finding roots of such a polynomial and, consequently, in factoring the polynomial. Before proving this result, observe that if $f(x)$ is a polynomial in **Q**[*x*], then $f(x)$ can be written as $f(x) = rg(x)$, where $r \in \mathbf{Q}$ and $g(x) \in \mathbf{Z}[x]$. For example, if $f(x) = \tfrac{1}{3}x^3 + \tfrac{7}{8}x + 5$, then $f(x) = \tfrac{1}{24}(8x^3 + 21x + 120)$. Moreover, the rational roots, if any exist, of $f(x)$ and $g(x) = 8x^3 + 21x + 120$ are exactly the same. Thus, if we can find the rational roots of the polynomial $g(x) \in \mathbf{Z}[x]$, then we have found the rational roots of $f(x) \in \mathbf{Q}[x]$.

6.3.8   Rational Root Theorem

Let $f(x) = a_0 + a_1 x + a_2 x^2 + \cdots + a_n x^n$ be a polynomial in $\mathbf{Z}[x]$, and suppose that $\frac{p}{q}$ is a nonzero rational number that has been reduced to lowest terms. If $\frac{p}{q}$ is a root of $f(x)$, then $p \mid a_0$ and $q \mid a_n$.

Proof. Let $\frac{p}{q}$ be a root of $f(x)$ that has been reduced to lowest terms. Then

$$a_0 + a_1 \frac{p}{q} + a_2 \left(\frac{p}{q}\right)^2 + \cdots + a_n \left(\frac{p}{q}\right)^n = 0.$$

Multiplying both sides of this equation by q^n produces

$$a_0 q^n + a_1 p q^{n-1} + a_2 p^2 q^{n-2} + \cdots + a_{n-1} p^{n-1} q + a_n p^n = 0,$$

so

$$a_0 q^n + a_1 p q^{n-1} + a_2 p^2 q^{n-2} + \cdots + a_{n-1} p^{n-1} q = -a_n p^n.$$

Factoring q from the terms on the left side of this equation shows that

$$q(a_0 q^{n-1} + a_1 p q^{n-2} + a_2 p^2 q^{n-3} + \cdots + a_{n-1} p^{n-1}) = -a_n p^n.$$

Hence, we see that $q \mid a_n p^n$. But $\frac{p}{q}$ is reduced to lowest terms, so p and q are relatively prime. Thus, $q \mid a_n p^n$ implies that $q \mid a_n$. A similar argument can be used to show that $p \mid a_0$. ∎

The Rational Root Theorem 6.3.8 provides a convenient method for describing the rational numbers that can be considered as candidates for roots of a polynomial $f(x) \in \mathbf{Z}[x]$. Any rational number $\frac{p}{q}$ that has been reduced to lowest terms and is a root of $f(x)$ must satisfy the condition that p divides the constant term of $f(x)$ and q divides the leading coefficient. This condition on p and q can be used to form the finite set of rational numbers, each of which has the possibility of being a root of $f(x)$. These numbers can then be tested to find the rational roots of $f(x)$. If it turns out that none of these rational numbers is a root of the polynomial, then the polynomial has no rational roots.

6.3.9   Examples

Find the rational roots of each of the following polynomials in $\mathbf{Q}[x]$.

1. $f(x) = x^3 + \frac{1}{2}x^2 + \frac{3}{2}x - 1$

Solution. First write $f(x)$ as $f(x) = \frac{1}{2}(2x^3 + x^2 + 3x - 2)$. Then the roots of $f(x)$ and of $g(x) = 2x^3 + x^2 + 3x - 2$ are exactly the same. The possible factors of the constant term of $g(x)$ are ± 1 and ± 2. Hence, $p \in \{\pm 1, \pm 2\}$. Likewise, the possible factors of the leading coefficient are ± 1 and ± 2, so $q \in \{\pm 1, \pm 2\}$. Hence, by the Rational Root Theorem 6.3.8, if $g(x)$ has a rational root $\frac{p}{q}$, then $\frac{p}{q} \in \left\{ \pm 1, \pm 2, \pm \frac{1}{2} \right\}$. First, 1 cannot be a root since the coefficients of $g(x)$ do not add to 0. It is also obvious that 2 is not a root of $g(x)$. Hence, we test $\frac{1}{2}$. Since $g\left(\frac{1}{2}\right) = 0$, $\frac{1}{2}$ is a root of $g(x)$, so $x - \frac{1}{2}$ is a factor of $g(x)$. Dividing $g(x)$ by $x - \frac{1}{2}$ produces a quotient of $2x^2 + 2x + 4$ and so $g(x) = \left(x - \frac{1}{2}\right)(2x^2 + 2x + 4)$. Use of the quadratic formula shows that the roots of $2x^2 + 2x + 4$ are nonreal numbers. Therefore, the only rational root of $g(x)$ is $\frac{1}{2}$, so the only rational root of $f(x)$ is $\frac{1}{2}$. In fact, $\frac{1}{2}$ is the only real root of $f(x)$. Hence, $f(x) = \frac{1}{2}\left(x - \frac{1}{2}\right)$ $(2x^2 + 2x + 4)$ is the factorization of $f(x)$ into irreducible factors in **R**[x].
The nonreal roots of $2x^2 + 2x + 4$ are $x = -\frac{1}{2} \pm \frac{\sqrt{7}}{2}i$, so
$$f(x) = \left(x - \frac{1}{2}\right)\left(x + \frac{1}{2} + \frac{\sqrt{7}}{2}i\right)\left(x + \frac{1}{2} - \frac{\sqrt{7}}{2}i\right)$$ is the factorization of $f(x)$ into a product of linear polynomials in **C** [x].

2. $f(x) = \frac{2}{3}x^4 + 2x^2 + \frac{4}{3}$

Solution. Since $f(x) = \frac{2}{3}(x^4 + 3x^2 + 2)$, the rational roots of $f(x)$ are exactly the same as the rational roots of $g(x) = x^4 + 3x^2 + 2$. Since all the powers of x in $g(x)$ are even, it is not possible for $x^4 + 2x^2 + 2$ to add to 0, no matter what rational number is substituted for x. Hence, $f(x)$ has no rational roots. For the same reason, $f(x)$ can have no real roots, so the four roots of $f(x)$ cannot be real numbers. Note that $f(x)$ factors as
$$f(x) = \frac{2}{3}(x^2 + 2)(x^2 + 1) = \frac{2}{3}(x + \sqrt{2}\,i)\,(x - \sqrt{2}\,i)(x + i)(x - i).$$
The first factorization of $f(x)$ is a product of irreducible quadratic polynomials in **R**[x], and the second is a product of linear polynomials in **C** [x].

The Rational Root Theorem 6.3.8 can also be used to show that certain real numbers are not rational numbers.

3. If s is a prime integer, show that $\sqrt{s}$ is not a rational number.

Solution. Consider the polynomial $f(x) = x^2 - s \in \mathbf{Z}[x]$. Factors of the constant term are ± 1 and $\pm s$, so $p \in \{\pm 1, \pm s\}$. Factors of the leading coefficient are ± 1 and so $q \in \{\pm 1\}$. Thus, by the Rational Root Theorem 6.3.8, a rational root $\frac{p}{q}$ of $f(x)$ must be an element of the set $\{\pm 1, \pm s\}$. Substitution into $f(x)$ shows that no element of this set is a root of $f(x)$. Hence, $f(x)$ has no rational roots. But $\sqrt{s}$ is a root of $f(x)$, so $\sqrt{s}$ cannot be a rational number.

Using the Rational Root Theorem 6.3.8 to find the rational roots of a polynomial in $\mathbf{Z}[x]$ works quite well, provided that the number of factors of the constant term and the leading coefficient are within reason. When the number of these factors is large and when the number of reduced ratios of these factors is also large, then a large number of possibilities may have to be tested before a rational root is found, if such a root exists at all. If a polynomial in $\mathbf{Z}[x]$ is irreducible in $\mathbf{Q}[x]$, then it can have no roots in $\mathbf{Q}$. Hence, it would be helpful to be able to test a given polynomial in $\mathbf{Z}[x]$ to see if it is irreducible in $\mathbf{Q}[x]$ before beginning a search for its rational roots. There is a test that works for some but (unfortunately) not very many polynomials in $\mathbf{Z}[x]$. It is known as **Eisenstein's Irreducibility Criterion.** The following two lemmas lay the groundwork necessary to establish this test.

6.3.10 Lemma

Let $f(x)$, $g(x)$, $h(x) \in \mathbf{Z}[x]$, and suppose that $f(x) = g(x)h(x)$. If p is a prime integer that divides every coefficient of $f(x)$, then p divides every coefficient of $g(x)$ or p divides every coefficient of $h(x)$.

Proof. Let

$$f(x) = a_0 + a_1 x + a_2 x^2 + \cdots + a_m x^m,$$
$$g(x) = b_0 + b_1 x + b_2 x^2 + \cdots + b_n x^n, \quad \text{and}$$
$$h(x) = c_0 + c_1 x + c_2 x^2 + \cdots + c_s x^s,$$

and suppose that the theorem is false. Then p divides every coefficient of $f(x)$, and there exist coefficients b_j and c_k of $g(x)$ and $h(x)$, respectively, such that $p \nmid b_j$ and

$p \nmid c_k$. Suppose also that j and k are the smallest nonnegative integers such that $p \nmid b_j$ and $p \nmid c_k$. Using the definition of multiplication of polynomials in $\mathbf{Z}[x]$, the coefficient of x^{j+k} in $g(x)h(x)$ is given by $b_0 c_{j+k} + \cdots + b_{j-1} c_{k+1} + b_j c_k + b_{j+1} c_{k-1} + \cdots + b_{j+k} c_0$. But $f(x) = g(x)h(x)$, so it follows that

$$a_{j+k} = b_0 c_{j+k} + \cdots + b_{j-1} c_{k+1} + b_j c_k + b_{j+1} c_{k-1} + \cdots + b_{j+k} c_0.$$

Hence,

$$b_j c_k = a_{j+k} - (b_0 c_{j+k} + \cdots + b_{j-1} c_{k+1} + b_{j+1} c_{k-1} + \cdots + b_{j+k} c_0). \qquad (**)$$

Since j is the smallest nonnegative integer such that $p \nmid b_j$, $p \mid b_0, p \mid b_1, \ldots, p \mid b_{j-1}$. Similarly, k is the smallest nonnegative integer such that $p \nmid c_k$, so $p \mid c_0, p \mid c_1, \ldots, p \mid c_{k-1}$. Since $p \mid a_{j+k}$, it follows that p divides the right side of equation $(**)$, so $p \mid b_j c_k$. But p is a prime, and consequently, $p \mid b_j$ or $p \mid c_k$, which is a contradiction. Thus, if a prime p divides every coefficient of $f(x)$, then p must divide every coefficient of $g(x)$ or every coefficient of $h(x)$. ∎

The following is an illustration of the technique that is used in the proof of the lemma immediately following the example.

6.3.11   Example

We are going to show that if a polynomial in $\mathbf{Z}[x]$ can be factored into a product of two polynomials in $\mathbf{Q}[x]$, then it can be factored into a product of two polynomials in $\mathbf{Z}[x]$. For example, the polynomial

$$f(x) = 3 - 5x - 57x^2 + 5x^3 + 150x^4$$

can be factored as $f(x) = g(x)h(x)$, where

$$g(x) = -\frac{9}{4} - \frac{3}{4}x + \frac{15}{2}x^2 \qquad \text{and} \qquad h(x) = -\frac{4}{3} + \frac{8}{3}x + 20x^2.$$

The least common multiple (see Exercise 14 of Problem Set 1.1) of the denominators of $g(x)$ and $h(x)$ are 4 and 3, respectively. Hence, $4g(x) = -9 - 3x + 30x^2$ and $3h(x) = -4 + 8x + 60x^2$ are polynomials in $\mathbf{Z}[x]$. Thus, we see that

$$2 \cdot 2 \cdot 3f(x) = (-9 - 3x + 30x^2)(-4 + 8x + 60x^2).$$

The coefficients of the second polynomial in this product are divisible by 4, so

$$3f(x) = (-9 - 3x + 30x^2)(-1 + 2x + 15x^2).$$

The coefficients of the first polynomial are divisible by 3. Consequently,

$$f(x) = (-3 - x + 10x^2)(-1 + 2x + 15x^2)$$

and so $f(x)$ can be factored into a product of two polynomials in $\mathbf{Z}[x]$.

6.3.12 Lemma

If $f(x)$ is a polynomial in $\mathbf{Z}[x]$ that can be factored as $f(x) = g(x)h(x)$ in $\mathbf{Q}[x]$, then there exist polynomials $g^*(x)$, $h^*(x) \in \mathbf{Z}[x]$ such that $f(x) = g^*(x)h^*(x)$. Moreover, $\deg(g(x)) = \deg(g^*(x))$ and $\deg(h(x)) = \deg(h^*(x))$.

Proof. Suppose that $f(x) = g(x)h(x)$, where $g(x)$, $h(x) \in \mathbf{Q}[x]$. If a and b are the least common multiples of the denominators of the coefficients of $g(x)$ and $h(x)$, respectively, then $g'(x) = ag(x)$ and $h'(x) = bh(x)$ have integral coefficients. Hence, $abf(x) = abg(x)h(x) = [ag(x)][bh(x)] = g'(x)h'(x)$. If p is a prime divisor of ab, then p divides every coefficient of $abf(x)$. By Lemma 6.3.10, p must divide every coefficient of either $g'(x)$ or $h'(x)$. If p divides every coefficient of $g'(x)$ and p is removed from every coefficient of $g'(x)$, the result is a polynomial with integral coefficients. The same observation holds if p divides every coefficient of $h'(x)$. Since ab factors into a finite number of primes, we can repeatedly divide both sides of $abf(x) = g(x)h(x)$ by the prime factors of ab and after a finite number of steps arrive at $f(x) = g^*(x)h^*(x)$, where $g^*(x)$ and $h^*(x)$ are polynomials in $\mathbf{Z}[x]$. It is obvious that the degrees of the polynomials $g(x)$ and $h(x)$ are not changed by these divisions. Hence, $\deg(g(x)) = \deg(g^*(x))$ and $\deg(h(x)) = \deg(h^*(x))$. ∎

We can now prove Eisenstein's Irreducibility Criterion mentioned earlier. **Ferdinand Gotthold Eisenstein (1823–1852)** was a German mathematician who is best known today for the following irreducibility theorem that bears his name.

6.3.13 Eisenstein's Irreducibility Criterion

Let $f(x) = a_0 + a_1x + \cdots + a_nx^n$ be a polynomial in $\mathbf{Z}[x]$ of positive degree n. If there exists a prime p such that

(i) $p \mid a_i$ for $i = 0, 1, 2, \ldots, n - 1$
(ii) $p \nmid a_n$ and
(iii) $p^2 \nmid a_0$,

then $f(x)$ is irreducible in $\mathbf{Q}[x]$.

Proof. If $f(x)$ is reducible in $\mathbf{Q}[x]$, then there exist polynomials $g(x)$ and $h(x)$ in $\mathbf{Q}[x]$ such that $f(x) = g(x)h(x)$. Moreover, $n = s + t$ with $1 \leq s, t \leq n$, where $\deg(g(x)) = s$ and $\deg(h(x)) = t$. Lemma 6.3.12 shows that there are $g^*(x)$ and $h^*(x)$ in $\mathbf{Z}[x]$ such that $f(x) = g^*(x)h^*(x)$ with $\deg(g(x)) = \deg(g^*(x))$ and $\deg(h(x)) = \deg(h^*(x))$. If

$$g^*(x) = b_0 + b_1 x + \cdots + b_s x^s \qquad \text{and}$$

$$h^*(x) = c_0 + c_1 x + \cdots + c_t x^t, \qquad \text{then}$$

$$a_0 + a_1 x + \cdots + a_n x^n = (b_0 + b_1 x + \cdots + b_s x^s)(c_0 + c_1 x + \cdots + c_t x^t). \qquad (*)$$

Suppose next that p is a prime that satisfies criteria (i) through (iii) of the theorem. Now $a_0 = b_0 c_0$, and since p is prime and $p \mid a_0$, p must divide either b_0 or c_0. We claim that p cannot divide both b_0 and c_0. If $p \mid b_0$ and $p \mid c_0$, then there are integers k_1 and k_2 such that $b_0 = pk_1$ and $c_0 = pk_2$. From this we see that $a_0 = b_0 c_0 = p^2 k_1 k_2$, so $p^2 \mid a_0$, which is a contradiction. Suppose that $p \mid b_0$ and $p \nmid c_0$. (A similar proof holds if $p \nmid b_0$ and $p \mid c_0$.) Since $a_n = b_s c_t$ and $p \nmid a_n$, $p \nmid b_s$ (and $p \nmid c_t$). Let k, $k \leq s < n$, be the smallest positive integer such that $p \nmid b_k$. Clearly such an integer exists since $p \nmid b_s$. Next, by equating the coefficient of x^k on both sides of the equation $(*)$, we have

$$a_k = b_0 c_k + b_1 c_{k-1} + \cdots + b_k c_0 \qquad \text{and thus}$$

$$b_k c_0 = a_k - b_0 c_k + b_1 c_{k-1} + \cdots + b_{k-1} c_1.$$

Since $p \mid a_k$ and k is the smallest positive integer such that $p \nmid b_k$, every term on the right side of this last equation is divisible by p. Hence, p divides $b_k c_0$, which is impossible since $p \nmid b_k$ and $p \nmid c_0$. Therefore, under the conditions stated in the theorem, $f(x)$ cannot be reducible in $\mathbf{Q}[x]$, so $f(x)$ is irreducible in $\mathbf{Q}[x]$. ∎

6.3.14   Examples

1. The polynomial $f(x) = 2 + 6x - 4x^2 + 3x^3$ is irreducible in $\mathbf{Q}[x]$ since the prime $p = 2$ divides 2, 6, and -4, but $2 \nmid 3$ and $2^2 \nmid 2$.
2. The polynomial $f(x) = 20 + 15x - 25x^2 + 35x^3 + 7x^4$ is irreducible in $\mathbf{Q}[x]$ since the prime $p = 5$ divides 20, 15, -25, and 35, but $5 \nmid 7$ and $5^2 \nmid 20$.
3. If n is a positive integer and p is a prime integer, then $f(x) = x^n - p$ is irreducible in $\mathbf{Q}[x]$. This follows since $p \mid p$ but $p \nmid 1$ and $p^2 \nmid p$. Hence, for every positive integer n, there are polynomials of degree n in $\mathbf{Z}[x]$ that are irreducible in $\mathbf{Q}[x]$.

Obviously, there are polynomials in $\mathbf{Z}[x]$ to which Eisenstein's Irreducibility Criterion cannot be applied. This criterion gives no information whatsoever concerning the irreducibility of $f(x) = 2 + 4x + 6x^2$ in $\mathbf{Q}[x]$ since no prime p exists such that

$p \mid 2$ and $p \mid 4$ but $p \nmid 6$ and $p^2 \nmid 2$. An application of the quadratic formula shows that the roots of $f(x)$ are complex numbers, so $f(x)$ is indeed irreducible in $\mathbf{Q}[x]$.

If Eisenstein's Criterion cannot be applied to a polynomial, it is sometimes the case that one can determine if a polynomial in $\mathbf{Z}[x]$ is irreducible in $\mathbf{Q}[x]$ by considering a polynomial in $\mathbf{Z}_p[x]$ for some prime p.

6.3.15   Definition

If $f(x) = a_0 + a_1 x + a_2 x^2 + \cdots + a_n x^n$ is a polynomial in $\mathbf{Z}[x]$ and p is a prime integer, then the polynomial

$$[f]_p(x) = [a_0] + [a_1]x + [a_2]x^2 + \cdots + [a_n]x^n \text{ in } \mathbf{Z}_p[x]$$

is referred to as the **polynomial in $\mathbf{Z}_p[x]$ corresponding to $f(x)$**.

6.3.16   Example

If $f(x) = 3 + 4x + 6x^2 + 8x^3$, then the polynomial corresponding to $f(x)$ in $\mathbf{Z}_3[x]$ is $[f]_3(x) = [3] + [4]x + [6]x^2 + [8]x^3 = x + [2]x^3$. The polynomial corresponding to $f(x)$ in $\mathbf{Z}_5[x]$ is $[f]_5(x) = [3] + [4]x + x^2 + [3]x^3$.

6.3.17   Theorem

Let $f(x) = a_0 + a_1 x + a_2 x^2 + \cdots + a_n x^n$ be a polynomial in $\mathbf{Z}[x]$, and suppose that p is a prime such that $p \nmid a_n$. If $[f]_p(x)$ is irreducible in $\mathbf{Z}_p[x]$, then $f(x)$ is irreducible in $\mathbf{Q}[x]$.

Proof. If $[f]_p(x)$ is irreducible in $\mathbf{Z}_p[x]$, then either $f(x)$ is irreducible in $\mathbf{Q}[x]$ or it is not. Suppose that $f(x)$ is reducible in $\mathbf{Q}[x]$. Then, by Lemma 6.3.12, there are polynomials $g(x), h(x) \in \mathbf{Z}[x]$, each with positive degree, such that $f(x) = g(x)h(x)$. Since a_n is the product of the leading coefficients of $g(x)$ and $h(x)$ and $p \nmid a_n$, p cannot divide the leading coefficient of $g(x)$ and p cannot divide the leading coefficient of $h(x)$. Therefore,

$$\deg([g]_p(x)) = \deg(g(x)) \qquad \text{and} \qquad \deg([h]_p(x)) = \deg(h(x)),$$

and it is not difficult to show that $[f]_p(x) = [g]_p(x)[h]_p(x)$. Hence, we have found polynomials $[g]_p(x), [h]_p(x) \in \mathbf{Z}_p[x]$, each with positive degree, that factor $[f]_p(x)$. But this contradicts the assumption that $[f]_p(x)$ is irreducible in $\mathbf{Z}_p[x]$. Thus, if $[f]_p(x)$ is irreducible in $\mathbf{Z}_p[x]$, then $f(x)$ must be irreducible in $\mathbf{Q}[x]$. ∎

6.3.18 | Examples

1. Consider the polynomial $f(x) = 9x^4 + 4x^3 - 3x + 7$. Since $2 \nmid 9$, we will try $p = 2$ for the prime of Theorem 6.3.17. The polynomial in $\mathbf{Z}_2[x]$ corresponding to $f(x)$ is $[f]_2(x) = x^4 + x + [1]$. Now $[f]_2([0]) = [1]$ and $[f]_2([1]) = [1]$, so $[f]_2(x)$ has no roots in $\mathbf{Z}_2$ and consequently no linear factors in $\mathbf{Z}_2[x]$. Hence, if $[f]_2(x)$ factors in $\mathbf{Z}_2[x]$, it must be the product of two quadratic polynomials in $\mathbf{Z}_2[x]$. The only quadratic polynomials in $\mathbf{Z}_2[x]$ are $x^2, x^2 + x, x^2 + [1]$, and $x^2 + x + [1]$. Long division can be used to show that none of these polynomials is a factor of $[f]_2(x)$, so $[f]_2(x)$ is irreducible in $\mathbf{Z}_2[x]$. Therefore, by Theorem 6.3.17, $f(x)$ is irreducible in $\mathbf{Q}[x]$.
2. Consider the polynomial $f(x) = x^4 + 10x^3 + 5x^2 + 4x + 7$. The polynomial in $\mathbf{Z}_2[x]$ corresponding to $f(x)$ is $[f]_2(x) = x^4 + x^2 + [1]$. Since $[f]_2([0]) = [1]$ and $[f]_2([1]) = [1]$, $[f]_2(x)$ has no roots in $\mathbf{Z}_2$, so $[f]_2(x)$ has no linear factors in $\mathbf{Z}_2[x]$. Consequently, if $[f]_2(x)$ factors in $\mathbf{Z}_2[x]$, then both factors must be of degree 2. As noted in Example 1, the only quadratic polynomials in $\mathbf{Z}_2[x]$ are $x^2, x^2 + x, x^2 + [1]$, and $x^2 + x + [1]$. Since $(x^2 + x + [1])^2 = [f]_2(x)$, the polynomial corresponding to $f(x)$ in $\mathbf{Z}_2[x]$ is reducible in $\mathbf{Z}_2[x]$. Hence, no information can be drawn about the irreducibility of $f(x)$ in $\mathbf{Q}[x]$ when using $p = 2$ for the prime of Theorem 6.3.17. The point of this example is that even though the polynomial in $\mathbf{Z}_2[x]$ corresponding to $f(x)$ is reducible in $\mathbf{Z}_2[x]$, this does not imply that $f(x)$ is reducible in $\mathbf{Q}[x]$. In fact, $f(x)$ is actually irreducible in $\mathbf{Q}[x]$. Read Theorem 6.3.17 and take careful note of this difference.

SUMMARY

General Facts.

1. If $f(x) \in R[x]$, then $c \in R$ is a root of $f(x)$ if and only if $x - c$ is a factor of $f(x)$.
2. If R is not a field and $f(x) \in R[x]$ is of degree n, then it is possible for $f(x)$ to have more than n roots in R.
3. If F is a field and $f(x) \in F[x]$ is of degree n, then $f(x)$ can have at most n roots in F.
4. If $f(x) \in F[x]$ is of degree 1, then $f(x)$ is irreducible in $F[x]$.

Polynomials in **Z**[x] *and* **Q**[x].

5. If a nonzero rational number $\frac{p}{q}$, reduced to lowest terms, is a root of $f(x) \in \mathbf{Z}[x]$, then p divides the constant term of $f(x)$ and q divides the leading coefficient.

6. If a prime p exists such that p divides the coefficients of $f(x) \in \mathbf{Z}[x]$ except for the leading coefficient and p^2 does not divide the constant term, then $f(x)$ is irreducible in $\mathbf{Q}[x]$.

Polynomials in $\mathbf{R}[x]$.

7. If $z \in \mathbf{C}$ is a root of $f(x) \in \mathbf{R}[x]$, then so is $\bar{z}$, the conjugate of z.

8. The number of nonreal roots of a polynomial $f(x) \in \mathbf{R}[x]$ is an even integer.

9. Any polynomial $f(x) \in \mathbf{R}[x]$ can be factored in $\mathbf{R}[x]$ into a product of linear and irreducible quadratic polynomials.

10. If $f(x) \in \mathbf{R}[x]$ is of odd degree, then $f(x)$ must have at least one root in $\mathbf{R}$.

Polynomials in $\mathbf{C}[x]$.

11. If $f(x) \in \mathbf{C}[x]$ is of degree n, then $f(x)$ has at most n roots in $\mathbf{C}$. If each root is counted according to its multiplicity, then the number of roots is exactly n.

12. If $f(x) \in \mathbf{C}[x]$ and $\deg(f(x)) = n$, then $f(x)$ can be factored in $\mathbf{C}[x]$ into a product $f(x) = a(x - c_1)(x - c_2) \cdots (x - c_n)$ of linear polynomials, where a is the leading coefficient of $f(x)$ and $c_1, c_2, \ldots, c_n$ are the n not necessarily distinct roots of $f(x)$ in $\mathbf{C}$.

HISTORICAL NOTES

Ferdinand Gotthold Eisenstein (1823–1852). Eisenstein was born in Berlin, Germany, in 1823, and he suffered from very poor health throughout his brief life. After reading an article by Hamilton on Abel's work, Eisenstein was stimulated to do research in mathematics. A student of Gauss, Eisenstein worked on a variety of topics that included quadratic and cubic forms, and he made contributions to matrix theory. During the latter years of his life, Eisenstein's health deteriorated until he was confined to bed. This, however, did not diminish his interest in mathematical research, and he continued to publish mathematics papers until the time of his death at the age of 29.

Problem Set 6.3

1. In each of the following, a root of the polynomial is given. Find all of the remaining roots of $f(x)$ in $\mathbf{C}$.

    (a) $f(x) = -9 + 18x - x^2 + 2x^3$, $c = \dfrac{1}{2}$

    (b) $f(x) = -15 + 17x - 7x^2 + x^3$, $c = 2 + i$

    (c) $f(x) = 351 - 162x + 66x^2 - 18x^3 + 3x^4$, $c = 3 - 2i$

    (d) $f(x) = -4 + 8x - 6x^2 - 4x^3 + 4x^4$, $c = \sqrt{2}$

2. Find a polynomial in **R**[x] with the given roots and given degree.

   (a) $c_1 = \frac{1}{2}$ and $c_2 = 2 - i$ of degree 3
   (b) $c_1 = 1 + \sqrt{2}\,i$ and $c_2 = \frac{1}{2} - \frac{2}{3}i$ of degree 4
   (c) $c_2 = 3 - 2i$, $c_3 = 2 + 3i$ of degree 5 and with any real root

3. Factor each of the following polynomials into a product of irreducible polynomials in **R**[x].

   (a) $f(x) = x^4 - 1$
   (b) $f(x) = x^3 - x^2 - 3x + 6$
   (c) $f(x) = x^4 + x^3 + 2x^2 + x + 1$

4. Prove that $f(x) \in$ **R**[x] is irreducible in **R**[x] if and only if $f(x)$ is a linear polynomial or $f(x) = ax^2 + bx + c$, $a \neq 0$, with $b^2 - 4ac < 0$.

5. In Exercise 10 of Problem Set 4.4, it was shown that a quadratic equation $ax^2 + bx + c = 0$ with its coefficients in **C** has solutions given by

$$x = \frac{-b + \sqrt[c]{b^2 - 4ac}}{2a},$$

   where $\sqrt[c]{b^2 - 4ac}$ denotes the complex square roots of $b^2 - 4ac$. Use this to find the roots of each of the following polynomials in **C**[x].

   (a) $f(x) = x^2 + ix + 1$
   (b) $f(x) = x^2 - x + 1$
   (c) $f(x) = x^2 + (2 - 3i)x - (5 + i)$

6. If $f(x) \in$ **R**[x] and $z = a + bi$ is a root of $f(x)$, then we know that $\bar{z} = a - bi$ is also a root of $f(x)$. Find a polynomial $f(x) \in$ **C**[x] with degree greater than 1 such that a complex number z is a root of $f(x)$ but $\bar{z}$ is not.

7. Use the Rational Root Theorem 6.3.8 to write each of the following polynomials in **Z**[x] as a product of irreducible polynomials in **Q**[x].

   (a) $f(x) = -2 - x + 2x^2 + x^3$
   (b) $f(x) = -4 + 13x^2 + 6x^3$
   (c) $f(x) = 7 + 33x + 25x^2 - 31x^3 + 6x^4$

8. Use Eisenstein's Irreducibility Criterion 6.3.13 to show that each of the following polynomials in **Z**[x] is irreducible in **Q**[x].

   (a) $f(x) = 6 + 4x^2 - 5x^4$
   (b) $f(x) = 7 + 35x - 21x^2 + 14x^3 + 2x^7$
   (c) $f(x) = 6 + 12x - 36x^2 + 3x^3$

9. Find a prime p such that for each of the following polynomials the corresponding polynomial $[f]_p[x]$ is irreducible in $\mathbf{Z}_p[x]$.

    (a) $f(x) = 7 - 3x + 4x^3 + 9x^4$
    (b) $f(x) = 6 + 4x + 6x^2 + 7x^4$
    (c) Can you conclude that the polynomials in (a) and (b) are irreducible in $\mathbf{Q}[x]$? Why?

10. Let $f(x) = a_0 + a_1x + a_2x^2 + \cdots + a_nx^n$ be a polynomial in $\mathbf{Z}[x]$, and suppose that p is a prime such that $p \mid a_1, p \mid a_2, \ldots, p \mid a_n$ but $p \nmid a_0$ and $p^2 \nmid a_n$. Prove that $f(x)$ is irreducible in $\mathbf{Q}[x]$. Hint: Let $x = \dfrac{1}{y}$, consider $g(y) = y^n f\left(\dfrac{1}{y}\right) \in \mathbf{Z}[y]$, and apply Eisenstein's Irreducibility Criterion to $g(y)$. Use this to show that $f(x) = 3 + 8x - 12x^3 + 2x^4$ is irreducible in $\mathbf{Q}[x]$.

11. If p is a prime, show that for any integer $n \geq 2$, $\sqrt[n]{p}$ is an irrational number.

TECHNOLOGY PROBLEMS

Use a computer algebra system to solve each of the following problems.

12. Factor each of the following polynomials into a product of linear and/or irreducible quadratic polynomials in $\mathbf{R}[x]$. Which of the polynomials can be factored into a product of linear factors in $\mathbf{Z}[x]$? Are any of the polynomials irreducible in $\mathbf{Z}[x]$, $\mathbf{Q}[x]$, or $\mathbf{R}[x]$?

    (a) $f(x) = 12x^6 + 41x^5 - 143x^4 - 251x^3 - 49x^2 + 30x$
    (b) $g(x) = x^5 + x^4 + x^3 + x^2 + x + 1$
    (c) $h(x) = x^3 + x^2 + x + 2$
    (d) $k(x) = x^7 + x^6 - 5x^5 - 5x^4 + 2x^3 + 2x^2 + 8x + 8$

13. Find the rational roots, if any, of each of the following polynomials.

    (a) $f(x) = 4x^4 - 20x^2 + 24$
    (b) $g(x) = 48x^5 + 100x^4 - 52x^3 - 127x^2 + x + 30$
    (c) $h(x) = 40x^5 + 32x^4 + 22x^3 - 16x^2 - 8x + 2$

14. The polynomial $f(x) = x^3 - (3 + \sqrt{2})x + 2 + \sqrt{2}$ is of odd degree and so must have at least one real root. Use your computer algebra system to find the real root(s) of $f(x)$.

■7

Modular Arithmetic in $F[x]$ and Unique Factorization Domains

*When F is a field, F[x] and **Z** share many properties that may not be enjoyed by integral domains in general. For example, both have a unique factorization property by which their elements can be expressed as a product of other elements in the domain that are in some sense unique. Both of these domains also have a division algorithm that, among other things, allows one to compute the greatest common divisor of a pair of their elements. A congruence-class arithmetic can also be developed for F[x] that is quite similar to the congruence-class arithmetic developed in Section 1.3 for the integers. We begin with congruence-class arithmetic in F[x].*

7.1   MODULAR ARITHMETIC IN F [x]

Section Overview. In this section we show that mod $f(x)$ is a congruence relation on $F[x]$, and the associated congruence class arithmetic is developed. It is shown that when $f(x)$ is irreducible in $F[x]$, $F[x] / (f(x))$ is a field that contains a subfield that is isomorphic to F. Throughout this section F is a field.

7.1.1   Definition

Let $f(x)$, $g(x)$, $h(x) \in F[x]$, and suppose that $f(x)$ is a nonzero polynomial. If $g(x) - h(x)$ is divisible by $f(x)$, then $g(x)$ is said to be **congruent to** $h(x)$ **modulo** $f(x)$. The expression $g(x) \equiv h(x)(\mathrm{mod}\ f(x))$ indicates that $g(x)$ is congruent to $h(x)$ modulo $f(x)$.

7.1.2   Examples

1. If $f(x) = 2x^2 + 1$, $g(x) = 6x^4 + 4x^2 - 3x + \frac{5}{2}$ and $h(x) = 2x^3 + \frac{3}{2}x^2 - 2x + \frac{11}{4}$, then $f(x)$, $g(x)$, $h(x) \in \mathbf{Q}[x]$ and $g(x) \equiv h(x)\ (\mathrm{mod}\ f(x))$ in $\mathbf{Q}[x]$ since $g(x) - h(x) =$
$$6x^4 - 2x^3 + \frac{5}{2}x^2 - x - \frac{1}{4} = (2x^2 + 1)\left(3x^2 - x - \frac{1}{4}\right).$$

2. If $f(x) = [2]x^2 + [1]$, $g(x) = [2]x^3 + x^2 + [4]$, and $h(x) = [3]x^2 + [4]x$ are in $\mathbf{Z}_5[x]$, then $g(x) \equiv h(x)(\mathrm{mod}\ f(x))$ in $\mathbf{Z}_5[x]$ since $g(x) - h(x) = [2]x^3 + [3]x^2 + x + [4] = ([2]x^2 + [1])(x + [4])$.

Recall that the modular relation mod n induces a congruence relation on $\mathbf{Z}$. The same is true of the relation mod $f(x)$ on $F[x]$. We first show that mod $f(x)$ is an equivalence relation on $F[x]$.

7.1.3   Theorem

If $f(x)$ is a nonzero polynomial in $F[x]$, then the relation mod $f(x)$ is an equivalence relation on $F[x]$.

Proof. It is obvious that the relation mod $f(x)$ on $F[x]$ is reflexive and symmetric. It is also transitive, for if $g(x)$, $h(x)$, and $k(x)$ are in $F[x]$ and such that $g(x) \equiv h(x)(\mathrm{mod}\ f(x))$ and $h(x) \equiv k(x)(\mathrm{mod}\ f(x))$, then $g(x) - k(x) = (g(x) - h(x)) + (h(x) - k(x))$. Since $f(x)$ divides each summand on the right side of this last equation, $f(x) \mid (g(x) - k(x))$. Thus, $g(x) \equiv k(x)\ (\mathrm{mod}\ f(x))$, so transitivity holds. Therefore, mod $f(x)$ is an equivalence relation on $F[x]$. ∎

7.1.4   Definition

If $f(x)$ is a nonzero polynomial in $F[x]$, the **equivalence class** determined by $g(x) \in F[x]$ and the relation mod $f(x)$ is the set $\overline{g(x)} = \{h(x) \in F[x] \mid g(x) \equiv h(x)(\mathrm{mod}\ f(x))\}$.

The equivalence class $\overline{g(x)}$ of Definition 7.1.4 can be expressed in other useful ways: $\overline{g(x)} = \{h(x) \in F[x] | g(x) = h(x) + a(x)f(x) \text{ for some } a(x) \in F[x]\}$ or as $\overline{g(x)} = \{h(x) | f(x) | (g(x) - h(x))\}$. Because of Theorem 0.2.3, the following corollary to Theorem 7.1.3 is immediate.

7.1.5   Corollary

The equivalence relation mod $f(x)$ on $F[x]$ has the following properties.

1. Either $\overline{g(x)} = \overline{h(x)}$ or $\overline{g(x)} \cap \overline{h(x)} = \varnothing$ for any $g(x), h(x) \in F[x]$.
2. For any $g(x), h(x) \in F[x]$, $g(x) \equiv h(x)(\bmod f(x))$ if and only if $\overline{g(x)} = \overline{h(x)}$.

In order for mod $f(x)$ to be a congruence relation on $F[x]$, this relation must preserve both addition and multiplication of polynomials in $F[x]$. The following theorem shows that this is indeed the case.

7.1.6   Theorem

If $f(x)$ is a nonzero polynomial in $F[x]$, then the equivalence relation mod $f(x)$ is a congruence relation on $F[x]$. That is, if $g_1(x), g_2(x), h_1(x), h_2(x) \in F[x]$ are such that $g_1(x) \equiv h_1(x) \ (\bmod f(x))$ and $g_2(x) \equiv h_2(x) \ (\bmod f(x))$, then

$$g_1(x) + g_2(x) \equiv h_1(x) + h_2(x)(\bmod f(x)) \qquad \text{and}$$
$$g_1(x)g_2(x) \equiv h_1(x)h_2(x)(\bmod f(x)).$$

Proof. Let $a_1(x), a_2(x) \in F[x]$ be such that (1) $g_1(x) - h_1(x) = a_1(x)f(x)$ and (2) $g_2(x) - h_2(x) = a_2(x)f(x)$. Adding equations (1) and (2) gives

$$g_1(x) - h_1(x) + g_2(x) - h_2(x) = a_1(x)f(x) + a_2(x)f(x) \qquad \text{and thus}$$
$$(g_1(x) + g_2(x)) - (h_1(x) + h_2(x)) = (a_1(x) + a_2(x))f(x).$$

Hence, $g_1(x) + g_2(x) \equiv h_1(x) + h_2(x) \ (\bmod f(x))$.

Next, multiply both sides of equation (1) by $g_2(x)$ and both sides of equation (2) by $h_1(x)$. This produces

$$g_1(x)g_2(x) - h_1(x)g_2(x) = a_1(x)f(x)g_2(x) \qquad \text{and}$$
$$h_1(x)g_2(x) - h_1(x)h_2(x) = h_1(x)a_2(x)f(x).$$

By adding the last two equations, we see that

$$g_1(x)_2g(x) - h_1(x)h_2(x) = (a_1(x)g_2(x) + h_1(x)a_2(x))f(x).$$

Therefore, $g_1(x)_2g(x) \equiv h_1(x)h_2(x) \ (\bmod f(x))$. ■

Recall that for the congruence relation mod n on the ring of integers, if $m \in \mathbf{Z}$, then $[m] = $ [remainder of the division of m by n]. Hence, there are $n - 1$ distinct congruence classes $[0], [1], \ldots, [n - 1]$ determined by the congruence relation mod n on $\mathbf{Z}$. A similar result holds for the congruence relation mod $f(x)$ on $F[x]$, with the focus of the proof shifting to the degree of the polynomials.

7.1.7   Lemma

If $f(x), g(x), h(x) \in F[x]$, $f(x) \neq 0$, then $g(x) \equiv h(x) (\mathrm{mod}\ f(x))$ if and only if $g(x)$ and $h(x)$ have the same remainder $r(x)$ when divided by $f(x)$. Furthermore, $\overline{r(x)} = \overline{g(x)} = \overline{h(x)}$.

Proof. If $g(x) \equiv h(x)\ (\mathrm{mod}\ f(x))$, there is an $a(x) \in F[x]$ such that $g(x) - h(x) = a(x)f(x)$. By the Division Algorithm for Polynomials 6.1.11, there exist $q_1(x), q_2(x), r_1(x), r_2(x) \in F[x]$ such that $g(x) = q_1(x)f(x) + r_1(x)$, where $r_1(x) = 0$ or $\deg(r_1(x)) < \deg(f(x))$, and $h(x) = q_2(x)f(x) + r_2(x)$, where $r_2(x) = 0$ or $\deg(r_2(x)) < \deg(f(x))$. Suppose that $r_1(x) \neq r_2(x)$. Then

$$a(x)f(x) = g(x) - h(x) = (q_1(x) - q_2(x))f(x) + (r_1(x) - r_2(x)),$$

which implies that $(a(x) - q_1(x) + q_2(x))f(x) = r_1(x) - r_2(x)$. Therefore, $f(x)$ divides $r_1(x) - r_2(x)$, but this cannot be the case since $\deg(r_1(x) - r_2(x)) < \deg(f(x))$. Hence, $r_1(x) = r_2(x)$, so $g(x)$ and $h(x)$ have the same remainder when divided by $f(x)$.

Conversely, suppose that $g(x)$ and $h(x)$ have the same remainder $r(x)$ when divided by $f(x)$. Then $g(x) = q(x)f(x) + r(x)$ for some $q(x) \in F[x]$, which implies that $g(x) - r(x) = q(x)f(x)$. Hence, $g(x) \equiv r(x)(\mathrm{mod}\ f(x))$. Similarly, $h(x) \equiv r(x)(\mathrm{mod}\ f(x))$, so because the equivalence relation mod $f(x)$ is symmetric and transitive, we see that $g(x) \equiv h(x)\ (\mathrm{mod}\ f(x))$.

If $g(x)$ and $h(x)$ have the same remainder $r(x)$ when divided by $f(x)$, we have just shown that $g(x) \equiv r(x)(\mathrm{mod}\ f(x))$, $h(x) \equiv r(x)(\mathrm{mod}\ f(x))$, and $g(x) \equiv h(x)(\mathrm{mod}\ f(x))$. It now follows from Corollary 7.1.5 that $\overline{r(x)} = \overline{g(x)} = \overline{h(x)}$. ∎

To simplify notation, the principal ideal $f(x)F[x] = \{f(x)g(x)\,|\,g(x) \in F(x)\}$ is denoted by $(f(x))$, and $F[x]/(f(x))$ denotes the set of all congruence classes modulo $f(x)$. Since the relation mod $f(x)$ on $F[x]$ is a congruence relation, Theorem 0.4.8 shows that addition and multiplication of congruence classes are well-defined binary operations on $F[x]/(f(x))$. The straightforward proof of the following theorem is left as an exercise.

7.1.8 Theorem

If $f(x)$ is a nonconstant polynomial in $F[x]$, then $F[x]/(f(x))$ is a commutative ring with identity under congruence-class addition and multiplication. Furthermore, $\phi : F \to F[x]/(f(x))$ defined by $\phi(a) = \overline{a}$ is an injective ring homomorphism that embeds F in $F[x]/(f(x))$.

7.1.9 Examples

1. Consider the polynomial ring $\mathbf{Z}_2[x]$, and let $f(x) = x^2 + x + [1]$. By Lemma 7.1.7, the distinct congruence classes in $\mathbf{Z}_2[x]/(f(x))$ are determined by the remainders of the division of polynomials in $\mathbf{Z}_2[x]$ by $f(x)$. Hence, there are four distinct congruence classes in $\mathbf{Z}_2[x]/(f(x))$ since a remainder of such a division must be the zero polynomial or it can have degree of at most 1. The only polynomials in $\mathbf{Z}_2[x]$ that satisfy these conditions are $[0]$, $[1]$, x, and $x + [1]$, so the congruence classes in $F[x]/(f(x))$ are $\overline{[0]}$, $\overline{[1]}$, $\overline{x}$, and $\overline{x + [1]}$. The addition and multiplication tables for the ring $\mathbf{Z}_2[x]/(f(x))$ are given below:

+	$\overline{[0]}$	$\overline{[1]}$	$\overline{x}$	$\overline{x + [1]}$
$\overline{[0]}$	$\overline{[0]}$	$\overline{[1]}$	$\overline{x}$	$\overline{x + [1]}$
$\overline{[1]}$	$\overline{[1]}$	$\overline{[0]}$	$\overline{x + [1]}$	$\overline{x}$
$\overline{x}$	$\overline{x}$	$\overline{x + [1]}$	$\overline{[0]}$	$\overline{[1]}$
$\overline{x + [1]}$	$\overline{x + [1]}$	$\overline{x}$	$\overline{[1]}$	$\overline{[0]}$

•	$\overline{[0]}$	$\overline{[1]}$	$\overline{x}$	$\overline{x + [1]}$
$\overline{[0]}$	$\overline{[0]}$	$\overline{[0]}$	$\overline{[0]}$	$\overline{[0]}$
$\overline{[1]}$	$\overline{[0]}$	$\overline{[1]}$	$\overline{x}$	$\overline{x + [1]}$
$\overline{x}$	$\overline{[0]}$	$\overline{x}$	$\overline{x + [1]}$	$\overline{[1]}$
$\overline{x + [1]}$	$\overline{[0]}$	$\overline{x + [1]}$	$\overline{[1]}$	$\overline{x}$

To construct the multiplication table, we have used the fact that the remainder of the division of x^2 by $x^2 + x + [1]$ in $\mathbf{Z}_2[x]$ is $x + [1]$. We also arrive at the same conclusion if we consider $\overline{x^2 + x + [1]} = \overline{[0]}$ in

$\mathbf{Z}_2[x] / (f(x))$. In $\mathbf{Z}_2[x] / (f(x))$, $\overline{x^2 + x + [1]} = \overline{[0]}$ implies that

$$\overline{x}^2 + \overline{x} + \overline{[1]} = \overline{[0]} \text{ or } \overline{x}^2 = -\overline{[1]}\,\overline{x} - \overline{[1]} = \overline{x} + \overline{[1]} = \overline{x + [1]}.$$

Hence,

$$\overline{x}^2 = \overline{x + [1]} \qquad \text{and thus}$$

$$\overline{x}(\overline{x + [1]}) = \overline{x}^2 + \overline{x} = \overline{x + [1]} + \overline{x} = \overline{[1]} \qquad \text{and}$$

$$(\overline{x + [1]})^2 = \overline{x^2 + [2]x + [1]} = \overline{x}^2 + \overline{[1]}$$

$$= \overline{x + [1]} + \overline{[1]} = \overline{x}.$$

In the preceding example, every nonzero element of $F[x] / (f(x))$ has a multiplicative inverse, so $F[x] / (f(x))$ is a field. The following example shows that this is not always the case. The difference is that in Example 1, $f(x)$ is irreducible in $\mathbf{Z}_2[x]$. ($f(x)$ has no roots in $\mathbf{Z}_2[x]$.) In the following example, we consider $\mathbf{Z}_2[x] / (g(x))$, where $g(x)$ is reducible in $\mathbf{Z}_2[x]$.

2. Consider $g(x) = x^2 + [1]$ in $\mathbf{Z}_2[x]$. $[0], [1], x,$ and $x + [1]$ are the only polynomials that can be the remainder of the division of a polynomial in $\mathbf{Z}_2[x]$ by $g(x)$. Consequently, the elements of $\mathbf{Z}_2[x] / (g(x))$ are $\overline{[0]}, \overline{[1]}, \overline{x},$ and $\overline{x + [1]}$. The addition table for $\mathbf{Z}_2[x] / (g(x))$ is the same as in Example 1. Since $\overline{x^2 + [1]} = \overline{[0]}$ in $\mathbf{Z}_2[x] / (g(x))$, we see that $\overline{x}^2 = -\overline{[1]} = \overline{[1]}$, so

$$\overline{x}(\overline{x + [1]}) = \overline{x}^2 + \overline{x} = \overline{x} + \overline{[1]} = \overline{x + [1]} \qquad \text{and}$$

$$(\overline{x + [1]})^2 = \overline{x}^2 + [2]\overline{x} + [1] = \overline{[1]} + \overline{[1]} = \overline{[0]}.$$

The multiplication table for $\mathbf{Z}^2[x] / (g(x))$ is given below.

$\bullet$	$\overline{[0]}$	$\overline{[1]}$	$\overline{x}$	$\overline{x + [1]}$
$\overline{[0]}$	$\overline{[0]}$	$\overline{[0]}$	$\overline{[0]}$	$\overline{[0]}$
$\overline{[1]}$	$\overline{[0]}$	$\overline{[1]}$	$\overline{x}$	$\overline{x + [1]}$
$\overline{x}$	$\overline{[0]}$	$\overline{x}$	$\overline{[1]}$	$\overline{x + [1]}$
$\overline{x + [1]}$	$\overline{[0]}$	$\overline{x + [1]}$	$\overline{x + [1]}$	$\overline{[0]}$

Inspection of the table shows that $\overline{x + [1]}$ does not have a multiplicative inverse in $\mathbf{Z}_2[x] / (g(x))$, so $\mathbf{Z}_2[x] / (g(x))$ is not a field. Note that $g(x)$ is reducible in $\mathbf{Z}_2[x]$. Indeed, $g(x) = (x + [1])^2 = x^2 + [1]$ in $\mathbf{Z}_2[x]$.

In Example 1, $\mathbf{Z}_2[x] / (f(x))$ is a field, while in Example 2, $\mathbf{Z}_2[x] / (g(x))$ is not. The interesting point is that $F[x] / (f(x))$ is a field if and only if $f(x)$ is irreducible in $F[x]$. The following theorem clarifies this situation. Compare the theorem with Corollary 3.2.17.

7.1.10   **Theorem**

If $f(x)$ is a polynomial in $F[x]$ of positive degree, then the following are equivalent:
1. The polynomial $f(x)$ is irreducible in $F[x]$.
2. The ring $F[x] / (f(x))$ is a field.
3. The ring $F[x] / (f(x))$ is an integral domain.

Proof

$1 \Rightarrow 2$.   We have already seen in Theorem 7.1.8 that $F[x] / (f(x))$ is a commutative ring with identity. Hence, to show that $F[x] / (f(x))$ is a field, we need only show that every nonzero element of $F[x] / (f(x))$ has a multiplicative inverse in $F[x] / (f(x))$. If $\overline{g(x)}$ is a nonzero element of $F[x] / (f(x))$, let $r(x)$ be the remainder of the division of $g(x)$ by $f(x)$. Then, by Lemma 7.1.7, $\overline{r(x)} = \overline{g(x)}$. Since $f(x)$ is irreducible in $F[x]$, $r(x)$ and $f(x)$ are relatively prime. Therefore, by Theorem 6.2.14, there are polynomials $a(x)$ and $b(x)$ in $F[x]$ such that $e = a(x)f(x) + b(x)r(x)$. Hence,

$$\bar{e} = \overline{a(x)f(x) + b(x)r(x)} = \overline{a(x)}\ \overline{f(x)} + \overline{b(x)}\ \overline{r(x)}.$$

But $\overline{f(x)} = \bar{0}$ in $F[x] / (f(x))$, so

$$\bar{e} = \overline{b(x)}\ \overline{r(x)} = \overline{b(x)}\ \overline{g(x)}.$$

Thus, every nonzero element of $F[x] / (f(x))$ has a multiplicative inverse in $F[x] / (f(x))$, and so $F[x] / (f(x))$ is a field.

$2 \Rightarrow 3$.   This is obvious since every field is an integral domain.

$3 \Rightarrow 1$.   Let $F[x] / (f(x))$ be an integral domain, and suppose that $f(x)$ is reducible in $F[x]$. Then $f(x)$ must have degree of at least 2, and there exist polynomials $g(x)$ and $h(x)$ in $F[x]$, each with positive degree, such that $f(x) = g(x)h(x)$. But then $\bar{0} = \overline{f(x)} = \overline{g(x)h(x)} = \overline{g(x)}\ \overline{h(x)}$. Now F is a field and so $\deg(f(x)) = \deg(g(x)) + \deg(h(x))$. Thus, $\deg(g(x)) < \deg(f(x))$ and $\deg(h(x)) < \deg(f(x))$. Hence, neither $g(x)$ nor $h(x)$ is divisible by $f(x)$, so $\overline{g(x)} \neq 0$ and $\overline{h(x)} \neq 0$. Therefore, $F[x] / (f(x))$ has zero-divisors when $f(x)$ is reducible in $F[x]$. But this cannot be the case since we have assumed that $F[x] / (f(x))$ is an integral domain. Therefore, if $F[x] / (f(x))$ is an integral domain, then $f(x)$ must be irreducible in $F[x]$. ∎

7.1.11 Corollary

If $f(x)$ is irreducible in $F[x]$, then $F[x]/(f(x))$ is a field that contains a subfield isomorphic to F.

Proof. This follows immediately from Theorems 7.1.8 and 7.1.10. ∎

7.1.12 Example

The polynomial $x^2 + 1$ is irreducible in $\mathbf{R}[x]$, so $\mathbf{R}[x]/(x^2 + 1)$ is a field that contains a subfield isomorphic to $\mathbf{R}$. We claim that $\mathbf{R}[x]/(x^2 + 1)$ is isomorphic to $\mathbf{C}$, the field of complex numbers. Since $\overline{x^2 + 1} = \overline{0}$ in $\mathbf{R}[x]/(f(x))$, it follows that $\overline{x}^2 = -\overline{1}$. Furthermore, $x^2 + 1$ is a polynomial of degree 2, so each equivalence class in $\mathbf{R}[x]/(x^2 + 1)$ can be represented by a polynomial with degree at most 1 or by the zero polynomial. The mapping $\phi : \mathbf{C} \to \mathbf{R}[x]/(x^2 + 1)$ defined by $\phi(a + bi) = \overline{a + bx}$ is a ring isomorphism, and the complex number i corresponds to the congruence class $\overline{x}$ under this mapping. Note that $i^2 = -1$ in $\mathbf{C}$ and $\overline{x}^2 = -\overline{1}$ in $\mathbf{R}[x]/(f(x))$.

Problem Set 7.1

1. For the given field F, let $f(x), g(x), h(x) \in F[x]$. Determine if $g(x) \equiv h(x) \pmod{f(x)}$.

   (a) $f(x) = x^2 + 1$, $g(x) = 2 - x + x^2 + x^3$, $h(x) = 2 - 2x + x^2$ in $\mathbf{Q}[x]$
   (b) $f(x) = x^2 + x$, $g(x) = 5 + 2x + 8x^2 + 5x^3 + x^4$, $h(x) = 5 - x + x^2$ in $\mathbf{Q}[x]$
   (c) $f(x) = [2]x + [1]$, $g(x) = [3]x^3 + [4]x + [2]$, $h(x) = [4]x^2 + x$ in $\mathbf{Z}_5[x]$
   (d) $f(x) = x + \sqrt{3}$
   $g(x) = 5 + 2\sqrt{3} + 2\sqrt{6}x + \left(2\sqrt{2} + \sqrt{3}\right)x^2 + x^3$
   $h(x) = 5 + \sqrt{3} - x$ in $\mathbf{R}[x]$

2. For each of the polynomials $f(x)$ in $F[x]$, construct addition and multiplication tables for $F[x]/(f(x))$, and determine if $F[x]/(f(x))$ is a field.

   (a) $f(x) = x^2 + x$ in $\mathbf{Z}_2[x]$
   (b) $f(x) = x^2$ in $\mathbf{Z}_2[x]$

(c) $f(x) = x^3 + x + [1]$ in $\mathbf{Z}_2[x]$

3. If $f(x)$ is a nonzero constant polynomial in $F[x]$, show that there is exactly one congruence class in $F[x] / (f(x))$.

4. If $a \in F$, describe the congruence classes in the field $F[x] / (x - a)$.

5. Let $f(x)$ be a polynomial in $\mathbf{Z}_3[x]$ of degree 4. How many distinct congruence classes are there in $\mathbf{Z}_3[x] / (f(x))$? If $f(x)$ is of degree n, how many distinct congruence classes are there in $\mathbf{Z}_p[x] / (f(x))$ if p is a prime?

6. If $g(x)$ and $f(x)$ are relatively prime in $F[x]$ and $g(x)h(x) \equiv g(x)z(x) \pmod{f(x)}$, show that $h(x) \equiv z(x) \pmod{f(x)}$.

7. Prove Theorem 7.1.8.

8. If $f(x)$ and $g(x)$ are relatively prime polynomials in $F[x]$, prove that $\overline{g(x)}$ has a multiplicative inverse in $F[x] / (f(x))$.

9. Show that the mapping ϕ of Example 7.1.12 is a well-defined ring isomorphism.

10. Explain why $\mathbf{Q}[x] / (x^2 - 2)$ is a field, and then prove that the fields $\mathbf{Q}[\sqrt{2}\,]$ and $\mathbf{Q}[x] / (x^2 - 2)$ are isomorphic.

11. When F is a field, prove that every ideal of $F[x]$ looks like $(f(x))$ for some $f(x) \in F[x]$; that is, prove that $F[x]$ is a principal ideal ring.

12. Show that $\mathbf{Z}_2[x] / (x^3 + x + 1)$ is a field, and then show that $\mathbf{Z}_2[x] / (x^3 + x + 1)$ contains the three roots of $x^3 + x + 1$.

TECHNOLOGY PROBLEMS

Use a computer algebra system to solve each of the following problems.

13. Determine which, if any, of the following modular relations are valid.
    (a) If $f(x) = x^3 + 2x - 5$,
    $\qquad g_1(x) = x^4 + 4x^3 + 2x^2 + 3x - 20$,
    $\qquad g_2(x) = x^6 + 4x^4 - 10x^3 + 4x^2 - 20x + 25$,
    $\qquad g_3(x) = x^5 - 6x^4 + 11x^3 - 17x^2 + 48x - 45$, and
    $\qquad g_4(x) = 6x^4 - 5x^3 - 102x^2 - 109x - 30$,
    $\qquad$ then $g_1(x)g_2(x) \equiv g_3(x)g_4(x) \pmod{f(x)}$ in $\mathbf{R}[x]$.
    (b) If $f(x) = [3]x^2 + [2]x + [1]$,
    $\qquad g_1(x) = [2]x^3 + [3]x + [1]$,
    $\qquad g_2(x) = x^5 + [3]x^3 + [2]x + [2]$,
    $\qquad g_3(x) = x^2 + [3]$, and
    $\qquad g_4(x) = x^3 + [2]x + [3]$
    $\qquad$ then $g_1(x)g_2(x) \equiv g_3(x)g_4(x) \pmod{f(x)}$ in $\mathbf{Z}_4[x]$.

(c) If $f(x) = 3x^2 + 2$,
$$g_1(x) = 2x^2 + 3,$$
$$g_2(x) = 3x^2 + 4x + 1,$$
$$g_3(x) = 2x^2 + x + 2, \qquad \text{and}$$
$$g_4(x) = 3x^2 + 2x + 4,$$
then $g_1(x)g_2(x) \equiv g_3(x)g_4(x) \pmod{f(x)}$ in $\mathbf{Z}_5[x]$.

14. (a) Show that $f(x) = x^3 + 2x^2 + 3$ is irreducible in $\mathbf{Z}_5[x]$.

 (b) Show that $\mathbf{Z}_5[x] / (f(x))$ contains all the roots of $f(x)$.

7.2   EUCLIDEAN DOMAINS, PRINCIPAL IDEAL DOMAINS, UNIQUE FACTORIZATION DOMAINS

Section Overview. When F is a field, $\mathbf{Z}$ and $F[x]$ have several properties in common that are not shared by integral domains in general. Both have a Euclidean Division Algorithm, both are principal ideal domains, and both have a unique factorization property by which their elements can be expressed as a product of elements in the domain that are in some sense unique. Because of the computational advantages these domains have over other integral domains, it is worthwhile to investigate integral domains that share similar properties. It is the purpose of this section to study such domains, namely, Euclidean Domains, Principal Ideal Domains, and Unique Factorization Domains. It will be shown that every Euclidean Domain is a Principal Ideal Domain and that every Principal Ideal Domain is a Unique Factorization Domain.

Many interesting concepts in ring theory result from an attempt to generalize properties of rings such as $\mathbf{Z}$ and the ring of polynomials $F[x]$ over a field F. For example, in the ring $F[x]$, the following two conditions are satisfied:

1. For all nonzero polynomials $f(x), g(x) \in F[x]$, $\deg(f(x)) \leq \deg(f(x)g(x))$.

2. If $f(x), g(x) \in F[x]$, $g(x) \neq 0$, then there are polynomials $q(x), r(x) \in F[x]$ such that $f(x) = g(x)q(x) + r(x)$, where $r(x) = 0$ or $\deg(r(x)) < \deg(g(x))$.

The desire to generalize these observations concerning the polynomial ring $F[x]$ to other integral domains leads to the concept of a Euclidean Domain.

7.2.1   Definition

If D is an integral domain, a mapping v of the nonzero elements of D to the set of nonnegative integers $\mathbf{N}_0$ is said to be a **Euclidean valuation** on D if the following two properties are satisfied.

> 1. For all nonzero $a, b \in D$, $v(a) \le v(ab)$.
> 2. If $a, b \in D$, $b \ne 0$, then there exist $q, r \in D$ such that $a = bq + r$, where $r = 0$ or $v(r) < v(b)$.
>
> An integral domain D together with a Euclidean valuation defined on D is said to be a **Euclidean Domain**.

7.2.2   **Examples**

1. The ring **Z** is a Euclidean Domain if the function v is defined by $v(a) = |a|$ for each nonzero integer a. Note that if $a, b \in$ **Z** are both nonzero, then $|a| \le |ab|$, so $v(a) \le v(ab)$. For any $a, b \in$ **Z**, $b \ne 0$, the Division Algorithm for the integers shows that there are integers q and r such that $a = bq + r$, with $r = 0$ or $0 < r < b$, so $r = 0$ or $v(r) < v(b)$.

2. For any field F, $F[x]$ is a Euclidean Domain under the valuation v defined by $v(f(x)) = \deg(f(x))$ for each nonzero polynomial $f(x)$ in $F[x]$. Properties of the degree function and the Division Algorithm for Polynomials 6.1.11 show that $F[x]$ is indeed a Euclidean Domain.

3. Every field F is a Euclidean Domain if v is defined on the nonzero elements of F by $v(a) = 1$. If $a, b \in F$, then obviously $v(a) \le v(ab)$, and if $b \ne 0$, then $a = b(b^{-1}a) + 0$.

4. The ring **Z**$[i] = \{a + bi \mid a, b \in$ **Z**$\}$ of **Gaussian integers** is an integral domain since it is a subring of the field **C** of complex numbers. The domain **Z**$[i]$ is a Euclidean Domain if v is defined on the nonzero elements of **Z**$[i]$ by $v(a + bi) = a^2 + b^2$. This follows because if $x = a + bi$ and $y = c + di$ are in **Z**$[i]$, then $a, b, c,$ and d are integers and

$$a^2 + b^2 \le (a^2 + b^2)(c^2 + d^2).$$

Hence, $v(x) \le v(xy)$, so the first condition of Definition 7.2.1 holds. Next, let's show that the second condition of Definition 7.2.1 holds. If $y \ne 0$, then y has a multiplicative inverse in **Q**$[i]$. Indeed,

$$y^{-1} = \frac{1}{c+di} = \frac{c-di}{(c+di)(c-di)} = \frac{c}{c^2+d^2} - \frac{d}{c^2+d^2}\,i \in \textbf{Q}[i].$$

From this it is easily verified that xy^{-1} is a complex number in **Q**$[i]$. Suppose $xy^{-1} = c' + d'i$, where $c', d' \in$ **Q**. Since c' and d' are rational numbers, each lies between two consecutive integers. Let m and n be the two integers that are closest to c' and d', respectively.

Then $|c' - m| \leq \frac{1}{2}$ and $|d' - n| \leq \frac{1}{2}$. Hence,

$$x = y(c' + d'i)$$
$$= y[(c' - m + m) + (d' - n + n)i]$$
$$= y[m + ni] + y[(c' - m) + (d' - n)i]$$
$$= yq + r,$$

where $q = m + ni$ and $r = y[(c' - m) + (d' - n)i]$. Since $q \in \mathbf{Z}[i]$, we will have shown that the second condition of Definition 7.2.1 holds if we can show that $r \in \mathbf{Z}[i]$ and either $r = 0$ or $v(r) < v(y)$. Since $r = x - yq$ and $x, y, q \in \mathbf{Z}[i]$, it follows immediately that $r \in \mathbf{Z}[i]$. Finally, note that

$$v(r) = v([c + di][(c' - m) + (d' - n)i])$$
$$= v([c(c' - m) - d(d' - n)] + [c(d' - n) + d(c' - m)]i)$$
$$= [c(c' - m) - d(d' - n)]^2 + [c(d' - n) + d(c' - m)]^2$$
$$= (c^2 + d^2)[(c' - m)^2 + (d' - n)^2]$$
$$= v(y)[(c' - m)^2 + (d' - n)^2].$$

But $(c' - m)^2 + (d' - n)^2 \leq \left(\frac{1}{2}\right)^2 + \left(\frac{1}{2}\right)^2$. Hence, if $r \neq 0$, then $v(r) \leq \frac{1}{2}v(y) < v(y)$. Therefore, $\mathbf{Z}[i]$ is a Euclidean Domain.

An example of what was developed in the preceding proof might be beneficial at this point. If you examine the proof carefully, you will see that it tells us how to find q and r. Suppose that $x = 3 + 4i$ and $y = 1 - 5i$. Then $y^{-1} = \frac{1}{26} + \frac{5}{26}i$ and $xy^{-1} = -\frac{17}{26} + \frac{19}{26}i$. Now the integers closest to $-\frac{17}{26}$ and $\frac{19}{26}$ are -1 and 1, respectively. Hence, $q = -1 + i$. Since $r = x - yq$, we see that $r = (3 + 4i) - (1 - 5i)(-1 + i) = -1 - 2i$. Note also that $v(r) = 5$ and $v(y) = 26$, which shows that $v(r)$ is less than one half of $v(y)$.

Recall that an ideal I in a commutative ring with identity is a **principal ideal** if I is generated by a single element, that is, if $I = (a) = \{ab \mid b \in R\}$. An integral domain in which every ideal is a principal ideal is called a **Principal Ideal Domain**. It was shown in Example 1 of 7.2.2 that $\mathbf{Z}$ is a Euclidean Domain, while Theorem 3.2.15 shows that $\mathbf{Z}$ is a Principal Ideal Domain. We now demonstrate that this is always the case. We prove that if an integral domain is a Euclidean Domain, then it is a Principal Ideal Domain.

7.2.3   Theorem

Every Euclidean Domain is a Principal Ideal Domain.

Proof.   Suppose that D is a Euclidean Domain with Euclidean valuation v, and let I be an ideal of D. We claim that I is a principal ideal. If $I = \{0\}$, there is nothing to prove since $I = (0)$. If $I \neq \{0\}$, then I contains at least one nonzero element. Let $S = \{v(a) \,|\, a \in I, a \neq 0\}$. Then S is a nonempty subset of the well-ordered set $\mathbf{N}_0$, so S has a smallest element. Let $b \in I$ be such that $v(b)$ is minimal in S. Then $b \neq 0$, so if $a \in I$, then there exist $q, r \in D$ such that $a = bq + r$, where either $r = 0$ or $v(r) < v(b)$. If $r \neq 0$, then $r = a - bq \in I$ since I is an ideal of D and $a, b \in I$. But then $v(r) < v(b)$ contradicts the minimality of $v(b)$. Thus, it must be the case that $r = 0$, so $a = bq \in (b)$. Hence, $I \subseteq (b)$. The reverse containment is obvious since $b \in I$, and so $I = (b)$. ∎

7.2.4   Corollary

If F is any field, then $F[x]$ is a Principal Ideal Domain.

Proof.   We have seen in Example 2 of 7.2.2 that $F[x]$ is a Euclidean Domain, so Theorem 7.2.3 shows that $F[x]$ is a Principal Ideal Domain. ∎

7.2.5   Theorem

The following are equivalent for any ring F.

1. F is a field.
2. $F[x]$ is a Euclidean Domain.
3. $F[x]$ is a Principal Ideal Domain.

Proof. $1 \Rightarrow 2$ is Example 2 of 7.2.2 and $2 \Rightarrow 3$ is Theorem 7.2.3. Hence, we need show only that $3 \Rightarrow 1$. If $a \in F$, $a \neq 0$, consider the set $I = (a, x) = \{ag(x) + xh(x) \,|\, g(x), h(x) \in F(x)\}$. It is not difficult to show that I is an ideal of $F[x]$. (I is the ideal generated by a and x.) Since $F[x]$ is a Principal Ideal Domain, there is an $f(x) \in F[x]$ such that $I = (f(x))$. Now $a, x \in I$, so there are $g(x), h(x) \in F[x]$ such that $f(x)g(x) = a$ and $f(x)h(x) = x$. From $\deg(f(x)) + \deg(g(x)) = \deg(f(x)g(x)) = \deg(a) = 0$, we see that $\deg(f(x)) = \deg(g(x)) = 0$. Thus, $f(x)$ and $g(x)$ are nonzero constant polynomials. If $f(x) = b$, then $bh(x) = x$, and if we let $x = e$ in this last equation, we see that $bc = e$, where $h(e) = c \in F$. Hence, b has a multiplicative inverse in F, and this implies that $I = (f(x)) = (b) = F[x]$. Therefore, $e \in I$, so $e = ag_1(x) + xh_1(x)$ for a pair of polynomials $g_1(x), h_1(x) \in F[x]$. If we now let $x = 0$, then $ad = e$, where $g_1(0) = d \in F$. Hence, $a \neq 0$ has a multiplicative inverse and so F is a field. ∎

7.2.6   Corollary

$\mathbf{Z}[x]$ is not a Principal Ideal Domain.

Proof. If $\mathbf{Z}[x]$ is a Principal Ideal Domain, Theorem 7.2.5 indicates that $\mathbf{Z}$ is a field, and we know this is not the case. Thus, $\mathbf{Z}[x]$ cannot be a Principal Ideal Domain. ∎

7.2.7   Example

As an additional illustration that $\mathbf{Z}[x]$ is not a Principal Ideal Domain, consider the ideal I composed of all the polynomials in $\mathbf{Z}[x]$ whose constant term is an integral multiple of 3. (You should show that I is an ideal in $\mathbf{Z}[x]$.) We claim that I is not a principal ideal. Suppose to the contrary that I is a principal ideal of $\mathbf{Z}[x]$. Then there is a polynomial $f(x)$ in $\mathbf{Z}[x]$ such that $I = (f(x))$. Since $h(x) = 3$ is in I, there is a polynomial $g(x)$ in $\mathbf{Z}[x]$ such that $3 = f(x)g(x)$. Thus, $0 = \deg(3) = \deg(f(x)) + \deg(g(x))$, so it follows that $\deg(f(x)) = 0$. Therefore, $f(x)$ is a constant polynomial that must look like $f(x) = 3k$ for some integer k. But $3 + x \in I$, so there must be a polynomial $p(x) \in \mathbf{Z}[x]$ such that $3 + x = f(x)p(x) = 3kp(x)$. But this is clearly impossible, so I cannot be a principal ideal. Thus, $\mathbf{Z}[x]$ is not a Principal Ideal Domain.

Theorem 7.2.3 shows that every Euclidean Domain is a Principal Ideal Domain. The converse does not necessarily hold.

7.2.8   Example

It is known that $\mathbf{Z}[\sqrt{-19}] = \{a + b\sqrt{-19} \mid a$ and b are both even or both odd integers} is a Principal Ideal Domain, but not a Euclidean Domain. We will not attempt to prove this result since a proof would take us too far afield from our purpose. We simply accept this result as valid and indicate that it shows that a Principal Ideal Domain need not be a Euclidean Domain. Additional information on this domain can be found in the article "A Principal Ideal Ring That Is Not a Euclidean Ring," published in 1973 in *Mathematics Magazine*, volume 46, pages 34–38.

In Definition 6.2.20, units, associates, irreducible elements, and reducible elements were defined for polynomial domains. These concepts are now extended to more general integral domains, along with an explanation of what it means for an element of an integral domain to be prime.

7.2.9   Definition

Let D be an integral domain. An element $u \in D$ is a **unit** in D if u has a multiplicative inverse in D. If $a, b \in D$ and there is a unit $u \in D$ such that $a = bu$, then a is said to be an **associate** of b. If $a = bu$ indicates that a is an associate of b, then b is an associate of a since $b = au^{-1}$. Hence, we often simply say that a and b are associates. A nonzero element of an integral domain D is said to be **irreducible** in D if it is not a unit in D and its only divisors in D are units and its associates. Equivalently, a nonzero, nonunit element $a \in D$ is irreducible in D if, whenever $a = bc$, either b or c is a unit in D. If a nonzero, nonunit element of D is not irreducible, then it is said to be a **reducible** element of D. A nonzero element p of an integral domain is said to be **prime** if p is not a unit and whenever $p \mid ab$, $p \mid a$ or $p \mid b$.

If p is a prime element of an integral domain, one can show by induction that if $p \mid a_1 a_2 \cdots a_n$, then p divides at least one of the a_i's. We leave the proof as an exercise.

7.2.10   Examples

1. In $\mathbf{R}[x]$, the polynomials $\frac{2}{3}x - \frac{4}{3}$ and $14x - 28$ are associates since $14x - 28 = 21\left(\frac{2}{3}x - \frac{4}{3}\right)$ and 21 is a unit in $\mathbf{R}[x]$. Similarly, $6x^2 + 18x + 45$ and $\frac{2}{5}x^2 + \frac{6}{5}x + 3$ are associates in $\mathbf{R}[x]$.

2. The definition of a prime element in an integral domain is taken from one of the properties of a prime integer (Corollary 1.1.16). Recall that an integer p is defined to be prime in $\mathbf{Z}$ if $p > 1$ and the only divisors of p are ± 1 and $\pm p$. When the integral domain under consideration is $\mathbf{Z}$, the definitions of a prime given in Definitions 1.1.12 and 7.2.9 do not determine exactly the same elements of $\mathbf{Z}$. If p is a prime in $\mathbf{Z}$ in the sense of Definition 1.1.12, then both p and $-p$ are prime in $\mathbf{Z}$ in the sense of Definition 7.2.9.

3. All nonzero elements of a field F are units, and every pair of nonzero elements are associates. Indeed, if $a, b \in F$ are both nonzero, then $b^{-1}a$ is a unit in F and $a = b(b^{-1}a)$. Since every nonzero element in a field F is a unit, there are no irreducible elements in F.

7.2.11   Lemma

If p is a prime element of an integral domain D, then p is irreducible in D.

Proof. Let p be a prime element of D, and suppose that $p = ab$. Then $p \mid ab$, so $p \mid a$ or $p \mid b$. If $p \mid a$, then $a = pk$ for some element k of D. But then $p = ab = (pk)b = p(kb)$, so $kb = e$ since D is an integral domain. Thus, b is a unit. Similarly, if $p \mid b$, then a is a unit, so in either case $p = ab$ implies that either a or b is a unit. Hence, p is irreducible. ∎

We have just seen that in an integral domain a prime element is also irreducible. The following example shows that the converse need not hold; that is, there exist integral domains and irreducible elements in these domains that are not prime.

7.2.12 Example

The ring $\mathbf{Z}[\sqrt{-5}]$ is a subring of the field $\mathbf{C}$ of complex numbers, so $\mathbf{Z}[\sqrt{-5}]$ is an integral domain. We claim that 3 is an irreducible element of this domain. First, note that 3 is not a unit in $\mathbf{Z}[\sqrt{-5}]$. If 3 were a unit, there would be an element $a + b\sqrt{-5} \in \mathbf{Z}[\sqrt{-5}]$ such that $3(a + b\sqrt{-5}) = 1$. This implies that $3a = 1$, which is impossible since a is an integer. Now suppose that 3 is reducible. Then there are elements $a + b\sqrt{-5}, c + d\sqrt{-5} \in \mathbf{Z}[\sqrt{-5}]$ such that $3 = (a + b\sqrt{-5})(c + d\sqrt{-5})$. Taking conjugates, we see that $3 = \overline{3} = \overline{(a + b\sqrt{-5})}\,\overline{(c + d\sqrt{-5})} = (a - b\sqrt{-5})(c - d\sqrt{-5})$, so we have

$$9 = 3 \cdot \overline{3}$$
$$= (a + b\sqrt{-5})(a - b\sqrt{-5})(c + d\sqrt{-5})(c - d\sqrt{-5})$$
$$= (a^2 + 5b^2)(c^2 + 5d^2).$$

Now 9 factors in $\mathbf{Z}$ as $(-3)(-3)$, $3 \cdot 3$, $(-1)(-9)$, or $1 \cdot 9$. Since $a^2 + 5b^2$ and $c^2 + 5d^2$ are positive integers, it must be the case that

$$a^2 + 5b^2 = 3 \quad \text{and} \quad c^2 + 5d^2 = 3 \qquad \text{or} \qquad (1)$$
$$a^2 + 5b^2 = 1 \quad \text{and} \quad c^2 + 5d^2 = 9 \qquad \text{or} \qquad (2)$$
$$a^2 + 5b^2 = 9 \quad \text{and} \quad c^2 + 5d^2 = 1. \qquad (3)$$

It is obvious that there are no integers that satisfy equations (1). Equations (2) are satisfied if and only if $b = 0$, $a = \pm 1$, $d = 0$, and $c = \pm 3$. But if $a = \pm 1$ and $b = 0$, then $a + b\sqrt{-5}$ is a unit. Similarly, if equations (3) are satisfied by integers, then $c = \pm 1$ and $d = 0$ so that $c + d\sqrt{-5}$ is a unit. Thus, 3 is irreducible in $\mathbf{Z}[\sqrt{-5}]$. Finally, we claim that 3 is not prime in $\mathbf{Z}[\sqrt{-5}]$. Since

$3 \mid 6$ and $6 = (1 - \sqrt{-5})(1 + \sqrt{-5})$, it must be the case that $3 \mid (1 - \sqrt{-5})(1 + \sqrt{-5})$. But if 3 is prime, then $3 \mid (1 - \sqrt{-5})$ or $3 \mid (1 + \sqrt{-5})$. If $3 \mid (1 - \sqrt{-5})$, then there is an element $c + d\sqrt{-5} \in \mathbf{Z}[\sqrt{-5}]$ such that $1 - \sqrt{-5} = 3(c + d\sqrt{-5}) = 3c + 3d\sqrt{-5}$. But this implies that $3c = 1$ for some integer c, and this is clearly impossible. Similarly, $3 \nmid (1 + \sqrt{-5})$, so 3 cannot be prime in $\mathbf{Z}[\sqrt{-5}]$.

In an integral domain, a prime element must be irreducible. However, the preceding example shows that the converse does not hold. The following theorem shows, however, that these two concepts are equivalent in a Principal Ideal Domain. To prove the theorem, we need the following lemma. The proof of the lemma is straightforward and is left as an exercise.

7.2.13   Lemma

If a and b are elements of an integral domain D, then $(a) = (b)$ if and only if a and b are associates.

7.2.14   Theorem

If D is a Principal Ideal Domain, then an element of D is prime if and only if it is irreducible.

Proof.   Let D be a Principal Ideal Domain. Because of Lemma 7.2.11, if p is a prime element of D, then p is irreducible. Hence, it remains only to prove that if p is irreducible, then p is prime. Let $p \in D$ be irreducible and suppose that $p \mid ab$. To show that p is prime, we must show that $p \mid a$ or $p \mid b$. Consider the ideal $I = \{px + by \mid x, y \in D\}$. Since D is a Principal Ideal Domain, there is an element $c \in D$ such that $(c) = I$. Since $p = pe + b0$ is in I, p can be expressed as $p = cd$ for some $d \in D$. But p is irreducible, so either c or d must be a unit. If c is a unit, it follows that $e \in I$ and so $I = D$. Hence, $e = px' + by'$ for some $x', y' \in D$. But then $a = apx' + aby'$, so $p \mid a$ since $p \mid apx'$ and $p \mid aby'$. Hence, when $p = cd$ and c is a unit, $p \mid a$. Now suppose that $p = cd$ and that d is a unit. Then by Lemma 7.2.13, $(p) = (c) = I$. Thus, $b = p0 + be \in I = (p)$, so $b = pk$ for some $k \in D$. Therefore $p \mid b$. Hence, $p \mid ab$ implies that $p \mid a$ or $p \mid b$, and thus p is prime. ■

7.2.15   Definition

An integral domain D is said to be a **Unique Factorization Domain** if each nonzero, nonunit element $a \in D$ can be expressed as a product $a = p_1 p_2 \cdots p_m$ of irreducible elements of D. Furthermore, if $a = q_1 q_2 \cdots q_n$

is another such factorization of a as a product of irreducible elements of D, then $m = n$, and after a suitable reindexing of the q_k's (or the p_k's if you like), p_k and q_k are associates for $k = 1, 2, ..., n$.

7.2.16 Examples

1. **Z** is a Unique Factorization Domain since every nonzero integer can be factored into a product of primes and ± 1's. This factorization is unique up to rearrangement of factors and the insertion of the units ± 1. For example, 12 can be factored as $12 = 2 \cdot 2 \cdot 3 = -2 \cdot 3 \cdot -2$. But after rearranging the factors as $2 \cdot 2 \cdot 3 = -2 \cdot -2 \cdot 3$, we see that the factors 2 and $-2 = -1 \cdot 2$ are associates since -1 is a unit in **Z**.

2. The ring **R**$[x]$ is a Unique Factorization Domain. For instance, $f(x) = x^2 - x - 6$ factors as $f(x) = (x - 3)(x + 2)$, and this factorization of $f(x)$ is unique up to associates. For example, $f(x) = (2x + 4)\left(\frac{1}{2}x - \frac{3}{2}\right)$, but $x + 2$ and $2x + 4 = 2(x + 2)$ are associates, as are $x - 3$ and $\frac{1}{2}x - \frac{3}{2} = \frac{1}{2}(x - 3)$, since 2 and $\frac{1}{2}$ are units in **R**.

3. There are integral domains that are *not* Unique Factorization Domains. For example, consider **Z**$[\sqrt{-6}]$. Then

$$5 \cdot 2 = 10 = \left(2 + \sqrt{-6}\right)\left(2 - \sqrt{-6}\right).$$

Using the same techniques as those used in Example 7.2.12, it can be shown that $2, 5, 2 + \sqrt{-6}$, and $2 - \sqrt{-6}$ are irreducible in **Z**$[\sqrt{-6}]$. Moreover, neither 2 nor 5 is an associate of either $2 + \sqrt{-6}$ or $2 - \sqrt{-6}$. Hence, **Z**$[\sqrt{-6}]$ is not a Unique Factorization Domain, and consequently, not every integral domain is a Unique Factorization Domain. A more detailed treatment of this example is presented in the next section.

It has been shown that every Euclidean Domain is a Principal Ideal Domain. Our task now is to show that every Principal Ideal Domain is a Unique Factorization Domain. To do this, we need to consider ascending chains of ideals in an integral domain and what it means for such a chain to have finite length. A ring (which may not be commutative) in which every ascending chain of right (left) ideals has finite length is said to be a right (left) **Noetherian ring**. These rings are named after **Emmy Noether** (1882–1935), who made significant advances in research in ring theory using ascending chains of right ideals. Her work played an important role in the development of ring theory as a major mathematical topic.

In the Principal Ideal Domain **Z**, one can show that if $(a_1) \subseteq (a_2) \subseteq \cdots$ is an **ascending chain of ideals** of **Z**, then there is an integer k such that $(a_1) \subseteq (a_2) \subseteq \cdots \subseteq (a_k) = (a_{k+1}) = (a_{k+2}) = \cdots$, that is, every ascending chain of ideals of **Z** must have finite length. For example,

$$(270) \subseteq (54) \subseteq (27) \subseteq (9) \subseteq (3),$$

and no proper ideal of **Z** can contain (3) since (3) is a maximal ideal of **Z** (Theorem 3.2.16). Hence, this chain of ideals of **Z** must terminate with (3) and so is of finite length. We now make these ideas more precise and then show that every ascending chain of ideals must have finite length not only in **Z**, but in every Principal Ideal Domain.

7.2.17   Definition

> If $I_1 \subseteq I_2 \subseteq I_3 \subseteq \cdots$ is an **ascending chain of ideals** in an integral domain D, then we say that the chain has **finite length** if there is a positive integer k such that $I_n = I_k$ for all $n \geq k$. An integral domain D is said to satisfy the **ascending chain condition** or to be a **Noetherian Domain** if every ascending chain of ideals of D has finite length.

7.2.18   Theorem

Every Principal Ideal Domain is a Noetherian Domain.

Proof. Let D be a Principal Ideal Domain, and suppose that $(a_1) \subseteq (a_2) \subseteq (a_3) \subseteq \cdots$ is an ascending chain of ideals of D. If $I = \cup_{i=1}^{\infty}(a_i)$, then I is an ideal of D. This follows easily, for if $a, b \in I$, then $x \in (a_m)$ and $y \in (a_n)$ for some positive integers m and n. If $m \leq n$, then $(a_m) \subseteq (a_n)$, so x and y are in (a_n). Hence, $x - y \in (a_n) \subseteq I$, and thus I is a subgroup of D. Similarly, if $a \in D$ and $x \in I$, then $x \in (a_n)$ for some positive integer n. Thus, $ax \in (a_n) \subseteq I$ and I is an ideal of D. Since D is a Principal Ideal Domain, there is an element $c \in D$ such that $(c) = I$. If k is the smallest positive integer such that $c \in (a_k)$, then $I \subseteq (a_k) \subseteq (a_{k+1}) \subseteq \cdots \subseteq I$. Therefore, $(a_n) = (a_k)$ for all $n \geq k$. ∎

The fact that a Principal Ideal Domain satisfies the ascending condition is precisely the property of these domains that allows us to prove that Principal Ideal Domains are Unique Factorization Domains. A key step in the proof depends on the following lemma.

7.2.19   Lemma

Let a be a reducible element of an integral domain D. If $a = a_1 a_2$, where neither a_1 nor a_2 is a unit, then $(a) \subset (a_1) \subset D$.

Proof. Suppose that $(a) = (a_1)$. Then $a_1 \in (a)$, so $a_1 = ab$ for some b in D. But then $a_1 = a_1 a_2 b$, and this implies that $a_2 b = e$, so a_2 is a unit. Therefore, if a_2 is not a unit, then $(a) \neq (a_1)$ and the first inclusion $(a) \subset (a_1)$ is indeed proper. Similarly, the second inclusion $(a_1) \subset D$ is proper since a_1 is not a unit. ∎

7.2.20	**Theorem**

Every Principal Ideal Domain is a Unique Factorization Domain.

Proof. Suppose that D is a Principal Ideal Domain, and let a be any nonzero nonunit in D. The first step in the proof is to show that a has at least one irreducible factor in D. If a is irreducible, then a is a product with one factor, and we are finished. If a is reducible, then $a = a_1 b_1$ for a pair of nonzero elements a_1, $b_1 \in D$, and neither a_1 nor b_1 is a unit in D. Lemma 7.2.19 shows that $(a) \subset (a_1)$. If a_1 is irreducible in D, we are done, so suppose that a_1 is reducible. Then $a_1 = a_2 b_2$ for a pair of nonzero elements a_2, b_2 in D, and neither a_2 nor b_2 is a unit. Invoking Lemma 7.2.19 again, we have $(a_1) \subset (a_2)$. If a_2 is irreducible, then $a = a_2 b_1 b_2$, and we are finished. If a_2 is reducible, then this process can be continued. If this process does not terminate after a finite number of steps, then we have an ascending chain

$$(a) \subset (a_1) \subset (a_2) \subset (a_3) \subset \cdots$$

of ideals of D that is not of finite length. But this contradicts Theorem 7.2.18, so such a chain must terminate. Hence, there is a positive integer k such that $(a_n) = (a_k)$ for all $n \geq k$, and a_k is irreducible in D. Since $a \in (a) \subset (a_k)$, $a = a_k b$ for some $b \in D$, and this shows that when D is a Principal Ideal Domain, every nonzero, nonunit element of D contains at least one irreducible factor.

The second step in the proof is to use the fact that every nonzero, nonunit element of D contains at least one irreducible factor to show that each nonzero, nonunit element of D can actually be written as a product of irreducible elements of D. Let a be a nonzero, nonunit element of D. If a is irreducible, we are finished. If not, we can write $a = c_1 d_1$, where c_1 is irreducible and d_1 is not a unit. If d_1 is irreducible, we are done. If d_1 is not irreducible, write $d_1 = c_2 d_2$, where c_2 is irreducible and d_2 is not a unit. If d_2 is irreducible, then $a = c_1 c_2 d_2$, and we are finished. If d_2 is not irreducible, then the process can be continued to obtain an ascending chain of ideals in D:

$$(a) \subseteq (c_1) \subseteq (c_2) \subseteq (c_3) \subseteq \cdots$$

We know that this chain must terminate since Principal Ideal Domains are Noetherian. If k is a positive integer such that $(c_n) = (c_k)$ for all $n \geq k$, then

$a = c_1 c_2 c_3 \cdots c_k d_k$, and d_k is not a unit. The element d_k must be irreducible, for otherwise the chain does not terminate at (c_k). Hence, every nonzero nonunit in D can be expressed as a product of irreducible elements of D.

Finally, it remains to show that such a factorization is unique up to associates and the order of the factors. Suppose that $a \in D$ has two factorizations

$$a = a_1 a_2 \cdots a_m = b_1 b_2 \cdots b_n,$$

where each a_i and each b_i is irreducible. We proceed by induction on m. If $m = 1$, then $a = a_1$ is irreducible, and it follows immediately that $n = 1$ and $a_1 = b_1$. Now make the induction hypothesis that when an element of D is expressed as a product with fewer than m irreducible factors, the factorization is unique up to associates and the order of the factors. If $a = a_1 a_2 \cdots a_m = b_1 b_2 \cdots b_n$, where each a_i and b_i is irreducible, then $a_1 \mid b_1 b_2 \cdots b_n$, and since a_1 is prime in D (Theorem 7.2.14), a_1 divides one of the b_i's. Suppose that our notation has been chosen so that $a_1 \mid b_1$. Then $b_1 = u a_1$, where u is a unit, since the only divisors of an irreducible element in D are associates and units. Hence,

$$a_1 a_2 \cdots a_m = a_1 (u b_2) \cdots b_n.$$

Canceling a_1 from both sides of this equation leads to

$$a_2 \cdots a_m = (u b_2) \cdots b_n, \qquad (*)$$

and the product on the left of this equation has fewer than m factors. Thus, by the induction hypothesis, this factorization is unique up to associates and the order of the factors. Thus, $m = n$, and after a suitable reindexing of the b_i's, each a_i is an associate of one of the factors on the right in equation $(*)$. This completes the proof. ∎

Thus, PID $\Rightarrow$ UFD, where PID and UFD mean Principal Ideal Domain and Unique Factorization Domain, respectively. If F is any field, Theorem 7.2.5 shows that $F[x]$ is a PID, so Theorem 7.2.20 shows that $F[x]$ is a UFD. This observation provides a proof for Theorem 6.2.22. The following example shows that UFD $\not\Rightarrow$ PID.

7.2.21 | Example

$\mathbf{Z}[x]$ is not a Principal Ideal Domain, as was shown in Corollary 7.2.6. $\mathbf{Z}[x]$, however, is a Unique Factorization Domain. To show this, we need the following two definitions. If $f(x)$ is in $\mathbf{Z}[x]$, then the greatest common divisor of the coefficients of $f(x)$, denoted by $c(f(x))$, is referred to as the **content** of $f(x)$. If $c(f(x)) = 1$, then $f(x)$ is said to be a **primitive polynomial**. Using Lemma 6.3.10, one can show that if $f(x) = f_1(x)f_2(x)\cdots f_n(x)$, then $f(x)$ is primitive if and only if each $f_i(x)$ is primitive.

To prove that $\mathbf{Z}[x]$ is a Unique Factorization Domain, we first establish that every polynomial $f(x) \in \mathbf{Z}[x]$ can be factored as $f(x) = a_1 a_2 \cdots a_k f_1(x)f_2(x)\cdots f_n(x)$, where each a_i is an irreducible polynomial in $\mathbf{Z}[x]$ of degree 0 and each $f_i(x)$ is a primitive irreducible polynomial. If $f(x)$ is a polynomial in $\mathbf{Z}[x]$ and $c(f(x)) = a$, then dividing a from the coefficients of $f(x)$ leaves a primitive polynomial $\bar{f}(x) \in \mathbf{Z}[x]$ and $f(x) = a\bar{f}(x)$. If a is factored into a product of primes as $a_1 a_2 \cdots a_k$, then each a_i is an irreducible polynomial of degree 0 in $\mathbf{Z}[x]$ and $f(x) = a_1 a_2 \cdots a_k \bar{f}(x)$. Hence, it suffices to show that every primitive polynomial in $\mathbf{Z}[x]$ can be factored into a product of primitive irreducible polynomials.

We proceed by using induction on the degree of the polynomial. If $f(x)$ is a primitive polynomial in $\mathbf{Z}[x]$ of degree 1, then $f(x)$ is irreducible in $\mathbf{Z}[x]$ since $c(f(x)) = 1$. Now make the induction hypothesis that each primitive polynomial of degree less than m can be factored into a product of primitive irreducible polynomials, and let $f(x)$ be a primitive polynomial in $\mathbf{Z}[x]$ of degree m. If $f(x)$ is irreducible, there is nothing to prove, so suppose that $f(x)$ is reducible in $\mathbf{Z}[x]$. Then there are polynomials $g(x), h(x) \in \mathbf{Z}[x]$ such that $f(x) = g(x)h(x)$. Furthermore, $g(x)$ and $h(x)$ must be primitive since $f(x)$ is primitive. Since $\deg(g(x)) < m$ and $\deg(h(x)) < m$, the induction hypothesis lets us factor both $g(x)$ and $h(x)$ into a product of primitive irreducible polynomials. This clearly implies that $f(x)$ is a product of primitive irreducible polynomials. Thus, it follows by induction that any primitive polynomial in $\mathbf{Z}[x]$ can be factored into a product of primitive irreducible polynomials. Therefore, any polynomial $f(x) \in \mathbf{Z}[x]$ can be factored as $f(x) = a_1 a_2 \cdots a_k f_1(x) \cdot f_2(x) \cdots f_n(x)$, where each a_i is an irreducible polynomial of degree 0 and each $f_i(x)$ is a primitive irreducible polynomial.

Next, we show uniqueness of this factorization of $f(x)$. If $f(x) = b_1 b_2 \cdots b_j g_1(x)g_2(x)\cdots g_m(x)$ is also a factorization of $f(x)$, where each b_i is an irreducible polynomial of degree 0 and each $g_i(x)$ is a primitive irreducible polynomial, we claim that $j = k$ and $m = n$, and after a suitable reindexing of the b_i's and the $g_i(x)$'s, $a_i = \pm b_i$ for $i = 1, 2, \ldots, k$ and $f_i(x) = \pm g_i(x)$ for $i = 1, 2, \ldots, n$. (Since the only units in $\mathbf{Z}[x]$ are ± 1, this just says that, after a suitable reindexing, a_i and b_i, and $f_i(x)$ and $g_i(x)$, are associates.) Since $c(f(x)) = a_1 a_2 \cdots a_k = b_1 b_2 \cdots b_j$ and the a_i's and b_i's are all primes, the Fundamental Theorem of Arithmetic 1.1.17 tells us immediately that $k = j$, and after a suitable reindexing of the b_i's, $a_i = \pm b_i$ for $i = 1, 2, \ldots, k$. Now $\mathbf{Z}[x]$ is an integral domain, so it follows from

$$a_1 a_2 \cdots a_k f_1(x) f_2(x) \cdots f_n(x) = b_1 b_2 \cdots b_j g_1(x) g_2(x) \cdots g_m(x)$$

that

$$f_1(x) f_2(x) \cdots f_n(x) = g_1(x) g_2(x) \cdots g_m(x).$$

Now $\mathbf{Z}[x] \subseteq \mathbf{Q}[x]$, so each of the polynomials $f_i(x)$ and $g_i(x)$ can be considered as a polynomial in $\mathbf{Q}[x]$, which is a PID by Corollary 7.2.4. Since each $f_i(x)$ is a primitive irreducible polynomial in $\mathbf{Z}[x]$, it follows that $f_i(x)$ is irreducible in $\mathbf{Q}[x]$. Hence, by Theorem 7.2.14, each $f_i(x)$ is prime in $\mathbf{Q}[x]$. Consequently, $f_1(x) f_2(x) \cdots f_n(x) = g_1(x) g_2(x) \cdots g_m(x)$ implies that $f_1(x) \mid g_1(x) g_2(x) \cdots g_m(x)$, so $f_1(x)$ divides one of the $g_i(x)$'s. Suppose that our notation has been chosen so that $f_1(x) \mid g_1(x)$. Since $g_1(x)$ is irreducible in $\mathbf{Q}[x]$, the only divisors of $g_1(x)$ are units and associates in $\mathbf{Q}[x]$. This observation and Lemma 6.3.12 show that $f_1(x) = \pm g_1(x)$. It now follows that $f_2(x) \cdots f_n(x) = \pm g_2(x) \cdots g_m(x)$. If $n \leq m$, repetition of this argument leads to $f_2(x) = \pm g_2(x), \ldots, f_n(x) = \pm g_n(x)$, and $1 = g_{n+1}(x) \cdots g_m(x)$. Therefore, $g_{n+1}(x) = \cdots = g_m(x) = \pm 1$. But 1 and -1 are units in $\mathbf{Z}[x]$, contradicting the assumption that $g_{n+1}(x), \ldots, g_m(x)$ are irreducible. Hence, $n = m$. A similar argument applies if $m \leq n$. We thus see that there are Unique Factorization Domains that are not Principal Ideal Domains.

We have seen that in an integral domain every prime element is irreducible and that the concepts prime and irreducible are equivalent in a Principal Ideal Domain. The preceding example shows that a Unique Factorization Domain need not be a Principal Ideal Domain. However, in a Unique Factorization Domain an element is still prime if and only if it is irreducible. We conclude this section with a proof of this fact.

7.2.22   Theorem

In a Unique Factorization Domain, an element is prime if and only if it is irreducible.

Proof. Since a prime element of any integral domain is irreducible, it suffices to prove that an irreducible element is prime in any Unique Factorization Domain. Let D be a Unique Factorization Domain, and suppose that p is an irreducible element that divides the product ab of elements of D. Then there is an element $c \in D$ such that $ab = pc$. D is a Unique Factorization Domain, so both a and b can be factored into a product of irreducible elements of D. Suppose that $a = a_1 a_2 \cdots a_m$ and $b = b_1 b_2 \cdots b_n$, where each a_i and each b_i is irreducible in D. Then $pc = a_1 a_2 \cdots a_m b_1 b_2 \cdots b_n$, and so p must be an associate of at least one of the a_i's or b_i's. If p is an associate of a_k, $1 \le k \le m$, then $a = a_1 \cdots a_{k-1}(pu)a_{k+1} \cdots a_m$, where u is a unit in D. Hence, $p \mid a$. Similarly, if p is an associate of one of the factors of b, then $p \mid b$. Thus, when p is an irreducible element of D and $p \mid ab$, then $p \mid a$ or $p \mid b$, so p is prime. ∎

We conclude with the observation (where ED means Euclidean Domain) that

$$\text{ED} \Rightarrow_{7.2.3} \text{PID} \Rightarrow_{7.2.20} \text{UFD} \qquad \text{but}$$

$$\text{UFD} \not\Rightarrow_{7.2.21} \text{PID} \not\Rightarrow_{7.2.8} \text{ED}.$$

A mnemonic device for remembering the implications ED $\Rightarrow$ PID $\Rightarrow$ UFD is that the implications are in alphabetical order. The subscripts on $\Rightarrow$ and $\not\Rightarrow$ indicate where a proof of the implication or a counterexample to the implication can be found. We have also seen that

prime $\Rightarrow_{7.2.11}$ irreducible in every integral domain,
irreducible need not $\Rightarrow_{7.2.12}$ prime in an integral domain,
prime $\Leftrightarrow_{7.2.14}$ irreducible in a PID, and
prime $\Leftrightarrow_{7.2.22}$ irreducible in a UFD.

HISTORICAL NOTES

Emmy Noether (1882–1935). Born in Erlangen, Germany, in 1882, Emmy Noether was the daughter of a professor of mathematics at the University of Erlangen. Although she initially studied to become a language teacher, her interest shifted to mathematics, and she began to audit mathematics courses at the University of Erlangen. At that time women were not permitted to enroll officially in German universities; however, she persisted and eventually was permitted to matriculate. Noether received her doctorate in mathematics from the University of Erlangen in 1907 for her work on ternary biquadratic forms. Since academic positions were not open to women, she served as an assistant to her father until David Hilbert invited her to Göttingen to work with him in his study of Einstein's general theory of relativity. Although she could not hold an academic position, she was able to teach courses that were listed under Hilbert's name. Noether finally obtained a position at the University of Göttingen in 1919, which in time provided her with a modest salary. Her reputation as a mathematician grew. Noether's life took a dramatic turn when the Nazis came to power and she, as a Jew, was forced to flee Germany. She obtained a position at Bryn Mawr College in Pennsylvania, where she remained until her death in 1935.

Emmy Noether is best known today for her pioneering research in ring theory using ascending chains of right (or left) ideals. Her work played an important role in the development of ring theory as a major mathematical topic. Today rings in which every ascending chain of right (left) ideals has finite length are known as right (left) **Noetherian rings** in her honor.

Problem Set 7.2

1. Find all the associates of the polynomial $x^2 + [3]$ in $\mathbf{Z}_5[x]$. What are the units in $\mathbf{Z}_5[x]$?

2. Prove that each of the following hold in an integral domain D.

   (a) If a and b are associates and b and c are associates, then a and c are associates.
   (b) a and b are associates if and only if $a \mid b$ and $b \mid a$.
   (c) a and b are associates if and only if $(a) = (b)$.

3. If D is a Euclidean Domain with Euclidean valuation v, prove the following:

   (a) $v(a) = v(-a)$ for all $a \in D$, $a \neq 0$.
   (b) $v(a) \geq v(e)$ for all $a \in D$, $a \neq 0$.
   (c) For any $a \neq 0$, $v(a) = v(e)$ if and only if a is a unit in D.
   (d) If n is an integer such that $v(e) + n \geq 0$, show that if v^* is defined on the nonzero elements of D by $v^*(a) = v(a) + n$, then v^* is a Euclidean valuation on D.

4. $\mathbf{Z}[i]$ is an integral domain, so $\mathbf{Z}[i]$ has a field of fractions (Section 4.1). If $Q(\mathbf{Z}[i])$ denotes the field of fractions of $\mathbf{Z}[i]$, prove that $Q(\mathbf{Z}[i])$ is isomorphic to $\mathbf{Q}[i]$.

5. Consider the result of Exercise 4. If n is a square free integer, is it always the case that the field of fractions $Q(\mathbf{Z}[\sqrt{n}])$ of $\mathbf{Z}[\sqrt{n}]$ is isomorphic to $\mathbf{Q}[\sqrt{n}]$? Justify your answer. Recall that an integer $n \neq 0, 1$ is square free if n is not divisible by the square of any integer $k \neq \pm 1$.

6. If I is a nonzero ideal of $\mathbf{Z}[i]$, show that the factor ring $\mathbf{Z}[i] / I$ is finite.

7. If D is a Principal Ideal Domain and $f : R \to S$ is a ring epimorphism, show that every ideal in S is a principal ideal. Must S also be an integral domain?

8. If p is prime in an integral domain D and $p \mid a_1 a_2 \cdots a_n$, show that p divides at least one of the a_i's for any integer $n \geq 1$.

9. Prove that an ideal (p) is a maximal ideal in a Principal Ideal Domain if and only if p is a prime element of the integral domain.

10. If D is a Euclidean Domain with valuation v, prove each of the following:

    (a) If a and b are associates, then $v(a) = v(b)$.
    (b) If $a \mid b$ and $v(a) = v(b)$, then a and b are associates.

11. Let D be a Euclidean Domain with Euclidean valuation v. If v is a constant function, prove that D is a field.

12. Use Lemma 6.3.10 to show that in $\mathbf{Z}[x]$ if $f(x) = f_1(x) f_2(x) \cdots f_n(x)$, then $f(x)$ is primitive if and only if each $f_i(x)$ is primitive.

13. Show that $\mathbf{Z}[\sqrt{10}]$ is not a Unique Factorization Domain by factoring 6 in two different ways.

14. Can one develop a Euclidean Algorithm for finding the greatest common divisor of a pair of nonzero elements of a Euclidean Domain?

7.3   NORMED DOMAINS

Section Overview. The purpose of this section is to take a closer look at integral domains of quadratic integers $\mathbf{Z}[\sqrt{n}]$. By defining a norm N on these domains, we see how the units and irreducible elements of these domains can often be determined. We also see that, in some cases, when N is restricted to the nonzero elements of $\mathbf{Z}[\sqrt{n}]$, N is a Euclidean valuation on $\mathbf{Z}[\sqrt{n}]$.

We saw in Example 4 of 7.2.2 that $\mathbf{Z}[i]$ is a Euclidean Domain, and Example 3 of 7.2.16 showed that $\mathbf{Z}[\sqrt{-6}]$ is not a Unique Factorization Domain. Hence, $\mathbf{Z}[\sqrt{-6}]$ cannot be a Euclidean Domain. (Remember that ED $\Rightarrow$ PID $\Rightarrow$ UFD.) Recall that an integer, different from 0 and 1, is square free if it is not divisible by the square of any integer $k \neq \pm 1$. If n is a square free integer and $\mathbf{Z}[\sqrt{n}] = \{a + b\sqrt{n} \mid a, b \in \mathbf{Z}\}$, then $\mathbf{Z}[\sqrt{n}]$ is an integral domain when addition and multiplication are defined on $\mathbf{Z}[\sqrt{n}]$ in the usual manner. This follows since $\mathbf{Z}[\sqrt{n}]$ is a subring of the field of complex numbers. The ring $\mathbf{Z}[\sqrt{n}]$ is said to be a **domain of quadratic integers.** Note that $\mathbf{Z} \subseteq \mathbf{Z}[\sqrt{n}]$ since for any $a \in \mathbf{Z}$, $a = a + 0\sqrt{n}$. Moreover, the multiplicative identity of $\mathbf{Z}[\sqrt{n}]$ is the integer 1. If $x = a + b\sqrt{n}$ is in $\mathbf{Z}[\sqrt{n}]$, $\bar{x} = a - b\sqrt{n}$ is often referred to as the **conjugate** of x.

7.3.1 | Definition

If D is an integral domain, then a mapping $N : D \to \mathbf{N}_0$ is said to be a **norm** on D if $N(x) = 0$ if and only if $x = 0$ and $N(xy) = N(x)N(y)$ for all $x, y \in D$.

If N is a norm on D and $x \in D$, then $N(x)$ is referred to as the norm of x. An integral domain D together with a norm N defined on D is said to be a **normed domain**.

Before considering examples of normed domains, let's compare the properties of a Euclidean valuation and a norm for an integral domain D. The set of nonzero elements of D is denoted by D^*.

EUCLIDEAN VALUATION

A function $v : D^* \to \mathbf{N}_0$ such that

1. for all $a, b \in D$, $v(a) \le v(ab)$, and
2. for all $a, b \in D$, $b \neq 0$, there exist q, $r \in D$ such that $a = bq + r$, where $r = 0$ or $v(r) < v(b)$.

NORM

A function $N : D \to \mathbf{N}_0$ such that

1. $N(x) = 0$ if and only if $x = 0$, and
2. $N(xy) = N(x)N(y)$ for all $x, y \in D$.

Let D be a normed domain with norm N. Since $N(x)$ and $N(y)$ are positive integers when $x \neq 0$ and $y \neq 0$, it follows that $N(x) \le N(x)N(y)$. Thus, a norm N, restricted to the nonzero elements of D, is a Euclidean valuation on D if for each a and b in D, $b \neq 0$, there are q, $r \in D$ such that $a = bq + r$, where $r = 0$ or $N(r) < N(b)$. Of course, when a norm N is a Euclidean valuation, then D together with N forms a Euclidean Domain. As shown in the following examples, a norm on a domain may or may not give rise to a Euclidean valuation, and a Euclidean valuation may or may not give rise to a norm.

7.3.2 | Examples

1. The integral domain $\mathbf{Z}$ is a normed domain with norm $N(x) = |x|$ for x in $\mathbf{Z}$. If N is restricted to the nonzero elements of $\mathbf{Z}$, then $\mathbf{Z}$ is a Euclidean Domain with Euclidean valuation $v = N$.
2. If F is a field, then F is a normed domain if N is defined on F by $N(x) = 1$ for each $x \neq 0$ and $N(0) = 0$. If N is restricted to the nonzero elements of F, then F is a Euclidean Domain with Euclidean valuation $v = N$.

In Examples 1 and 2, can you show that property 2 of a Euclidean valuation is satisfied?

3. If F is a field, then $F[x]$ is a Euclidean Domain under the mapping defined on $F[x]$ by $v(f(x)) = \deg(f(x))$ for all $f(x) \neq 0$. If v is extended to $F[x]$ by setting $v(0) = 0$, then $N = v$ is *not* a norm on $F[x]$. This follows because if $f(x) \neq 0$ and $g(x) \neq 0$, then $N(f(x)g(x)) \neq N(f(x))N(g(x))$. In fact, $N(f(x)g(x)) = \deg(f(x)g(x)) = \deg(f(x)) + \deg(g(x)) = N(f(x)) + N(g(x))$. Thus, $F[x]$ does not become a normed domain under this extension.

Examples 1 and 2 show that in some cases a normed domain D is also a Euclidean Domain when the norm is restricted to the nonzero elements of D. Conversely, Example 3 shows that a Euclidean Domain D may not be a normed domain if the Euclidean valuation v is extended to D by setting $v(0) = 0$. For a Euclidean Domain, there is no assurance that $v(xy) = v(x)v(y)$ for all $x, y \in D$. So, in general, an extended Euclidean valuation is not a norm. Since we are interested in domains of quadratic integers, there is a useful norm that can be defined on $\mathbf{Z}[\sqrt{n}]$.

7.3.3   Definition

Define N on $\mathbf{Z}[\sqrt{n}]$ by

$$N : \mathbf{Z}[\sqrt{n}] \to \mathbf{N}_0, \text{ where } N(a + b\sqrt{n}) = |a^2 - nb^2|.$$

The following theorem shows that N is indeed a norm on $\mathbf{Z}[\sqrt{n}]$. It also establishes several properties of N that are often helpful in determining the units and irreducible elements of $\mathbf{Z}[\sqrt{n}]$. The proof of the following theorem is a straightforward computation, so it is left as an exercise.

7.3.4   Theorem

If n is a square free integer, and N is defined on $\mathbf{Z}[\sqrt{n}]$ as in Definition 7.3.3, then:

1. If $x \in \mathbf{Z}[\sqrt{n}]$, then $N(x) = 0$ if and only if $x = 0$.
2. For all $x, y \in \mathbf{Z}[\sqrt{n}]$, $N(xy) = N(x)N(y)$.
3. For any $x \in \mathbf{Z}[\sqrt{n}]$, $N(x) = N(\bar{x})$.
4. If $x \in \mathbf{Z}[\sqrt{n}]$, then x is a unit in $\mathbf{Z}[\sqrt{n}]$ if and only if $N(x) = 1$.
5. If $N(x)$ is a prime integer, then x is an irreducible element of $\mathbf{Z}[\sqrt{n}]$.

We thus see that $\mathbf{Z}[\sqrt{n}]$, together with the norm in Definition 7.3.3, is a normed domain. Recall that we saw in Lemma 7.2.11 that prime $\Rightarrow$ irreducible in every integral domain, while Theorems 7.2.14 and 7.2.22 show that prime $\Leftrightarrow$ irreducible when we are working either in a Principal Ideal Domain or a Unique Factorization Domain. The first of the following examples shows how one can go about factoring an element in a ring of quadratic integers into a product of irreducible elements, and the second example shows that there are integral domains that are not Unique Factorization Domains. Since a Euclidean Domain must be a Unique Factorization Domain, this also shows that there are integral domains that are not Euclidean Domains. The third example provides two additional illustrations of domains of quadratic integers. One of these domains is a Euclidean Domain, and the other is not.

7.3.5   Examples

1. It was proven in Example 4 of 7.2.2 that $\mathbf{Z}[i]$ is a Euclidean Domain under the Euclidean valuation $v(a + bi) = N(a + bi)$ for all $a + bi \neq 0$ in $\mathbf{Z}[i]$. In this case $\mathbf{Z}[i]$ is also a normed domain if we set $v(0) = 0$. Since $N(a + bi) = a^2 + b^2 = 1$ if and only if $a = \pm 1$ and $b = 0$ or $a = 0$ and $b = \pm 1$, part 4 of Theorem 7.3.4 shows that the units of $\mathbf{Z}[i]$ are ± 1 and $\pm i$. If $a^2 + b^2 = N(a + bi) = p$ for some prime p, then $a + bi$ is an irreducible element of $\mathbf{Z}[i]$. For example, $1 + i$, $1 - i$, $2 + i$, etc., are irreducible elements in $\mathbf{Z}[i]$. Since $\mathbf{Z}[i]$ is a Euclidean Domain, it is a Unique Factorization Domain, so how does one go about factoring an element of $\mathbf{Z}[i]$ into a product of irreducibles? If $x \in \mathbf{Z}[i]$ and $x = x_1 x_2 \cdots x_m$ is a factorization of x into a product of irreducible elements of $\mathbf{Z}[i]$, then one can argue by induction that

    $$N(x) = N(x_1)N(x_2)\cdots N(x_m).$$

    If x_i is an irreducible factor of x, then $N(x_i) \mid N(x)$. This fact is often helpful in factoring an element of $\mathbf{Z}[i]$ into a product of irreducible elements. As an illustration, suppose that we are to factor $4 + 3i$. Since $N(4 + 3i) = 4^2 + 3^2 = 25$, the norm of any factor of $4 + 3i$ must be 1, 5, or 25. If $N(a + bi) = 1$, then $a + bi$ is a unit, so $a + bi$ cannot be an irreducible factor of $4 + 3i$. If $N(a + bi) = 5$, then $a^2 + b^2 = 5$ has solutions $a = \pm 2$ and $b = \pm 1$ or $a = \pm 1$ and $b = \pm 2$. If we let $x_1 = 2 + i$ and $x_2 = 1 - 2i$, then $x_1 x_2 = (2 - i)(1 + 2i) = (2 + 2) + (4 - 1)i = 4 + 3i$, so this is a factorization of $4 + 3i$ into a product of irreducible elements of $\mathbf{Z}[i]$. Note that both $2 - i$ and $1 + 2i$ are irreducible in $\mathbf{Z}[i]$, since the norm of each is a prime integer. We also see that $4 + 3i = -i(1 + 2i)(1 + 2i)$, but since $-i(1 + 2i) = 2 - i$ and $-i$ is a unit in $\mathbf{Z}[i]$, the elements $1 + 2i$ and $2 - i$ are associates. $4 + 3i$ can also be factored as $i(2 - i)(2 - i)$, but in this case i is a unit, and $i(2 - i) = 1 + 2i$ shows again that $1 + 2i$ and $2 - i$ are associates. This confirms that the factorization of $4 + 3i$ in $\mathbf{Z}[i]$ is unique up to associates, and we know that this must be the case since $\mathbf{Z}[i]$ is a Unique Factorization Domain.

2. In Example 3 of 7.2.16, we made the observation that $\mathbf{Z}[\sqrt{-6}]$ is not a Unique Factorization Domain, but the details were deferred to this section. We now show that this is indeed the case. In an integral domain, Lemma 7.2.11 shows that a prime element is always irreducible, and Theorem 7.2.22 shows that the converse holds in a Unique Factorization Domain. Hence, we can show that $\mathbf{Z}[\sqrt{-6}]$ is not a Unique Factorization Domain by finding an element of $\mathbf{Z}[\sqrt{-6}]$ that is irreducible but not prime. For this, consider $5 + 0\sqrt{-6}$. We claim that $5 + 0\sqrt{-6}$ is irreducible but not prime. Suppose that $5 = (a + b\sqrt{-6})(c + d\sqrt{-6})$. If we can show that either $a + b\sqrt{-6}$ or $c + d\sqrt{-6}$ is a unit in $\mathbf{Z}[\sqrt{-6}]$, then 5 is irreducible. Using the norm function of Definition 7.3.3,

$$25 = N(5) = N(a + b\sqrt{-6})N(c + d\sqrt{-6})$$
$$= (a^2 + 6b^2)(c^2 + 6d^2).$$

Thus, $(a^2 + 6b^2)(c^2 + 6d^2)$ is a factorization of 25 by positive integers, and the only factorizations of 25 by positive integers are $5 \cdot 5$ and $1 \cdot 25$. Suppose that

$$N(a + b\sqrt{-6}) = a^2 + 6b^2 = 5 \qquad \text{and}$$
$$N(c + d\sqrt{-6}) = c^2 + 6d^2 = 5. \qquad (*)$$

In equations $(*)$ if $b \neq 0$, then $a^2 + 6b^2 \geq 6$, so the equation $5 = a^2 + 6b^2$ has no solution in the integers unless $b = 0$. But if $b = 0$, $a^2 = 5$ has no solution in the integers since $\sqrt{5}$ is an irrational number. Similar observations hold for the equation $5 = c^2 + 6d^2$, so the equations $(*)$ have no solutions in the integers. Therefore, it cannot be the case that $N(a + b\sqrt{-6}) = 5$ and $N(c + d\sqrt{-6}) = 5$. The possibilities left are that $N(a + b\sqrt{-6}) = 1$ and $N(c + d\sqrt{-6}) = 25$ or that $N(a + b\sqrt{-6}) = 25$ and $N(c + d\sqrt{-6}) = 1$. Hence, we see that either $N(a + b\sqrt{-6})$ or $N(c + d\sqrt{-6})$ must be 1, so part 4 of Theorem 7.3.4 shows that either $a + b\sqrt{-6}$ or $c + d\sqrt{-6}$ is a unit.

Therefore,

$$5 = (a + b\sqrt{-6})(c + d\sqrt{-6})$$

implies that one of the factors is a unit, so 5 is irreducible in $\mathbf{Z}[\sqrt{-6}]$. Finally, let's show that 5 is not prime in $\mathbf{Z}[\sqrt{-6}]$. Now $5 \mid 15$ and

$$15 = (3 + \sqrt{-6})(3 - \sqrt{-6}),$$

so if 5 is prime, then it must be the case that $5 \mid (3 + \sqrt{-6})$ or $5 \mid (3 - \sqrt{-6})$. If $5 \mid (3 + \sqrt{-6})$, there is an element $a + b\sqrt{-6}$ in $\mathbf{Z}[\sqrt{-6}]$ such that $3 + \sqrt{-6} = 5(a + b\sqrt{-6}) = 5a + 5b\sqrt{-6}$. But this implies that $5a = 3$, and since a is an integer, this is clearly impossible. Thus, $5 \nmid ((3 + \sqrt{-6})$. Similarly, $5 \nmid (3 - \sqrt{-6})$, so 5 is not prime in $\mathbf{Z}[\sqrt{-6}]$. Thus, 5 is an irreducible element of $\mathbf{Z}[\sqrt{-6}]$ that is not prime. Since an element x in a Unique Factorization Domain is prime if and only if it is irreducible, $\mathbf{Z}[\sqrt{-6}]$ is not a Unique Factorization Domain. This also shows that it is impossible to define a Euclidean valuation v on $\mathbf{Z}[\sqrt{-6}]$. If it were possible, then $\mathbf{Z}[\sqrt{-6}]$ would be a Euclidean Domain, and we know that this would imply that $\mathbf{Z}[\sqrt{-6}]$ is a Unique Factorization Domain.

3. The domain $\mathbf{Z}[\sqrt{-2}]$ is a Euclidean Domain, while $\mathbf{Z}[\sqrt{-3}]$ is not. The details are left as exercises.

Example 2 above shows that there are normed domains that are not Unique Factorization Domains. In a Unique Factorization Domain D, we know that every element can be expressed as a product of irreducible elements of D. But what happens with factorization in a normed domain that is not a Unique Factorization Domain? Can every element of such a domain be expressed as a product of irreducible elements with only the uniqueness property of factoring failing to hold? The following theorem and Example 7.3.8 show that this is exactly what happens.

7.3.6   **Theorem**

If D is a normed domain with norm N, then every nonzero element of D that is not a unit can be expressed as a product of irreducible elements of D.

Proof.   Let S be the set of all nonzero, nonunit elements of D that cannot be expressed as a product of irreducible elements. The proof will be complete if we can show that $S = \emptyset$. If $S \neq \emptyset$, the set $N(S) = \{N(x) \mid x \in S\}$ is a nonempty subset of N_0. Since N_0 is well-ordered, $N(S)$ has a smallest element. Hence, there is an $x \in S$ such that $N(x) \leq N(y)$ for all y in S. Now x is not irreducible, for if it were, x would be a product of irreducible elements with one factor and so could not be in S. Thus, x is reducible, so there are nonunits $a, b \in D$ such that $x = ab$. Now at least one of a and b is not the product of irreducible elements, for if both a and b are the product of irreducible elements of D, then x is a product of irreducible elements of D. Hence, either a or b must be in S. Since a and b are nonunits, $N(a) > 1$ and $N(b) > 1$, and if $a \in S$, it follows that $N(a) < N(a)N(b) = N(ab) = N(x)$. But this contradicts the minimality of $N(x)$ in $N(S)$. A similar argument holds if $b \in S$. Thus, $S = \emptyset$, so every nonzero, nonunit in D can be expressed as a product of irreducible elements of D. ∎

7.3.7   **Corollary**

If n is a square free integer, then every nonzero, nonunit element of the normed domain $\mathbf{Z}[\sqrt{n}]$ can be expressed as a product of irreducible elements.

7.3.8   **Example**

In Example 2 of 7.3.5 we saw that $\mathbf{Z}[\sqrt{-6}]$ is not a Unique Factorization Domain. Hence, $\mathbf{Z}[\sqrt{-6}]$ is a normed domain that is not a Unique Factorization Domain. This is demonstrated by the fact that the integer $15 \in \mathbf{Z}[\sqrt{-6}]$ can be factored in two distinct ways as a product of irreducible elements of $\mathbf{Z}[\sqrt{-6}]$:

$$3 \cdot 5 = 15 = (3 + \sqrt{-6})(3 - \sqrt{-6}).$$

Problem Set 7.3

All quadratic domains are to have the norm N as defined in Definition 7.3.3.

1. Is 5 an irreducible element in $\mathbf{Z}[\sqrt{2}]$? In $\mathbf{Z}[i]$?

2. Prove Theorem 7.3.4.

3. If $x = x_1 x_2 \cdots x_m$ in $\mathbf{Z}[\sqrt{n}]$, prove that $N(x) = N(x_1)N(x_2) \cdots N(x_m)$.

4. Factor $11 + 7i$ into a product of irreducible elements of $\mathbf{Z}[i]$.

5. Find all the units of the normed domain $\mathbf{Z}[\sqrt{-3}]$.

6. Factor 8 as a product of two irreducible elements and then as a product of three irreducible elements in the normed domain $\mathbf{Z}[\sqrt{-7}]$. Explain why $\mathbf{Z}[\sqrt{-7}]$ is not a Euclidean Domain when the norm N is restricted to the nonzero elements of $\mathbf{Z}[\sqrt{-7}]$.

7. (a) Explain why Example 7.2.12 shows that $\mathbf{Z}[\sqrt{-5}]$ is not a Unique Factorization Domain.
    (b) Part 5 of Theorem 7.3.4 shows that in $\mathbf{Z}[\sqrt{n}]$ if $N(x)$ is a prime integer, then x is irreducible. Show that the converse is false by showing that $1 + \sqrt{-5}$ is irreducible in $\mathbf{Z}[\sqrt{-5}]$ but that $N(1 + \sqrt{-5})$ is not a prime integer.

8. Show that $\mathbf{Z}[\sqrt{-2}]$ is a Euclidean Domain.

9. Show that $\mathbf{Z}[\sqrt{-3}]$ is not a Euclidean Domain.

10. In $\mathbf{Z}[i]$ find $a + bi$ and $c + di$ such that

$$3 + 7i = (1 + 2i)(a + bi) + (c + di),$$

where either $c + di = 0$ or $N(c + di) < 5$.

11. (a) Show that the normed domain $\mathbf{Z}[\sqrt{-2}]$ is a Euclidean Domain if $v(a + b\sqrt{-2}) = N(a + b\sqrt{-2})$ for every $a + b\sqrt{-2} \neq 0$ in $\mathbf{Z}[\sqrt{-2}]$.
    (b) Show that each of $2 + \sqrt{-2}, 2 - \sqrt{-2}, 5 + \sqrt{-2}$, and $11 - 7\sqrt{-2}$ is irreducible in $\mathbf{Z}[\sqrt{-2}]$.
    (c) Does $(2 + \sqrt{-2})(11 - 7\sqrt{-2}) = (2 - \sqrt{-2})(5 + \sqrt{-2})$ contradict unique factorization in $\mathbf{Z}[\sqrt{-2}]$?

Use a computer algebra system to solve the following problem.

12. Factor each of the following into a product of irreducible elements in $\mathbf{Z}[i]$, the ring of Gaussian integers.

(a) $120 + 64i$

(b) $-1792 - 5888i$

(c) $-1{,}474{,}944 + 855{,}808i$

8

Field Extensions

*If F is a field and f(x) is a nonconstant polynomial in F[x], then we know from the Factor Theorem 6.2.5 that c ∈ F is root of f(x) if and only if x − c is a factor of f(x). Thus, the problem of factoring a polynomial is very closely linked to finding the roots of f(x), whether the roots lie in F or in some field extension of F. (Recall that a field E is a **field extension** of a field F if F is a subfield of E.) A polynomial f(x) ∈ F[x] may or may not have a root in F. For example, f(x) = x² − 2 in **Q**[x] does not have a root in **Q**. The polynomial f(x), however, does have a root in the field **Q**[$\sqrt{2}$]. (See Exercise 2 of Problem Set 3.1.) In fact, both of the roots −$\sqrt{2}$ and $\sqrt{2}$ of f(x) are in **Q**[$\sqrt{2}$]. Since **Q** is a subfield of **Q**[$\sqrt{2}$], we have found a field extension of **Q** that contains the roots of f(x). Consequently, f(x) factors as f(x) = (x −$\sqrt{2}$)(x + $\sqrt{2}$) in the polynomial ring **Q**[$\sqrt{2}$][x]. Since f(x) factors as a product of linear factors over the field **Q**[$\sqrt{2}$] and since it can be shown that **Q**[$\sqrt{2}$] is the smallest field that contains the roots of f(x), we call **Q**[$\sqrt{2}$] the **splitting field** of f(x) over **Q**.*

*The field **C** has the property that every element of **C** is the root of a nonzero polynomial in **R**[x]. A field extension E of F with the property that every element of E is the root of a nonzero polynomial in F[x] is what we later refer to as an **algebraic field extension**. Another nice property of the complex numbers, given by the Fundamental Theorem of Algebra, is that **C** contains all the roots of every polynomial in **C**[x]. A field F that has the property that every polynomial in F[x] has a root in F (and consequently, all of its roots are in F) is said to be **algebraically closed**. These properties of the field of complex numbers bring up an interesting and parallel question for an arbitrary field: Does an algebraic field extension E of F exist such that E is algebraically closed? If an algebraic field extension E of F exists such that E is algebraically closed, then E is said to be the **algebraic closure** of F. Algebraic closures do exist; for example, as we have just seen, **C** is the algebraic closure of **R**.*

The purpose of this chapter is to investigate the existence of a splitting field for a polynomial and the existence of an algebraic closure for a field. Before beginning, we need to develop several concepts that play an important role in subsequent sections.

8.1   BASES OF FIELD EXTENSIONS

Section Overview. In this section, basic machinery is developed that will be used to investigate field extensions. The concepts presented in this section are fundamental to understanding the material in the sections that follow. So make sure you master the concepts of linear combinations, linearly independent and linearly dependent sets, span, and basis.

When E is a field extension of F, the additive identity and the multiplicative identity of F coincide with the additive identity and the multiplicative identity of E, respectively. The notation E/F indicates that E is a field extension of F, and to simplify language, we simply say that E/F is a field extension. Frequently occurring examples of field extensions are $\mathbf{R}/\mathbf{Q}$, $\mathbf{C}/\mathbf{R}$, and $\mathbf{C}/\mathbf{Q}$.

8.1.1   Definition

If E/F is a field extension and $S = \{x_1, x_2, \ldots, x_n\}$ is a subset of E, then an expression of the form

$$c_1x_1 + c_2x_2 + \cdots + c_nx_n,$$

where $c_1, c_2, \ldots, c_n \in F$, is said to be a **linear combination** of the elements of S. The c_i's are referred to as the **coefficients** of the x_i's, and if there is a need, F will be called the **coefficient field**. The set of all linear combinations of elements of S is said to be the **span** of S. (In passing we mention that E is actually a **vector space** over F. In this setting, elements of E are referred to as **vectors**, and the elements of F are called **scalars**. However, we delay discussion of this terminology until a later chapter, where vector spaces are studied in more detail.) If every element of E can be expressed as a linear combination of the elements of S, then S is said to **span** E. Hence, if S spans E, then for every $x \in E$, there are elements $c_1, c_2, \ldots, c_n$ in F such that

$$x = c_1x_1 + c_2x_2 + \cdots + c_nx_n.$$

If S has the property that $c_1 x_1 + c_2 x_2 + \cdots + c_n x_n = 0$ implies that $c_1 = c_2 = \cdots = c_n = 0$, then S is said to be a **linearly independent** set of elements of E. If one can find a linear combination $c_1 x_1 + c_2 x_2 + \cdots + c_n x_n = 0$ with at least one of the c_i's $\neq 0$, then S is said to be a **linearly dependent** set in E.

An important observation concerning a linearly independent set $S = \{x_1, x_2, \ldots, x_n\}$ is that the coefficients $c_1, c_2, \ldots, c_n$ in any expression for x are unique. This follows, for if

$$b_1 x_1 + b_2 x_2 + \cdots + b_n x_n = x = c_1 x_1 + c_2 x_2 + \cdots + c_n x_n, \qquad \text{then}$$

$$b_1 x_1 + b_2 x_2 + \cdots + b_n x_n - c_1 x_1 - c_2 x_2 - \cdots - c_n x_n = 0 \qquad \text{and thus}$$

$$(b_1 - c_1) x_1 + (b_2 - c_2) x_2 + \cdots + (b_n - c_n) x_n = 0.$$

But S is linearly independent, so $b_1 - c_1 = 0$, $b_2 - c_2 = 0$, ..., $b_n - c_n = 0$. Hence, $b_1 = c_1$, $b_2 = c_2$, ..., $b_n = c_n$, thus showing that the coefficients are indeed unique. A linearly independent subset of E that spans E is said to be a **basis** of the field extension E/F. The nice thing about a basis for a field extension E/F is that every element of E can be determined from the basis elements, in the sense that each element of E has a unique expression as a linear combination of the basis elements with the coefficients taken from F. The following theorem shows that there are several ways that a set can fail to be linearly independent.

8.1.2   Theorem

Let E/F be a field extension, and suppose that $S = \{x_1, x_2, \ldots, x_n\}$ is a subset of E.

1. If $0 \in S$, then S is linearly dependent.
2. If the elements of S are not distinct, then S is linearly dependent.
3. If any element of S can be expressed as a linear combination of the remaining elements in S, then S is linearly dependent.

Proof

1. This follows easily because, for example, if $x_1 = 0$, then for any $c \in F$, $c \neq 0$, $c x_1 + 0 \cdot x_2 + \cdots + 0 \cdot x_n = c0 = 0$. Thus, we have constructed a linear combination of the elements of S that is equal to zero and not all of the coefficients are zero. Thus, S is linearly dependent.
2. Suppose that $x_i = x_k$, where $1 \leq i, k \leq n$, $i \neq k$. Then we see that $0 x_1 + \cdots + 0 x_{i-1} + e x_i + 0 x_{i+1} + \cdots + 0 x_{k-1} + (-e) x_k + 0 x_{k+1} + \cdots + 0 x_n = 0$ and that not all of the coefficients are 0. Thus, S is linearly dependent.

3. Suppose that $x_k \in S$ is a linear combination of the remaining elements of S. Then there are constants $c_1, c_2, \ldots, c_{k-1}, c_{k+1}, \ldots, c_n$ such that

$$x_k = c_1 x_1 + \cdots + c_{k-1} x_{k-1} + c_{k+1} x_{k+1} + \cdots + c_n x_n \quad \text{and thus}$$

$$c_1 x_1 + \cdots + c_{k-1} x_{k-1} + (-e) x_k + c_{k+1} x_{k+1} + \cdots + c_n x_n = 0.$$

Since $-e \neq 0$, we have found a linear combination of the x_i's that is equal to 0 but not all of the coefficients are zero. Thus, S is linearly dependent. ∎

8.1.3   Examples

1. Consider the field extension $\mathbf{C}/\mathbf{R}$. The set $B = \{1, i\}$ is a basis for $\mathbf{C}$. This follows since any complex number z can be written as a linear combination of 1 and i. That is, $z = a1 + bi$, where $a, b \in \mathbf{R}$. Furthermore, $a1 + bi = 0$ if and only if $a = b = 0$, so B is linearly independent. There are subsets of $\mathbf{C}$ that span $\mathbf{C}$ that are not linearly independent. For example, the set $S = \{1, 1 + i, 1 - i\}$ spans $\mathbf{C}$, but this set is not linearly independent. To show that S spans $\mathbf{C}$, it must be shown that for any complex number $a + bi$ there are $c_1, c_2, c_3 \in \mathbf{R}$ such that

$$c_1 1 + c_2(1 + i) + c_3(1 - i) = a + bi \quad \text{or}$$
$$(c_1 + c_2 + c_3) + (c_2 - c_3)i = a + bi.$$

Hence, $c_1 + c_2 + c_3 = a$ and $c_2 - c_3 = b$, so $c_1 = a + b - 2c_2$ and $c_3 = c_2 - b$. Since a and b are given, a value can be assigned to c_2 to obtain values for c_1 and c_3. Hence, the set S spans $\mathbf{C}$, but S is not linearly independent since the coefficients $c_1, c_2,$ and c_3 are not unique. The coefficient field also plays an important role in determining a basis of a field extension. For example, $B = \{1, i\}$ is a basis of the field extension $\mathbf{C}/\mathbf{R}$, but B is *not* a basis for the field extension $\mathbf{C}/\mathbf{Q}$. If B were a basis for the field extension $\mathbf{C}/\mathbf{Q}$, then for each complex number $a + bi$, $c_1, c_2 \in \mathbf{Q}$ would exist such that $c_1 1 + c_2 i = a + bi$. Can you find elements $c_1, c_2 \in \mathbf{Q}$ such that $c_1 1 + c_2 i = \sqrt{2} + i$?

2. If F is a field, then F/F is a field extension with basis $B = \{e\}$ since every $x \in F$ can be written as $x = xe$, and $xe = 0$ clearly implies $x = 0$.

3. Since $x^2 + 2$ is irreducible in $\mathbf{R}[x]$, Corollary 7.1.11 shows that $\mathbf{R}[x]/(x^2 + 2)$ is a field extension of $\mathbf{R}$. Every element of $\mathbf{R}[x]/(x^2 + 2)$ is of the form $\overline{a + bx} = a\overline{1} + b\overline{x}$, where $a, b \in \mathbf{R}$, and it follows that $B = \{\overline{1}, \overline{x}\}$ is a basis for $\mathbf{R}[x]/(x^2 + 2)$ relative to the coefficient field $\mathbf{R}$.

4. There can be more than one basis of a given field extension. In Example 1, $B = \{1, i\}$ was shown to be a basis for the field extension

C/**R**. We claim that $B' = \{1 - i, 3i\}$ is also a basis for **C**/**R**. Let $a + bi$ be an arbitrary element of **C**. If $c_1(1 - i) + c_2(3i) = a + bi$, then it must be the case that $c_1 + (-c_1 + 3c_2)i = a + bi$. Hence, $c_1 = a$ and $-c_1 + 3c_2 = b$, so $c_2 = \frac{a+b}{3}$. Since a and b are elements of **R**, the equations

$$c_1 = a \quad \text{and} \quad c_2 = \frac{a + b}{3}$$

show that $c_1, c_2 \in$ **R** can always be found such that $c_1(1 - i) + c_2(3i) = a + bi$. Hence, B' spans **C**. If $c_1(1 - i) + c_2(3i) = 0$, then $c_1 = 0$ and $-c_1 + 3c_2 = 0$, so it must also be the case that $c_2 = 0$. Hence, B' is linearly independent, and so B' is a basis for the field extension **C**/**R**. Thus, the field extension **C**/**R** has more than one basis. Observe that $3 + 4i \in$ **C** and that relative to the basis B of **C**/**R**, $3 + 4i = 3 \cdot 1 + 4i$, so relative to B, the coefficients are $c_1 = 3$ and $c_2 = 4$. With respect to the second basis B' of **C**/**R**, $c_1 = 3$ and $c_2 = (3 + 4)/3 = \frac{7}{3}$. Note that $3(1 - i) + \frac{7}{3}(3i) = 3 - 3i + 7i = 3 + 4i$.

An important observation that you should make in this example is that given a basis B of a field extension E/F, for any $x \in E$, x can be expressed as a linear combination of the elements of B with the coefficients taken from F. Moreover, these coefficients are unique, but only relative to B. The important point is that when the basis is changed and x is expressed as a linear combination of the "new" basis elements, the coefficients may change.

If E/F is a field extension and $S = \{x_1, x_2, \ldots, x_n\}$ is a subset of E, it is often convenient to think of the elements of S as being ordered as they are listed in S. With this in mind, if k is such that $2 \le k \le n$, the elements x_1, $x_2, \ldots, x_{k-1}$ can be thought of as **preceding** x_k, and if k is such that $1 \le k \le n - 1$, the elements $x_{k+1}, x_{k+2}, \ldots, x_n$ can be thought of as **following** x_k.

8.1.4   Lemma

Let E/F be a field extension. Then a subset $S = \{x_1, x_2, \ldots, x_n\}$ of nonzero elements of E is linearly dependent if and only if there is an element $x_k \in S$, $2 \le k \le n$, such that x_k can be expressed as a linear combination of the elements of S that precede x_k.

Proof. Let k be an integer such that $2 \le k \le n$, and suppose that $x_k \in S$ can be expressed as a linear combination of $x_1, x_2, \ldots, x_{k-1}$. Then there are elements $c_1, c_2, \ldots, c_{k-1} \in F$, not all of which can be zero, such that

$$x_k = c_1 x_1 + c_2 x_2 + \cdots + c_{k-1} x_{k-1}.$$

But then

$$c_1 x_1 + c_2 x_2 + \cdots + c_{k-1} x_{k-1} + (-e) x_k + 0 x_{k+1} + \cdots + 0 x_n = 0,$$

so S is linearly dependent. Conversely, suppose that S is linearly dependent. Then there exist $c_1, c_2, \ldots, c_n \in F$, not all of which can be zero, such that

$$c_1 x_1 + c_2 x_2 + \cdots + c_n x_n = 0.$$

If k is the smallest subscript such that $c_k \neq 0$, c_k^{-1} exists since F is a field. If $k = 1$, then $c_1 x_1 = 0$ implies that $x_1 = c_1^{-1} c_1 x_1 = c_1^{-1} 0 = 0$, and we know this cannot be the case since the elements of S are nonzero elements of E. Thus, $k \geq 2$. But then

$$c_1 x_1 + c_2 x_2 + \cdots + c_k x_k = 0, \qquad \text{so}$$

$$c_k x_k = -c_1 x_1 - c_2 x_2 - \cdots - c_{k-1} x_{k-1} \qquad \text{or}$$

$$x_k = -c_k^{-1} c_1 x_1 - c_k^{-1} c_2 x_2 - \cdots - c_k^{-1} c_{k-1} x_{k-1}.$$

Therefore, an element of S exists that can be expressed as a linear combination of the elements of S that precede it. ∎

8.1.5   Corollary

If E/F is a field extension and $S = \{x_1, x_2, \ldots, x_n\}$ is a subset of nonzero elements of E, then S is linearly independent if and only if no element of S can be written as a linear combination of the elements that precede it.

Using a technique similar to that used in the proof of Lemma 8.1.4, it can also be shown that if each element of $S = \{x_1, x_2, \ldots, x_n\}$ is nonzero, then S is linearly dependent if and only if there is an element of S that can be written as a linear combination of elements of S that follow it. From this we see that S is linearly independent if and only if no element of S can be written as a linear combination of the elements of S that follow it.

In Example 4 of 8.1.3, we saw that a basis of a field extension E/F need not be unique. However, the number of elements in a basis for a field extension is unique. (We will deal only with field extensions E/F in which the number of elements in a basis is finite.) Before proving that the number of elements in a basis of a field extension is unique, the technique used in the proof of the theorem is illustrated through an example. If you study the example carefully, you should be able to work your way through the proof of the theorem. The technique used in the example and in the proof of the theorem can be described as a **replacement procedure**.

8.1.6   Example

Suppose that $B = \{x_1, x_2, x_3\}$ and $B' = \{y_1, y_2\}$ are bases for the field extension E/F. Since B' is a basis for E, x_1 can be written as a linear combination of the elements of B'. Theorem 8.1.2 indicates that the set $\{x_1, y_1, y_2\}$ is linearly dependent, so Lemma 8.1.4 tells us that there is an element of $\{x_1, y_1, y_2\}$ that can be written as a linear combination of the elements that precede it. Suppose this element is y_2. Then $y_2 = c_1 x_1 + c_2 y_1$, where $c_1, c_2 \in F$. We claim that $\{x_1, y_1\}$ spans E. If $x \in E$, then x can be written as a linear combination of the elements of B', so there are elements $c_1', c_2' \in F$ such that

$$
\begin{aligned}
x &= c_1'\, y_1 + c_2'\, y_2 \\
&= c_1'\, x_1 + c_2'\, (c_1 x_1 + c_2 y_1) \\
&= (c_1' + c_2'\, c_1) x_1 + c_2'\, c_2 y_1
\end{aligned}
$$

Hence, the set $\{x_1, y_1\}$ does indeed span E. In this process, we have replaced $y_2 \in B'$ with $x_1 \in B$ and obtained a set $\{x_1, y_1\}$ that continues to span E. Since $\{x_1, y_1\}$ spans E, $x_2 \in B$ can be written as a linear combination of the elements from $\{x_1, y_1\}$, so the set $\{x_1, x_2, y_1\}$ is linearly dependent. Thus, there is an element of $\{x_1, x_2, y_1\}$ that can be written as a linear combination of the elements that precede it. This cannot be x_2, for if it were x_2, this would imply that B is not a linearly independent set. Hence, it must be y_1. If $y_1 = c_1'' x_1 + c_2'' x_2$, then it follows easily that $\{x_1, x_2\}$ spans E. (In this step, we have replaced $y_1 \in B'$ in the spanning set $\{x_1, y_1\}$ with $x_2 \in B$.) But if $\{x_1, x_2\}$ spans E, then x_3 can be written as a linear combination of x_1 and x_2, and this contradicts the fact that B is a linearly independent set. Hence, if the field extension E/F has a basis with two elements, it cannot have a basis with three elements. An easy extension of this argument shows that E/F cannot have a basis with more than two elements.

8.1.7   Theorem

If a field extension E/F has a basis with n elements, then every basis of E/F has n elements.

Proof.   Suppose that $B = \{x_1, x_2, \ldots, x_m\}$ and $B' = \{y_1, y_2, \ldots, y_n\}$ are bases of E. Since B' is a basis for E, every element of E can be written as a linear combination of the elements of B'. In particular, x_1 can be so expressed, so the set $\{x_1, y_1, y_2, \ldots, y_n\}$ is linearly dependent. Hence, by Lemma 8.1.4, there is an integer j such that y_j can be written as a linear combination of $x_1, y_1, y_2, \ldots, y_{j-1}$. (It may be the case that $j = n$, but the following argument still works if the appropriate notational changes are made.) Thus, it follows that y_j can be removed from the set $\{x_1, y_1, y_2, \ldots, y_n\}$ and $\{x_1, y_1, y_2, \ldots, y_{j-1}, y_{j+1}, \ldots, y_n\}$ will still span E. This latter set will span E since any element of E can be written as a linear combination of the y_i's in B', and in such a linear combination y_j can be replaced by a linear combination of the elements in $\{x_1, y_1, y_2, \ldots, y_{j-1}\}$. Next, note that x_2 can be written as a linear combination of the elements in

$$\{x_1, y_1, y_2, \ldots, y_{j-1}, y_{j+1}, \ldots, y_n\},$$

so the set

$$\{x_1, x_2, y_1, y_2, \ldots, y_{j-1}, y_{j+1}, \ldots, y_n\}$$

is linearly dependent. Using Lemma 8.1.4 and the same argument as before, an element y_k of this set can be removed and the resulting set

$$\{x_1, x_2, y_1, y_2, \ldots, y_{j-1}, y_{j+1}, \ldots, y_{k-1}, y_{k+1}, \ldots, y_n\}$$

will still span E. If $m > n$ and the process of adding an x_i, removing a y_i, and producing a set that still spans E is continued, then all of y_i's will be exhausted before all of the x_i's are used. The last step that adds x_n and removes a y_i produces the set $\{x_1, x_2, \ldots, x_n\}$, that spans E. But then x_{n+1} is a linear combination of the elements in $\{x_1, x_2, \ldots, x_n\}$, and this contradicts the linear independence of B. Thus, it must be the case that $m \leq n$. If $n > m$, reversing the roles of B and B' in this argument shows that $n \leq m$. Hence, $m = n$, so the number of elements in a basis of the field extension E/F is unique. ∎

8.1.8   Definition

> If a field extension E/F has a basis with a finite number of elements, then E/F is said to be a **finite dimensional field extension**. The unique number, denoted by $[E : F]$, of elements in a finite basis of E is said to be the **dimension** of E over F.

8.1.9   Theorem

> If E/K and K/F are finite dimensional field extensions, then E/F is a finite dimensional field extension and $[E : K][K : F] = [E : F]$.

Proof. Obviously, E/F is a field extension since $F \subseteq K \subseteq E$ and when F is a subfield of K and K is a subfield of E, then F is a subfield of E. Now suppose that $[E : K] = m$ and $[K : F] = n$. Let $B = \{x_1, x_2, \ldots, x_m\}$ and $B' = \{y_1, y_2, \ldots, y_n\}$ be bases of the field extensions E/K and K/F, respectively. We claim that $B'' = \{x_i y_j \mid 1 \leq i \leq m,\ 1 \leq j \leq n\}$ is a basis of E/F. The set B'' is nothing more than the set of all possible products of an element of B and an element of B'. First, note that $x_i y_j \neq 0$ for any i and j since x_i and x_j are elements of E and a field has no zero-divisors. It is also the case that no two elements of B'' can be equal. If $x_i y_j = x_s y_t$, then $x_i y_j - x_s y_t = 0$, and we have a linear combination of the x_i's that is equal to zero in which some of the coefficients are nonzero. This says that B is not a linearly independent set, and this is a clear contradiction. Hence, B'' has mn elements. If we can show that B'' is a basis for the field extension E/F, then the proof will be complete.

What we are going to prove is that

$$\left. \begin{array}{l} E \text{ has } B \text{ as a basis over } K \\ K \text{ has } B' \text{ as a basis over } F \end{array} \right\} \Rightarrow \left\{ \begin{array}{l} E \text{ has } B'' \text{ as} \\ \text{a basis over } F. \end{array} \right.$$

The first step is to show that B'' spans E. If $x \in E$, then since B is a basis for E/K, there are elements $c_1, c_2, \ldots, c_n \in K$ such that

$$x = c_1 x_1 + c_2 x_2 + \cdots + c_m x_m. \tag{1}$$

But B' is a basis for K/F, so we see that

$$\begin{aligned} c_1 &= f_{11} y_1 + f_{12} y_2 + \cdots + f_{1n} y_n, \\ c_2 &= f_{21} y_1 + f_{22} y_2 + \cdots + f_{2n} y_n, \\ &\vdots \\ c_m &= f_{m1} y_1 + f_{m2} y_2 + \cdots + f_{mn} y_n, \end{aligned} \tag{2}$$

where each $f_{ij} \in F$. If the right side of each of the equations in (2) is substituted for the corresponding c_i in (1) and "multiplied out," then a little thought shows that the result is a linear combination of the $x_i y_j$'s with coefficients in F. Thus, B'' spans E relative to F. Finally, we need to show that B'' is linearly independent relative to F. Suppose that we make the substitutions just suggested but do not "multiply out" the result. If this is equal to zero, then

$$\begin{aligned} &(f_{11} y_1 + f_{12} y_2 + \cdots + f_{1n} y_n) x_1 + \\ &(f_{21} y_1 + f_{22} y_2 + \cdots + f_{2n} y_n) x_2 + \\ &\qquad\qquad \vdots \\ &(f_{m1} y_1 + f_{m2} y_2 + \cdots + f_{mn} y_n) x_m = 0. \end{aligned}$$

Now B is linearly independent with respect to K, and each coefficient of an x_i is actually a c_i that is in K. Thus,

$$f_{11}y_1 + f_{12}y_2 + \cdots + f_{1n}y_n = 0$$
$$f_{21}y_1 + f_{22}y_2 + \cdots + f_{2n}y_n = 0$$
$$\vdots$$
$$f_{m1}y_1 + f_{m2}y_2 + \cdots + f_{mn}y_n = 0$$

It now follows that each $f_{ij} = 0$ since B' is linearly independent with respect to F. Thus, B'' is linearly independent with respect to F, so B'' is a basis for the field extension E/F. This also shows that $[E:K][K:F] = [E:F]$ since B'' is a basis with mn elements. ∎

The proof of Theorem 8.1.9 establishes the useful fact that if E/K and K/F are finite dimensional field extensions with bases $B = \{x_1, x_2, \ldots, x_m\}$ and $B' = \{y_1, y_2, \ldots, y_n\}$, respectively, then $B'' = \{x_i y_j \mid x_i \in B \text{ and } y_j \in B' \text{ for } i = 1, 2, \ldots, m \text{ and } j = 1, 2, \ldots, n\}$ is a basis of E/F. Thus, we can obtain a basis for E/F by forming all possible products of an element of B and an element of B'.

| 8.1.10 | **Example** |

The set $B = \{1, \sqrt{2}\}$ is a basis of the field extension $\mathbf{Q}[\sqrt{2}]/\mathbf{Q}$. Now let $\mathbf{Q}[\sqrt{2}][i] = \{x + yi \mid x, y \in \mathbf{Q}[\sqrt{2}]\}$. It can be shown that $\mathbf{Q}[\sqrt{2}][i]$ is a field that contains $\mathbf{Q}[\sqrt{2}]$ as a subfield. A basis for the field extension $\mathbf{Q}[\sqrt{2}][i]/\mathbf{Q}[\sqrt{2}]$ is $B' = \{1, i\}$. Now $\mathbf{Q} \subseteq \mathbf{Q}[\sqrt{2}] \subseteq \mathbf{Q}[\sqrt{2}][i]$, so we have a field extension $\mathbf{Q}[\sqrt{2}][i]/\mathbf{Q}$. The proof of Theorem 8.1.9 shows that a basis B'' for $\mathbf{Q}[\sqrt{2}][i]/\mathbf{Q}$ is the set of all possible products of an element of B and an element of B'. Hence, $B'' = \{1, \sqrt{2}, i, \sqrt{2}\,i\}$ is a basis for $\mathbf{Q}[\sqrt{2}][i]/\mathbf{Q}$.

If E/F and K/F are field extensions of F and $f : E \longrightarrow K$ is a function such that $f(c) = c$ for all $c \in F$, then we say that f **leaves each element of F fixed** or that f **leaves F fixed elementwise**.

| 8.1.11 | **Theorem** |

Let E/F and K/F be finite dimensional field extensions of F. If there is a ring isomorphism $f : E \longrightarrow K$ that leaves F fixed elementwise, then $[E:F] = [K:F]$.

Proof.   Let $B = \{x_1, x_2, \ldots, x_m\}$ be a basis for E/F, and suppose that a ring isomorphism $f: E \longrightarrow K$ exists that leaves each element of F fixed. We claim that $B' = \{f(x_1), f(x_2), \ldots, f(x_n)\}$ is a basis for the field extension K/F. If $c_1 f(x_1) + c_2 f(x_2) + \cdots + c_n f(x_n) = 0$, then $f(c_1)f(x_1) + f(c_2)f(x_2) + \cdots + f(c_n)f(x_n) = 0$ since $f(c_i) = c_i$ for $i = 1, 2, \ldots, n$. Using the fact that f is a ring isomorphism gives

$$f(c_1 x_1) + f(c_2 x_2) + \cdots + f(c_n x_n) = 0 \quad \text{or}$$

$$f(c_1 x_1 + c_2 x_2 + \cdots + c_n x_n) = f(0) = 0.$$

But f is an injective function, so $c_1 x_1 + c_2 x_2 + \cdots + c_n x_n = 0$. Hence, we see that $c_1 = c_2 = \cdots = c_n$ since B is linearly independent. Thus, B' is linearly independent. It remains only to show that B' spans K. If $y \in K$, since f is a surjective function, there is an $x \in E$ such that $f(x) = y$. Now B spans E, and so $x = c_1 x_1 + c_2 x_2 + \cdots + c_n x_n$, where $c_1, c_2, \ldots, c_n \in F$. Hence,

$$
\begin{aligned}
y &= f(x) \\
&= f(c_1 x_1 + c_2 x_2 + \cdots + c_n x_n) \\
&= f(c_1 x_1) + f(c_2 x_2) + \cdots + f(c_n x_n) \\
&= f(c_1)f(x_1) + f(c_2)f(x_2) + \cdots + f(c_n)f(x_n) \\
&= c_1 f(x_1) + c_2 f(x_2) + \cdots + c_n f(x_n)
\end{aligned}
$$

Thus, B' spans K, so B' is a basis for K/F. Since B and B' both have n elements, this shows that $[E : F] = [K : F]$. ∎

Problem Set 8.1

1. Determine whether each of the following is a basis of the given field extension.

   (a) $\{2, 1 + i, 3 - 2i\}$, **C**/**R**

   (b) $\left\{\dfrac{1}{2}, \dfrac{4}{3}i\right\}$, **C**/**R**

   (c) $\left\{\dfrac{3}{4}, \dfrac{2}{3} + \dfrac{1}{2}i\right\}$, **C**/**R**

   (d) $\{1, \sqrt{2}\}$, **R**/**Q**

   (e) $\{5\}$, **Q**/**Q**

   (f) $\{\sqrt{2}, \sqrt{3} + i, \sqrt{3} - i\}$, **C**/**R**

2. The polynomial $f(x) = x^3 + x^2 + x + 2$ is irreducible in $\mathbf{Q}[x]$, so $\mathbf{Q}[x]/(f(x))$ is a field that extends $\mathbf{Q}$. Find a basis for this extension.

3. Use the argument given in Example 8.1.6 to show that if a field extension E/F has a basis with two elements, then it is impossible for E/F to have a basis $B = \{x_1, x_2, \ldots, x_n\}$ with $n \geq 3$.

4. If $\{x_1, x_2, x_3\}$ is a basis for a field extension E/F, is $\{x_1, x_1 + x_2, x_1 + x_2 + x_3\}$ also a basis for E/F?

5. If $B = \{x_1, x_2, \ldots, x_n\}$ is a basis of a field extension E/F, prove that every nonempty subset of B is linearly independent but that a proper subset of B cannot be a basis of the field extension.

6. Prove that $\{1, x\}$ is a linearly independent subset of the field extension $\mathbf{R}/\mathbf{Q}$ if and only if x is an irrational number.

7. If $B = \{x_1, x_2, \ldots, x_n\}$ is a basis of a field extension E/F and c is a nonzero element of F, prove that $cB = \{cx_1, cx_2, \ldots, cx_n\}$ is also a basis of E/F.

8. If $B = \{x_1, x_2, \ldots, x_n\}$ is a linearly independent subset of the field extension E/F and $x \in E$ is not a linear combination of the x_i's, prove that the set $B \cup \{x\} = \{x, x_1, x_2, \ldots, x_n\}$ is also linearly independent.

9. If E/F is a finite dimensional field extension, prove that $[E : F] = 1$ if and only if $E = F$.

10. Let E/F be a finite dimensional field extension. If S is a finite subset of nonzero elements of E that spans E, prove that S contains a basis of E/F.

11. Let E/F be a finite dimensional field extension and S a linearly independent subset of E. Prove there is a basis B of E such that $S \subseteq B$. Conclude that any linearly independent subset of E can be extended to a basis of E. Hint: See Exercise 8.

12. Let E/F be a finite dimensional field extension. Prove that the following are equivalent:

    (a) $B = \{x_1, x_2, \ldots, x_n\}$ is a basis for E/F.
    (b) B is a minimal spanning set for E over F. That is, a set $S \subseteq E$ cannot be found that spans E and $S \subset B$.
    (c) B is a maximal linearly independent subset of E relative to F. That is, a linearly independent subset S of E cannot be found such that $B \subset S$.

13. If E/F is a field extension, then a field K such that $F \subseteq K \subseteq E$ is said to be an **intermediate field** of the field extension. If $K \neq E$, then K is said to be a **proper intermediate field**. For a finite dimensional field extension E/F, show that if $[E : F] = p$ is a prime, then F is the only proper intermediate field.

8.2   SIMPLE FIELD EXTENSIONS

Section Overview. If F is a field and $f(x) \in F[x]$ has a root c in a field extension E of F, how does one go about determining whether a smallest field extension of F exists that contains c? Moreover, if such a field extension does exist, can its elements be described in a satisfactory way? And in what sense, if any, is this extension field of F unique? It is the purpose of this section to investigate these questions. Important concepts you should master are simple field extension, algebraic element, and minimal polynomial.

If $f(x)$ is an irreducible polynomial in $F[x]$ of degree 2 or greater, then $f(x)$ cannot have a root in F because if $c \in F$ is a root of $f(x)$, then we know that $x - c$ is a factor of $f(x)$. However, it may be the case that $f(x)$ has a root in some field extension of F. As we saw in the previous section, $f(x) = x^2 - 2$ does not have a root in $\mathbf{Q}$, but $f(x)$ does have a root in $\mathbf{R}$, namely, $\sqrt{2}$ (or $-\sqrt{2}$). We also saw in the previous section that $\mathbf{R}$ is not the smallest extension field of $\mathbf{Q}$ that contains $\sqrt{2}$. Indeed, $\sqrt{2} \in \mathbf{Q}[\sqrt{2}]$, and $\mathbf{Q}[\sqrt{2}]$ is an extension field of $\mathbf{Q}$ contained in $\mathbf{R}$. The structure of $\mathbf{Q}[\sqrt{2}]$ is explicitly known since its elements are of the form $a + b\sqrt{2}$, where a and b are in $\mathbf{Q}$. The question now becomes: Is $\mathbf{Q}[\sqrt{2}]$ the smallest field extension of $\mathbf{Q}$ that contains $\sqrt{2}$? As we shall see, the answer is yes. To answer this in a more general context, we investigate the situation where $f(x) \in F[x]$ has a root c in a field extension E of F. Our goal is to determine whether a smallest field extension of F exists that contains c. Before turning our attention to this question, there are several important points that need to be made.

Let E/F be a field extension, and suppose that $c \in E$.

Point 1: The intersection of all the subrings of E that contain F and c is the polynomial ring $F[c]$. If $\{R_\alpha\}_{\alpha \in \Delta}$ is the family of subrings of E that contain F and c, then we claim that $\cap_{\alpha \in \Delta} R_\alpha = F[c]$, the polynomial ring in c with coefficients in F. If x and y are in $\cap_{\alpha \in \Delta} R_\alpha$, then $x, y \in R_\alpha$ for each $\alpha \in \Delta$, so both $x - y$ and xy are in R_α for each $\alpha \in \Delta$. Hence, $x - y, xy \in \cap_{\alpha \in \Delta} R_\alpha$, and this shows that $\cap_{\alpha \in \Delta} R_\alpha$ is a subring of E. $F[c]$ is a subring of E that contains F and c, so $F[c]$ is one of the rings in the family $\{R_\alpha\}_{\alpha \in \Delta}$. Hence, $\cap_{\alpha \in \Delta} R_\alpha \subseteq F[c]$. Now $\cap_{\alpha \in \Delta} R_\alpha$ contains F and c, so polynomials in c of the form $a_0 + a_1 c + \cdots + a_n c^n$ are in $\cap_{\alpha \in \Delta} R_\alpha$ since $\cap_{\alpha \in \Delta} R_\alpha$ is closed under addition and multiplication. Therefore $F[c] \subseteq \cap_{\alpha \in \Delta} R_\alpha$, and thus $\cap_{\alpha \in \Delta} R_\alpha = F[c]$.

Point 2: The intersection $F(c)$ of all the subfields of E that contain both F and c is the smallest subfield of E that contains $F[c]$. Let $\{E_\alpha\}_{\alpha \in \Delta}$ be the family of subfields of E that contain F and c. Then $F[c] \subseteq E_\alpha$ for each $\alpha \in \Delta$, and by the same method used in point 1, it can be shown that $F(c) = \cap_{\alpha \in \Delta} E_\alpha$ is a subring of E that contains $F[c]$. $F(c)$ is a subfield of E because if $x \in F(c)$, $x \neq 0$, then $x \in E_\alpha$ for each $\alpha \in \Delta$, so x^{-1} is in each E_α. Hence, $x^{-1} \in F(c)$ and thus $F(c)$ is a field. $F(c)$ is the smallest subfield of E that contains $F[c]$. To see this, let E' be a subfield of E that contains $F[c]$, and suppose that $E' \subseteq F(c)$. Since E' is a member of the family $\{E_\alpha\}_{\alpha \in \Delta}$, $F(c) = \cap_{\alpha \in \Delta} E_\alpha \subseteq E'$, so $F(c) = E'$. Notice the distinction between the brackets [ ] in $F[c]$ and the parentheses ( ) in $F(c)$. $F[c]$ is the *ring of polynomials* in c while $F(c)$ denotes the *smallest subfield of E that contains $F[c]$.*

Point 3: $F(c)$ is isomorphic to the field of fractions of $F[c]$. Since $F[c] \subseteq E$ and since E is a field, the polynomial ring $F[c]$ is an integral domain, so $F[c]$ has a field of fractions $Q(F[c])$. Elements of $Q(F[c])$ look like congruence classes $\frac{f(c)}{g(c)}$, where $f(c), g(c) \in F[c]$ with $g(c) \neq 0$. Recall that two congruence classes $\frac{f(c)}{g(c)}$ and $\frac{h(c)}{k(c)}$ are equal if and only if $f(c)k(c) = g(c)h(c)$. The map $\phi : Q(F[c]) \longrightarrow E$ defined by $\phi\left(\frac{f(c)}{g(c)}\right) = f(c)g(c)^{-1}$ is well-defined because if $\frac{f(c)}{g(c)} = \frac{h(c)}{k(c)}$, then $f(c)k(c) = g(c)h(c)$, so $f(c)g(c)^{-1} = h(c)k(c)^{-1}$ in E. Hence, $\phi\left(\frac{f(c)}{g(c)}\right) = \phi\left(\frac{h(c)}{k(c)}\right)$. It is straightforward to show that ϕ is a ring monomorphism, so E contains a copy $\overline{Q(F[c])}$ of $Q(F[c])$. The field $\overline{Q(F[c])}$ is a subfield of E that contains F and c, so it follows immediately that $F(c) \subseteq \overline{Q(F[c])}$ because $F(c)$ is the intersection of all the subfields of E that contain F and c. If $f(c)g(c)^{-1}$ is in $\overline{Q(F[c])}$, then $f(c)$, $g(c) \in F[c] \subseteq F(c)$ and $g(c) \neq 0$. But $F(c)$ is a field, so $f(c)g(c)^{-1} \in F(c)$. Thus, $\overline{Q(F[c])} \subseteq F(c)$, so $F(c) = \overline{Q(F[c])}$. Since $Q(F[c])$ embeds in every field E' that contains $F[c]$ (just use the map ϕ given earlier with E replaced by E'), $F(c)$ is the smallest field extension of F that contains $F[c]$, not only in E but in every field E' that contains $F[c]$. Finally, $F(c)$ is unique up to isomorphism in the sense that it is isomorphic to the field of fractions of $F[c]$.

With the notation of points 1, 2, and 3, we have established the following theorem.

8.2.1 Theorem

Let E/F be a field extension, and suppose that $c \in E$.

1. The smallest subring of E that contains F and c is $F[c]$.
2. The smallest subfield of E that contains $F[c]$ is $F(c)$.
3. The field $F(c)$ is isomorphic to the field of fractions of $F[c]$, so $F(c)$ is unique up to isomorphism.

With this theorem in mind, let's return to an investigation of the problem of trying to find the smallest field extension of $\mathbf{Q}$ that contains a root of the polynomial $x^2 - 2 \in \mathbf{Q}[x]$. Since $\sqrt{2} \in \mathbf{R}$ is a root of $f(x)$, there is a field extension that contains a root of $f(x)$, namely, $\mathbf{R}$. Theorem 8.2.1 shows that $\mathbf{Q}(\sqrt{2})$, the smallest field extension that contains $\mathbf{Q}[\sqrt{2}]$, is isomorphic to the field of fractions of $\mathbf{Q}[\sqrt{2}]$. Next let's look at the polynomial ring $\mathbf{Q}[\sqrt{2}]$. Notice that a polynomial in $\mathbf{Q}[\sqrt{2}]$ such as

$$f(\sqrt{2}) = \tfrac{1}{2} + 2\sqrt{2} + \tfrac{2}{3}(\sqrt{2})^2 - \tfrac{5}{3}(\sqrt{2})^3 + (\sqrt{2})^4$$

reduces to

$$f(\sqrt{2}) = \tfrac{1}{2} + 2\sqrt{2} + \tfrac{4}{3} - \tfrac{10}{3}\sqrt{2} + 4$$

$$= \tfrac{35}{6} - \tfrac{4}{3}\sqrt{2}.$$

For a general polynomial $f(\sqrt{2}) = a_0 + a_1\sqrt{2} + a_2(\sqrt{2})^2 + \cdots + a_n(\sqrt{2})^n$ in $\mathbf{Q}[\sqrt{2}]$, let $a_m(\sqrt{2})^m$ be any term of $f(\sqrt{2})$. Then m is either an even or an odd integer, so

$$a_m(\sqrt{2})^m = \begin{cases} a_m 2^k & \text{if } m = 2k, \quad k \text{ an integer.} \\ a_m 2^k \sqrt{2} & \text{if } m = 2k+1, \ k \text{ an integer.} \end{cases}$$

This shows that every polynomial in $\mathbf{Q}[\sqrt{2}]$ reduces to $f(\sqrt{2}) = a + b\sqrt{2}$, where $a, b \in \mathbf{Q}$, so $\mathbf{Q}[\sqrt{2}] = \{a + b\sqrt{2} \mid a, b \in \mathbf{Q}\}$. Remember, this is how $\mathbf{Q}[\sqrt{2}]$ was actually defined in earlier chapters. Since $\mathbf{Q}(\sqrt{2})$ is a field of fractions of $\mathbf{Q}[\sqrt{2}]$, every element of $\mathbf{Q}(\sqrt{2})$ looks like a congruence class $\frac{a+b\sqrt{2}}{c+d\sqrt{2}}$, where $a, b, c, d \in \mathbf{Q}$ and $c + d\sqrt{2} \neq 0$. If the numerator and the denominator of $\frac{a+b\sqrt{2}}{c+d\sqrt{2}}$ are multiplied by $c - d\sqrt{2}$, we see that

$$\frac{a+b\sqrt{2}}{c+d\sqrt{2}} = \frac{(a+b\sqrt{2})(c-d\sqrt{2})}{(c+d\sqrt{2})(c-d\sqrt{2})} = \frac{ac-2bd}{c^2-2d^2} + \frac{bc-ad}{c^2-2d^2}\sqrt{2}. \qquad (*)$$

Since at least one of c and d is nonzero and since $\sqrt{2}$ is not a rational number, it can be shown that $\frac{ac-2bd}{c^2-2d^2}$ and $\frac{bc-ad}{c^2-2d^2}$ are in $\mathbf{Q}$. Hece, equation (*) shows that $\frac{a+b\sqrt{2}}{c+d\sqrt{2}} \in$ $\mathbf{Q}[\sqrt{2}]$, so we have $\mathbf{Q}(\sqrt{2}) \subseteq \mathbf{Q}[\sqrt{2}]$. Therefore, $\mathbf{Q}[\sqrt{2}] = \mathbf{Q}(\sqrt{2})$, so $\mathbf{Q}[\sqrt{2}]$ is actually a field. Not only is $\mathbf{Q}[\sqrt{2}]$ a field, it is the smallest field extension of $\mathbf{Q}$ that contains $\sqrt{2}$, and this is exactly the field we are looking for! The polynomial $f(x) = x^2 - 2$ is irreducible in $\mathbf{Q}[x]$, but $f(x)$ factors as $f(x) = (x - \sqrt{2})(x + \sqrt{2})$ in $\mathbf{Q}[\sqrt{2}][x]$. Because $f(x)$ factors into a product of linear factors in the polynomial ring $\mathbf{Q}[\sqrt{2}][x]$, we say that $f(x)$ **splits** over $\mathbf{Q}[\sqrt{2}]$. As we shall see, this example actually tells much of the story of what is to follow.

8.2.2   **Definition**

Suppose that E/F is a field extension, and let c be an element of E. The subfield $F(c)$ is said to be the field obtained from F by **adjoining the element c to F.** The field $F(c)$ is also referred to as a **simple field extension of F.** The term *simple* indicates that a single element of E has been adjoined to F. An element $c \in E$ is said to be **algebraic** over F if there is a nonzero polynomial $f(x) \in F[x]$ such that c is a root of $f(x)$. If such a polynomial does not exist, then c is **transcendental** over F.

8.2.3   **Examples**

1. The complex number $3i \in \mathbf{C}$ is algebraic over $\mathbf{R}$ since $3i$ is the root of $x^2 + 9 \in \mathbf{R}[x]$.
2. The irrational number $-\sqrt{5} \in \mathbf{R}$ is algebraic over $\mathbf{Q}$ since $-\sqrt{5}$ is a root of $x^2 - 5 \in \mathbf{Q}[x]$.
3. Every element c of a field F is algebraic over F since c is a root of $x - c \in F[x]$.
4. It is known that π and the base of the natural logarithmic function e are both transcendental over $\mathbf{Q}$, although neither of these facts has an easy proof. The fact that e is transcendental over $\mathbf{Q}$ was proven by **Charles Hermite** (1822–1901) in 1873, and **Ferdinand von Lindemann** (1852–1939) did the same for π in 1882. It is still unknown whether or not $\pi + e$ is transcendental over $\mathbf{Q}$.

It is sometimes the case that $F[c] = F(c)$, and in these instances $F(c)$ is the preferred notation. We have already seen that $\mathbf{Q}[\sqrt{2}] = \mathbf{Q}(\sqrt{2})$. The following example gives another instance where these two seemingly quite different rings $F[c]$ and $F(c)$ are equal.

| 8.2.4 | **Example** |

The complex number i is algebraic over $\mathbf{Q}$ since i is the root of the polynomial $x^2 + 1 \in \mathbf{Q}[x]$. Since $i^n = \pm 1$ when n is an even integer and $i^n = \pm i$ when n is an odd integer, any polynomial $f(i)$ in $\mathbf{Q}[i]$ reduces to the form $a + bi$ where $a, b \in \mathbf{Q}$. Elements of $\mathbf{Q}(i)$ look like $\frac{a + bi}{c + di}$, where $c + di \neq 0$. Since

$$\frac{a + bi}{c + di} = \frac{(a + bi)(c - di)}{(c + di)(c - di)} = \frac{ac + bd}{c^2 + d^2} + \frac{bc - ad}{c^2 + d^2}i \in \mathbf{Q}[i],$$

it follows that $\mathbf{Q}[i] = \mathbf{Q}(i)$. The polynomial $g(x) = x^2 + 1$ does not factor in $\mathbf{Q}[x]$ but does factor as $g(x) = (x + i)(x - i)$ in the polynomial ring $\mathbf{Q}(i)[x]$. Hence, $g(x)$ splits over $\mathbf{Q}(i)$. Hence, $\mathbf{Q}(i)$ is the smallest field extension of $\mathbf{Q}$ over which $g(x)$ factors into a product of linear factors. Notice that $\mathbf{Q}(i)$ is not a subfield of $\mathbf{R}$, but rather it is a subfield of $\mathbf{C}$.

The technique we have used to show that $\mathbf{Q}[\sqrt{2}] = \mathbf{Q}(\sqrt{2})$ and that $\mathbf{Q}[i] = \mathbf{Q}(i)$ involves determining what the reduced form of the polynomials in $\mathbf{Q}[\sqrt{2}]$ and $\mathbf{Q}[i]$ look like. In more complicated situations, this method can become quite cumbersome. For example, the real number $1 + \sqrt{1 + \sqrt{3}}$ is algebraic over $\mathbf{Q}$ since $1 + \sqrt{1 + \sqrt{3}}$ is a root of $f(x) = x^4 - 4x^3 + 4x^2 - 3$. Using the method described earlier, the first step in showing that $\mathbf{Q}[1 + \sqrt{1 + \sqrt{3}}] = \mathbf{Q}(1 + \sqrt{1 + \sqrt{3}})$ is to determine how polynomials in $1 + \sqrt{1 + \sqrt{3}}$ with coefficients in $\mathbf{Q}$ can be simplified, and this can be somewhat tedious. Fortunately, there is a better method based on the idea of a **minimal polynomial**. The following theorem lays the groundwork for this concept.

| 8.2.5 | **Theorem** |

Let E/F be a field extension, and suppose that $c \in E$. Then c is algebraic over F if and only if there is a unique monic irreducible polynomial $m(x) \in F[x]$ that has c as a root. Moreover, $m(x)$ satisfies the following two conditions.

1. There is no polynomial $g(x) \in F[x]$, $g(x) \neq 0$, that has c as a root such that $\deg(g(x)) < \deg(m(x))$.

2. If $g(x) \in F[x]$, $g(x) \neq 0$, has c as a root, then $m(x) \mid g(x)$.

Proof. Let $c \in E$ be algebraic over F. Then there is a nonconstant polynomial $f(x) \in F[x]$ that has c as a root, so the set of nonconstant polynomials in $F[x]$ that have c as a root is nonempty. Since $\mathbf{N}$ is well-ordered, there is a polynomial in $F[x]$ of minimal degree that has c as a root. (Can you explain why?) If $f(x)$ is such a polynomial, then $\deg(f(x)) \geq 1$. If $g(x) \in F[x]$ has c for a root, then $\deg(f(x)) \leq \deg(g(x))$. Thus, by the Division Algorithm for Polynomials, there are $q(x), r(x) \in F[x]$ such that $g(x) = q(x)f(x) + r(x)$, where $r(x) = 0$ or $\deg(r(x)) < \deg(f(x))$. Since $r(x) = g(x) - q(x)f(x)$ and $r(c) = g(c) - q(c)f(c) = 0$, it follows that $r(x) = 0$. (If $r(x) \neq 0$, then $f(x)$ is not a polynomial in $F[x]$ of minimal degree that has c as a root.) Hence, $g(x) = q(x)f(x)$ and thus $f(x) \mid g(x)$.

Next, we claim that $f(x)$ is irreducible in $F[x]$. If not, there are polynomials $g(x)$, $h(x) \in F[x]$ such that $f(x) = g(x)h(x)$ with $1 \leq \deg(g(x)) < \deg(f(x))$ and $1 \leq \deg(h(x)) < \deg(f(x))$. Since $0 = f(c) = g(c)h(c) \in E$ and since a field has no zero-divisors, either $g(c) = 0$ or $h(c) = 0$. But this contradicts the fact that $f(x)$ is a polynomial in $F[x]$ of minimal degree that has c as a root. Thus, $f(x)$ is irreducible in $F[x]$.

If a is the leading coefficient of $f(x)$ and we let $m(x) = a^{-1}f(x)$, then $m(x)$ is a monic irreducible polynomial in $F[x]$ that has c as a root. From the way $f(x)$ was chosen, $m(x)$ has the smallest degree of any nonzero polynomial in $F[x]$ that has c as a root. Furthermore, since $f(x) \mid g(x)$ for each $g(x)$ in $F[x]$ that has c as a root, we see that $m(x)$ also divides each such $g(x)$. Finally, if $m'(x)$ is another such monic irreducible polynomial in $F[x]$, then $m(x) \mid m'(x)$ and $m'(x) \mid m(x)$. But because $m(x)$ and $m'(x)$ are both monic and irreducible, $m(x) = m'(x)$. Therefore, $m(x)$ is unique.

Conversely, if there is a monic irreducible polynomial in $F[x]$ that has c as a root, then it follows immediately that c is algebraic over F. ∎

8.2.6   **Definition**

If E/F is a field extension and $c \in E$ is algebraic over F, then the unique monic irreducible polynomial $m(x)$ that has c as a root is called the **minimal polynomial in $F[x]$ of c.** The degree of $m(x)$ is the **degree of c over F.**

One point you should remember from Theorem 8.2.5 is that when c is algebraic over F, there is only one monic irreducible polynomial in $F[x]$ that has c as a root. Consequently, *when you find a monic irreducible polynomial in $F[x]$ that has c as a root, this polynomial is the minimal polynomial in $F[x]$ of c.*

8.2.7 | Examples

1. The complex number i is algebraic over $\mathbf{Q}$ since it is the root of $f(x) = x^2 + 1 \in \mathbf{Q}[x]$. The polynomial $f(x)$ is monic and irreducible in $\mathbf{Q}[x]$, so $f(x)$ is the minimal polynomial in $\mathbf{Q}[x]$ of i. Since $\deg(f(x)) = 2$, the degree of i over $\mathbf{Q}$ is 2.

2. The polynomial $f(x) = x^4 - 4x^3 + 4x^2 - 3 \in \mathbf{Q}[x]$ factors as $(x^2 - 2x - \sqrt{3})(x^2 - 2x + \sqrt{3})$, and neither factor is in $\mathbf{Q}[x]$. The Rational Root Theorem 6.3.8 shows that $f(x)$ has no rational roots. Hence, $f(x)$ is irreducible in $\mathbf{Q}[x]$. The real number $1 + \sqrt{1 + \sqrt{3}}$ is algebraic over $\mathbf{Q}$ since it is a root of $f(x)$. Since $f(x)$ is monic and irreducible in $\mathbf{Q}[x]$, $f(x)$ is the minimal polynomial in $\mathbf{Q}[x]$ of $1 + \sqrt{1 + \sqrt{3}}$. Note also that $\deg(f(x)) = 4$, so the degree of $1 + \sqrt{1 + \sqrt{3}}$ over $\mathbf{Q}$ is 4.

We now come to an important theorem. It shows the role that the minimal polynomial plays in determining the structure of the field extension $F(c)/F$ when c is algebraic over F. It tells us exactly what $F(c)$ looks like.

8.2.8 | Theorem

Let E/F be a field extension, and suppose that $c \in E$. If c is algebraic over F with minimal polynomial $m(x) \in F[x]$ of degree n, then

1. $F[c] = F(c)$ and $F(c)$ is isomorphic to $F[x]/(m(x))$,
2. $B = \{e, c, c^2, \ldots, c^{n-1}\}$ is a basis for the field extension $F(c)/F$, and
3. $[F(c) : F] = n$.

Proof

1. The mapping $\phi : F[x] \longrightarrow F[c]$ given by $\phi(f(x)) = f(c)$ is easily shown to be a well-defined ring epimorphism. If $K = \ker \phi$, then $f(x) \in K$, if and only if $f(c) = 0$. Hence, $m(x) \in K$, and $m(x)$ is of minimal degree with regard to the polynomials in K. Thus, it follows from Theorem 8.2.5 that $m(x) \mid f(x)$ for each $f(x) \in K$, so $K = (m(x))$. Since $m(x)$ is irreducible in $F[x]$, Theorem 7.1.10 shows that $F[x]/(m(x))$ is a field. By the First Isomorphism Theorem for Rings 3.3.12, we know that

$$\phi^* : F[x]/(m(x)) \longrightarrow F[c] \text{ defined by } \phi^*(f(x) + (m[x])) = f(c)$$

is a ring isomorphism from $F[x]/(m(x))$ to $F[c]$. Thus, $F[c]$ is a field, so $F[c] = F(c)$.

2. By Lemma 7.1.7, every element $f(x) + (m(x)) \in F[x] / (m(x))$ can be expressed as $r(x) + (m(x))$, where $r(x)$ is the remainder of the division of $f(x)$ by $m(x)$. Thus, $r(x) = a_0 + a_1 x + a_2 x^2 + \cdots + a_{n-1} x^{n-1}$, where some or all of the a_i's may be zero. The isomorphism

$$\phi^* : F[x] / (m(x)) \longrightarrow F[c] = F(c)$$

is such that $\phi^*(f(x) + (m[x])) = r(c) = a_0 e + a_1 c + a_2 c^2 + \cdots + a_{n-1} c^{n-1}$. Therefore, every element of $F(c)$ can be expressed as a linear combination of the elements in $\{e, c, c^2, \ldots, c^{n-1}\}$ with the coefficients in F, so this set spans $F(c)$. If $a_0 e + a_1 c + a_2 c^2 + \cdots + a_{n-1} c^{n-1} = 0$, then c is a root of the polynomial $g(x) = a_0 + a_1 x + a_2 x^2 + \cdots + a_{n-1} x^{n-1}$ in $F[x]$. But this is impossible since $\deg(g(x)) < \deg(m(x))$ and $m(x)$ is of minimal degree among the nonzero polynomials in $F[x]$ that have c as a root. Thus, $g(x)$ must be the zero polynomial, and this implies that $a_0 = a_1 = a_2 = \cdots = a_{n-1} = 0$. Therefore the set $\{e, c, c^2, \ldots, c^{n-1}\}$ is linearly independent with respect to F, so $\{e, c, c^2, \ldots, c^{n-1}\}$ is a basis for the field extension $F(c) / F$.

3. This follows immediately from part 2 since the field extension $F(c) / F$ has a basis with n elements. ∎

It is important to point out that Theorem 8.2.8 shows that since $F(c)$ is isomorphic to the field $F[x] / (m(x))$ when c is algebraic over F, the structure of $F(c)$ does not depend on the field extension E, but only on $F[x]$ and the polynomial $m(x)$. When we say that c is algebraic or transcendental over F, we are assuming that c lies in some field extension E of F. Nothing whatsoever need be known about E. The field extension E serves only as a field, where it can be shown that $f(c) = 0$.

8.2.9   Examples

1. The complex number $i \in \mathbf{C}$ is algebraic over $\mathbf{R}$, and the minimal polynomial of i in $\mathbf{R}[x]$ is $m(x) = x^2 + 1$. Since $\deg(m(x)) = 2$, $\mathbf{R}(i)$ has a basis with two elements. According to Theorem 8.2.8, $B = \{1, i\}$ is a basis for $\mathbf{R}(i)$. Since B also spans $\mathbf{C}$, $\mathbf{R}(i) = \mathbf{C}$.

2. The irrational number $\sqrt{3} \in \mathbf{R}$ is algebraic over $\mathbf{Q}$, and the minimal polynomial in $\mathbf{Q}[x]$ of $\sqrt{3}$ is $m(x) = x^2 - 3$. Theorem 8.2.8 indicates that $B = \{1, \sqrt{3}\}$ is a basis for $\mathbf{Q}(\sqrt{3})$, so $\mathbf{Q}(\sqrt{3}) = \{a1 + b\sqrt{3} \mid a, b \in \mathbf{Q}\}$. If $a + b\sqrt{3} \neq 0$ is in $\mathbf{Q}(\sqrt{3})$, to find the multiplicative inverse of $a + b\sqrt{3}$ in $\mathbf{Q}(\sqrt{3})$, simply perform the following computations.

$$\frac{1}{a + b\sqrt{3}} = \frac{a - b\sqrt{3}}{\left(a + b\sqrt{3}\right)\left(a - b\sqrt{3}\right)} = \frac{a}{a^2 - 3b^2} - \frac{b}{a^2 - 3b^2}\sqrt{3}$$

It is not difficult to show that since at least one of a and b must be a nonzero rational number and $\sqrt{3}$ is an irrational number,

$$\frac{a}{a^2 - 3b^2} - \frac{b}{a^2 - 3b^2}\sqrt{3} \in \mathbf{Q}(\sqrt{3}).$$

3. The polynomial $m(x) = x^4 - 2x^2 - 2$ has $\sqrt{1 + \sqrt{3}}$ as a root, so $\sqrt{1 + \sqrt{3}}$ is algebraic over $\mathbf{Q}$, and $m(x)$ is the minimal polynomial in $\mathbf{Q}[x]$ of $\sqrt{1 + \sqrt{3}}$. Since $\deg(m(x)) = 4$, the field extension $\mathbf{Q}(\sqrt{1 + \sqrt{3}})/\mathbf{Q}$ has a basis with four elements. Because of Theorem 8.2.8,

$$\left\{ 1, \sqrt{1 + \sqrt{3}}, \left(\sqrt{1 + \sqrt{3}}\right)^2, \left(\sqrt{1 + \sqrt{3}}\right)^3 \right\} \quad \text{or}$$

$$\left\{ 1, \sqrt{1 + \sqrt{3}}, 1 + \sqrt{3}, \left(1 + \sqrt{3}\right)\left(\sqrt{1 + \sqrt{3}}\right) \right\}$$

is a basis for $\mathbf{Q}(\sqrt{1 + \sqrt{3}})/\mathbf{Q}$. Elements of the field $\mathbf{Q}(\sqrt{1 + \sqrt{3}})$ are of the form

$$a + b\sqrt{1 + \sqrt{3}} + c(1 + \sqrt{3}) + d(1 + \sqrt{3})(\sqrt{1 + \sqrt{3}}),$$

where $a, b, c, d \in \mathbf{Q}$. Multiplicative inverses of nonzero elements of $\mathbf{Q}(\sqrt{1 + \sqrt{3}})$ may take a bit of work to compute. However, they can be found. A general method for finding inverses in a simple field extension is presented in the exercises.

8.2.10 Theorem

The following hold for any field extension E/F.

1. An element $c \in E$ is algebraic over F if and only if $F[c] = F(c)$.
2. An element $c \in E$ is transcendental over F if and only if $F[c] \subset F(c)$.
3. The field extension $F(c)/F$ is finite dimensional if and only if $c \in E$ is algebraic over F.

4. If $c_1, c_2 \in E$ are algebraic over F and have the same minimal polynomial in $F[x]$, then $F(c_1)$ and $F(c_2)$ are isomorphic fields.

Proof

1. Suppose that $c \in E$ is algebraic over F. Then Theorem 8.2.8 shows that $F[c] = F(c)$. Conversely, suppose that $F[c] = F(c)$. If $c = 0$, then c is the root of the polynomial $f(x) = x \in F[x]$, so c is algebraic over F. If $c \neq 0$, then c has a multiplicative inverse in the field $F(c)$. Hence, $c^{-1} \in F(c)$ is a polynomial in c. If $c^{-1} = a_0 + a_1 c + \cdots + a_n c^n$, then $e = a_0 c + a_1 c^2 + \cdots + a_n c^{n+1}$, and so $-e + a_0 c + a_1 c^2 + \cdots + a_n c^{n+1} = 0$. Thus, c is a root of the polynomial $f(x) = -e + a_0 x + a_1 x^2 + \cdots + a_n x^{n+1} \in F[x]$, so c is algebraic over F.
2. This follows immediately from part 1.
3. This is a consequence of part 1 and Theorem 8.2.8.
4. If $m(x) \in F[x]$ is the minimal polynomial for $c_1, c_2, \in E$, then, by Theorem 8.2.8, $F(c_1)$ and $F(c_2)$ are both isomorphic to $F[x] / (m(x))$ and hence to each other. ∎

8.2.11   Corollary

If $c_1, c_2, \ldots, c_n$ are algebraic over F with the same minimal polynomial in $F[x]$, then the fields $F(c_1), F(c_2), \ldots, F(c_n)$ are isomorphic.

When c_1 and c_2 are algebraic over F with the same minimal polynomial in $F[x]$, Theorem 8.2.10 shows that $F(c_1)$ and $F(c_2)$ are isomorphic simple field extensions of F. It is sometimes the case that not only are $F(c_1)$ and $F(c_2)$ isomorphic, but $F(c_1) = F(c_2)$. The first of the following two examples illustrates this fact, while the second example shows that this does not always have to be the case.

8.2.12   Examples

1. The monic polynomial $m(x) = x^2 - 7$ is clearly irreducible in $\mathbf{Q}[x]$. Thus, $m(x)$ is the minimal polynomial in $\mathbf{Q}[x]$ of $\sqrt{7}$ and $-\sqrt{7}$, so $\mathbf{Q}(\sqrt{7})$ and $\mathbf{Q}(-\sqrt{7})$ are isomorphic. Since $\mathbf{Q}(\sqrt{7})$ and $\mathbf{Q}(-\sqrt{7})$ have $\{1, \sqrt{7}\}$ and $\{1, -\sqrt{7}\}$ as bases, respectively, every element of $\mathbf{Q}(\sqrt{7})$ is of the form $a + b\sqrt{7}$, where $a, b \in \mathbf{Q}$, and every element of $\mathbf{Q}(-\sqrt{7})$ looks like $a + b(-\sqrt{7})$ with $a, b \in \mathbf{Q}$. Hence, $a + b\sqrt{7} \in \mathbf{Q}(\sqrt{7})$ implies that $a + (-b)(-\sqrt{7}) \in \mathbf{Q}(-\sqrt{7})$, and $a + b(-\sqrt{7}) \in \mathbf{Q}(-\sqrt{7})$ implies that $a + (-b)\sqrt{7} \in \mathbf{Q}(\sqrt{7})$. Therefore $\mathbf{Q}(\sqrt{7}) = \mathbf{Q}(-\sqrt{7})$.

2.   The polynomial $m(x) = x^3 - 5$ is irreducible in $\mathbf{Q}[x]$. Since $m(x)$ is monic, $m(x)$ is the minimal polynomial in $\mathbf{Q}[x]$ of $\sqrt[3]{5}$. If we set $\omega = -\frac{1}{2} + \frac{\sqrt{3}}{2}i$, then $\omega^3 = 1$. From this we see that $m(\omega\sqrt[3]{5}) = (\omega\sqrt[3]{5})^3 - 5 = \omega^3(\sqrt[3]{5})^3 - 5 = 5 - 5 = 0$, so $\omega\sqrt[3]{5}$ is also a root of $m(x)$. Thus, $\mathbf{Q}(\sqrt[3]{5})$ and $\mathbf{Q}(\omega\sqrt[3]{5})$ are isomorphic. Furthermore, there is an isomorphism $\phi : \mathbf{Q}(\sqrt[3]{5}) \longrightarrow \mathbf{Q}(\omega\sqrt[3]{5})$ such that $\phi(\sqrt[3]{5}) = \omega\sqrt[3]{5}$. Let's see if we can find this isomorphism. Since $\deg(m(x)) = 3$, $\mathbf{Q}(\sqrt[3]{5})$ has a basis with three elements, and by Theorem 8.2.8, $\{1, \sqrt[3]{5}, (\sqrt[3]{5})^2\} = \{1, \sqrt[3]{5}, \sqrt[3]{25}\}$ is such a basis. Each element of $\mathbf{Q}(\sqrt[3]{5})$ is of the form $a + b\sqrt[3]{5} + c\sqrt[3]{25}$, where $a, b, c \in \mathbf{Q}$. Similarly, $\{1, \omega\sqrt[3]{5}, (\omega\sqrt[3]{5})^2\} = \{1, \omega\sqrt[3]{5}, \omega^2\sqrt[3]{25}\}$ is a basis for $\mathbf{Q}(\omega\sqrt[3]{5})$, and elements of $\mathbf{Q}(\omega\sqrt[3]{5})$ look like $a + b\omega\sqrt[3]{5} + c\omega^2\sqrt[3]{25}$, with a, b and c in $\mathbf{Q}$. Let $\phi : \mathbf{Q}(\sqrt[3]{5}) \longrightarrow \mathbf{Q}(\omega\sqrt[3]{5})$ be defined by $\phi(a + b\sqrt[3]{5} + b\sqrt[3]{25}) = a + b\omega\sqrt[3]{5} + c\omega^2\sqrt[3]{25}$, and note that if $a = 0, b = 1$, and $c = 0$, then $\phi(\sqrt[3]{5}) = \omega\sqrt[3]{5}$. We leave it as an exercise to show that ϕ is a well-defined ring isomorphism. Since $\omega\sqrt[3]{5}$ is not a real number, $\omega\sqrt[3]{5} \notin \mathbf{Q}(\sqrt[3]{5}) \subseteq \mathbf{R}$, so $\mathbf{Q}(\sqrt[3]{5}) \neq \mathbf{Q}(\omega\sqrt[3]{5})$.

Theorem 8.2.10 shows that if c_1 and c_2 have the same minimal polynomial in $F[x]$, then $F(c_1)$ and $F(c_2)$ are isomorphic. We conclude this section with a theorem that shows that this is true in more general circumstances. If $\phi : F \longrightarrow F'$ is a ring isomorphism and $f(x) = a_0 + a_1x + \cdots + a_nx^n$ is a polynomial in $F[x]$, then for the purpose of the following proof, we refer to

$$f^\phi(x) = \phi(a_0) + \phi(a_1)x + \cdots + \phi(a_n)x^n$$

as the **corresponding polynomial** in $F'[x]$. The polynomial $f^\phi(x)$ in $F'[x]$ is nothing more than the polynomial $f(x)$ with ϕ applied to its coefficients. The inverse mapping $\phi^{-1} : F' \longrightarrow F$ is also a ring isomorphism (Problem Set 3.3, Exercise 5), so for each polynomial $g(x) = b_0 + b_1x + \cdots + b_nx^n$ in $F'[x]$, there is a corresponding polynomial

$$g^{\phi^{-1}}(x) = \phi^{-1}(b_0) + \phi^{-1}(b_1)x + \cdots + \phi^{-1}(b_n)x^n \text{ in } F[x].$$

To prove the theorem, we need the following lemma. The proof of the lemma is left as an exercise.

8.2.13   Lemma

If F and F' are fields and $\phi : F \longrightarrow F'$ is an isomorphism, then $\psi : F[x] \longrightarrow F'[x]$ defined by $\psi(f(x)) = f^\phi(x)$ is an isomorphism, and for any $f(x) \in F[x]$, $\deg(f(x)) = \deg(f^\phi(x))$.

Theorem

Let F and F' be fields, and suppose that $\phi : F \longrightarrow F'$ is a ring isomorphism. If c is algebraic over F with minimal polynomial $m(x) = a_0 + a_1 x + \cdots + a_n x^n \in F[x]$, then the corresponding polynomial $m^\phi(x)$ is monic and irreducible in $F'[x]$. If c' is a root of $m^\phi(x)$, then ϕ can be extended to an isomorphism $\psi : F(c) \longrightarrow F'(c')$ of fields such that $\psi(c) = c'$.

Proof. Let c be algebraic over F, and suppose that $m(x)$ is the minimal polynomial in $F[x]$ of c. If the corresponding polynomial $m^\phi(x)$ is reducible in $F'[x]$, then there exist polynomials $f(x)$ and $g(x)$ in $F'[x]$ that each have degree of at least 1 such that $m^\phi(x) = f(x)g(x)$. Because of Lemma 8.2.13, the ring isomorphism $\phi^{-1} : F' \longrightarrow F$ applied to $f(x)$ and $g(x)$ produces corresponding polynomials $f^{\phi^{-1}}(x), g^{\phi^{-1}}(x) \in F[x]$ such that $m(x) = f^{\phi^{-1}}(x)g^{\phi^{-1}}(x)$ and $\deg(f^{\phi^{-1}}(x)) = \deg(f(x))$ and $\deg(g^{\phi^{-1}}(x)) = \deg(g(x))$. From this it follows that if $m^\phi(x)$ is reducible in $F'[x]$, then $m(x)$ is reducible in $F[x]$. Since this contradicts the assumption that $m(x)$ is irreducible in $F[x]$, $m^\phi(x)$ must be irreducible in $F'[x]$. It also follows that because $m(x)$ is monic, $m^\phi(x)$ is monic as well. Therefore, if c' is a root of $m^\phi(x)$, then $m^\phi(x)$ is the minimal polynomial in $F'[x]$ of c'. Theorem 8.2.8 indicates that $F(c)$ is isomorphic to $F[x]/(m(x))$. Every element of $F(c)$ can be written as a polynomial $r(c)$, where $\deg(r(c)) < \deg(m(x))$. So if $\eta : F(c) \longrightarrow F[x]/(m(x))$ is defined by $\eta(r(c)) = r(x) + (m(x))$, then η is a ring isomorphism. Similarly, if we define $\eta' : F'[x]/(m^\phi(x)) \longrightarrow F'(c)$ by $\eta'(r'(x) + (m^\phi(x))) = r'(c)$, where $\deg(r'(x)) < \deg(m^\phi(x))$, then η' is a ring isomorphism. It also follows that $\phi : F[x]/(m(x)) \longrightarrow F'[x]/(m^\phi(x))$, where $\phi(f(x) + (m(x))) = f^\phi(x) + (m^\phi(x))$, is a ring isomorphism. We thus have the following sequence of isomorphisms:

$$F(c) \xrightarrow{\eta} F[x]/(m(x)) \xrightarrow{\phi} F'[x]/(m^\phi(x)) \xrightarrow{\eta'} F'(c') \qquad (*)$$
$$r(c) \longmapsto r(x) + (m(x)) \longmapsto r^\phi(x) + (m^\phi(x)) \longmapsto r^\phi(c').$$

Since $r(c)$ is a polynomial in c such that $\deg(r(c)) < \deg(m(x))$, if $r(c) = c$, then $r(x) = x$, so $c \longmapsto x + (m(x)) \longmapsto x + (m^\phi(x)) \longmapsto c'$. The composition $\eta'\phi\eta$ of the mappings in $(*)$ produces an isomorphism $\psi : F(c) \longrightarrow F'(c')$ such that $\psi(c) = c'$. ∎

HISTORICAL NOTES

Charles Hermite (1822–1901). Charles Hermite was born on December 24, 1822, in Dieuze, France. He became one of the leading French mathematicians of the second half of the 19th century. Although his academic performance as measured by formal examinations was, at best, indifferent, Hermite was a brilliant and creative mathematician who achieved a worldwide reputation with his work. His mathematical production yielded in 1873 a proof that e, the base of the natural logarithmic function, is a transcendental number. As a professor at the Sorbonne, a position that he held for some 27 years, he shaped the development of many of the fine young mathematicians of the period.

Ferdinand von Lindemann (1852–1939). Ferdinand von Lindemann was born in Hannover, Germany, in 1852, and he studied mathematics at Erlangen, where he received a Ph.D. in 1873 for his dissertation on non-Euclidean geometry. After serving as a professor at the University of Königsberg from 1883 until 1893, he accepted a chair at the University of Munich and remained there for the rest of his career. Today Lindemann is best known for his proof that π is a transcendental number, a proof that laid to rest the classical problem of Greek mathematics: squaring a circle. The problem, stated simply, is this: given a circle, using only a straightedge and compass, construct a square with area equal to that of the circle. Lindemann's result laid the groundwork for the proof that such constructions are impossible.

Problem Set   8.2

1. Show that each of the following is algebraic over $\mathbf{Q}$, and find the minimal polynomial in $\mathbf{Q}[x]$ of each.

   (a) $2 - 3i$
   (b) $1 + \sqrt[3]{5}$
   (c) $\sqrt{2 + \sqrt{3}}$
   (d) $\sqrt{1 + i}$
   (e) $\sqrt{3} + \sqrt{2}i$
   (f) $\sqrt{3 + i\sqrt{2}}$

2. Find the minimal polynomial of $\sqrt{3} + i$ in $\mathbf{Q}[x]$ and then in $\mathbf{R}[x]$.

3. Find a basis for each of the following field extensions.

   (a) $\mathbf{Q}(\sqrt{3})/\mathbf{Q}$
   (b) $\mathbf{Q}(\sqrt[3]{7})/\mathbf{Q}$
   (c) $\mathbf{Q}(\sqrt[4]{2})/\mathbf{Q}$
   (d) $\mathbf{Q}(\sqrt{2} + \sqrt{3})/\mathbf{Q}$

4. Find $[\mathbf{Q}(\sqrt{3} - \sqrt{5}) : \mathbf{Q}]$ and $[\mathbf{Q}(\sqrt{2} + \sqrt{10}) : \mathbf{Q}]$.

5. Determine whether the polynomial $x^2 - 5$ is irreducible in $\mathbf{Q}(\sqrt{2})[x]$.

6. If p is a prime integer, show that $\mathbf{Q}(\sqrt{p})$ and $\mathbf{Q}(-\sqrt{p})$ are isomorphic field extensions of $\mathbf{Q}$. Is it true that $\mathbf{Q}(\sqrt{p}) = \mathbf{Q}(-\sqrt{p})$?

7. If p is a prime integer, show that $\mathbf{Q}(\sqrt[3]{p})$ and $\mathbf{Q}(\omega\sqrt[3]{p})$ are isomorphic but that $\mathbf{Q}(\sqrt[3]{p}) \neq \mathbf{Q}(\omega\sqrt[3]{p})$, where $\omega = -\frac{1}{2} + \frac{\sqrt{3}}{2}i$.

8. (a) Prove that $\mathbf{Q}(\sqrt{3})$ cannot be isomorphic to $\mathbf{Q}(\sqrt{5})$.
    (b) Prove that if p and q are distinct prime integers, then $\mathbf{Q}(\sqrt{p})$ and $\mathbf{Q}(\sqrt{q})$ are not isomorphic.

9. (a) Show that $2 + 5\sqrt{2}$ is algebraic over $\mathbf{Q}$ and that $\mathbf{Q}(2 + 5\sqrt{2}) = \mathbf{Q}(\sqrt{2})$.
    (b) Does $\mathbf{Q}(5 - i) = \mathbf{Q}(1 + i)$? And does $\mathbf{Q}(1 + i) = \mathbf{Q}(i)$?
    (c) Let E/F be an extension field such that $[E : F] = 2$, and suppose that $c \in E$ is algebraic over F. If $a, b \in F, b \neq 0$, prove that $F(a + bc) = F(c)$. We say that a and b are **absorbed by the field** F. Show that if $c \in F$, then $F = F(c)$.

10. Prove Lemma 8.2.13.

11. Suppose that E/F is a field extension. If K is the set of all elements of E that are algebraic over F, prove that K is a subfield of E.

12. Show that the map ϕ used in the proof of Theorem 8.2.14 is a well-defined ring isomorphism. Show also that the mapping $\psi : F(c) \longrightarrow F'(c')$ that extends $\phi : F \longrightarrow F'$ is unique.

13. Let E/F be a field extension, and suppose that $\mathcal{A}$ is the set of all elements of E that are algebraic over F. Define the relation R on $\mathcal{A}$ by aRb if and only if there is a polynomial $f(x) \in F[x]$ such that a and b are roots of $f(x)$. Prove that R is an equivalence relation on $\mathcal{A}$.

14. Let E/F be a finite dimensional field extension. Let $c \in E$ be algebraic over F with minimal polynomial $m(x) \in F[x]$. Prove that $\deg(m(x))$ divides $[E : F]$.

15. Let F be a field with char $F \neq 2$, and suppose that E/F is a field extension such that $[E : F] = 2$. If $c \in E$ but $c \notin F$ is algebraic over F, then c is the root of a minimal polynomial $m(x) \in F[x]$ of degree at most 2. If $m(x) = ax + b$, where a, $b \in F, a \neq 0$, then $ac + b = 0$, so $c = -a^{-1}b$ is in F, which is a contradiction. Thus, $\deg(m(x)) = 2$, so $m(x) = x^2 + bx + d$, where $b, d \in F$. Show that

$$c = \frac{-b \pm \sqrt{b^2 - 4d}}{2e},$$

where e is the identity of F. At this point you should decide why it was assumed that char $F \neq 2$. Now $\sqrt{b^2 - 4d} \notin F$, for if it were in F, it would follow that $c \in F$. Hence, $b^2 - 4d$ is not a perfect square in F. Show that $F(c) = F(\sqrt{b^2 - 4d})$. Conclude that if F is a field with char $F \neq 2$, then any field extension E/F such that $[E : F] = 2$ must be of the form $F(\sqrt{c})$, where $c \in F$ is not a perfect square in F. Such simple field extensions of F are sometimes called **quadratic field extensions** of F. For example, $\mathbf{Q}(\sqrt{2})$ is a quadratic field extension of $\mathbf{Q}$.

16.   If E/F is a field extension and $c \in E$ is transcendental over F, prove that the rings $F[x]$ and $F[c]$ are isomorphic.

17.   **A method for computing the multiplicative inverse of a nonzero element in a simple field extension.** If c is algebraic over F with minimal polynomial $m(x)$, then Theorem 8.2.8 shows that $F(c)$ is isomorphic to the field $F[x]/(m(x))$. If $\deg(m(x)) = n$, then $B = \{e, c, c^2, \ldots, c^{n-1}\}$ is a basis for the field extension $F(c)/F$. Thus, every element of $F(c)$ has the form $a_0 + a_1c + a_2c^2 + \cdots + a_{n-1}c^{n-1}$, so every element of $F(c)$ is a polynomial in c of degree at most $n - 1$. If $f(c)$ is an element in $F(c)$, then $f(x) + (m(x))$ is the corresponding element of $F[x]/(m(x))$. We can compute the multiplicative inverse of a nonzero element of $F(c)$ by finding the multiplicative inverse of the corresponding element in $F[x]/(m(x))$ and then looking back to $F(c)$. If $f(c) \in F(c)$, $f(c) \neq 0$, then $f(x) + (m(x)) \neq 0$ in $F[x]/(m(x))$, and $f(x)$ is such that $\deg(f[x]) < \deg(m[x])$. Since $m(x)$ is irreducible in $F[x]$, $f(x)$ and $m(x)$ are relatively prime in $F[x]$. Consequently, by the Euclidean Algorithm for Polynomials, there are polynomials $a(x), b(x)$ in $F[x]$ such that $a(x)f(x) + b(x)m(x) = e$. Now $m(c) = 0$, so $a(c)f(c) = e$ in $F(c)$. Hence, $a(c) \in F(c)$ is the multiplicative inverse of $f(c) \in F(c)$.

Example. Find the multiplicative inverse of $1 + \sqrt[3]{3} - \sqrt[3]{9} \in \mathbf{Q}(\sqrt[3]{3})$.

Solution. The minimal polynomial in $\mathbf{Q}[x]$ of $\sqrt[3]{3}$ is $m(x) = x^3 - 3$. If $f(\sqrt[3]{3}) = 1 + \sqrt[3]{3} - \sqrt[3]{9}$, the corresponding element in $\mathbf{Q}[x]/(x^3 - 3)$ is $f(x) + (m(x)) = 1 + x - x^2 + (x^3 - 3)$. Now $f(x) = 1 + x - x^2$ and $m(x)$ are relatively prime in $\mathbf{Q}[x]$, so $\gcd(f(x), m(x)) = 1$. Using the Division Algorithm for Polynomials, we see that

$$x^3 - 3 = (-x - 1)(-x^2 + x + 1) + (2x + 2) \qquad \text{and}$$
$$-x^2 + x + 1 = -\tfrac{1}{2}x(2x - 2) + 1.$$

Hence, $m(x) = -(x + 1)f(x) + (2x - 2)$ and $f(x) = -\tfrac{1}{2}x(2x - 2) + 1$. Solving for 1 and $2x - 2$ gives

$$1 = f(x) + \tfrac{1}{2}x(2x - 2) \qquad \text{and}$$
$$2x - 2 = m(x) + (x + 1)f(x).$$

Therefore, by back-substitution, we arrive at

$$1 = f(x) + \tfrac{1}{2}x[m(x) + (x + 1)f(x)]$$

$$= \left(\tfrac{1}{2}x^2 + \tfrac{1}{2}x + 1\right)f(x) + \tfrac{1}{2}xm(x) \text{ and}$$

since $m(\sqrt[3]{3}) = 0$, $1 = \left(\frac{1}{2}\sqrt[3]{9} + \frac{1}{2}\sqrt[3]{3} + 1\right)(f(\sqrt[3]{3}))$ in $\mathbf{Q}(\sqrt[3]{3})$. The multiplicative inverse in $\mathbf{Q}(\sqrt[3]{3})$ of $1 + \sqrt[3]{3} - \sqrt[3]{9}$ is $1 + \frac{1}{2}\sqrt[3]{3} + \frac{1}{2}\sqrt[3]{9}$. You should compute $(1 + \sqrt[3]{3} - \sqrt[3]{9})\left(1 + \frac{1}{2}\sqrt[3]{3} + \frac{1}{2}\sqrt[3]{9}\right)$ to show that the result is indeed 1.

Use the method just demonstrated to find the multiplicative inverse of each of the following elements of the given field extension of $\mathbf{Q}$.

(a) $1 + 2\sqrt{3} \in \mathbf{Q}(\sqrt{3})$

(b) $3 + 2\sqrt{2}\,i \in \mathbf{Q}(\sqrt{2}\,i)$

(c) $2 - \sqrt[3]{2} + \sqrt[3]{4} \in \mathbf{Q}(\sqrt[3]{2})$

TECHNOLOGY PROBLEMS

Use a computer algebra system to solve the following problem.

18. Find the multiplicative inverse of each of the following in the given field extension of $\mathbf{Q}$, and verify each result.

(a) $\frac{1}{2} - \frac{3}{4}\sqrt[3]{3} + 24\sqrt[3]{9} \in \mathbf{Q}(\sqrt[3]{3})$

(b) $\frac{7}{5} + 3(\sqrt{1 + \sqrt{2}}) - \frac{3}{11}(\sqrt{1 + \sqrt{2}})^2 + 2(\sqrt{1 + \sqrt{2}})^3 \in \mathbf{Q}(\sqrt{1 + \sqrt{2}})$

(c) $5(\sqrt{2} + \sqrt{3}) - \frac{5}{8}(\sqrt{2} + \sqrt{3})^2 \in \mathbf{Q}(\sqrt{2} + \sqrt{3})$

8.3   ALGEBRAIC FIELD EXTENSIONS

Section Overview. When $E\,/\,F$ is a field extension, in Section 8.2 a single element $c \in E$ was adjoined to a field F to obtain the smallest field extension $F(c)$ of F that contains c. We now show that if c is algebraic over F, then every element of $F(c)$ is algebraic over F. Field extensions $E\,/\,F$ such that every element of E is algebraic over F are called algebraic field extensions. Make sure you understand iterated field extensions. This procedure is used in the next section to compute the splitting field of a polynomial $f(x) \in F[x]$.

8.3.1   Theorem

If $E\,/\,F$ is a finite dimensional field extension, then every element of E is algebraic over F.

Proof. Suppose $[E : F] = n$, let $a \in E$, and consider $S = \{e, a, a^2, \ldots, a^n\}$. Since S contains $n + 1$ elements, Exercise 11 of Problem Set 8.1 shows that S cannot be linearly independent. Hence, S is linearly dependent, so there exist $c_0, c_1, c_2, \ldots, c_n \in F$, at least one of which is nonzero, such that

$$c_0 e + c_1 a + c_2 a^2 + \cdots + c_n a^n = 0.$$

Therefore, a is a root of the nonzero polynomial

$$f(x) = c_0 + c_1 x + c_2 x^2 + \cdots + c_n x^n \in F[x],$$

so a is algebraic over F. ∎

8.3.2 Corollary

If c is algebraic over F, then every element of $F(c)$ is algebraic over F.

Proof. By Theorem 8.2.8, $[F(c) : F] = n$ for some positive integer n, so the result follows immediately from Theorem 8.3.1. ∎

8.3.3 Example

Corollary 8.3.2 indicates that every element of $\mathbf{Q}(\sqrt{2})$ is algebraic over $\mathbf{Q}$. In fact, if $a + b\sqrt{2} \in \mathbf{Q}(\sqrt{2})$, then $a + b\sqrt{2}$ is a root of the polynomial

$$
\begin{aligned}
f(x) &= (x - a + b\sqrt{2})(x - a - b\sqrt{2}) \\
&= (x - a)^2 - 2b^2 \\
&= x^2 - 2ax + (a^2 - 2b^2),
\end{aligned}
$$

and $f(x)$ is in $\mathbf{Q}[x]$.

The result of Theorem 8.3.1 suggests the following definition.

8.3.4 Definition

If E/F is a field extension such that every element of E is algebraic over F, then E is said to be an **algebraic field extension** of F.

Because of Corollary 8.3.2, we see that if c is algebraic over F, $F(c)$ is an algebraic field extension of F.

We now turn our attention to **iterated field extensions**. Suppose that E/F is a field extension, and let $c_1, c_2, \ldots, c_n \in E$. As shown in Section 8.2, c_1 can be adjoined to F to obtain the smallest field $F(c_1)$ such that $F[c_1] \subseteq F(c_1)$. Since $F \subseteq F[c_1]$, we have that $F \subseteq F(c_1) \subseteq E$. Similarly, c_2 can be adjoined to $F(c_1)$ to obtain the smallest field $F(c_1)(c_2) = F(c_1, c_2)$ that contains $F(c_1)[c_2]$. Since $F(c_1) \subseteq F(c_1)[c_2]$, we see that $F \subseteq F(c_1) \subseteq F(c_1, c_2)$. Continuing this process, a **chain** or **tower of field extensions**

$$F \subseteq F(c_1) \subseteq F(c_1, c_2) \subseteq F(c_1, c_2, c_3) \subseteq \cdots \subseteq F(c_1, c_2, \ldots, c_n)$$

of F can be constructed, where $F(c_1, c_2, \ldots, c_n) = F(c_1, c_2, \ldots, c_{n-1})(c_n)$. At each step in constructing the tower of field extensions, $F(c_1, c_2, \ldots, c_{k-1})(c_k)$ is the field of fractions of the integral domain $F(c_1, c_2, \ldots, c_{k-1}))[c_k]$, and, as such, $F(c_1, c_2, \ldots, c_{k-1})(c_k)$ is unique up to isomorphism. From these observations, it follows that $F(c_1, c_2, \ldots, c_n)$ is the smallest field extension of F containing $c_1, c_2, \ldots, c_n$ and that $F(c_1, c_2, \ldots, c_n)$ is unique up to isomorphism. We say that the field $F(c_1, c_2, \ldots, c_n)$ has been **obtained from F by adjoining the elements** $c_1, c_2, \ldots, c_n \in E$ to F.

8.3.5 | **Example**

The elements $\frac{3}{4} + 2\sqrt{3}, 6 - 5i \in \mathbf{C}$ are easily shown to be algebraic over $\mathbf{Q}$. If we adjoin $\frac{3}{4} + 2\sqrt{3}$ to $\mathbf{Q}$, we obtain the field $\mathbf{Q}(\sqrt{3}) = \mathbf{Q}\left(\frac{3}{4} + 2\sqrt{3}\right)$ since $\frac{3}{4}$ and 2 are absorbed by $\mathbf{Q}$. (See Problem Set 8.2, Exercise 9.) If we adjoin $6 - 5i$ to $\mathbf{Q}(\sqrt{3})$, we obtain the field $\mathbf{Q}(\sqrt{3}, i)$ since 6 and -5 are absorbed by $\mathbf{Q}(\sqrt{3})$. Hence, $\mathbf{Q}\left(\frac{3}{4} + 2\sqrt{3}, 6 - 5i\right) = \mathbf{Q}(\sqrt{3}, i)$ is the smallest extension field of $\mathbf{Q}$ that contains $\frac{3}{4} + 2\sqrt{3}$ and $6 - 5i$. Let's see whether we can find a basis for the field extension $\mathbf{Q}(\sqrt{3}, i)/\mathbf{Q}$. If we can find a basis, then we will know exactly what the elements of $\mathbf{Q}(\sqrt{3}, i)/\mathbf{Q}$ will look like. The minimal polynomial in $\mathbf{Q}[x]$ of $\sqrt{3}$ is $m(x) = x^2 - 3$, so $\mathbf{Q}(\sqrt{3})/\mathbf{Q}$ has a basis with two elements since $\deg(m(x)) = 2$. Theorem 8.2.8 indicates that $B = \{1, \sqrt{3}\}$ is such a basis.

We claim that the minimal polynomial in $\mathbf{Q}(\sqrt{3})[x]$ of i is $m'(x) = x^2 + 1$. Since $\mathbf{Q}(\sqrt{3})[x] \subseteq \mathbf{R}[x]$ and since $m'(x)$ is irreducible in $\mathbf{R}[x]$, $m'(x)$ must be irreducible in $\mathbf{Q}(\sqrt{3})[x]$. Thus, $m'(x)$ is the minimal polynomial in $\mathbf{Q}(\sqrt{3})[x]$ of i. Therefore, $B' = \{1, i\}$ is a basis for the field extension $\mathbf{Q}(\sqrt{3}, i)/\mathbf{Q}(\sqrt{3})$. The proof of Theorem 8.1.9 indicates that a basis of $\mathbf{Q}(\sqrt{3}, i)/\mathbf{Q}$ can be obtained by forming all possible products of an element of B and an element of B'. Thus, $B'' = \{1, \sqrt{3}, i, \sqrt{3}\,i\}$ is a basis for the field extension $\mathbf{Q}(\sqrt{3}, i)/\mathbf{Q}$, so every element of $\mathbf{Q}(\sqrt{3}, i)/\mathbf{Q}$ looks like $a1 + b\sqrt{3} + ci + d\sqrt{3}\,i$, where $a, b, c, d \in \mathbf{Q}$. $\mathbf{Q}(\sqrt{3}, i)/\mathbf{Q}$ is the smallest field extension of $\mathbf{Q}$ that contains $\frac{3}{4} + 2\sqrt{3}$ and $6 - 5i$. Notice also that

$$[\mathbf{Q}(\sqrt{3}, i) : \mathbf{Q}] = 2 \cdot 2 = [\mathbf{Q}(\sqrt{3}, i) : \mathbf{Q}(\sqrt{3})][\mathbf{Q}(\sqrt{3}) : \mathbf{Q}].$$

If c is algebraic over F, then c is algebraic over every field extension of F. This follows, for if c is algebraic over F, then c is the root of a nonzero polynomial in $F[x]$. If E is *any* field extension of F, $F \subseteq E$ implies that $F[x] \subseteq E[x]$, so c is a root of a nonzero polynomial in $E[x]$. Hence, if each of the elements $c_1, c_2, \ldots, c_n$ is algebraic over F, then each of these elements is algebraic over $F(c_1, c_2, \ldots, c_k)$ for $k = 1, 2, \ldots, n$. This observation leads to the following theorem.

8.3.6   Theorem

If each of the elements $c_1, c_2, \ldots, c_n$ is algebraic over F, then $F(c_1, c_2, \ldots, c_n)$ is an algebraic field extension of F.

Proof.   Suppose that each of the elements $c_1, c_2, \ldots, c_n$ is algebraic over F, and consider the tower of field extensions

$$F \subseteq F(c_1) \subseteq F(c_1, c_2) \subseteq F(c_1, c_2, c_3) \subseteq \cdots \subseteq F(c_1, c_2, \ldots, c_n).$$

To simplify notation, let $F_0 = F, F_1 = F(c_1), \ldots, F_n = F(c_1, c_2, \ldots, c_n)$ so that

$$F_0 \subseteq F_1 \subseteq F_2 \subseteq \cdots \subseteq F_n.$$

Since c_k is algebraic over F_{k-1}, it follows from Theorem 8.2.8 that $[F_k : F_{k-1}]$ is determined by the degree of the minimal polynomial in $F_{k-1}[x]$ of c_k. Hence, $[F_k : F_{k-1}]$ is finite for $k = 1, 2, \ldots, n$. Now Theorem 8.1.9 indicates that

$$[F_2 : F_0] = [F_2 : F_1][F_1 : F_0].$$

If both sides of this equation are multiplied by $[F_3 : F_2]$, using Theorem 8.1.9 again shows that

$$[F_3 : F_0] = [F_3 : F_2][F_2 : F_0] = [F_3 : F_2][F_2 : F_1][F_1 : F_0].$$

Next, multiply both sides of this last equation by $[F_4 : F_3]$. Then

$$[F_4 : F_0] = [F_4 : F_3][F_3 : F_0] = [F_4 : F_3][F_3 : F_2][F_2 : F_1][F_1 : F_0].$$

At the nth step in this process, we arrive at

$$[F_n : F_0] = [F_n : F_{n-1}] \cdots [F_4 : F_3][F_3 : F_2][F_2 : F_1][F_1 : F_0].$$

Because each factor on the right side of this last equation is a positive integer, $[F_n : F]$ is a positive integer. Theorem 8.3.1 now shows that F_n is an algebraic field extension of F. ■

The proof of Theorem 8.3.6 provides us with the following result. The notation established in the proof of Theorem 8.3.6 is used to state the corollary more compactly.

8.3.7   Corollary

$F(c_1, c_2, \ldots, c_n)/F$ is a finite dimensional field extension whenever $c_1, c_2, \ldots, c_n$ are algebraic over F and

$$[F_n : F_0] = [F_n : F_{n-1}] \cdots [F_4 : F_3][F_3 : F_2][F_2 : F_1][F_1 : F_0].$$

Let E/F be a field extension, and suppose that each of $c_1, c_2, \ldots, c_n \in E$ is algebraic over F. Theorem 8.3.6 shows that $F(c_1, c_2, \ldots, c_n)$ is an algebraic field extension of F. This shows that to determine if each element of $F(c_1, c_2, \ldots, c_n)$ is algebraic over F, one need only show that each of the elements $c_1, c_2, \ldots, c_n$ is algebraic over F.

Adjoining an element $c \in F$ to F produces F again since, in this case, F is the smallest field that contains both F and c. It is also the case that adjoining the root of a polynomial to a field F can produce a field that contains additional roots of the polynomial. Example 1 of 8.2.12 demonstrates this point, while Example 2 of 8.2.12 shows that this does not always have to be the case. Hence, we see that (1) if $c \in F$, then $F(c) = F$ and (2) adjoining a root of a polynomial to a field may produce a field that contains additional roots of the polynomial. These two facts show that when the elements $c_1, c_2, \ldots, c_n$ are algebraic over F, not all the terms in the tower of field extensions

$$F \subseteq F(c_1) \subseteq F(c_1, c_2) \subseteq F(c_1, c_2, c_3) \subseteq \cdots \subseteq F(c_1, c_2, \ldots, c_n)$$

need be distinct. This is illustrated in the following example.

8.3.8   Example

The elements $\sqrt{3}$, $-\sqrt{3}$, i, $-i$ are all algebraic over $\mathbf{Q}$ since each is a root of $f(x) = x^4 - 2x^2 - 3 \in \mathbf{Q}[x]$. These four roots of $f(x)$ produce the following tower of fields:

$$\mathbf{Q} \subseteq \mathbf{Q}(\sqrt{3}) \subseteq \mathbf{Q}(\sqrt{3}, -\sqrt{3}) \subseteq \mathbf{Q}(\sqrt{3}, -\sqrt{3}, i) \subseteq \mathbf{Q}(\sqrt{3}, -\sqrt{3}, i, -i).$$

But $-\sqrt{3} \in \mathbf{Q}(\sqrt{3})$, so $\mathbf{Q}(\sqrt{3}, -\sqrt{3}) = \mathbf{Q}(\sqrt{3})$. We see that $\mathbf{Q}(\sqrt{3}, -\sqrt{3}, i) = \mathbf{Q}(\sqrt{3}, i)$ and $\mathbf{Q}(\sqrt{3}, -\sqrt{3}, i, -i) = \mathbf{Q}(\sqrt{3}, i)$ since $-i$ is in $\mathbf{Q}(\sqrt{3}, i)$. Therefore the tower of extension fields reduces to

$$\mathbf{Q} \subseteq \mathbf{Q}(\sqrt{3}) \subseteq \mathbf{Q}(\sqrt{3}, i).$$

Consequently, not all the terms in the original tower of field extensions are distinct. Corollary 8.3.7 shows that

$$[\mathbf{Q}(\sqrt{3}, i):\mathbf{Q}] = [\mathbf{Q}(\sqrt{3}, i):\mathbf{Q}(\sqrt{3})][\mathbf{Q}(\sqrt{3}):\mathbf{Q}].$$

It was shown in Example 8.3.5 that $\{1, i\}$ is a basis for $\mathbf{Q}(\sqrt{3}, i)/\mathbf{Q}(\sqrt{3})$ and that $\{1, \sqrt{3}\}$ is a basis for $\mathbf{Q}(\sqrt{3})/\mathbf{Q}$. Hence, $[\mathbf{Q}(\sqrt{3}, i): \mathbf{Q}] = 2 \cdot 2 = 4$. It was also shown in Example 8.3.5 that $\{1, \sqrt{3}, i, \sqrt{3}i\}$ is a basis for the field extension $\mathbf{Q}(\sqrt{3}, i)/\mathbf{Q}$.

Finally, note that since $\sqrt{3}, -\sqrt{3}, i, -i \in \mathbf{Q}(\sqrt{3}, i), f(x) = x^4 - 2x^2 - 3$ factors completely in $\mathbf{Q}(\sqrt{3}, i)[x]$ as $f(x) = (x + \sqrt{3})(x - \sqrt{3})(x + i)(x - i)$. The field $\mathbf{Q}(\sqrt{3}, i)$ is the smallest field extension of $\mathbf{Q}$ that contains all the roots of $f(x)$. Subsequently, $\mathbf{Q}(\sqrt{3}, i)$ is referred to as a **splitting field** of $f(x)$.

We conclude this section with the following result dealing with the transitivity of algebraic field extensions.

8.3.9   Theorem

If E/K and K/F are algebraic field extensions, then E/F is an algebraic field extension.

Proof. Suppose that $c \in E$. Since c is algebraic over K, there is a polynomial $f(x) = a_0 + a_1x + a_2x^2 + \cdots + a_nx^n$ in $K[x]$ such that $f(c) = 0$. Consider the field $F(a_0, a_1, \ldots, a_n)$. Since $a_0, a_1, \ldots, a_n \in K$ are algebraic over F, Corollary 8.3.7 shows that $F(a_0, a_1, \ldots, a_n)/F$ is a finite dimensional field extension. Also note that c is a root of the polynomial $f(x) \in F(a_0, a_1, \ldots, a_n)[x]$, so c is algebraic over $F(a_0, a_1, \ldots, a_n)$. Hence, $F(a_0, a_1, \ldots, a_n, c) = F(a_0, a_1, \ldots, a_n)(c)$ is a finite dimensional extension of $F(a_0, a_1, \ldots, a_n)$, and $F(a_0, a_1, \ldots, a_n)$ is a finite dimensional field extension of F. But

$$[F(a_0, a_1, \ldots, a_n, c) : F] = [F(a_0, a_1, \ldots, a_n, c) : F(a_0, a_1, \ldots, a_n)][F(a_0, a_1, \ldots, a_n) : F],$$

and the two factors on the right side in this equation are positive integers. Therefore, $F(a_0, a_1, \ldots, a_n, c)/F$ is a finite dimensional field extension and thus is an algebraic field extension by Theorem 8.3.1. Since $c \in F(a_0, a_1, \ldots, a_n, c)$, c is algebraic over F. This shows that E/F is an algebraic field extension since c was chosen arbitrarily from E.   ■

Problem Set 8.3

1. Show that $f(x) = x^2 - 7$ is irreducible in $\mathbf{Q}(\sqrt{3})[x]$.

2. Show that $\mathbf{Q}(\sqrt{2} + \sqrt{3}) = \mathbf{Q}(\sqrt{2}, \sqrt{3})$. Compute $[\mathbf{Q}(\sqrt{2} + \sqrt{3}) : \mathbf{Q}]$.

Both $\sqrt{3}$ and $\sqrt{5}i$ are algebraic over $\mathbf{Q}$, so $\mathbf{Q}(\sqrt{3}, \sqrt{5}i)$ is an algebraic field extension of $\mathbf{Q}$. Find each of the following in Problems 3 though 7.

3. A basis for $\mathbf{Q}(\sqrt{3})/\mathbf{Q}$ and $[\mathbf{Q}(\sqrt{3}) : \mathbf{Q}]$

4. The minimal polynomial in $\mathbf{Q}(\sqrt{3})[x]$ of $\sqrt{5}i$

5. A basis for $\mathbf{Q}(\sqrt{3}, \sqrt{5}i)/\mathbf{Q}(\sqrt{3})$ and $[\mathbf{Q}(\sqrt{3}, \sqrt{5}i) : \mathbf{Q}(\sqrt{3})]$

6. A basis for $\mathbf{Q}(\sqrt{3}, \sqrt{5}i)/\mathbf{Q}$

7. $[\mathbf{Q}(\sqrt{3}, \sqrt{5}i) : \mathbf{Q}]$

8. The real numbers $\sqrt{2}$, $\sqrt{5}$, and $\sqrt{10}$ are algebraic over $\mathbf{Q}$. Are all of the terms in the tower of extension fields determined by these elements distinct?

9. The real numbers $\sqrt{2}$, $\sqrt{3}$, and $\sqrt{5}$ are algebraic over $\mathbf{Q}$. Are all of the terms in the tower of extension fields determined by these elements distinct? Compute $[\mathbf{Q}(\sqrt{2}, \sqrt{3}, \sqrt{5} : \mathbf{Q}]$, and find a basis for the field extension $\mathbf{Q}(\sqrt{2}, \sqrt{3}, \sqrt{5})/\mathbf{Q}$.

10. Find the minimal polynomial in $\mathbf{Q}[x]$ for $\sqrt{3 + \sqrt[3]{2}}$. Determine $[\mathbf{Q}(\sqrt{3 + \sqrt[3]{2}}) : \mathbf{Q}]$, and find a basis for $\mathbf{Q}(\sqrt{3 + \sqrt[3]{2}})/\mathbf{Q}$.

11. If c_1 and c_2 are algebraic over F, explain why $c_1 + c_2, c_1 - c_2, c_1 c_2$ are algebraic over F and, if $c_1 \neq 0$, why c_1^{-1} is algebraic over F.

8.4   SPLITTING FIELDS; ALGEBRAICALLY CLOSED FIELDS

Section Overview. If E/F is a field extension and $c \in E$ is algebraic over F, then, by definition, there is a polynomial $f(x)$ in $F[x]$ that has c

as a root. In the previous section the smallest field extension $F(c)$ of F that contains c was constructed. One purpose of this section is to consider this from a different point of view. We start with a polynomial $f(x) \in F[x]$ and ask whether an extension field E of F can be found that contains a root of $f(x)$. As we shall see, such a field can be found, and this fact will be used to construct a smallest field extension of F that contains all the roots of $f(x)$. Another goal of this section is to address a more general question. Can a field extension of F be found that contains the roots of every polynomial in $F[x]$? The field $\mathbf{C}$ of complex numbers has this property with respect to $\mathbf{R}$ since every polynomial in $\mathbf{R}[x]$ has all of its root in $\mathbf{C}$. In fact, every polynomial in $\mathbf{C}[x]$ has all of its roots in $\mathbf{C}$. In this section it is shown that a similar situation holds for any field F. We show that if F is a field, then there is an algebraic field extension E of F such that every polynomial in $E[x]$ has all of its roots in E.

If F is a field and $f(x)$ is a nonconstant polynomial in $F[x]$, is it possible to construct a field extension of F that contains a root of $f(x)$? If such a field can be constructed, can such a field be found that is in some sense minimal? If so, in what sense, if any, is this minimal field unique? We now turn our attention to an investigation of these questions. If $f(x) \in F[x]$, since $F[x]$ is a Unique Factorization Domain, $f(x)$ can be factored as $f(x) = f_1(x)f_2(x) \cdots f_n(x)$, where each $f_i(x)$ is an irreducible polynomial in $F[x]$. If $c \in E$ is a root of $f(x)$, then $f_1(c)f_2(c) \cdots f_n(c) = f(c) = 0$ in E, and since E is a field, E is free of zero-divisors, so $f_i(c) = 0$ for at least one of $i = 1, 2, \ldots, n$. Conversely, if $c \in E$ is a root of at least one of the irreducible factors of $f(x)$, then c is a root of $f(x)$. Hence, c is a root of $f(x)$ if and only if c is a root of one or more of the irreducible factors of $f(x)$. Therefore, in our search for a field that contains a root of $f(x)$, we can begin by assuming that $f(x)$ is irreducible. The following theorem was first proven by **Leopold Kronecker** (1823–1891).

8.4.1   Kronecker's Theorem

If F is a field and $f(x)$ is an irreducible polynomial in $F[x]$, then a field extension E/F exists such that E contains a root of $f(x)$.

Proof. Let $f(x) = a_0 + a_1x + a_2x^2 + \cdots + a_nx^n$. Since $f(x)$ is irreducible in $F[x]$, Theorem 7.1.10 shows that $E = F[x]/(f(x))$ is a field that contains a copy of F. We can identify F with its copy in E and consider F to be a subfield of E. Let $\bar{x} = x + (f(x))$. Then $\bar{x} \in E$, and we claim that $\bar{x}$ is a root of $f(x)$. Now

$$f(\bar{x}) = a_0\bar{x}^0 + a_1\bar{x} + a_2\bar{x}^2 + \cdots + a_n\bar{x}^n,$$

and we also see that

$$\overline{x}^k = (x + (f(x)))^k = x^k + (f(x))$$

for any nonnegative integer k. Therefore,

$$f(\overline{x}) = a_0(e + (f(x))) + a_1(x + (f(x))) + a_2(x^2 + (f(x))) + \cdots + a_n(x^n + (f(x)))$$

$$= a_0 + a_1 x + a_2 x^2 + \cdots + a_n x^n + (a_0 + a_1 + a_2 + \cdots + a_n)(f(x))$$

$$= f(x) + a(f(x)) \quad \text{where } a = a_0 + a_1 + a_2 + \cdots + a_n.$$

All the coefficients of $f(x)$ cannot add to 0, for if they did, the identity e of F would be a root, and $x - e$ would be a factor of $f(x)$. But then $f(x)$ would not be irreducible in $F[x]$, contrary to our assumption. Hence, $a \neq 0$, so since a has a multiplicative inverse in F, $a(f(x)) = (f(x))$. Therefore,

$$f(\overline{x}) = f(x) + (f(x)) = 0 + (f(x)) = \overline{0},$$

which shows that $\overline{x} \in E$ is a root of $f(x)$.

8.4.2   Corollary

If F is a field and $f(x)$ is a nonconstant polynomial in $F[x]$, then there is a field extension of F that contains a root of $f(x)$.

Proof. Since $F[x]$ is a Unique Factorization Domain, $f(x)$ can be factored as $f(x) = f_1(x)f_2(x) \cdots f_n(x)$, where $f_k(x)$ is irreducible in $F[x]$ for $k = 1, 2, \ldots, n$. Hence, if k is an integer such that $1 \leq k \leq n$, then, by Theorem 8.4.1, there is a field extension of F that contains a root of $f_k(x)$. Since a root of $f_k(x)$ is also a root of $f(x)$, there is a field extension of F that contains a root of $f(x)$. ∎

8.4.3   Example

Consider the polynomial $g(x) = x^4 + 5x^2 + 6$ that factors as $g(x) = (x^2 + 2)(x^2 + 3)$. Both factors are irreducible in $\mathbf{Q}[x]$, so if we choose the factor $f(x) = x^2 + 2$, then $\mathbf{Q}[x]/(f(x))$ is a field such that $\mathbf{Q} \subseteq \mathbf{Q}[x]/(f(x))$. The proof of Theorem 8.4.1 shows that $\overline{x} = x + (f(x))$ is a root of $f(x)$. Indeed, $f(\overline{x}) = \overline{x}^2 + 2 = (x + (f(x)))^2 + 2 = x^2 + (f(x)) + 2 = x^2 + 2 + (f(x)) = 0 + (f(x))$ in $\mathbf{Q}[x]/(f(x))$. Thus, $\mathbf{Q}[x]/(f(x))$ is a field extension of $\mathbf{Q}$ that contains a root of $f(x)$. Since $\overline{x}$ is a root of $f(x)$ and $f(x)$ is a factor of $g(x)$, $\overline{x}$ is also a root of $g(x)$. Therefore, $\mathbf{Q}[x]/(f(x))$ is a field extension of $\mathbf{Q}$ that contains a root of $g(x)$. We could just as well have chosen $f(x)$ to be the irreducible polynomial $f(x) = x^2 + 3$ to produce the field extension.

Corollary 8.4.2 shows that for any nonconstant polynomial $f(x) \in F[x]$, we can produce a field extension of F that contains at least one root of $f(x)$. Can a field extension of F be found that contains all the roots of $f(x)$? If so, can a smallest such field be found?

8.4.4   Definition

Let F be a field, and suppose that $f(x) \in F[x]$ is a nonconstant polynomial of degree n. If E/F is a field extension, then $f(x)$ is said to **split** over E if there exist not necessarily distinct $c_1, c_2, \ldots, c_n \in E$ such that $f(x) = a(x - c_1)(x - c_2) \cdots (x - c_n)$, where a is the leading coefficient of $f(x)$. If $\mathcal{S}/F$ is a field extension and $f(x) \in F[x]$ splits over $\mathcal{S}$ but over no intermediate field properly contained in $\mathcal{S}$, then $\mathcal{S}$ is said to be a **splitting field** for $f(x)$. Recall that if E/F is a field extension and K is a field such that $F \subseteq K \subseteq E$, then K is an **intermediate field** of the field extension.

8.4.5   Theorem

Suppose that E/F is a field extension, and let $f(x) \in F[x]$ be a nonconstant polynomial of degree n. If $f(x)$ splits over E as

$$f(x) = a(x - c_1)(x - c_2) \cdots (x - c_n),$$

then $\mathcal{S} = F(c_1, c_2, \ldots, c_n)$ is a splitting field for $f(x)$.

Proof.   Since $c_1, c_2, \ldots, c_n \in F(c_1, c_2, \ldots, c_n)$ are the roots of $f(x)$, $f(x)$ splits over $F(c_1, c_2, \ldots, c_n)$. Now suppose that $F \subseteq F' \subseteq F(c_1, c_2, \ldots, c_n)$ and that $f(x)$ splits over F'. Since $E[x]$ is a Unique Factorization Domain and $F'[x] \subseteq E[x]$, there is only one way that $f(x)$ can split over F', and that is as $f(x) = a(x - c_1)(x - c_2) \cdots (x - c_n)$. Hence, $c_1, c_2, \ldots, c_n \in F'$, so $F(c_1, c_2, \ldots, c_n) \subseteq F'$ since $F(c_1, c_2, \ldots, c_n)$ is the smallest subfield of E that contains all the roots of $f(x)$. We have shown that $F' = F(c_1, c_2, \ldots, c_n)$, so $\mathcal{S} = F(c_1, c_2, \ldots, c_n)$ is a splitting field for $f(x)$. ∎

The preceding theorem shows that if $f(x) \in F[x]$ splits over some field extension E of F, then $f(x)$ has a splitting field. The following theorem shows that E can be dispensed with altogether.

8.4.6   Theorem

If F is a field and $f(x)$ is a nonconstant polynomial in $F[x]$, then a splitting field for $f(x)$ exists.

Proof. We first prove by induction that for *any* field F, if $f(x)$ is a nonconstant polynomial in $F[x]$, then a field extension E/F exists such that $f(x)$ splits over E. Once this is done, Theorem 8.4.5 can be invoked to show that $f(x)$ has a splitting field. Let F be any field. If $f(x)$ is a polynomial in $F[x]$ of degree 1, then $f(x) = ax + b$, where a and b are in F with $a \neq 0$. Hence, $f(x) = a(x - (-a^{-1}b))$, so $f(x)$ splits over F. Hence, we can let $E = F$. Now make the induction hypothesis that, for any field F, a polynomial in $F[x]$ of degree $k \geq 1$ splits over some field extension E of F. If $f(x)$ is a polynomial in $F[x]$ of degree $k + 1$, then, by Corollary 8.4.2, there is a field extension E_1/F such that E_1 contains a root c_1 of $f(x)$. Hence, $f(x)$ factors as $f(x) = (x - c_1)g(x)$ in $E_1[x]$. But $\deg(g(x)) = k$ and E_1 is a field, so the induction hypothesis indicates a field extension E/E_1 exists such that E contains the roots $c_2, \ldots, c_{k+1}$ of $g(x)$ and $g(x) = a(x - c_2) \cdots (x - c_{k+1})$ in $E[x]$, where a is the leading coefficient of $g(x)$. Now $F \subseteq E_1 \subseteq E$, so E contains all the roots $c_1, c_2, \ldots, c_{k+1}$ of $f(x)$ and thus $f(x)$ splits over E as

$$f(x) = a(x - c_1)(x - c_2) \cdots (x - c_{k+1}).$$

It follows by induction that if $f(x)$ is a nonconstant polynomial in $F[x]$, then a field extension E/F exists such that $f(x)$ splits over E. Therefore, by Theorem 8.4.5, there is a splitting field S for $f(x)$.

8.4.7 | Example

Find a splitting field S of $f(x) = x^4 + 1 \in \mathbf{Q}[x]$, and compute $[S : \mathbf{Q}]$.

Solution. We know by the Fundamental Theorem of Algebra that there is a $z \in \mathbf{C}$ such that $z^4 + 1 = 0$ and that z is a complex root of -1. Since $-1 = \cos \pi + i \sin \pi$, Theorem 4.4.6 shows that there are four 4th roots of -1 that are given by $u_k = \cos(\frac{\pi + 2k\pi}{4}) + i \sin(\frac{\pi + 2k\pi}{4})$ for $k = 0, 1, 2, 3$. Therefore,

$$u_0 = \cos\left(\frac{\pi}{4}\right) + i\sin\left(\frac{\pi}{4}\right) = \frac{\sqrt{2}}{2} + \frac{\sqrt{2}}{2}i,$$

$$u_1 = \cos\left(\frac{3\pi}{4}\right) + i\sin\left(\frac{3\pi}{4}\right) = -\frac{\sqrt{2}}{2} + \frac{\sqrt{2}}{2}i,$$

$$u_2 = \cos\left(\frac{5\pi}{4}\right) + i\sin\left(\frac{5\pi}{4}\right) = -\frac{\sqrt{2}}{2} + \frac{\sqrt{2}}{2}i, \quad \text{and}$$

$$u_3 = \cos\left(\frac{7\pi}{4}\right) + i\sin\left(\frac{7\pi}{4}\right) = \frac{\sqrt{2}}{2} + \frac{\sqrt{2}}{2}i.$$

are the roots of $f(x)$ in $\mathbf{C}$. Consider the field extension $\mathbf{Q}(\sqrt{2}, i)/\mathbf{Q}$. A basis for the field extension $\mathbf{Q}(\sqrt{2})/\mathbf{Q}$ is clearly $B = \{1, \sqrt{2}\}$. Using the same techniques as in Example 8.3.5, it can be shown that the minimal polynomial in $\mathbf{Q}(\sqrt{2})[x]$ of i is $m(x) = x^2 + 1$. The set $B' = \{1, i\}$ is a basis for the field extension $\mathbf{Q}(\sqrt{2}, i)/\mathbf{Q}(\sqrt{2})$, so a basis B'' for the field extension $\mathbf{Q}(\sqrt{2}, i)/\mathbf{Q}$ can be found by forming all possible products of an element from B and an element of B'. Hence, $B'' = \{1, \sqrt{2}, i, \sqrt{2}i\}$ is a basis of $\mathbf{Q}(\sqrt{2}, i)/\mathbf{Q}$. Therefore, every element of $\mathbf{Q}(\sqrt{2}, i)$ can be written as

$$a + b\sqrt{2} + ci + d\sqrt{2}i,$$

where $a, b, c, d, \in \mathbf{Q}$. By choosing $a, b, c,$ and d correctly, it is easy to show that $u_0, u_1, u_2, u_3, \in \mathbf{Q}(\sqrt{2},i)$, so $\mathbf{Q}(\sqrt{2},i)$ is the splitting field $\mathcal{S}$ for $f(x)$. Since $\mathbf{Q}(\sqrt{2},i)/\mathbf{Q}$ has a basis with four elements, $[\mathcal{S}:\mathbf{Q}] = 4$.

The only remaining property of splitting fields that we need to prove is that the splitting field $\mathcal{S}$ for a polynomial $f(x) \in F[x]$ is unique up to isomorphism.

Recall that if $f(x) = a_0 + a_1 x + \cdots + a_n x^n$ is a polynomial in $F[x]$ and $\phi : F \to F'$ is a ring isomorphism, then

$$f^\phi(x) = \phi(a_0) + \phi(a_1)x + \cdots + \phi(a_n)x^n$$

is the **corresponding polynomial** in $F'[x]$. Furthermore, if we define $\psi : F[x] \to F'[x]$ by $\psi(f(x)) = f^\phi(x)$, then Lemma 8.2.13 shows that ψ is a ring isomorphism.

8.4.8 Theorem

Let F and F' be fields, and suppose that $\phi : F \to F'$ is a ring isomorphism. If $\mathcal{S}$ is a splitting field over F for a nonconstant polynomial $f(x) \in F[x]$ and $\mathcal{S}'$ is a splitting field over F' for the corresponding polynomial $f^\phi(x) \in F'[x]$, then ϕ can be extended to a ring isomorphism $\overline{\phi} : \mathcal{S} \to \mathcal{S}'$.

Proof. We use induction on the degree of $f(x)$. If F is any field and $f(x)$ is any polynomial in $F[x]$ of degree 1, then $f(x) = ax + b$, where $a, b \in F$ and $a \neq 0$. We can factor $f(x)$ as $f(x) = a(x - (-a^{-1}b))$, so $f(x)$ splits over F. Since $\phi(a)$, $\phi(b) \in F'[x]$ and $\phi(a) \neq 0$, the polynomial $f^\phi(x)$ in $F'[x]$ corresponding to $f(x)$ factors as $f^\phi(x) = \phi(a)(x -(-\phi(a)^{-1}\phi(b)))$, so $f^\phi(x)$ splits over F'. Hence, we have $\mathcal{S} = F, \mathcal{S}' = F',$ and $\overline{\phi} = \phi$, so the theorem is true for any field F and any polynomial in $F[x]$ of degree 1. Now make the induction hypothesis that the theorem is true for any field F and any polynomial in $F[x]$ of degree $k \geq 1$, and let $f(x) \in F[x]$ have degree $k + 1$. If $m(x)$ is a monic irreducible factor of $f(x)$, suppose that $c_1 \in \mathcal{S}$ is a root of $m(x)$. Such a root of $m(x)$ exists, since any root of $m(x)$ is a root of $f(x)$ and all the roots of $f(x)$ are in $\mathcal{S}$. It follows that $m^\phi(x)$ is a monic irreducible factor of $f^\phi(x)$ in $F'[x]$, so if c_1' is a root of $m^\phi(x)$, then

$c_1' \in \mathcal{S}'$. Consequently, Theorem 8.2.14 indicates that ϕ can be extended to an isomorphism $\phi_1 : F(c_1) \to F'(c_1')$. Moreover, $f(x)$ factors as $f(x) = (x - c_1)g(x)$ in $F(c_1)[x]$, $f^\phi(x)$ factors as $f^\phi(x) = (x - c_1')g^\phi(x)$ in $F'(c_1')[x]$, and this shows that $\deg(g(x)) = \deg(g^\phi(x)) = k$. Now $\mathcal{S}$ is a splitting field of $g(x)$ over $F(c_1)$, and $\mathcal{S}'$ is a splitting field for $g^\phi(x)$ over $F'(c_1')$, so because $\deg(g(x)) = k$, the induction hypothesis tells us that the isomorphism ϕ_1 can be extended to an isomorphism $\overline{\phi} : \mathcal{S} \to \mathcal{S}'$. ∎

8.4.9 Corollary

If $f(x) \in F[x]$ is a nonconstant polynomial, then any two splitting fields for $f(x)$ are isomorphic.

Proof. The proof follows immediately. Just let $F' = F$ in the preceding proof, and take $\phi : F \to F$ to be the identity mapping. ∎

8.4.10 Example

The polynomial $f(x) = x^2 + 1 \in \mathbf{Q}[x]$ has $\pm i$ as roots. Thus, $\mathbf{Q}(i)$ and $\mathbf{Q}(-i)$ are splitting fields for $f(x)$. The mapping $\phi : \mathbf{Q}(i) \to \mathbf{Q}(-i)$ defined by $\phi(a + bi) = a - bi$ is easily shown to be a ring isomorphism.

Since any two splitting fields of a nonconstant polynomial $f(x)$ are isomorphic, it is now legitimate to speak of *the* splitting field of $f(x)$.

We now turn our attention to a much more general question. Given a field F, does a field extension of F exist that contains all the roots of *every* polynomial in $F[x]$? Furthermore, if such a field does exist, in what sense, if any, is it unique?

8.4.11 Definition

A field F is said to be **algebraically closed** if every nonconstant polynomial in $F[x]$ has a root in F. A field extension E/F is said to be an **algebraic closure** of F provided that

1. E is algebraically closed and
2. E is an algebraic field extension of F.

The following theorem establishes several interesting properties of algebraically closed fields.

8.4.12	**Theorem**

The following are equivalent for any field F.

1. F is algebraically closed.
2. Every irreducible polynomial in $F[x]$ has degree 1.
3. Every nonconstant polynomial in $F[x]$ splits into a product of its leading coefficient and linear factors of the form $x - c$, where $c \in F$.
4. If E/F is an algebraic field extension, then $E = F$.

Proof

$1 \Rightarrow 2$. Suppose that $f(x)$ is irreducible in $F[x]$. Since F is algebraically closed, $f(x)$ has a root $c \in F$. Thus, $f(x)$ factors as $f(x) = (x - c)g(x)$, where $g(x) \in F[x]$. Since $f(x)$ is irreducible in $F[x]$, $g(x)$ must be a unit in $F[x]$. But the only units in $F[x]$ are the constant polynomials, so there is an $a \in F$, $a \neq 0$, such that $g(x) = a$. Hence, $f(x) = a(x - c)$, so $\deg(f(x)) = 1$.

$2 \Rightarrow 3$. Let $f(x) \in F[x]$. Since $F[x]$ is a Unique Factorization Domain, $f(x) = f_1(x)f_2(x) \cdots f_n(x)$, where each $f_i(x)$ is irreducible in $F[x]$. Since statement 2 holds, each $f_1(x)$ has degree 1. Thus, $f_i(x) = a_i(x - c_i)$, where $a_i, c_i \in F$ for $i = 1, 2, \ldots, n$. Therefore, $f(x) = a(x - c_1)(x - c_2) \cdots (x - c_n)$, where $a = a_1 a_2 \cdots a_n$ is the leading coefficient of $f(x)$.

$3 \Rightarrow 4$. Suppose that E/F is an algebraic extension. If $c \in E$, then c is algebraic over F. Let $m(x)$ be the minimal polynomial in $F[x]$ of c. Since $m(x)$ is irreducible in $F[x]$, it follows from statement 3 that $m(x)$ must have degree 1. If $m(x) = x + a$, where $a \in F$, then $0 = m(c) = c + a$. Thus, $c = -a$ is in F, so $E = F$.

$4 \Rightarrow 1$. Suppose that $f(x)$ is a nonconstant polynomial in $F[x]$. By Corollary 8.4.2, there is a field extension E/F such that E contains a root c of $f(x)$. We know from Corollary 8.3.2 that $F(c)/F$ is an algebraic field extension. Hence, by statement 4, $F(c) = F$, so $c \in F$. Therefore, F is algebraically closed. ∎

The following two theorems are important. The first establishes a fundamental relation between algebraic field extensions and algebraically closed fields. The second shows that every field F has an algebraic closure that is unique up to isomorphism. With regard to the isomorphism, it actually shows much more. It shows that an isomorphism can be found between any two algebraic closures of F that leave each element of F fixed. Unfortunately, a proof of either theorem would take us too far afield from our purpose, so these theorems are stated without proof.

8.4.13   Theorem

Suppose that E/F is an algebraic field extension. If K is an algebraically closed field and $\phi : F \to K$ is a ring monomorphism, there is a ring monomorphism $\overline{\phi} : E \to K$ that extends ϕ to E.

8.4.14   Theorem

If F is a field, then there exists a field extension $\overline{F}/F$ such that $\overline{F}$ is the algebraic closure of F. Furthermore, if $\overline{F}$ and $\overline{F}^*$ are algebraic closures of F, then there exists a ring isomorphism $\phi : \overline{F} \to \overline{F}^*$ such that $\phi(a) = a$ for all $a \in F$.

Recall that at the conclusion of Section 4.4, the basic structures of our number systems were summarized. This summary was incomplete since, at that point, all of the properties that characterize $\mathbf{C}$ were not at hand. The following theorem completes the characterization of $\mathbf{C}$ promised in Chapter 4.

8.4.15   Theorem

The field of complex numbers $\mathbf{C}$ is the algebraic closure of the field $\mathbf{R}$ of real numbers and, as such, is unique up to isomorphism.

Proof.   The Fundamental Theorem of Algebra 6.3.1 shows that $\mathbf{C}$ is algebraically closed. $\mathbf{C/R}$ is an algebraic field extension because if $z = a + bi \in \mathbf{C}$, then z is a root of $f(x) = x^2 - 2ax + (a^2 + b^2) \in \mathbf{R}[x]$. Therefore, $\mathbf{C}$ is the algebraic closure of $\mathbf{R}$. Theorem 8.4.14 now shows that $\mathbf{C}$ is unique up to isomorphism. ∎

8.4.16   Corollary

$\mathbf{C}$ contains the algebraic closure of each of its subfields.

In particular, $\mathbf{C}$ contains the algebraic closure $\overline{\mathbf{Q}}$ of $\mathbf{Q}$. The field $\overline{\mathbf{Q}}$ is the set of all complex numbers that are algebraic over $\mathbf{Q}$. In the literature, $\overline{\mathbf{Q}}$ is often referred to as the **field of algebraic numbers**. It was mentioned earlier that π and e, where e is the base of the natural logarithmic function, are not in $\overline{\mathbf{Q}}$. The irrational numbers π and e are transcendental over $\mathbf{Q}$.

HISTORICAL NOTES

Leopold Kronecker (1823–1891). Leopold Kronecker was born in 1823 in what is now Legnica, Poland. The son of well-to-do parents, Kronecker received an excellent private education before entering the local gymnasium, where he received special instruction from the algebraist Kummer. It was under the guidance of Kummer that Kronecker became interested in mathematics, and his doctoral dissertation, accepted by the University of Berlin in 1845, was grounded in Kummer's work in the theory of numbers. After completing his degree, Kronecker assumed control of family business affairs and attained great financial success. Throughout his eight-year business career, he retained his interest in mathematics, pursuing it largely as a hobby but maintaining contacts with Kummer and other mathematicians. In 1861, Kronecker was elected to the Berlin Academy and took advantage of his privilege as a member to lecture at the University of Berlin. Financially independent, Kronecker held no official position until 1883, when Kummer retired from the University of Berlin and Kronecker was appointed to fill his chair.

Kronecker held a rather limited view of mathematics. He objected to the use of irrational numbers and denied the existence of transcendental numbers. He believed that all mathematics should be based on relationships among integers and that all proofs should be completed in a finite number of steps using relationships that themselves were derived in a finite number of steps from integers. In Kronecker's view, "God created the integers, all else is the work of man."

Problem Set 8.4

1. Show that $f(x) = x^2 - 5$ and $g(x) = x^2 - 2x - 4$ are irreducible in $\mathbf{Q}[x]$. Prove that $\mathbf{Q}(\sqrt{5})$ is the splitting field for each of these polynomials. Conclude that it is possible for two distinct polynomials in $\mathbf{Q}[x]$ to have the same splitting field.

2. Find the splitting field $\mathcal{S}$ for each of the following polynomials in $\mathbf{Q}[x]$. Find a basis for $\mathcal{S}/\mathbf{Q}$ and compute $[\mathcal{S} : \mathbf{Q}]$.

   (a) $f(x) = x^4 - 6x^2 - 7$
   (b) $f(x) = x^4 - 13x^2 + 36$
   (c) $f(x) = x^4 - 7x^2 + 10$

3. Find the splitting field $\mathcal{S}$ for each of the following. Compute $[\mathcal{S} : \mathbf{Z}_2]$, and find a basis for the field extension $\mathcal{S}/\mathbf{Z}_2$.

   (a) $f(x) = x^2 + [1] \in \mathbf{Z}_2[x]$
   (b) $f(x) = x^2 + x + [1] \in \mathbf{Z}_2[x]$

4. Find the splitting field $\mathcal{S}$ for $f(x) = x^2 + 2\sqrt{2} + 3 \in \mathbf{Q}(\sqrt{2})[x]$, and compute $[\mathcal{S} : \mathbf{Q}]$.

5. Suppose that $f(x)$, $g(x)$, $h(x) \in F[x]$ are nonconstant polynomials. If $f(x) = g(x)h(x)$ and $\mathcal{S}$ is the splitting field of $f(x)$, show that $\mathcal{S}$ contains the splitting field for $g(x)$ and $h(x)$. What happens if either $g(x)$ or $h(x)$ is a constant polynomial?

6. Suppose that $\mathcal{S}$ is the splitting field for a nonconstant polynomial $f(x)$ in $F[x]$. If E is a field such that $F \subseteq E \subseteq \mathcal{S}$, prove that $\mathcal{S}$ is the splitting field for $f(x)$ as a polynomial in $F[x]$ if and only if $\mathcal{S}$ is the splitting field for $f(x)$ as a polynomial in $E[x]$.

7. Let $f(x) \in F[x]$ be a nonconstant polynomial of degree n. If $\mathcal{S}$ is the splitting field for $f(x)$, show that $[\mathcal{S} : F] \leq n!$. Hint: Use induction on the degree of $f(x)$.

8. Let F be a field. Prove that for any positive integer n, if $f_1(x)$, $f_2(x)$, ..., $f_n(x) \in F[x]$ are nonconstant polynomials, then there exists a field extension E/F such that E contains a root of $f_i(x)$ for $i = 1, 2, \ldots, n$. Hint: Use induction and Theorem 8.4.1.

9. Explain why neither **R** nor **C** is algebraic over **Q**.

10. Prove that **C** contains the algebraic closure of each of its subfields. Are there subfields F of **C** such that $\bar{F} \neq$ **C**?

11. If $F_0 \subseteq F_1 \subseteq \cdots \subseteq F_n \subseteq \cdots$ is a tower of fields, prove that $F = \cup_{n \geq 0} F_n$ is a field.

12. For any field F, $F \subseteq \bar{F} \subseteq \bar{\bar{F}} \subseteq \bar{\bar{\bar{F}}} \subseteq \cdots$. Show that this process of taking the algebraic closure of a field stops after the first step; that is, show that $\bar{F} = \bar{\bar{F}}$.

13. If E/F is an algebraic extension such that every $f(x) \in F[x]$ splits over E, prove that $E = \bar{F}$.

8.5   GEOMETRIC CONSTRUCTIONS AND THE FAMOUS PROBLEMS OF ANTIQUITY

Section Overview. Three classical problems that were of great interest to Greek mathematicians were trisecting an angle, doubling a cube, and squaring a circle. In this section we use some of the material developed in the previous sections on field extensions to show that these problems cannot always be accomplished using only a straightedge and compass.

There are geometric constructions that cannot be completed successfully using only a straightedge and compass. Among these impossible constructions are three constructions that are referred to as the three famous problems of antiquity: trisecting an angle, doubling a cube, and squaring a circle.

Geometric constructions are usually first encountered in elementary geometry. A geometric construction means a construction that can be completed successfully using only a straightedge and a compass. By a straightedge we do not mean a ruler that can be used to measure distances, but rather a straightedge on which you are not permitted to place any marks and that is used only for drawing straight lines. The compass is used to draw circles and arcs of circles and as agent for transferring lengths of line segments. You have probably performed several of the following constructions in high school geometry, in which case you are familiar with some of the constructions that can be completed successfully using a straightedge and compass.

1. Draw a line through two distinct points.
2. Transfer the length of a line segment onto another line.
3. Bisect an angle.
4. Construct the perpendicular bisector of a line segment.
5. Construct an equilateral triangle whose sides are of given length.
6. Construct an isosceles triangle with base of a given length and sides of a given length.
7. Construct a triangle with 30-, 60-, and 90-degree angles.
8. Construct a circle with a given center and of a given radius.
9. Given a line and point not on the line, construct a line through the given point parallel to the given line.
10. Given a line segment, construct a square whose sides are equal in length to the given line segment.
11. Construct a perpendicular to a given line at a given point on the line.
12. Given a line and a point not on the line, construct a perpendicular from the given point to the given line.

The three famous problems of antiquity were first considered by the Greeks, who were interested in mathematics not only for its practical value, but also for the creative and challenging ideas it presented.

The Three Famous Problems of Antiquity

The Problem of Trisecting an Angle.   Is it always possible to trisect a given angle?

The Problem of Doubling a Cube.   Is it possible to construct a cube whose volume is twice that of a given cube?

The Problem of Squaring a Circle.   Can one construct a square whose area is equal to the area of a given circle?

Over the years, many general methods of geometric construction for each of these problems have been attempted, only to be shown to be invalid. Even today, when we know that these constructions cannot be completed successfully, a (necessarily fallacious) method for trisecting an angle, doubling a cube, or squaring a circle appears from time to time. It will now be shown that it is impossible to develop methods for successfully performing geometric constructions for the three problems of antiquity.

Constructible Numbers

We begin our discussion of constructible numbers with the assumption that we are given a line segment that has been assigned a length of 1 unit. All other lengths of line segments must be obtained from this line segment by geometric constructions using only a straightedge and compass.

8.5.1   **Definition**

A positive real number a is said to be a **constructible length** if a line segment of length a can be obtained from a finite number of straightedge and compass constructions beginning with a line segment that has been assigned a length of 1 unit. A point is taken to be a line segment of length 0, so 0 is a constructible real number. A real number a is said to be a **constructible number** if $|a|$, the absolute value of a, is a constructible length. An angle that measures α degrees is said to be a **constructible angle** if it can be constructed in a finite number of steps using a straightedge and compass.

Obviously, any integer n is a constructible number since 0 is constructible and if $n \neq 0$, then a line segment of length $|n|$ can be obtained by using a compass to transfer $|n|$ lengths of the line segment of length 1 onto a line, placing these line segments end-to-end. Recall from high school geometry that *corresponding parts of similar triangles are proportional*. This fact allows us to establish an equality of ratios between corresponding parts of similar triangles. For example, from the similar triangles in the figure,

we can establish the equality $\frac{x}{6} = \frac{3}{4}$. This property of similar triangles allows us to prove that the set of all constructible numbers is a subfield of the field of real numbers **R**.

8.5.2	**Theorem**

If **K** is the set of all constructible real numbers, then **K** is a subfield of **R**.

Proof. It is easily shown that if a and b are constructible numbers, then $a + b$ and $a - b$ are also constructible. Thus, since 0 is a constructible number, we now need to show that if a and b are nonzero real numbers, then ab and $\frac{a}{b}$ are constructible. Since a and b are constructible, we assume that two line segments of length $|a|$ and length $|b|$ have been obtained from a line segment with length 1 by a finite number of straightedge and compass constructions. To show that ab is constructible, first draw two half-lines emanating from a point P, and use a compass to transfer the line segment of length $|a|$ to the figure below. Denote this line segment by PA. Make a similar transfer of the line segments of length 1 and length $|b|$ to the figure, and denote these line segments by PU and PB, respectively. Draw a line through the points A and U, and then construct a line through B parallel to this line. Label the point of intersection of the half-line through P and A and the line through B as C. The triangles $\triangle PAU$ and $\triangle PCB$ are similar, so it follows that if c denotes the length of the line segment from P to C, then $\frac{c}{|b|} = \frac{|a|}{1}$. Hence, $c = |a|\,|b| = |ab|$, so ab is a constructible number.

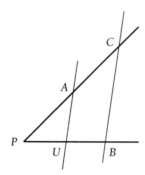

To show that $\frac{a}{b}$ is a constructible number, proceed in a similar manner, and construct the figure below. However, this time, first draw the line through the points A and B, and then construct a line through U that is parallel to the line through the points A and B. Denote the point of intersection of the line through U with the half-line through P and A by C. Then the triangles $\triangle PCU$ and $\triangle PAB$ are similar, so if c denotes the length of the line segment from P to C, then $\frac{c}{1} = \frac{|a|}{|b|}$ or $c = \left|\frac{a}{b}\right|$. Hence, $\frac{a}{b}$ is a constructible number. The other field properties follow since $\mathbf{K} \subseteq \mathbf{R}$, so $\mathbf{K}$ is a field. ∎

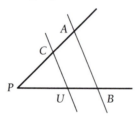

8.5.3   Collorary

The field of rational numbers $\mathbf{Q}$ is a subfield of $\mathbf{K}$.

Proof. This follows from the fact that every integer is a constructible number. Hence, if m and n are integers, then $m + n$, $m - n$, mn, and $\frac{m}{n}$, $n \neq 0$, are constructible. Thus, $\mathbf{Q} \subseteq \mathbf{K}$. ∎

8.5.4   Definition

Let F be a subfield of $\mathbf{R}$. The set of all points $(a, b) \in F \times F$ is called the **plane of** F. A straight line with an equation of the form $ax + by + c = 0$, where $a, b, c \in F$, is said to be a **line in** F, and a circle with equation $x^2 + y^2 + ax + by + c = 0$, where $a, b, c \in F$, is a **circle in** F. A line segment is said to be in F if its endpoints are in the plane of F. (Line segment refers to the usual concept of a line segment in $\mathbf{R}^2$.)

8.5.5   Lemma

Any line passing through two points in the plane of F is a line in F, and any circle with its center in the plane of F and its radius in F is a circle in F.

Proof. If (a_1, b_1) and (a_2, b_2) are points in the plane of F, then the equation of the line through these two points is given by

$$y = \frac{b_2 - b_1}{a_2 - a_1}(x - a_1) + b_1.$$

This equation can be reduced to an equation of the form $ax + by + c = 0$ with $a, b, c \in F$. If (a, b) is a point in the plane of F and $r \in F$, then the equation of the circle with center at (a, b) and radius r is

$$(x - a)^2 + (y - b)^2 = r^2.$$

If this equation is expanded and simplified, then the equation produced is also seen to be an equation of a circle with the required form. ∎

8.5.6   Lemma

Let F be a subfield of $\mathbf{R}$. The point(s) of intersection of lines and circles in F lie in the plane of $F(\sqrt{s})$ for some $s \in F$, $s \geq 0$.

Proof. Consider two lines in F, $a_1 x + b_1 y + c_1 = 0$ and $a_2 x + b_2 y + c_2 = 0$, that are not parallel. Solving for x and y, we see that

$$x = \frac{b_1 c_2 - b_2 c_1}{a_1 b_2 - a_2 b_1} \quad \text{and} \quad y = \frac{a_2 c_1 - a_1 c_2}{a_1 b_2 - a_2 b_1}.$$

Since the coefficients of the lines are in F, it follows that x and y are in F, so (x, y) lies in the plane of $F = F(\sqrt{0})$. If $x^2 + y^2 + dx + ey + f = 0$ and $ax + by + c = 0$ are equations of a circle and a line in F, respectively, and $b \neq 0$, then $y = -\frac{ax + c}{b}$. Substituting this expression for y into the equation of the circle and simplifying produces a quadratic equation with its coefficients in F of the form

$$x^2 + gx + h = 0.$$

The quadratic formula yields the solutions for x:

$$x = \frac{-g \pm \sqrt{g^2 - 4h}}{2}.$$

Since we are assuming that the line intersects the circle, $s = g^2 - 4h \geq 0$ (the line might be tangent to the circle, in which case $s = 0$), so it follows that both solutions $x = \frac{-g \pm \sqrt{s}}{2}$ lie in $F(\sqrt{s})$. Substituting these values of x into $y = -\frac{ax+c}{b} \in F$ produces values of y that are in $F(\sqrt{s})$. Thus, the points of intersection lie in the plane of $F(\sqrt{s})$. A similar result holds if $a \neq 0$. Finally, if $x^2 + y^2 + a_1 x + b_1 y + c_1 = 0$ and $x^2 + y^2 + a_2 x + b_2 y + c_2 = 0$ are equations of distinct intersecting circles in F, subtracting these equations gives $(a_1 - a_2)x + (b_1 - b_2)y + (c_1 - c_2) = 0$, and this is a line in F. Solving this equation of a line with the equation of either of the circles produces the points of intersection of the circles, and we have just shown that the points of intersection of a line in F with a circle in F produces points that lie in the plane of $F(\sqrt{s})$ for some $s \in F, s \geq 0$. ∎

8.5.7   Lemma

If $a \geq 0$ is a constructible number, then so is $\sqrt{a}$.

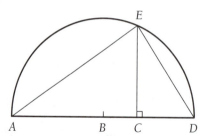

Proof. If $a = 0$, there is nothing to prove, so suppose that $a \neq 0$. Since a and 1 are constructible numbers, $a + 1$ is constructible, so we can construct a semicircle with diameter of length $a + 1$ using only a straightedge and compass. Thus, the figure above can be constructed in a finite number of steps using a straightedge and compass. The center of the semicircle is at B and has radius $\frac{a+1}{2}$. The length of the line segment from C to D is 1 unit. Facts from elementary geometry show that the triangles $\triangle ACE$ and $\triangle ECD$ are similar. Since the length of the line segment from A to C is a, it follows that if we let the length of the line segment from C to E be c, then $\frac{c}{a} = \frac{1}{c}$. Hence, $c^2 = a$, so $c = \sqrt{a}$. Thus $\sqrt{a}$ is a constructible number. ∎

8.5.8   Lemma

If F is a subfield of $\mathbf{R}$ and a line segment in F has length a, then $a \in F(\sqrt{s})$ for some $s \in F, s \geq 0$.

Proof. If $a = 0$, then the line segment has length 0, and $a \in F = F(\sqrt{0})$. If $a > 0$, then (a_1, b_1) and (a_2, b_2), the endpoints of the line segment, lie in the plane of F, so $a = \sqrt{(a_1 - a_2)^2 + (b_1 - b_2)^2}$. Thus, $a = \sqrt{s}$, where $s = (a_1 - a_2)^2 + (b_1 - b_2)^2 > 0$. Moreover, $s \in F$ since $a_1, a_2, b_1, b_2 \in F$. Hence, $a \in F(\sqrt{s})$. ■

8.5.9   Theorem

If a is a constructible number, then a is algebraic over $\mathbf{Q}$, and the degree of the minimal polynomial in $\mathbf{Q}[x]$ of a is a power of 2.

Proof. If a is a constructible number, then a line segment of length $|a|$ can be constructed in a finite number of steps using a straightedge and compass. Each step of the construction involves intersections of lines and circles and transfer of lengths. We start in the plane of $\mathbf{Q}$ and, to simplify notation, let $\mathbf{Q}_k = \mathbf{Q}(\sqrt{a_1}, \dots, \sqrt{a_k})$, for $k = 1, 2, \dots, n$. By Lemma 8.5.6, at the first step of the construction, the points of intersection of the lines and circles lie in the plane of $\mathbf{Q}_1$ for some rational number $a_1 \geq 0$. Note that the points might lie in the plane of $\mathbf{Q}$, in which case $a_1 = 0$. Hence, $[\mathbf{Q}_1 : \mathbf{Q}] = 1$ or 2. At the next step of this construction, we can consider the lines and circles to be in $\mathbf{Q}_1$. Again, by Lemma 8.5.6, the points of intersection of these lines and circles lie in the plane of $\mathbf{Q}_1$ or in the plane of $\mathbf{Q}(\sqrt{a_1})(\sqrt{a_2}) = \mathbf{Q}_2$ for some $a_2 \geq 0$, $a_2 \in \mathbf{Q}_1$. Again, we see that $[\mathbf{Q}_2 : \mathbf{Q}_1] = 1$ or 2. Since a is a constructible number, after $n - 1$ steps, we arrive at a line segment of length $|a|$, and at this step, the endpoints of this segment lie in the plane of $\mathbf{Q}_{n-1}$ for some $a_{n-1} \geq 0$, $a_{n-1} \in \mathbf{Q}_{n-2}$. As in the previous cases, $[\mathbf{Q}_{n-1} : \mathbf{Q}_{n-2}] = 1$ or 2. But we now see from Lemma 8.5.8 that $|a|$ is in $\mathbf{Q}(\sqrt{a_1}, \dots, \sqrt{a_{n-1}})(\sqrt{a_n}) = \mathbf{Q}_n$ for some $a_n \geq 0$ in $\mathbf{Q}_{n-1}$. Since $\mathbf{Q}_n$ is a field, it follows that $a \in \mathbf{Q}_n$. Furthermore, $[\mathbf{Q}_n : \mathbf{Q}_{n-1}] = 1$ or 2. Corollary 8.3.7 now shows that

$$[\mathbf{Q}_n : \mathbf{Q}] = [\mathbf{Q}_n : \mathbf{Q}_{n-1}][\mathbf{Q}_{n-1} : \mathbf{Q}_{n-2}] \cdots [\mathbf{Q}_2 : \mathbf{Q}_1][\mathbf{Q}_1 : \mathbf{Q}].$$

Since each factor in this product is either 1 or 2, $[\mathbf{Q}_n : \mathbf{Q}] = 2^m$ for some integer m, $0 \leq m \leq n$. This indicates that $\mathbf{Q}_n$ is a finite dimensional field extension of $\mathbf{Q}$, and thus, by Theorem 8.3.1, $\mathbf{Q}_n$ is an algebraic extension of $\mathbf{Q}$. Since a is an element of $\mathbf{Q}_n$, we have that a is algebraic over $\mathbf{Q}$. Note also that $\mathbf{Q}(a)$ is a subfield of $\mathbf{Q}_n$, so

$$2^m = [\mathbf{Q}_n : \mathbf{Q}] = [\mathbf{Q}_n : \mathbf{Q}(a)][\mathbf{Q}(a) : \mathbf{Q}].$$

Hence, $[\mathbf{Q}(a) : \mathbf{Q}]$ divides 2^m, and this indicates that $[\mathbf{Q}(a) : \mathbf{Q}]$ is a power of 2. Thus, the minimal polynomial in $\mathbf{Q}[x]$ of a must have a degree that is a power of 2. ■

We need the following two observations before the questions posed by the three problems of antiquity can be settled.

8.5.10   Lemma

For any angle of α degrees, $\cos 3\alpha \equiv 4 \cos^3 \alpha - 3 \cos \alpha$. Furthermore, if an angle of 20 degrees is a constructible angle, then $\cos 20°$ is a constructible number.

Proof. Recall that for any angles of α and β degrees,

$$\sin (\alpha + \beta) \equiv \sin \alpha \cos \beta + \sin \beta \cos \alpha,$$
$$\cos (\alpha + \beta) \equiv \cos \alpha \cos \beta - \sin \alpha \sin \beta, \quad \text{and}$$
$$\sin^2 \alpha + \cos^2 \alpha \equiv 1.$$

By letting $\alpha = \beta$, it follows from $\cos(\alpha + \beta) \equiv \cos \alpha \cos \beta - \sin \alpha \sin \beta$ that $\cos 2\alpha \equiv \cos^2\alpha - \sin^2\beta \equiv 2 \cos^2\alpha - 1$. Similarly, we see that the identity for $\sin(\alpha + \beta)$ gives $\sin 2\alpha \equiv 2 \sin \alpha \cos \alpha$ when $\alpha = \beta$. Hence,

$$\cos 3\alpha \equiv \cos(2\alpha + \alpha)$$
$$\equiv \cos 2\alpha \cos \alpha - \sin 2\alpha \sin \alpha$$
$$\equiv (2\cos^2\alpha - 1)\cos\alpha - (2\sin\alpha\cos\alpha)\sin\alpha$$
$$\equiv 2\cos^3\alpha - \cos\alpha - 2\sin^2\alpha\cos\alpha$$
$$\equiv 2\cos^3\alpha - \cos\alpha - 2(1 - \cos^2\alpha)\cos\alpha$$
$$\equiv 2\cos^3\alpha - \cos\alpha - 2\cos\alpha + 2\cos^3\alpha$$
$$\equiv 4\cos^3\alpha - 3\cos\alpha.$$

Finally, suppose that we can construct an angle of 20 degrees using a straight-edge and compass. From the point P, using a compass, transfer the unit length

to the terminal side of the angle by placing one end of the segment at the point P. Label the other endpoint of the segment as U. From the point U construct a perpendicular to the initial side of the angle to obtain the right triangle $\triangle PAU$. Then $\cos 20°$ is the length of the line segment from P to A. Hence, if an angle of 20 degrees is constructible, then $\cos 20°$ is a constructible number. ∎

One conclusion that can be drawn from Lemma 8.5.10 is that if $\cos 20°$ is not a constructible number, then an angle of 20 degrees is not a constructible angle.

8.5.11   Theorem

It is not possible to find general methods of geometric construction with a straightedge and compass for trisecting a given angle, for doubling a cube, or for squaring a circle.

Proof. To show that it is not possible to find a general method of geometric construction with a straightedge and compass for trisecting a given angle, it suffices to find an angle that cannot be trisected. A similar observation holds for the other two problems of antiquity.

If it is possible to always trisect a given angle, then an angle of 20 degrees can be constructed by trisecting an angle of 60 degrees. Note that an angle of 60 degrees is a constructible angle since it can be constructed by constructing an equilateral triangle. Consequently, by Lemma 8.5.10, $\cos 20°$ would be a constructible number. Since $\cos 60° = \frac{1}{2}$, we see from Lemma 8.5.10 that $\frac{1}{2} = \cos(3 \cdot 20°) = 4\cos^3 20° - 3\cos 20°$, so $\cos 20°$ is a root of the polynomial $f(x)$ $4x^3 - 3x - \frac{1}{2} \in \mathbf{Q}[x]$. The polynomial $8x^3 - 6x - 1$ has no rational roots, so $f(x)$ is irreducible in $\mathbf{Q}[x]$. Hence, $\cos 20°$ has a minimal polynomial of degree 3. Theorem 8.5.9 now shows that $\cos 20°$ cannot be a constructible number since 3 is not a power of 2. Thus, an angle of 20 degrees cannot be constructed, so it is impossible to trisect an angle of 60 degrees.

To show that it is not always possible to double a cube, construct a cube with each side of length 1. If we can double this cube, then we must be able to construct a cube with side of length s such that $s^3 = 2$. Thus, s must be a root of the polynomial $f(x) = x^3 - 2$. Since $f(x)$ has no rational roots, $f(x)$ is irreducible in $\mathbf{Q}[x]$, so s has a minimal polynomial in $\mathbf{Q}[x]$ of degree 3, and 3 is not a power of 2. Hence, s is not a constructible number, so it is impossible to duplicate a cube with a side of length 1.

Finally, construct a circle with radius 1 and center at $(0,0)$. If we can always square a circle, then we would be able to construct a square with area equal to π, the area of our circle. If s is the length of the side of the square, then $s^2 = \pi$, so $s = \sqrt{\pi}$ would be a constructible number. But Theorem 8.5.9 indicates that $\sqrt{\pi}$ would be algebraic over $\mathbf{Q}$, and this in turn would imply that π is algebraic over $\mathbf{Q}$. But this is impossible since we know that **Ferdinand von Lindemann** (1852–1939) proved that π is transcendental over $\mathbf{Q}$. (Information on the life of Lindemann can be found in the Historical Notes at the end of Section 8.2.) ■

Construction Problems.   In the Historical Notes at the end of Section 1.3 it was pointed out that **Carl Friedrich Gauss** (1777–1855) at the age of 19 proved that it was possible to construct a regular 17-gon using only a straightedge and compass. This problem had defied solution for over 2000 years. Not only did Gauss prove that it was possible, but he actually was able to construct a regular 17-gon. A theorem due to Gauss asserts that if $p \neq 2$ is a prime, then a regular p-gon is constructible using only a straightedge and compass if and only if p is of the form $p = 2^{2^n} + 1$ for some integer $n \geq 0$. Consequently, it is impossible to construct a regular 7-gon, a regular 11-gon, and a regular 13-gon since 7, 11, and 13 are not primes of the correct form.

Primes of the form $2^{2^n} + 1$ are known as **Fermat primes**. The values of $n = 0$, 1, 2, 3, and 4 give the primes 3, 5, 17, 257, and 65,537. The values $n = 5$ and $n = 6$ give $4,294,967,297 = (641)(6,700,417)$ and $18,446,744,073,709,551,617 = (274,177)(67,280,421,310,721)$, respectively, neither of which is prime.

HISTORICAL NOTES

It is difficult to give an accurate date for when each of the three famous problems of antiquity first appeared, but it is known that these problems were of considerable interest to Greek mathematicians. The problem of squaring a circle appeared in mathematical writing as early as 1650 B.C., while the problem of doubling the cube dates from around 430 B.C. Trisecting an angle is probably the least well-known of these problems, doubling a cube was the most famous in ancient Greek times, and squaring a circle became more famous in recent times. Amateur mathematicians were especially interested in the problem of squaring a circle. Even though we now know that each of these problems is impossible to solve using only a straightedge and compass, even to this day amateur mathematicians continue to claim that they have found a valid construction for solving one or more of these problems. The problem of squaring a circle became so popular among amateur mathematicians and the number of attempted solutions so numerous that in 1775 the Paris Academy passed a resolution indicating that no further attempted solutions would be examined. The Royal Society in London passed a similar resolution a few years later in an effort to stem the flow of "proofs" that a circle could be squared. Of course, the underpinnings for the proof that a circle cannot be squared rests on Lindemann's proof that π is a transcendental number.

Problem Set 8.5

All constructions are to be carried out using only a straightedge and compass.

1. If a and b are constructible numbers, show that $a + b$ and $a - b$ are constructible.

2. Show that an angle of 90 degrees can be trisected. If n is a positive integer, can an angle of $n \cdot 90$ degrees always be trisected?

3. Prove that the following are equivalent for an angle of α degrees.
   (a) An angle of α degrees is a constructible angle.
   (b) $\cos \alpha$ is a constructible number.
   (c) $\sin \alpha$ is a constructible number.

4. Prove that an angle of 40 degrees is not a constructible angle..

5. Prove that it is possible to double a square. That is, given a square, it is possible using only a straightedge and compass to construct a square with twice the area of the given square.

6. (a) Prove that an angle of 72 degrees is a constructible angle. Hint: Draw an isosceles triangle with sides of length a and 1, with the side of length 1 opposite a 36-degree angle. Next, draw a bisector of one of the 72-degree angles, and then use similar triangles to show that $a = \frac{1 + \sqrt{5}}{2}$. Note the use of *draw* as opposed to *construct*. Now explain why an angle of 72 degrees is constructible.

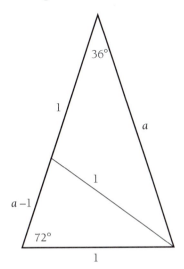

   (b) Prove that it is possible to construct a regular pentagon.

7. Suppose that angles of α and β degrees are constructible angles.

   (a) Prove that an angle of $\alpha + \beta$ degrees is a constructible angle. Hint: Consider $\sin(\alpha + \beta)$ and Exercise 2.
   (b) *If* m *and* n *are nonnegative integers, prove that an angle of* $m\alpha + n\beta$ *degrees is a constructible angle.*
   (c) Prove that it is not possible to construct angles of 1 and of 2 degrees.

8. If n is a positive integer, a **regular** *n*-gon is an n-sided figure with n sides all having equal length, as shown in the figure. Exactly one of the following n-gons is not constructible using a straightedge and compass. Find the one that is not constructible and explain why. Explain why all the other n-gons are constructible.

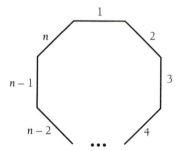

   (a) regular 6-gon (hexagon)
   (b) regular 8-gon (octagon)
   (c) regular 9-gon (nonagon)
   (d) regular 10-gon (decagon)
   (e) regular 12-gon (dodecagon)

9. Prove that it is impossible to find a general method of geometric construction to construct a cube whose volume is 3 times that of a given cube. Does the statement remain true if 3 is replaced by a prime integer p?

9

An Introduction to Galois Theory

The problem of finding the roots of a polynomial dates back to around 400 B.C. with the attempt of the Babylonians to find the roots of quadratic polynomials. The Babylonians did not understand the notion of an equation, so their methods amounted to algorithmic techniques. The problems they solved would, in modern notation, give rise to quadratic polynomials. Around 300 B.C. Euclid developed a geometric method for finding certain lengths, and his method also amounted to finding the roots of a quadratic polynomial. Additional contributions for finding the roots of a quadratic polynomial were made by Hindu mathematicians around A.D. 600 and by Arabic mathematicians around A.D. 800. Arabic mathematicians were able to compute the roots of quadratic polynomials, and the technique they used was essentially what we recognize today as the quadratic formula. It was not until around 1145 that a book appeared in Europe that gave a complete solution of the quadratic equation. This led mathematicians to turn their attention to polynomials of degree 3 and higher. They attempted to find a formula that would yield the roots of all polynomials of a given degree. As we shall see, this is always possible for polynomials of degree 2, 3, and 4 but impossible for polynomials of higher degree.

9.1   CLASSICAL FORMULAS

Section Overview. This section investigates polynomials for which a formula can be developed that will yield the roots of the polynomial. For the sake of simplicity, only polynomials in $\mathbf{R}[x]$ are investigated. Techniques are developed for finding the roots of quadratic, cubic, and quartic polynomials in $\mathbf{R}[x]$.

9.1.1   Quadratic Polynomials

As you well know, the roots of the quadratic polynomial

$$f(x) = ax^2 + bx + c \quad a,b,c \in \mathbf{R}, \ a \neq 0,$$

are given by the quadratic formula

$$x = \frac{-b \pm \sqrt{b^2 - 4ac}}{2a},$$

the derivation of which can be found in any college algebra book. The usual method for developing the quadratic formula is to complete the square. We use a different method that illustrates the technique that is used to develop formulas for cubic and quartic polynomials in $\mathbf{R}[x]$. Notice that the quadratic formula is expressed in terms of the coefficients of the quadratic polynomial.

9.1.2   Definition

A polynomial $f(x) \in \mathbf{R}[x]$ of degree n is said to be **reduced** if it has no x^{n-1} term, that is, if $f(x)$ is a polynomial of the form

$$f(x) = a_n x^n + a_{n-2} x^{n-2} + \cdots + a_1 x + a_0.$$

9.1.3   Example

The polynomials $2x^3 + 6x - 8$ and $x^4 + 2x^2 - 1$ are reduced, while $3x^5 + 5x^4 - 6x + 7$ is not.

The following lemma shows that each polynomial has a corresponding reduced polynomial and that a root of the reduced polynomial gives rise to a root of the original polynomial. The proof is a straightforward computation, so it is left as an exercise.

9.1.4   Lemma

If $f(y) = a_n y^n + a_{n-1} y^{n-1} + \cdots + a_1 y + a_0$ is a polynomial of degree n, then the substitution

$$y = x - \frac{a_{n-1}}{n a_n}$$

produces a reduced polynomial

$$\bar{f}(x) = f\left(x - \frac{a_{n-1}}{na_n}\right).$$

Furthermore, if u is a root of $\bar{f}(x)$, then $u - \frac{a_{n-1}}{na_n}$ is a root of $f(y)$.

Now let's return to the development of the quadratic formula. Consider the quadratic polynomial $f(y) = ay^2 + by + c$, $a \neq 0$, and replace y by $x - \frac{b}{2a}$. This produces

$$\bar{f}(x) = a\left(x - \frac{b}{2a}\right)^2 + b\left(x - \frac{b}{2a}\right) + c$$

$$= ax^2 - bx + \frac{b^2}{4a} + bx - \frac{b^2}{2a} + c$$

$$= ax^2 + \left(c - \frac{b^2}{4a}\right).$$

We see that replacing y by $x - \frac{b}{2a}$ does indeed produce a reduced polynomial, as was asserted by Lemma 9.1.4. Setting $\bar{f}(x)$ equal to 0 and solving for x yields the roots

$$u = \pm\frac{\sqrt{b^2 - 4ac}}{2a}$$

of $\bar{f}(x)$. By Lemma 9.1.4, the roots of $f(y)$ are given by

$$y = u - \frac{b}{2a} = -\frac{b}{2a} \pm \frac{\sqrt{b^2 - 4ac}}{2a}.$$

As we will see later, the development of the quadratic formula is useful in finding the roots of the cubic and quartic polynomials. It is interesting that the discovery of the quadratic formula forced mathematicians to confront the problem of taking the square root of negative numbers. They had the tendency to dismiss such numbers as meaningless. Over time, the resolution of this difficulty played an important role in the development of complex numbers.

9.1.5   Cubic Polynomials

We claim that a formula can be found that yields the roots of a cubic polynomial in $\mathbf{R}[x]$. Consider the cubic polynomial $f(y) = ay^3 + by^2 + cy + d$, and note that since $f(y) = a\left(y^3 + \dfrac{b}{a}y^2 + \dfrac{c}{a}y + \dfrac{d}{a}\right)$, u is a root of $f(y)$ if and only if u is a root of $y^3 + \dfrac{b}{a}y^2 + \dfrac{c}{a}y + \dfrac{d}{a}$. Hence, we can assume that $f(y)$ is monic, so we begin by considering a polynomial of the form

$$f(y) = y^3 + by^2 + cy + d.$$

Next, make the substitution $y = x - \dfrac{b}{3}$ in order to obtain the associated reduced polynomial $\bar{f}(x)$.

$$\bar{f}(x) = \left(x - \frac{b}{3}\right)^3 + b\left(x - \frac{b}{3}\right)^2 + c\left(x - \frac{b}{3}\right) + d$$

$$= x^3 + \left(c - \frac{b^2}{3}\right)x + \left(d - \frac{bc}{3} + \frac{2b^3}{27}\right)$$

$$= x^3 + mx - n,$$

where

$$m = c - \frac{b^2}{3} \quad \text{and} \quad n = -d + \frac{bc}{3} - \frac{2b^3}{27}.$$

If we can find the roots of $\bar{f}(x)$ in terms of m and n, we can find the roots of $f(y)$ in terms of b, c, and d. We now give a solution of $x^3 + mx = n$ presented by **Gerome Cardan** (1501–1576). (Information on the life of Cardan can be found in the Historical Notes at the end of Section 4.4.) Cardan was an Italian mathematician who published a solution of $x^3 + mx = n$ in 1545 in his book *Ars Magna*. (Be sure to read the Historical Notes at the end of this section for insight into the controversy that developed around the discovery of the solution of the cubic equation.) If $(u - v)^3 + 3uv(u - v)$ is expanded and simplified, the result is $u^3 - v^3$, so by comparing

$$(u - v)^3 \quad + \ 3uv(u - v) = u^3 - v^3 \qquad \text{with}$$
$$\downarrow \qquad\qquad \downarrow \quad \downarrow \qquad\qquad \downarrow$$
$$x^3 \quad\ + \ m \quad x \quad = \quad n,$$

we see that if u and v are solutions of the equations $3uv = m$ and $u^3 - v^3 = n$, then $x = u - v$ is a solution of the equation $x^3 + mx = n$. Now $3uv = m$ gives $v = \dfrac{m}{3u}$, and substituting this into $u^3 - v^3 = n$ gives

$$u^3 - \frac{m^3}{27u^3} = n, \quad \text{or} \quad u^6 - nu^3 - \frac{m^3}{27} = 0,$$

which is a quadratic equation in u^3. Applying the quadratic formula, we arrive at

$$u^3 = \frac{n \pm \sqrt{n^2 + \frac{4m^3}{27}}}{2} = \frac{n}{2} \pm \sqrt{\left(\frac{n}{2}\right)^2 + \left(\frac{m}{3}\right)^3},$$

and from this we can find u. If we choose the plus sign in the last equation, then

$$u = \sqrt[3]{\frac{n}{2} + \sqrt{\left(\frac{n}{2}\right)^2 + \left(\frac{m}{3}\right)^3}}.$$

If we solve $3uv = m$ for u, then $u = \frac{m}{3v}$, so $u^3 - v^3 = n$ produces

$$\frac{m^3}{27v^3} - v^3 = n, \quad \text{or} \quad v^6 + nv^3 - \frac{m^3}{27} = 0.$$

Following exactly the same procedure as we did for u, we see that

$$v = \sqrt[3]{-\frac{n}{2} + \sqrt{\left(\frac{n}{2}\right)^2 + \left(\frac{m}{3}\right)^3}},$$

so

$$x = u - v = \sqrt[3]{\frac{n}{2} + \sqrt{\left(\frac{n}{2}\right)^2 + \left(\frac{m}{3}\right)^3}} - \sqrt[3]{-\frac{n}{2} + \sqrt{\left(\frac{n}{2}\right)^2 + \left(\frac{m}{3}\right)^3}}$$

is one root of $x^3 + mx - n$. If ω is the complex cube root of unity $-\frac{1}{2} + \frac{\sqrt{3}}{2}i$, then it can be shown that $x = \omega u - \omega^2 v$ and $x = \omega^2 u - \omega v$ are also roots of $x^3 + mx - n$. (See Exercise 3.) We can conclude that if u and v are chosen to be such that $3uv = m$ and $u^3 - v^3 = n$, then the roots of the cubic polynomial $x^3 + mx - n$ are given by the **cubic formula**:

$$x = u - v = \sqrt[3]{\frac{n}{2} + \sqrt{\left(\frac{n}{2}\right)^2 + \left(\frac{m}{3}\right)^3}} - \sqrt[3]{-\frac{n}{2} + \sqrt{\left(\frac{n}{2}\right)^2 + \left(\frac{m}{3}\right)^3}}$$

$$x = \omega \sqrt[3]{\frac{n}{2} + \sqrt{\left(\frac{n}{2}\right)^2 + \left(\frac{m}{3}\right)^3}} - \omega^2 \sqrt[3]{-\frac{n}{2} + \sqrt{\left(\frac{n}{2}\right)^2 + \left(\frac{m}{3}\right)^3}}$$

$$x = \omega^2 \sqrt[3]{\frac{n}{2} + \sqrt{\left(\frac{n}{2}\right)^2 + \left(\frac{m}{3}\right)^3}} - \omega \sqrt[3]{-\frac{n}{2} + \sqrt{\left(\frac{n}{2}\right)^2 + \left(\frac{m}{3}\right)^3}}$$

No claim is made that the cubic formula will produce the roots of a cubic polynomial in their simplest form. The important point is that a formula, expressed in terms of the coefficients of the polynomial, can be found that will yield its roots.

9.1.6   Examples

1. Consider the cubic equation $x^3 + 6x = 20$. $x = 2$ is clearly a root of this equation. Let's see what result will be produced by the preceding formulas. Comparing $x^3 + mx = n$ with $x^3 + 6x = 20$, we see that $m = 6$ and $n = 20$. Hence, $\frac{m}{3} = 2$ and $\frac{n}{2} = 10$. From this it follows that

    $$u^3 = 10 + 6\sqrt{3} \quad \text{and} \quad v^3 = -10 + 6\sqrt{3}.$$

    Hence,

    $$x = \sqrt[3]{10 + 6\sqrt{3}} - \sqrt[3]{-10 + 6\sqrt{3}}$$

    is a root of the cubic polynomial $x^3 + 6x - 20$. It follows that

    $$\sqrt[3]{10 + 6\sqrt{3}} - \sqrt[3]{-10 + 6\sqrt{3}} = 2,$$

    but this is by no means obvious. (You should show this.) The remaining roots can be found by using the cubic formula or the quadratic formula; the latter formula in this case is the most efficient. Since $x^3 + 6x - 20$ factors as $(x - 2)(x^2 + 2x + 10)$, an application of the quadratic formula to $x^2 + 2x + 10$ yields the roots $x = -1 + 3i$ and $x = -1 - 3i$.

2. Finding the roots of a cubic equation by using the cubic formula forced mathematicians to deal with the issue of complex numbers. They could dismiss the nonreal roots of a quadratic equation arising from an application of the quadratic formula as meaningless, but complex numbers often arise in the cubic formula for a cubic equation that has real roots. For example, the cubic polynomial $f(x) = x^3 - 6x - 4$ has three real roots, namely, $x = -2$, $x = 1 + \sqrt{3}$, and $x = 1 - \sqrt{3}$. Yet an application of the cubic formula to $x^3 - 6x = 4$ produces $u = \sqrt[3]{2 + 2i}$ and $v = \sqrt[3]{-2 + 2i}$. Since $3uv = 3\sqrt[3]{2 + 2i}\sqrt[3]{-2 + 2i} = 3\sqrt[3]{(2 + 2i)(-2 + 2i)} = 3\sqrt[3]{-8} = -6$ and $u^3 - v^3 = 4$, $x = u - v = \sqrt[3]{2 + 2i} - \sqrt[3]{-2 + 2i}$ is a root of $f(x)$. Therefore, $\sqrt[3]{2 + 2i} - \sqrt[3]{-2 + 2i}$ must be a real number. Mathematicians who worked with the cubic formula were confronted with

the fact that the difference of two quantities that were previously dismissed as meaningless could be a real number. As was pointed out earlier, the resolution of this difficulty played an important role in the development of complex numbers.

9.1.7   Quartic Polynomials

Around 1545 a method of finding the roots of a quartic polynomial was discovered by **Lodovico Ferrari** (1522–1565), who was a student of Cardan. It was due to Cardan's encouragement that Ferrari worked on quartic polynomials. However, an effective technique was discovered by **René Descartes** (1596–1650) in 1637, and his method is the one presented here. If we can find the roots of a general monic reduced quartic polynomial, then we can find the roots of a general quartic polynomial. Do you see why? Consider

$$f(x) = x^4 + bx^2 + cx + d,$$

and set

$$x^4 + bx^2 + cx + d = (x^2 + Ax + B)(x^2 - Ax + C).$$

Then

$$x^4 + bx^2 + cx + d = x^4 + (-A^2 + B + C)x^2 + (-AB + AC)x + BC.$$

Equating coefficients produces

$$-A^2 + B + C = b, \tag{1}$$

$$-AB + AC = c, \qquad \text{and} \tag{2}$$

$$BC = d. \tag{3}$$

Solving equations (1) and (2) for B and C yields

$$B = \frac{1}{2}\left(A^2 + b - \frac{c}{A}\right), \ A \neq 0 \qquad \text{and}$$

$$C = \frac{1}{2}\left(A^2 + b + \frac{c}{A}\right), \ A \neq 0.$$

Substituting these values for B and C into equation (3) produces

$$\left(A^2 + b - \frac{c}{A}\right)\left(A^2 + b + \frac{c}{A}\right) = 4d,$$

which gives

$$A^6 + 2bA^4 + (b^2 - 4d)A^2 - c^2 = 0.$$

This last equation is a cubic in A^2, and we know how to find the roots of a cubic polynomial. This will yield A^2, and from this we can easily find A. With this done, we can return to equations (1) and (2) to compute B and C. If $A = 0$, just solve equations (1) and (3) for B and C. Once A, B, and C are known, we can consider

$$x^4 + bx^2 + cx + d = (x^2 + Ax + B)(x^2 - Ax + C)$$

and apply the quadratic formula to $x^2 + Ax + B$ and $x^2 - Ax + C$. This yields the roots of $f(x) = x^4 + bx^2 + cx + d$. Actually producing a formula for the roots of $f(x)$, as was done for quadratic and cubic polynomials, is cumbersome, so we will be satisfied with Descartes' technique for finding the roots of a quartic polynomial outlined earlier. However, it is possible to find a formula, expressed in terms of the coefficients of the quartic polynomial, that yields its roots.

9.1.8   Example

Let's use Descartes' method to find the roots of $f(x) = x^4 + 5x^2 - 4$. Comparing this with $f(x) = x^4 + bx^2 + cx + d$, we see that $b = 5, c = 0$, and $d = -4$. Hence, the equation

$$A^6 + 2bA^4 + (b^2 - 4d)A^2 - c^2 = 0$$

of 9.1.7 becomes

$$A^6 + 10A^4 + 41A^2 = 0.$$

This last equation obviously has $A = 0$ as a solution, so the equations

$$-A^2 + B + C = b, \tag{1}$$

$$-AB + AC = c, \quad \text{and} \tag{2}$$

$$BC = d \tag{3}$$

of 9.1.7 reduce to

$$B + C = 5, \tag{4}$$

$$BC = -4. \tag{5}$$

Solving equations (4) and (5) for B produces $B = \frac{5 \pm \sqrt{41}}{2}$. If $B = \frac{5 + \sqrt{41}}{2}$, then $C = \frac{5 - \sqrt{41}}{2}$, and if $B = \frac{5 - \sqrt{41}}{2}$, then $C = \frac{5 + \sqrt{41}}{2}$. Hence, we see that

$$x^4 + 5x^2 - 4 = \left(x^2 + \frac{5 + \sqrt{41}}{2}\right)\left(x^2 + \frac{5 - \sqrt{41}}{2}\right).$$

From this it follows that the roots of $f(x) = x^4 + 5x^2 - 4$ are given by

$$x = \pm i \sqrt{\frac{5 + \sqrt{41}}{2}} \quad \text{and} \quad x = \pm \sqrt{\frac{-5 + \sqrt{41}}{2}}.$$

You might have noticed that $x^4 + 5x^2 - 4$ is a quadratic in x^2, so the quadratic formula can be applied to show that $x^2 = -\frac{5 \pm \sqrt{41}}{2}$; thus, by taking square roots we arrive at the same result as with Descartes' method. A more interesting and much more tedious problem is encountered if one tries to compute the roots of a quartic polynomial such as $f(x) = x^4 + 4x - 3$ using Descartes' method. You should compute the roots of $x^4 + 4x - 3$ using his method.

Historical Notes

As noted in the introduction to this chapter, the problem of finding the roots of a polynomial dates back to the ancient culture of the Babylonians who, around 400 B.C., were attempting to find the roots of quadratic polynomials. Euclid's geometrical approach around 300 B.C. and the contributions of Hindu and Arabic mathematicians provided later mathematicians with the tools to solve such equations. However, not until A.D. 1145 did the first book appear in Europe that gave a complete solution of quadratic equations. Formulas for the roots of cubic and quartic polynomials were found in the 16th century, born amid the furor of controversy and great rivalry among Italian mathematicians of the day.

Credit for solving the cubic equation probably should go to **Scipione del Ferro** (1465–1526), who was a lecturer in arithmetic and geometry at the University of Bologna. Del Ferro could solve only cubic equations of the form $x^3 + mx = n$, but, as we know now, that is all that is required since every cubic equation can be reduced to one of this form. On his deathbed, del Ferro confided his solution to one of his students, who used it in a competition with **Nicolo Fontana** (1499–1557), also known as **Tartaglia**, which means "the stammerer." In the competition Tartaglia utilized his own method for the solution of cubics and handily won the contest. As his fame spread among Italian mathematicians, Tartaglia's accomplishments came to the attention of **Gerome Cardan** (1501–1576) in Milan, who, with much persuasive flattery, convinced Tartaglia to reveal his method; he did so on the condition that Cardan not publish it. Cardan kept his promise to Tartaglia and did not include the results in his book *Practica Arithmeticae*. Over the next several years Cardan worked out a method for finding the roots of cubic polynomials on his own, and his student **Lodovico Ferrari** (1522–1565) discovered a method of finding the roots of quartic polynomials. After Cardan and Ferrari had an opportunity to inspect the papers of del Ferro and verify that del Ferro had indeed provided the initial solution to the cubic equation, Cardan published the methods for solving cubic and quartic equations in his most important book, the *Ars Magna*. Although Cardan gave full credit to Tartaglia's solutions, Tartaglia was not satisfied and furiously challenged Cardan to a debate. It was, however, Ferrari who took up the gauntlet and defeated Tartaglia, a victory that greatly enhanced Ferrari's reputation. He was appointed to the lucrative position of tax assessor for the governor of Milan and subsequently to a professorship in mathematics at the University of Bologna. Unfortunately, he would not long enjoy his success; he died at age 43, reportedly of arsenic poisoning administered by his sister. To this day, the formula that gives the roots of a cubic polynomial is usually attributed to Cardan even though it was originally discovered by del Ferro.

René Descartes (1596–1650). René Descartes, born near Tours, France, in 1596, entered the Jesuit college of La Flèche in Anjou at the age of eight. As his health was frail, he received permission to rest in bed as late as he pleased in the mornings. The time was used in meditation and study, a practice he continued until near the end of his life. After taking degrees in law at the University of Poitiers in Paris, Descartes traveled around Europe, served as a gentleman-soldier, and in 1628 finally settled in Holland, where he remained for the next 20 years, pursuing his studies in philosophy and mathematics. In 1637 Descartes published his *Discourse on the Method of Rightly Conducting the Reason*, a major philosophical work. As a demonstration of the method, he included three treatises on optics, meteorology, and geometry. The third, *La Géométrie*, which applies algebra to geometry, in essence outlined analytical geometry. In 1649 Queen Christina of Sweden persuaded Descartes to come to Stockholm to serve as her private tutor in philosophy and to help her establish a Royal Swedish Academy of Sciences. The rigors of the Swedish winter, coupled with Christina's insistence that their tutoring sessions be held at 5:00 A.M. in the icy library of her palace, soon worked its fatal effect; Descartes succumbed to an inflammation of the lungs, dying at the age of 54.

Problem Set 9.1

1. Prove Lemma 9.1.4.

2. Show that $\sqrt[3]{10 + 6\sqrt{3}} - \sqrt[3]{-10 + 6\sqrt{3}} = 2$.

3. Consider the cubic equation $x^3 + mx = n$. We know from Cardan's solution that if u and v are chosen to be such that $3uv = m$ and $u^3 - v^3 = n$, then $x = u - v$ is a solution of $x^3 + mx = n$. If u and v are chosen in this way and ω is the complex cube root of unity $-\frac{1}{2} + \frac{\sqrt{3}}{2}i$, prove that $x = \omega u - \omega^2 v$ and $x = \omega^2 u - \omega v$ are also solutions of this equation.

4. Use methods of this section to compute the roots of each of the following polynomials.

   (a) $f(x) = x^3 - 3x + 1$

   (b) $f(x) = x^3 + x - 2$

   (c) $f(x) = x^4 - 5x^2 + 6$

5. If r_1, r_2, r_3 are the roots of a monic cubic polynomial $f(x)$, then the expansion of $(x - r_1)(x - r_2)(x - r_3)$ produces $f(x)$. A similar observation holds for the roots of a quartic polynomial. Show that the polynomials in parts (b) and (c) of Exercise 4 can be constructed from the roots that were computed by methods of this section.

6. Show that the roots of a general monic cubic polynomial can be expressed in terms of its coefficients.

7. How can the situation in Exercise 6 be handled if the cubic polynomial is not monic?

9.2   SOLVABLE GROUPS

Section Overview. The goal of this section is to lay the groundwork for answering the question of whether or not the roots of every polynomial of degree 5 or higher can be found by means of a formula. We have already seen in the previous section that this is always possible for polynomials of degree 2, 3, and 4. One concept that plays an important role in answering this classical question about polynomials is that of a

solvable group. A solvable group derives its name from its close connection to finding the roots of a polynomial.

The roots of the quadratic polynomial $f(x) = ax^2 + bx + c \in \mathbf{R}[x]$, $a \neq 0$,
are given by the quadratic formula

$$x = \frac{-b \pm \sqrt{b^2 - 4ac}}{2a}.$$

This formula can be expressed in terms of the coefficients of $f(x)$ using only the
field operations of addition, subtraction, multiplication, and division and the
operation of extraction of roots. When a formula for finding the roots of a class
of polynomials over a field F can be expressed in terms of the coefficients of the
polynomials using only the field operations and extraction of roots, the polynomials in the class are said to be **solvable by radicals.** Whether or not it is possible to solve a class of polynomials by radicals is an age-old problem in
mathematics. By the end of the 16th century, the question had been answered
for polynomials of degree 2, 3, and 4, but not much progress was made on the
problem for polynomials of degree 5 and higher for the next 300 years. Then,
early in the 19th century, **Paolo Ruffini (1765–1822)** and **Niels Abel (1802–
1829)**, working independently, were able to prove that not all quintic polynomials can be solved by radicals. (See the Historical Notes at the end of Section
2.1 for information on Abel and at the end of this section for Ruffini.) Although
this was a major step in the right direction, it left open the question regarding
which polynomials of degree 5 and higher can be solved by radicals. To answer
this question, we first need to investigate the concept of a solvable group.

9.2.1   **Definition**

A group G is said to be **solvable** if there is a chain of subgroups $\{e\} = G_n \subseteq
G_{n-1} \subseteq \cdots \subseteq G_1 \subseteq G_0 = G$ of G such that the following two conditions
hold:

1. G_k is normal in G_{k-1} for $k = 1, 2, \ldots, n$.
2. G_{k-1}/G_k is an abelian group for $k = 1, 2, \ldots, n$.

A chain of subgroups $\{e\} = G_n \subseteq G_{n-1} \subseteq \cdots \subseteq G_1 \subseteq G_0 = G$ that satisfies these conditions is referred to as a **solvable series** for G.

We are particularly interested in the solvability of $\mathbf{S}_n$ for $n \geq 1$, where $\mathbf{S}_n$
denotes the symmetric group on n letters investigated in Section 5.1. These
groups arise in connection with solving polynomials by radicals. Note that any

abelian group G is solvable since $\{e\} = G_1 \subseteq G_0 = G$ is a solvable series for G. Since $\mathbf{S}_1$ and $\mathbf{S}_2$ contain 1 and 2 elements, respectively, $\mathbf{S}_1$ and $\mathbf{S}_2$ are easily shown to be abelian, so both of these groups are solvable. The following two examples show that $\mathbf{S}_3$ and $\mathbf{S}_4$ are solvable groups.

9.2.2 | **Examples**

1. $\mathbf{S}_3$ is solvable. To see this, consider $\mathbf{S}_3$ and the chain of subgroups

$$\{e\} = G_2 \subseteq G_1 = \{e, (1, 2, 3), (1, 3, 2)\} \subseteq G_0 = \mathbf{S}_3,$$

   where $e = (1)$. The number of elements in $\mathbf{S}_3$ is $3! = 6$. Since there are three elements in G_1, it follows that $[\mathbf{S}_3 : G_1] = 2$, so from Exercise 17 of Problem Set 2.2, G_1 is a normal subgroup of $\mathbf{S}_3$. The fact that $[\mathbf{S}_3 : G_1] = 2$ indicates that there are two elements in $\mathbf{S}_3/G_1$, so Exercise 20 of Problem Set 2.1 shows that G_1 and $\mathbf{S}_3/G_1$ are abelian. Thus, $\mathbf{S}_3$ is a solvable group.

2. $\mathbf{S}_4$ is solvable. Consider the chain of subgroups $\{e\} \subseteq \{e, (1, 2)(3, 4)\} \subseteq \{e, (1, 2)(3, 4), (1, 3)(2, 4), (1, 4)(2, 3)\} \subseteq \mathbf{A}_4 \subseteq \mathbf{S}_4$. $\mathbf{A}_4$ is the alternating group of all even permutations in $\mathbf{S}_4$. Exercise 7 of Problem Set 5.1 shows that $\mathbf{A}_4$ is a normal subgroup of $\mathbf{S}_4$. Since the number of elements in $\mathbf{A}_4$ is $\frac{4!}{2} = 12$ (Problem Set 5.1, Exercise 8), $[\mathbf{S}_4 : \mathbf{A}_4] = 2$, so $\mathbf{S}_4/\mathbf{A}_4$ is abelian (again by Exercise 20 of Problem Set 2.1). One can show that

$$\{e, (1, 2)(3, 4)\} \quad \text{and}$$
$$\{e, (1, 2)(3, 4), (1, 3)(2, 4), (1, 4)(2, 3)\}$$

   are normal subgroups of

$$\{e, (1, 2)(3, 4), (1, 3)(2, 4), (1, 4)(2, 3)\} \quad \text{and} \quad \mathbf{A}_4,$$

   respectively, and that

$$\{e, (1, 2)(3, 4), (1, 3)(2, 4), (1, 4)(2, 3)\}/\{e, (1, 2)(3, 4)\} \quad \text{and}$$
$$\mathbf{A}_4/\{e, (1, 2)(3, 4), (1, 3)(2, 4), (1, 4)(2, 3)\}$$

   are abelian. (The details are left as exercises.) Thus, $\mathbf{S}_4$ is solvable.

We now prove several properties of solvable groups that will be needed in our investigation of solving polynomials by radicals.

Theorem

If G is a solvable group, then every homomorphic image of G is solvable.

Proof.   Suppose that $f: G \rightarrow G'$ is a group epimorphism so that G' is a homomorphic image of G. If G is solvable, we claim that G' is also solvable. Let $\{e\} = G_n \subseteq G_{n-1} \subseteq \cdots \subseteq G_1 \subseteq G_0 = G$ be a solvable series for G. Then $\{e'\} \subseteq f(G_n) \subseteq f(G_{n-1}) \subseteq \cdots \subseteq f(G_1) \subseteq f(G_0) = G'$ is a chain of subgroups of G'. If we can show that this chain is a solvable series for G', then G' is a solvable group. First, note that if $x' \in f(G_{i-1})$, then there is an $x \in G_{i-1}$ such that $f(x) = x'$. Moreover, if $g' \in f(G_i)$, then there is a g in G_i such that $f(g) = g'$. Hence, $x'g'x'^{-1} = f(x)f(g)f(x)^{-1} = f(x)f(g)f(x^{-1}) = f(xgx^{-1})$. But G_i is normal in G_{i-1}, so, by Theorem 2.2.15, $xgx^{-1} \in G_i$ and thus $f(xgx^{-1}) \in f(G_i)$. Therefore, $x'g'x'^{-1} \in f(G_i)$, so $f(G_i)$ is normal in $f(G_{i-1})$. Next, we need to show that $f(G_{i-1})/f(G_i)$ is abelian for $i = 1$, $2, \ldots, n$. Define $\phi: G_{i-1} \rightarrow f(G_{i-1})/f(G_i)$ by $\phi(x) = f(x)f(G_i)$. Now $G_i \subseteq G_{i-1}$ and so if $x \in G_i$, then $\phi(x) = f(x)f(G_i) = f(xG_i) = f(G_i)$, and this is the identity of $f(G_{i-1})/f(G_i)$. Thus, $x \in \ker \phi$, so we have shown that $G_i \subseteq \ker \phi$. Hence, we have group homomorphisms

$$G_{i-1}/G_i \longrightarrow G_{i-1}/\ker \phi \longrightarrow f(G_{i-1})/f(G_i)$$

with the first map being an epimorphism and the second an isomorphism. Therefore, $f(G_{i-1})/f(G_i)$ is a homomorphic image of G_{i-1}/G_i, which is abelian. Since a homomorphic image of an abelian group must be abelian, we see that $f(G_{i-1})/f(G_i)$ is abelian. Consequently, G' is a solvable group. ∎

Theorem

If G is a solvable group, then every subgroup of G is solvable.

Proof.   Suppose that H is a subgroup of G. Let's show that H is solvable. Let $\{e\} = G_n \subseteq G_{n-1} \subseteq \cdots \subseteq G_1 \subseteq G_0 = G$ be a solvable series for G. To simplify notation, let $K_i = G_i \cap H$ for $i = 0, 1, \ldots, n$, and consider the chain $\{e\} = K_n \subseteq K_{n-1} \subseteq \cdots \subseteq K_1 \subseteq K_0 = H$. First, we claim that K_i is a normal subgroup of K_{i-1} for $i = 1$, $2, \ldots, n$. If x is in K_{i-1} and g is in K_i, then $xgx^{-1} \in G_i$ since G_i is a normal subgroup of G_{i-1}. Furthermore, $xgx^{-1} \in H$ since H is a subgroup of G. Thus, xgx^{-1} is in $G_i \cap H = K_i$, so, by Theorem 2.2.15, K_i is a normal subgroup of K_{i-1}. Finally, we need to show that K_{i-1}/K_i is abelian. First note that $K_i = G_i \cap H = G_i \cap G_{i-1} \cap H = K_{i-1} \cap G_i$, so it follows that $K_{i-1}/K_i = K_{i-1}/(K_{i-1} \cap G_i)$. But by the Third Isomorphism Theorem for Groups 2.3.15, $K_{i-1}/(K_{i-1} \cap G_i)$ is isomorphic to $K_{i-1}G_i/G_i$ and so K_{i-1}/K_i and $K_{i-1}G_i/G_i$ are isomorphic groups. Now $K_{i-1}G_i/G_i$ is an abelian group since it is a subgroup of the abelian group G_{i-1}/G_i. Thus, K_{i-1}/K_i must be abelian, and so $\{e\} = K_n \subseteq K_{n-1} \subseteq \cdots \subseteq K_1 \subseteq K_0 = H$ is a solvable series for H. Therefore, H is a solvable group. ∎

Our goal now is to show that $\mathbf{S}_n$ is not solvable when $n \geq 5$. To do this, we need the following definition. Recall that if A is a subset of a group G, then the intersection of the family of subgroups of G that contain A is a subgroup of G denoted by $\langle A \rangle$. The group $\langle A \rangle$ is said to be the **subgroup of G generated by A**. (See Exercise 19 of Problem Set 2.2.)

9.2.5 Definition

Let G be a group, and suppose that $x, y \in G$. The **commutator** of x and y is the element $xyx^{-1}y^{-1}$. If $A = \{xyx^{-1}y^{-1} \mid x, y \in G\}$, then the group $G^d = \langle A \rangle$ is called the **derived** or **commutator subgroup** of G. Now set $G^{[1]} = G^d$ and inductively define $G^{[k+1]} = (G^{[k]})^d$. The group $G^{[n]}$ is said to be the **nth derived** or **nth commutator subgroup** of G.

9.2.6 Theorem

A group G is abelian if and only if $G^d = \{e\}$.

Proof. Let $A = \{xyx^{-1}y^{-1} \mid x, y \in G\}$. If $G^d = \{e\}$, then $A \subseteq G^d = \{e\}$ indicates that $xyx^{-1}y^{-1} = e$ for any $x, y \in G$. Thus, $xy = yx$, so G is abelian. Conversely, if G is abelian, then $xy = yx$ implies that $xyx^{-1}y^{-1} = e$ for all $x, y \in G$. Hence, $A \subseteq \{e\}$. But G^d is the intersection of all the subgroups of G that contain A and $\{e\}$ is one of these groups, so it must be the case that $G^d \subseteq \{e\}$. Thus, $G^d = \{e\}$. ∎

Theorem 9.2.6 demonstrates that the derived subgroup of G is, in some sense, a measure of how "far" a nonabelian group G is from being abelian. Since $\{e\} \subseteq G^d \subseteq G$, it shows that the "closer" G^d is to $\{e\}$, the closer G is to being an abelian group. On the other hand, the closer G^d is to G, the further G is from being an abelian group.

9.2.7 Theorem

If H is a subgroup of G, then $G^d \subseteq H$ if and only if H is a normal subgroup of G and G/H is abelian.

Proof. Suppose that $G^d \subseteq H$. If $h \in H$ and $x \in G$, we need to show that $xhx^{-1} \in H$. Now $xhx^{-1}h^{-1} \in G^d \subseteq H$, so $xhx^{-1} = (xhx^{-1}h^{-1})h \in H$. Thus, H is a normal subgroup of G. Next, let's show that G/H is abelian. If xH and $yH \in G/H$, then $(xH)(yH)(xH)^{-1}(yH)^{-1} = (xH)(yH)(x^{-1}H)(y^{-1}H) = xyx^{-1}y^{-1}H$. Since $xyx^{-1}y^{-1} \in G^d \subseteq H$, it follows that $xyx^{-1}y^{-1}H = H$. Hence, $(xH)(yH)(xH)^{-1}(yH)^{-1} = H$, and so $(xH)(yH) = H(yH)(xH) = (Hy)H(xH) = (yH^2)(xH) = (yH)(xH)$. Therefore, G/H is abelian.

Conversely, suppose that H is a normal subgroup of G and that G/H is abelian. Then for any $x, y \in G$, $(xH)(yH) = (yH)(xH)$, so $xyx^{-1}y^{-1} \in H$. Thus, the set $A = \{xyx^{-1}y^{-1} \mid x, y \in G\} \subseteq H$, and this implies that $G^d \subseteq H$. ∎

The next theorem relates the derived subgroups of G to the solvability of G and provides us with the tool necessary to show that $\mathbf{S}_n$ is not solvable for any integer $n \geq 5$. In the proof of this theorem, we use the fact that if G_1 and G_2 are subgroups of a group G such that $G_1 \subseteq G_2$, then $G_1^d \subseteq G_2^d$. You will be asked to prove this in the exercises.

9.2.8   Theorem

A group G is solvable if and only if $G^{[n]} = \{e\}$ for some positive integer n.

Proof. If $G^{[n]} = \{e\}$, where n is a positive integer, it follows directly from Theorem 9.2.7 that $\{e\} = G^{[n]} \subseteq \cdots \subseteq G^{[2]} \subseteq G^{[1]} \subseteq G$ is a solvable series for G. Hence, G is solvable. Conversely, suppose that G is solvable, and let $\{e\} = G_n \subseteq G_{n-1} \subseteq \cdots \subseteq G_1 \subseteq G_0 = G$ be a solvable series for G. Since G_i is a normal subgroup of G_{i-1} and G_{i-1}/G_i is abelian, it follows from Theorem 9.2.7 that $G_{i-1}^d \subseteq G_i$ for $i = 1, 2, \ldots, n$. Hence,

$$G^{[1]} \subseteq G_0^d \subseteq G_1,$$

$$G^{[2]} \subseteq G_1^d \subseteq G_2,$$

$$G^{[3]} \subseteq G_2^d \subseteq G_3,$$

$$\vdots$$

$$G^{[n]} \subseteq G_{n-1}^d \subseteq G_n = \{e\}.$$

Consequently, $G^{[n]} = \{e\}$, and this completes the proof. ∎

We now have at hand all the mathematical machinery necessary to prove that $\mathbf{S}_n$ is not solvable for $n \geq 5$.

9.2.9   **Theorem**

For any integer $n \geq 5$, the symmetric group $\mathbf{S}_n$ is not solvable.

Proof.   The proof is by induction. Let $\zeta = (a_1, a_2, a_3)$ be any 3-cycle in $\mathbf{S}_n$. Since $n \geq 5$, there are integers a_4 and a_5 in $\{1, 2, ..., n\}$ such that a_1, a_2, a_3, a_4, and a_5 are distinct. If $\alpha = (a_1, a_2, a_4)$ and $\beta = (a_1, a_3, a_5)$, then

$$\zeta = (a_1, a_2, a_3)$$
$$= (a_1, a_2, a_4)(a_1, a_3, a_5)(a_1, a_4, a_2)(a_1, a_5, a_3)$$
$$= \alpha\beta\alpha^{-1}\beta^{-1} \in \mathbf{S}_n^{[1]}.$$

Hence, $\mathbf{S}_n^{[1]}$ contains every 3-cycle of $\mathbf{S}_n$. Now suppose that $\mathbf{S}_n^{[k]}$ contains every 3-cycle of $\mathbf{S}_n$. Then, by an argument similar to the preceding one, $\mathbf{S}_n^{[k+1]} = (\mathbf{S}_n^{[k]})^d$ contains every 3-cycle of $\mathbf{S}_n$. It follows by induction that $\mathbf{S}_n^{[k]}$ contains every 3-cycle of $\mathbf{S}_n$ for each integer $k = 1, 2, ...$. But this implies that it is not possible for $\mathbf{S}_n^{[k]} = \{e\}$ for any positive integer k. Thus, by Theorem 9.2.8, $\mathbf{S}_n$ is not solvable for any integer $n \geq 5$. ∎

We have shown that $\mathbf{S}_n$ is a solvable group if $1 \leq n \leq 4$ and that $\mathbf{S}_n$ is not solvable for $n \geq 5$. At this point it is not apparent how solvable groups are connected to finding the roots of polynomials. The ties between solvable groups and polynomials are developed later, so for now, be satisfied in the knowledge that there is a connection.

HISTORICAL NOTES

Paolo Ruffini (1765–1822). Paolo Ruffini was born in Valentano, Italy, in 1765, and he received his education in mathematics and medicine at the University of Modena. Ruffini taught analysis at Modena while still a student and in 1791 was appointed to a position in mathematics there. He eventually lost the professorship because of his refusal, on religious grounds, to swear allegiance to the Cisalpine Republic (which included Modena) that Napoleon Bonaparte had established in the aftermath of the French Revolution. Ruffini was able to support himself through the practice of medicine, while continuing his mathematical research on fifth-degree equations. In 1799 his results were published in a treatise in which he declared that quintic polynomials could not be solved by radicals. The proof of this was a remarkable piece of work in mathematics with the exception of one flaw. Ruffini's proof was difficult to understand, and other mathematicians failed to acknowledge his result. Although he spent a great deal of time corresponding with mathematicians, trying to have his result accepted, no one wrote him to point out the error in his proof; thus, he did not have a chance to correct it. Ruffini also unsuccessfully sought verification of his results from the Institute in Paris and from the Royal Society. Some years later, in the mid-1820s, Niels Abel published a complete proof that quintic polynomials cannot be solved by radicals. While Abel is usually credited with establishing this result, Ruffini had laid the groundwork.

Problem Set 9.2

1. Complete Example 2 of 9.2.2 by showing each of the following:

    (a) $\{e,(1,2)(3,4)\}$ and $\{e,(1,2)(3,4), (1,3)(2,4), (1,4)(2,3)\}$ are normal subgroups of $\{e,(1,2)(3,4), (1,3)(2,4), (1,4)(2,3)\}$ and $\mathbf{A}_4$, respectively.
    (b) Show that

    $$\{e, (1, 2)(3, 4), (1, 3)(2, 4), (1, 4)(2, 3)\}/\{e, (1, 2)(3, 4)\}$$

    is an abelian group.

2. If G_1 and G_2 are subgroups of a group G such that $G_1 \subseteq G_2$, prove that $G_1^d \subseteq G_2^d$.

3. If G is a solvable group and $H \neq \{e\}$ is a subgroup of G, prove that $H^d \neq H$.

4. Prove that if H is a normal subgroup of a group G and if H and G/H are solvable, then G is a solvable group. Hint: Let $\{e\} = H_n \subseteq H_{n-1} \subseteq \cdots \subseteq H_1 \subseteq H_0 = H$ be a solvable series for H, and let $\{H\} = \{eH\} = \overline{K}_m \subseteq \overline{K}_{m-1} \subseteq \cdots \subseteq \overline{K}_1 \subseteq \overline{K}_0 = G/H$ be a solvable series for G/H. Use the Correspondence Theorem for Groups 2.3.16 to show that there are subgroups K_i of G, for $i = 0, 1, \ldots, m$, each of which contains H, such that $\overline{K}_i = K_i/H$ for each i. A careful reading of the Correspondence Theorem 2.3.16 will yield the result.

9.3   SOLVABILITY BY RADICALS

Section Overview. In the first section of this chapter, it was shown that polynomials of degree 2, 3, and 4 can be solved by radicals. One purpose of this section is to show that this is, in general, not possible for polynomials of degree 5 or higher. There are specific polynomials, however, of degree 5 and higher that can be solved by radicals. Another purpose of this section is to establish criteria that can be used to determine exactly which of these polynomials can be solved by radicals.

As we saw in Section 9.2, a result by Ruffini and Abel ruled out the possibility of finding a formula that would yield the roots of every quintic polynomial. However, there are specific polynomials of degree 5 that can be solved by radicals. For instance, the polynomial $f(x) = x^5 - 3x^4 - 2x^3 + 8x^2 - 4$ has 1, $\pm\sqrt{2}$, and $1\pm\sqrt{3}$ as it roots. The result by Ruffini and Abel was incomplete in the sense that it did not establish criteria by which one could determine the quintic polynomials that can be solved by radicals. This question was settled by **Evariste Galois** (1811–1832) who, shortly before his death at the age of 20, outlined the basis of a remarkable theory that shows exactly which polynomials can be solved by radicals. This theory has been so fruitful that it is now called Galois theory in his honor. The basic idea of Galois was to investigate a group of permutations of the roots of a polynomial. In the 1920s **Emil Artin** (1898–1962) realized that the theory of Galois could be developed more elegantly using groups of automorphisms of fields, and this led to what we now consider to be the modern approach to Galois theory. Modern Galois theory is a theory that results from the interplay among polynomials, groups, and fields. It is Artin's approach that is presented here, but this should not be construed as diminishing the genius of Galois. The two approaches are equivalent. The central idea is that for a field extension E/F, a certain type of automorphism of E induces a permutation of the roots of a polynomial in $F[x]$.

9.3.1   Definition

If F is a field, then a ring isomorphism $\alpha : F \rightarrow F$ is said to be an **automorphism** of F. The set of all automorphisms of F is denoted by Aut(F). If E/F is a field extension, then an automorphism $\alpha : E \rightarrow E$ such that $\alpha(a) = a$ for all $a \in F$ is said to **leave each element of F fixed**. We also say that α **leaves F fixed elementwise** when $\alpha(a) = a$ for each $a \in F$. Gal(E/F) denotes the set of all automorphisms of E that leave each element of F fixed. Gal(E/F) $\neq \varnothing$ since the identity automorphism $\epsilon : E \rightarrow E$ is in Gal(E/F). Throughout this section, ϵ always denotes an identity automorphism.

9.3.2   Examples

1. The mapping $\alpha : \mathbf{C} \rightarrow \mathbf{C}$ that is defined by $\alpha(a + bi) = a - bi$ is easily shown to be a ring isomorphism, so $\alpha \in$ Aut($\mathbf{C}$). Since $\alpha(a) = a$ for

each $a \in \mathbf{R}$, α leaves $\mathbf{R}$ fixed elementwise, so $\alpha \in \mathrm{Gal}(\mathbf{C}/\mathbf{R})$. In fact, it can be shown that $\mathrm{Gal}(\mathbf{C}/\mathbf{R}) = \{\epsilon, \alpha\}$. (See Exercise 11, Problem Set 3.3.)

2. The mapping $\beta: \mathbf{Q}(\sqrt{2}) \to \mathbf{Q}(\sqrt{2})$ given by $\beta(a + b\sqrt{2}) = a - b\sqrt{2}$ is a ring isomorphism, so $\beta \in \mathrm{Aut}(\mathbf{Q}(\sqrt{2}))$. Since $\beta(a) = a$ for each $a \in \mathbf{Q}$, α leaves $\mathbf{Q}$ fixed elementwise. Hence, $\beta \in \mathrm{Gal}(\mathbf{Q}(\sqrt{2})/\mathbf{Q}) = \{\epsilon, \beta\}$.

You will be asked to prove the following theorem in the exercises.

9.3.3   Theorem

If F is a field, then $\mathrm{Aut}(F)$ is a group under the operation of composition of functions. Moreover, if E/F is a field extension, then $\mathrm{Gal}(E/F)$ is a subgroup of $\mathrm{Aut}(E)$.

When E/F is a field extension, the group $\mathrm{Gal}(E/F)$ is the **Galois group** of the field extension. If $\mathcal{S}$ is the splitting field of $f(x) \in F[x]$, then $\mathrm{Gal}(\mathcal{S}/F)$ is said to be the **Galois group of the polynomial** $f(x)$. The following theorem shows the connection between elements of $\mathrm{Gal}(E/F)$ and the roots of polynomials in $F[x]$.

9.3.4   Theorem

Let E/F be a field extension. If $\alpha \in \mathrm{Gal}(E/F)$ and $c \in E$ is a root of a polynomial $f(x) \in F[x]$, then $\alpha(c)$ is also a root of $f(x)$.

Proof. Suppose that $f(x) = a_0 + a_1x + \cdots + a_nx^n$ is a polynomial in $F[x]$ that has a root c in a field extension E of F. Then $f(c) = 0$, so we have

$$a_0 + a_1c + \cdots + a_nc^n = 0.$$

If $\alpha \in \mathrm{Gal}(E/F)$ and if α is applied to the preceding equation, then since α preserves addition and multiplication, we see that

$$\alpha(a_0) + \alpha(a_1c) + \cdots + \alpha(a_nc^n) = \alpha(0) = 0 \qquad \text{and}$$

$$\alpha(a_0) + \alpha(a_1)\alpha(c) + \cdots + \alpha(a_n)\alpha(c)^n = 0.$$

But $\alpha(a_k) = a_k$ for $k = 0, 1, \ldots, n$, so

$$a_0 + a_1\alpha(c) + \cdots + a_n\alpha(c)^n = 0$$

Hence, $\alpha(c)$ is a root of $f(x)$. ∎

If E/F is a field extension and $c \in E$ is algebraic over F, then c is the root of a monic irreducible polynomial $m(x) \in F[x]$, the minimal polynomial in $F[x]$ of c. If $\alpha \in \mathrm{Gal}(E/F)$, then the preceding theorem shows that $\alpha(c)$ is also a root of $m(x)$. Hence, α maps roots of $m(x)$ to roots of $m(x)$. It is legitimate to ask if the converse holds. That is, if $c, c' \in E$ are roots of a monic irreducible polynomial $m(x) \in F[x]$, is there an $\alpha \in \mathrm{Gal}(E/F)$ such that $\alpha(c) = c'$? If E is the splitting field $\mathcal{S}$ of some polynomial in $F[x]$, the answer is yes.

9.3.5   Theorem

If $\mathcal{S}$ is the splitting field of some polynomial $f(x) \in F[x]$ and $c, c' \in \mathcal{S}$, then there is an $\alpha \in \mathrm{Gal}(\mathcal{S}/F)$ such that $\alpha(c) = c'$ if and only if c and c' have the same minimal polynomial in $F[x]$.

Proof. Suppose that c and c' have the same minimal polynomial $m(x)$ in $F[x]$. Let's show that there is an $\alpha \in \mathrm{Gal}(\mathcal{S}/F)$ such that $\alpha(c) = c'$. By Theorem 8.2.14, $F(c)$ and $F(c')$ are isomorphic by an isomorphism ψ, with $\psi(c) = c'$, that extends the identity isomorphism $\epsilon : F \to F$. Note that ψ leaves each element of F fixed since if $a \in F$, then $\psi(a) = \epsilon(a) = a$. Since $F \subseteq F(c)$, it follows that $F[x] \subseteq F(c)[x]$, so $f(x) \in F(c)[x]$. If we adjoin the roots of $f(x)$ to F or to $F(c)$, we arrive at the same splitting field $\mathcal{S}$ of $f(x)$. A similar observation holds for $F(c')$, so Theorem 8.4.8 and its Corollary 8.4.9 show that ψ can be extended to an isomorphism $\alpha : \mathcal{S} \to \mathcal{S}$. Moreover, α leaves each element of F fixed since ψ leaves each element of F fixed. Furthermore, $\alpha(c) = \psi(c) = c'$. This shows that there is an α in $\mathrm{Gal}(\mathcal{S}/F)$ such that $\alpha(c) = c'$.

Conversely, suppose that there is an $\alpha \in \mathrm{Gal}(\mathcal{S}/F)$ such that $\alpha(c) = c'$. Let's show that c and c' have the same minimal polynomial in $F[x]$. If $m(x)$ is the minimal polynomial in $F[x]$ of c, then c is a root of $m(x)$. Theorem 9.3.4 shows that $\alpha(c) = c'$ is a root of $m(x)$. Since there is exactly one monic irreducible polynomial in $F[x]$ that has c' as a root, $m(x)$ must also be the minimal polynomial in $F[x]$ of c'. $\blacksquare$

The next result is helpful in determining the distinct automorphisms of a finite dimensional field extension.

9.3.6   Theorem

Let $E = F(c_1, c_2, \ldots, c_n)$ be a finite dimensional extension of F. Then $\alpha \in \mathrm{Gal}(E/F)$ is completely determined by the values it takes on $c_1, c_2, \ldots, c_n$. That is, if $\alpha, \beta \in \mathrm{Gal}(E/F)$ and $\alpha(c_k) = \beta(c_k)$ for $k = 1, 2, \ldots, n$, then $\alpha = \beta$.

Proof. Let $\alpha, \beta \in \mathrm{Gal}(E/F)$ be such that $\alpha(c_k) = \beta(c_k)$ for $k = 1, 2, \ldots, n$. Then $\delta = \beta^{-1} \circ \alpha$ is such that $\delta(c_k) = \beta^{-1} \circ \alpha(c_k) = \beta^{-1}(\alpha(c_k)) = \beta^{-1}(\beta(c_k)) = c_k$ for each k. We claim that δ is the identity automorphism $\epsilon : E \to E$. By Theorem 8.2.8, $\{e, c_1, c_1^2, \ldots, c_1^{n-1}\}$ is a basis for $F(c_1)$, so for any a in $F(c_1)$, there are $a_0, a_1, \ldots, a_{n-1} \in F$ such that $a = a_0 + a_1 c_1 + \cdots + a_{n-1} c_1^{n-1}$. It follows that

$$\begin{aligned}
\delta(a) &= \delta(a_0 + a_1 c_1 + \cdots + a_{n-1} c_1^{n-1}) \\
&= \delta(a_0) + \delta(a_1)\delta(c_1) + \cdots + \delta(a_{n-1})\delta(c_1)^{n-1} \\
&= a_0 + a_1 c_1 + \cdots + a_{n-1} c_1^{n-1} \\
&= a.
\end{aligned}$$

Consequently, δ is the identity map on $F(c_1)$. Repetition of the same argument with F replaced by $F(c_1)$ and c_1 by c_2 shows that δ is the identity automorphism on $F(c_1)(c_2) = F(c_1, c_2)$. A similar repetition shows that δ is the identity automorphism on $F(c_1, c_2, c_3)$, and after $n-1$ repetitions we arrive at the conclusion that δ is the identity automorphism on E. Hence, we have $\beta^{-1} \circ \alpha = \epsilon$, and this clearly implies that $\alpha = \beta$. ∎

9.3.7   Examples

1. Let's compute the Galois group of $f(x) = x^2 - 2 \in \mathbf{Q}[x]$. Now $\mathcal{S} = \mathbf{Q}(\sqrt{2})$ is the splitting field of the polynomial $f(x)$, so, by Theorem 9.3.6, any $\alpha \in \mathrm{Gal}(\mathcal{S}/\mathbf{Q})$ is determined by its action on $\sqrt{2}$. Since $\sqrt{2}$ is a root of the minimal polynomial $f(x) = x^2 - 2 \in \mathbf{Q}[x]$, Theorem 9.3.5 indicates that $\alpha(\sqrt{2})$ must also be a root of $m(x)$. Hence, $\alpha(\sqrt{2}) = \sqrt{2}$ or $\alpha(\sqrt{2}) = -\sqrt{2}$, so there are two automorphisms in $\mathrm{Gal}(\mathcal{S}/\mathbf{Q})$. One automorphism is given by $\sqrt{2} \longmapsto \sqrt{2}$ and the other by $\sqrt{2} \longmapsto -\sqrt{2}$. Since $\mathbf{Q}(\sqrt{2}) = \{a + b\sqrt{2} \mid a, b \in \mathbf{Q}\}$, it follows that the two automorphisms ϵ and α in $\mathrm{Gal}(\mathcal{S}/\mathbf{Q})$ are defined by

    $$\epsilon(a + b\sqrt{2}) = a + b\sqrt{2} \qquad \text{and} \qquad \alpha(a + b\sqrt{2}) = a - b\sqrt{2}.$$

    Finally, observe that if ϵ and α are restricted to the set $R = \{\sqrt{2}, -\sqrt{2}\}$ of roots of $f(x)$, then these restricted maps, also denoted by ϵ and α, are permutations of R. Using the two-row notation for permutations, ϵ and α are the permutations $\begin{pmatrix} \sqrt{2} & -\sqrt{2} \\ \sqrt{2} & -\sqrt{2} \end{pmatrix}$ and $\begin{pmatrix} \sqrt{2} & -\sqrt{2} \\ -\sqrt{2} & \sqrt{2} \end{pmatrix}$, respectively. If $\sqrt{2}$ is replaced by 1 and $-\sqrt{2}$ is replaced by 2 in these permutations, then this establishes a mapping $\phi : \text{Gal}(\mathcal{S}/\mathbf{Q}) \rightarrow \mathbf{S}_2$ given by $\epsilon \longmapsto \begin{pmatrix} 1 & 2 \\ 1 & 2 \end{pmatrix} = (1)$ and $\alpha \longmapsto \begin{pmatrix} 1 & 2 \\ 2 & 1 \end{pmatrix} = (1, 2)$. It follows that ϕ is a group isomorphism, so $\text{Gal}(\mathcal{S}/\mathbf{Q})$ and $\mathbf{S}_2$ are isomorphic groups. Since $\mathcal{S}$ has $\{1, \sqrt{2}\}$ as a basis, $[\mathcal{S} : \mathbf{Q}] = 2 = |\text{Gal}(\mathcal{S}/\mathbf{Q})|$, the order of $\text{Gal}(\mathcal{S}/\mathbf{Q})$.

2. For this example we compute the Galois group of polynomial $g(x) = (x^2 - 2)(x^2 + 3) \in \mathbf{Q}[x]$. The splitting field of $g(x)$ is $\mathcal{S} = \mathbf{Q}(\sqrt{2}, \sqrt{3}i)$, so, by Theorem 9.3.6, an automorphism in $\text{Gal}(\mathcal{S}/\mathbf{Q})$ is determined by the values it takes on $\sqrt{2}$ and $\sqrt{3}i$. If α is an automorphism in $\text{Gal}(\mathcal{S}/\mathbf{Q})$, then $\alpha(\sqrt{2}) = \sqrt{2}$ or $\alpha(\sqrt{2}) = -\sqrt{2}$ since $\sqrt{2}$ and $-\sqrt{2}$ have $m(x) = x^2 - 2 \in \mathbf{Q}[x]$ for a minimal polynomial. Likewise, $\alpha(\sqrt{3}i) = \sqrt{3}i$ or $\alpha(\sqrt{3}i) = -\sqrt{3}i$ since $\sqrt{3}i$ and $-\sqrt{3}i$ are roots of $m(x) = x^2 + 3 \in \mathbf{Q}[x]$. Theorem 9.3.5 shows that it cannot be the case that $\alpha(\sqrt{2})$ is either of $\pm\sqrt{3}i$ since $\sqrt{2}$ and neither of $\pm\sqrt{3}i$ are roots of the same minimal polynomial in $\mathbf{Q}[x]$. Similarly, we cannot have $\alpha(\sqrt{3}i) = \sqrt{2}$ or $\alpha(\sqrt{3}i) = -\sqrt{2}$. Therefore, there are four automorphisms in $\text{Gal}(\mathcal{S}/\mathbf{Q})$:

$$\epsilon : \begin{cases} \sqrt{2} \longmapsto \sqrt{2} \\ \sqrt{3}i \longmapsto \sqrt{3}i \end{cases} \qquad \alpha : \begin{cases} \sqrt{2} \longmapsto -\sqrt{2} \\ \sqrt{3} \longmapsto \sqrt{3}i \end{cases}$$

$$\beta : \begin{cases} \sqrt{2} \longmapsto -\sqrt{2} \\ \sqrt{3}i \longmapsto -\sqrt{3}i \end{cases} \qquad \gamma : \begin{cases} \sqrt{2} \longmapsto -\sqrt{2} \\ \sqrt{3}i \longmapsto -\sqrt{3}i \end{cases}$$

Since $\mathcal{S} = \mathbf{Q}(\sqrt{2}, \sqrt{3}i) = \{a + b\sqrt{2} + c\sqrt{3}i + d\sqrt{2}\sqrt{3}i \mid a, b, c, d \in \mathbf{Q}\}$, it follows that the automorphisms $\epsilon, \alpha, \beta, \gamma : \mathcal{S} \rightarrow \mathcal{S}$ are

$$\epsilon(a + b\sqrt{2} + c\sqrt{3}i + d\sqrt{2}\sqrt{3}i) = a + b\sqrt{2} + c\sqrt{3}i + d\sqrt{2}\sqrt{3}i,$$
$$\alpha(a + b\sqrt{2} + c\sqrt{3}i + d\sqrt{2}\sqrt{3}i) = a - b\sqrt{2} + c\sqrt{3}i - d\sqrt{2}\sqrt{3}i,$$
$$\beta(a + b\sqrt{2} + c\sqrt{3}i + d\sqrt{2}\sqrt{3}i) = a + b\sqrt{2} - c\sqrt{3}i - d\sqrt{2}\sqrt{3}i, \quad \text{and)}$$
$$\gamma(a + b\sqrt{2} + c\sqrt{3}i + d\sqrt{2}\sqrt{3}i) = a - b\sqrt{2} - c\sqrt{3}i + d\sqrt{2}\sqrt{3}i.$$

As in Example 1, each of the automorphisms $\epsilon, \alpha, \beta, \gamma \in \text{Gal}(\mathcal{S}/\mathbf{Q})$ can be viewed as a permutation of the set $R = \{\sqrt{2}, -\sqrt{2}, \sqrt{3}i, -\sqrt{3}i\}$ of roots of the polynomial $g(x) = (x^2 - 2)(x^2 + 3)$. Thus, $\epsilon, \alpha, \beta,$ and γ can be represented as

$$\epsilon: \begin{pmatrix} \sqrt{2} & -\sqrt{2} & \sqrt{3}i & -\sqrt{3}i \\ \sqrt{2} & -\sqrt{2} & \sqrt{3}i & -\sqrt{3}i \end{pmatrix}, \qquad \alpha: \begin{pmatrix} \sqrt{2} & -\sqrt{2} & \sqrt{3}i & -\sqrt{3}i \\ -\sqrt{2} & \sqrt{2} & \sqrt{3}i & -\sqrt{3}i \end{pmatrix},$$

$$\beta: \begin{pmatrix} \sqrt{2} & -\sqrt{2} & \sqrt{3}i & -\sqrt{3}i \\ \sqrt{2} & -\sqrt{2} & -\sqrt{3}i & \sqrt{3}i \end{pmatrix}, \quad \text{and} \quad \gamma: \begin{pmatrix} \sqrt{2} & -\sqrt{2} & \sqrt{3}i & -\sqrt{3}i \\ -\sqrt{2} & \sqrt{2} & -\sqrt{3}i & \sqrt{3}i \end{pmatrix}.$$

Replacing $\sqrt{2}$ by 1, $-\sqrt{2}$ by 2, $\sqrt{3}i$ by 3, and $-\sqrt{3}i$ by 4 in these permutations establishes a group homomorphism $\phi : \mathrm{Gal}(\mathcal{S}/\mathbf{Q}) \to S_4$ given by

$$\epsilon \mapsto \begin{pmatrix} 1\ 2\ 3\ 4 \\ 1\ 2\ 3\ 4 \end{pmatrix} = (1), \qquad \alpha \mapsto \begin{pmatrix} 1\ 2\ 3\ 4 \\ 2\ 1\ 3\ 4 \end{pmatrix} = (1,2),$$

$$\beta \mapsto \begin{pmatrix} 1\ 2\ 3\ 4 \\ 1\ 2\ 4\ 3 \end{pmatrix} = (3,4), \quad \text{and} \quad \gamma \mapsto \begin{pmatrix} 1\ 2\ 3\ 4 \\ 2\ 1\ 4\ 3 \end{pmatrix} = (1,2)(3,4).$$

It can be shown that ϕ is an injective function, so $\mathrm{Gal}(\mathcal{S}/\mathbf{Q})$ is isomorphic to a subgroup of S_4. Since S_4 has $4! = 24$ elements, $\mathrm{Gal}(\mathcal{S}/\mathbf{Q})$ is isomorphic to a proper subgroup of S_4. Notice that $\mathcal{S}$ has $\{1, \sqrt{2}, \sqrt{3}i, \sqrt{2}\sqrt{3}i\}$ as a basis, so $[\mathcal{S}:\mathbf{Q}] = 4 = |\mathrm{Gal}(\mathcal{S}/\mathbf{Q})|$.

3. Let $f(x) = (x^2 - 2)(x^2 + 3)(x^2 + 3) \in \mathbf{Q}[x]$. The splitting field $\mathcal{S}$ for $f(x)$ is the same as for the polynomial $g(x) = (x^2 - 2)(x^2 + 3)$ of Example 2 since the repeated factor $x^2 + 3$ in $f(x)$ adds no "new" roots to the list of roots of $g(x)$. Hence, the Galois group of $f(x)$ and $g(x)$ are equal. Even though $\deg(f(x)) = 6$, $[\mathcal{S}:\mathbf{Q}] = 4 = |\mathrm{Gal}(\mathcal{S}/\mathbf{Q})|$.

4. Consider the field extension $\mathbf{Q}(\sqrt[3]{2})/\mathbf{Q}$. Let's compute $\mathrm{Gal}(\mathbf{Q}(\sqrt[3]{2})/\mathbf{Q})$. Since $\sqrt[3]{2}$ is a root of the monic irreducible polynomial $m(x) = x^3 - 2 \in \mathbf{Q}[x]$, $m(x) = (x - \sqrt[3]{2})(x^2 + \sqrt[3]{2}x + \sqrt[3]{4})$. An application of the quadratic formula to the quadratic factor of $m(x)$ shows that $m(x)$ has two nonreal roots. If $\alpha \in \mathrm{Gal}(\mathbf{Q}(\sqrt[3]{2})/\mathbf{Q})$, then α is an automorphism of $\mathbf{Q}(\sqrt[3]{2})$ that leaves the elements of $\mathbf{Q}$ fixed. Since α permutes the roots of $m(x)$ that lie in $\mathbf{Q}(\sqrt[3]{2})$, the only possibility is that $\alpha(\sqrt[3]{2}) = \sqrt[3]{2}$ because the two nonreal roots of $m(x)$ are not in $\mathbf{Q}(\sqrt[3]{2})$. Hence, $\mathrm{Gal}(\mathbf{Q}(\sqrt[3]{2})/\mathbf{Q}) = \{\epsilon\}$. Since $\deg(m(x)) = 3$, $[(\mathbf{Q}(\sqrt[3]{2}):\mathbf{Q}] = 3$, so $|\mathrm{Gal}(\mathbf{Q}(\sqrt[3]{2})/\mathbf{Q})| < [(\mathbf{Q}(\sqrt[3]{2}):\mathbf{Q}]$.

Recall that when F is a field, $F[x]$ is a Unique Factorization Domain. If $f(x)$ is a nonconstant polynomial in $F[x]$, then $f(x)$ can be factored (uniquely up to associates and the order of the factors) in $F[x]$ into a product of not necessarily distinct irreducible factors:

$$f(x) = a f_1(x) f_2(x) \cdots f_n(x),$$

where a is the leading coefficient of $f(x)$. If $\mathcal{S}$ is the splitting field of $f(x)$, then $f(x)$ can also be factored in $\mathcal{S}[x]$ as

$$f(x) = a(x - c_1)^{n_1} (x - c_2)^{n_2} \cdots (x - c_k)^{n_k},$$

where $c_1, c_2, \ldots, c_k$ are the distinct roots of $f(x)$ in $\mathcal{S}$. If $n_i = 1$, then c_i is a **simple root** of $f(x)$, and if $n_i > 1$, then c_i is said to be a **repeated root** or a **multiple root** of $f(x)$ of multiplicity n_i.

9.3.8 Examples

1. The polynomial $f(x) = x^4 - x^2 - 6 \in \mathbf{Q}[x]$ factors as $f(x) = (x^2 + 2)(x^2 - 3)$ in $\mathbf{Q}[x]$ and as $f(x) = (x - \sqrt{2}i)(x + \sqrt{2}i)(x - \sqrt{3})(x + \sqrt{3})$ in $\mathcal{S}[x]$, where $\mathcal{S} = \mathbf{Q}(\sqrt{2}i, \sqrt{3})$. Each of the roots of $f(x)$ is a simple root.
2. The polynomial $f(x) = x^3 - 3x + 2 \in \mathbf{Q}[x]$ factors as $f(x) = (x - 1)^2(x + 2)$ in $\mathbf{Q}[x]$. The splitting field for $f(x)$ is $\mathcal{S} = \mathbf{Q}$. Thus, 1 is a double root and -2 is a simple root of $f(x)$.

9.3.9 Definition

For a field F, an irreducible polynomial $f(x) \in F[x]$ of degree n is said to be **separable** if $f(x)$ has n distinct roots in its splitting field. A polynomial in $F[x]$ is said to be **separable** if each of its irreducible factors is separable. A polynomial that is not separable is said to be **inseparable**.

9.3.10 Examples

1. It may be the case that an irreducible polynomial in $F[x]$ of degree n has fewer than n distinct roots in its splitting field. For example, consider the field $\mathbf{Z}_2$, and let a be an element that is transcendental over $\mathbf{Z}_2$. Adjoin a

and a^2 to $\mathbf{Z}_2$, and note that $\mathbf{Z}_2(a^2) \subseteq \mathbf{Z}_2(a)$. This follows, for if $f(x) = b_0 + b_1(a^2) + b_2(a^2)^2 + \cdots + b_n(a^2)^n$ is a polynomial in $\mathbf{Z}_2[a^2]$, then $f(x) = b_0 + b_1 a^2 + b_2 a^4 + \cdots + b_n a^{2n} \in \mathbf{Z}_2[a]$. Thus, a quotient of polynomials in a^2 is a quotient of polynomials in a. Hence, $\mathbf{Z}_2(a^2) \subseteq \mathbf{Z}_2(a)$. Since $a \notin \mathbf{Z}_2(a^2)$, $x^2 - a^2$ has no roots in $\mathbf{Z}_2(a^2)$ and so is irreducible in $\mathbf{Z}_2(a^2)[x]$. But $(x - a)(x - a) = x^2 - 2ax + a^2 = x^2 + a^2 = x^2 - a^2$ since $\mathbf{Z}_2$ has characteristic 2 and $a^2 = -a^2$ in $\mathbf{Z}_2(a)$. Hence, $x^2 - a^2$ splits over $\mathbf{Z}_2(a)$, and $a \in \mathbf{Z}_2(a)$ is a double root of $x^2 - a^2$. The field $\mathbf{Z}_2(a)$ is the splitting field for $x^2 - a^2$ as a polynomial in $\mathbf{Z}_2(a^2)[x]$. Therefore, $x^2 - a^2$ is an irreducible polynomial of degree 2 in $\mathbf{Z}_2(a^2)[x]$ that has only one distinct root in $\mathbf{Z}_2(a)$. The fact that $x^2 - a^2$ has a as double root shows that it is an inseparable polynomial.

2. The polynomial $f(x) = (x^2 + 2)(x^2 + 2)(x^2 - 3)$ is separable in $\mathbf{Q}[x]$, with splitting field $\mathbf{Q}(\sqrt{2}i, \sqrt{3})$. Each of the irreducible factors $x^2 + 2$ and $x^2 - 3$ has no repeated roots. Notice also that the degree of the irreducible factor $x^2 + 2$ is equal to the number of distinct roots $\pm\sqrt{2}i$ of $x^2 + 2$. A similar observation holds for $x^2 - 3$. The fact that $f(x)$ has two factors of $x^2 + 2$ does not disqualify it from being separable. The requirement for separability is for each individual irreducible factor to have no repeated roots, and $f(x)$ satisfies this condition.

Example 1 illustrates an important point. The number of distinct roots of an irreducible polynomial in its splitting field can be less than the degree of the irreducible polynomial. This can happen when the irreducible polynomial is inseparable, but not when it is separable.

Remember that one of our goals is to establish the result of Ruffini and Abel that indicates that quintic polynomials are not solvable by radicals. To do this, we need the ability to count the automorphisms of a Galois group $\mathrm{Gal}(\mathcal{S}/F)$, where $\mathcal{S}$ is the splitting field of a separable polynomial in $F[x]$. This ability is used later when an example is given of a polynomial of degree 5 that cannot be solved by radicals.

9.3.11   Theorem

For any separable polynomial in $F[x]$ with splitting field $\mathcal{S}$, the number of extensions of an automorphism $\alpha : F \to F$ to $\mathcal{S}$ is exactly $[\mathcal{S} : F]$.

Proof. The proof is by induction on the integer $[\mathcal{S} : F]$. Let $f(x)$ be any separable polynomial in $F[x]$, and suppose that $\mathcal{S}$ is the splitting field of $f(x)$. If $[\mathcal{S} : F] = 1$, then $\mathcal{S} = F$, and it is trivial that $\alpha : F \rightarrow F$ has only $1 = [F : F]$ extension to F. Hence, we have the result when $[\mathcal{S} : F] = 1$. Next, suppose that $[\mathcal{S} : F] > 1$, and make the induction hypothesis that for any field K and any separable polynomial $g(x) \in K[x]$ with splitting field $\mathcal{S}_K$ such that $[\mathcal{S}_K : K] < [\mathcal{S} : F]$, the number of extensions of an automorphism $\beta : K \rightarrow K$ to $\mathcal{S}_K$ is exactly $[\mathcal{S}_K, K]$. Since $[\mathcal{S} : F] > 1$, $f(x)$ must have at least one monic irreducible factor with degree at least 2. If $m(x)$ is such a factor of $f(x)$, then $m(x)$ is the minimal polynomial of each of its roots. If c and c' are roots of $m(x)$, then Theorem 8.2.14 shows that there is a ring isomorphism $\beta : F(c) \rightarrow F(c')$, with $\beta(c) = c'$, that extends $\alpha : F \rightarrow F$. Furthermore, β is unique by Exercise 12 of Problem Set 8.2. Since $f(x)$ is separable, $m(x)$ must also be separable, so if $\deg(m(x)) = n$, there are exactly n of these extensions of α, one for each distinct root c' of $m(x)$. Since $\mathcal{S}$ is the splitting field for $f(x)$ when $f(x)$ is viewed as a polynomial in either $F(c)[x]$ or $F(c')[x]$, Theorem 8.4.8 shows that β can be extended to an automorphism $\bar{\beta}$ of $\mathcal{S}$. Suppose that β can be extended to an automorphism of $\mathcal{S}$ in m ways. The following diagram illustrates this situation.

$$\begin{array}{ccc}
 & \bar{\beta} & \\
\mathcal{S} & \rightarrow & \mathcal{S} \\
\uparrow & & m \text{ ways} \\
 & \beta & \\
F(c) & \rightarrow & F(c') \\
\uparrow & & n \text{ ways} \\
 & \alpha & \\
F & \rightarrow & F
\end{array}$$

Since α can be extended to $F(c)$ in n ways and β can be extended to $\mathcal{S}$ in m ways, then by the First Principle of Counting given in Example 5.1.5, α can be extended to $\mathcal{S}$ in mn ways. Since $m(x)$ is irreducible, Theorem 8.2.8 shows that $[F(c) : F] = n \geq 2$. Now Corollary 8.3.7 shows that $[\mathcal{S} : F] = [\mathcal{S} : F(c)][F(c) : F]$, so it must be the case that $[\mathcal{S} : F(c)] < [\mathcal{S} : F]$. The induction hypothesis tells us that the number m of extensions of β to $\mathcal{S}$ is exactly $[\mathcal{S} : F(c)]$. Therefore, the number of extensions of α to $\mathcal{S}$ is $mn = [\mathcal{S} : F(c)][F(c) : F] = [\mathcal{S} : F]$. ∎

9.3.12 Corollary

For any separable polynomial $f(x) \in F[x]$ with splitting field $\mathcal{S}$, $|\mathrm{Gal}(\mathcal{S}/F)| = [\mathcal{S} : F]$.

Proof. Choose $\alpha = \epsilon$, the identity automorphism of F, for the automorphism of Theorem 9.3.11. The set of extensions of ϵ to $\mathcal{S}$ is $\text{Gal}(\mathcal{S}/F)$, and Theorem 9.3.11 indicates that the number of these extensions is $[\mathcal{S}:F]$. Therefore, $|\text{Gal}(\mathcal{S}/F)| = [\mathcal{S}:F]$. ∎

Examples 1 and 2 of 9.3.7 are particular illustrations of Corollary 9.3.12. Now let's make the connection between automorphisms in $\text{Gal}(\mathcal{S}/F)$, where $\mathcal{S}$ is the splitting field of some separable polynomial in $F[x]$ and the permutations in $\mathbf{S}_n$, the symmetric group on n letters.

9.3.13   Theorem

If $f(x)$ is a separable polynomial in $F[x]$ with splitting field $\mathcal{S}$, then the Galois group $\text{Gal}(\mathcal{S}/F)$ of $f(x)$ is isomorphic to a subgroup of $\mathbf{S}_n$ for some positive integer n.

Proof. Suppose that $R = \{c_1, c_2, \ldots, c_n\}$ is the set of distinct roots of $f(x)$. If $\mathcal{S}/F$ is the splitting field of $f(x)$ and $\alpha \in \text{Gal}(\mathcal{S}/F)$, then $\alpha : \mathcal{S} \to \mathcal{S}$ is an automorphism, so α is a bijection. The restriction $\alpha|_R$ of α to R produces a mapping $\alpha|_R : R \to R$, whose image is the set $\{\alpha(c_1), \alpha(c_2), \ldots, \alpha(c_n)\}$. Theorem 9.3.4 shows that $\{\alpha(c_1), \alpha(c_2), \ldots, \alpha(c_n)\} \subseteq R$, and since α is an injection, $\{\alpha(c_1), \alpha(c_2), \ldots, \alpha(c_n)\} = R$. Therefore, α induces a permutation of the set of roots of $f(x)$.

If c_1 is replaced by 1, c_2 by 2, $\ldots$, and c_n by n, then for each $\alpha \in \text{Gal}(\mathcal{S}/F)$, there is a corresponding permutation θ_α in $\mathbf{S}_n$. If $\phi : \text{Gal}(\mathcal{S}/F) \to \mathbf{S}_n$ is defined by $\phi(\alpha) = \theta_\alpha$, it is not difficult to show that ϕ is a well-defined group homomorphism. If α and β in $\text{Gal}(\mathcal{S}/F)$ are such that $\phi(\alpha) = \phi(\beta)$, then $\theta_\alpha = \theta_\beta$, so it must be the case that $\alpha(c_k) = \beta(c_k)$ for $k = 1, 2, \ldots, n$. Theorem 9.3.6 shows that $\alpha = \beta$, so ϕ is an injection. Hence, $\text{Gal}(\mathcal{S}/F)$ is isomorphic to a subgroup of $\mathbf{S}_n$. Notice that the integer n is the number of distinct roots of $f(x)$. ∎

We can now state Galois' theorem on the solvability of polynomials by radicals. Galois' theorem is not proved here since quite a bit of mathematics would have to be developed before a proof could be offered. However, enough mathematics has been developed to make the statement of the theorem meaningful. The theorem is used to illustrate the result of Ruffini and Abel that indicates that polynomials in $\mathbf{Q}[x]$ of degree 5 exist that cannot be solved by radicals. Since the polynomials we consider have their coefficients in a field with characteristic 0, we now assume that the field F has characteristic 0. This assumption simplifies matters, and the following lemma allows us to apply what we have proven for separable polynomials to polynomials over a field of characteristic 0. You will be asked to prove the lemma as an exercise.

9.3.14   Lemma

If F is a field of characteristic 0, then every irreducible polynomial in $F[x]$ is separable.

This lemma actually tells us that over a field of characteristic 0, every polynomial $f(x)$ is separable. This follows since a polynomial is a product of irreducible polynomials, and this lemma indicates that each irreducible factor has no repeated roots. Thus, $f(x)$ is separable.

Recall that the theorem of Ruffini and Abel showed that it is not possible to find a formula that can be used to find the roots of all polynomials of degree 5. Their result left open the question of exactly which polynomials of degree 5 are solvable by radicals. The following remarkable theorem of Galois settles this question.

9.3.15   Galois' Theorem

If F is a field of characteristic 0, then a polynomial $f(x) \in F[x]$ with degree ≥ 1 is solvable by radicals if and only if the Galois group of $f(x)$ is a solvable group.

We also need the following theorem due to **Augustin-Louis Cauchy** (1789–1857) to prove the result of Ruffini and Abel. (Information on Cauchy can be found in the Historical Notes at the end of Section 5.1.) A proof of Cauchy's theorem can be found in "Another Proof of Cauchy's Group Theorem," published in 1959 in the *American Mathematical Monthly*, volume 66, page 119. The commutative version of this theorem is Lemma 5.4.12.

9.3.16   Cauchy's Theorem

If the order of a finite group G is divisible by a prime p, then G contains an element of order p.

Finally, we are in a position to prove the theorem of Ruffini and Abel dealing with the solvability of polynomials in $\mathbf{Q}[x]$ of degree 5. Notice the variety of theorems that are brought into play when proving this result.

9.3.17   Ruffini and Abel's Theorem

There exists a quintic polynomial in $\mathbf{Q}[x]$ that is not solvable by radicals.

Proof. Consider $f(x) = 3x^5 - 15x + 5 \in \mathbf{Q}[x]$. Calculus can be used to show that the graph of $f(x)$ must look something like the figure below. Hence, $f(x)$ has three real roots that are approximately -1.56912, 0.334167, and 1.39682. The other two roots of $f(x)$ cannot be real numbers. Since $5|5$, $5|15$, $5 \nmid 3$, and $5^2 \nmid 5$, $f(x)$ is irreducible in $\mathbf{Q}[x]$ by Eisenstein's Criterion 6.3.13. If c is a real root of $f(x)$, then by Theorem 8.2.8, $[F(c) : F] = \deg(f(x)) = 5$. Corollary 8.3.7 shows that $[\mathcal{S} : F] = [\mathcal{S} : F(c)]\ [F(c) : F] = [\mathcal{S} : F(c)]5$, where $\mathcal{S}$ is the splitting field of $f(x)$. By Corollary 9.3.12, $|\mathrm{Gal}(\mathcal{S}/F)| = [\mathcal{S} : F]$, so 5 divides $|\mathrm{Gal}(\mathcal{S}/F)|$. Therefore, by Theorem 9.3.16, $\mathrm{Gal}(\mathcal{S}/F)$ contains an element of order 5. Invoking Theorem 9.3.13, we see that $\mathrm{Gal}(\mathcal{S}/F)$ is isomorphic to a subgroup G of $\mathbf{S}_5$, and this means that G contains a 5-cycle. Since G and $\mathrm{Gal}(\mathcal{S}/F)$ are isomorphic groups, $\mathrm{Gal}(\mathcal{S}/F)$ is solvable if and only if G is solvable. Theorem 9.2.9 shows that $\mathbf{S}_5$ is not solvable, so it follows from Galois' theorem that $f(x)$ is not solvable by radicals if we can show that $G = \mathbf{S}_5$. Since the two nonreal roots of $f(x)$ are roots of the same minimal polynomial in $\mathbf{Q}[x]$, Theorem 9.3.5 shows that there is automorphism in $\mathrm{Gal}(\mathcal{S}/\mathbf{Q})$ that permutes these two roots

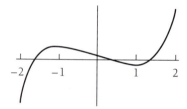

and maps each real root to itself. Thus, G also contains a 2-cycle. But it can be shown that any subgroup of $\mathbf{S}_5$ that contains a 2-cycle and a 5-cycle must equal $\mathbf{S}_5$. (See Exercise 10.) Hence, $G = \mathbf{S}_5$, and the proof is complete. ∎

What we have shown is not the complete story. It can be shown using the concept of a symmetric function that for any field F with characteristic 0 and any positive integer $n \geq 5$, there is a polynomial of degree n in $F[x]$ whose Galois group is isomorphic to $\mathbf{S}_n$. Thus, when F is a field of characteristic 0, Galois' theorem shows that a general polynomial in $F[x]$ of degree $n \geq 5$ cannot be solved by radicals. We know from Section 9.2 that the groups $\mathbf{S}_2$, $\mathbf{S}_3$, and $\mathbf{S}_4$ are solvable groups. Since $\mathrm{Gal}(\mathcal{S}/F)$ is isomorphic to a subgroup of $\mathbf{S}_n$ and since subgroups of solvable groups are solvable, Galois' theorem shows that all polynomials of degree 2, 3, and 4 over a field F with characteristic 0 are solvable by radicals. Galois' theorem provides an additional confirmation of the conclusions of Section 9.1. Formulas for computing the roots of a polynomial of degree 2, 3, and 4 were actually found in Section 9.1, while Galois' theorem shows only that such formulas exist.

HISTORICAL NOTES

Evariste Galois (1811–1832). Evariste Galois was born in Bourg-la-Reine near Paris, France. His parents were well-educated, and his mother was his teacher for the first 12 years of his life. She provided him with an excellent grounding in the classics, but it was mathematics that truly won his interest. When he began his formal schooling, Galois disliked textbook mathematics and went straight to the writings of mathematicians such as Legendre, Lagrange, and Abel. He took the competitive examinations for entrance to the Ecole Polytechnique without adequate preparation and failed. A second attempt also met with failure. (During the course of the second examination, Galois threw an eraser at one of the examiners.) In 1829 he entered the Ecole Normale Supérieure to prepare for a teaching career.

In March of 1829 Galois published his first mathematics paper, followed shortly by the submission of articles dealing with the algebraic solutions of equations to the Académie des Sciences. Cauchy was assigned to report on the work but somehow misplaced the paper. A second misfortune of a similar nature occurred in 1830 when Galois submitted papers to the Académie in competition for the Grand Prize in Mathematics. Again, his manuscript was misplaced, and he was not considered for the prize. Frustrated by these disasters, the hot-tempered young Galois threw himself into the turbulent French political environment. An ardent republican, Galois was tried and found innocent of threatening the king's life. In 1831 he was again arrested, and this time sentenced to six months' imprisonment. While in detention, he revised his work and studied elliptic functions. When the cholera epidemic of 1832 swept through Paris, Galois was transferred to a nursing home, where he resumed his work and wrote several essays on the philosophy of mathematics. In May of 1832, shortly after his release, Galois was challenged to a duel, the cause of which has been attributed to an unfortunate love affair and his republican politics. The night before the duel he desperately scrawled a manuscript outlining his principal mathematical results and requested that they be sent to Jacobi or Gauss for review. (In 1843 it was announced that a concise solution had been found among Galois' papers for the problem of solving a polynomial of prime degree by radicals.) In the duel by pistols the following day, Galois was badly wounded and died, at the age of 20, on May 31, 1832.

Emil Artin (1898–1962). Emil Artin was born in Vienna, Austria, near the close of the 19th century. Although Artin was not Jewish, he fell victim to Nazi policies that cost him his position at the University of Hamburg. He moved to the United States and spent the next 20 years of his life in academic appointments at Notre Dame (1937–1938), Indiana University (1938–1946), and Princeton University (1946–1958). In 1958 he was appointed to a position at the University of Hamburg, and he returned to Germany. Artin made many contributions to mathematics, but he is probably best known for his work with the minimum condition on descending chains of right ideals in a ring. Such rings today are known as right **Artinian rings** in his honor. (Recall that Emy Noether worked with the maximum condition on ascending chains of right ideals in a ring. Rings that satisfy this condition are known today as Noetherian rings. Information on the life of Noether can be found in the Historical Notes at the end of Section 7.2.) Artin's work includes contributions to the theory of associative algebras, Galois theory, semi-simple algebras, finite simple groups, and many other areas of mathematics. Among Artin's books are *Geometric Algebra*, published in 1957, and *Class Field Theory*, which appeared in 1961.

Problem Set 9.3

1. Prove each of the following.

   (a) If F is a field, then $\text{Aut}(F)$ is a group under the operation of composition of functions.
   (b) If E/F is a field extension, then $\text{Gal}(E/F)$ is a subgroup of $\text{Aut}(E)$.

2. Determine the splitting field $\mathcal{S}$ of each of the polynomials in $\mathbf{Q}[x]$, and compute the Galois group $\text{Gal}(\mathcal{S}/\mathbf{Q})$.

   (a) $f(x) = x^2 + 3$
   (b) $f(x) = x^4 + 5x^2 + 6$
   (c) $f(x) = x^4 + 8x^2 + 7$
   (d) $f(x) = x^4 + x^3 + 7x^2 + 5x + 10$

3. For the polynomials of Exercise 2, $\text{Gal}(\mathcal{S}/\mathbf{Q})$ is isomorphic to a subgroup of $\mathbf{S}_n$ for an appropriate value of n. In each case, find n and determine the group of permutations in $\mathbf{S}_n$ that is isomorphic to $\text{Gal}(\mathcal{S}/\mathbf{Q})$.

4. Determine the Galois group of the field extensions $\mathbf{Q}(\sqrt{3}, i)/\mathbf{Q}$ and $\mathbf{Q}(\sqrt[3]{5}, i)/\mathbf{Q}$.

5. Show that the polynomials $f(x) = x^2 - 5$, $g(x) = x^4 - 10x^2 + 25$, and $f(x) = x^2 - 5x - 5$ all have the same Galois group.

6. Argue that the Galois group of the polynomial $f(x) = x^5 - 1$ contains exactly four automorphisms.

7. Prove that the Galois group of an irreducible quadratic polynomial in $\mathbf{Q}[x]$ is isomorphic to the additive group $\mathbf{Z}_2$.

8. Prove Lemma 9.3.14. Hint: Show that each of the following hold:

    (a) For any field F, if $f(x)$ is an irreducible polynomial in $F[x]$, then $f(x)$ is separable if and only if $f(x)$ and its formal derivative $D[f(x)]$ are relatively prime.

    (b) For any field F, if $f(x)$ is an irreducible polynomial in $F[x]$, then $f(x)$ is separable if and only if $D[f(x)] \neq 0$.

    (c) If F is a field with characteristic 0 and $f(x) \in F[x]$ is irreducible, then $D[f(x)] \neq 0$.

    See Exercise 18 of Problem Set 6.1 and Exercise 12 of Problem Set 6.2.

9. (a) Show that the mappings $\phi : \mathrm{Gal}(\mathcal{S}/\mathbf{Q}) \to \mathbf{S}_4$ of Example 2 of 9.3.7 is a well-defined group monomorphism.

    (b) Show that the mapping $\phi : \mathrm{Gal}(\mathcal{S}/F) \to \mathbf{S}_n$ defined by $\phi(\alpha) = \theta_\alpha$ of Theorem 9.3.13 is a well-defined group monomorphism.

10. If G is a subgroup of $\mathbf{S}_5$ that contains a permutation of length 2 and one of length 5, prove that $G = \mathbf{S}_5$.

11. Show that the Galois group of the polynomial $f(x) = 2x^5 - 10x + 5$ is isomorphic to $\mathbf{S}_5$. Is $f(x)$ solvable by radicals?

12. Find the Galois group of the polynomial $f(x) = x^3 - 2x + 1$, and show that it is isomorphic to a subgroup of $\mathbf{S}_3$. Is $f(x)$ solvable by radicals? If so, solve $f(x)$ by radicals.

13. Let F be a field, and suppose that H is a subgroup of $\mathrm{Aut}(F)$. Prove that $F_H = \{a \in F \mid \alpha(a) = a$ for all $\alpha \in H\}$ is a subfield of F called the **fixed field** of H with respect to F.

14. Show that if E/F is a field extension, then $F \subseteq E_{\mathrm{Gal}(E/F)}$, where $E_{\mathrm{Gal}(E/F)}$ is defined in Exercise 13. Find an example where the containment is proper.

9.4   THE FUNDAMENTAL THEOREM OF GALOIS THEORY

Section Overview. In this section the Fundamental Theorem of Galois Theory is presented and discussed briefly. This theorem, which is at the center of Galois theory, is the result of an interplay among polynomials, groups, and fields. The Fundamental Theorem is not presented in its full generality. What is presented, however, will give you the flavor of Galois theory, and hopefully it will enable you to appreciate the genius of Galois.

As previously indicated, Galois theory results from certain connections between polynomials, groups, and fields. Before we begin to develop these connections, recall that if E/F is a field extension, then a field K such that $F \subseteq K \subseteq E$ is said to be an **intermediate field** of the extension E/F. (See Exercise 13 of Problem Set 8.1.)

9.4.1   Theorem

Let E be a field, and suppose that H is a subgroup of $\mathrm{Aut}(E)$. Then $E_H = \{a \in E \mid \alpha(a) = a \text{ for all } \alpha \in H\}$ is a subfield of E.

Proof. We first show that E_H is a subgroup of E. Note that $E_H \neq \phi$ since $e \in E_H$. If $a, b \in E_H$ and $\alpha \in H$, then $\alpha(a - b) = \alpha(a) - \alpha(b) = a - b$, so $a - b \in E_H$. Hence, E_H is a subgroup of E. Likewise, $\alpha(ab) = \alpha(a)\alpha(b) = ab$ shows that $ab \in E_H$, so E_H is a subring of E. If $a \in E_H$, $a \neq 0$, then a has a multiplicative inverse a^{-1} in E. But then $e = \alpha(e) = \alpha(aa^{-1}) = \alpha(a)\alpha(a^{-1}) = a\alpha(a^{-1})$. Hence, by the uniqueness of multiplicative inverses, we have $a^{-1} = \alpha(a^{-1})$, so $a^{-1} \in E_H$. ■

The field E_H of Theorem 9.4.1 is called the **fixed field** of H with respect to E or **the subfield of E corresponding to H.**

9.4.2   Corollary

If E/F is a field extension and H is a subgroup of the Galois group $\mathrm{Gal}(E/F)$, then there is an intermediate field E_H corresponding to H.

Proof. Theorem 9.4.1 shows that $E_H \subseteq E$ and $F \subseteq E_H$ since every automorphism in $\mathrm{Gal}(E/F)$ leaves each element of F fixed. ■

9.4.3   Examples

1. The field extension $\mathcal{S} = \mathbf{Q}(\sqrt{2})/\mathbf{Q}$ is the splitting field of the polynomial $f(x) = x^2 - 2$. The Galois group $\mathrm{Gal}(\mathcal{S}/\mathbf{Q})$ has two automorphisms ϵ and α, where ϵ is the identity automorphism

determined by $\sqrt{2} \longmapsto \sqrt{2}$, and α is the automorphism determined by $\sqrt{2} \longmapsto -\sqrt{2}$. Hence, $\text{Gal}(\mathcal{S}/\mathbf{Q}) = \{\epsilon, \alpha\}$ has only two subgroups, $H_1 = \{\epsilon\}$ and $H_2 = \{\epsilon, \alpha\}$. The fixed field of H_1 is $\mathbf{Q}(\sqrt{2})$, and the fixed field of H_2 is $\mathbf{Q}$. Note that the correspondence is order-reversing:

$$H_1 \subseteq H_2$$
$$\downarrow \quad \downarrow$$
$$\mathcal{S} \supseteq \mathbf{Q}$$

If $\mathcal{G}$ denotes the set of all subgroups of $\text{Gal}(\mathcal{S}/\mathbf{Q})$ and $\mathcal{F}$ is the set of all intermediate fields, then the mapping $\phi : \mathcal{G} \to \mathcal{F}$ is said to be a **Galois correspondence**. In this case, the Galois correspondence is a bijection.

2. The field extension $\mathcal{S} = \mathbf{Q}(\sqrt{2}, \sqrt{3})$ is the splitting field of the polynomial $g(x) = (x^2 - 2)(x^2 - 3)$. In this case, the Galois group of $\mathcal{S}/\mathbf{Q}$ is $\text{Gal}(\mathcal{S}/\mathbf{Q}) = \{\epsilon, \alpha, \beta, \gamma\}$, where ϵ, α, β, and γ are determined by

$$\epsilon : \begin{cases} \sqrt{2} \longmapsto \sqrt{2} \\ \sqrt{3} \longmapsto \sqrt{3} \end{cases}, \qquad \alpha : \begin{cases} \sqrt{2} \longmapsto -\sqrt{2} \\ \sqrt{3} \longmapsto \sqrt{3} \end{cases},$$

$$\beta : \begin{cases} \sqrt{2} \longmapsto \sqrt{2} \\ \sqrt{3} \longmapsto -\sqrt{3} \end{cases}, \quad \text{and} \quad \gamma : \begin{cases} \sqrt{2} \longmapsto -\sqrt{2} \\ \sqrt{3} \longmapsto -\sqrt{3} \end{cases}.$$

The subgroups of $\text{Gal}(\mathcal{S}/\mathbf{Q})$ are $H_1 = \{\epsilon\}$, $H_2 = \{\epsilon, \alpha\}$, $H_3 = \{\epsilon, \beta\}$, $H_4 = \{\epsilon, \gamma\}$, and $H_5 = \{\epsilon, \alpha, \beta, \gamma\}$. The fixed fields of H_1, H_2, H_3, H_4, and H_5 are $\mathcal{S}, \mathbf{Q}(\sqrt{3}), \mathbf{Q}(\sqrt{2}), \mathbf{Q}(\sqrt{6})$, and $\mathbf{Q}$, respectively. Again we see that the Galois correspondence is order-reversing:

$$H_1 \subseteq \quad H_2 \quad \subseteq H_5 \qquad H_1 \subseteq \quad H_3 \quad \subseteq H_5 \qquad H_1 \subseteq \quad H_4 \quad \subseteq H_5$$
$$\downarrow \quad \downarrow \quad \downarrow \qquad\qquad \downarrow \quad \downarrow \quad \downarrow \qquad\qquad \downarrow \quad \downarrow \quad \downarrow$$
$$\mathcal{S} \supseteq \mathbf{Q}(\sqrt{3}) \supseteq \mathbf{Q} \qquad \mathcal{S} \supseteq \mathbf{Q}(\sqrt{2}) \supseteq \mathbf{Q} \qquad \mathcal{S} \supseteq \mathbf{Q}(\sqrt{6}) \supseteq \mathbf{Q}$$

The lattices of subgroups and corresponding subfields are

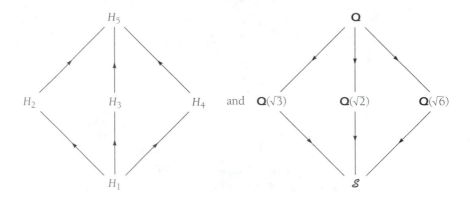

where the arrows indicate the direction of containment. Since $\mathbf{Q}$, $\mathbf{Q}(\sqrt{2})$, $\mathbf{Q}(\sqrt{3})$, $\mathbf{Q}(\sqrt{6})$, and $\mathcal{S} = \mathbf{Q}(\sqrt{2}, \sqrt{3})$ are all of the intermediate fields, we see, as in Example 1, that the Galois correspondence between the subgroups of $\mathrm{Gal}(\mathcal{S}/\mathbf{Q})$ and the intermediate fields of the field extension $\mathcal{S}/\mathbf{Q}$ is an order-reversing, one-to-one correspondence.

We have just seen that the set $\mathcal{G}$ of all subgroups of $\mathrm{Gal}(\mathcal{S}/\mathbf{Q})$ determines the set $\mathcal{F}$ of all intermediate fields of $\mathcal{S}/\mathbf{Q}$ and that the correspondence is bijective. Now let's look at the reverse question. Does the set $\mathcal{F}$ of all intermediate fields of the field extension $\mathcal{S}/\mathbf{Q}$ determine the set of all subgroups $\mathcal{G}$ of $\mathrm{Gal}(\mathcal{S}/\mathbf{Q})$? Since we easily see that

$$\mathrm{Gal}(\mathcal{S}/\mathbf{Q}) = \{\epsilon, \alpha, \beta, \gamma\} = H_5, \quad \mathrm{Gal}(\mathcal{S}/\mathbf{Q}(\sqrt{2})) = \{\epsilon, \beta\} = H_3,$$

$$\mathrm{Gal}(\mathcal{S}/\mathbf{Q}(\sqrt{3})) = \{\epsilon, \alpha\} = H_2, \quad \mathrm{Gal}(\mathcal{S}/\mathbf{Q}(\sqrt{6})) = \{\epsilon, \gamma\} = H_4,$$

$$\text{and} \quad \mathrm{Gal}(\mathcal{S}/\mathcal{S}) = \{\epsilon\} = H_1,$$

the answer in this case is clearly Yes, and this also establishes a one-to-one correspondence.

3. For this example, consider the field extension $\mathbf{Q}(\sqrt{5}, \sqrt[3]{2})/\mathbf{Q}$. In this case, $\sqrt{5}$ is a root of $x^2 - 5$, and $\mathbf{Q}(\sqrt{5}, \sqrt[3]{2})$ contains both roots of this polynomial. We also see that $\sqrt[3]{2}$ is a root of $x^3 - 2$ but that $\mathbf{Q}(\sqrt{5}, \sqrt[3]{2})$ does not contain the two nonreal roots of $x^3 - 2$. Hence, $\mathbf{Q}(\sqrt{5}, \sqrt[3]{2})$ is not the splitting field for $f(x) = (x^2 - 5)(x^3 - 2)$. Since $x^3 - 2$ is the minimal polynomial in $\mathbf{Q}[x]$ of $\sqrt[3]{2}$, $\mathbf{Q}(\sqrt{5}, \sqrt[3]{2})$ is not the

splitting field for any polynomial in $\mathbf{Q}[x]$. This follows since any polynomial in $\mathbf{Q}[x]$ that has $\sqrt[3]{2}$ as a root would have to have $x^3 - 2$ as a factor. The intermediate fields of the field extension $\mathbf{Q}(\sqrt{5}, \sqrt[3]{2})\,/\,\mathbf{Q}$ are

$$\mathbf{Q}, \qquad \mathbf{Q}(\sqrt{5}), \qquad \mathbf{Q}(\sqrt[3]{2}), \qquad \text{and} \qquad \mathbf{Q}(\sqrt{5}, \sqrt[3]{2}).$$

Let's see if we can determine $\mathrm{Gal}(\mathbf{Q}(\sqrt{5}, \sqrt[3]{2})/\mathbf{Q})$. If α is any automorphism in $\mathrm{Gal}(\mathbf{Q}(\sqrt{5}, \sqrt[3]{2}))$, then by Theorem 9.3.6, α is completely determined by its action on $\sqrt{5}$ and $\sqrt[3]{2}$. Furthermore, $\alpha(\sqrt{5})$ must be a root of the same minimal polynomial in $\mathbf{Q}[x]$ as $\sqrt{5}$. Since $\pm\sqrt{5}$ are both in $\mathbf{Q}(\sqrt{5}, \sqrt[3]{2})$, $\alpha(\sqrt{5}) = \sqrt{5}$ or $\alpha(\sqrt{5}) = -\sqrt{5}$. Likewise, $\alpha(\sqrt[3]{2})$ can take only one value in $\mathbf{Q}(\sqrt{5}, \sqrt[3]{2})$, and that is $\sqrt[3]{2}$. Thus, there are only two automorphisms in $\mathrm{Gal}(\mathbf{Q}(\sqrt{5}, \sqrt[3]{2})/\mathbf{Q})$, and they are determined by

$$\epsilon : \begin{cases} \sqrt{5} \longmapsto \sqrt{5} \\ \sqrt[3]{2} \longmapsto \sqrt[3]{2} \end{cases} \quad \text{and} \quad \alpha : \begin{cases} \sqrt{5} \longmapsto -\sqrt{5} \\ \sqrt[3]{2} \longmapsto \sqrt[3]{2} \end{cases}.$$

Hence, $\mathrm{Gal}(\mathbf{Q}(\sqrt{5}, \sqrt[3]{2})/\mathbf{Q}) = \{\epsilon, \alpha\}$, so the only subgroups are $H_1 = \{\epsilon\}$ and $H_2 = \{\epsilon, \alpha\}$. The intermediate field determined by H_1 is $\mathbf{Q}(\sqrt{5}, \sqrt[3]{2})$, and the intermediate field determined by H_2 is $\mathbf{Q}(\sqrt[3]{2})$. In this case we see that the Galois correspondence fails to be a one-to-one correspondence. As we shall see, this happens because $\mathbf{Q}(\sqrt{5}, \sqrt[3]{2})$ is not the splitting field of a polynomial in $\mathbf{Q}[x]$.

   In Examples 1 and 2, there is a one-to-one correspondence between the subgroups of the Galois group $\mathrm{Gal}(\mathcal{S}\,/\,\mathbf{Q})$ and the intermediate fields of $\mathcal{S}\,/\,\mathbf{Q}$. This happened because in both cases $\mathcal{S}$ is the splitting field of a polynomial in $\mathbf{Q}[x]$. In Example 3, we saw a situation where the Galois correspondence failed to be a one-to-one correspondence. This happened because in the field extension $\mathbf{Q}(\sqrt{5}, \sqrt[3]{2})\,/\,\mathbf{Q}, \mathbf{Q}(\sqrt{5}, \sqrt[3]{2})$ is not the splitting field of any polynomial in $\mathbf{Q}[x]$.

   A finite dimensional field extension $E\,/\,F$ is said to be a **separable extension** if the minimal polynomial in $F[x]$ of each $c \in E$ is a separable polynomial. (Recall that Theorem 8.3.1 indicates that every element of E is algebraic over F when $E\,/\,F$ is finite dimensional.) The field E is a **normal extension** of F if whenever an irreducible polynomial $f(x) \in F[x]$ has a root in E, then every root of $f(x)$ is in E.

The Fundamental Theorem of Galois Theory is often stated in terms of a finite dimensional, separable, normal extension E/F. But when F has characteristic 0, this just says that E is the splitting field of some polynomial in $F[x]$. Because of this, we assume that the field F has characteristic 0 even though this assumption eliminates finite fields from consideration. We now state the Fundamental Theorem of Galois Theory over a field of characteristic 0. Examples 1 and 2 of 9.4.3 serve as illustrations of the theorem.

9.4.4   Fundamental Theorem of Galois Theory

Let F be a field of characteristic 0, and suppose that $\mathcal{S}$ is the splitting field of some polynomial in $F[x]$. If $\mathcal{G}$ is the set of all subgroups of $\mathrm{Gal}(\mathcal{S}/F)$ and $\mathcal{F}$ is the set of all intermediate fields of $\mathcal{S}/F$, then the following hold:

1. The Galois correspondence $\phi : \mathcal{G} \to \mathcal{F}$ defined by $\phi(H) = \mathcal{S}_H$ is an order-reversing, one-to-one correspondence with inverse mapping $\psi : \mathcal{F} \to \mathcal{G}$ given by $\psi(K) = \mathrm{Gal}(\mathcal{S}/K)$.
2. If K is an intermediate field of the field extension $\mathcal{S}/F$, then $|\mathrm{Gal}(\mathcal{S}/K)| = [\mathcal{S} : K]$ and $[K : F] = [\mathrm{Gal}(\mathcal{S}/F) : \mathrm{Gal}(\mathcal{S}/K)]$.
3. An intermediate field K is the splitting field of a polynomial in $F[x]$ if and only if $\mathrm{Gal}(\mathcal{S}/K)$ is a normal subgroup of $\mathrm{Gal}(\mathcal{S}/F)$, and when this is the case, the group $\mathrm{Gal}(K/F)$ is isomorphic to the factor group $\mathrm{Gal}(\mathcal{S}/F)/\mathrm{Gal}(\mathcal{S}/K)$.

9.4.5   Example

Consider the polynomial $f(x) = x^3 - 2 \in \mathbf{Q}[x]$. One root of $f(x)$ is clearly $\sqrt[3]{2}$, and $f(x) = (x - \sqrt[3]{2})(x^2 + \sqrt[3]{2}x + \sqrt[3]{4})$. If the quadratic formula is applied to the quadratic factor, the result is that $\omega \sqrt[3]{2}$ and $\omega^2 \sqrt[3]{2}$ are the other two roots of $f(x)$, where $\omega = -\frac{1}{2} + \frac{\sqrt{3}}{2}i$ is a complex cube root of unity. The splitting field of $f(x)$ is $\mathcal{S} = \mathbf{Q}(\sqrt[3]{2}, \omega \sqrt[3]{2}, \omega^2 \sqrt[3]{2})$, and $\omega \in \mathcal{S}$ since $\omega = \frac{1}{2}(\sqrt[3]{2})^2 \omega \sqrt[3]{2} \in \mathcal{S}$. From this it follows that the splitting field is $\mathcal{S} = \mathbf{Q}(\sqrt[3]{2}, \omega)$ and that the intermediate fields are

$$\mathbf{Q}, \quad \mathbf{Q}(\omega), \quad \mathbf{Q}(\sqrt[3]{2}), \quad \mathbf{Q}(\omega \sqrt[3]{2}), \quad \mathbf{Q}(\omega^2 \sqrt[3]{2}), \quad \text{and} \quad \mathcal{S}.$$

Since the minimal polynomial in $\mathbf{Q}[x]$ of ω is $m(x) = x^2 + x + 1$ and since $f(x)$ is the minimal polynomial in $\mathbf{Q}[x]$ of $\sqrt[3]{2}$, it follows that $[\mathcal{S} : \mathbf{Q}] =$

$(\deg(m(x)))(\deg(f(x))) = 2 \cdot 3 = 6$. Hence, $|\mathrm{Gal}(\mathcal{S}/\mathbf{Q})| = 6$, so there are exactly six automorphisms of $\mathcal{S}$ that leave $\mathbf{Q}$ fixed elementwise. Since $f(x)$ has three distinct roots, $\mathrm{Gal}(\mathcal{S}/\mathbf{Q})$ is isomorphic to a subgroup of $\mathbf{S}_3$. But $\mathbf{S}_3$ also has $3! = 6$ elements, so $\mathrm{Gal}(\mathcal{S}/\mathbf{Q})$ and $\mathbf{S}_3$ must be isomorphic. The automorphisms in $\mathrm{Gal}(\mathcal{S}/\mathbf{Q})$ are determined by the following permutations of the roots of $f(x)$.

$$\epsilon: \begin{cases} \sqrt[3]{2} \longmapsto \sqrt[3]{2} \\ \omega\sqrt[3]{2} \longmapsto \omega\sqrt[3]{2} \\ \omega^2\sqrt[3]{2} \longmapsto \omega^2\sqrt[3]{2} \end{cases} \quad \alpha_1: \begin{cases} \sqrt[3]{2} \longmapsto \omega\sqrt[3]{2} \\ \omega\sqrt[3]{2} \longmapsto \sqrt[3]{2} \\ \omega^2\sqrt[3]{2} \longmapsto \omega^2\sqrt[3]{2} \end{cases} \quad \alpha_2: \begin{cases} \sqrt[3]{2} \longmapsto \sqrt[3]{2} \\ \omega\sqrt[3]{2} \longmapsto \omega^2\sqrt[3]{2} \\ \omega^2\sqrt[3]{2} \longmapsto \omega\sqrt[3]{2} \end{cases}$$

$$\alpha_3: \begin{cases} \sqrt[3]{2} \longmapsto \omega^2\sqrt[3]{2} \\ \omega\sqrt[3]{2} \longmapsto \omega\sqrt[3]{2} \\ \omega^2\sqrt[3]{2} \longmapsto \sqrt[3]{2} \end{cases} \quad \alpha_4: \begin{cases} \sqrt[3]{2} \longmapsto \omega\sqrt[3]{2} \\ \omega\sqrt[3]{2} \longmapsto \omega^2\sqrt[3]{2} \\ \omega^2\sqrt[3]{2} \longmapsto \sqrt[3]{2} \end{cases} \quad \alpha_5: \begin{cases} \sqrt[3]{2} \longmapsto \omega^2\sqrt[3]{2} \\ \omega\sqrt[3]{2} \longmapsto \sqrt[3]{2} \\ \omega^2\sqrt[3]{2} \longmapsto \omega\sqrt[3]{2} \end{cases}$$

If $\sqrt[3]{2}$ is replaced by 1, $\omega\sqrt[3]{2}$ by 2, and $\omega^2\sqrt[3]{2}$ by 3 and permutations of $\{1, 2, 3\}$ are written as cycles, then we have the following correspondence between the automorphisms in $\mathrm{Gal}(\mathcal{S}/\mathbf{Q})$ and elements of $\mathbf{S}_3$.

$$\epsilon \leftrightarrow (1) \qquad \alpha_3 \leftrightarrow (1, 3)$$
$$\alpha_1 \leftrightarrow (1, 2) \qquad \alpha_4 \leftrightarrow (1, 2, 3)$$
$$\alpha_2 \leftrightarrow (2, 3) \qquad \alpha_5 \leftrightarrow (1, 3, 2)$$

The subgroups of $\mathbf{S}_3$ are $\{(1)\}, \{(1),(1,2)\}, \{(1),(2,3)\}, \{(1),(1,3)\}, \{(1), (1,2,3), (1,3,2)\}$, and $\mathbf{S}_3$, and the corresponding subgroups of $\mathrm{Gal}(\mathcal{S}/\mathbf{Q})$ are $H_1 = \{\epsilon\}$, $H_2 = \{\epsilon, \alpha_1\}$, $H_3 = \{\epsilon, \alpha_2\}$, $H_4 = \{\epsilon, \alpha_3\}$, $H_5 = \{\epsilon, \alpha_4, \alpha_5\}$, and $\mathrm{Gal}(\mathcal{S}/\mathbf{Q})$. The intermediate fields determined by these subgroups are given in the following correspondence table.

$$\mathcal{G} \leftrightarrow \mathcal{F}$$
$$H_1 \leftrightarrow S$$
$$H_2 \leftrightarrow \mathbf{Q}(\omega^2\sqrt[3]{2})$$
$$H_3 \leftrightarrow \mathbf{Q}(\sqrt[3]{2})$$
$$H_4 \leftrightarrow \mathbf{Q}(\omega\sqrt[3]{2})$$

$$\mathcal{G} \leftrightarrow \mathcal{F}$$
$$H_5 \leftrightarrow \mathbf{Q}(\omega)$$
$$\mathrm{Gal}(\mathcal{S}/\mathbf{Q}) \leftrightarrow \mathbf{Q}$$

Since $\omega = \frac{1}{2}(\sqrt[3]{2})^2\omega\sqrt[3]{2}$ and

$$
\begin{aligned}
\alpha_5(\omega) &= \alpha_5\!\left(\frac{1}{2}(\sqrt[3]{2})^2\omega \,\sqrt[3]{2} \right) \\
&= \frac{1}{2}\alpha_5((\sqrt[3]{2})^2)\alpha_5(\omega \,\sqrt[3]{2}) \\
&= \frac{1}{2}\omega^4\sqrt[3]{4}\,\sqrt[3]{2} \\
&= \omega^4 \\
&= \omega,
\end{aligned}
$$

α_5 fixes ω. Similarly, $\alpha_4(\omega) = \omega$, so the group H_5 fixes $\mathbf{Q}(\omega)$ elementwise. Since $\mathbf{Q}(\sqrt[3]{2})$ is not the splitting field of any polynomial in $\mathbf{Q}[x]$, H_3 is not a normal subgroup of $\mathrm{Gal}(\mathcal{S}/\mathbf{Q})$. Moreover, $\mathbf{Q}(\omega)$ is the splitting field for the polynomial $m(x) = x^2 + x + 1 \in \mathbf{Q}[x]$, so the group H_5 is a normal subgroup of $\mathrm{Gal}(\mathcal{S}/\mathbf{Q})$. The lattices of subgroups and corresponding intermediate fields are given below. The arrows indicate the direction of inclusion.

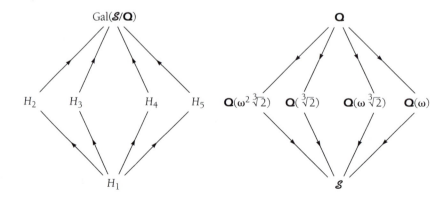

Problem Set 9.4

1. If possible, find a polynomial in $\mathbf{Q}[x]$ for which each of the following is its splitting field $\mathcal{S}$. When possible, give the Galois correspondence between the subgroups of $\mathrm{Gal}(\mathcal{S}/\mathbf{Q})$ and the intermediate fields of $\mathcal{S}/\mathbf{Q}$.

    (a) $\mathbf{Q}(\sqrt{5})$

    (b) $\mathbf{Q}(\sqrt{2}\,i)$

    (c) $\mathbf{Q}(\sqrt{2},\sqrt{3}\,i)$

    (d) $\mathbf{Q}(\sqrt[3]{5},\sqrt{3})$

2. (a) Show that $\omega = -\frac{1}{2} + \frac{\sqrt{3}}{2}i$ is a cube root of unity. That is, show that $\omega^3 = 1$.

    (b) Show that $\mathbf{Q}(\omega)$ is the splitting field for the polynomial $f(x) = x^2 + x + 1$. Hint: Show that ω and $\omega^2 = \overline{\omega}$, the complex conjugate of ω, are the two roots of $f(x)$.

3. Show that $\mathcal{S} = \mathbf{Q}(\sqrt[4]{2}, i)$ is the splitting field for the polynomial $f(x) = x^4 - 2 \in \mathbf{Q}(x)$ and that $|\mathrm{Gal}(\mathcal{S}/\mathbf{Q})| = 8$.

4. Find the splitting field $\mathcal{S}$ for the polynomial $f(x) = x^2 + 1 \in \mathbf{R}[x]$. Compute the subgroups of $\mathrm{Gal}(\mathcal{S}/\mathbf{R})$, and exhibit the Galois correspondence between the subgroups of $\mathrm{Gal}(\mathcal{S}/\mathbf{R})$ and the intermediate fields of the extension $\mathcal{S}/\mathbf{R}$.

5. Determine the Galois group of $f(x) = (x^2 - 2)(x^2 - 3)(x^2 - 5) \in \mathbf{Q}[x]$. Exhibit the Galois correspondence between the subgroups of $\mathrm{Gal}(\mathcal{S}/\mathbf{Q})$ and the intermediate fields of $\mathcal{S}/\mathbf{Q}$.

6. Suppose that $\mathcal{S}$ is the splitting field of $f(x) = x^4 + 1 \in \mathbf{Q}[x]$. Compute $\mathrm{Gal}(\mathcal{S}/\mathbf{Q})$, and exhibit the Galois correspondence between the subgroups of $\mathrm{Gal}(\mathcal{S}/\mathbf{Q})$ and the intermediate fields of $\mathcal{S}/\mathbf{Q}$.

7. Show that the Galois group of the polynomial $f(x) = x^4 - 5x^2 + 6$ is isomorphic to the group $\mathbf{Z}_2 \times \mathbf{Z}_2$.

8. If $E/\mathbf{Q}$ is a field extension and $\alpha \in \mathrm{Aut}(E)$, prove that α must fix $\mathbf{Q}$ elementwise. Hint: If $\frac{p}{q} \in \mathbf{Q}$, show that $q\alpha\left(\frac{p}{q}\right) = p$.

9. Theorem 9.4.1 shows that when E is a field and H is a subgroup of $\text{Aut}(E)$, then $E_H = \{a \in E \mid \alpha(a) = a \text{ for all } \alpha \in H\}$ is an intermediate field of the field extension E / F. If K is an intermediate field of a field extension E / F, prove that $\text{Gal}(E / K)$ is a subgroup of $\text{Gal}(E / F)$.

10

Vector Spaces

In the previous chapters, several basic structures of abstract algebra were studied. Groups, rings, integral domains, and fields, along with examples of each, were considered. In this chapter an important algebraic structure known as a vector space is investigated. A vector space is formed from two of the algebraic structures that we have already studied, namely, additive abelian groups and fields. Vector spaces enjoy a wide variety of uses, not only in mathematics but in many areas of science and engineering. The theory of these spaces is quite extensive, so we provide only an introduction.

Much of what is presented in this chapter frequently appears in undergraduate texts on linear algebra, often without emphasis on proof. For this reason, it is beneficial to revisit this material with attention to the details.

10.1   VECTOR SPACES

Section Overview. In this section, the definition of a vector space is motivated by considering certain shared properties of two algebraic structures that we have previously studied, namely, $\mathbf{R}^3$ and $M_2(\mathbf{R})$. Pay particular attention to the concepts of a subspace of a vector space and a basis of a vector space. Be sure to note that the number of elements in a basis of a vector space is unique.

As a motivation for the definition of a vector space, we investigate one algebraic structure that can be formed from the set of 3-tuples in $\mathbf{R}^3$, and then we look at certain algebraic properties of $M_2(\mathbf{R})$. The shared properties of the algebraic structures presented in these two examples provide a model for the definition of a vector space.

10.1.1   Examples

1. Consider the set $\mathbf{R}^3$ of all 3-tuples $a = (a_1, a_2, a_3)$ whose entries are real numbers. For example, $(2, \frac{5}{6}, \sqrt{2})$ is an element of $\mathbf{R}^3$. We know that $\mathbf{R}^3$ is an additive abelian group under componentwise addition. The additive identity is $(0, 0, 0)$, and the additive inverse of $a = (a_1, a_2, a_3)$ is $-a = (-a_1, -a_2, -a_3)$. Each element of $\mathbf{R}^3$ can be multiplied by a real number c if we define multiplication of an element of $\mathbf{R}^3$ by an element of $\mathbf{R}$ as

$$ca = c(a_1, a_2, a_3) = (ca_1, ca_2, ca_3).$$

For example,

$$3\left(2, \tfrac{5}{6}, \sqrt{2}\right) = \left(3 \cdot 2, 3 \cdot \tfrac{5}{6}, 3\sqrt{2}\right)$$
$$= \left(6, \tfrac{5}{2}, 3\sqrt{2}\right).$$

If $a, b \in \mathbf{R}^3$ and $c, c_1, c_2 \in \mathbf{R}$, then it is not difficult to show that

$$c(a + b) = ca + cb, \tag{1}$$

$$(c_1 c_2)a = c_1(c_2 a), \tag{2}$$

$$(c_1 + c_2)a = c_1 a + c_2 a, \tag{3}$$

$$1a = a. \tag{4}$$

For example, to show equation (1), let $a = (a_1, a_2, a_3)$ and $b = (b_1, b_2, b_3)$. Then

$$c(a + b) = c((a_1, a_2, a_3) + (b_1, b_2, b_3))$$
$$= c(a_1 + b_1, a_2 + b_2, a_3 + b_3)$$
$$= (c(a_1 + b_1), c(a_2 + b_2), c(a_3 + b_3))$$
$$= (ca_1 + cb_1, ca_2 + cb_2, ca_3 + cb_3)$$
$$= (ca_1, ca_2, ca_3) + (cb_1, cb_2, cb_3)$$

$$= c(a_1, a_2, a_3) + c(b_1, b_2, b_3))$$

$$= ca + cb.$$

In general, for each integer $n \geq 2$, $\mathbf{R}^n$ is an additive abelian group under componentwise addition. The additive identity is $(0, 0, \ldots, 0)$, and the additive inverse of $a = (a_1, a_2, \ldots, a_n)$ is $-a = (-a_1, -a_2, \ldots, -a_n)$. Elements of $\mathbf{R}^n$ can be multiplied by elements of $\mathbf{R}$ if multiplication is defined by

$$ca = c(a_1, a_2, \ldots, a_n) = (ca_1, ca_2, \ldots, ca_n).$$

Moreover, $\mathbf{R}^n$ together with this multiplication satisfies the conditions in equations (1) through (4) given earlier.

2. Consider the additive abelian group $M_2(\mathbf{R})$ under ordinary matrix addition.

The additive identity of this group is $\begin{pmatrix} 0 & 0 \\ 0 & 0 \end{pmatrix}$, and the additive inverse of

$A = \begin{pmatrix} a_{11} & a_{12} \\ a_{21} & a_{22} \end{pmatrix}$ is $-A = \begin{pmatrix} -a_{11} & -a_{12} \\ -a_{21} & -a_{22} \end{pmatrix}$. Each element of $M_2(\mathbf{R})$ can be

multiplied by a real number if we define multiplication by

$$cA = c\begin{pmatrix} a_{11} & a_{12} \\ a_{21} & a_{22} \end{pmatrix} = \begin{pmatrix} ca_{11} & ca_{12} \\ ca_{21} & ca_{22} \end{pmatrix}.$$

As an illustration, if $c = \frac{1}{2}$ and $A = \begin{pmatrix} 2 & 4\sqrt{3} \\ 1 & \pi \end{pmatrix}$, then

$$\frac{1}{2}\begin{pmatrix} 2 & 4\sqrt{3} \\ 1 & \pi \end{pmatrix} = \begin{pmatrix} \frac{1}{2} \cdot 2 & \frac{1}{2} \cdot 4\sqrt{3} \\ \frac{1}{2} \cdot 1 & \frac{1}{2} \cdot \pi \end{pmatrix} = \begin{pmatrix} 1 & 2\sqrt{3} \\ \frac{1}{2} & \frac{\pi}{2} \end{pmatrix}.$$

As in Example 1, it can be shown that for any $A, B \in M_2(\mathbf{R})$ and $c, c_1, c_2 \in \mathbf{R}$ that

$$c(A + B) = cA + cB, \tag{1}$$

$$(c_1 c_2)A = c_1(c_2 A), \tag{2}$$

$$(c_1 + c_2)A = c_1 A + c_2 A, \tag{3}$$

$$1A = A. \tag{4}$$

The two preceding examples have the following properties in common: (1) both are additive abelian groups, and (2) an external multiplication defined on the elements of the group by elements of **R** satisfies the list of properties in equations (1) through (4) given in both examples. These examples are illustrations of a very fundamental and important algebraic structure known as a vector space.

10.1.2   Definition

If V is an additive abelian group and F is a field with identity e, then V is said to be a **vector space** over F if there is a **scalar multiplication** $F \times V \to V$ given by $(c, v) \longmapsto cv$ that satisfies the following properties:

1. $(c_1 + c_2)v = c_1v + c_2v$ for all $c_1, c_2 \in F$ and all $v \in V$.
2. $c(v_1 + v_2) = cv_1 + cv_2$ for all $c \in F$ and all $v_1, v_2 \in V$.
3. $(c_1c_2)v = c_1(c_2v)$ for all $c_1, c_2 \in F$ and all $v \in V$.
4. $ev = v$ for all $v \in V$.

Elements of V are called **vectors**, elements of F are **scalars**.

One point should be mentioned before proceeding. The symbol 0 is used to denote both the zero vector (the additive identity of V) and the additive identity of the field F. The context of the discussion can be used to determine the element that 0 represents.

10.1.3   Additional Examples of Vector Spaces

1. The field of complex numbers **C** is a vector space over the field **R** of real numbers. The additive abelian group of **C** is the additive abelian group of the vector space. The vectors of this vector space are complex numbers, and the scalars are real numbers. If $z = a + bi \in$ **C** and $c \in$ **R**, scalar multiplication is defined by

$$cz = c(a + bi) = ca + cbi.$$

2. If F is a field, then $F[x]$ is an additive abelian group under ordinary addition of polynomials. If $f(x) = a_0 + a_1x + \cdots + a_nx^n \in F[x], c \in F,$ and scalar multiplication is defined by $cf(x) = ca_0 + ca_1x + \cdots + ca_nx^n$, then $F[x]$ is a vector space over F. The vectors are the polynomials in $F[x]$, and the scalars are elements of F.

3. Suppose that n is a nonnegative integer, and let F be a field. If $P_n(F)$ denotes the set of all polynomials in $F[x]$ with degree $\leq n$ together with the zero polynomial, then $P_n(F)$ is an additive abelian group under ordinary polynomial addition. If $f(x) = a_0 + a_1 x + \cdots + a_n x^n \in P_n(F)$ and $c \in F$, let scalar multiplication be defined by $cf(x) = ca_0 + ca_1 x + \cdots + ca_n x^n$. Then $P_n(F)$ is a vector space over F. In this example, the vectors are polynomials in $P_n(F)$, and the scalars are the elements of F.

4. If n is a positive integer and F is a field, then the set of all n-tuples F^n with their entries from F is an additive abelian group. If scalar multiplication is defined by

$$cu = c(u_1, u_2, \ldots, u_n) = (cu_1, cu_2, \ldots, cu_n),$$

    then F^n is a vector space over F. The vectors are the n-tuples, and the scalars are the elements of F. Example 1 of 10.1.1 is a special case of this example; just let $F = \mathbf{R}$. A vector in F^n can be viewed as a **row vector**.

5. If $\overline{F^n}$ denotes the set of all transposes of the row vectors in F^n, then $\overline{F^n}$ is also a vector space over F under the operations defined as follows: if

$$u = \begin{pmatrix} u_1 \\ u_2 \\ \vdots \\ u_n \end{pmatrix}, \ v = \begin{pmatrix} v_1 \\ v_2 \\ \vdots \\ v_n \end{pmatrix} \in \overline{F^n}$$

    and $c \in F$, then

$$u + v = \begin{pmatrix} u_1 + v_1 \\ u_2 + v_2 \\ \vdots \\ u_n + v_n \end{pmatrix} \ \text{and} \ c \begin{pmatrix} u_1 \\ u_2 \\ \vdots \\ u_n \end{pmatrix} = \begin{pmatrix} cu_1 \\ cu_2 \\ \vdots \\ cu_n \end{pmatrix}.$$

    The vectors in $\overline{F^n}$ are called **column vectors**.

6. If n is a positive integer and F is a field, then the set $M_{m \times n}(F)$ of all $m \times n$ matrices with entries from F is an additive abelian group under matrix addition. If scalar multiplication is defined by

$$cA = c \begin{pmatrix} a_{11} & a_{12} & \cdots & a_{1n} \\ a_{21} & a_{22} & \cdots & a_{2n} \\ \vdots & \vdots & \vdots & \vdots \\ a_{m1} & a_{m2} & a_{m3} & a_{mn} \end{pmatrix} = \begin{pmatrix} ca_{11} & ca_{12} & \cdots & ca_{1n} \\ ca_{21} & ca_{22} & \cdots & ca_{2n} \\ \vdots & \vdots & \vdots & \vdots \\ ca_{m1} & ca_{m2} & \cdots & ca_{mn} \end{pmatrix},$$

then $M_{m \times n}(F)$ is a vector space over F. If $m = n = 2$ and $F = \mathbf{R}$, this yields Example 2 of 10.1.1. In the present example, the vectors are the $m \times n$ matrices, and the scalars are the field elements.

7. If E/F is a field extension, then F is a subfield of E. The field E can be viewed as a vector space over F. The additive abelian group of E is the additive abelian group for the vector space, and if $x \in E$, then x is thought of as a vector, and $c \in F$ is a scalar in F. The product cx in E can be viewed as the scalar multiplication. In particular, F itself is a vector space over F. (Do you see why?)

8. If F is a field, let $F[[x]]$ denote the set of all formal power series in x with their coefficients in F. Elements of $F[[x]]$ look like

$$f(x) = a_0 + a_1 x + a_2 x^2 + \cdots + a_n x^n + a_{n+1} x^{n+1} + \cdots.$$

If elements of $F[[x]]$ are added by adding corresponding coefficients, then $F[[x]]$ is an additive abelian group. Moreover, if scalar multiplication is defined by

$$cf(x) = ca_0 + ca_1 x + ca_2 x^2 + \cdots + ca_n x^n + ca_{n+1} x^{n+1} + \cdots$$

then $F[[x]]$ is a vector space over F. The vectors are the formal power series, and the scalars are the field elements.

9. We have seen in Example 1 of 3.1.9 that $q(\mathbf{R})$, the set of all real quaternions, is a division ring. This is also a vector space over $\mathbf{R}$ if addition of quaternions is defined in the usual way and if scalar multiplication is defined by $r(a + bi + cj + dk) = ra + rbi + rcj + rdk$. The set of all rational quaternions $q(\mathbf{Q})$ is also a vector space over $\mathbf{Q}$ if addition and scalar multiplication are defined in a similar fashion.

Recall that in Section 8.1 the concept of a basis of a field extension was developed. A similar development can be carried out for vector spaces.

10.1.4   Definition

If V is a vector space over a field F and S is a set of vectors of V, then an expression of the form

$$c_1 v_1 + c_2 v_2 + \cdots + c_n v_n,$$

where $v_1, v_2, \ldots, v_n \in S$ and $c_1, c_2, \ldots, c_n \in F$, is said to be a (finite) **linear combination** of the elements of S. The c_i's are referred to as the **coefficients** of the v_i's. If every element of V can be expressed as a linear combination of the elements in S, then S is said to **span** V. (A set S of vectors that spans V can contain an "infinite number" of vectors. The condition that must hold is that every vector in V can be expressed as a linear combination of a *finite* number of vectors from S. The vectors in S that are used in such expressions can vary as the vector v varies.) If S has the property that for any $v_1, v_2, \ldots, v_n \in S$,

$$c_1 v_1 + c_2 v_2 + \cdots + c_n v_n = 0$$

implies that $c_1 = c_2 = \cdots = c_n = 0$, then S is said to be a **linearly independent** set of vectors of V. To simplify language, we often simply say that S is linearly independent. If a linear combination of elements of S exists such that $c_1 v_1 + c_2 v_2 + \cdots + c_n v_n = 0$ with at least one of the c_i's $\neq 0$, then S is said to be **linearly dependent**. A linearly independent set of vectors of V that spans V is a **basis** for V.

It can be shown that *every vector space has a basis*, but a complete proof of this result is beyond the scope of this text. At this point, we assume that every vector space has a basis. Later in this section, a partial development of this result, omitting the proof of one key theorem, is given.

There are vector spaces that have finite bases, and vector spaces that do not. Our attention is confined primarily to vector spaces that have finite bases. However, examples of vector spaces that do not have a finite basis are given. The proof of the following theorem is an exercise.

10.1.5   Theorem

Let V be a vector space over a field F, and suppose that $S = \{v_1, v_2, \ldots, v_n\}$ is a set of vectors in V that spans V. Then S is a basis for V if and only if in each expression $v = c_1 v_1 + c_2 v_2 + \cdots + c_n v_n$ for the vectors $v \in V$, the coefficients are unique.

Example 4 of 8.1.3 shows that a vector space can have more than one basis. An important observation that should be made from Example 4 is that given a basis B of a vector space over a field F, then for any $v \in V$, v can be expressed as a linear combination of the elements of B with the coefficients taken from F. Moreover, these coefficients are unique, but only relative to B. The point is that when the basis is changed and v is expressed as a linear combination of the "new" basis elements, the coefficients may change. Hence, uniqueness of coefficients holds only relative to a given basis.

At this point it might be a good idea to reread the material in Section 8.1 on bases of field extensions. The proofs presented there of Theorem 8.2.1, Lemma 8.1.4, and Theorem 8.1.7 should give you an idea of how one can go about proving each of the following two theorems. (If E/F is field extension, then E is a vector space over F, so just replace E with V.) The first of these two theorems points out two ways that a set of vectors can fail to be linearly independent.

10.1.6   Theorem

Let V be a vector space over a field F, and suppose that $S = \{v_1, v_1, \ldots, v_n\}$ is a nonempty subset of V.

1. If $0 \in S$, then S is linearly dependent.
2. If the elements of S are not distinct, then S is linearly dependent.

As was pointed out earlier, Example 4 of 8.1.3 shows that a vector space can have more than one basis. Examples 1 and 3 of 10.1.9 provide additional verification of this fact. The following important theorem shows, however, that the number of elements in a basis of a vector space is unique.

10.1.7   Theorem

If vector space V over a field F has a basis with n elements, then every basis of V has n elements.

10.1.8   Definition

If a vector space V over a field F has a basis B with a finite number of elements, then V is said to be a **finite dimensional vector space**, otherwise V is said to be an **infinite dimensional vector space**. If V is a finite dimensional vector space over a field F with basis B, then the unique number of elements of B, denoted by $\dim_F V$, is said to be the **dimension** of V.

10.1.9   Examples

1. The field of complex numbers $\mathbf{C}$ is a vector space over $\mathbf{R}$. A basis for this vector space is the set of vectors $B = \{1, i\}$. Thus, $\dim_{\mathbf{R}} \mathbf{C} = 2$, and

every basis of this vector space has two elements. A basis for vector space is *not* unique. For example, $B' = \{1, 2 - i\}$ is also a basis for **C** as a vector space over **R**.

2. If F is a field, then $B = \{e, x, x^2, \ldots, x^n, \ldots\}$ is a basis for the vector space $F[x]$. If $f(x) \in F[x]$, then $f(x)$ is a linear combination of a finite number of elements of B. This is *not* a finite dimensional vector space. This basis is referred to as the **standard basis** for $F[x]$.

3. If F is a field and n is a nonnegative integer, then $B = \{e, x, x^2, \ldots, x^n\}$ is the **standard basis** for the vector space $P_n(F)$, the vector space of all polynomials of degree $\leq n$ together with the zero polynomial. In this case, $\dim_F P_n(F) = n + 1$. For example, $B = \{1, x, x^2\}$ is the standard basis for $P_2(\mathbf{R})$. $B' = \{1, 1 - x, x + x^2\}$ is also a basis for $P_2(\mathbf{R})$. This is another demonstration that the basis of vector space is *not* unique, but observe that each basis B and B' of $P_2(\mathbf{R})$ has three elements.

4. If n is a nonnegative integer and F is a field with identity e, then F^n is a vector space over F. If

$$e_1 = (e, 0, 0, \ldots, 0),$$

$$e_2 = (0, e, 0, \ldots, 0),$$

$$e_3 = (0, 0, e, \ldots, 0), \qquad \text{and}$$

$$\vdots$$

$$e_n = (0, 0, 0, \ldots, e),$$

then $B = \{e_1, e_2, e_3, \ldots, e_n\}$ is a basis for F^n. Note that $\dim_F F^n = n$. The set of vectors B is the **standard basis** for F^n. If each e_i is transposed to a column vector, the result is the standard basis for the vector space $\overline{F^n}$. (See Example 5 of 10.1.3.)

5. The set of 2×2 matrices $M_2(\mathbf{Q})$ is a vector space over $\mathbf{Q}$. Let

$$E_{11} = \begin{pmatrix} 1 & 0 \\ 0 & 0 \end{pmatrix}, E_{12} = \begin{pmatrix} 0 & 1 \\ 0 & 0 \end{pmatrix}, E_{21} = \begin{pmatrix} 0 & 0 \\ 1 & 0 \end{pmatrix}, \text{ and } E_{22} = \begin{pmatrix} 0 & 0 \\ 0 & 1 \end{pmatrix}.$$

If $\begin{pmatrix} a & b \\ c & d \end{pmatrix} \in M_2(\mathbf{Q})$, then $\begin{pmatrix} a & b \\ c & d \end{pmatrix} = aE_{11} + bE_{12} + cE_{21} + dE_{22}$. Hence, the set of vectors $B = \{E_{11}, E_{12}, E_{21}, E_{22}\}$ spans $M_2(\mathbf{Q})$. If

$$aE_{11} + bE_{12} + cE_{21} + dE_{22} = \begin{pmatrix} 0 & 0 \\ 0 & 0 \end{pmatrix},$$

then

$$a\begin{pmatrix} 1 & 0 \\ 0 & 0 \end{pmatrix} + b\begin{pmatrix} 0 & 1 \\ 0 & 0 \end{pmatrix} + c\begin{pmatrix} 0 & 0 \\ 1 & 0 \end{pmatrix} + d\begin{pmatrix} 0 & 0 \\ 0 & 1 \end{pmatrix} = \begin{pmatrix} 0 & 0 \\ 0 & 0 \end{pmatrix}$$

gives

$$\begin{pmatrix} a & 0 \\ 0 & 0 \end{pmatrix} + \begin{pmatrix} 0 & b \\ 0 & 0 \end{pmatrix} + \begin{pmatrix} 0 & 0 \\ c & 0 \end{pmatrix} + \begin{pmatrix} 0 & 0 \\ 0 & d \end{pmatrix} = \begin{pmatrix} 0 & 0 \\ 0 & 0 \end{pmatrix},$$

so

$$\begin{pmatrix} a & b \\ c & d \end{pmatrix} = \begin{pmatrix} 0 & 0 \\ 0 & 0 \end{pmatrix}.$$

Hence, $a = b = c = d = 0$, which shows that B is linearly independent. Therefore, B is a basis for $M_2(\mathbf{Q})$. Note that $\dim_{\mathbf{Q}} M_2(\mathbf{Q}) = 4$.

In general, if $M_{m \times n}(F)$ is the set of $m \times n$ matrices over a field F, then $M_{m \times n}(F)$ is a vector space over F. If we let E_{ij} be the $m \times n$ matrix with e in the (i, j)th position and with 0 in the other positions, it follows that

$$(a_{ij}) = a_{11}E_{11} + \cdots + a_{1n}E_{1n} + a_{21}E_{21} + \cdots + a_{2n}E_{2n}$$

$$+ a_{m1}E_{m1} + \cdots + a_{mn}E_{mn}$$

for any matrix $(a_{ij}) \in M_{m \times n}(F)$. Hence, we see that the set B of vectors $\{E_{ij}\}$ spans $M_{m \times n}(F)$. It follows also that B is linearly independent, so B is a basis for $M_{m \times n}(F)$. For this example, $\dim_F M_{m \times n}(F) = mn$, and B is the **standard basis** for this vector space.

It was mentioned earlier that every vector space has a basis, but it may be difficult to display a concrete basis of a given vector space. This is the case with the two following examples.

6. Consider the field extension $\mathbf{R}/\mathbf{Q}$. $\mathbf{R}$ is a vector space over $\mathbf{Q}$, but a specific basis for this vector space is yet to be discovered.
7. The set of formal power series $F[[x]]$ with their coefficients in a field F is a vector space over F. This is another example of a vector space for which no one has been able to find a specific basis, even though we know that such a basis exists. Every element of $F[[x]]$ looks like

$$f(x) = a_0 + a_1 x + a_2 x^2 + \cdots + a_n x^n + a_{n+1}x^{n+1} + \cdots$$

but do not be misled to think that $B = \{e, x, x^2, \ldots, x^n, \ldots\}$ is a basis for this space. The formal power series $f(x)$ is not a linear combination of a finite number of elements from B.

If V is a vector space over a field F, then a nonempty subset U of V is a **subspace** of V if U is a vector space when the operations of addition and scalar multiplication are restricted to U.

10.1.10   Theorem

Let U be a nonempty subset of a vector space V over a field F. Then U is a subspace of V if and only if U is closed under addition and scalar multiplication.

Proof. If U is a subspace of V, then U is closed under addition and scalar multiplication. This follows since U is a vector space under the same operations of addition and scalar multiplication defined on V. Thus, these are binary operations on U and consequently take their values in U.

Conversely, suppose that U is closed under addition and scalar multiplication. If $u, u' \in U$, then $u - u' = u + (-e)u' \in U$, so it follows from Theorem 2.2.3 that U is a subgroup of the additive group of V. (You will be asked to show that $(-e)v = -v$ for all $v \in V$ in the exercises.) The scalar multiplication properties of a vector space are inherited from V; that is, these properties hold in U since they hold in V. For example, if u and u' are vectors in U and $c \in F$ are such that $c(u + u') \neq cu + cu'$ in U, then $c(u + u') \neq cu + cu'$ in V since $U \subseteq V$. But this is a clear contradiction, so it must be the case that $c(u + u') = cu + cu'$. Likewise for the other scalar multiplication properties. ∎

10.1.11   Examples

1. Every vector space V over a field F has at least two subspaces, namely, the zero subspace $\{0\}$ and V itself.
2. With the notation established in Examples 2, 3, and 8 of 10.1.3, we see that $P_n(F)$ is a subspace of $F[x]$. Both $P_n(F)$ and $F[x]$ are subspaces of $F[[x]]$.
3. Each of the following is a subspace of the vector space $M_2(\mathbf{R})$.

$$U_1 = \left\{ \begin{pmatrix} a & 0 \\ 0 & 0 \end{pmatrix} \middle| a \in \mathbf{R} \right\}$$

$$U_2 = \left\{ \begin{pmatrix} a & 0 \\ 0 & b \end{pmatrix} \middle| a, b \in \mathbf{R} \right\}$$

$$U_3 = \left\{ \begin{pmatrix} a & a+b \\ a+b & b \end{pmatrix} \middle| a, b \in \mathbf{R} \right\}$$

The subspaces of $M_2(\mathbf{R})$ are numerous. Can you list three others? Is

$$U = \left\{ \begin{pmatrix} a & b \\ c & d \end{pmatrix} \middle| \; a, b, c, d \in \mathbf{Q} \right\} \text{ a subspace of } M_2(\mathbf{R})?$$

The proof of the following theorem is straightforward, so it is left as an exercise. (See Theorem 2.2.5.)

10.1.12   Theorem

If V is a vector space over a field F, and $\{U_\alpha\}_{\alpha \in \Delta}$ is a family of subspaces of V, then $\cap_{\alpha \in \Delta} U_\alpha$ is a subspace of V.

10.1.13   Definition

Let S be a possibly empty set of vectors in a vector space V over a field F. If $\{U_\alpha\}_{\alpha \in \Delta}$ is the set of all subspaces of V that contain S, then the subspace $\langle S \rangle = \cap_{\alpha \in \Delta} U_\alpha$ of V is said to be the **subspace of V generated by** S. If $\langle S \rangle = V$, then S is said to be a **set of generators** of V.

When S is a set of vectors of V, the family $\{U_\alpha\}_{\alpha \in \Delta}$ of Definition 10.1.13 is nonempty since V is a member of $\{U_\alpha\}_{\alpha \in \Delta}$. Note also that if $S = \varnothing$, then every subspace of V contains S, even the zero subspace $\{0\}$. Since $\{0\}$ is contained in every subspace of V, it follows that $\langle \varnothing \rangle = \{0\}$. Thus, $\varnothing$ is a set of generators for the vector space $\{0\}$ even though $\varnothing$ contains no elements. The subspace $\langle S \rangle$ can be described much more precisely. The following lemma shows that if $S \neq \varnothing$, then $\langle S \rangle$ is just the set of all (finite) linear combinations of elements of S.

10.1.14   Lemma

If V is a vector space over a field F and S is a nonempty set of vectors in V, then $\langle S \rangle$ is the set of all linear combinations of elements of S.

Proof. If $\{U_\alpha\}_{\alpha \in \Delta}$ is the family of subspaces of V that contain S, then $\langle S \rangle = \cap_{\alpha \in \Delta} U_\alpha$. Let LC be the set of all linear combinations of elements of S. If $u = c_1 u_1 + c_2 u_2 + \cdots + c_m u_m$ and $u' = c_1' u_1' + c_2' u_2' + \cdots + c_n' u_n'$ are elements of LC, then

$$u - u' = c_1 u_1 + c_2 u_2 + \cdots + c_m u_m + (-c_1') u_1' + (-c_2') u_2' + \cdots + (-c_n') u_n'$$

is a linear combination of elements of S, so $u - u' \in$ LC. It follows from Theorem 2.2.3 that LC is a subgroup of V. If $c \in F$, then

$$cu = c(c_1 u_1 + c_2 u_2 + \cdots + c_m u_m) = (cc_1)u_1 + (cc_2)u_2 + \cdots + (cc_m)u_m$$

shows that LC is closed under scalar multiplication. Thus, LC is a subspace of V. Since $u = eu \in$ LC for each $u \in S$, we see that $S \subseteq$ LC. Hence, LC is a subspace of V that contains S, so LC is a member of the family $\{U_\alpha\}_{\alpha \in \Delta}$. Therefore, $\langle S \rangle = \cap_{\alpha \in \Delta} U_\alpha \subseteq$ LC.

Conversely, suppose that $u = c_1 u_1 + c_2 u_2 + \cdots + c_m u_m \in$ LC, where $u_i \in S$ for $i = 1, 2, \ldots, m$. But $S \subseteq U_\alpha$ for each $\alpha \in \Delta$, so since each U_α is closed under addition and scalar multiplication, it follows that $u \in U_\alpha$ for each $\alpha \in \Delta$. Hence, $u \in \cap_{\alpha \in \Delta} U_\alpha = \langle S \rangle$, so we have LC $\subseteq \langle S \rangle$. Consequently, $\langle S \rangle =$ LC. ∎

Let V be a vector space over a field F, and S a nonempty set of vectors of V. If S is linearly independent, then S is said to be a **maximal set of linearly independent vectors** in V if whenever S' is a set of linearly independent set of vectors of V such that $S \subseteq S'$, then $S = S'$. If $\langle S \rangle = V$, then S is said to be a **minimal set of generators** of V if whenever S' is a set of vectors in V such that $S' \subseteq S$ and $\langle S' \rangle = V$, then $S = S'$. The following theorem is often useful when finding a basis for a vector space. The vector space in the theorem is not required to be finite dimensional.

10.1.15   Theorem

If V is a vector space over a field F and S is a nonempty set of vectors of V, then the following are equivalent.

1. S is a basis for V.
2. S is a maximal set of linearly independent vectors in V.
3. S is a minimal set of generators of V.

Proof

$1 \Rightarrow 2$. Suppose that S is a basis of V and that S is not a maximal set of linearly independent vectors in V. Then there is a set S' of linearly independent vectors of V such that $S \subset S'$. If $v \in S' \setminus S$, then v can be expressed as a linear combination of elements of S. If $v_1, v_2, \ldots, v_n \in S$ and $c_1, c_2, \ldots, c_n \in F$ are such that $v = c_1 v_1 + c_2 v_2 + \cdots + c_n v_n$, then $c_1 v_1 + c_2 v_2 + \cdots + c_n v_n + (-e)v = 0$ is a linear combination of elements of S', and all of the coefficients are not zero. Hence, S' is not linearly independent, and this is a contradiction. Thus, if S is a basis for V, then S must be a maximal linearly independent set of vectors in V.

$2 \Rightarrow 3$. Suppose that S is a maximal set of linearly independent vectors in V but that S is not a minimal set of generators of V. Then there exists a set S' of vectors of V such that $S' \subset S$ and $\langle S' \rangle = V$. If $v \in S \setminus S'$, then $v \in V = \langle S' \rangle$, so, by Lemma 10.1.14, v can be written as a linear combination of elements of S'. If $v_1, v_2, \ldots, v_n \in S'$ and $c_1, c_2, \ldots, c_n \in F$ are such that $v = c_1 v_1 + c_2 v_2 + \cdots + c_n v_n$, then $c_1 v_1 + c_2 v_2 + \cdots + c_n v_n + (-e)v = 0$ is a linear combination of elements of S, and all of the coefficients are not zero. This contradicts the fact that the set S is linearly independent. Hence, if S is a maximal set of linearly independent vectors of V, then S must be a minimal set of generators of V.

$3 \Rightarrow 1$. Suppose that S is a minimal set of generators of V. Then $\langle S \rangle = V$, so S spans V. Therefore, it remains only to show that S is linearly independent. If S is not linearly independent, then there are vectors $v_1, v_2, \ldots, v_n \in S$ and $c_1, c_2, \ldots, c_n \in F$ such that $c_1 v_1 + c_2 v_2 + \cdots + c_n v_n = 0$, and at least one of the coefficients is nonzero. Suppose that our notation has been chosen so that $c_1 \neq 0$. Then $v_1 = c_1^{-1}(-c_2)v_2 + \cdots + c_1^{-1}(-c_n)v_n$, so in any linear combination of elements of S that has a summand containing v_1, v_1 can be replaced by $c_1^{-1}(-c_2)v_2 + \cdots + c_1^{-1}(-c_n)v_n$. This observation shows that $\langle S \setminus \{v_1\} \rangle = V$, so S is not a minimal set of generators for V. Therefore, any minimal set of generators of V must be linearly independent, so such a set of generators of V is a basis of V. ∎

We mentioned earlier that a proof would not be given for the fact that every vector space has a basis. The key to proving this fact is the following theorem, which cannot be proven without developing advanced topics. A corollary to this theorem is the fact that every vector space has a basis. We can prove the corollary.

10.1.16   Theorem

If V is a vector space over a field F and S is a set of linearly independent vectors of V, then there is a basis B of V such that $S \subseteq B$.

10.1.17   Corollary

If V is a vector space over a field F, then V has a basis.

Proof. If V is the zero vector space $\{0\}$, then we have seen that $\langle \emptyset \rangle = \{0\}$, so $\emptyset$ spans V. Note that $\emptyset$ is a linearly independent set of vectors in V. Why? Because if $\emptyset$ is linearly dependent, then for some positive integer n, vectors $v_1, v_2, \ldots, v_n$ in $\emptyset$ and scalars $c_1, c_2, \ldots, c_n$, not all of which are zero, would have to exist such that $c_1v_1 + c_2v_2 + \cdots + c_nv_n = 0$. Since this clearly cannot happen, $\varnothing$ is a set of linearly independent vectors in V. Thus, $\varnothing$ is a basis for the vector space $\{0\}$. If $V \neq \{0\}$, let $v \in V$, $v \neq 0$. Then the set $\{v\}$ is a linearly independent set of vectors in V. Thus, by Theorem 10.1.16, there is a basis B for V such that $\{v\} \subseteq B$, and this completes the proof. ■

HISTORICAL NOTES

The beginnings of the concept of a vector space can be traced back to **Hermann Grassmann** (1809–1877) and **Giuseppe Peano** (1858–1932). Grassmann, who was born in what is now Szczecin, Poland, was a brilliant mathematician whose work centered around expressing geometric ideas symbolically. The theme of Grassmann's work can be generalized to spaces of dimension higher than 3, and today his work, along with the contributions of other mathematicians, is known as the theory of exterior algebras. The idea of dimension and subspace appear in his work, and he proved the well-known formula

$$\dim(U_1 + U_2) = \dim U_1 + \dim U_2 - \dim(U_1 \cap U_2),$$

where U_1 and U_2 are subspaces of a vector space V. Unfortunately, the importance of Grassmann's work was not recognized during his lifetime. At the age of 53, Grassmann became disillusioned by the lack of interest in his work, and he turned to Sanskrit studies. (Sanskrit is the classical literary language of India.) He wrote a Sanskrit dictionary that is still in use today.

The first mathematician actually to state the abstract definition of a vector space was Giuseppe Peano, who was born in Cuneo, Italy. Peano's work was along the same lines as that of Grassmann, and he gave the definition of what he called a linear system. Peano's linear system was very close to what we today call a vector space. In this system, Peano defined addition and scalar multiplication of vectors. Peano's work, like that of Grassmann, went unnoticed by the mathematical community, and it wasn't until around 1918 that his definition of a vector space became a part of standard mathematical terminology. Peano is best known for his set of axioms describing the set of positive integers, which are referred to as Peano's axioms in his honor.

Problem Set 10.1

1. If V is a vector space over a field F with identity e, prove each of the following.

    (a) $0v = 0$ for all $v \in V$. In the expression $0v$, 0 is the additive identity of F, and the 0 on the right in the equation $0v = 0$ is the zero vector.
    (b) $(-e)v = -v$ for all $v \in V$.

2. Prove Theorem 10.1.5.

3. Which of the following subsets of $\mathbf{R}^4$ are subspaces of $\mathbf{R}^4$, where we are considering $\mathbf{R}^4$ as a vector space over $\mathbf{R}$?

   (a) $U_1 = \{(a_1, a_2, a_3, a_4) \mid a_2 \geq 0\}$
   (b) $U_2 = \{(a_1, a_2, a_3, a_4) \mid a_1 + a_2 = a_3\}$
   (c) $U_3 = \{(a_1, a_2, a_3, a_4) \mid a_2 a_4 = 0\}$

4. Determine which, if any, of the following are a subspace of the vector space $F[x]$, where we are considering $F[x]$ as a vector space over F.

   (a) $U_1 = \{f(x) \in F[x] \mid f(x) = f(-x)\}$
   (b) $U_2 = \{f(x) \in F[x] \mid f(-x) = -f(x)\}$
   (c) $U_3 = \{f(x) \in F[x] \mid f(0) = f(1) = 0\}$

5. Show that the set of vectors $S = \{(2, 1, 0), (-1, 0, 1)\}$ is linearly independent in $\mathbf{R}^3$, where we are considering $\mathbf{R}^3$ to be a vector space over $\mathbf{R}$. Find an additional vector (a_1, a_2, a_3) in $\mathbf{R}^3$ such that $B = \{(2, 1, 0), (-1, 0, 1), (a_1, a_2, a_3)\}$ is a basis of $\mathbf{R}^3$.

6. If we consider $\mathbf{R}^3$ as a vector space over $\mathbf{R}$, prove that $U = \{(x, y, z) \mid x - 3y + z = 0\}$ is a subspace of $\mathbf{R}^3$, and find a basis for U.

7. If $\{u, v, w\}$ is a linearly independent set of vectors in a vector space V over a field F, decide if $\{u + v, v + w, u + w\}$ is linearly independent.

8. If v is a nonzero vector in a vector space V over a field F, prove that $\{v\}$ is a linearly independent set of vectors.

9. If V is a finite dimensional vector space over a field F and $\dim_F V = n$, explain why any set of vectors of V containing more than n elements must be linearly dependent.

10. If V is a vector space over a field F and $\{U_\alpha\}_{\alpha \in \Delta}$ is a family of subspaces of V, prove that $\cap_{\alpha \in \Delta} U_\alpha$ is a subspace of V.

11. Prove that the set U of all solutions (x, y, z) of the system of equations

$$x + y - z = 0$$
$$3x - y - z = 0$$
$$x - 5y + z = 0$$

is a subspace of $\mathbf{R}^3$. Find a basis for U.

12. Prove that the set of vectors $B = \{e_1, e_2, e_3, \ldots, e_n\}$ of Example 4 of 10.1.9 is a basis of the vector space F^n.

13. (a) Prove that the set of vectors $B = \{E_{11}, E_{12}, E_{21}, E_{22}\}$ of Example 5 of 10.1.9 is a basis for the vector space $M_2(\mathbf{Q})$.

 (b) Prove that the set of vectors $B = \{E_{ij}\}$ of Example 5 of 10.1.9 is a basis for the vector space $M_{m \times n}(F)$.

14. Prove Theorem 10.1.6.

15. Suppose that V is a vector space over F and that S is a finite subset of V such that $\langle S \rangle = V$. Prove that V is finite dimensional over F. Hint: If S is not a minimal set of generators of V, then there is a proper subset S_1 of S such that $\langle S_1 \rangle = V$. If S_1 is a minimal set of generators of V, then S_1 is a basis for V. If S_1 is not a minimal set of generators of V, then there is a proper subset S_2 of S_1 such that $\langle S_2 \rangle = V$. Continuing in this way, a descending chain $S \supset S_1 \supset S_2 \supset \cdots$ of sets of generators of V can be constructed. Argue that there must exist an integer k such that $S_k = S_{k+1} = S_{k+2} = \cdots$ and consequently that S_k is a basis for V. This is an argument that every finitely generated vector space has a basis.

16. Let V be a vector space over a field F, and suppose that U_1 and U_2 are subspaces of V. Prove that $U_1 + U_2 = \{u_1 + u_2 \mid u_1 \in U_1, u_1 \in U_2\}$ is a subspace of V.

17. If U_1 and U_2 are subspaces of a vector space V over a field F such that $V = U_1 + U_2$ and $U_1 \cap U_2 = \{0\}$, then we say that V is the **direct sum** of U_1 and U_2. This is denoted by $V = U_1 \oplus U_2$. (See Exercise 16.)

 (a) If $V = U_1 \oplus U_2$, prove that every vector $v \in V$ has a unique expression as $v = u_1 + u_2$, where $u_1 \in U_1$ and $u_2 \in U_2$.

 (b) If U_1 is a subspace of V, prove that there is a subspace U_2 of V such that $V = U_1 \oplus U_2$. The subspace U_2 is the **complement** of U_1 in V. Hint: The vector space U_1 has a basis B_1, so there is a basis B for V such that $B_1 \subseteq B$. Consider $B_2 = B \setminus B_1$.

18. Let $\{U_i\}_{i=1}^n$ be a family of subspaces of a vector space V such that
$V = U_1 + U_2 + \cdots + U_n$ and $U_i \cap (U_1 + \cdots + U_{i-1} + U_{i+1} + \cdots + U_n) = \{0\}$
for each i, where $U_1 + U_2 + \cdots + U_n = \{u_1 + u_2 + \cdots + u_n \mid u_i \in U_i$ for $i = 1, 2, \ldots, n\}$. Under these conditions, we say that V is the (internal) **direct sum** of the family of subspaces $\{U_i\}_{i=1}^n$. The notation $\oplus_{i=1}^n U_i = U_1 \oplus U_2 \oplus \cdots \oplus U_n$ indicates that V is the direct sum of the family of subspaces $\{U_i\}_{i=1}^n$.

 If V is a finite dimensional vector space, prove that V is the direct sum $\oplus_{i=1}^n U_i$ of a finite family $\{U_i\}_{i=1}^n$ of 1-dimensional subspaces and that each $v \in V$ has a unique expression as $v = u_1 + u_2 + \cdots + u_n$, where $u_i \in U_i$ for $i = 1, 2, \ldots, n$. Hint: Let $B = \{v_1, v_2, \ldots, v_n\}$ be a basis for V, and consider the subspaces $U_i = Fv_i = \{cv_i \mid c \in F\}$ for $i = 1, 2, \ldots, n$.

19. Consider the vector space $\mathbf{R}^4$ over $\mathbf{R}$. If $U_1 = \langle\{(1, 1, 0, 0), (0, 0, 1, 1)\}\rangle$ and $U_2 = \langle\{(0, 0, 1, 1), (1, 0, 0, 1)\}\rangle$, show that

$$\dim_{\mathbf{R}}(U_1 + U_2) = \dim_{\mathbf{R}} U_1 + \dim_{\mathbf{R}} U_2 - \dim_{\mathbf{R}}(U_1 \cap U_2).$$

20. If V is a finite dimensional vector space over a field F and U_1 and U_2 are subspaces of V, prove that

$$\dim_F(U_1 + U_2) = \dim_F U_1 + \dim_F U_2 - \dim_F(U_1 \cap U_2).$$

21. If V is a finite dimensional vector space and U is a subspace of V, prove that U must also be finite dimensional and $\dim_F U \le \dim_F V$.

TECHNOLOGY PROBLEMS

Use a computer algebra system to solve each of the following problems.

22. Determine whether the following set of vectors is a basis for $\mathbf{R}^7$.

$$S = \{(1, 1, 1, 0, 1, 1, 0), (0, 1, 1, 0, 0, 1, 0), (0, 0, 1, 1, 0, 1, 1),$$
$$(0, 0, 0, 0, 1, 1, 1), (1, 1, 1, 1, 1, 0, 0), (0, 0, 0, 0, 0, 1, 1),$$
$$(0, 0, 0, 1, 1, 0, 0)\}$$

23. Determine if

$$S = \{1, 1 - x, x + x^2, x + x^2 + x^3, x^4 + x^5, x^5\}$$

is a basis for the vector space $P_5(\mathbf{R})$.

24. Determine if the set of vectors

$$S = \{(1, 1, 0), (1, 1, 1), (0, 1, 0), (0, 1, 1), (1, 2, 3)\}$$

spans $\mathbf{R}^3$. Show that S is a linearly dependent set of vectors.

10.2   LINEAR TRANSFORMATIONS

Section Overview. In this section, the concept of a linear transformation from a vector space U to a vector space V is established, and the idea of vector space isomorphism is investigated. One important result of the section is that a finite dimensional vector space over a field F is isomorphic to F^n, and this has the effect of setting up a coordinate system for the vector space. It is also shown that when the vector spaces are finite dimensional, a matrix that corresponds to a linear transformation $\phi : U \to V$ exists. This matrix determines a linear transformation for the coordinate space of U to the coordinate space of V. Finally, commutative diagrams are introduced as a visual device for tracing the effect of the composition of linear transformations. Make sure you understand how to use these diagrams. Such diagrams are a valuable tool for keeping track of the effect of sequences of compositions of linear transformations.

We saw in Chapters 2 and 3 that it is sometimes possible to establish a one-to-one correspondence between the elements of two groups or two rings in such a way that the algebraic properties of the two groups or rings are transferred by the correspondence. This led to the concept of group and ring homomorphism. A similar situation holds for vector spaces. It is sometimes the case that a correspondence can be established between two vector spaces in such a way that algebraic properties of the vector spaces are transferred through the correspondence. We give two examples that serve as a model for the definition of a linear transformation.

10.2.1   Examples

1. Consider the vector spaces $P_2(\mathbf{R})$ and $\mathbf{R}^3$. If $a_0 + a_1 x + a_2 x^2$ is a vector in $P_2(\mathbf{R})$, then since $B = \{e, x, x^2\}$ is a basis of $P_2(\mathbf{R})$, the coefficients of $a_0 + a_1 x + a_2 x^2$ are unique, so a well-defined mapping from $P_2(\mathbf{R})$ to $\mathbf{R}^3$ is given by $a_0 + a_1 x + a_2 x^2 \longmapsto (a_0, a_1, a_2)$. Conversely, any vector (a_0, a_1, a_2) in $\mathbf{R}^3$ uniquely determines a vector $a_0 + a_1 x + a_2 x^2$ in $P_2(\mathbf{R})$, so we have a well-defined mapping $(a_0, a_1, a_2) \longmapsto a_0 + a_1 x + a_2 x^2$. These mappings are clearly inverses of each other, so Theorem 0.3.7 indicates that each of these mappings is a bijection. Because of this, we have a one-to-one correspondence between the vectors of $P_2(\mathbf{R})$ and the vectors of $\mathbf{R}^3$. If $f(x) = a_0 + a_1 x + a_2 x^2$, $g(x) = b_0 + b_1 x + b_2 x^2$, and $c \in \mathbf{R}$, it follows that the sums and scalar products of corresponding vectors correspond.

$$
\begin{array}{ccl}
P_2(\mathbf{R}) & \longleftrightarrow & \mathbf{R}^3 \\
f(x) & \longleftrightarrow & (a_0, a_1, a_2) \\
g(x) & \longleftrightarrow & (b_0, b_1, b_2) \\
f(x) + g(x) & \longleftrightarrow & (a_0 + b_0, a_1 + b_1, a_2 + b_2) \\
cf(x) & \longleftrightarrow & (ca_0, ca_1, ca_2)
\end{array}
$$

If the mapping $a_0 + a_1x + a_2x^2 \longmapsto (a_0, a_1, a_2)$ is denoted by ϕ, then

$$\phi(f(x) + g(x)) = \phi((a_0 + a_1x + a_2x^2) + (b_0 + b_1x + b_2x^2))$$
$$= \phi((a_0 + b_0) + (a_1 + b_1)x + (a_2 + b_2)x^2)$$
$$= (a_0 + b_0, a_1 + b_1, a_2 + b_2)$$
$$= (a_0, a_1, a_2) + (b_0, b_1, b_2)$$
$$= \phi(f(x)) + \phi(g(x))$$

and

$$\phi(cf(x)) = \phi(c(a_0 + a_1x + a_2x^2))$$
$$= \phi((ca_0) + (ca_1)x + (ca_2)x^2)$$
$$= (ca_0, ca_1, ca_2)$$
$$= c(a_0, a_1, a_2)$$
$$= c\phi(f(x)).$$

More generally, if F is a field, let $f(x) = a_0 + a_1x + a_2x^2 + \cdots + a_nx^n$ be a polynomial in $P_n(F)$. Then the mapping $\phi : P_n(F) \to F^{n+1}$ defined by $\phi(f(x)) = (a_0, a_1, a_2, \ldots, a_n)$ is a bijective function with the properties

$$\phi(f(x) + g(x)) = \phi(f(x)) + \phi(g(x)) \qquad \text{and}$$
$$\phi(cf(x)) = c\phi(f(x))$$

for all $f(x), g(x) \in P_n(F)$ and $c \in F$.

2. Consider the vector space $M_2(\mathbf{R})$. If $A = \begin{pmatrix} a_{11} & a_{12} \\ a_{21} & a_{22} \end{pmatrix}$ is a vector in $M_2(\mathbf{R})$, then we have a mapping $\phi : M_2(\mathbf{R}) \to \mathbf{R}^4$ defined by $\phi\left(\begin{pmatrix} a_{11} & a_{12} \\ a_{21} & a_{22} \end{pmatrix} \right) = (a_{11}, a_{12}, a_{21}, a_{22})$ and a mapping $\psi : \mathbf{R}^4 \to M_2(\mathbf{R})$ given by $\psi((a, b, c, d)) = \begin{pmatrix} a & b \\ c & d \end{pmatrix}$. It follows that $\psi \circ \phi = 1_{M_2(\mathbf{R})}$ and $\phi \circ \psi = 1_{\mathbf{R}^4}$, so ϕ and ψ are bijective functions. Thus, there is a one-to-one correspondence between the vectors of $M_2(\mathbf{R})$ and those of $\mathbf{R}^4$. If $A = \begin{pmatrix} a_{11} & a_{12} \\ a_{21} & a_{22} \end{pmatrix}$ and $B = \begin{pmatrix} b_{11} & b_{12} \\ b_{21} & b_{22} \end{pmatrix}$ are vectors in $M_2(\mathbf{R})$ and $c \in \mathbf{R}$, then

$$\phi(A + B) = \phi\left(\begin{pmatrix} a_{11} & a_{12} \\ a_{21} & a_{22} \end{pmatrix} + \begin{pmatrix} b_{11} & b_{12} \\ b_{21} & b_{22} \end{pmatrix}\right)$$

$$= \phi\left(\begin{pmatrix} a_{11} + b_{11} & a_{12} + b_{12} \\ a_{21} + b_{21} & a_{22} + b_{22} \end{pmatrix}\right)$$

$$= (a_{11} + b_{11}, a_{12} + b_{12}, a_{21} + b_{21}, a_{22} + b_{22})$$

$$= (a_{11}, a_{12}, a_{21}, a_{22}) + (b_{11}, b_{12}, b_{21}, b_{22})$$

$$= \phi\left(\begin{pmatrix} a_{11} & a_{12} \\ a_{21} & a_{22} \end{pmatrix}\right) + \phi\left(\begin{pmatrix} b_{11} & b_{12} \\ b_{21} & b_{22} \end{pmatrix}\right)$$

$$= \phi(A) + \phi(B)$$

and

$$\phi(cA) = \phi\left(c\begin{pmatrix} a_{11} & a_{12} \\ a_{21} & a_{22} \end{pmatrix}\right)$$

$$= \phi\left(\begin{pmatrix} ca_{11} & ca_{12} \\ ca_{21} & ca_{22} \end{pmatrix}\right)$$

$$= (ca_{11}, ca_{12}, ca_{21}, ca_{22})$$

$$= c(a_{11}, a_{12}, a_{21}, a_{22})$$

$$= c\phi\left(\begin{pmatrix} a_{11} & a_{12} \\ a_{21} & a_{22} \end{pmatrix}\right)$$

$$= c\phi(A).$$

In a similar manner, it can be shown that if $A, B \in M_n(F)$ and $c \in F$, then $\phi(A + B) = \phi(A) + \phi(B)$ and $\phi(cA) = c\phi(A)$, where ϕ is given by

$$\begin{pmatrix} a_{11} & a_{12} & \cdots & a_{1n} \\ a_{21} & a_{22} & \cdots & a_{2n} \\ \vdots & \vdots & \vdots & \vdots \\ a_{n1} & a_{n2} & \cdots & a_{nn} \end{pmatrix} \longmapsto (a_{11}, \ldots, a_{1n}, \ldots, a_{n1}, \ldots, a_{nn})$$

The two examples of 10.2.1 provide a model for the following definition.

10.2.2   Definition

Let U and V be vector spaces over the same field F. A bijective function $\phi : U \to V$ such that

$$\phi(u_1 + u_2) = \phi(u_1) + \phi(u_2) \quad \text{and}$$

$$\phi(cu) = c\phi(u)$$

for all $u, u_1, u_2 \in U$ and $c \in F$ is said to be a **vector space isomorphism**. In this case, we say that U and V are **isomorphic vector spaces**. A mapping $\phi : U \to V$ such that $\phi(u_1 + u_2) = \phi(u_1) + \phi(u_2)$ and $\phi(cu) = c\phi(u)$ for all $u, u_1, u_2 \in U$ and all $c \in F$ is said to be a **vector space homomorphism** or a **linear transformation**. We say that a vector space homomorphism preserves addition and scalar multiplication. If a linear transformation is surjective, it is a **vector space epimorphism**, and if it is injective, then it is a **vector space monomorphism**. The **kernel K** of a linear transformation $\phi : U \to V$ is defined to be $K = \{u \in U \mid \phi(u) = 0\}$. The kernel of ϕ is often denoted by ker ϕ.

10.2.3   Examples

1. If $F[x]$ is the set of all polynomials with their coefficient in F, then $F[x]$ is a vector space over F. Suppose also that $D[f(x)]$ denotes the formal derivative of $f(x) \in F[x]$. (See Exercise 18, Problem Set 6.1.) If $f(x)$, $g(x) \in F[x]$ and $c \in F$, then $D : F[x] \to F[x]$ since the derivative of a polynomial in $F[x]$ is a polynomial in $F[x]$. Moreover,

$$D[f(x) + g(x)] = D[f(x)] + D[g(x)] \quad \text{and}$$
$$D[cf(x)] = cD[f(x)].$$

Thus, D is a linear transformation from $F[x]$ to $F[x]$. The kernel of D is the set of all constant polynomials in $F[x]$.

2. Let $V = \{f : [a, b] \to \mathbf{R} \mid f \text{ is a continuous function}\}$. Then V is a vector space over $\mathbf{R}$ under function addition and multiplication of a function by a real number. The field of real numbers $\mathbf{R}$ is a vector space over itself. Addition of vectors is the addition of real numbers, where the real numbers being added are viewed as the vectors of the space. Scalar multiplication is the multiplication ca of a real number by a real number. The first real number c in the product ca is considered to be a scalar, and the second real number a is viewed as a vector. It is not

difficult to show that all of the vector space properties are satisfied so that **R** is indeed a vector space over itself.

If we let $I(f) = \int_a^b f(x)\, dx$, then

$$I(f+g) = \int_a^b (f(x) + g(x))\, dx = \int_a^b f(x)\, dx + \int_a^b g(x)\, dx = I(f) + I(g)$$

and $\quad I(cf) = \int_a^b cf(x)\, dx = c \int_a^b f(x)\, dx = cI(f).$

Hence, I is a linear transformation from the vector space V to the vector space **R**. The kernel of I is the set of all functions in V such that

$$\int_b^a f(x)\, dx = 0.$$

3. Consider the usual vector spaces $\mathbf{R}^3$ and $\mathbf{R}^2$, and let $\phi : \mathbf{R}^3 \to \mathbf{R}^2$ be defined by $\phi((a_1, a_2, a_3)) = (a_1 + a_2, a_1 + a_3)$. We claim that ϕ is a linear transformation. If $a = (a_1, a_2, a_3)$, $b = (b_1, b_2, b_3) \in \mathbf{R}^3$ and $c \in \mathbf{R}$, then $a + b = (a_1, a_2, a_3) + (b_1, b_2, b_3) = (a_1 + b_1, a_2 + b_2, a_3 + b_3)$ and $c(a_1, a_2, a_3) = (ca_1, ca_2, ca_3)$. From this we see that

$$\phi(a + b) = (a_1 + b_1 + a_2 + b_2, a_1 + b_1 + a_3 + b_3)$$
$$= (a_1 + a_2, a_1 + a_3) + (b_1 + b_2, b_1 + b_3)$$
$$= \phi(a) + \phi(b)$$

and

$$\phi(ca) = (ca_1 + ca_2, ca_1 + ca_3)$$
$$= c(a_1 + a_2, a_1 + a_3)$$
$$= c\phi(a).$$

Hence, ϕ is a linear transformation from $\mathbf{R}^3$ to $\mathbf{R}^2$. The kernel of ϕ is the set of all (a_1, a_2, a_3) such that $(a_1 + a_2, a_1 + a_3) = (0, 0)$. Thus, $a_1 + a_2 = 0$ and $a_1 + a_3 = 0$, so $a_2 = -a_1$ and $a_3 = -a_1$. From this it follows that ker $\phi = \{c(1, -1, -1) \mid c \in \mathbf{R}\}$.

4. Consider the 2×3 matrix $\begin{pmatrix} c_{11} & c_{12} & c_{13} \\ c_{21} & c_{22} & c_{23} \end{pmatrix}$ over **R**, and define

$\phi : \mathbf{R}^2 \to \mathbf{R}^3$ by

$$\phi((a_1, a_2)) = (a_1, a_2)\begin{pmatrix} c_{11} & c_{12} & c_{13} \\ c_{21} & c_{22} & c_{23} \end{pmatrix}$$

$$= (a_1 c_{11} + a_2 c_{21}, a_1 c_{12} + a_2 c_{22}, a_1 c_{13} + a_2 c_{23})$$

Then ϕ is a linear transformation from the vector space $\mathbf{R}^2$ to the vector space $\mathbf{R}^3$. Indeed, if $a = (a_1, a_2)$ and $b = (b_1, b_2)$ are vectors in $\mathbf{R}^2$ and $c \in \mathbf{R}$, then $a + b = (a_1 + b_1, a_2 + b_2)$ and $ca = (ca_1, ca_2)$. Hence,

$$\phi(a + b) = (a_1 + b_1, a_2 + b_2)\begin{pmatrix} c_{11} & c_{12} & c_{13} \\ c_{21} & c_{22} & c_{23} \end{pmatrix}$$

$$= ((a_1 + b_1)c_{11} + (a_2 + b_2)c_{21},$$
$$(a_1 + b_1)c_{12} + (a_2 + b_2)c_{22}, (a_1 + b_1)c_{13} + (a_2 + b_2)c_{23})$$

$$= (a_1 c_{11} + a_2 c_{21}, a_1 c_{12} + a_2 c_{22}, a_1 c_{13} + a_2 c_{23}) +$$
$$(b_1 c_{11} + b_2 c_{21}, b_1 c_{12} + b_2 c_{22}, b_1 c_{13} + b_2 c_{23})$$

$$= (a_1, a_2)\begin{pmatrix} c_{11} & c_{12} & c_{13} \\ c_{21} & c_{22} & c_{23} \end{pmatrix} + (b_1, b_2)\begin{pmatrix} c_{11} & c_{12} & c_{13} \\ c_{21} & c_{22} & c_{23} \end{pmatrix}$$

$$= \phi(a) + \phi(b)$$

and

$$\phi(ca) = (ca_1, ca_2)\begin{pmatrix} c_{11} & c_{12} & c_{13} \\ c_{21} & c_{22} & c_{23} \end{pmatrix}$$

$$= ((ca_1)c_{11} + (ca_2)c_{21}, (ca_1)c_{12} + (ca_2)c_{22}, (ca_1)c_{13} + (ca_2)c_{23})$$
$$= c(a_1 c_{11} + a_2 c_{21}, a_1 c_{12} + a_2 c_{22}, a_1 c_{13} + a_2 c_{23})$$
$$= c\phi(a).$$

More generally, if A is a matrix in $M_{m \times n}(\mathbf{R})$, then $\phi : \mathbf{R}^m \to \mathbf{R}^n$ defined by $\phi(a) = aA$ is a linear transformation, where $a = (a_1, a_2, \ldots, a_m) \in \mathbf{R}^m$. This is also true if $\mathbf{R}$ is replaced by an arbitrary field F.

5. If $A = \begin{pmatrix} c_{11} & c_{12} & c_{13} \\ c_{21} & c_{22} & c_{23} \end{pmatrix}$ is a matrix in $M_{2 \times 3}(\mathbf{R})$, then

$$\phi\left(\begin{pmatrix} a_1 \\ a_2 \\ a_3 \end{pmatrix}\right) = \begin{pmatrix} c_{11} & c_{12} & c_{13} \\ c_{21} & c_{22} & c_{23} \end{pmatrix}\begin{pmatrix} a_1 \\ a_2 \\ a_3 \end{pmatrix}$$

defines a linear transformation from $\overline{\mathbf{R}}^3$ to $\overline{\mathbf{R}}^2$, where $\overline{\mathbf{R}}^n$, $n \geq 2$, is the vector space of column vectors given in Example 5 of 10.1.3.

You should make an important observation from this example and Example 4. When we are dealing with row vectors, they are written on the left of the matrix that determines the linear transformation, and when we are dealing with column vectors, they are written on the right of the matrix.

Now let A be an $n \times n$ square matrix. Then A can be used to define a linear transformation from $\mathbf{R}^n$ to $\mathbf{R}^n$. If the transpose of A is used to define a linear mapping from $\overline{\mathbf{R}}^n$ to $\overline{\mathbf{R}}^n$, then these two mappings are, up to transpose, the same mapping. For example, consider the matrix

$$A = \begin{pmatrix} c_{11} & c_{12} \\ c_{21} & c_{22} \end{pmatrix} \in M_2(\mathbf{R}), \text{ and let } \phi : \mathbf{R}^2 \to \mathbf{R}^2 \text{ be such that}$$

$\phi(a) = aA$ for all $a \in \mathbf{R}^2$. Then

$$\phi((a_1, a_2)) = (a_1, a_2)\begin{pmatrix} c_{11} & c_{12} \\ c_{21} & c_{22} \end{pmatrix} = (a_1 c_{11} + a_2 c_{21}, a_1 c_{12} + a_2 c_{22}). \quad (1)$$

The linear transformation $\overline{\phi} : \overline{\mathbf{R}}^2 \to \overline{\mathbf{R}}^2$ determined by A^t is given by

$$\overline{\phi}\left(\begin{pmatrix} a_1 \\ a_2 \end{pmatrix}\right) = \begin{pmatrix} c_{11} & c_{21} \\ c_{12} & c_{22} \end{pmatrix}\begin{pmatrix} a_1 \\ a_2 \end{pmatrix} = \begin{pmatrix} a_1 c_{11} + a_2 c_{21} \\ a_1 c_{12} + a_2 c_{22} \end{pmatrix}. \quad (2)$$

Observe that the result of $\overline{\phi}\left(\begin{pmatrix} a_1 \\ a_2 \end{pmatrix}\right)$ is the transpose of $\phi((a_1, a_2))$ so that $\phi((a_1, a_2))^t = \overline{\phi}\left(\begin{pmatrix} a_1 \\ a_2 \end{pmatrix}\right)$. This is what we meant when we indicated that the mappings are, up to transpose, the same mapping. This is always the case for a square matrix $A \in M_n(F)$ over a field F since for any $a \in F^n$,

$\phi(a)^t = (aA)^t = A^t a^t = \overline{\phi}(a^t)$. (See Exercise 5 of Problem Set 0.5.)

Compare the first of the following two theorems with Theorem 2.2.16 and the second with Theorems 2.3.5 and 3.3.10. An investigation of the proofs of Theorems 2.2.16, 2.3.5, and 3.3.10 will give you an idea of how to construct a proof of each of the following two theorems about vector spaces.

10.2.4   Theorem

If V is a vector space over F and U is a subspace of V, then the set of all cosets V/U is a vector space over F under the operations $(v + U) + (v' + U) = (v + v') + U$ and $c(v + U) = cv + U$.

10.2.5   Theorem

If U and V are vector spaces over F and $\phi : U \rightarrow V$ is a linear transformation, then the following hold.

1. The kernel K of ϕ is a subspace of U.
2. The linear transformation ϕ is a vector space monomorphism if and only if $K = \{0\}$.
3. If ϕ is an epimorphism, then the vector spaces U/K and V are isomorphic.
4. Suppose that $\dim_F U = n$. If $B = \{k_1, k_2, \ldots, k_m\}$ is a basis of K and $B' = \{k_1, k_2, \ldots, k_m, u_1, u_2, \ldots, u_t\}$ is a basis of U that contains B, then $n = m + t$ and $\bar{B} = \{u_i + K \mid i = 1, 2, \ldots, t\}$ is a basis for U/K.

Linear transformations play an important role in the study of vector spaces. The following theorem gives one connection between isomorphism and finite dimensional vector spaces.

10.2.6   Theorem

If V is a vector space over F and $\dim_F V = n$, then V and F^n are isomorphic.

Proof. Since $\dim_F V = n$, V has a basis with n elements. Suppose that $B = \{v_1, v_2, \ldots, v_n\}$ is a basis for V. If $v \in V$, then there are scalars $c_1, c_2, \ldots, c_n$ in F such that $v = c_1 v_1 + c_2 v_2 + \cdots + c_n v_n$. Moreover, we know from Theorem 10.1.5 that the coefficients $c_1, c_2, \ldots, c_n$ are unique. Thus, we have a well-defined mapping $\zeta : V \rightarrow F^n$ given by $\zeta(v) = (c_1, c_2, \ldots, c_n)$. We claim that ζ is a vector space isomorphism. Let's first show that ζ preserves addition and multiplication. If $v = c_1 v_1 + c_2 v_2 + \cdots + c_n v_n$ and $v' = c_1' v_1 + c_2' v_2 + \cdots + c_n' v_n$ are in V and $c \in F$, then $v + v' = (c_1 + c_1')v_1 + (c_2 + c_2')v_2 + \cdots + (c_n + c_n')v_n$ and $kv = (kc_1)v_1 + (kc_2)v_2 + \cdots + (kc_n)v_n$. Hence,

$$\zeta(v + v') = (c_1 + c_1', c_2 + c_2', ..., c_n + c_n')$$

$$= (c_1, c_2, ..., c_n) + (c_1', c_2', ..., c_n')$$

$$= \zeta(v) + \zeta(v')$$

and

$$\zeta(cv) = (cc_1, cc_2, ..., cc_n)$$

$$= c(c_1, c_2, ..., c_n)$$

$$= c\zeta(v).$$

Therefore, ζ preserves addition and scalar multiplication.

Finally, we need to show that ζ is a bijection. If $v \in \ker \zeta$ and $v = c_1v_1 + c_2v_2 + \cdots + c_nv_n$, then $\zeta(v) = (0, 0, ..., 0)$ and so $(c_1, c_2,..., c_n) = (0, 0, ..., 0)$. Hence, $c_1 = c_2 = \cdots = c_n = 0$, so $\ker \zeta = \{0\}$. Hence, it follows from Theorem 10.2.5 that ζ is injective. Finally, if $(c_1, c_2, ..., c_n) \in F^n$, then $v = c_1v_1 + c_2v_2 + \cdots + c_nv_n \in V$ and $\zeta(v) = (c_1, c_2, ..., c_n)$, so ζ is a surjective mapping. Therefore, ζ is a vector space isomorphism. ∎

10.2.7 Corollary

If U and V are finite dimensional vector spaces over the same field F, then U and V are isomorphic if and only if $\dim_F U = \dim_F V$.

The following theorem shows that a linear transformation is completely determined by the values it takes on basis elements.

10.2.8 Theorem

Let U and V be vector spaces over the same field F, and suppose that $\phi : U \rightarrow V$ is a linear transformation. If B is a basis for U, then ϕ is uniquely determined by the set of vectors $\phi(B) = \{\phi(u) \mid u \in B\}$ in V. That is, ϕ is determined by the values it takes on the basis elements of U.

Proof. If $u \in U$, then u can be written as a linear combination of vectors in B. Thus, there are scalars $c_1, c_2, ..., c_n \in F$ and vectors $u_1, u_2, ..., u_n \in B$ such that $u = c_1u_1 + c_2u_2 + \cdots + c_nu_n$. Hence,

$$\phi(u) = \phi(c_1 u_1 + c_2 u_2 + \cdots + c_n u_n)$$
$$= \phi(c_1 u_1) + \phi(c_2 u_2) + \cdots + \phi(c_n u_n)$$
$$= c_1 \phi(u_1) + c_2 \phi(u_2) + \cdots + c_n \phi(u_n),$$

which shows that $\phi(u)$ is determined by the vectors in $\phi(B)$. If $\psi : U \to V$ is another linear transformation that takes the same values on basis elements of U, then we see that

$$\phi(u) = \phi(c_1 u_1 + c_2 u_2 + \cdots + c_n u_n)$$
$$= c_1 \phi(u_1) + c_2 \phi(u_2) + \cdots + c_n \phi(u_n)$$
$$= c_1 \psi(u_1) + c_2 \psi(u_2) + \ldots + c_n \psi(u_n)$$
$$= \psi(c_1 u_1 + c_2 u_2 + \cdots + c_n u_n)$$
$$= \psi(u)$$

for each $u \in U$, so $\phi = \psi$. This shows that ϕ is uniquely determined by the values it takes on the basis elements of U. ∎

Suppose that U and V are vector spaces over F, and let B be a basis of U. Let $\phi : B \to V$ be a well-defined function, and suppose that $u \in U$. If $u = c_1 u_1 + c_2 u_2 + \cdots + c_n u_n$, where $u_1, u_2, \ldots, u_n \in B$ and $c_1, c_2, \ldots, c_n \in F$, and if $\bar{\phi} : U \to V$ is defined by $\bar{\phi}(u) = c_1 \phi(u_1) + c_2 \phi(u_2) + \cdots + c_n \phi(u_n)$, then $\bar{\phi}$ is a linear transformation. It is the usual practice to also denote $\bar{\phi}$ by ϕ, and we say that ϕ has been **extended linearly** (from B) to U. (You will be asked in the exercises to show that ϕ is a linear transformation.) Thus, a linear transformation $\phi : U \to V$ determines a function $\phi : B \to V$, and conversely, any function $\phi : B \to V$ determines a linear transformation $\phi : U \to V$. Now let U and V be vector spaces over F, and suppose that U is finite dimensional with basis $B = \{u_1, u_2, \ldots, u_n\}$. If $v_1, v_2, \ldots, v_n$ are n not necessarily distinct vectors in V, then there is *exactly one* linear transformation $\phi : U \to V$ such that $\phi(u_i) = v_i$ for $i = 1, 2, \ldots, n$. Indeed, we can define the function $\phi : B \to V$ by $\phi(u_i) = v_i$ for each i and extend ϕ linearly to U.

Corollary 10.2.7 shows that if U and V are finite dimensional vector spaces over F, then U and V are isomorphic if and only if U and V have the same dimension. However, as shown by the following example, two vector spaces over the same field do not have to be finite dimensional in order to be isomorphic.

10.2.9 | Example

The set $\mathbf{R}_0[x]$ of all polynomials with their coefficients in $\mathbf{R}$, each of which has zero constant term, is a vector space over $\mathbf{R}$ under the operations of polynomial addition and multiplication of a polynomial by a real number. It's easy to see that $\mathbf{R}_0[x]$ is an infinite dimensional vector space over $\mathbf{R}$ with basis $B = \{x, x^2, \ldots, x^n, \ldots\}$. Furthermore, $\mathbf{R}_0[x]$ is a proper subspace of $\mathbf{R}[x]$, the vector space of all polynomials over $\mathbf{R}$. Define $D : \mathbf{R}_0[x] \to \mathbf{R}[x]$ by $D[f(x)] = f'(x)$, where $f'(x)$ is the formal derivative of $f(x)$. Example 1 of 10.2.3 shows that D is a linear transformation. The mapping $I : \mathbf{R}[x] \to \mathbf{R}_0[x]$ defined by $I(f(x)) = \int f(x)\, dx$, where the constant of integration is always chosen to be zero, is a linear transformation that is an inverse function for D. Since I is an inverse function for D, Theorem 0.3.7 shows that D is a bijection. Hence, D is a vector space isomorphism. This is an interesting example. It shows that it is possible for an infinite dimensional vector space to be isomorphic to a proper subspace of itself, something that is not possible for finite dimensional vector spaces.

10.2.10   Theorem

If U and V are finite dimensional vector spaces over F and $\phi : U \to V$ is a linear transformation that is an epimorphism, then $\dim_F U = \dim_F K + \dim_F V$, where $K = \ker \phi$.

Proof. Suppose that $\dim_F U = n$ and $\dim_F K = m$. Then, by Exercise 21 of Problem Set 10.1, $m \le n$. If $B = \{k_1, k_2, \ldots, k_m\}$ is a basis for K, then Theorem 10.1.16 indicates that there is a basis B' of U such that $B \subseteq B'$. If $B' = \{k_1, k_2, \ldots, k_m, u_1, u_2, \ldots, u_t\}$, then $n = m + t$, and Theorem 10.2.5 shows that the set of cosets $\{u_i + K \mid i = 1, 2, \ldots, t\}$ is a basis of U/K. Since ϕ is an epimorphism, Theorem 10.2.5 also shows that U/K and V are isomorphic. It now follows from Corollary 10.2.7 that $\dim_F (U/K) = \dim_F V = t$. Thus, $\dim_F U = n = m + t = \dim_F K + \dim_F V$. ∎

If A is a matrix in $M_{m \times n}(\mathbf{R})$, then $\phi : \mathbf{R}^m \to \mathbf{R}^n$ defined by $\phi(a) = aA$ is a linear transformation, where $a = (a_1, a_2, \ldots, a_m) \in \mathbf{R}^m$. Example 4 of 10.2.3 is a particular illustration of such a linear transformation. The converse also holds; that is, if $\phi : \mathbf{R}^m \to \mathbf{R}^n$ is a linear transformation, then there is a matrix $A \in M_{m \times n}(\mathbf{R})$ such that $\phi(a) = aA$ for each vector $a \in \mathbf{R}^m$. How one can go about finding such a matrix A is demonstrated in the following example.

10.2.11   **Example**

Consider the linear transformation $\phi : \mathbf{R}^3 \to \mathbf{R}^2$ that is given by
$\phi((a_1, a_2, a_3)) = (a_1 + a_2, a_1 + a_3)$. This was shown to be a linear
transformation in Example 3 of 10.2.3. Consider the bases $B = \{e_1 = (1, 0, 0),$
$e_2 = (0, 1, 0), e_3 = (0, 0, 1)\}$ and $B' = \{(e_1' = (1, 0), e_2' = (0, 1)\}$ of $\mathbf{R}^3$ and $\mathbf{R}^2$,
respectively. Since a linear transformation is determined by the values it takes on
basis elements, we first compute $\phi(e_1)$, $\phi(e_2)$, and $\phi(e_3)$.

$$\phi(e_1) = (1 + 0, 1 + 0) = (1, 1) = 1e_1' + 1e_2'$$

$$\phi(e_2) = (0 + 1, 0 + 0) = (1, 0) = 1e_1' + 0e_2'$$

$$\phi(e_3) = (0 + 0, 0 + 1) = (0, 1) = 0e_1' + 1e_2'$$

Notice that we have selected a particular order for writing the preceding array.
Everything has been written in increasing order of the subscripts of the e_i's and
e_i's that name the basis elements of B and B', respectively. The column
containing $\phi(e_1)$, $\phi(e_2)$, and $\phi(e_3)$ has been written in increasing order of
subscripts as we go down the column, and the subscripts of the basis elements
of B' have been written in an increasing order as we read each line in the array
from left to right. When a particular order is chosen for the basis elements of a
basis for a vector space V, then the basis is said to be an **ordered basis**. The matrix
corresponding to ϕ depends on the order chosen for the basis elements. If the
order of the basis elements of either B or B' is changed, then the matrix
corresponding to ϕ may also change. Under the order chosen for the preceding
bases B and B', the coefficient matrix of the expressions in e_1' and e_2' is

$$\begin{pmatrix} 1 & 1 \\ 1 & 0 \\ 0 & 1 \end{pmatrix}.$$

Observe that

$$\phi((a_1, a_2, a_3)) = (a_1, a_2, a_3) \begin{pmatrix} 1 & 1 \\ 1 & 0 \\ 0 & 1 \end{pmatrix} = (a_1 + a_2, a_1 + a_3),$$

so let

$$A_\phi(B, B') = \begin{pmatrix} 1 & 1 \\ 1 & 0 \\ 0 & 1 \end{pmatrix}.$$

As mentioned earlier, the matrix $A_\phi(B, B')$ depends on the order chosen for the elements in the bases B and B'. For example, suppose we change the subscripts on the vectors in B and write $\bar{B} = \{\bar{e}_1 = (0, 0, 1),$ $\bar{e}_2 = (1, 0, 0), \bar{e}_3 = (0, 1, 0)\}$. Then under this "new" ordering of B,

$$\phi(\bar{e}_1) = (0 + 0, 0 + 1) = (0, 1) = 0e'_1 + 1e_2,$$

$$\phi(\bar{e}_2) = (1 + 0, 1 + 0) = (1, 1) = 1e'_1 + 1e'_2, \qquad \text{and}$$

$$\phi(\bar{e}_3) = (0 + 1, 0 + 0) = (1, 0) = 1e'_1 + 0e'_2, \qquad \text{so}$$

$$A_\phi = (\bar{B}, B') = \begin{pmatrix} 0 & 1 \\ 1 & 1 \\ 1 & 0 \end{pmatrix}.$$

Notice that

$$(a_1, a_2, a_3) \begin{pmatrix} 0 & 1 \\ 1 & 1 \\ 1 & 0 \end{pmatrix} = (a_2 + a_3, a_1 + a_2),$$

so this product does not yield $\phi((a_1, a_2, a_3)) = (a_1 + a_2, a_1 + a_3)$. The important point to observe is that since the order of the basis has been changed, the order of the components of the vector are also changed in the same manner to (a_3, a_1, a_2), so

$$(a_3, a_1, a_2) \begin{pmatrix} 0 & 1 \\ 1 & 1 \\ 1 & 0 \end{pmatrix} = (a_1 + a_2, a_1 + a_3).$$

In general, if F is a field and $\phi : F^m \to F^n$ is a linear transformation, then for each pair of ordered bases B and B' of F^m and F^n, respectively, there is a matrix $A_\phi(B, B') \in M_{m \times n}(F)$. The matrix $A_\phi(B, B')$ clearly depends on the bases chosen for F^m and F^n and on the order chosen for the elements of B and B'. We refer to this matrix as **the matrix corresponding to ϕ** (relative to the ordered bases B and B'). Once the bases of F^m and F^n have been chosen and the order on these bases has been set, the matrix corresponding to ϕ is unique. This follows from the fact that ϕ is uniquely determined by the values it takes on basis vectors in B and from the fact that the coefficients of any linear combination of elements of B' are unique. The matrix corresponding to ϕ can be found in the same manner demonstrated in Example 10.2.11. We simplify notation and denote the matrix $A_\phi(B, B')$ by A_ϕ when the ordered bases B and B' are understood. *The notation A_ϕ is often used in two different ways. It serves to denote the matrix corresponding to ϕ, and it also denotes the R-linear mapping $A_\phi : F^m \to F^n$ defined by $(a_i) \longmapsto (a_i)A_\phi$.* The context of the discussion indicates how the notation is being used.

When V is a finite dimensional vector space over F of dimension n, Theorem 10.2.6 shows that V and F^n are isomorphic. If $B = \{v_1, v_2, \ldots, v_n\}$ is an ordered basis of V, then each $v \in V$ can be written as $v = c_1 v_1 + c_2 v_2 + \cdots + c_n v_n$, and the coefficients $c_1, c_2, \ldots, c_n \in F$ are unique. The vector $(c_1, c_2, \ldots, c_n)$ in F^n is often referred to as the **coordinate vector** of v, and F^n is the **coordinate space** for V. This, in effect, sets up a coordinate system for the vector space V with $(c_1, c_2, \ldots, c_n)$ acting as the coordinates of the vector v. (The summands in the expression $v = c_1 v_1 + c_2 v_2 + \cdots + c_n v_n$ should follow the same order as the order of the basis vectors so that the components of the coordinate vector will also follow this order.) The vector space isomorphism $\zeta : V \to F^n$ of Theorem 10.2.6 defined by $v \longmapsto (c_1, c_2, \ldots, c_n)$ is said to be the **coordinate map** of V relative to B. This leads to the following theorem, which is illustrated by examples.

10.2.12   Theorem

Let U and V be finite dimensional vector spaces over the same field F. Suppose also that $\dim_F U = m$, $\dim_F V = n$, and B and B' are ordered bases of U and V, respectively. If $\phi : U \to V$ is a linear transformation and A_ϕ is the matrix corresponding to ϕ, then $\phi = \zeta_V^{-1} \circ A_\phi \circ \zeta_U$, where $\zeta_U : U \to \mathbf{R}^m$ and $\zeta_V : V \to \mathbf{R}^n$ are the coordinate maps of U and V.

A technique that is often used to clarify the situation described in Theorem 10.2.12 is to display a diagram such as the one that follows.

The order of application of the functions ζ_U, A_ϕ, and ζ_V^{-1} can be determined by following the arrows around the diagram, a technique known as **chasing the diagram**. Since $\phi = \zeta_V^{-1} \circ A_\phi \circ \zeta_U$, we see that mapping an element of U to V via ϕ produces the same result as mapping the element from U to F^m by ζ_U, then mapping this result from F^m to F^n by A_ϕ, and finally mapping the result in F^n from F^n to V by ζ_V^{-1}. Since taking these two paths from U to V

always produces the same element of V, we say that **the diagram is commutative**. The use of such diagrams as a visual display of the order of composition of functions is a useful tool in many areas of mathematics.

10.2.13 Examples

1. Consider the vector spaces $P_4(\mathbf{R})$ and $P_3(\mathbf{R})$. Then $D : P_4(\mathbf{R}) \to P_3(\mathbf{R})$ defined by $D[f(x)] = f'(x)$, the derivative of $f(x)$, is a linear transformation. The sets of vectors $B = \{1, x, x^2, x^3, x^4\}$ and $B' = \{1, x, x^2, x^3\}$ are bases of $P_4(\mathbf{R})$ and $P_3(\mathbf{R})$, respectively. Let's compute the matrix of D relative to B and B'. The bases B and B' are ordered from left to right as they are written. This is the **standard ordering** on the bases B and B'.

$$D[1] \; = 0 \quad = 0 \cdot 1 + 0x + 0x^2 + 0x^3$$
$$D[x] \; = 1 \quad = 1 \cdot 1 + 0x + 0x^2 + 0x^3$$
$$D[x^2] = 2x \; = 0 \cdot 1 + 2x + 0x^2 + 0x^3$$
$$D[x^3] = 3x^2 = 0 \cdot 1 + 0x + 3x^2 + 0x^3$$
$$D[x^4] = 4x^3 = 0 \cdot 1 + 0x + 0x^2 + 4x^3$$

The coefficient matrix of the expressions in the elements of B' is the matrix that corresponds to D. Hence,

$$A_D = \begin{pmatrix} 0 & 0 & 0 & 0 \\ 1 & 0 & 0 & 0 \\ 0 & 2 & 0 & 0 \\ 0 & 0 & 3 & 0 \\ 0 & 0 & 0 & 4 \end{pmatrix}.$$

We claim that the diagram

is commutative. If $f(x) = a_0 + a_1x + a_2x^2 + a_3x^3 + a_4x^4$ is a polynomial in $P_4(\mathbf{R})$, then

$$A_D \circ \zeta_{P_4(\mathbf{R})}(f(x)) = (a_0, a_1, a_2, a_3, a_4) \begin{pmatrix} 0 & 0 & 0 & 0 \\ 1 & 0 & 0 & 0 \\ 0 & 2 & 0 & 0 \\ 0 & 0 & 3 & 0 \\ 0 & 0 & 0 & 4 \end{pmatrix}$$

$$= (a_1, 2a_2, 3a_3, 4a_4)$$

and $(\zeta_{P_3(\mathbf{R})}^{-1}((a_1, 2a_2, 3a_3, 4a_4)) = a_1 + 2a_2 x + 3a_3 x^2 + 4a_4 x^3 = D[f(x)]$.

Thus, $D = \zeta_{P_3(\mathbf{R})}^{-1} \circ A_D \circ \zeta_{P_4(\mathbf{R})}$, so the diagram is indeed commutative.

2. Consider $\phi : P_3(\mathbf{R}) \to P_2(\mathbf{R})$ defined by $\phi(a_0 + a_1 x + a_2 x^2 + a_3 x^3) = \frac{3}{4} a_0 + \left(a_1 + \frac{1}{2} a_2 \right) x + \left(a_2 - \frac{2}{3} a_3 \right) x^2$. One can show that ϕ is a linear transformation, so let's compute the matrix that corresponds to ϕ relative to the standard ordered bases for $P_3(\mathbf{R})$ and $P_2(\mathbf{R})$, respectively. Since

$$\phi(1) = \tfrac{3}{4} \cdot 1 + (0 + 0)x + (0 - 0)x^2 = \tfrac{3}{4} \cdot 1 + 0x + 0x^2,$$

$$\phi(x) = 0 \cdot 1 + (1 + 0)x + (0 - 0)x^2 = 0 \cdot 1 + 1x + 0x^2,$$

$$\phi(x^2) = 0 \cdot 1 + \left(0 + \tfrac{1}{2} \right)x + (1 - 0)x^2 = 0 \cdot 1 + \tfrac{1}{2}x + 1x^2, \qquad \text{and}$$

$$\phi(x^3) = 0 \cdot 1 + (0 + 0)x + \left(0 - \tfrac{2}{3} \right)x^2 = 0 \cdot 1 + 0x + \left(-\tfrac{2}{3} \right)x^2,$$

the coefficient matrix of the expressions in the basis elements of B' is

$$A_\phi = \begin{pmatrix} \frac{3}{4} & 0 & 0 \\ 0 & 1 & 0 \\ 0 & \frac{1}{2} & 1 \\ 0 & 0 & -\frac{2}{3} \end{pmatrix}.$$

The coordinate vector in $\mathbf{R}^4$ of $f(x) = a_0 + a_1 x + a_2 x^2 + a_3 x^3 \in P_3(\mathbf{R})$ is $a = (a_0, a_1, a_2, a_3)$, and

$$aA_\phi = (a_0, a_1, a_2, a_3) \begin{pmatrix} \frac{3}{4} & 0 & 0 \\ 0 & 1 & 0 \\ 0 & \frac{1}{2} & 1 \\ 0 & 0 & -\frac{2}{3} \end{pmatrix} = \left(\tfrac{3}{4} a_0, a_1 + \tfrac{1}{2} a_2, a_2 - \tfrac{2}{3} a_3 \right)$$

It follows that the diagram

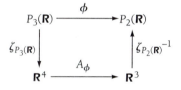

is commutative.

We conclude this section with an observation about the set of all linear transformations from a vector space U to a vector space V.

10.2.14 Theorem

If U and V are vector spaces over a field F and $\mathcal{L}(U,V)$ is the set of all linear transformations from U to V, then $\mathcal{L}(U,V)$ is a vector space over F under the operations of function addition and multiplication of a function by a scalar. Moreover, if U and V are finite dimensional vector spaces with dimensions m and n, respectively, then $\dim_F \mathcal{L}(U,V) = mn$.

Proof. (Sketched) It is straightforward to show that $\mathcal{L}(U,V)$ is a vector space over F under the operations of function addition and the multiplication of a linear transformation in $\mathcal{L}(U,V)$ by an element of F. If ϕ is a linear transformation in $\mathcal{L}(U,V)$, we have seen that there is a unique matrix $A_\phi \in M_{m \times n}(F)$ corresponding to ϕ relative to a pair of fixed ordered bases B and B' of $\mathcal{L}(U, V)$ and $M_{m \times n}(F)$, respectively. Since the basis $B' = \{E_{ij} \mid E_{ij}$ is the $m \times n$ matrix over F with e in the $(i, j)th$ position and zeros elsewhere$\}$ of $M_{m \times n}(\mathbf{R})$ contains mn elements, $\dim_F M_{m \times n}(F) = mn$. If we define $\theta : \mathcal{L}(U, V) \to M_{m \times n}(F)$ by $\theta(\phi) = A_\phi$, then θ is a vector space isomorphism. Hence, $\dim_F \mathcal{L}(U, V) = mn$ since isomorphic vector spaces have the same dimension. ∎

Problem Set 10.2

1. Prove that the transformation $\phi : \mathbf{R}^4 \to \mathbf{R}^2$ defined by $\phi((a_1, a_2, a_3, a_4)) = (a_1 + a_2, a_3 + a_4)$ is linear, and find the images of the vectors $(-1, 3, 7, 2)$ and $(0, 2, -3, -1)$.

2. Decide whether each of the following transformations is linear and give reasons for your conclusion.

(a) $\phi : \mathbf{R}^3 \to \mathbf{R}^2 : (a_1, a_2, a_3) \longmapsto (a_1 - a_2, a_2 - a_3)$

(b) $\phi : \mathbf{R}^3 \to \mathbf{R}^3 : (a_1, a_2, a_3) \longmapsto (2a_1, a_2 + 3, 4a_3)$

(c) $\phi : \mathbf{R}^3 \to P_{10}(\mathbf{R}) : (a_1, a_2, a_3) \longmapsto a_1 x^2 + a_2 x^4 + a_3 x^6$

3. Compute the kernel of each of the transformations in Problem 2 that is linear. Find a basis for each kernel.

4. Each of the following is a linear transformation. Find the matrix that corresponds to the linear transformation relative to the given bases.

   (a) $\phi : \mathbf{R}^4 \to \mathbf{R}^3 : (a_1, a_2, a_3, a_4) \longmapsto (2a_1 + 3a_2, a_2 - 4a_3, a_3 + 3a_4)$
   relative to the standard ordered bases of $\mathbf{R}^4$ and $\mathbf{R}^3$

   (b) $I : P_3(\mathbf{R}) \to P_4(\mathbf{R}) : f(x) \longmapsto \int f(x)dx$, where the constant of integration is always chosen to be zero, relative to the standard ordered bases for $P_3(\mathbf{R})$ and $P_4(\mathbf{R})$

   (c) $D : P_4(\mathbf{R}) \to P_3(\mathbf{R}) : f(x) \longmapsto f'(x)$, where $f'(x)$ is the derivative of the polynomial $f(x)$, relative to the bases $B = \{1, 1 + x, x^2, x + x^3, x^4\}$ and $B' = \{1, x, x^2, x + x^3\}$ of $P_4(\mathbf{R})$ and $P_3(\mathbf{R})$, respectively, if B and B' are ordered from left to right as they are written

5. Under the conditions of part (c) of Exercise 4, show that the diagram

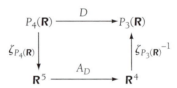

is commutative.

6. Let U and V be finite dimensional vector spaces over F with bases B and B', respectively. If $\phi : U \to V$ is a linear transformation, show by example that the matrix corresponding to ϕ depends on the ordering of the elements in B and B'.

7. If U and V are vector spaces over a field F, $\phi : U \to V$ is a linear transformation, and $\lambda \in F$, prove that the set of all vectors $u \in U$ such that $\phi(u) = \lambda u$ is a subspace of U. If v is a fixed nonzero vector in V, is the set of all $u \in U$ such that $\phi(u) = v$ a subspace of U? Show how you arrived at your conclusion.

8. If U and V are vector spaces over a field F and $\phi : U \to V$ is a linear transformation, an element $u \in U$ is said to be fixed by ϕ if $\phi(u) = u$. Prove or disprove that the set of all points that are fixed by ϕ is a subspace of U.

9. Prove Theorem 10.2.4.

10. Prove Theorem 10.2.5.

11. Let U and V be finite dimensional vector spaces over a field F, and suppose that $\dim_F U = \dim_F V$. If $\phi : U \to V$ is a linear transformation, prove that the following are equivalent:

    (a) ϕ is an isomorphism.
    (b) ϕ is an injective mapping.
    (c) ϕ is a surjective mapping.

12. If $\phi : M_3(\mathbf{R}) \to M_3(\mathbf{R})$ is defined by

$$\begin{pmatrix} a_{11} & a_{12} & a_{13} \\ a_{21} & a_{22} & a_{23} \\ a_{31} & a_{32} & a_{33} \end{pmatrix} \longmapsto \begin{pmatrix} a_{11} & 0 & 0 \\ a_{21} & a_{22} & 0 \\ a_{31} & a_{32} & a_{33} \end{pmatrix},$$

    show that ϕ is a linear transformation, and find the matrix corresponding to ϕ relative to the ordered basis $\{E_{11}, E_{12}, E_{13}, \ldots, E_{31}, E_{32}, E_{33}\}$ of $M_3(\mathbf{R})$. What is the kernel of ϕ?

13. Suppose that U and V are vector spaces over F, and let B be a basis of U. Let $\phi : B \to V$ be a well-defined function, and suppose that $u \in U$. If $u = c_1 u_1 + c_2 u_2 + \cdots + c_n u_n$, where $u_1, u_2, \ldots, u_n \in B$, $c_1, c_2, \ldots, c_n \in F$, and $\overline{\phi} : U \to V$ is defined by $\overline{\phi}(u) = c_1 \phi(u_1) + c_2 \phi(u_2) + \cdots + c_n \phi(u_n)$, prove that $\overline{\phi}$ is a linear transformation.

14. If $\phi : U \to V$ is a vector space isomorphism, prove that a set of vectors B of U is a basis for U if and only if $\phi(B) = \{\phi(u) \mid u \in B\}$ is a basis of V.

15. If $\phi : U \to V$ is a vector space isomorphism, prove that the inverse function $\phi^{-1} : V \to U$ is a vector space isomorphism.

16. Consider the linear transformation $\phi : \overline{\mathbf{R}}^3 \to \overline{\mathbf{R}}^2$ that is defined by

$$\phi \left(\begin{pmatrix} a_1 \\ a_2 \\ a_3 \end{pmatrix} \right) = \begin{pmatrix} a_1 - a_3 \\ a_2 - a_3 \end{pmatrix}.$$

    Find the matrix that corresponds to ϕ relative to the standard ordered bases of $\overline{\mathbf{R}}^3$ and $\overline{\mathbf{R}}^2$. Hint: Follow the same procedure as in Example 10.2.11, and transpose the result.

10.3   INNER PRODUCT SPACES

Section Overview. This section provides a brief introduction to inner product spaces. Examples are provided of several different inner products on a vector space. Properties of the dot product on $\mathbf{R}^n$ are used as a model to define an inner product on a vector space. This leads to the definition of a norm of a vector and to the concept of the distance between two vectors. Several important properties of inner product spaces are treated in the exercises of this section.

If $a = (a_1, a_2)$, $b = (b_1, b_2) \in \mathbf{R}^2$, then the **dot product** of the vectors a and b is defined by

$$a \bullet b = a_1 b_1 + a_2 b_2.$$

Thus, the dot product of two vectors a and b is a scalar, and for this reason $a \bullet b$ is sometimes referred to as the **scalar product** of a and b. As an example, if $a = (1, -1)$ and $b = \left(3, \frac{1}{2}\right)$, then $a \bullet b = 1 \cdot 3 + (-1)\frac{1}{2} = 3 - \frac{1}{2} = \frac{5}{2}$. Notice that $a \bullet a = a_1^2 + a_2^2$, so $a \bullet a \geq 0$ and $a \bullet a = 0$ if and only if a is the zero vector in $\mathbf{R}^2$. The dot product is a function $\bullet$ from $\mathbf{R}^2 \times \mathbf{R}^2$ to $\mathbf{R}$ with the following properties:

1. $a \bullet b = b \bullet a$ for all $a, b \in \mathbf{R}^2$.
2. $(ka) \bullet b = k(a \bullet b)$ for all $a, b \in \mathbf{R}^2$ and all $k \in \mathbf{R}$.
3. $(a + b) \bullet c = a \bullet b + a \bullet c$ for all $a, b, c \in \mathbf{R}^2$.
4. $a \bullet a \geq 0$ and $a \bullet a = 0$ if and only if a is the zero vector in $\mathbf{R}^2$.

If we set

$$\|a\| = \sqrt{a \bullet a} = \sqrt{a_1^2 + a_2^2},$$

then $\|a\|$ is the distance from the origin to the point with coordinates (a_1, a_2) in the (x, y)-plane. We refer to $\|a\|$ as the **norm** of the vector a. Notice that $a - b = (a_1 - b_2, a_2 - b_2)$, so

$$\|a - b\| = \sqrt{(a - b) \bullet (a - b)} = \sqrt{(a_1 - b_1)^2 + (a_2 - b_2)^2}$$

Hence, $\|a - b\|$ is the distance in the (x, y)-plane between the points with coordinates (a_1, a_2) and (b_1, b_2). When $a = (a_1, a_2)$ and $b = (b_1, b_2)$ are viewed as vectors in the vector space $\mathbf{R}^2$, we say that $\|a - b\|$ is the **distance between the vectors** a and b. More generally, if $a = (a_1, a_2, \ldots, a_n)$ and $b = (b_1, b_2, \ldots, b_n)$ are vectors in $\mathbf{R}^n$, $n \geq 2$, then the dot product of a and b is defined by

$$a \bullet b = a_1 b_1 + a_2 b_2 + \cdots + a_n b_n.$$

For example, if $n = 5$, $a = \left(-2, \frac{1}{2}, 0, 3, \frac{2}{3} \right)$, and $b = (1, -1, 3, 2, 3)$, then $a \bullet b = $

$-2 \cdot 1 + \frac{1}{2}(-1) + 0 \cdot 3 + 3 \cdot 2 + \frac{2}{3} \cdot 3 = -2 - \frac{1}{2} + 0 + 6 + 2 = \frac{11}{2}$. The norm of

a is

$$\|a\| = \sqrt{a \bullet a} = \sqrt{a_1^2 + a_2^2 + \cdots + a_n^2}, \qquad \text{and}$$

$$\|a - b\| = \sqrt{(a_1 - b_1)^2 + (a_2 - b_2)^2 + \cdots + (a_n - b_n)^2}$$

is the distance between the vectors a and b. It is not difficult to show that this scalar product on $\mathbf{R}^n$ satisfies conditions 1, 2, 3, and 4 given earlier.

The dot product just defined on the vector space $\mathbf{R}^n$ is a special case of a more general type of scalar product that can often be defined on a vector space. We use the properties of the dot product on $\mathbf{R}^n$ as a model for the definition of an inner product on a vector space. At first, we restrict our attention to vector spaces over $\mathbf{R}$.

10.3.1   Definition

If V is a vector space over $\mathbf{R}$, then a function $(\cdot \mid \cdot)$ from $V \times V$ to $\mathbf{R}$ is said to be an **inner product** on V if the following conditions are satisfied:

1. $(u \mid v) = (v \mid u)$ for all $u, v \in V$.
2. $(ku \mid v) = k(u \mid v)$ for all $u, v \in V$ and $k \in \mathbf{R}$.
3. $(u + v \mid w) = (u \mid w) + (v \mid w)$ for all $u, v, w \in V$.
4. $(v \mid v) \geq 0$ for all $v \in V$, and $(v \mid v) = 0$ if and only if $v = 0$.

If $(\cdot \mid \cdot)$ is an inner product on a vector space V, then we define the **norm** of a vector $v \in V$ by $\|v\| = \sqrt{(v \mid v)}$, and the **distance between two vectors** $u, v \in V$ is $\|u - v\|$. A vector space V together with an inner product is said to be an **inner product space**.

Because of the symmetry of the inner product provided by condition 1 of the preceding definition, we immediately have the following properties:

2′. $(u \mid kv) = k(u \mid v)$ for all $u, v \in V$ and $k \in \mathbf{R}$.
3′. $(w \mid u + v) = (w \mid u) + (w \mid v)$ for all $u, v, w \in V$.

The proof of these properties is left as an exercise.

10.3.2   Examples

1. The dot product $\bullet$ defined on $\mathbf{R}^n$, $n \geq 2$, by $(a \mid b) = a \bullet b = a_1 b_1 + a_2 b_2 + \cdots + a_n b_n$ for all $a = (a_1, a_2, \ldots, a_n)$, $b = (b_1, b_2, \ldots, b_n) \in \mathbf{R}^n$

is an inner product on $\mathbf{R}^n$. This is the **standard inner product on $\mathbf{R}^n$**. The vector space $\mathbf{R}^n$ together with this inner product is often referred to as **Euclidean n-space**.

2. If $a = (a_1, a_2, a_3), b = (b_1, b_2, b_3) \in \mathbf{R}^3$, then $(a \mid b) = a_1 b_1 + 5a_2 b_2 + 2a_3 b_3$ is an inner product on $\mathbf{R}^3$. The norm of a is given by
    $\|a\| = \sqrt{a_1^2 + 5a_2^2 + 2a_3^2}$, and the distance between a and b is
    $\|a - b\| = \sqrt{(a_1 - b_1)^2 + 5(a_2 - b_2)^2 + 2(a_3 - b_3)^2}$. For example, if $a =$
    $(0, 1, 2)$ and $b = (1, -2, 1)$, then $\|a\| = \sqrt{0^2 + 5 \cdot 1^2 + 2 \cdot 2^2} = \sqrt{13}$ and
    $\|a - b\| = \sqrt{(0 - 1)^2 + 5(1 + 2)^2 + 2(1 - 2)^2} = \sqrt{48}$.

3. The function $(f \mid g) = \displaystyle\int_0^1 f(x)g(x)\, dx$ is the **standard inner product** on
    $P_n(\mathbf{R})$. Suppose that $n = 2$, and let $f(x) = 1 + 2x$. Since
    $$(f \mid f) = \int_0^1 (1 + 2x)^2 dx = \int_0^1 (1 + 4x + 4x^2)\, dx = \left(x + 2x^2 + \tfrac{4}{3}x^3 \right)\Big|_0^1 =$$
    $1 + 2 + \frac{4}{3} = \frac{13}{3}$, the norm of $f(x)$ is $\|f\| = \sqrt{(f \mid f)} = \sqrt{\tfrac{13}{3}}$. Furthermore,
    if $f(x) = 3x^2$ and $g(x) = 3 + 4x^2$, then $(f - g \mid f - g) =$
    $$\int_0^1 (3x^2 - 3 - 4x^2)\, dx = \int_0^1 (-3 - x^2)\, dx = \int_0^1 (9 + 6x^2 + x^4)\, dx =$$
    $\left(9x + 2x^3 + \tfrac{1}{5}x^5 \right)\Big|_0^1 = \frac{56}{5}$, so the distance between the vectors $f(x)$ and
    $g(x)$ is $\|f - g\| = \sqrt{(f - g \mid f - g)} = \sqrt{\tfrac{56}{5}}$.

4. If a and b are real numbers such that $a < b$, let $V = \{ f : [a, b] \to \mathbf{R} \mid f$
    is a continuous function$\}$. Then V is a vector space over $\mathbf{R}$ under addition
    of functions and multiplication of a function by a real number. The
    function $(f \mid g) = \displaystyle\int_a^b f(x)g(x)\, dx$ from $V \times V$ to $\mathbf{R}$ defines an inner
    product on the vector space V over $\mathbf{R}$.

If V is a vector space over $\mathbf{C}$, the field of complex numbers, then an inner product can also be defined on V.

10.3.3   Definition

If V is a vector space over $\mathbf{C}$, then an **inner product** on V is a function $(\cdot \mid \cdot) : V \times V \to \mathbf{C}$ that satisfies the following conditions:

1. $(u|v) = \overline{(v|u)}$, the complex conjugate of $(v \mid u)$, for all $u, v \in V$.
2. $(ku \mid v) = k(u \mid v)$ for all $u, v \in V$ and $k \in \mathbf{C}$.
3. $(u + v \mid w) = (u \mid v) + (u \mid w)$ for all $u, v, w \in V$.
4. $(u \mid u) \geq 0$, and $(u \mid u) = 0$ if and only if $u = 0$.

Notice that because of condition 1, $(u \mid u) = \overline{(u|u)}$, so $(u \mid u)$ is a real number. Hence, the inequality $(u \mid u) \geq 0$ given in condition 4 makes sense.

10.3.4   Example

Consider the vector space $\mathbf{C}^2$ over $\mathbf{C}$, and define $(\cdot \mid \cdot) : \mathbf{C}^2 \times \mathbf{C}^2 \to \mathbf{C}$ by $(u|v) = u_1 \overline{v_1} + u_2 \overline{v_2}$, where $u = (u_1, u_2)$ and $v = (v_1, v_2)$ are vectors in $\mathbf{C}^2$. Then $(\cdot \mid \cdot)$ is an inner product on $\mathbf{C}^2$. For example, if $u = (3 - 2i, 1 + 2i)$ and $v = (2 + i, 1 - i)$, then

$$(u|v) = (3 - 2i)\overline{(2 + i)} + (1 + 2i)\overline{(1 - i)}$$
$$= (3 - 2i)(2 - i) + (1 + 2i)(1 + i)$$
$$= 3 - 4i.$$

If the function $(\cdot \mid \cdot) : \mathbf{C}^n \times \mathbf{C}^n \to \mathbf{C}$ is defined by $(u \mid v) = u_1 \overline{v_1} + u_2 \overline{v_2} + \cdots + u_n \overline{v_n}$, where $u = (u_1, u_2, \ldots, u_n)$ and $v = (v_1, v_2, \ldots, v_n)$ are vectors in $\mathbf{C}^n$, then $(\cdot \mid \cdot)$ is an inner product on $\mathbf{C}^n$. This is the **standard inner product** on $\mathbf{C}^n$. ∎

Problem Set 10.3

1. If $(\cdot \mid \cdot)$ is the standard inner product on $\mathbf{R}^3$, compute each of the following.
   (a) The norm of the vector $a = \left(\frac{1}{2}, -2, \frac{3}{4} \right)$
   (b) The distance between the vectors $a = \left(1, \frac{2}{3}, 1 \right)$ and $b = \left(-1, 2, \frac{1}{2} \right)$

2. Let V be a vector space over F, where $F = \mathbf{R}$ or $F = \mathbf{C}$. If $(\cdot \mid \cdot)$ is such that $(u|v) = 0$ for every pair of vectors $u, v \in V$, is $(\cdot \mid \cdot)$ an inner product on V?

3. Determine if the function $(\cdot \mid \cdot)$ is an inner product on the given vector space.
   (a) If $a = (a_1, a_2)$, $b = (b_1, b_2) \in \mathbf{R}^2$, then $(a \mid b) = a_1b_1 - 4a_2b_2$.
   (b) If $a = (a_1, a_2, a_3)$, $b = (b_1, b_2, b_3) \in \mathbf{R}^3$, then $(a \mid b) = a_1b_1 + a_2b_3 + a_3b_2$.
   (c) If $f(x), g(x) \in P_2(\mathbf{R})$, let $(f \mid g) = \int_0^1 x^2 f(x)g(x)\,dx$.

4. If $(a \mid b) = (a_1, a_2)\begin{pmatrix} 1 & -4 \\ 3 & 2 \end{pmatrix}\begin{pmatrix} b_1 \\ b_2 \end{pmatrix}$ for any two vectors $a = (a_1, a_2)$ and $b = (b_1, b_2)$ in $\mathbf{R}^2$, determine whether $(\cdot \mid \cdot)$ is an inner product on $\mathbf{R}^2$.

5. Let V be a finite dimensional vector space over F with inner product $(\cdot \mid \cdot)$ and basis $B = \{v_1, v_2, \ldots, v_n\}$. Then the matrix $(c_{ij}) \in M_n(F)$ defined by $c_{ij} = (v_i \mid v_j)$, $1 \le i, j \le n$, is the **matrix of the inner product** relative to the ordered basis B.
   (a) If $(f \mid g) = \int_0^1 f(x)g(x)\,dx$ is the standard inner product on $P_1(\mathbf{R})$, find the matrix M of this inner product corresponding to the standard ordered basis $B = \{v_1 = 1, v_2 = x\}$ of $P_1(\mathbf{R})$. Hint: Compute $c_{11} = (v_1 \mid v_1)$, $c_{12} = (v_1 \mid v_2)$, etc.
   (b) If $f(x) = 1 + 2x$ and $g(x) = 3 + 4x$ are vectors in $P_1(\mathbf{R})$, show that
   $$(f \mid g) = (1, 2)M\begin{pmatrix} 3 \\ 4 \end{pmatrix}.$$
   If $f(x) = a_0 + a_1x$ and $g(x) = b_0 + b_1(x)$ are vectors in $P_1(\mathbf{R})$, show that
   $$(f \mid g) = (a_0, a_1)M\begin{pmatrix} b_0 \\ b_1 \end{pmatrix} = (b_0, b_1)M\begin{pmatrix} a_0 \\ a_1 \end{pmatrix}.$$

6. Show that the function $(\cdot \mid \cdot)$ given in Examples 1 and 3 of 10.3.2 is indeed an inner product on the given vector space.

7. If $(\cdot \mid \cdot)$ is an inner product on a vector space V over $\mathbf{R}$, show that each of the following holds.
   (a) $(u \mid kv) = k(u \mid v)$ for all $u, v \in V$ and $k \in \mathbf{R}$.
   (b) $(w \mid u + v) = (w \mid u) + (w \mid v)$ for all $u, v, w \in V$.
   (c) $(k_1u_1 + k_2u_2 \mid v) = k_1(u_1 \mid v) + k_2(u_2 \mid v)$ for all $u_1, u_2, v \in V$ and all $k_1, k_2 \in \mathbf{R}$.
   (d) $(u \mid k_1v_1 + k_2v_2) = k_1(u \mid v_1) + k_2(u \mid v_2)$ for all $u, v_1, v_2 \in V$ and all $k_1, k_2 \in \mathbf{R}$.

8. If $(\cdot \mid \cdot)$ is an inner product on a vector space V over $\mathbf{C}$, show that each of the following holds.

    (a) $(u \mid kv) = \bar{k}\,(u \mid v)$ for all $u, v \in V$ and $k \in \mathbf{C}$.

    (b) $(w \mid u + v) = (w \mid u) + (w \mid v)$ for all $u, v, w \in V$.

    (c) $(k_1 u_1 + k_2 u_2 \mid v) = k_1(u_1 \mid v) + k_2(u_2 \mid v)$ for all $u_1, u_2, v \in V$ and all $k_1, k_2 \in \mathbf{C}$.

    (d) $(u \mid k_1 v_1 + k_2 v_2) = \bar{k}_1(u \mid v_1) + \bar{k}_2(u \mid v_2)$ for all $u, v_1, v_2 \in V$ and all $k_1, k_2 \in \mathbf{C}$.

9. Let $(f \mid g) = \displaystyle\int_a^b f(x)g(x)\,dx$, where $f(x), g(x) \in P_n(\mathbf{R})$, $n \geq 1$, and

    suppose that a and b are real numbers such that $a < b$. Then $(\cdot \mid \cdot)$ is an inner product on $P_n(\mathbf{R})$.

    (a) If $a = 1$, $b = 3$, and $n \geq 2$, compute the norm of $f(x) = 1 + 2x - x^2$.
    (b) If $a = 1$, $b = 3$, and $n \geq 2$, compute the distance between the vectors $f(x) = 1 + x^2$ and $g(x) = 3x + 2x^2$.
    (c) If $d\,(f, g)$ denotes the distance between the vectors $f(x)$ and $g(x)$, show that

    $$d(f, g) = \left[\int_a^b (f(x)g(x))^2\,dx \right]^{1/2}.$$

    (d) If $a \leq b$, is $(\cdot \mid \cdot)$ always an inner product?

10. Let V be a vector space over $\mathbf{R}$ with inner product $(\cdot \mid \cdot)$. Prove that each of the following holds.

    (a) If $u \in V$ and $(u \mid 0) = 0$, then $u = 0$.
    (b) If $u \in V$ is such that $(v \mid u) = 0$ for all $v \in V$, then $u = 0$.
    (c) If B is a basis of V and u is a vector in V such that $(v \mid u) = 0$ for every $v \in B$, then $u = 0$.

11. Let V be a finite dimensional vector space over $\mathbf{R}$ of dimension n, and suppose that $(\cdot \mid \cdot)$ is an inner product defined on V. Suppose also that $B = \{v_1, v_2, \ldots, v_n\}$ is a set of vectors in V such that

    $$(v_i \mid v_j) = \begin{cases} 1 \text{ if } i = j \\ 0 \text{ if } i \neq j \end{cases}.$$

(a) Prove that B is a basis for V.

(b) If $v \in V$, prove that $v = (v \mid v_1)v_1 + (v \mid v_2)v_2 + \cdots + (v \mid v_n)v_n$.

12. If V is a vector space over $\mathbf{R}$ with inner product $(\cdot \mid \cdot)$, show that

$|(u \mid v)| \le \|u\| \|v\|$ for all vectors $u, v \in V$. This is known as **Schwarz's inequality**.

Hint: The polynomial $f(x) = ax^2 - 2bx + c, a > 0$, has minimum value

$c - \frac{b^2}{a}$. Hence, if $f(x) \ge 0$, then $b^2 < ac$. Schwarz's inequality is trivially true if

$u = 0$ or $v = 0$. If both are nonzero, let x be a nonzero scalar and consider

$$(xu - v \mid xu - v) = (u \mid u)x^2 - 2(u \mid v)x + (v \mid v)$$

Herman Schwarz (1843–1921) was educated in Berlin and taught in Berlin, Göttingen, and Zurich. His great concern for students occupied much of his time, and because of this, he was not widely published. However, the methods Schwarz developed for solving problems proved to be of lasting importance.

13. If V is a vector space over $\mathbf{R}$ with inner product $(\cdot \mid \cdot)$, we see from Schwarz's ine-

quality that $\dfrac{|(u \mid v)|}{\|u\| \|v\|} \le 1$. Thus, $-1 \le \dfrac{(u \mid v)}{\|u\| \|v\|} \le 1$, so we can define the cosine of

the angle θ between two vectors $u, v \in V$ as

$$\cos \theta = \frac{(u \mid v)}{\|u\| \|v\|} .$$

(a) If $(\cdot \mid \cdot)$ is the standard inner product on the vector space $\mathbf{R}^3$, find the

cosine of the angle between the vectors $u = \left(1, \frac{1}{2}, \frac{3}{4}\right)$ and $v = \left(2, 1, \frac{5}{3}\right)$.

(b) If $(f \mid g) = \displaystyle\int_0^1 f(x)g(x)\, dx$ is the standard inner product on $P_3(\mathbf{R})$, find

the cosine of the angle between the vectors $f(x) = 2x + x^3$ and

$g(x) = 1 + x^2$.

11

Modules

This chapter concludes our investigations of the basic structures of abstract algebra with a brief discussion of modules. Since every vector space is a module, vector spaces could be presented as part of this chapter. However, a separate presentation more clearly demonstrates the similarities and the differences between the two concepts. As with vector spaces, the theory of modules is quite extensive, so we provide only an introduction.

11.1   MODULES

Section Overview. A module is just a modest generalization of a vector space. You will notice that the definition of a module is almost exactly the same as the definition of a vector space. One difference is that in the definition of a vector space, the scalars are taken from a field, while for a module, the scalars belong to an arbitrary ring. Another important distinction is that the external multiplication of elements of the module by ring elements must be designated as operating on the left or on the right—a condition that is not necessary for vector spaces. Since modules and vector spaces are very closely connected, you should compare the results presented in this chapter with the material in Chapter 10 on vector spaces.

The following definition of a module shows that every vector space is a module. All one has to do to arrive at the definition of a vector space is to replace the ring R with a field F. We assume throughout this chapter that R is a ring with identity that may not be commutative.

11.1.1   Definition

If R is a ring, an additive abelian group M is said to be a **left R-module** if there is an external multiplication $R \times M \longrightarrow M : (a, x) \longmapsto ax$ between elements of R and elements of M that satisfies the following conditions:

1. $(a + b)x = ax + bx$ for all $x \in M$ and all $a, b \in R$.
2. $a(x + y) = ax + ay$ for all $x, y \in M$ and all $a \in R$.
3. $(ab)x = a(bx)$ for all $x \in M$ and all $a, b \in R$.
4. $ex = x$ for all $x \in M$.

Right R-modules are defined symmetrically; just place all of the ring elements in conditions 1 through 4 in the definition on the right of the elements of M. Elements of a module are *not* usually referred to as vectors unless the module under consideration is actually a vector space. However, for convenience of language, we continue to refer to ring elements as scalars and to the external multiplication of elements of a module by ring elements as **scalar multiplication**. (This terminology is nonstandard and not entirely correct since a scalar is a quantity that can be described by a number. But then, is an element of an arbitrary field a scalar?) We also refer to this multiplication as scalar multiplication on M when R is understood. A subgroup N of a left R-module M is a **submodule** of M if, when the scalar multiplication on M is restricted to N, N is a left R-module.

The first important point that should be made is that, in general, if M is a left R-module, then M *cannot* be made into a right R-module by switching sides with scalar multiplication. That is, if M is a left R-module, then M cannot be made into a right R-module by defining $x * a = ax$ for all $a \in R$ and $x \in M$. Conditions 1, 2, and 4 in the definition of a left R-module carry over, but the offending property is condition 3 of the definition. To see the difficulty, suppose that defining $x * a = ax$ *does* make M into a right R-module. Then for $a, b \in R$ and $x \in M$, $(ab)x = x * (ab) = (x * a) * b = b(x * a) = b(ax) = (ba)x$. But if the ring is not commutative, it may be the case that $ab \neq ba$, so there is no reason to believe that $(ab)x = (ba)x$. In fact, examples of left R-modules abound where we can find

$a, b \in R$ and $x \in M$ such that $(ab)x \neq (ba)x$. Of course, if the ring is commutative, this difficulty disappears, and a left R-module is a right R-module, and vice versa. For this reason, it is immaterial whether or not the scalars for a vector space operate on the left of vectors or on the right. We can switch the side of scalar multiplication with impunity when dealing with vector spaces. This also holds when working with modules over a commutative ring but, as we have just seen, not when the ring is noncommutative.

Examples of modules are numerous. Since every vector space is a module, each of the examples given in Section 10.1.3 is a module. In fact, in many of these examples, if the field is replaced by a ring R with identity that is not a field, the result is a left R-module that is not a vector space. This is true of Examples 2 through 6 and Example 8 of Section 10.1.3. You should revisit these examples and make sure you understand why each becomes a left R-module when the field is replaced by a general ring with identity. The following are additional examples of modules. In general, we deal with left R-modules. Because of the symmetry between the definitions of a left R-module and a right R-module, any result we prove or demonstrate for left R-modules also holds for right R-modules.

11.1.2 Examples

1. If R is a ring, then R is both a left R-module and a right R-module. The scalar multiplication of elements of R by ring elements is just the multiplication defined on R.

2. The set $\overline{R}^2$ of all 2×1 column matrices with entries from a ring R is an additive abelian group under componentwise addition defined by

$$\begin{pmatrix} a_1 \\ a_2 \end{pmatrix} + \begin{pmatrix} b_1 \\ b_2 \end{pmatrix} = \begin{pmatrix} a_1 + b_1 \\ a_2 + b_2 \end{pmatrix}.$$

One way to turn the group $\overline{R}^2$ into a left $M_2(R)$-module is to define scalar multiplication on $\overline{R}^2$ as

$$\begin{pmatrix} c_{11} & c_{12} \\ c_{21} & c_{22} \end{pmatrix} \begin{pmatrix} a_1 \\ a_2 \end{pmatrix} = \begin{pmatrix} c_{11}a_1 + c_{12}a_2 \\ c_{21}a_1 + c_{22}a_2 \end{pmatrix}.$$

In general, for any integer $n \geq 2$, $\overline{R}^n$ is a left $M_n(R)$-module if scalar multiplication is defined on $\overline{R}^n$ by following the example of the definition of multiplication of elements of $\overline{R}^2$ by elements of $M_2(R)$.

3. The set R^2 of all ordered pairs (a_1, a_2) is an additive abelian group under componentwise addition. This group is a right $M_2(R)$-module if scalar multiplication is defined on R^2 as

$$(a_1, a_2) \begin{pmatrix} c_{11} & c_{12} \\ c_{21} & c_{22} \end{pmatrix} = (a_1 c_{11} + a_2 c_{21}, \ a_1 c_{12} + a_2 c_{22}).$$

   If $n \geq 2$ is an integer and R^n is made into an additive abelian group by defining addition on R^n componentwise, then R^n is a right $M_n(R)$-module if scalar multiplication is defined on R^n in the obvious way.

4. If $n \geq 2$ is an integer and $\{M_i\}_{i=1}^{n}$ is a family of left R-modules, then $\times_{i=1}^{n} M_i = M_1 \times M_2 \times \cdots \times M_n = \{(x_1, x_2, \ldots, x_n) \mid x_i \in M_i \text{ for } i = 1, 2, \ldots, n\}$ is an additive abelian group if addition is defined on $\times_{i=1}^{n} M_i$ by

$$(x_1, x_2, \ldots, x_n) + (y_1, y_2, \ldots, y_n) = (x_1 + y_1, x_2 + y_2, \ldots, x_n + y_n).$$

   If scalar multiplication of elements of $\times_{i=1}^{n} M_i$ by ring elements is defined as

$$a(x_1, x_2, \ldots, x_n) = (ax_1, ax_2, \ldots, ax_n),$$

   then $\times_{i=1}^{n} M_i$ is a left R-module. The module $\times_{i=1}^{n} M_i$ is the **direct product** of the family $\{M_i\}_{i=1}^{n}$. In particular, $R^n = R \times R \times \cdots \times R$ (n factors) is a left R-module for each positive integer $n \geq 2$.

5. If G is an additive abelian group, then G is a **Z**-module. The scalar multiplication defined on G by integers is given by nx. The notation nx represents an integer n times a group element x. (See "Exponents and Multiples in a Group" in Section 2.1 for the details.) Since **Z** is a commutative ring, $nx = xn$, so any additive abelian group is both a left and a right **Z**-module.

6. Let G be an additive abelian group, and suppose that $E = \text{End}(G)$ is the set of all group homomorphisms $f: G \longrightarrow G$. Then E is a ring with identity under function addition and function composition. The identity of E is the identity mapping $1_G: G \longrightarrow G: x \longmapsto x$. The additive abelian group G is a left E-module under the operation defined by $f * x = f(x)$ for all $f \in E$ and $x \in G$.

7. Let $\{M_i\}_{i \geq 1}$ be the family of left R-modules, and consider the set $\times_{i=1}^{\infty} M_i$ of all sequences $(x_1, x_2, x_3, \ldots)$ such that $x_i \in M$ for each i. If addition is defined on $\times_{i=1}^{\infty} M_i$ as

$$(x_1, x_2, x_3, \ldots) + (y_1, y_2, y_3, \ldots) = (x_1 + y_1, x_2 + y_2, x_3 + y_3, \ldots),$$

   then $\times_{i=1}^{\infty} M_i$ is an additive abelian group. The additive identity is $(0_1, 0_2, 0_3, \ldots)$, where 0_i is the additive identity of M_i for $i = 1, 2, 3, \ldots$

and the additive inverse of $(x_1, x_2, x_3, \ldots)$ is $(-x_1, -x_2, -x_3, \ldots)$. If scalar multiplication is defined on $\times_{i=1}^{\infty} M_i$ as

$$a(x_1, x_2, x_3, \ldots) = (ax_1, ax_2, ax_3, \ldots),$$

then $\times_{i=1}^{\infty} M_i$ is a left R-module. The module $\times_{i=1}^{\infty} M_i$ is the **direct product** of the family $\{M_i\}_{i \geq 1}$.

8. Let $\times_{i=1}^{\infty} M_i$ be the left R-module of Example 7, and suppose that $\oplus_{i=1}^{\infty} M_i$ is the set of all elements $(x_1, x_2, x_3, \ldots)$ in $\times_{i=1}^{\infty} M_i$ such that at most a finite number of the x_i's are nonzero. Then $\oplus_{i=1}^{\infty} M_i$ is a left R-module under componentwise addition and the same scalar multiplication defined on $\times_{i=1}^{\infty} M_i$. In fact, $\oplus_{i=1}^{\infty} M_i$ is a submodule of $\times_{i=1}^{\infty} M_i$. A proof of this is delayed until techniques have been developed that will enable us to quickly show that a nonempty subset of a module is a submodule. The module $\oplus_{i=1}^{\infty} M_i$ is said to be the **direct sum** of the family $\{M_i\}_{i \geq 1}$.

The following theorem presents results that hold for every module. The proof is straightforward, so it appears in the exercises. We use subscripts in part 1 of the theorem to add clarity. The notation 0_M denotes the additive identity of the module M, and 0_R denotes the additive identity of the ring.

11.1.3   Theorem

The following hold for every left R-module M.

1. For any $a \in R$ and each $x \in M$, $0_R x = 0_M$ and $a 0_M = 0_M$.
2. For any $a \in R$ and each $x \in M$, $(-a)x = a(-x) = -(ax)$.

In Definition 11.1.1, a submodule N of a left R-module M was defined to be a subgroup of M that is a left R-module under the scalar multiplication defined on M. The following theorem provides a quick method for determining whether a nonempty subset is a submodule of M.

11.1.4   Theorem

If N is a nonempty subset of a left R-module M, then the following are equivalent:

1. The set N is a submodule of M.
2. For all $x, y \in N$ and $a \in R$, $x + y$ and ax belong to N.
3. For all $x, y \in N$ and $a \in R$, $x - y$ and ax belong to N.

Proof

1 $\Rightarrow$ 2. Since N is a submodule of M, N is a subgroup of the additive group of M, and N is closed under the scalar multiplication defined on M. Hence, we see that N is closed under addition and that $ax \in N$ for each $x \in N$ and all $a \in R$.

2 $\Rightarrow$ 3. Let $x, y \in N$. Since $ax \in N$ for all $x \in N$ and each $a \in R$, we see, in particular, that $(-e)y \in N$. Therefore, $x + (-e)y \in N$ since N is closed under addition. Since $N \subseteq M$, $x + (-e)y \in M$, so, by part 2 of Theorem 11.1.3, we see that $x + (-e)y = x + [-(ey)] = x + (-y) = x - y$ in M. (Why couldn't we just do this in N? Because at this point we don't know that N is a module, and this is part of the assumption of Theorem 11.1.3.) Since the same equality holds in N as in M, it must be the case that $x + (-e)y = x - y$ in N. Therefore, $x - y \in N$.

3 $\Rightarrow$ 1. Since N is closed under subtraction, N is a subgroup of the additive group of M. We also have that $ax \in N$ for each $a \in R$ and all $x \in N$. Each of conditions 1 through 4 of Definition 11.1.1 must hold in N since they hold in M; that is, these properties are inherited from M. To see why this is the case, suppose, for example, that we can find $x, y \in N$ and $a \in R$ such that $a(x + y) \neq ax + ay$. Then since $N \subseteq M$, we have found $x, y \in M$ and $a \in R$ such that $a(x + y) \neq ax + ay$. But we know this cannot be the case since M is a left R-module. ∎

Theorem 11.1.4 makes the following theorem easy to prove. We provide a proof of part 1 and leave the proof of part 2 as an exercise.

11.1.5   Theorem

Let M be a left R-module.

1. If $N_1, N_2, \ldots, N_n$ are submodules of M, then $N = N_1 + N_2 + \cdots + N_n = \{x_1 + x_2 + \cdots + x_n \mid x_1 \in N_1, x_2 \in N_2, \ldots, x_n \in N_n\}$ is a submodule of M.

2. If $\{M_\alpha\}_{\alpha \in \Delta}$ is a family of submodules of M, then $\cap_{\alpha \in \Delta} M_\alpha$ is a submodule of M.

Proof.   First note that $N \neq \varnothing$ since $0_1 + 0_2 + \cdots + 0_n$ is in N, where $0_1 = 0_2 = \cdots = 0_n = 0$, the additive identity of M. Suppose that $x_1 + x_2 + \cdots + x_n$ and $y_1 + y_2 + \cdots + y_n$ are elements of N. Since M is an additive *abelian* group, the sum $(x_1 + x_2 + \cdots + x_n) + (y_1 + y_2 + \cdots + y_n)$ can be written as $(x_1 + y_1) + (x_2 + y_2) + \cdots + (x_n + y_n)$, and this is an element of N because

$x_i + y_i \in N_i$ for $i = 1, 2, \ldots, n$. Why? Because each N_i is a submodule of M and as such, by Theorem 11.1.4, is closed under addition. Finally, if $a \in R$, then $a(x_1 + x_2 + \cdots + x_n) = ax_1 + ax_2 + \cdots + ax_n \in N$ since $ax_i \in N_i$ for $i = 1, 2, \ldots, n$. Theorem 11.1.4 now shows that N is a submodule of M. ∎

11.1.6   Examples

1. We have seen in Example 1 of 11.1.2 that R is a left R-module and a right R-module. The R-submodules of R, when R is viewed as a left R-module, are the left ideals of R. Similarly, if R is viewed as a right R-module, the submodules of R are the right ideals of R.

2. If G is an additive abelian group, then Example 5 of 11.1.2 shows that G is a **Z**-module. The submodules of G are the subgroups of G.

3. If $x \in M$, then $Rx = \{ax \mid a \in R\}$ is a submodule of M. The module Rx is referred to as the cyclic submodule of M generated by x. If there is an $x \in M$ such that $Rx = M$, then M is said to be a **cyclic module**. If x_1, $x_2, \ldots, x_n \in M$, then Theorem 11.1.5 shows that $Rx_1 + Rx_2 + \cdots + Rx_n$ is a submodule of M. It is the **submodule generated by the set** $\{x_1, x_2, \ldots, x_n\}$.

4. If $M = \{0\}$, then M is the **zero module**. A nonzero left R-module M always has at least two submodules, namely, the zero module and M itself. If $M \neq \{0\}$ and if $\{0\}$ and M are the only submodules of M, then M is said to be a **simple** left R-module. Notice that a simple module must be cyclic. This follows because if $x \in M, x \neq 0$, then Rx is a nonzero submodule of M, so it must be the case that $Rx = M$.

5. If M_1 and M_2 are left R-modules, then $M_1 \times M_2$ is a left R-module under componentwise addition and the scalar multiplication on $M_1 \times M_2$ defined in Example 4 of 11.1.2. Both $M_1 \times \{0\}$ and $\{0\} \times M_2$ are submodules of $M_1 \times M_2$, but neither M_1 nor M_2 is a submodule of $M_1 \times M_2$. In general, if $\{M_i\}_{i=1}^{n}$ is a family of left R-modules, then $\{0\} \times \cdots \{0\} \times M_i \times \{0\} \times \cdots \times \{0\}$ is a submodule of the left R-module $\times_{i=1}^{n} M_i$ for $i = 1, 2, \ldots, n$.

6. Consider the left $M_2(R)$-module $\overline{R}^2$ of Example 2 of 11.1.2, and let $R = \mathbf{Q}$. Then $\overline{\mathbf{Q}}^2$ is a left $M_2(\mathbf{Q})$-module. If we let $N_1 = \left\{ \begin{pmatrix} a \\ 0 \end{pmatrix} \middle| a \in \mathbf{Q} \right\}$ and $N_2 = \left\{ \begin{pmatrix} 0 \\ b \end{pmatrix} \middle| b \in \mathbf{Q} \right\}$, then neither N_1 nor N_2 is a submodule of $\overline{\mathbf{Q}}^2$. Both are subgroups of the additive group $\overline{\mathbf{Q}}^2$, but neither is closed under the scalar multiplication defined on $\overline{\mathbf{Q}}^2$. For example,

$$\begin{pmatrix} 1 & 1 \\ \frac{1}{2} & -1 \end{pmatrix}\begin{pmatrix} 2 \\ 0 \end{pmatrix} = \begin{pmatrix} 2 \\ 1 \end{pmatrix} \notin N_1 \quad \text{and} \quad \begin{pmatrix} 1 & 1 \\ \frac{1}{2} & -1 \end{pmatrix}\begin{pmatrix} 0 \\ 2 \end{pmatrix} = \begin{pmatrix} 2 \\ -2 \end{pmatrix} \notin N_2.$$

7. For this example, we return to Example 8 of 11.1.2, where it was indicated but not shown that $\oplus_{i=1}^{\infty} M_i$ is a submodule of $\times_{i=1}^{\infty} M_i$. This now follows easily from Theorem 11.1.4. First, note that the set $\oplus_{i=1}^{\infty} M_i$ is nonempty since $(0_1, 0_2, 0_3, \ldots)$ is in $\oplus_{i=1}^{\infty} M_i$. It also follows that the sum of any two elements of $\oplus_{i=1}^{\infty} M_i$ can have at most a finite number of nonzero components, and the same is true of an element of $\oplus_{i=1}^{\infty} M_i$ times a ring element. Hence, $\oplus_{i=1}^{\infty} M_i$ is closed under addition and under multiplication of elements of $\oplus_{i=1}^{\infty} M_i$ by ring elements. Therefore, $\oplus_{i=1}^{\infty} M_i$ is indeed a submodule of $\times_{i=1}^{\infty} M_i$.

If N is a normal subgroup of a group G, we saw in Section 2.2 that we can form the factor group G/N. Similarly, when I is an ideal of a ring R, we can form the factor ring R/I. An analogous situation holds for modules; if N is a submodule of a left R-module M, we can form the **factor module** M/N.

11.1.7   Theorem

If N is a submodule of a left R-module M, then the set of all cosets $M/N = \{x + N \mid x \in M\}$ can be made into a left R-module by defining

$$(x + N) + (y + N) = (x + y) + N \qquad \text{and}$$
$$a(x + N) = ax + N$$

for all $x + N, y + N \in M/N$ and all $a \in R$.

Proof. Since N is a subgroup of the additive abelian group M, N is a normal subgroup of M. The results of Section 2.2 show that M/N is an additive abelian group under coset addition. It remains then only to show that if we set $a(x + N) = ax + N$ for all $x + N \in M/N$ and $a \in R$, this multiplication of elements of M/N by ring elements is well-defined and makes M/N into a left R-module. To show that this operation is well-defined, suppose that $x + N$ and $y + N$ are cosets in M/N such that $x + N = y + N$, and let $a, b \in R$ be such that $a = b$. Since $x + N = y + N$, $x - y \in N$, so $x - y = z$ for some $z \in N$. But multiplication of elements of M by ring elements is well-defined, so $a(x - y) = az$, which gives $ax = ay + az$. Since scalar multiplication on M is well-defined, we see that $ax = by + bz$, and this gives $ax - by = bz \in N$. Therefore, $ax + N = by + N$ or $a(x + N) = b(y + N)$, and this tells us that the operation is well-defined.

The fact that conditions 1 through 4 of Definition 11.1.1 hold follows directly from the fact that these properties hold for the left R-module M. For example, to show that condition 1 holds, suppose $a, b \in R$ and that $x + N$ is a coset in M / N. Then

$$(a + b)(x + N) = (a + b)x + N$$
$$= (ax + bx) + N$$
$$= (ax + N) + (bx + N)$$
$$= a(x + N) + b(x + N).$$

The proofs of conditions 2, 3, and 4 are just as straightforward. ∎

11.1.8   Examples

1. The set $M_2(R)$ of all 2×2 matrices with entries from the ring R is a left R-module under matrix addition and scalar multiplication defined by

$$a \begin{pmatrix} c_{11} & c_{12} \\ c_{21} & c_{22} \end{pmatrix} = \begin{pmatrix} ac_{11} & ac_{12} \\ ac_{21} & ac_{22} \end{pmatrix}.$$

The set $N = \left\{ \begin{pmatrix} c_{11} & 0 \\ c_{21} & 0 \end{pmatrix} \middle| c_{11}, c_{21} \in R \right\}$ is a submodule of $M_2(R)$. Cosets in the factor module $M_2(R) / N$ look like

$$\begin{pmatrix} c_{11} & c_{12} \\ c_{21} & c_{22} \end{pmatrix} + N = \begin{pmatrix} 0 & c_{12} \\ 0 & c_{22} \end{pmatrix} + \begin{pmatrix} c_{11} & 0 \\ c_{21} & 0 \end{pmatrix} + N$$

$$= \begin{pmatrix} 0 & c_{12} \\ 0 & c_{22} \end{pmatrix} + N.$$

2. The set $P_4(R) = \{a_0 + a_1x + a_2x^2 + a_3x^3 + a_4x^4 \mid a_i \in R \text{ for } i = 0, 1, 2, 3, 4\}$ is a left R-module under polynomial addition and scalar multiplication of elements of $P_4(R)$ by ring elements $a \in R$ given by

$$a(a_0 + a_1x + a_2x^2 + a_3x^3 + a_4x^4) =$$
$$(aa_0) + (aa_1)x + (aa_2)x^2 + (aa_3)x^3 + (aa_4)x^4.$$

The set $N = \{a_0 + a_2x^2 + a_4x^4 \mid a_0, a_2, a_4 \in R\}$ is a submodule of $P_4(R)$. Cosets in the factor module $P_4(R) / N$ can be written in the form $(a_1x + a_3x^3) + N$. This follows since

$$a_0 + a_1 x + a_2 x^2 + a_3 x^3 + a_4 x^4 + N = (a_1 x + a_3 x^3) + (a_0 + a_2 x^2 + a_4 x^4) + N$$
$$= (a_1 x + a_3 x^3) + N.$$

3. If M_1 and M_2 are left R-modules, then $M_1 \times M_2$ is a left R-module under the operations established in Example 4 of 11.1.2. Both $M_1 \times \{0\}$ and $\{0\} \times M_2$ are submodules of $M_1 \times M_2$. You should determine what the cosets look like in the factor modules $(M_1 \times M_2) / (\{0\} \times M_2)$ and $(M_1 \times M_2) / (M_1 \times \{0\})$.

As with groups, rings, and vector spaces, there are mappings between left R-modules that preserve the algebraic structure of the modules. Before we can consider examples, a formal definition of these mappings is required.

11.1.9   Definition

Let M and N be left R-modules. A function $f : M \longrightarrow N$ such that

$$f(x + y) = f(x) + f(y) \qquad \text{and}$$
$$f(ax) = af(x)$$

for all $x, y \in M$ and $a \in R$ is said to be an **R-linear mapping**. Such mappings are also sometimes referred to as **R-homomorphisms** or **R-module homomorphisms**. An R-module homomorphism that is injective is an **R-monomorphism**, and one that is surjective is an **R-epimorphism**. If $f : M \longrightarrow N$ is an R-monomorphism, then we say that M **embeds** in N, and if $f : M \longrightarrow N$ is an R-epimorphism, then N is a **homomorphic image** of M. An R-linear mapping that is both injective and surjective is said to be an **R-module isomorphism**. If $f : M \longrightarrow N$ is an R-linear isomorphism, then we say that N is a **copy** of M. The set $\ker f = \{x \in M \mid f(x) = 0\}$ is the **kernel** of the R-linear mapping $f : M \longrightarrow N$, and $f(M) = \{f(x) \mid x \in M\}$ is the **image** of f. The image of f is denoted by $\operatorname{Im} f$ or by $f(M)$.

Notice the similarity between the definition of an R-linear mapping and Definition 10.2.2 of a linear transformation. Actually, a linear transformation is just a special case of an R-linear mapping. When the ring is a field, the modules involved are vector spaces, and an R-linear mapping is a linear transformation.

11.1.10 | Examples

1. If M is a left R-module, then for any submodule N of M, the mapping $\eta : M \longrightarrow M/N : x \longmapsto x + N$ is an R-epimorphism. It is referred to as the **canonical R-epimorphism** from M to the factor module M/N.

2. If M_1 and M_2 are left R-modules, we have seen that $M_1 \times M_2$ is a left R-module. The mappings $k_1 : M_1 \longrightarrow M_1 \times M_2 : x \longrightarrow (x, 0)$ and $k_2 : M_2 \longrightarrow M_1 \times M_2 : y \longrightarrow (0, y)$ are both R-monomorphisms called the **first** and **second canonical injections**, respectively. It is a straight forward proof to show that $f_1 : M_1 \longrightarrow M_1 \times \{0\}$ defined by $f_1(x) = (x,0)$ and $f_2 : M_2 \longrightarrow \{0\} \times M_2$ given by $f_2(y) = (0, y)$ are both R-isomorphisms. Hence, $M_1 \times M_2$ contains submodules that are isomorphic to M_1 and M_2.

    Similarly, the mappings $pr_1 : M_1 \times M_2 \longrightarrow M_1 : (x, y) \longmapsto x$ and $pr_2 : M_1 \times M_2 \longrightarrow M_2 : (x, y) \longmapsto y$ are R-epimorphisms. These mappings are called the **first** and **second canonical projections**, respectively.

3. If $R[x]$ is the additive group of polynomials in x with their coefficients on R, then $R[x]$ is a left R-module under polynomial addition and under scalar multiplication defined on $R[x]$ by

    $$a(a_0 + a_1 x + a_2 x^2 + \cdots + a_n x^n) = (aa_0) + (aa_1)x + (aa_2)x^2 + \cdots + (aa_n)x^n.$$

    If $D[f(x)]$ denotes the formal derivative of a polynomial $f(x) \in R[x]$, then

    $$D[f(x) + g(x)] = D[f(x)] + D[g(x)] \qquad \text{and}$$
    $$D[af(x)] = aD[f(x)]$$

    for all $f(x), g(x) \in R[x]$ and $a \in R$. Hence, D is an R-linear mapping from $R[x]$ to $R[x]$. (See Exercise 18, Problem Set 6.1.)

4. Let M_1 and M_2 be left R-modules such that $M_1 \neq M_2$. Then $M_1 \times M_2$ and $M_2 \times M_1$ are isomorphic left R-modules, but $M_1 \times M_2 \neq M_2 \times M_1$. The mapping $f : M_1 \times M_2 \longrightarrow M_2 \times M_1$, defined by $f((x_1, x_2)) = (x_2, x_1)$, is easily shown to be an R-module isomorphism.

5. The set $M_2(R)$ of 2×2 matrices over R is a left R-module under the operations of matrix addition and scalar multiplication defined on

    $M_2(R)$ by $a \begin{pmatrix} a_{11} & a_{12} \\ a_{21} & a_{22} \end{pmatrix} = \begin{pmatrix} aa_{11} & aa_{12} \\ aa_{21} & aa_{22} \end{pmatrix}$ We also see from Example 4 of

    11.1.2 that R^4 is a left R-module. The mapping given by

    $$\begin{pmatrix} a_{11} & a_{12} \\ a_{21} & a_{22} \end{pmatrix} \longmapsto (a_{11}, a_{12}, a_{21}, a_{22})$$

from $M_2(R)$ to R^4 is an R-linear isomorphism. Hence, $M_2(R)$ and R^4 are isomorphic as left R-modules. In general, for each integer $n \geq 2$, the left R-modules $M_n(R)$ and R^{n^2} are isomorphic.

The proof of the following theorem is left as an exercise. A proof can be fashioned after those given for similar properties that hold for group and ring homomorphisms.

11.1.11 Theorem

Let $f : M \longrightarrow N$ be an R-linear mapping.

1. The set $\ker f$ is a submodule of M, and f is injective if and only if $\ker f = \{0\}$.
2. If X is a submodule of M, $f(X) = \{f(x) \mid x \in X\}$ is a submodule of N.
3. If Y is a submodule of N, $f^{-1}(Y) = \{x \in M \mid f(x) \in Y\}$ is a submodule of M.

Each of the following theorems is analogous to a theorem for groups given in Section 2.3. Each is also analogous to a theorem for rings presented in Section 3.3. You should compare the following theorems with Theorems 2.3.11, 2.3.14, 2.3.15, and 2.3.16 and with Theorems 3.3.12, 3.3.15, 3.3.16, and 3.3.17. As you read the following theorems, fill in the details of each proof by providing an argument for each statement that is left unproven. Be sure to show that each mapping is well-defined.

11.1.12 First Isomorphism Theorem for Modules

If $f : M \longrightarrow N$ is an R-linear epimorphism, then $M / \ker f$ and N are isomorphic.

Proof. If $\phi : M / \ker f \longrightarrow N$ is defined by $\phi(x + \ker f) = f(x)$, then ϕ is well-defined and such that

$$
\begin{aligned}
\phi((x + \ker f) + (y + \ker f)) &= \phi((x + y) + \ker f) \\
&= f(x + y) \\
&= f(x) + f(y) \\
&= \phi(x + \ker f) + \phi(y + \ker f).
\end{aligned}
$$

In addition, for any $a \in R$,

$$\phi(a(x + \ker f)) = \phi(ax + \ker f)$$
$$= f(ax)$$
$$= af(x)$$
$$= a\phi(x + \ker f).$$

Hence, ϕ is an R-linear mapping that is obviously surjective. If $\phi(x + \ker f) = 0$, then $f(x) = 0$. Thus, $x \in \ker f$, which indicates that $x + \ker f = 0$. Therefore, $\ker \phi = \{0\}$, so ϕ is injective. ∎

11.1.13   Corollary

If $f : M \longrightarrow N$ is an R-linear mapping, then $M / \ker f$ and $f(M)$ are isomorphic.

11.1.14   Second Isomorphism Theorem for Modules

If L and N are submodules of a left R-module M, then $L/(L \cap N)$ and $(L + N)/N$ are isomorphic left R-modules.

Proof.   If $f : L \longrightarrow (L + N)/N$ is given by $f(x) = x + N$, then f is a well-defined R-linear epimorphism. Theorem 11.1.12 indicates that $L/\ker f$ and $(L + N)/N$ are isomorphic. Since $\ker f = L \cap N$, this proves the theorem. ∎

11.1.15   Third Isomorphism Theorem for Modules

If L and N are submodules of a left R-module M with $L \subseteq N$, then $(M/L)/(N/L)$ and M/N are isomorphic.

Proof.   Let $f : M/L \longrightarrow M/N$ be defined by $f(x + L) = x + N$. Since $L \subseteq N$, f is well-defined. Furthermore, it's easy to show that f is an R-epimorphism with $\ker f = N/L$. Theorem 11.1.12 now shows that $(M/L)/(N/L)$ and M/N are isomorphic. ∎

11.1.16   Correspondence Theorem for Modules

If $f : M \longrightarrow N$ is an R-epimorphism, then there is a one-to-one correspondence between the submodules of M that contain $\ker f$ and the submodules of N.

Proof.   The proof is left as an exercise. (See Theorem 2.3.16.) ∎

Problem Set 11.1

1. Consider the set M of all 2×2 matrices of the form $\begin{pmatrix} a & b \\ b & a \end{pmatrix}$, where $a, b \in R$. Decide whether M is a left R-module if addition is matrix addition and scalar multiplication of elements of M by ring elements is defined as $c \begin{pmatrix} a & b \\ b & a \end{pmatrix} = \begin{pmatrix} ca & -b \\ -b & ca \end{pmatrix}$. If M is a left R-module, give a proof. If M is not a left R-module, show why not.

2. Consider the set M of all 3×3 matrices with entries from R. Make M into a left R-module by defining addition on M as matrix addition and scalar multiplication by

    $$a \begin{pmatrix} a_{11} & a_{12} & a_{13} \\ a_{21} & a_{22} & a_{23} \\ a_{31} & a_{32} & a_{33} \end{pmatrix} = \begin{pmatrix} aa_{11} & aa_{12} & aa_{13} \\ aa_{21} & aa_{22} & aa_{23} \\ aa_{31} & aa_{32} & aa_{33} \end{pmatrix}$$

    (a) If N is the set of all matrices in $M_3(R)$ of the form

    $$\begin{pmatrix} a_{11} & 0 & 0 \\ 0 & a_{22} & 0 \\ 0 & 0 & a_{33} \end{pmatrix},$$

    decide if N is a submodule of $M_3(R)$. If N is a submodule of $M_3(R)$, show why, and if N is not a submodule of $M_3(R)$, show why not.

    (b) If N is the set of all matrices in $M_3(R)$ of the form

    $$\begin{pmatrix} e & a_{12} & a_{13} \\ a_{21} & e & a_{23} \\ a_{31} & a_{32} & e \end{pmatrix},$$

    decide if N is a submodule of $M_3(R)$. If N is a submodule of $M_3(R)$, show why, and if N is not a submodule of $M_3(R)$, show why not.

3. Read the discussion following Definition 11.1.1, which demonstrates that a left R-module M cannot be made into a right R-module by defining $x * a = ax$ for all $a \in R$ and $x \in M$. Find a left R-module M and elements $a, b \in R$ and $x \in M$ such that $(ab)x \neq (ba)x$. Hint: See Example 2 of 11.1.2, and let $R = \mathbf{R}$.

4. Prove Theorem 11.1.3.

5. Show that there are left R-modules M in which $ax = 0$, where both $a \in R$ and $x \in M$ are nonzero. Prove that this is not possible in a vector space.

6. Let N be a submodule of a left R-module M, and suppose that $x + N$ and $y + N$ are cosets of N in M.

    (a) Prove that $x + N = y + N$ if and only if $x - y \in N$.
    (b) Prove that either $(x + N) \cap (y + N) = \varnothing$ or $x + N = y + N$.

7. Prove part 2 of Theorem 11.1.5.

8. If M is a left R-module, prove that $A = \{a \in R \mid ax = 0 \text{ for all } x \in M\}$ is an ideal of R. The ideal A is referred to as the **annihilator** of M.

9. In Example 3 of 11.1.6, if $x_1, x_2, \ldots, x_n$ are elements of a left R-module M, then $Rx_1 + Rx_2 + \cdots + Rx_n$ is the submodule of M generated by the set $\{x_1, x_2, \ldots, x_n\}$. Prove that $Rx_1 + Rx_2 + \cdots + Rx_n$ is the intersection of all the submodules of M that contain the set $\{x_1, x_2, \ldots, x_n\}$.

10. In the proof of Theorem 11.1.7, it was shown that condition 1 in Definition 11.1.1 of a left R-module holds for the factor module M / N. Show that conditions 2, 3, and 4 also hold.

11. Show that the mapping $\eta : M \longrightarrow M / N$ of Example 1 of 11.1.10 is a well-defined R-module epimorphism.

12. Let k_1, k_2, pr_1, and pr_2 be the R-linear mapping defined in Example 2 of 11.1.10.

    (a) Show that each of these mappings is well-defined and R-linear.
    (b) Show that $k_1 \circ pr_1 + k_2 \circ pr_2 = 1_{M_1 \times M_2}$ where $1_{M_1 \times M_2}$ is the identity mapping $M_1 \times M_2 \longrightarrow M_1 \times M_2 : (x, y) \longmapsto (x, y)$.
    (c) Show that $pr_1 \circ k_1$ and $pr_2 \circ k_2$ are the identity mappings $M_1 \longrightarrow M_1$ and $M_2 \longrightarrow M_2$, respectively.
    (d) Show that if $pr_1 \circ k_1 \times pr_2 \circ k_2 : M_1 \times M_2 \longrightarrow M_1 \times M_2$ is defined by $pr_1 \circ k_1 \times pr_2 \circ k_2((x, y)) = (pr_1 \circ k_1(x), pr_2 \circ k_2(y))$, then $pr_1 \circ k_1 \times pr_2 \circ k_2 = 1_{M_1 \times M_2}$.

13. If $M_1 \subseteq M_2 \subseteq \cdots \subseteq M_n \subseteq \cdots$ is an ascending chain of submodules of a left R-module M, prove that $\cup_{i \geq 1} M_i$ is a submodule of M.

14. Let N be a submodule of a left R-module M, and suppose that K is a left ideal of R.

    (a) Show that KN is a submodule of M, where KN is the set of all finite sums of the form $a_1 x_1 + a_2 x_2 + \cdots + a_n x_n$ with $x_i \in N$ and $a_i \in K$ for $i = 1, 2, \ldots, n$. The integer n is not fixed; a finite sum can contain any finite number of summands.

(b) Show also that if K is an ideal of R, then M/KM is an R/K-module if addition and scalar multiplication are defined on M/KM by

$$(x + KM) + (y + KM) = (x + y) + KM \qquad \text{and}$$
$$(a + K)(x + KM) = ax + KM.$$

15. Prove that the mapping $D : R[x] \longrightarrow R[x]$ given in Example 3 of 11.1.10 is a well-defined R-module homomorphism.

16. Show that the mapping given in Example 4 of 11.1.10 is a well-defined R-module isomorphism.

17. Show that the mapping $\begin{pmatrix} a_{11} & a_{12} \\ a_{21} & a_{22} \end{pmatrix} \longmapsto (a_{11}, a_{12}, a_{21}, a_{22})$ from $M_2(R)$ to R^4

given in Example 5 of 11.1.10 is an R-module isomorphism.

18. Prove Theorem 11.1.11.

19. If $f : R \longrightarrow S$ is a ring homomorphism such that $f(e_R) = e_s$ and if M is a left S-module, show that M can be made into a left R-module by defining scalar multiplication $*$ of elements $x \in M$ by ring elements $a \in R$ as $a * x = f(a)x$.

20. Let M and N be left R-modules, and suppose that $f : M \longrightarrow N$ is an R-linear mapping. If X is a submodule of M such that $X \subseteq \ker f$, prove that the mapping $f^* : M/X \longrightarrow N$ defined by $f^*(x + X) = f(x)$ is a well-defined R-linear mapping. We say that f^* is the **mapping induced by** f.

21. (a) If M is a left R-module and $x \in M$, show that $(0 : x) = \{a \in R \mid ax = 0\}$ is a left ideal of R.
    (b) If M is a left R-module and $x \in M$, show that $R/(0 : x)$ is a left R-module.
    (c) If M is a left R-module and there is an $x \in M$ such that $Rx = M$, then M is said to be a **cyclic module**. If $Rx = M$, prove that M and $R/(0 : x)$ are isomorphic left R-modules. Hint: Show that $f : R \longrightarrow Rx$ defined by $f(a) = ax$ is an R-linear epimorphism. What is the kernel of f?

22. (a) If $f : M \longrightarrow N$ and $g : N \longrightarrow X$ are R-linear mappings, show that $g \circ f : M \longrightarrow X$ is an R-linear mapping.
    (b) If $f : M \longrightarrow N$ is an R-linear isomorphism, then f is a bijective function with an inverse function $f^{-1} : N \longrightarrow M$. Prove that f^{-1} is a well-defined R-linear isomorphism.

23. If M_1 and M_2 are left R-modules, it was pointed out in Example 5 of 11.1.6 that $\{0\} \times M_2$ and $M_1 \times \{0\}$ are submodules of $M_1 \times M_2$. Prove that $(M_1 \times M_2)/(\{0\} \times M_2)$ and M_1 are isomorphic and that $(M_1 \times M_2)/(M_1 \times \{0\})$ and M_2 are isomorphic. Hint: Consider the first

and second canonical projections pr_1 and pr_2, respectively, of Example 2 of 11.1.10, and use Theorem 11.1.12.

24. If $n \geq 2$ is an integer, prove that if α is any permutation in $\mathbf{S}_n$ and $\{M_i\}_{i=1}^{n}$ is a family of left R-modules, then $M_1 \times M_2 \times \cdots \times M_n$ and $M_{\alpha(1)} \times M_{\alpha(2)} \times \cdots \times M_{\alpha(n)}$ are isomorphic. Hint: See Example 4 of 11.1.10 and Exercise 16.

25. Prove Theorem 11.1.16.

11.2   MODULES OF QUOTIENTS

Section Overview. The method of construction used in Section 4.1 to develop the field of fractions of an integral domain will now be generalized to modules over a general commutative ring with identity. Be sure to compare what is developed in this section with the procedure presented in Section 4.1. You will find that the construction of the field of fractions of an integral domain is a special case of the theory developed here when the module under consideration is an integral domain viewed as a module over itself. Exercise 10 of Problem Set 4.1 is also relevant. You may find it helpful to redo that exercise before reading this section.

When R is an integral domain, the field of fractions of R is just the set of all fractions $\frac{a}{b}$ where $a, b \in R$, $b \neq 0$. These fractions can be added and multiplied in a manner that is analogous with adding and multiplying rational numbers. In this section we consider a similar situation where the quotients $\frac{x}{s}$ are such that x is an element of a module and s is an element taken from a certain type of subset S of R. We will see that it is possible to make this set of ratios into an R-module. In addition, when the module is a ring viewed as a module over itself, these quotients can be made into a ring.

We now assume that R is a commutative ring with identity. Since R is commutative, an R-module M is both a left and a right R-module, and we can write $xa = ax$ for all $a \in R$ and $x \in M$. Because of this, there is no need to distinguish the side of scalar multiplication, so M is referred to simply as an R-module.

If $Q(D)$ is the field of fractions of an integral domain D, then addition and multiplication of elements of $Q(D)$ are defined as

$$\frac{a}{b} + \frac{c}{d} = \frac{ad + bc}{bd} \qquad \text{and} \qquad \frac{a}{b} \cdot \frac{c}{d} = \frac{ac}{bd}.$$

The set D^* of nonzero elements of D serves as the denominators of the fractions in $Q(D)$. If we are to generalize the procedure for constructing the field of fractions of D to modules, several properties possessed by D^* must be observed. First, $b, d \in D^*$ implies that $bd \neq 0$, so $bd \in D^*$, that is, D^* is closed under multiplication. Notice also that $e \in D^*$ and, of course, $0 \notin D^*$. As we shall see, the elements s of any nonempty subset S of R that have these same properties can serve as the denominators of quotients $\frac{x}{s}$, where x is an element from an R-module M. These properties of the set of nonzero elements of an integral domain lead to the definition of a multiplicatively closed set in a commutative ring R.

11.2.1   Definition

A nonempty subset S of R is said to be a **multiplicatively closed set** in R if $e \in S, 0 \notin S$, and $st \in S$, whenever $s, t \in S$.

11.2.2   Examples

1. If $a \in R$ is such that $a^n \neq 0$ for any integer $n \geq 0$, then $S = \{a^n \mid n \in \mathbf{Z}, n \geq 0\}$ is a multiplicatively closed set in R. This follows because if $a^m, a^n \in S$, then $a^m a^n = a^{m+n} \in S$. Since $a^0 = e, e \in S$.

2. If $R[x]$ is the ring of polynomials in x over R, then the set $S = \{e, x, x^2, x^3, \dots\}$ is a multiplicatively closed set in $R[x]$. If $R[[x]]$ is the ring of formal power series in x over R, then S is also a multiplicatively closed set in $R[[x]]$.

3. The set $S = \{e\}$ is clearly closed under multiplication.

4. A nonzero element $a \in R$ is said to be a **regular element** of R if a is not a zero-divisor. We claim that if S is the set of all regular elements of R, then S is a multiplicatively closed set in R. First, note that if $a, b \in S$, then $ab \neq 0$ since both a and b are nonzero and neither a nor b is a zero-divisor. Next, we assert that ab cannot be a zero-divisor in R. To see this, suppose that ab is a zero-divisor. Then there is a $c \in R, c \neq 0$, such that $(ab)c = 0$. But $a(bc) = (ab)c = 0$, so since $bc \neq 0$, a is a zero-divisor. This implies that $a \notin S$, and we have a contradiction. Hence, if $a, b \in S$, then ab is a regular element of R and thus $ab \in S$.

5. If D is an integral domain, then D^*, the set of nonzero elements of D, is multiplicatively closed. This is just a special case of Example 4 since D^* is the set of nonzero-divisors of D.

6. Recall that an ideal P of R is said to be a prime ideal of R, if whenever $ab \in P$, either $a \in P$ or $b \in P$. If P is a prime ideal of R, then we claim that $S = R \setminus P$ is a multiplicatively closed set in R. Indeed, let $a, b \in S$. Then either $ab \in S$ or $ab \notin S$. If $ab \notin S$, then $ab \in P$, so either $a \in P$ or $b \in P$ since P is prime. If $a \in P$, then $a \notin S$, and if $b \in P$, then $b \notin S$. Hence, we have arrived at the contradictory statement that if $a, b \in S$ and $ab \notin S$, then either $a \notin S$ or $b \notin S$. Therefore, if $a, b \in S$, it must be the case that $ab \in S$, so S is multiplicatively closed.

In Lemma 4.1.1 for an integral domain D with nonzero elements D^*, the relation E was defined on $D \times D^*$ by $(a, b)E(c, d)$ if and only if $ad = bc$ where $(a, b), (c, d) \in D \times D^*$, and it was shown that E is an equivalence relation. Remember that this just says that the product of the two outside members of the ordered pairs in $(a, b)E(c, d)$ is equal to the product of the inside members as indicated in the following figure.

Now suppose that M is an R-module, and let S be a multiplicatively closed set in R. Define the relation E on $S \times M$ by $(s, x)E(t, y)$ if and only if $sy = tx$. Again, we are just setting the product of two outside members in the ordered pairs in $(s, x)E(t, y)$ equal to the product of the two inside members. Since R is commutative, M is a left R-module as well as a right R-module, and we can switch sides with scalar multiplication at will. For this reason, we could just as well carry out our development with $M \times S$ in place of $S \times M$. However, since we are dealing primarily with left R-modules, $S \times M$ is the preferred notation. Now let's see whether we can prove that $(s, x)E\ (t, y)$ if and only if $sy = tx$ is an equivalence relation on $S \times M$. That E is reflexive and symmetric follows directly from the fact that R is commutative. So it remains to be shown that E is transitive. Let $(s, x), (t, y), (u, z) \in S \times M$ be such that $(s, x)E(t, y)$ and $(t, y)E(u, z)$. Then (1) $sy = tx$ and (2) $tz = uy$. Multiplying both sides of equation (1) by u and both sides of equation (2) by s gives $suy = tux$ and $tsz = suy$, which in turn implies that $tsz = tux$ or $t(sz - ux) = 0$. In Section 4.1, we assumed that $M = D$ was an integral domain and that $S = D^*$. In that case, $t \in D^*$, so t is not a zero-divisor in D. Hence, $t(sz - ux) = 0$ implies that $sz = ux$ in D, so $(s, x)E(u, z)$. Therefore, when the module under consideration is an integral domain, viewed as a module over itself, E is transitive.

Returning to $S \times M$, we have $t \in S$, so $t \neq 0$. Since it is possible that $t \neq 0$ and $sz - ux \neq 0$ and yet $t(sz - ux) = 0$ in M (see Exercise 5 of Problem Set 11.1.), the technique used for showing that E is transitive on $D \times D^*$ cannot be used to show that E is transitive on $S \times M$. It appears that we have run into an obstruction for developing a module of quotients. However, there is a way around this difficulty. The trick is to take the obstruction into consideration when defining the equivalence relation on $S \times M$. This is exactly what is done in the following lemma. Although the "new" definition of E on $S \times M$ makes the proofs of the following lemmas and theorems somewhat more technical, the beauty is that it clears the way for the development of a module of quotients.

We now assume that S is a multiplicatively closed set in R and that M is an R-module.

11.2.3   Lemma

The relation E, defined on $S \times M$ by $(s, x)E(t, y)$ if and only if there is a $u \in S$ such that $u(sy - tx) = 0$, is an equivalence relation on $S \times M$.

Proof.   If $(s, x) \in S \times M$, then $sx = sx$, so we let $u = e$. Hence, $(s, x)E(s, x)$, so E is reflexive. Likewise, if $(s, x)E(t, y)$ with $(s, x), (t, y) \in S \times M$, then there is a $u \in S$ such that $u(sy - tx) = 0$. Hence, $u(tx - sy) = 0$, so $(t, y)E(s, x)$. Thus, E is symmetric. Finally, suppose that $(s, x), (t, y), (u, z) \in S \times M$ are such that $(s, x)E(t, y)$ and $(t, y)E(u, z)$. Then there are $w, w' \in S$ such that

$$wsy - wtx = w(sy - tx) = 0 \qquad \text{and} \tag{1}$$

$$w'tz - w'uy = w'(tz - uy) = 0. \tag{2}$$

Multiplying both sides of equation (1) by $w'u$ and both sides of equation (2) by ws produces

$$ww'suy - ww'tux = 0 \qquad \text{and} \tag{3}$$

$$ww'stz - ww'suy = 0. \tag{4}$$

Adding equations (3) and (4) gives $ww't(sz) - ww't(ux) = 0$ or $ww't$ $(sz - ux) = 0$. But $ww't \in S$, so $(s, x)E(u, z)$. This shows that E is transitive, and the proof is finished. ■

The notation $\frac{x}{s}$ is used to denote the equivalence class determined by the ordered pair $(s, x) \in S \times M$ under the equivalence relation E defined on $S \times M$ in Lemma 11.2.3. We let $S^{-1}M$ denote the set of all such equivalence classes. Our goal is to show that $S^{-1}M$ can be made into an R-module.

In the process of constructing the field of fractions of an integral domain D, addition of fractions in $Q(D)$ was defined by $\frac{a}{b} + \frac{c}{d} = \frac{ad+bc}{bd}$. A similar procedure can be used to turn $S^{-1}M$ into an additive abelian group. The first step is to show that the equivalence relation E of Lemma 11.2.3 is a congruence relation on $S \times M$ if addition is defined on $S \times M$ in an appropriate manner. After this, Theorem 0.4.8 can be invoked to make $S^{-1}M$ into an additive abelian group.

11.2.4   **Lemma**

If addition is defined on $S \times M$ by $(s, x) + (t, y) = (st, sy + tx)$ for all $(s, x), (t, y) \in S \times M$, then this addition is well-defined, and the equivalence relation E of Lemma 11.2.3 is a congruence relation on $S \times M$ with respect to this operation. Furthermore, if addition is defined on $S^{-1}M$ by $\frac{x}{s} + \frac{y}{t} = \frac{sy + tx}{st}$, for all $\frac{x}{t}, \frac{y}{t} \in S^{-1}M$, then $S^{-1}M$ is an additive abelian group under this operation.

Proof. We show first that the binary operation defined on $S \times M$ is well-defined. To show the operations are well-defined, suppose that $(s, x) = (s', x')$ and $(t, y) = (t', y')$. It follows from the definition of equality of ordered pairs that $s = s'$, $x = x'$, $t = t'$, and $y = y'$. Since addition and scalar multiplication are well-defined on M, $sy + tx = s'y' + t'x'$ and $st = s't'$. Hence, $(st, sy + tx) = (s't', s'y' + t'x')$, so $(s, x) + (t, y) = (s', x') + (t', y')$. Therefore, the operation is well-defined.

Finally, we need to show that E is a congruence relation on $S \times M$ with respect to this operation. If $(s, x)E(s', x')$ and $(t, y)E(t', y')$, it must be shown that

$$((s,x) + (t,y))E((s',x') + (t',y')),$$

which is the same as showing that

$$(st, sy + tx)E(s't', s'y' + t'x').$$

If $(s, x)E(s', x')$ and $(t, y)E(t', y')$, then there are $w, w' \in S$ such that

$$wsx' - ws'x = w(sx' - s'x) = 0$$

$$w'ty' - w't'y = w'(ty' - t'y) = 0$$

or

$$wsx' = ws'x \tag{1}$$

$$w't y' = w't'y. \tag{2}$$

Multiplying both sides of equation (1) by $w'tt'$ and both sides of equation (2) by wss' gives

$$ww'stt'x' = ww's'tt'x, \tag{3}$$

$$ww'ss'ty' = ww'ss't'y. \tag{4}$$

Adding equations (3) and (4) produces

$$ww'stt'x' + ww'ss'ty' = ww's'tt'x + ww'ss't'y \qquad \text{or}$$

$$ww'(st(s'y' + t'x')) = ww'(s't'(sy + tx)),$$

and this gives

$$ww'(s't'(sy + tx) - st(s'y' + t'x')) = 0.$$

Since $ww' \in S$, we have $(st, sy + tx)E(s't', s'y' + t'x')$, so E is a congruence relation on $S \times M$. If we now invoke Theorem 0.4.8, the congruence-class addition given by $\frac{x}{s} + \frac{y}{t} = \frac{sy + tx}{st}$ is well-defined. This makes $S^{-1}M$ into an additive abelian group with additive identity $\frac{0}{e}$, and the additive inverse of $\frac{x}{s}$ is $-\frac{x}{s} = \frac{-x}{s}$. The proof of the fact that addition is associative is left as an exercise. ∎

Remember that the goal is to make $S^{-1}M$ into a left R-module. It was just shown that $S^{-1}M$ can be turned into an additive abelian group, so it remains to be shown that a scalar multiplication can be defined on $S^{-1}M$ that satisfies conditions 1 through 4 of Definition 11.1.1. *Throughout the remainder of the section,* E *denotes the congruence relation of* Lemmas 11.2.3 *and* 11.2.4.

11.2.5   Theorem

Let M be an R-module, and suppose that S is a multiplicatively closed set in R. If addition is defined on $S^{-1}M$ as in Lemma 11.2.4 and scalar multiplication is defined on $S^{-1}M$ by $a \cdot \frac{x}{s} = \frac{ax}{s}$ for all $a \in R$ and $\frac{x}{s} \in S^{-1}M$, then $S^{-1}M$ is an R-module.

Proof.   Lemma 11.2.4 shows that $S^{-1}M$ is an additive abelian group if addition is defined on $S^{-1}M$ by $\frac{x}{s} + \frac{y}{t} = \frac{sy + tx}{st}$ for all $\frac{x}{s}, \frac{y}{t} \in S^{-1}M$, so define scalar multiplication on $S^{-1}M$ by $a \cdot \frac{x}{s} = \frac{ax}{s}$ for $a \in R$ and $\frac{x}{s} \in S^{-1}M$. The first requirement is to show that this operation is well-defined. Suppose that $a = a'$ and $\frac{x}{s} = \frac{x'}{s'}$.

Then $(s, x)E(s', x')$ implies that there is a $u \in S$ such that $u(sx' - s'x) = 0$. Thus, we see that $u(s(ax') - s'(ax)) = a[u(sx' - s'x)] = 0$. But M is an R-module, so the scalar multiplication defined on M is well-defined. Hence, $ax' = a'x'$, so $u(s(a'x') - s'(ax)) = 0$. It follows that $(s, ax)E(s', a'x')$, so $\frac{ax}{s} = \frac{a'x'}{s'}$, or $a \cdot \frac{x}{s} = a' \cdot \frac{x'}{s'}$. Therefore, scalar multiplication is well-defined.

Finally, we need to show that this makes $S^{-1}M$ into an R-module. We prove condition 1 of Definition 11.1.1 and leave the proof of conditions 2, 3, and 4 as exercises. If $a, b \in R$ and $\frac{x}{s} \in S^{-1}M$, then

$$(a + b) \cdot \frac{x}{s} = \frac{(a + b)x}{s} = \frac{s(ax + b)x}{s^2} = \frac{sax + sbx}{s^2} = \frac{ax}{s} + \frac{bx}{s} = a \cdot \frac{x}{s} + b \cdot \frac{y}{s},$$

so we have proved condition 1. ∎

11.2.6 Definition

Let M be an R-module, and suppose that S is a multiplicatively closed set in R. If E is the congruence relation on $S \times M$ of Lemmas 11.2.3 and 11.2.4, we say that the congruence class $\frac{x}{s}$ is a **fraction** or **quotient** in $S^{-1}M$. The left R-module $S^{-1}M$ is referred to as a **module of quotients** of M.

We saw in Section 4.1 that an integral domain D embeds in its field of fractions. The following theorem shows that this property does not necessarily carry over to modules.

11.2.7 Theorem

If S is a multiplicatively closed set in R and M is an R-module, then the set $t(M) = \{x \in M \mid$ there is an $s \in S$ such that $sx = 0\}$ is a submodule of M. Furthermore, $M/t(M)$ embeds in $S^{-1}M$.

Proof. Because of Lemma 11.1.4, to show that $t(M)$ is a submodule of M, it suffices to show that if $x, y \in t(M)$ and $a \in R$, then $x + y \in t(M)$ and $ax \in t(M)$. If $x, y \in S$, then there are $s, t \in S$ such that $sx = 0$ and $ty = 0$. From this we see that $st(x + y) = (st)x + (st)y = t(sx) + s(ty) = s0 + t0 = 0$ and that $s(ax) = a(sx) = a0 = 0$. Therefore, $t(M)$ is indeed a submodule of M. Note that $t(M) \neq \emptyset$ since $0 \in t(M)$. Hence, we can form the factor module $M/t(M)$, so we define the mapping $f : M/t(M) \longrightarrow S^{-1}M$ by $f(x + t(M)) = \frac{x}{e}$. We claim that f is a well-defined injective R-linear mapping.

First, we show that f is well-defined. If $x + t(M) = y + t(M)$, then $y - x \in t(M)$, so there is an $s \in S$ such that $s(ey - ex) = s(y - x) = 0$. Thus, $(e,x)E(e,y)$, so $\frac{x}{e} = \frac{y}{e}$, which gives $f(x + t(M)) = f(y + t(M))$.

Second, we show that f is R-linear. This follows because for any $x + t(M)$, $y + t(M) \in M/t(M)$ and $a \in R$,

$$f(x + t(M) + y + t(M)) = f((x + y) + t(M))$$
$$= \frac{x + y}{e}$$
$$= \frac{x}{e} + \frac{y}{e}$$
$$= f(x + t(M)) + f(y + t(M))$$

and

$$f(a(x + t(M))) = f(ax + t(M))$$
$$= \frac{ax}{e}$$
$$= a \cdot \frac{x}{e}$$
$$= af(x + t(M)).$$

Finally, we need to show that f is an injection. Since f is R-linear, in light of Theorem 11.1.11, it suffices to show that $\ker f = \{0\}$. If $x + t(M)$ is in $\ker f$, then $f(x + t(M)) = \frac{0}{e}$, the additive identity of $S^{-1}M$. From this we see that $\frac{x}{e} = \frac{0}{e}$, so $(e, x)E(e, 0)$. This implies there is a $u \in S$ such that $u(e0 - xe) = 0$, so there is a $u \in S$ such that $ux = 0$. Hence, x is in $t(M)$, and from this it follows that $x + t(M) = 0 + t(M)$, the additive identity of $M/t(M)$. Therefore, $\ker f = \{0\}$. ∎

11.2.8   Corollary

If $t(M) = 0$, then the module M embeds in $S^{-1}M$.

The submodule $t(M)$ of M developed in Theorem 11.2.7 is said to be the **S-torsion submodule** of M relative to the multiplicatively closed set S in R. An element $x \in t(M)$ is said to be an **S-torsion element** of M. The prefix S is dropped from S-torsion when S is understood.

The following lemma gives additional properties of elements of modules of quotients. The proof of each part of the lemma is straightforward.

11.2.9   Lemma

The following hold in $S^{-1}M$.

1.   If $\frac{x}{s} \in S^{-1}M$, then $\frac{x}{s} = \frac{tx}{ts}$ for any $t \in S$.

2.   If $\frac{x}{s} \in S^{-1}M$, then $\frac{x}{s} = \frac{0}{e}$ if and only if there is $t \in S$ such that $tx = 0$.

3.   For any $s \in S$, $\frac{0}{s} = \frac{0}{e}$.

11.2.10   Examples

1.   Let D be an integral domain, and suppose that D^* is the multiplicatively closed set of all nonzero elements of D. If D is viewed as a D-module, the congruence relation defined in Lemma 11.2.3 reduces to the congruence relation defined on $D \times D^*$ in Section 4.1 since u can be taken to be the identity e. Moreover, $D^{*-1}D$ can be made into a ring by defining $\frac{a}{b} \cdot \frac{c}{d} = \frac{ac}{bd}$, and we have $D^{*-1}D = Q(D)$, the field of fractions of D. In this case, $t(D) = 0$ and D embeds in $Q(D)$.

2.   If addition and scalar multiplication are defined on $\mathbf{Z}[\sqrt{2}]$ by $(x_1 + y_1\sqrt{2}) + (x_2 + y_2\sqrt{2}) = (x_1 + x_2) + (y_1 + y_2)\sqrt{2}$ and if $n(x_1 + y_1\sqrt{2}) = nx_1 + ny_1\sqrt{2}$ for all $x_1 + y_1\sqrt{2}, x_2 + y_2\sqrt{2} \in \mathbf{Z}[\sqrt{2}]$ and $n \in \mathbf{Z}$, then $\mathbf{Z}[\sqrt{2}]$ is a $\mathbf{Z}$-module. The set of nonzero integers $\mathbf{Z}^*$ is a multiplicatively closed set in $\mathbf{Z}$. If we form the quotient module $\mathbf{Z}^{*-1}\mathbf{Z}[\sqrt{2}]$, then elements of $\mathbf{Z}^{*-1}\mathbf{Z}[\sqrt{2}]$ look like congruence classes $\frac{x + y\sqrt{2}}{n}$, $n \neq 0$. It is not difficult to show that $t(\mathbf{Z}[\sqrt{2}]) = 0$, so $\mathbf{Z}[\sqrt{2}]$ embeds in $\mathbf{Z}^{*-1}\mathbf{Z}[\sqrt{2}]$ via the mapping defined by $x + y\sqrt{2} \longmapsto \frac{x + y\sqrt{2}}{1}$.

3.   The additive abelian group $\mathbf{Z}_6$ is a $\mathbf{Z}$-module, and $\mathbf{Z}^*$ is a multiplicatively closed set in $\mathbf{Z}$. Hence, we can form the module of quotients $\mathbf{Z}^{*-1}\mathbf{Z}_6$. Elements of $\mathbf{Z}^{*-1}\mathbf{Z}_6$ look like $\frac{[x]}{m}$, where $[x] \in \mathbf{Z}_6$ and $m \in \mathbf{Z}^*$. Now $\frac{[x]}{m} = \frac{[y]}{n}$ in $\mathbf{Z}^{*-1}\mathbf{Z}_6$ if and only if $(m, [x])E(n, [y])$ if and only if there is element $k \in \mathbf{Z}^*$ such that $k(m[y] - n[x]) = [0]$. Since we can let $k = 6$, it follows that there is only one congruence class in $\mathbf{Z}^{*-1}\mathbf{Z}_6$, and this must be $\frac{[0]}{1}$. Hence, $\mathbf{Z}^{*-1}\mathbf{Z}_6$ is the zero $\mathbf{Z}$-module.

4. If the ring $\mathbf{Z}_4$ is viewed as a module over itself, $S = \{[1], [3]\}$ is a multiplicatively closed set in $\mathbf{Z}_4$. If we form the module of quotients $S^{-1}\mathbf{Z}_4$, elements of $S^{-1}\mathbf{Z}_4$ are of the form $\frac{[x]}{[m]}$, where $[x] \in \mathbf{Z}_4$ and $[m] \in S$. Hence,

$$S^{-1}\mathbf{Z}_4 = \left\{\frac{[0]}{[1]}, \frac{[1]}{[1]}, \frac{[2]}{[1]}, \frac{[3]}{[1]}, \frac{[0]}{[3]}, \frac{[1]}{[3]}, \frac{[2]}{[3]}, \frac{[3]}{[3]}\right\}.$$

But we immediately see that $\frac{[0]}{[1]} = \frac{[0]}{[3]}$ and $\frac{[1]}{[1]} = \frac{[3]}{[3]}$, so $S^{-1}\mathbf{Z}_4$ reduces to

$$S^{-1}\mathbf{Z}_4 = \left\{\frac{[0]}{[1]}, \frac{[1]}{[1]}, \frac{[2]}{[1]}, \frac{[3]}{[1]}, \frac{[1]}{[3]}, \frac{[2]}{[3]}\right\}.$$

These elements of $S^{-1}\mathbf{Z}_4$ are still not distinct. It will be the case that $\frac{[3]}{[1]} = \frac{[1]}{[3]}$ if we can find $[k] \in S$ such that $[k]([3][3] - [1][1]) = [0]$. But $[3][3] - [1][1] = [8] = [0]$ in $\mathbf{Z}_4$, so this is true for any $[k] \in S$. Using similar reasoning, we see that $\frac{[2]}{[1]} = \frac{[2]}{[3]}$, so $S^{-1}\mathbf{Z}_4$ can be further reduced to

$$S^{-1}\mathbf{Z}_4 = \left\{\frac{[0]}{[1]}, \frac{[1]}{[1]}, \frac{[2]}{[1]}, \frac{[3]}{[1]}\right\}.$$

The elements in $S^{-1}\mathbf{Z}_4$ are now distinct. It is also the case that there are no nonzero torsion elements in $\mathbf{Z}_4$ since $[1][x] \neq [0]$ and $[3][x] \neq [0]$ for $x = 1, 2,$ and 3. Therefore, $\mathbf{Z}_4$ embeds in $S^{-1}\mathbf{Z}_4$. The mapping is the obvious one: $[x]$ in $\mathbf{Z}_4$ maps to $\frac{[x]}{[1]}$ in $S^{-1}\mathbf{Z}_4$. In fact, this mapping is an isomorphism. Inspect the preceding work carefully. Can you determine the property that each element of S possesses that ensures that $S^{-1}\mathbf{Z}_4$ and $\mathbf{Z}_4$ are isomorphic?

5. If R is a ring and P is a prime ideal of R, then we saw in Example 6 of 11.2.2 that $S = R \setminus P$ is a multiplicatively closed set in R. If R is viewed as a module over itself, then we can form the module of quotients $S^{-1}R$. The R-module $S^{-1}R$ can be made into a ring by defining $\frac{x}{s} \cdot \frac{y}{t} = \frac{xy}{st}$ for all $\frac{x}{s}, \frac{y}{t} \in S^{-1}R$. This ring has many interesting properties, one of which is that it has a unique maximal ideal. A commutative ring R that has a

unique maximal ideal is said to be a **local ring**. The ring $S^{-1}R$ is said to be the **localization** of R at the prime ideal P. The torsion submodule $t(R)$ is an ideal of R, and the R-linear mapping $f : R/t(R) \longrightarrow S^{-1}R$ given by $f(a + t(R)) = \frac{a}{e}$ is a ring monomorphism.

6. If S is the set of all regular elements of R, then from Example 4 of 11.2.2 we know that S is a multiplicatively closed set in R. The left R-module $S^{-1}R$ can be made into a ring by defining multiplication as $\frac{x}{s} \cdot \frac{y}{t} = \frac{xy}{st}$ for all $\frac{x}{s}, \frac{y}{t} \in S^{-1}R$. This ring is often denoted by $Q_{cl}(R)$ and is called the **classical ring of quotients** of R. The torsion submodule of the left R-module R is $\{0\}$, and there is a ring monomorphism from R to $Q_{cl}(R)$ defined by $x \longmapsto \frac{x}{e}$.

Problem Set 11.2

In each of the following problems, S is a multiplicatively closed set in R, and the modules are R-modules, where R is a commutative ring with identity.

1. Show that the operation of addition defined on $S^{-1}M$ in Lemma 11.2.4 is associative.

2. Complete the proof of Theorem 11.2.5.

3. Prove Lemma 11.2.9.

4. (a) If S_1 and S_2 are multiplicatively closed sets in R, is $S_1 \cap S_2$ a multiplicatively closed set in R?
    (b) If $\{S_i\}_{i \geq 1}$ is a family of multiplicatively closed sets in R, must $\cap_{i \geq 1} S_i$ be a multiplicatively closed set in R?
    (c) If $S_1 \subseteq S_2 \subseteq S_3 \subseteq \cdots$ is an ascending chain of multiplicatively closed sets in R, argue that $\cup_{i \geq 1} S_i$ is a multiplicatively closed set in R.

5. The set $S = \{e, x, x^2, x^3, \cdots\}$ is a multiplicatively closed set in the ring $R[[x]]$ of formal power series over R. Determine what the elements of $S^{-1}R[[x]]$ look like, and show that each such element can be written as $a_0 x^{-n} + a_1 x^{-n+1} + \cdots + a_{n-1} x^{-1} + a_n + a_{n+1} x + a_{n+2} x^2 + \cdots$ for an appropriately chosen value of the integer n. See Example 7 of 10.1.9.

6. If S is a multiplicatively closed set in R such that every element of S has a multiplicative inverse in R and if M is an R-module, prove that $S^{-1}M$ and M are isomorphic R-modules. Compare this result with Example 4 of 11.2.10.

7. (a) If P is a prime ideal of R and $S = R \setminus P$, show that $S^{-1}R$ has a unique maximal ideal. Hint: Consider $S^{-1}P = \{\frac{x}{s} \mid x \in P \text{ and } s \in S\}$.

    (b) If I is an ideal of R, is $S^{-1}I = \{\frac{x}{s} \mid x \in I \text{ and } s \in S\}$ an ideal of $S^{-1}R$? Give reasons for your answer.

8. (a) If $f : M \longrightarrow N$ is an R-linear mapping, show that f can be used to define an R-linear mapping $S^{-1}f : S^{-1}M \longrightarrow S^{-1}N$. Hint: Consider $(S^{-1}f)\left(\frac{x}{s}\right) = \frac{f(x)}{s}$.

    (b) If $f, g : M \longrightarrow N$ are R-linear mappings, does $S^{-1}(f + g) = S^{-1}f + S^{-1}g$?

    (c) If $f : M \longrightarrow N$ and $g : N \longrightarrow X$, does $S^{-1}(g \circ f) = S^{-1}g \circ S^{-1}f$?

9. Prove that the mapping $f : R \longrightarrow Q_{\mathrm{cl}}(R)$ defined in Example 6 of 11.2.10 is a ring monomorphism.

10. Explain why the torsion submodule $t(R)$ of R of Example 5 of 11.2.10 is an ideal of R.

11. Let S be a multiplicatively closed set in R, and suppose that $f : M \longrightarrow N$ is an R-linear mapping.

    (a) If f is injective, must $S^{-1}f$ be injective? Give reasons for your response.

    (b) If f is surjective, must $S^{-1}f$ also be surjective? Give reasons for your answer.

12. Let $f : R \longrightarrow R'$ be a ring homomorphism such that $f(e) = e'$.

    (a) If S is a multiplicatively closed set in R, is $f(S)$ a multiplicatively closed set in R'?

    (b) If S' is a multiplicatively closed set in R', is $f^{-1}(S')$ a multiplicatively closed set in R?

11.3   FREE MODULES

Section Overview. In this section we investigate modules that behave somewhat like vector spaces. Every vector space has a basis, but this is not true for modules. A module that does have a basis is said to be a free module. You should compare the properties of free modules presented in this section with the analogous properties of vector spaces and note the similarities and the differences.

There are modules that behave somewhat like vector spaces in that they have a basis. The definition of a basis of a left R-module is exactly the same as that for a vector space, with the exception that the scalars are taken from a ring that may not be a field and the ring may not be commutative. If B is a set of elements of a left R-module M, then B is said to **span** M if for each $x \in M$, there are elements $x_1, x_2, \ldots, x_n \in B$ and ring elements $a_1, a_2, \ldots, a_n$ such that $x = a_1 x_1 + a_2 x_2 + \cdots + a_n x_n$. The set B is said to be **linearly independent** if whenever $x_1, x_2, \ldots, x_n \in B$ and $a_1, a_2, \ldots, a_n \in R$ are such that $a_1 x_1 + a_2 x_2 + \cdots + a_n x_n = 0$, then $a_1 = a_2 = \cdots = a_n = 0$. If B is linearly independent and spans M, then B is said to be a **basis** for M. A module that has a basis is said to be a **free module**. Since every vector space has a basis, every vector space is a free module. For a vector space, we have seen that the number of elements in a basis is unique. This is not true, in general, for modules, and this fact is one of the major differences between modules and vector spaces. Subsequently, an example will be given of a module that has a basis with one element and a basis with two elements. In this section we do *not* assume that the ring is commutative.

11.3.1   Examples

1. If addition is defined on $R^n = R \times R \times \cdots \times R$ (n factors) componentwise and scalar multiplication is given by

    $$a(a_1, a_2, \ldots, a_n) = (aa_1, aa_2, \ldots, aa_n),$$

    then R^n is a free left R-module. A basis for R^n is the set $B = \{e_1, e_2, \ldots, e_n\}$ where

    $$e_1 = (e, 0, 0, \ldots, 0)$$
    $$e_2 = (0, e, 0, \ldots, 0)$$
    $$\vdots$$
    $$e_n = (0, 0, 0, \ldots, e)$$

    If $(a_1, a_2, \ldots, a_n) \in R^n$, then $(a_1, a_2, \ldots, a_n) = a_1 e_1 + a_2 e_2 + \cdots + a_n e_n$, and $a_1 e_1 + a_2 e_2 + \cdots + a_n e_n = 0$ clearly implies that $(a_1, a_2, \ldots, a_n) = 0 = (0, 0, 0, \ldots, 0)$, which in turn gives $a_1 = a_2 = \cdots = a_n = 0$. We say that the left R-module R^n is the **direct product** of n factors of R.

2. Let $R_i = R$ for $i = 1, 2, 3, \ldots$, and suppose that $\oplus_{i=1}^{\infty} R_i$ is the set of all sequences $(a_1, a_2, a_3, \ldots, a_n, \ldots)$ where at most a finite number of the a_i's are nonzero. If addition is componentwise and scalar multiplication is defined by

    $$a(a_1, a_2, a_3, \ldots, a_n, \ldots) = (aa_1, aa_2, aa_3, \ldots, aa_n, \ldots),$$

then $\oplus_{i=1}^{\infty} R_i$ is a free left R-module. A basis for $\oplus_{i=1}^{\infty} R_i$ is the set $B = \{e_1, e_2, \ldots, e_n, \ldots\}$, where

$$e_1 = (e, 0, 0, \ldots, 0, \ldots)$$
$$e_2 = (0, e, 0, \ldots, 0, \ldots)$$
$$\vdots$$
$$e_n = (0, 0, 0, \ldots, e, \ldots)$$
$$\vdots$$

It is not difficult to show that B is a linearly independent set of elements of $\oplus_{i=1}^{\infty} R_i$ that spans $\oplus_{i=1}^{\infty} R_i$.

3. Let $\{R_i\}_{i \geq 1}$ be such that $R_i = R$, for $i = 1, 2, \ldots$, and suppose that $\times_{i \geq 1} R_i$ is the set of all sequences of the form $(a_1, a_2, \ldots, a_n, \ldots)$ where $a_i \in R_i$ for $i = 1, 2, \ldots$. The set $\times_{i \geq 1} R_i$ can be made into a left R-module by defining addition componentwise and by defining scalar multiplication by $a(a_1, a_2, \ldots, a_n, \ldots) = (aa_1, aa_2, \ldots, aa_n, \ldots)$. If the elements of the set $S = \{e_1, e_2, \ldots, e_n, \ldots\}$ are defined as in Example 2, then S is *not* a basis for the left R-module $\times_{i \geq 1} R_i$. Why? Because every element of $\times_{i \geq 1} R_i$ cannot be expressed as a linear combination of a *finite number* of elements from S, which is a requirement if S is to be a basis. The module $\times_{i \geq 1} R_i$ contains elements $(a_1, a_2, \ldots, a_n, \ldots)$, where $a_i \neq 0$ for each i. For such an element, we cannot write

$$(a_1, a_2, \ldots, a_n, \ldots) = a_1 e_1 + a_2 e_2 + \cdots + a_n e_n + \cdots.$$

The right side of this equation is an infinite series, and we have no way of telling whether the series converges to $(a_1, a_2, \ldots, a_n, \ldots)$ without additional structure on the module. This is in direct contrast to elements $(a_1, a_2, \ldots, a_n, \ldots)$ of $\oplus_{i=1}^{\infty} R_i$, where at most a finite number of the a_i's can be nonzero. For an element in $\oplus_{i=1}^{\infty} R_i$ we *can* write

$$(a_1, a_2, \ldots, a_n, \ldots) = a_1 e_1 + a_2 e_2 + \cdots + a_n e_n + \cdots$$

since the sum on the right side of this equation is actually a finite sum.

In Example 1 the module R^n has a basis with a finite number of elements, while in Example 2 the basis of $\oplus_{i=1}^{\infty} R_i$ contains an "infinite number" of elements. Examples 2 and 3 bring out an interesting point. Example 8 of 11.1.2 shows that $\oplus_{i=1}^{\infty} R_i$ is a submodule of $\times_{i \geq 1} R_i$. Hence, it is possible for a module that is not free to have submodules that are free.

The proof of the following theorem is straightforward, so we leave it as an exercise. You should compare this theorem with Theorems 10.1.5 and 10.2.6.

11.3.2 | Theorem

Let M be a left R-module, and suppose that $B = \{x_1, x_2, \ldots, x_n\}$ spans M.

1. The set B is a basis for M if and only if for each $x \in M$, the coefficients in the expression $x = a_1x_1 + a_2x_2 + \cdots + a_nx_n$ are unique.
2. If B is a basis for M, then R^n and M are isomorphic.

Since the set $\{x_1, x_2, \ldots, x_n\}$ of Theorem 11.3.2 spans M, $M = Rx_1 + Rx_2 + \cdots + Rx_n$. The isomorphism between R^n and M of part 2 of this theorem should now be obvious. It is the mapping from R^n to M given by $(a_1, a_2, \ldots, a_n) \longmapsto a_1x_1 + a_2x_2 + \cdots + a_nx_n$.

Example 3 of 11.3.1 shows that it is possible for a module that is not free to have a submodule that is free. We now show that it is possible for a free module to have submodules that are not free.

11.3.3 | Examples

1. A submodule of a free module need not be free. To see this, consider the $\mathbf{Z}_6$-module $\mathbf{Z}_2 \times \mathbf{Z}_3$, where addition is defined componentwise and scalar multiplication is given by $[k]_6([x]_2, [y]_3) = ([kx]_2, [ky]_3)$. We claim that $B = \{([1]_2, [1]_3)\}$ is a basis for $\mathbf{Z}_2 \times \mathbf{Z}_3$. Since

$$[0]_6([1]_2, [1]_3) = ([0 \cdot 1]_2, [0 \cdot 1]_3) = ([0]_2, [0]_3),$$
$$[4]_6([1]_2, [1]_3) = ([4 \cdot 1]_2, [4 \cdot 1]_3) = ([0]_2, [1]_3),$$
$$[2]_6([1]_2, [1]_3) = ([2 \cdot 1]_2, [2 \cdot 1]_3) = ([0]_2, [2]_3),$$
$$[3]_6([1]_2, [1]_3) = ([3 \cdot 1]_2, [3 \cdot 1]_3) = ([1]_2, [0]_3),$$
$$[1]_6([1]_2, [1]_3) = ([1 \cdot 1]_2, [1 \cdot 1]_3) = ([1]_2, [1]_3), \quad \text{and}$$
$$[5]_6([1]_2, [1]_3) = ([5 \cdot 1]_2, [5 \cdot 1]_3) = ([1]_2, [2]_3),$$

we see that B spans $\mathbf{Z}_2 \times \mathbf{Z}_3$. Now suppose that $[k]_6([1]_2, [1]_3) = ([0]_2, [0]_3)$. Then $[k]_2 = [0]_2$ and $[k]_3 = [0]_3$, so $2 \mid k$ and $3 \mid k$. Hence, $6 \mid k$, and consequently $[k]_6 = [0]_6$. Therefore, B is a linearly independent set that spans $\mathbf{Z}_2 \times \mathbf{Z}_3$ and so is a basis for $\mathbf{Z}_2 \times \mathbf{Z}_3$. Both $\mathbf{Z}_2 \times \{[0]_3\}$ and $\{[0]_2\} \times \mathbf{Z}_3$ are submodules of $\mathbf{Z}_2 \times \mathbf{Z}_3$, but neither can be a free $\mathbf{Z}_6$-module. For example, $\mathbf{Z}_2 \times \{[0]_3\}$ has two elements, so if $\mathbf{Z}_2 \times \{[0]_3\}$ were a free $\mathbf{Z}_6$-module, Theorem 11.3.2 shows that $\mathbf{Z}_2 \times \{[0]_3\}$ and $\mathbf{Z}_6^n$ would be isomorphic for some positive integer n. But $\mathbf{Z}_6^n$ has at least six elements, so this cannot be the case. A similar observation holds for $\{[0]_2\} \times \mathbf{Z}_3$. Thus, submodules of free modules need not be free. This also demonstrates that there are modules that are not free. The $\mathbf{Z}_6$-module $\mathbf{Z}_2 \times \{[0]_3\}$ is not a free $\mathbf{Z}_6$-module.

2. Example 1 shows that there are modules that are not free. An additional illustration of this fact is given by the $\mathbf{Z}$-module $\mathbf{Z}_8$. If $\mathbf{Z}_8$ is a free $\mathbf{Z}$-module, then $\mathbf{Z}_8$ and $\mathbf{Z}^n$ are isomorphic for some positive integer n. But this is impossible since $\mathbf{Z}_8$ contains eight elements, while $\mathbf{Z}^n$ has an "infinite number" of elements. Since every abelian group is a $\mathbf{Z}$-module, a similar argument shows that a finite abelian group cannot be a free $\mathbf{Z}$-module. In fact, if a ring R contains an "infinite number" of elements, then a finite left R-module M cannot be free. Because of this observation, we see that it is not possible for a vector space to contain a finite number of vectors over a field F with an "infinite number" of scalars. If a field F contains an "infinite number" of elements, then so must any vector space over F.

It is important to point out that if the ring is not a field and if a free left R-module has a basis with n elements, we *do not* say that the dimension of M is n, as we do with vector spaces. The integer n may not be unique. On this note, we close our brief discussion of free modules with an example of a module that has a basis with one element and a basis with two elements. The following notation and lemma are required to present the example.

If M is a free left R-module with basis $B = \{x_1, x_2, \ldots, x_n, \ldots\}$, then for each $x \in M$ there are scalars $a_1, a_2, \ldots, a_n, \ldots \in R$, where at most a finite number of the a_i's are nonzero, such that

$$x = a_1x_1 + a_2x_2 + \cdots + a_nx_n + \cdots = \sum a_ix_i.$$

The notation $\sum a_ix_i$ enables us to write expressions for elements of M as linear combinations of basis elements more compactly. A point to remember is that $\sum a_ix_i$ is actually a finite sum since at most a finite number of the a_i's can be nonzero. With this notation, if $x = \sum a_ix_i$ and $y = \sum b_ix_i$ are elements of M and $a \in R$, addition and scalar multiplication are given by $x + y = \sum a_ix_i + \sum b_ix_i = \sum (a_i + b_i)x_i$ and $ax = a(\sum a_ix_i) = \sum (aa_i)x_i$.

11.3.4   Lemma

Let M be a free left R-module with basis $B = \{x_i, x_2, \ldots, x_n, \ldots\}$. If $f : B \longrightarrow M$ is a function, then $f^* : M \longrightarrow M$ given by $f^*(x) = \sum a_i f(x_i)$ is an R-linear mapping.

Proof. If $x = \sum a_i x_i$ and $y = \sum b_i x_i$ are elements of M and $a \in R$, then

$f^*(x + y) = f^*(\sum a_i x_i + b_i x_i) = f^*(\sum (a_i + b_i)x_i) = \sum (a_i + b_i)f(x_i) =$

$\sum a_i f(x_i) + \sum b_i f(x_i) = f^*(x) + f^*(y)$ and $f^*(ax) = f^*(a \sum a_i x_i) =$

$f^*(\sum (aa_i)x_i) = \sum (aa_i)f(x_i) = a \sum a_i f(x_i) = a f^*(\sum a_i x_i) = a f^*(x)$. ∎

When M is a free left R-module, Lemma 11.3.3 shows that in order to define a linear mapping from M to M, it suffices to define the function on basis elements of M. One can obtain a function $f : B \longrightarrow M$ by selecting a set $\{y_i\}_{i \geq 1}$ of elements of M, all of which need not be distinct, and defining $f : B \longrightarrow M$ by $f(x_i) = y_i$ for $i = 1, 2, 3, \ldots$. When f^* has been obtained from f as in Lemma 11.3.3, we say that f has been **extended linearly** to M. It is also common practice to denote f^* by f, and we follow this practice. The idea of extending a mapping defined on basis elements to a linear mapping was also touched upon briefly in Section 10.2.

11.3.5   Example

Let R be any ring with identity, and suppose that $M = \bigoplus_{i=1}^{\infty} R_i$ is the left R-module of Example 2 of 11.3.1. If $E = \text{End}_R(M)$ is the set of all R-linear mappings from M to M, then E is a ring under addition and composition of functions. The identity for E is the identity mapping $1_E : E \longrightarrow E : x \longmapsto x$. If E is viewed as a left E-module over itself, we claim that E has a basis with one element and a basis with two elements. First, note that the set $\{1_E\}$ is a basis for E, since any function f in E can be written as $f = f \circ 1_E$, and if $f \circ 1_E = 0$, then $f = 0$. Next, we show that E has a basis with two elements. Recall from Example 2 of 11.3.1 that the set $B = \{e_1, e_2, \ldots, e_n, \ldots\}$ is a basis for M, where

$$e_1 = (e, 0, 0, \ldots, 0, \ldots)$$
$$e_2 = (0, e, 0, \ldots, 0, \ldots)$$
$$\vdots$$
$$e_n = (0, 0, 0, \ldots, e, \ldots)$$
$$\vdots$$

To construct a basis for E with two elements, we define two functions on the basis elements of M and then extend each linearly to M. Toward this end, let

$$g(e_1) = 0 \qquad\qquad h(e_1) = e_1$$
$$g(e_2) = e_1 \qquad\qquad h(e_2) = 0$$
$$g(e_3) = 0 \qquad\qquad h(e_3) = e_2$$
$$g(e_4) = e_2 \qquad\qquad h(e_4) = 0$$
$$\vdots \qquad\qquad\qquad \vdots$$
$$g(e_{2k-1}) = 0 \qquad\qquad h(e_{2k-1}) = e_k$$
$$g(e_{2k}) = e_k \qquad\qquad h(e_{2k}) = 0$$
$$\vdots \qquad\qquad\qquad \vdots$$

Briefly, g maps basis elements with an odd subscript to 0 and basis elements with an even subscript $2k$ to the basis element with subscript k. A similar description holds for h : h maps basis elements with even subscripts to 0 and basis elements with an odd subscript $2k - 1$ to the basis element with subscript k. Now that g and h have been defined on basis elements of M, each can be extended linearly to M to obtain R-linear mappings $g, h \in E$. We claim that the set $\{g,h\}$ is a basis for E as a left E-module. Two properties must be shown. We must show that if $f \in E$, then there are scalars $f_1, f_2 \in E$ such that $f = f_1 \circ g + f_2 \circ h$ and that if $g_1, h_1 \in E$ are such that $g_1 \circ g + h_1 \circ h = 0$, then $g_1 = h_1 = 0$. If the first condition holds, then $\{g, h\}$ spans E, and the second indicates that the set $\{g, h\}$ is linearly independent. To show that B spans E, suppose that $f \in E$, define $f_1, f_2 : B \longrightarrow M$ by $f_1(e_k) = f(e_{2k})$ and $f_2(e_k) = f(e_{2k-1})$ for $k = 1, 2, 3, \ldots$, and extend each of these functions linearly to M. If $x \in M$, then $x = \sum a_i e_i$ where at most a finite number of the a_is are nonzero. Since

$$x = a_1 e_1 + a_2 e_2 + a_3 e_3 + a_4 e_4 + \cdots$$
$$= (a_2 e_2 + a_4 e_4 + \cdots) + (a_1 e_1 + a_3 e_3 + \cdots),$$

we see that x can be rewritten as

$$x = \sum a_{2k} e_{2k} + \sum a_{2k-1} e_{2k-1}.$$

Let's consider the action of $f_1 \circ g + f_2 \circ h$ on $\sum a_{2k} e_{2k}$ and $\sum a_{2k-1} e_{2k-1}$:

$$(f_1 \circ g + f_2 \circ h)(\sum a_{2k} e_{2k}) = \sum (f_1 \circ g + f_2 \circ h)(a_{2k} e_{2k})$$
$$= \sum a_{2k}[(f_1 \circ g)(e_{2k}) + (f_2 \circ h)(e_{2k})]$$
$$= \sum a_{2k}[f_1(e_k) + f_2(0)]$$
$$= \sum a_{2k} f(e_{2k})$$
$$= f(\sum a_{2k} e_{2k}).$$

Similarly, $(f_1 \circ g + f_2 \circ h)(\sum a_{2k-1}e_{2k-1}) = f(\sum a_{2k-1}e_{2k-1})$, so we have

$$(f_1 \circ g + f_2 \circ h)(x) = (f_1 \circ g + f_2 \circ h)(\sum a_{2k}e_{2k} + \sum a_{2k-1}e_{2k-1})$$
$$= (f_1 \circ g + f_2 \circ h)(\sum a_{2k}e_{2k}) + (f_1 \circ g + f_2 \circ h)(\sum a_{2k-1}e_{2k-1})$$
$$= f(\sum a_{2k}e_{2k}) + f(\sum a_{2k-1}e_{2k-1})$$
$$= f(\sum a_{2k}e_{2k} + \sum a_{2k-1}e_{2k-1})$$
$$= f(x).$$

Therefore, $f = f_1 \circ g + f_2 \circ h$, so $\{g, h\}$ spans E. Finally, we need to show that $\{g, h\}$ is linearly independent. For this, let $g_1, h_1 \in E$, and suppose that
$$g_1 \circ g + h_1 \circ h = 0.$$

Then

$$0 = (g_1 \circ g + h_1 \circ h)(e_{2k}) = (g_1 \circ g)(e_{2k}) + (h_1 \circ h)(e_{2k})$$
$$= g_1(e_k) + h_1(0)$$
$$= g_1(e_k) \qquad \text{for } k = 1, 2, 3, \dots.$$

Hence, $g_1 = 0$ since it is zero on basis elements of M. Similarly, $h_1 = 0$, so the set $\{g, h\}$ is linearly independent.

Since $\{1_E\}$ and $\{g, h\}$ are bases of E when E is viewed as a left E-module, the number of elements in a basis for a module need not be unique. In contrast to the ring E, there are rings R for which the number of elements in a basis of a free module is unique. Such rings are said to have the **Independent Basis Number** property or that they are IBN rings. Every field is an IBN ring, and more generally, it can be shown that any commutative ring with identity is an IBN ring.

Problem Set 11.3

1. Prove Theorem 11.3.2.

2. Show that the set B as defined in Example 2 of 11.3.1 is a basis of the left R-module $\bigoplus_{i=1}^{\infty} R_i$.

3. Let $S = \{x_1, x_2, \dots, x_n\}$ be a possibly empty set of elements of a left R-module M.

   (a) Prove that the intersection of the submodules of M that contain S is the smallest submodule $\langle S \rangle$ of M that contains S.
   (b) Prove that $\langle \varnothing \rangle = \{0\}$.
   (c) Prove that $\langle S \rangle = Rx_1 + Rx_2 + \cdots + Rx_n$.

4. If R is a ring with identity, then $M_2(R)$ is a left R-module under matrix addition and scalar multiplication given by

$$a \begin{pmatrix} a_{11} & a_{12} \\ a_{21} & a_{22} \end{pmatrix} = \begin{pmatrix} aa_{11} & aa_{12} \\ aa_{21} & aa_{22} \end{pmatrix}.$$

Show that $M_2(R)$ is a free left R-module.

5. If addition is defined componentwise and scalar multiplication is defined as in Example 1 of 11.3.3, determine which, if any, of the following modules are free modules, and give a basis for those that are free.

   (a) $\mathbf{Z}_2 \times \mathbf{Z}_5$ as a $\mathbf{Z}_{10}$-module
   (b) $\mathbf{Z}_2 \times \mathbf{Z}_4$ as a $\mathbf{Z}_8$-module
   (c) $\mathbf{Z}_2 \times \mathbf{Z}_2 \times \mathbf{Z}_3$ as a $\mathbf{Z}_6$-module

6. Let $f, g : M \longrightarrow N$ be R-linear mappings. If M is a free R-module, prove that f and g agree on a basis of M if and only if $f = g$.

7. Let M and N be left R-modules, and suppose that $f : M \longrightarrow N$ is an R-linear mapping. If M is free with basis B, prove that N is free with basis $f(B) = \{f(x) \mid x \in B\}$ if and only if f is an R-isomorphism.

8. If M is a free left R-module with basis $B = \{x_1, x_2, \ldots, x_n\}$, prove that the mapping $f : R^n \longmapsto M$ defined by

$$f((a_1, a_2, \ldots, a_n)) = a_1 x_1 + a_2 x_2 + \cdots + a_n x_n$$

is an isomorphism.

9. Return to Example 11.3.4 and show, using the notation and definitions provided there, that $(f_1 \circ g + f_2 \circ h)(\sum a_{2k-1} \, e_{2k-1}) = f(\sum a_{2k-1} \, e_{2k-1})$.

10. If R is a ring, prove that it is possible for R and R^2 to be isomorphic as left R-modules. Hint: Consider Example 11.3.4 and let $R = E$.

11. If R is a ring, prove that it is possible for R and R^n to be isomorphic as left R-modules for each positive integer n. Hint: Consider Exercise 10 and use induction.

12. Let E be the ring of Example 11.3.4.

    (a) Use a vector space argument to show that the ring E cannot be a field. Hint: See Theorem 10.1.7.
    (b) Prove that E is not a field by showing that there are nonzero elements in E that do not have an inverse under the operation of function composition.

Appendix

Answers and Hints to Selected Problems

Problem Set 0.1

1. $\mathscr{P}(A) = \{\varnothing, \{a\}, \{b\}, \{c\}, \{a, b\}, \{a, c\}, \{b, c\}, A\}$

5. Suppose that $A \subseteq B$. If $x \in \mathbf{C}_x B$, then $x \in X$ but $x \notin B$. Since $A \subseteq B$, $x \notin B \Rightarrow x \notin A$. Hence, $x \in X$ and $x \notin A$, which tells us that $x \in \mathbf{C}_x A$. $\therefore \mathbf{C}_x B \subseteq \mathbf{C}_x A$. Conversely, if $\mathbf{C}_x B \subseteq \mathbf{C}_x A$ and $x \in A$, then $x \notin \mathbf{C}_x A$. But $\mathbf{C}_x B \subseteq \mathbf{C}_x A$, so $x \notin \mathbf{C}_x B$. Hence, $x \in B$ and thus $A \subseteq B$.

9. (a) $A \triangle B = \{1, 2, 6, 7\}$

13. (a) $[1, \infty)$ (c) $\{0\}$

Problem Set 0.2

1. (a) $\mathrm{Dom}(R_1) = \{3, 5, 7\}$ and $\mathrm{Rng}(R_1) = \{2, 4, 6\}$
   (b) $\mathrm{Dom}(R_2) = \{1, 3, 5, 7\}$ and $\mathrm{Rng}(R_2) = \{2, 4, 6, 8\}$
   (c) $\mathrm{Dom}(R_3) = \{1, 7\}$ and $\mathrm{Rng}(R_3) = \{2, 4, 6, 8\}$

5. $E = \{(1, 1), (2, 2), (1, 2), (2, 1), (3, 3), (4, 4), (3, 4), (4, 3), (5, 5), (6, 6), (5, 6), (6, 5)\}$. The equivalence classes are A_1, A_2, and A_3.

9. If $(a, b) \in A$, then $ab = ba$, so R is reflexive.

   If $(a, b)R(c, d)$, then $ad = bc$, so $cb = da$ and R is symmetric.

If $(a, b)R(c, d)$ and $(c, d)R(e, f)$, then (1) $ad = bc$ and (2) $cf = de$. Hence, $adf = bcf$ and $cfb = deb$, so $d(af - be) = 0$. This implies that $af = be$. Therefore, $(a, b)R(e, f)$, so R is transitive.

Problem Set 0.3

1. (a) $h\left(\dfrac{p}{q}\right) = \dfrac{4p}{q} + 1$ is well-defined.

    (b) $f\left(\dfrac{p}{q}\right) = 2p - 3q$ is not well-defined.

5. (a) $f(x) = \dfrac{x+2}{3}$ is an inverse function for f.

    (b) $f(x) = \sqrt[3]{x-1}$ is an inverse function for f.

9. If $g \circ f(x) = g \circ f(y)$, then $g(f(x)) = g(f(y))$, so $f(x) = f(y)$ since g is injective. But f is injective, so $f(x) = f(y)$ implies that $x = y$. If f and g are surjective and $z \in C$, then there is a $y \in B$ such that $g(y) = z$. Also, there is an $x \in A$ such that $f(x) = y$. Hence, $g \circ f(x) = g(f(x)) = g(y) = z$, so $g \circ f$ is surjective.

Problem Set 0.4

1. Not closed   5.   Closed

9. Commutative but not associative. Does not have an identity.

13. (a) $\begin{pmatrix} \dfrac{-1}{3} & \dfrac{4}{3} \\ \dfrac{5}{9} & \dfrac{-2}{9} \end{pmatrix}$ (b) Does not have an inverse.

     (c) $\begin{pmatrix} \dfrac{342}{119{,}505} & \dfrac{21}{119{,}505} \\ \dfrac{121}{119{,}505} & \dfrac{-342}{119{,}505} \end{pmatrix}$

Problem Set 0.5

1. (a) $A + D = \begin{pmatrix} -1 & 1 & 5 \\ 1 & 1 & 1 \\ -4 & 2 & 1 \\ 6 & 0 & 0 \end{pmatrix}$ (c) $A(BC) = \begin{pmatrix} -23 & 8 \\ 23 & -7 \\ -31 & 7 \\ 22 & -3 \end{pmatrix}$

and $(AB)C = A(BC)$

(e) $B + C = DNE$

5.   (a) Yes, $(A + B)^t = A^t + B^t$ (b) $(AB)^t = B^t A^t$, but $(AB)^t \neq A^t B^t$

Problem Set 1.1

3.   Let $n = 4, x = 2$, and $y = 6$.

5.   (a) True   (b) True   (c) True   (d) False

7.   (a) $\gcd(12345, 5040) = 15, a = -89$, and $b = 218$.
     (b) $\gcd(85672, 242040) = 8, a = 7289$, and $b = -2580$.

9.   Let a and b be integers such that $1 = a(xy) + bz$. Then $1 = (ay)x + bz$, so 1 is the smallest positive integer that can be written as a linear combination of x and z. Hence $\gcd(x, z) = 1$. Likewise $1 = (ax)y + bz$, so $\gcd(y, z) = 1$.

11.   Let $d = \gcd(x, y)$ and $d' = \gcd(nx, ny)$. Then d and d' are the smallest positive integers that can be written as a linear combination of x and y and of nx and ny, respectively. Suppose that $a, b, s,$ and t are integers such that $d = ax + by$ and $d' = s(nx) + t(ny)$. Since $d' = (sx + ty)n$, we see that $n|d'$. Let k be a positive integer such that $d' = nk$. Now $nd = a(nx) + b(nx)$, so it must be the case that $kn = d' \leq nd$, so $k \leq d$. Since $nk = (sx + ty)n$ implies that $k = sx + ty$ and since d is the smallest positive integer that can be written in this form, $k = d$. Thus $d' = nd$, so $\gcd(nx, ny) = n \gcd(x, y)$.

Problem Set 1.2

1.   If $n = 1$, then $5^n - 1 = 5^1 - 1 = 4$, which is clearly divisible by 4. Now make the induction hypothesis that $4|(5^k - 1)$.

5.   If $n = 1$, then $\sum_{i=1}^{1} i^3 = 1 = \left[\frac{1(1 + 1)}{2}\right]^2$, so make the induction hypothesis that $\sum_{i=1}^{k} i^3 = \left[\frac{k(k + 1)}{2}\right]^2$.

9.   If $n = 2$, then $\sum_{i=1}^{2} \frac{1}{\sqrt{i}} = 1 + \frac{1}{\sqrt{2}}$. Now $\sqrt{2} > 1$, so $\sqrt{2} + 1 > 2$. Hence
$1 + \frac{1}{\sqrt{2}} = \frac{\sqrt{2} + 1}{\sqrt{2}} > \frac{2}{\sqrt{2}} = \sqrt{2}$. Thus $\sum_{i=1}^{2} \frac{1}{\sqrt{i}} > \sqrt{2}$, so make the induction hypothesis that $\sum_{i=1}^{k} \frac{1}{\sqrt{i}} > \sqrt{k}$.

13. Let #A denote the number of elements in the finite set A. If $n = 1$, then S has exactly one element, so the only subsets of S are $\emptyset$ and S. Hence #$\mathscr{P}(S) = 2 = 2^1$ elements. Now make the induction hypothesis that if S has k elements, then #$\mathscr{P}(S) = 2^k$, and consider the set $S' = \{x_1, x_2, \ldots, x_k, x_{k+1}\}$ with $k + 1$ distinct elements.

17. The proof is by induction. If $n = 1$, then $(a + b)^1 = a + b = \binom{1}{0}ab^0 + \binom{1}{1}a^0b^1$, so make the induction hypothesis that

$$(x + y)^k = \binom{k}{0}x^k y^0 + \binom{k}{1}x^{k-1}y + \binom{k}{2}x^{k-2}y^2 + \cdots + \binom{k}{r}x^{k-r}y^r + \cdots +$$

$$+ \binom{k}{k-1}xy^{k-1} + \binom{k}{k}x^0 y^k,$$ and make use of the result of Exercise 16 and

$$\binom{k}{k} = \binom{k+1}{0} \text{ and } \binom{k}{k} = \binom{k+1}{k+1}.$$ The result will follow by induction.

Problem Set 1.3

1. (a) [1]   (b) [3]   (c) [3]   (d) [2]   (e) Yes

7. (a) $x = 0$ and $x = 2$ are solutions.
    (b) $x = 300$, $x = 613$, and $x = 926$ are solutions.
    (c) $x = 6$ is a solution.
    (d) $x = 1$ and $x = 8$ are solutions.

11. (a) [2], [3], [4], [6], [8], [9], and [10].
     (b) [5], [7], [10], [14], [15], [20], [21], [25], [28], [30]
     (c) There are no zero-divisors in $\mathbf{Z}_{31}$.

15. (a) [0] and [2]   (b) [0] and [6]
     (c) [0], [3], [6], [9], [12], [15], [18], [21], and [24]

Problem Set 2.1

1. The set $2\mathbf{Z}$ is an abelian group under addition of integers.

5. The set $\{[1],[5],[7],[11]\}$ is an abelian group under multiplication modulo 12.

9. (a) This follows from the fact that $\mathbf{N}$ is well-ordered.

(b) Let q and r be integers such that $m = n_0 q + r$, where $0 \le r < n_0$. Note that $q \ne 0$ since $n_0 < m$. Then $e = x^m = x^{n_0 q + r} = x^{n_0 q} x^r = \left(x^{n_0}\right)^q x^r = e^q x^r = e x^r = x^r$. But n_0 is the smallest positive integer such that $x^{n_0} = e$, so $r = 0$. Hence, $m = n_0 q$, so $n_0 \mid m$.

13. (a) If $x \in G$, then $x_L^{-1} x = e_L$, so $x_L^{-1} x x_L^{-1} = e_L x_L^{-1} = x_L^{-1}$. Since every element of G has a left inverse and since $x_L^{-1} \in G$, there is a $y \in G$ such that $y x_L^{-1} = e_L$. Therefore $y x_L^{-1} x x_L^{-1} = y x_L^{-1}$, so $e_L x x_L^{-1} = e_L$. Hence $x x_L^{-1} = e_L$. Consequently, if we now let $x^{-1} = x_L^{-1}$, then $x^{-1} x = x x^{-1} = e_L$. Finally, we show that e_L is also a right identity for G. Since $e_L x = x$ and $e_L = x^{-1} x$, we see that $x e_L = x(x^{-1} x) = (x x^{-1}) x = e_L x = x$. Therefore $e_L x = x e_L = x$, so $e = e_L$ is an identity for G. Thus, G is a group.

15. If $(x, y),(z, w) \in G$, then $x, y, z, w \in \mathbf{R}$, so $xz, yz + w \in \mathbf{R}$ and so $(xz, yz + w) \in G$. Hence, the operation is closed. The operation is associative since $(a, b)[(x, y)(z, w)] = (a, b)(xz, yz + w) = (axz, bxz + yz + w)$ and $[(a, b)(x, y)][(z, w)] = (ax, bx + y)(z, w) = (axz, bxz + yz + w)$. The identity is $(1, 0)$, and the inverse of (x, y) is $(x^{-1}, -x^{-1}y)$. Thus, G is a group. G is not abelian.

17. The operation is closed since $\begin{pmatrix} 1 & x \\ 0 & 1 \end{pmatrix}\begin{pmatrix} 1 & y \\ 0 & 1 \end{pmatrix} = \begin{pmatrix} 1 & x+y \\ 0 & 1 \end{pmatrix}$ and $x + y$ is in $\mathbf{Q}$. If

$\begin{pmatrix} 1 & x \\ 0 & 1 \end{pmatrix}, \begin{pmatrix} 1 & y \\ 0 & 1 \end{pmatrix}, \begin{pmatrix} 1 & z \\ 0 & 1 \end{pmatrix} \in G$, then $\begin{pmatrix} 1 & x \\ 0 & 1 \end{pmatrix}\left[\begin{pmatrix} 1 & y \\ 0 & 1 \end{pmatrix}\begin{pmatrix} 1 & z \\ 0 & 1 \end{pmatrix}\right] =$

$\begin{pmatrix} 1 & x \\ 0 & 1 \end{pmatrix}\begin{pmatrix} 1 & y+z \\ 0 & 1 \end{pmatrix} = \begin{pmatrix} 1 & x+(y+z) \\ 0 & 1 \end{pmatrix} = \begin{pmatrix} 1 & (x+y)+z \\ 0 & 1 \end{pmatrix} =$

$\begin{pmatrix} 1 & x+y \\ 0 & 1 \end{pmatrix}\begin{pmatrix} 1 & z \\ 0 & 1 \end{pmatrix} = \left[\begin{pmatrix} 1 & x \\ 0 & 1 \end{pmatrix}\begin{pmatrix} 1 & y \\ 0 & 1 \end{pmatrix}\right]\begin{pmatrix} 1 & z \\ 0 & 1 \end{pmatrix}$ so the operation is associative.

The identity is $\begin{pmatrix} 1 & 0 \\ 0 & 1 \end{pmatrix}$ and $\begin{pmatrix} 1 & x \\ 0 & 1 \end{pmatrix}^{-1} = \begin{pmatrix} 1 & -x \\ 0 & 1 \end{pmatrix}$. Finally,

$\begin{pmatrix} 1 & x \\ 0 & 1 \end{pmatrix}\begin{pmatrix} 1 & y \\ 0 & 1 \end{pmatrix} = \begin{pmatrix} 1 & x+y \\ 0 & 1 \end{pmatrix} = \begin{pmatrix} 1 & y+x \\ 0 & 1 \end{pmatrix} = \begin{pmatrix} 1 & y \\ 0 & 1 \end{pmatrix}\begin{pmatrix} 1 & x \\ 0 & 1 \end{pmatrix}$, and so the operation is commutative. Hence, G is an abelian group.

23. If $x, y \in G$, then $xy \in G$, so $xyxy = (xy)^2 = e = ee = x^2 y^2 = xxyy$.
Multiplying both sides of $xyxy = xxyy$ on the left by x^{-1} and on the right by
y^{-1}, we see that $yx = xy$, and so G is an abelian group.

Problem Set 2.2

3. (b)

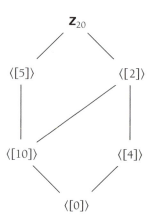

7. (c) Since $ad - bc = 1$, $\begin{pmatrix} a & b \\ c & d \end{pmatrix}^{-1} = \begin{pmatrix} d & -b \\ -c & a \end{pmatrix}$. Hence, if $\begin{pmatrix} a & b \\ c & d \end{pmatrix}, \begin{pmatrix} x & y \\ z & w \end{pmatrix} \in H_3$,

then $\begin{pmatrix} a & b \\ c & d \end{pmatrix}\begin{pmatrix} x & y \\ z & w \end{pmatrix}^{-1} = \begin{pmatrix} a & b \\ c & d \end{pmatrix}\begin{pmatrix} w & -y \\ -z & x \end{pmatrix} = \begin{pmatrix} aw - bz & -ay + bx \\ cw - dz & -cy + dx \end{pmatrix}$. But

$(aw - bz)(-cy + dx) - (-ay + bx)(cw - dz) = -acyw + adxw + bcyz - bdxz + acyw - adyz - bcxw + bdxz = adxw + bcyz - adyz - bcxw = (ad - bc)(xw - yz) = 1 \cdot 1 = 1$.

Hence, $\begin{pmatrix} a & b \\ c & d \end{pmatrix}\begin{pmatrix} x & y \\ z & w \end{pmatrix}^{-1} \in H_3$.

(d) H_1 is a subgroup of H_2, but H_2 is not a subgroup of H_3.

13. If $a + bi, c + di \in H$, then $(a + bi)(c + di)^{-1} = (a + bi)\left(\dfrac{c}{c^2 + d^2} - \dfrac{d}{c^2 + d^2}i \right) =$

$\dfrac{ac + bd}{c^2 + d^2} + \dfrac{bc - ad}{c^2 + d^2}i = (ac + bd) + (bc - ad)i$ since $c^2 + d^2 = 1$. Now

$(ac + bd)^2 + (bc - ad)^2 = a^2c^2 + 2abcd + b^2d^2 + b^2c^2 - 2abcd + a^2d^2 = a^2c^2 + b^2d^2 + b^2c^2 + a^2d^2 = (a^2 + b^2)(c^2 + d^2) = 1 \cdot 1 = 1$. Hence

$(a + bi)(c + di)^{-1}$ is in H, so H is a subgroup of $\mathbf{C}^*$.

15. For any finite group G, let $\#G$ denote the number of elements of G. We see from the proof of Lagrange's Theorem that when G is finite, $\#G$ equals the index of a subgroup H in G times the number of elements in H. Hence, $\#G = [G:H_2](\#H_2)$ and $\#H_2 = [H_2:H_1](\#H_1)$. Therefore $\#G = [G:H_2][H_2:H_1](\#H_1)$. But $\#G = [G:H_1](\#H_1)$, so $[G:H_1](\#H_1) = [G:H_2][H_2:H_1](\#H_1)$. Hence $[G:H_1] = [G:H_2][H_2:H_1]$.

17. The answer is yes. Suppose that the distinct left cosets are $eH = H$ and xH. Consider the right coset Hx, and note that either $Hx = H$ or $Hx \neq H$. If $Hx = H$, show that this implies that $xH = H$, which is a contradiction. Hence, H and Hx are the distinct right cosets of H in G. Since $H \cap xH = \varnothing$ and $H \cap Hx = \varnothing$, $G = H \cup xH$ and $G = H \cup Hx$. The result follows from the observation that $Hx = G \setminus H = xH$.

Problem Set 2.3

3. h is a group homomorphism.

7. $g \circ f(xy) = g(f(xy)) = g(f(x)f(y)) = g(f(x))g(f(y)) = (g \circ f(x))(g \circ f(y))$

11. Suppose that $G = \{e_G, x\}$ and $H = \{e_H, y\}$, and show that the mapping $f: G \to H$ defined by $f(e_G) = e_H$ and $f(x) = y$ is a group isomorphism. Finally, suppose that $G = \{e_G, x, x'\}$ and $H = \{e_H, y, y'\}$. First, observe that in G, we cannot have $xx' = x$, for if so, then $x' = e_G$ and G would not have three elements. Similarly, $x'x \neq x$, $xx' \neq x'$, and $x'x \neq x'$. Hence, in G, $xx' = x'x = e_G$, $x^2 = x'$, and $x'^2 = x$. Likewise for H. Therefore, the multiplication tables for G and H are

$\bullet$	e_G	x	x'
e_G	e_G	x	x'
x	x	x'	e_G
x'	x'	e_G	x

and

$\bullet$	e_H	y	y'
e_H	e_H	y	y'
y	y	y'	e_H
y'	y'	e_H	y

The mapping $f: G \to H$ defined by $f(e_G) = e_H$, $f(x) = y$, and $f(x') = y'$ is a group isomorphism.

13. If $f(x) = ax + b$ and $g(x) = cx + d \in G$, then $f \circ g(x) = f(g(x)) = f(cx + d) = a(cx + d) + b = (ac)x + (ad + b) = a'x + b'$, where $a' = ac$ and $b' = ad + b$. Since $a \neq 0$ and $c \neq 0$, $ac \neq 0$, so $f \circ g(x) \in G$. Hence, G is closed under the operation. The operation is associative since function composition is an associative operation. It is easy to show that the identity

of G is the mapping $I(x) = ex + 0 = x$ and that if $f(x) = ax + b \in G$, then $f^{-1}(x) = \frac{1}{a}x - \frac{b}{a}$ is the inverse of $f(x)$. Therefore, G is a group.

15. In order for the operation to be closed, we must have $x \circ y \neq 1$ for $x, y \in H$. If $x \circ y = 1$, then $x + y - xy = 1$, where $x \neq 1$ and $y \neq 1$. Now $x + y - xy = x + (1 - x)y$, so if $x + y - xy = 1$, then $x + (1 - x)y = 1$, so $(1 - x)y = 1 - x$. But since $x \neq 1$, this implies that $y = 1$, which is a contradiction. Hence, $x \circ y \neq 1$, so the operation on H is closed. The operation is associative and $e_H = 0$. Moreover, if $x \in H$, then $x^{-1} = \frac{-x}{1-x} \in H$, so H has inverses. Thus, H is a group. If $x, y \in \mathbf{Q}^*$, then $f(x) \circ f(y) = (1 - x) \circ (1 - y) = (1 - x) + (1 - y) - (1 - x)(1 - y) = 1 - x + 1 - y - 1 + y + x - xy = 1 - xy = f(xy)$. Hence, f is a group homomorphism. If $x \in H$, then $x \neq 1$, so $1 - x \neq 0$ and thus $1 - x \in \mathbf{Q}^*$. Since $f(1 - x) = x$, f is a surjective mapping. Finally, if $f(x) = f(y)$, then $1 - x = 1 - y$, so $x = y$. Hence, f is injective, and f is an isomorphism.

21. (a) Since $\ker f = 18\mathbf{Z}$, the correspondence between the subgroups of $\mathbf{Z}$ that contain $18\mathbf{Z}$ and the subgroups of $\mathbf{Z}_{18}$ is $2\mathbf{Z} \leftrightarrow \langle[2]\rangle$, $3\mathbf{Z} \leftrightarrow \langle[3]\rangle$, $6\mathbf{Z} \leftrightarrow \langle[6]\rangle$, $9\mathbf{Z} \leftrightarrow \langle[9]\rangle$, $18\mathbf{Z} \leftrightarrow \langle[0]\rangle$, and $\mathbf{Z} \leftrightarrow \mathbf{Z}_{18}$.
    (c) Since $\ker f^* = \langle[6]\rangle$, the correspondence between the subgroups of $\mathbf{Z}_{24}$ that contain $\langle[6]\rangle$ and the subgroups of $f^*(\mathbf{Z}_{24})$ is $\langle[2]\rangle \leftrightarrow \langle[8]\rangle$, $\langle[3]\rangle \leftrightarrow \langle[12]\rangle$, $\langle[6]\rangle \leftrightarrow \langle[0]\rangle$, and $\mathbf{Z}_{24} \leftrightarrow f^*(\mathbf{Z}_{24}) = \langle[4]\rangle$.

Problem Set 3.1

1. $x = c - b$ is a solution. If s and s' are solutions, then $s + b = s' + b$ implies that $s = s'$. If R is a field and $a \neq 0$, then $a^{-1}(b - c)$ is a solution. If s and s' are solutions, then $as + b = as' + b$, so $as = as'$. Hence, $a^{-1}as = a^{-1}as'$ and thus $s = s'$.

3. If $a \neq 0$ and $ax = ay$, then $0 = ax - ay = a(x - y)$. But the ring has no zero-divisors, and $a \neq 0$. Hence, $x - y = 0$ or $x = y$. Similarly, if $xa = ya$, then $x = y$.

9. In the first table, there is no additive identity. For the second table, a is the additive identity. However, b does not have an additive inverse.

11. It follows from $(x + x)^2 = x + x$ and $(x + x)^2 = x + x + x + x$ that $x + x = 0$. Hence $x = -x$ for every $x \in R$. Next, note that $(x + y)^2 = x + y$ and $(x + y)^2 = x + xy + yx + y$. Hence $x + y = x + xy + yx + y$, so $xy + yx = 0$. Thus, $xy = -(yx) = (-y)x = yx$, so the ring is commutative.

13. A field cannot have zero-divisors since if $xy = 0$ in a field and $x \neq 0$, then $y = x^{-1}xy = x^{-1}0 = 0$. Therefore, if R is a Boolean ring that is a field, Exercise 12 indicates that R cannot contain more than two elements. Since a field must contain at least two elements, we see that R contains exactly two elements. Since a field has an additive identity and a multiplicative identity, $R = \{0, e\}$.

17.

$$\begin{pmatrix} a & b \\ -b & a \end{pmatrix}\begin{pmatrix} c & d \\ -d & c \end{pmatrix} = \begin{pmatrix} ac - bd & ad + bc \\ -bc - ad & -bd + ac \end{pmatrix} = \begin{pmatrix} c & d \\ -d & c \end{pmatrix}\begin{pmatrix} a & b \\ -b & a \end{pmatrix}$$

and

$$\begin{pmatrix} a & b \\ -b & a \end{pmatrix}\begin{pmatrix} \dfrac{a}{a^2+b^2} & \dfrac{-b}{a^2+b^2} \\ \dfrac{b}{a^2+b^2} & \dfrac{a}{a^2+b^2} \end{pmatrix} = \begin{pmatrix} \dfrac{a^2+b^2}{a^2+b^2} & \dfrac{-ab+ba}{a^2+b^2} \\ \dfrac{-ba+ab}{a^2+b^2} & \dfrac{(-b)^2+a^2}{a^2+b^2} \end{pmatrix} = \begin{pmatrix} 1 & 0 \\ 0 & 1 \end{pmatrix}$$

23. (a) $f(x) + g(x) = \frac{5}{2} + 4x - \frac{3}{2}x^2 - 8x^3$ and $f(x)g(x) = 4 + 8x - 16x^3$

(b) $f(x) + g(x) = [2] + [3]x + [5]x^2 + [3]x^3 + [2]x^4$ and

$f(x)g(x) = [4]x^2 + [3]x^3 + [3]x^4 + [4]x^6$

(c) $f(x) + g(x) = \begin{pmatrix} 3 & 2 \\ -1 & 1 \end{pmatrix} + \begin{pmatrix} -2 & 1 \\ -1 & 1 \end{pmatrix}x + \begin{pmatrix} 1 & -3 \\ 1 & 7 \end{pmatrix}x^2$ and

$f(x)g(x) = \begin{pmatrix} -8 & 5 \\ 7 & -4 \end{pmatrix}x + \begin{pmatrix} 3 & 2 \\ -1 & 1 \end{pmatrix}\begin{pmatrix} 0 & 1 \\ 0 & 4 \end{pmatrix}x^2 + \begin{pmatrix} -2 & 1 \\ -4 & 3 \end{pmatrix}x^3 + \begin{pmatrix} 0 & -3 \\ 0 & 7 \end{pmatrix}x^4$

Problem Set 3.2

3. The multiplication and addition tables for S are

+	[0]	[2]	[4]	[6]	[8]
[0]	[0]	[2]	[4]	[6]	[8]
[2]	[2]	[4]	[6]	[8]	[0]
[4]	[4]	[6]	[8]	[0]	[2]
[6]	[6]	[8]	[0]	[2]	[4]
[8]	[8]	[0]	[2]	[4]	[6]

•	[0]	[2]	[4]	[6]	[8]
[0]	[0]	[0]	[0]	[0]	[0]
[2]	[0]	[4]	[8]	[2]	[6]
[4]	[0]	[8]	[6]	[4]	[2]
[6]	[0]	[2]	[4]	[6]	[8]
[8]	[0]	[6]	[2]	[8]	[4]

Note that $[2][6] = [2], [4][6] = [4]$, and $[8][6] = [8]$, so $[6]$ is an identity for S. Since $[1]$ is the identity for $\mathbf{Z}_{10}$ and $[1] \neq [6]$, it is possible for a subring S of a ring R with identity to have an identity that is different from that of R.

5.   Note that $S_1 \cap S_2 \neq \emptyset$ since $0 \in S_1 \cap S_2$. If $x, y \in S_1 \cap S_2$, then $x, y \in S_1$ and $x, y \in S_2$. Hence, $x - y$, $xy \in S_1$ and $x - y$, $xy \in S_2$ since S_1 and S_2 are subrings of R. Therefore, $x - y$, $xy \in S_1 \cap S_2$, so $S_1 \cap S_2$ is a subring of R. Now generalize this for $\{S_\alpha\}_{\alpha \in \Delta}$.

11.   The multiplicative identity of $M_2(\mathbf{R})$ is $\begin{pmatrix} 1 & 0 \\ 0 & 1 \end{pmatrix}$. The matrix $\begin{pmatrix} 1 & 2 \\ 2 & 4 \end{pmatrix}$ does not have a multiplicative inverse since $\det\begin{pmatrix} 1 & 2 \\ 2 & 4 \end{pmatrix} = 0$. Therefore, $M_2(\mathbf{R})$ is not a division ring. Now suppose that I is a nonzero ideal of $M_2(\mathbf{R})$. If $\begin{pmatrix} a & b \\ c & d \end{pmatrix}$ is a nonzero element of I, then $a, b, c,$ or $d \neq 0$. Suppose that $a \neq 0$. A similar proof holds if $b, c,$ or $d \neq 0$. Now

$$\begin{pmatrix} 1 & 0 \\ 0 & 0 \end{pmatrix}\begin{pmatrix} a & b \\ c & d \end{pmatrix}\begin{pmatrix} a^{-1} & 0 \\ 0 & 0 \end{pmatrix} = \begin{pmatrix} 1 & 0 \\ 0 & 0 \end{pmatrix}\begin{pmatrix} 1 & 0 \\ ca^{-1} & 0 \end{pmatrix} = \begin{pmatrix} 1 & 0 \\ 0 & 0 \end{pmatrix} \in I \text{ since } I \text{ is an ideal.}$$

Also $\begin{pmatrix} 0 & 0 \\ 1 & 0 \end{pmatrix}\begin{pmatrix} 1 & 0 \\ 0 & 0 \end{pmatrix}\begin{pmatrix} 0 & 1 \\ 0 & 0 \end{pmatrix} = \begin{pmatrix} 0 & 0 \\ 1 & 0 \end{pmatrix}\begin{pmatrix} 0 & 1 \\ 0 & 0 \end{pmatrix} = \begin{pmatrix} 0 & 0 \\ 0 & 1 \end{pmatrix} \in I$, again since I is an

ideal. Hence, $\begin{pmatrix} 1 & 0 \\ 0 & 1 \end{pmatrix} = \begin{pmatrix} 1 & 0 \\ 0 & 0 \end{pmatrix} + \begin{pmatrix} 0 & 0 \\ 0 & 1 \end{pmatrix} \in I$, so $I = M_2(\mathbf{R})$.

23.   Let $\frac{a}{b}, \frac{c}{d} \in R$. Then $\frac{a}{b} - \frac{c}{d} = \frac{ad - bc}{bd}$, so since 5 is prime, $5 \nmid b$ and $5 \nmid d$ imply that $5 \nmid bd$. Hence, $\frac{a}{b} - \frac{c}{d} \in R$. Similarly, $\left(\frac{a}{b}\right)\left(\frac{c}{d}\right) = \frac{ac}{bd} \in R$, so R is a subring of $\mathbf{Q}$. Hence, R is a commutative ring. R has an identity since $\frac{1}{1} \in R$. Therefore, R is a commutative ring with identity. If $\frac{a}{b}, \frac{c}{d} \in I$, then $\frac{a}{b} - \frac{c}{d} = \frac{ad - bc}{bd}$ and $5 \mid (ad - bc)$ since $5 \mid a$ and $5 \mid c$. If $\frac{a}{b} \in I$ and $\frac{c}{d} \in R$, then $\left(\frac{a}{b}\right)\left(\frac{c}{d}\right) = \frac{ac}{bd}$ and $5 \mid ac$ since $5 \mid a$. Hence, I is an ideal of R. Now consider the factor ring R/I.

Elements of R/I look like $\frac{a}{b} + I$, where $5 \nmid b$. Using the division algorithm, there are integers q and r, $0 \leq r < 5$, such that $a = 5q + r$. Hence, $\frac{a}{b} + I = \frac{5q + r}{b} + I = \frac{5q}{b} + \frac{r}{b} + I = \frac{r}{b} + I$. Hence, every element of R/I can be written as $\frac{r}{b} + I$, where $5 \nmid b$ and $0 \leq r < 5$. If $\frac{r}{b} + I$ is a nonzero element of R/I, then $r \neq 0$, so $\frac{b}{r} + I \in R$ since $5 \nmid r$. Thus, $\left(\frac{r}{b} + I\right)\left(\frac{b}{r} + I\right) = \left(\frac{r}{b}\right)\left(\frac{b}{r}\right) + I = \frac{1}{1} + I$, and so R/I is a field. A similar result holds if 5 is replaced by a prime p.

Problem Set 3.3

3. f is a ring homomorphism, and $\ker f = \left\{ \begin{pmatrix} 0 & y \\ 0 & 0 \end{pmatrix} \middle| y \in \mathbf{Z} \right\}$.

5. Let $y, y' \in S$. Since f is surjective, there exist $x, x' \in R$ such that $f(x) = y$ and $f(x) = y'$. Furthermore, since f is injective, $f^{-1}(y) = x$ and $f^{-1}(y') = x'$. Now $f(x + x') = f(x) + f(x') = y + y'$, so again, since f is injective, $f^{-1}(y + y') = x + x' = f^{-1}(y) + f^{-1}(y')$. We also have that $f(xx') = f(x)f(x') = yy'$, and so $f^{-1}(yy') = xx' = f^{-1}(y)f^{-1}(y')$. Hence, f^{-1} is a ring homomorphism. If $f^{-1}(y) = f^{-1}(y')$, then $f \circ f^{-1}(y) = f \circ f^{-1}(y')$, which implies that $y = y'$. Hence, f^{-1} is injective. Finally, if $x \in R$, then $f(x) = y \in S$ and $f^{-1}(y) = f^{-1} \circ f(x) = x$, so f^{-1} is surjective. Thus, f^{-1} is a ring isomorphism.

11. Use the fact that $f(i)^2 = f(i^2) = f(-1) = -1$ to show that $f(x + yi) = x + yi$ and $f(x + yi) = x - yi$ are the only ring isomorphisms $f : \mathbf{C} \to \mathbf{C}$ that leave elements of $\mathbf{R}$ fixed.

13. $f(x + y) = (0, x + y) = (0, x) + (0, y) = f(x) + f(y)$ and $f(xy) = (0, xy) = (0 \cdot 0, 0 \cdot y + x \cdot 0 + xy) = (0, x)(0, y) = f(x)f(y)$. If $f(x) = f(y)$, then $(0, x) = (0, y)$, so $x = y$. Finally, if $(0, x) \in R'$, then $f(x) = (0, x)$. This shows that f is a ring isomorphism.

15. If $f(e_R) \neq e_S$, then $f(e_R) - e_S \neq 0$. So $f(e_R)(f(e_R) - e_S) = f(e_R)f(e_R) - f(e_R)e_S = f(e_R{}^2) - f(e_R) = 0$. Hence, $f(e_R)$ will be a zero-divisor in S provided that $f(e_R) \neq 0$. If $f(e_R) = 0$, then for any $x \in R$, $f(x) = f(xe_R) = f(x)f(e_R) = f(x)0 = 0$, which implies that f is the zero map, and this is not the case. Hence, $f(e_R) \neq 0$, so $f(e_R)$ is a zero-divisor in S.

17. (a) Since R is commutative and f is an epimorphism, S is a commutative ring. (See the proof of Exercise 16.) Next, suppose that P is a prime ideal of S. Then $f^{-1}(P)$ is an ideal of R, so let $ab \in f^{-1}(P)$. Then $f(a)f(b) = f(ab) \in P$, so either $f(a)$ or $f(b)$ is in P since P is prime. If $f(a) \in P$, then $a \in f^{-1}(P)$, and if $f(b) \in P$, then $b \in f^{-1}(P)$. Hence, $f^{-1}(P)$ is a prime ideal of R since $ab \in f^{-1}(P)$ implies that either a or b is in $f^{-1}(P)$.

    (b) Suppose that I is an ideal of R such that $f^{-1}(M) \subseteq I \subseteq R$. Since f is an epimorphism, $M = f(f^{-1}(M)) \subseteq f(I) \subseteq f(R) = S$. Therefore, since M is a maximal ideal of S, it must be the case that either $M = f(I)$ or $f(I) = S$. Since f is an epimorphism, if $M = f(I)$, then $I = f^{-1}(M)$, and if $f(I) = S$, then $I = f^{-1}(S) = R$. Hence, when I is an ideal of R such that $f^{-1}(M) \subseteq I \subseteq R$, either $f^{-1}(M) = I$ or $I = R$. Thus, $f^{-1}(M)$ is a maximal ideal of R.

Problem Set 4.1

7. Suppose that F is a proper subfield of $\mathbf{Z}_p$. Since $\mathbf{Z}_p$ has order p, F must be finite, and the additive subgroup of F is a proper subgroup of the additive subgroup of $\mathbf{Z}_p$. Hence, by Lagrange's Theorem, the order of F must divide the order of $\mathbf{Z}_p$, which is clearly impossible. Now let F be a subfield of $\mathbf{Q}$. We know that the identity of $\mathbf{Q}$ is the identity of each subfield of $\mathbf{Q}$, so $1 \in F$. It follows by induction that $n \in F$ for each $n \in \mathbf{N}$, so $\mathbf{N} \subset F$. Since F contains the additive inverse of each of its elements, $-n \in F$ for each $n \in \mathbf{N}$. Also, since F is closed under addition, $0 = n + (-n) \in F$, so we see that $\mathbf{Z} \subset F$. If $\frac{a}{b}$ is any rational number, then $a, b \in \mathbf{Z} \subset F$ with $b \neq 0$. Now b has an inverse b^{-1} in F and an inverse $\frac{1}{b}$ in $\mathbf{Q}$. Since the inverse of b is unique in $\mathbf{Q}$, it must be the case that $b^{-1} = \frac{1}{b}$. Thus, $\frac{a}{b} = a \cdot \frac{1}{b} \in F$, so $\mathbf{Q} \subseteq F$. Therefore, $F = \mathbf{Q}$.

9. (a) If $\frac{a}{b}, \frac{c}{d} \in D$, define addition on D by $\frac{a}{b} + \frac{c}{d} = \frac{ad+bc}{bd}$ and $\frac{a}{b} \cdot \frac{c}{d} = \frac{ac}{bd}$. These operations are well-defined since addition and multiplication of rational numbers are well-defined. The operations are also closed since 5 is prime and since $5 \nmid b$ and $5 \nmid d$ imply that $5 \nmid bd$. Note also since $5 \nmid 1$, $a = \frac{a}{1} \in D$ for each $a \in \mathbf{Z}$. Hence, $\mathbf{Z} \subseteq D \subseteq \mathbf{Q}$. It is easy to show that D is a commutative ring with identity 1 and that D is an integral domain since it is a subring of a field and therefore must be free of zero-divisors. Hence, $Q(D)$ exists. From the construct of $Q(D)$, we see that elements of $Q(D)$ look like $\left(\frac{a}{b}\right) / \left(\frac{c}{d}\right)$, where $\left(\frac{c}{d}\right) \neq 0$ in D. Now $\left(\frac{a}{b}\right) / \left(\frac{c}{d}\right) = \frac{ad}{bc} \in \mathbf{Q}$, so it follows that $Q(D) \subseteq \mathbf{Q}$. Conversely, if $\frac{a}{b} \in \mathbf{Q}$, then $\frac{a}{b} = \left(\frac{a}{1}\right) / \left(\frac{b}{1}\right) \in Q(D)$, so $\mathbf{Q} \subseteq Q(D)$. Hence, $Q(D) = \mathbf{Q}$.

11. $Q(2\mathbf{Z}) = \left\{ \frac{2m}{2n} \mid m,n \in \mathbf{Z} \text{ with } n \neq 0 \right\}$, $Q(\mathbf{Z}) = \mathbf{Q} = \left\{ \frac{m}{n} \mid m, n \in \mathbf{Z} \text{ with } n \neq 0 \right\}$, and the mapping $f: Q(2\mathbf{Z}) \rightarrow \mathbf{Q}: \frac{2m}{2n} \mapsto \frac{m}{n}$ is a ring isomorphism.

Problem Set 4.2

1. Suppose that R is an integral domain. If $xy = xz$, then $x(y - z) = 0$. But $x \neq 0$ and R has no zero-divisors, so it must be the case that $y - z = 0$.

Thus, $y = z$. Conversely, suppose that if $xy = xz$ and $x \neq 0$, then $y = z$. To show that R is an integral domain, we must show that R is free of zero-divisors. If $xy = 0$ and $x \neq 0$, then $xy = x \cdot 0$ implies that $y = 0$, and this completes the proof.

7. In any ordered integral domain, $x^2 \geq 0$ for all $x \in D$. Since $e > 0$, $x^2 + e > 0$. Thus, by the trichotomy property of ordered integral domains, $x^2 + e \neq 0$ for all $x \in D$.

9. Let $f : \mathbf{Z} \longrightarrow D$ be defined by $f(x) = \begin{pmatrix} x & 0 \\ 0 & x \end{pmatrix}$, and show that f is a well-defined ring isomorphism.

13. (a) If x and y are rational numbers and $x < y$, then $x + x < x + y$, so $2x < x + y$. Hence, $x < \frac{x+y}{2}$. Likewise, $x < y$ implies that $x + y < y + y$, so $\frac{x+y}{2} < y$. Thus, $x < \frac{x+y}{2} < y$, and $\frac{x+y}{2}$ is a rational number.

  (b) By part (a) there is a rational number a_1 such that $x < a_1 < y$. Similarly, there is a rational number a_2 such that $a_1 < a_2 < y$. Hence, we have $x < a_1 < a_2 < y$. Repeating this process n times produces $x < a_1 < a_2 < \cdots < a_n < y$.

  (c) If s is the smallest positive rational number, then $s \leq x$ for every rational number x. But $0 < s$ and 0 is a rational number, by part (a) there is a rational number a such that $0 < a < s$, which is a contradiction. Thus, there is no smallest rational number.

Problem Set 4.3

3. (b) If $x > 0$ is such that $x^2 < 2$, let $y = \frac{x(x^2+6)}{3x^2+2}$. Then $y - x = \frac{2x(2-x^2)}{3x^2+2}$.
   Now x, $2 - x^2$, and $3x^2 + 2$ are all positive, so $y - x > 0$. Hence $x < y$.
   We also see that $y^2 - 2 = \frac{(x^2-2)^3}{(3x^2+2)^2} < 0$, and so $y^2 < 2$, which shows
   that $y \in r$. Hence, r has no largest element. Now we need to show that r is closed downward. If $x \in r$ and $0 \leq y < x$, then $y^2 < x^2 < 2$, so $y \in r$. If $x < 0$ and $y < x$, then $y < 0$, and so $y \in r$. Thus, r has no smallest element, so r is a cut.

7. Since $x < y$, $0 < y - x$. Thus, $0 < \frac{y-x}{\sqrt{2}}$, so $x < x + \frac{y-x}{\sqrt{2}}$. We also
   see that since $1 < \sqrt{2}$, $y - x < (y-x)\sqrt{2}$, which implies that $\frac{y-x}{\sqrt{2}} < y - x$.

Therefore, $x + \frac{y-x}{\sqrt{2}} < y$, so $x < x + \frac{y-x}{\sqrt{2}} < y$. The answer is yes. To see that, suppose that r and r' are real numbers. Then, because of Theorem 4.3.18, we can find a rational number x such that $r < x < r'$. Likewise, we can find a rational number y such that $r < x < y < r'$. Hence, $r < x < x + \frac{y-x}{\sqrt{2}} < y < r'$. Now $\sqrt{2}$ is irrational, so $\frac{1}{\sqrt{2}}$ must be irrational by part (e) of Exercise 4. Thus, by part (d) of Exercise 4, $\frac{y-x}{\sqrt{2}}$ is irrational. Therefore, $x + \frac{y-x}{\sqrt{2}}$ is irrational by part (c) of Exercise 4.

11. Since $\bar{0} = \{x \in \mathbf{Q} \mid x < 0\} \subseteq \{x \in \mathbf{Q} \mid x < 1\} = \bar{1}, \bar{1} > \bar{0}$. If r is a cut and $r > \bar{0}$, then $\bar{1} \, r = \{x_1 x_2 \mid x_1 \in \bar{1}, x_1 \geq 0, \text{ and } x_2 \in r, x_2 \geq 0\} \cup \{x \in \mathbf{Q} \mid x < 0\}$. If $x_1 x_2 \in \{x_1 x_2 \mid x_1 \in \bar{1}, x_1 \geq 0, \text{ and } x_2 \in r, x_2 \geq 0\}$, then $0 \leq x_1 < 1$, so $0 \leq x_1 x_2 < x_2$. Hence, $x_1 x_2 \in r$ since r is closed downward. Furthermore, if $x \in \{x \in \mathbf{Q} \mid x < 0\}$, then $x < x_2$, so $x \in r$, again since r is closed downward. Thus $\bar{1} r \subseteq r$. Conversely, if $x \in r$, then either $x < 0$ or $x \geq 0$. If $x < 0$, then $x \in \{x \in \mathbf{Q} \mid x < 0\}$, so $x \in \bar{1}r$. If $x = 0$, then $x = 0 \cdot 0 \in \bar{1}r$, so suppose that $x > 0$. Then $-\frac{x}{x} = -1 \in \bar{1}$ and $-x < 0 < x$, and so $-x \in r$. Thus, $x = \left(-\frac{x}{x}\right)(-x)$ is in $\bar{1}r$. In any case, $r \subseteq \bar{1}r$, so $r = \bar{1}r$. A similar proof can be used to show that $r\bar{1} = r$.

Problem Set 4.4

1. (c) $\left|3 - \sqrt{2}i\right| = \sqrt{3^2 + (-\sqrt{2})^2} = \sqrt{9 + 2} = \sqrt{11}$

 (d) $\dfrac{(5-i)(2-i)}{(2+i)(2-i)} - \dfrac{(1-5i)(6-7i)}{(6+7i)(6-7i)} = \dfrac{(10-1)-(5+2)i}{4+1} -$

 $\dfrac{(6-35)-(7+30)i}{36+49} = \dfrac{9-7i}{5} - \dfrac{-29-37i}{85} = \dfrac{153-119i+29+37i}{85} =$

 $\dfrac{182-82i}{85} = \dfrac{182}{85} - \dfrac{82}{85}i.$

3. (a) 6   (c) -8

5. Use induction.

9. (a) $u_0 = \cos\frac{\pi}{2} + i\sin\frac{\pi}{2} = i,$

   $u_1 = \cos\frac{7\pi}{6} + i\sin\frac{7\pi}{6} = -\frac{\sqrt{3}}{2} - \frac{1}{2}i,$ and

   $u_2 = \cos\frac{11\pi}{6} + i\sin\frac{11\pi}{6} = \frac{\sqrt{3}}{2} - \frac{1}{2}i$

   (c) $u_0 = \sqrt[4]{2}(\cos\frac{5\pi}{12} + i\sin\frac{5\pi}{12}) \cong 0.307789 + 1.14869i,$

   $u_1 = \sqrt[4]{2}(\cos\frac{11\pi}{12} + i\sin\frac{11\pi}{12}) \cong -1.14869 + 0307789i,$

   $u_2 = \sqrt[4]{2}(\cos\frac{17\pi}{12} + i\sin\frac{17\pi}{12}) \cong -0.307789 - 1.14869i,$ and

   $u_3 = \sqrt[4]{2}(\cos\frac{23\pi}{12} + i\sin\frac{23\pi}{12}) \cong 1.14869 - 0.307789i$

11. (a) Let $u = r^{-1}(\cos\theta - i\sin\theta)$. Since multiplicative inverses are unique, it suffices to show that $zu = 1$. Now $zu = r^{-1}(\cos\theta - i\sin\theta) r(\cos\theta + i\sin\theta) = rr^{-1}(\cos^2\theta + \sin^2\theta) = 1$. Hence, $z^{-1} = u$.

17. Show that $f: \mathbf{C} \longrightarrow q(\mathbf{R})$ defined by $f(a + bi) = a + bi + 0i + 0k$ is a well-defined ring monomorphism.

Problem Set 5.1

1. (a) $\alpha\beta = \begin{pmatrix} 1\ 2\ 3\ 4\ 5 \\ 5\ 3\ 4\ 1\ 2 \end{pmatrix}$ (c) $\alpha^2 = \begin{pmatrix} 1\ 2\ 3\ 4\ 5 \\ 3\ 2\ 1\ 5\ 4 \end{pmatrix}$

3. (a) $(1, 3, 5)(2, 4, 6)$ (c) $(1,4,3)$

5. (a) Odd   (c) Even

7. Suppose that $\alpha, \beta \in \mathbf{S}_n$. If α and β are both even, then $\alpha\beta$ is even, so $f(\alpha\beta) = [0] = [0] + [0] = f(\alpha) + f(\beta)$. If α and β are both odd, then $\alpha\beta$ is even, and $f(\alpha\beta) = [0] = [1] + [1] = f(\alpha) + f(\beta)$. If α is odd and β is even, then $\alpha\beta$ is odd, and $f(\alpha\beta) = [1] = [1] + [0] = f(\alpha) + f(\beta)$, and similarly if α is even and β is odd. Therefore, f is a group homomorphism. Note that $f(\alpha) = [0]$ if and only if $\alpha \in \mathbf{A}_n$, the set of all even permutations in $\mathbf{S}_n$. This shows that $\ker f = \mathbf{A}_n$ is a normal subgroup of $\mathbf{S}_n$ since the kernels of group homomorphisms are normal subgroups.

9. (a) 5   (c) 3

11. $\mathbf{A}_3 = \{R_0, R_{120}, R_{360}\}$

Problem Set 5.2

1. If G is a cyclic group, then $G = \langle x \rangle$ for some $x \in G$. Hence, if $y, y' \in G$, then there are integers m and n such that $y = x^m$ and $y' = x^n$. Hence, $yy' = x^m x^n = x^{m+n} = x^{n+m} = x^n x^m = y'y$.

3. The cyclic subgroups of $\mathbf{Z}$ look like $\langle n \rangle$, where $n \in \mathbf{Z}$.

5. (a) Suppose that $\mathbf{R}$ is an additive cyclic group. Then there is an element $x \in \mathbf{R}$, $x \neq 0$, such that $\mathbf{R} = \langle x \rangle = \{nx \mid n \in \mathbf{Z}\}$. Since $\mathbf{Z}$ is closed under addition, x cannot be an integer, for if it were, then $\mathbf{R} = \langle x \rangle \subseteq \mathbf{Z}$, which is clearly impossible. Likewise, if x is a rational number, then $\mathbf{R} = \langle x \rangle \subseteq \mathbf{Q}$, again an impossibility. Hence, if the additive group $\mathbf{R}$ has a generator x, x has to be an irrational number. If x is irrational, then $\mathbf{R} = \langle x \rangle$ implies that there is an integer n such that $nx = 1$. But then $x = \frac{1}{n}$, so x is rational, a contradiction. Hence, $\mathbf{R}$ cannot be an additive cyclic group.

7. If x is a generator for $\mathbf{Z}$ and $x \neq \pm 1$, then there is an integer n such that $nx = 1$. But then $x = \frac{1}{n}$, so x is not an integer unless $n = \pm 1$. But then $x = \pm 1$, which is a contradiction. Thus, the only generators of $\mathbf{Z}$ are 1 and -1.

9. (a) Suppose that $\gcd(m, n) = 1$. Then $\mathbf{o}(x^m) = \dfrac{n}{\gcd(m, n)} = n$. Hence $\langle x^m \rangle = G$. Conversely, suppose that $\langle x^m \rangle = G$. Then it must be the case that $\mathbf{o}(x^m) = n$. Consequently, $\dfrac{n}{\gcd(m, n)} = n$, which implies that $\gcd(m, n) = 1$.

    (b) $[1], [3], [5], [7], [9]$

11. Consider $\mathbf{Z}_n$.

13. Suppose that $\mathbf{o}(x) = m$ and $\mathbf{o}(y) = n$, and let $k = \text{lcm}(\mathbf{o}(x), \mathbf{o}(y))$. Now both m and n divide k, and so there are integers a and b such that $k = am$ and $k = bn$. Hence, $(xy)^k = x^k y^k = x^{ma} y^{nb} = (x^m)^a (y^n)^b = e$. Thus, $\mathbf{o}(xy)$ divides k, so $\mathbf{o}(xy) \leq \text{lcm}(\mathbf{o}(x), \mathbf{o}(y))$. Conversely, if $\mathbf{o}(xy) = s$, then $(xy)^s = e$, and so $x^s y^s = e$. Hence, $y^s = x^{-s} \in \langle x \rangle$. Thus, $y^s \in \langle x \rangle \cap \langle y \rangle = \{e\}$, so $n \mid \mathbf{o}(xy)$. Similarly, $m \mid \mathbf{o}(xy)$, so it must be the case that $\text{lcm}(\mathbf{o}(x), \mathbf{o}(y))$ divides $\mathbf{o}(xy)$. Hence, $\text{lcm}(\mathbf{o}(x), \mathbf{o}(y)) \leq \mathbf{o}(xy)$. Therefore, $\mathbf{o}(xy) = \text{lcm}(\mathbf{o}(x), \mathbf{o}(y))$.

15. Suppose that $|G| = |H| = n$. Then $G = \{e, x, x^2, \ldots, x^{n-1}\}$ and $H = \{e, y, y^2, \ldots, y^{n-1}\}$. Show that $f : G \longrightarrow H$ defined by $f(x^k) = y^k$ is a well-defined group isomorphism.

Problem Set 5.3

1. If $G_1 \times G_2$ is abelian, then $(x, y)(x, y') = (x', y')(x, y)$ for all $(x, y), (x', y') \in G_1 \times G_2$. Thus, $(xx', yy') = (x'x, y'y)$, so it follows that $xx' = x'x$ and $yy' = y'y$ for all $x, x' \in G_1$ and all $y, y' \in G_2$. Thus, G_1 and G_2 are abelian. Conversely, if G_1 and G_2 are abelian and $(x, y), (x', y') \in G_1 \times G_2$, then $(x, y)(x', y') = (xx', yy') = (x'x, y'y) = (x', y')(x, y)$, and so $G_1 \times G_2$ is abelian.

3. (a) By Theorem 5.2.13, it suffices to show that $\mathbf{Z}_m \times \mathbf{Z}_n$ is cyclic and has order mn. Clearly, $|\mathbf{Z}_m \times \mathbf{Z}_n| = mn$. We claim that $\mathbf{Z}_m \times \mathbf{Z}_n = \langle ([1]_m, [1]_n) \rangle$. If $\mathbf{o}([1]_m, [1]_n) = s$, then $(s[1]_m, s[1]_n) = s([1]_m, [1]_n) = ([0]_m, [0]_n)$, and so $s[1]_m = [0]_m$ and $s[1]_n = [0]_n$. Since $\mathbf{o}([1]_m) = m$ and $\mathbf{o}([1]_n) = n$, Theorem 5.2.8 implies that $m \mid s$ and $n \mid s$. Hence, there are integers k_1 and k_2 such that $s = mk_1$ and $s = nk_2$. Now, by Theorem 1.1.7, there are integers a and b such that $1 = ma + nb$. Thus, $s = mas + nbs = mnak_2 + mnbk_1 = mn(ak_2 + bk_1)$, so $mn \mid s$. Consequently, $mn \leq s$. Note also that $mn([1]_m, [1]_n) = (nm[1]_m, mn[1]_n) = ([0]_m, [0]_n)$, so again by Theorem 5.2.8, $s \mid mn$. Hence, $s \leq mn$, so $mn = s$. Therefore, we have $|\mathbf{Z}_m \times \mathbf{Z}_n| = \mathbf{o}(([1]_m, [1]_n)$, so $\mathbf{Z}_m \times \mathbf{Z}_n = \langle ([1]_m, [1]_n) \rangle$.

    (b) Since $m \mid mn$ and $n \mid mn$, by Theorem 5.2.10, G has a cyclic subgroup G_1 of order m and a cyclic subgroup G_2 of order n. Theorem 5.2.13 shows that G_1 is isomorphic to $\mathbf{Z}_m$ and that G_2 is isomorphic to $\mathbf{Z}_n$.

5. Suppose that $f : G_1 \longrightarrow H_1$ and $g : G_2 \longrightarrow H_2$ are group isomorphisms. Show that $\phi : G_1 \times G_2 \longrightarrow H_1 \times H_2$ defined by $\phi(x, y) = (f(x), g(y))$ is a group isomorphism.

7. Let $f : (G_1/H_1) \times (G_2/H_2) \longrightarrow G/(H_1H_2)$ be defined by $f((xH_1, yH_2)) = xyH_1H_2$. Show that f is a group isomorphism.

Problem Set 5.4

3. (d)

Elementary divisors	$\longrightarrow$	Group	Invariant factors	$\longrightarrow$	Group
2, 2, 2, 3, 5, 5		$\mathbf{Z}_2 \times \mathbf{Z}_2 \times \mathbf{Z}_2 \times \mathbf{Z}_3 \times \mathbf{Z}_5 \times \mathbf{Z}_5$	2, 10, 30		$\mathbf{Z}_2 \times \mathbf{Z}_{10} \times \mathbf{Z}_{30}$
$2, 2^2, 3, 5, 5$		$\mathbf{Z}_2 \times \mathbf{Z}_4 \times \mathbf{Z}_3 \times \mathbf{Z}_5 \times \mathbf{Z}_5$	10, 60		$\mathbf{Z}_{10} \times \mathbf{Z}_{60}$
$2^3, 3, 5, 5$		$\mathbf{Z}_8 \times \mathbf{Z}_3 \times \mathbf{Z}_5 \times \mathbf{Z}_5$	5, 120		$\mathbf{Z}_5 \times \mathbf{Z}_{120}$

$2, 2, 2, 3, 5^2$	$\mathbf{Z}_2 \times \mathbf{Z}_2 \times \mathbf{Z}_2 \times \mathbf{Z}_3 \times \mathbf{Z}_{25}$	$2, 2, 150$	$\mathbf{Z}_2 \times \mathbf{Z}_2 \times \mathbf{Z}_{150}$
$2, 2^2, 3, 5^2$	$\mathbf{Z}_2 \times \mathbf{Z}_4 \times \mathbf{Z}_3 \times \mathbf{Z}_{25}$	$2, 300$	$\mathbf{Z}_2 \times \mathbf{Z}_{300}$
$2^3, 3, 5^2$	$\mathbf{Z}_8 \times \mathbf{Z}_3 \times \mathbf{Z}_{25}$	600	$\mathbf{Z}_{600}$

5. (d)

By elementary divisors   By invariant factors

$\mathbf{Z}_2 \times \mathbf{Z}_2 \times \mathbf{Z}_2 \times \mathbf{Z}_4 \times \mathbf{Z}_3 \times \mathbf{Z}_3 \times \mathbf{Z}_3 \times \mathbf{Z}_5 \times \mathbf{Z}_{25} \times \mathbf{Z}_{125}$ $\mathbf{Z}_2 \times \mathbf{Z}_{30} \times \mathbf{Z}_{150} \times \mathbf{Z}_{1500}$

9. If $x \in G$, then $px \in pG$, so $pG \neq \varnothing$. Finally, if $px, py \in pG$, where $x, y \in G$, then $px - py = p(x - y) \in pG$ since $x - y \in G$. If G is a p-group and $px \in pG$, where $x \in G$, then x has order that is a power of p. Suppose that $\mathbf{o}(x) = p^n$. Then $p^{n-1}(px) = p^n x = 0$, so $\mathbf{o}(px) = p^{n-1}$. Hence, pG is a p-subgroup of G.

13. Since G is cyclic, let $x \in G$ such that $G = \langle x \rangle$. Then $\mathbf{o}(x) = p^k$. If t is an integer such that $0 \leq t \leq k$, then by Theorem 5.2.8, $\mathbf{o}(p^{k-t}x) = \dfrac{p^k}{\gcd(p^{k-t}, p^k)} = \dfrac{p^k}{p^{k-t}} = pt$. Hence, $p^{k-t}x$ generates a subgroup of G of order p^t.

Problem Set 6.1

1. (a) (2) holds   (c) (1) holds

3. $\mathbf{Z}[\sqrt[3]{2}] = \{a + b\sqrt[3]{2} + c\sqrt[3]{4} \mid a, b, c \in \mathbf{Z}\}$.

7. Use induction.

9. There are six polynomials in $\mathbf{Z}_3[x]$ of degree 1 and $m(n)(n-1)$ polynomials in $\mathbf{Z}_n[x]$ of degree m.

11. Let $f(x) = a_0 + a_1 x + \cdots + a_m x^m$ and $g(x) = b_0 + b_1 x + \cdots + b_n x^n$ be polynomials in $R[x]$, where R is commutative. To show that $f(x)g(x) = g(x)f(x)$, it suffices to show that the general kth term of $f(x)g(x)$ is equal to the general kth term of $g(x)f(x)$. The kth term of $f(x)g(x)$ is $c_k = a_0 b_k + a_1 b_{k-1} + \cdots + a_k b_0$. Since addition and multiplication are commutative in R, the terms of c_k can be rearranged to give $c'_k = b_0 a_k + b_1 a_{k-1} + \cdots + b_k a_0$, which is the kth term of the product $g(x)f(x)$.

15. If $F[x]$ is a field, then $x \in F[x]$ has a multiplicative inverse in $F[x]$. Let $a_0 + a_1x + \cdots + a_nx^n \in F[x]$ be the multiplicative inverse in $F[x]$ of x. Then $a_0x + a_1x^2 + \cdots + a_nx^{n+1} = (a_0 + a_1x + \cdots + a_nx^n)x = e$, which is clearly impossible. Thus, x does not have a multiplicative inverse in $F[x]$, so $F[x]$ is not a field. To complete the exercise, we need to show that $f(x) \in F[x]$ has a multiplicative inverse in $F[x]$ if and only if $f(x)$ is a nonzero constant polynomial. If $f(x)$ has a multiplicative inverse in $F[x]$, then there is a $g(x) \in F[x]$ such that $f(x)g(x) = e$. This implies that $f(x) \neq 0$ since if $f(x) = 0$, then $f(x)g(x) = 0$ and we would have $0 = e$, which is not possible in a field. We also have $\deg(f(x)) + \deg(g(x)) = \deg(f(x)g(x)) = \deg(e) = 0$, so $\deg(f(x)) = \deg(g(x)) = 0$. Therefore, $f(x)$ is a constant polynomial. Conversely, if $f(x) = a \neq 0$, then $a \in F$, so a^{-1} exists in F. Hence, $f(x)g(x) = e$ where $g(x) = a^{-1}$.

Problem Set 6.2

3. $[5](x - [1])(x - [2]) = [5](x - [4])(x - [5]) = [5]x^2 + [3]x + [4]$. Furthermore, $[1],[2]$ and $[3],[4]$ are the only pairs of roots c_1, c_2 such that $f(x) = [5](x - c_1)(x - c_2)$.

5. Suppose that $c(x) = \gcd(a(x), b(x))$ with $c(x) \neq e$. Then $\deg(c(x)) \geq 1$, and there exist $k_1(x), k_2(x) \in F[x]$ such that $a(x) = c(x)k_1(x)$ and $b(x) = c(x)k_2(x)$. But this implies that $f(x) = d(x)c(x)k_1(x)$ and $g(x) = d(x)c(x)k_2(x)$. Hence, $d(x)c(x) \mid f(x)$, $d(x)c(x) \mid g(x)$, and $\deg(d(x)c(x)) > \deg(d(x))$, which is impossible since $d(x)$ is a divisor of $f(x)$ and $g(x)$ of minimal degree. Hence, it must be the case that $\gcd(a(x), b(x)) = e$.

9. Consider $ax^2 + bx + c = 0$. Since $a \neq 0$, a^{-1}exists in F. To simplify notation, let's agree to write a^{-1} as $\frac{1}{a}$. Then $x^2 + \frac{b}{a}x = -\frac{c}{a}$. Now F does not have characteristic 2, so $2e \neq 0$. Thus, $(2e)a \neq 0$ since F has no zero-divisors. But $(2e)a = 2(ea) = 2a$, so $2a \neq 0$. Hence, $\frac{1}{2a}$ exists in F, as does $\left(\frac{1}{2a}\right)^2 = \frac{1}{4a^2}$. Therefore, we see that

$$x^2 + \frac{b}{a}x + \frac{b^2}{4a^2} = \frac{b^2}{4a^2} - \frac{c}{a} = \frac{b^2 - 4ac}{4a^2}.$$

If $b^2 - 4ac$ is a perfect square in F, then there is an $s \in F$ such that $s^2 = b^2 - 4ac$. Hence,

$$\left(x + \frac{b}{a}\right)^2 = \frac{s^2}{4a^2} = \left(\frac{s}{2a}\right)^2,$$

which implies that

$$x + \frac{b}{a} = \pm\frac{s}{2a}, \quad \text{or} \quad x = \frac{-b \pm s}{2a} = \frac{-b \pm \sqrt{b^2 - 4ac}}{2a}.$$

11. (a) The roots are [4] and [3].   (c) The roots are [3] and [2].

13. Let F have more than n elements, and suppose that $f(c) = 0$ for all $c \in F$. If $c_1, c_2, \ldots, c_n$ are n distinct elements of F, then each c_i is a root of $f(x)$. If a is the leading coefficient of $f(x)$, then we know that $f(x)$ can be factored as $f(x) = a(x - c_1)(x - c_2) \cdots (x - c_n)$. If c is an element of F such that $c \neq c_i$ for $i = 1, 2, \ldots, n$, then $f(c) = 0$. But this implies that $a(c - c_1)(c - c_2) \cdots (c - c_n) = 0$, so F has zero-divisors, which is impossible since F is a field. Thus, it cannot be the case that $f(c) = 0$, and so $f(c) \neq 0$.

Problem Set 6.3

1. (c) $\bar{c} = 3 + 2i$ is also a root of $f(x)$, so $(x - 3 + 2i)(x - 3 - 2i) = x^2 - 6x + 13$ is a factor of $f(x)$. Dividing $f(x)$ by $x^2 - 6x + 13$ gives $3x^2 + 27$. Hence, $c = 3i$ and $\bar{c} = -3i$ are roots of $f(x)$.

5. (a) $x = \frac{-i + \sqrt[c]{i^2 - 4(1)(1)}}{2(1)} = \frac{-i + \sqrt[c]{-5}}{2}$. The complex square roots of -5 are $\sqrt{5}i$ and $-\sqrt{5}i$. Hence, we see that $x = \frac{(-1 + \sqrt{5})i}{2}$ and $x = \frac{(-1 - \sqrt{5})i}{2}$ are the roots of $f(x)$.

7. (c) The rational roots of $f(x)$, if any exist, must be elements of the set $\left\{ \pm 1, \pm 7, \pm\frac{1}{2}, \pm\frac{7}{2}, \pm\frac{1}{3}, \pm\frac{7}{3}, \pm\frac{1}{6}, \pm\frac{7}{6} \right\}$. The rational roots are $x = -\frac{1}{3}$ and $x = \frac{7}{2}$, and $f(x) = 6(x + \frac{1}{3})(x - \frac{7}{2})(x^2 - 2x - 1)$.

9. (a) Since $2 \nmid 9$, let's try 2 for the prime. The polynomial in $\mathbf{Z}_2[x]$ corresponding to $f(x)$ is $[f]_2[x] = [1] + x + x^4$. Since neither [0] nor [1] is a root of $[f]_2[x]$, if $[f]_2[x]$ factors in $\mathbf{Z}_2[x]$, it must be the product of two quadratic polynomials in $\mathbf{Z}_2[x]$. The only quadratic polynomials in $\mathbf{Z}_2[x]$ are $x^2, x^2 + x, x^2 + [1]$, and $x^2 + x + [1]$. Long division can be used to show that none of these polynomials divide $[f]_2[x]$. Hence, $[f]_2[x]$ is irreducible in $\mathbf{Z}_2[x]$.

11. Consider the polynomial $f(x) = x^n - p$ in $\mathbf{Z}[x]$.

Problem Set 7.1

1. (a) $g(x) \equiv h(x) \pmod{f(x)}$
    (c) $g(x)$ and $h(x)$ are not congruent modulo $f(x)$.

3. Suppose that $f(x) = a$, where a is a nonzero element of F. Since F is a field, a^{-1} exists in F. If $g(x) \in F[x]$ and b_k is a coefficient of $g(x)$, then $b_k = a(a^{-1}b_k)$, so $a \mid b_k$. Hence, a divides every coefficient of $g(x)$, and so a divides every coefficient of $g(x) - 0$, where 0 is the zero polynomial. Thus, $f(x) \mid (g(x) - 0)$, so $g(x) \equiv 0 \pmod{f(x)}$. Therefore, there is only one congruence class, and this is the class determined by the zero polynomial.

5. There are 81 congruence classes in $\mathbf{Z}_3[x] / (f(x))$ and p^{n+1} congruence classes in $\mathbf{Z}_p[x] / (f(x))$ when $\deg(f(x)) = n$.

7. By Theorem 6.1.3, $F[x]$ is a commutative ring with identity, and $(f(x))$ is an ideal of $F[x]$. Hence, $F[x] / (f(x))$ is just a factor ring of $F[x]$ that is commutative and has an identity. (See Theorem 3.2.7 for the details.) Thus, it remains only to show that ϕ is a well-defined ring monomorphism. Suppose that $a = b$, then $\bar{a} = a + (f(x)) = b + (f(x)) = \bar{b}$ in $F[x] / (f(x))$. Thus, $\phi(a) = \phi(b)$, so ϕ is well-defined. ϕ is a ring homomorphism because if $a, b \in F$, then $\phi(a + b) = \overline{a+b} = \bar{a} + \bar{b} = \phi(a) + \phi(b)$ and $\phi(ab) = \overline{ab} = \bar{a}\,\bar{b} = \phi(a)\phi(b)$. Finally, if $\phi(a) = \phi(b)$, then $a + (f(x)) = \bar{a} = \bar{b} = b + (f(x))$. Hence, $a - b \in (f(x))$, which implies that $\bar{a} - \bar{b} = \overline{a-b} = \bar{0}$ in $F[x] / (f(x))$, so $\bar{a} = \bar{b}$.

9. The mapping $\phi : \mathbf{C} \longrightarrow \mathbf{R}[x] / (x^2 + 1)$ is defined by $\phi(a + bi) = \overline{a + bx}$. First note that in $\mathbf{R}[x] / (x^2 + 1))$, $\overline{x^2 + 1} = \bar{0}$, so it follows that $\overline{x^2} = \overline{-1}$ in $\mathbf{R}[x] / (x^2 + 1)$. If $a + bi = c + di$, then $a = c$ and $b = d$. Hence, $\overline{a + bx} = \overline{c + dx}$, which implies that $\overline{a + bx} = \overline{c + dx}$, so $\phi(a + bi) = \phi(c + di)$. Therefore, ϕ is well-defined. ϕ is a ring homomorphism since $\phi((a + bi) + (c + di)) = \phi((a + c) + (b + d)i) = \overline{(a + c) + (b + d)x} = \overline{(a + bx) + (c + dx)} = \overline{a + bx} + \overline{c + dx} = \phi(a + bi) + \phi(c + di)$ and $\phi(a + bi)\phi(c + di) = (\overline{a + bx})(\overline{c + dx}) = \overline{(a + bx)(c + dx)} = \overline{ac + (ad + bc)x + bdx^2} = \overline{ac} + \overline{(ad + bc)x} + \overline{bd\,x^2} = \overline{ac} + \overline{(ad + bc)x} + \overline{bd}\,\overline{-1} = \overline{ac - bd} + \overline{(ad + bc)x} = \overline{(ac - bd) + (ad + bc)x} =$

$\phi((ac - bd) + (ad + bc)i) = \phi((a + bi)(c + di))$. ϕ is clearly surjective, for if $\overline{a + bx} \in \mathbf{R}[x]/(x^2 + 1)$, then $a, b \in \mathbf{R}$, so $a + bi \in \mathbf{C}$ is such that $\phi(a + bi) = \overline{a + bx}$. If $\phi(a + bi) = \overline{0}$, then $\overline{a + bx} = \overline{0}$, and it follows that $a + bx = a + bx - 0 \in (x^2 + 1)$. But this implies that $x^2 + 1$ divides $a + bx$, which is impossible unless $a + bx$ is the zero polynomial. Thus, $a = b = 0$, so $a + bi = 0$. Therefore, ker $\phi = \{0\}$, and so ϕ is injective.

11. Let I be an ideal of $F[x]$. If $I = \{0\}$, then $I = (0)$, so suppose that $I \neq \{0\}$. If I contains a nonzero constant polynomial, say $f(x) = a$, then $f(x)^{-1} = a^{-1}$ exists in $F[x]$ and $e = aa^{-1} \in I$, so $I = F[x]$. Hence, $I = (e)$. Finally, suppose that $I \neq \{0\}$ and that I contains no nonzero constant polynomials. Let $f(x)$ be a polynomial in I of minimal positive degree. Such a polynomial exists since $\mathbf{N}$ is well-ordered. If $g(x) \in I$, then there are polynomials $q(x)$, $r(x) \in F[x]$ such that $g(x) = f(x)q(x) + r(x)$, where $r(x) = 0$ or $\deg(r(x)) < \deg(f(x))$. Since $r(x) = g(x) - f(x)q(x) \in I$, it must be the case that $r(x) = 0$, for otherwise $r(x)$ will be a polynomial in I with degree smaller than that of $f(x)$. Therefore, $g(x) = f(x)q(x)$, so $I \subseteq (f(x))$. Since the reverse containment is obvious, we have $I = (f(x))$.

Problem Set 7.2

1. If $f(x)$ is an associate of $x^2 + [3]$, there must be a unit $u(x)$ in $\mathbf{Z}_5[x]$ such that $f(x) = u(x)(x^2 + [5])$. We first need to find the units in $\mathbf{Z}_5[x]$. If $u(x)$ is a unit in $\mathbf{Z}_5[x]$, then there is a $v(x)$ in $\mathbf{Z}_5[x]$ such that $u(x)v(x) = [1]$. Since $\mathbf{Z}_5$ is a field, $\deg(u(x)) + \deg(v(x)) = \deg(u(x)v(x)) = \deg([1]) = 0$. Therefore, $\deg(u(x)) = 0$, so $u(x)$ is a constant. Hence, the units in $\mathbf{Z}_5[x]$ are $[1]$, $[2]$, $[3]$, and $[4]$. Thus, the associates of $x^2 + [3]$ are $x^2 + [3]$, $[2](x^2 + [3]) = [2]x^2 + [1]$, $[3](x^2 + [3]) = [3]x^2 + [4]$, and $[4](x^2 + [3]) = [4]x^2 + [2]$.

3. (a) The technique of proof is to use the fact that $v(ab) \geq v(a)$ for all nonzero a and b in D and to make special choices for a and b. $v(-a) = v(-ea) \geq v(a)$ and $v(a) = v((-e)(-a)) \geq v(-a)$. Hence, $v(-a) \geq v(a)$ and $v(a) \geq v(-a)$ and so $v(a) = v(-a)$.

    (c) Suppose that $v(a) = v(e)$. Since D is a Euclidean Domain, there exist q and r in D such that $e = aq + r$, where $r = 0$ or $v(r) < v(a) = v(e)$. By part (b), it is impossible for $v(r) < v(e)$, so $r = 0$. Hence, $aq = e$, and so a is a unit. Conversely, suppose that a is a unit. Then there is an element b in D such that $ab = e$. Hence, we see that $v(e) = v(ab) \geq v(a)$. But by part (b), it is impossible for $v(e) > v(a)$. Therefore, $v(e) = v(a)$.

7. If I is an ideal of S, $f^{-1}(I)$ is an ideal of R. Since R is a Principal Ideal Domain, there is an $a \in R$ such that $f^{-1}(I) = (a)$. We claim that $(f(a)) = I$. If $y \in I$, then there is an $x \in R$ such that $f(x) = y$. Hence, $x \in f^{-1}(I)$, so $x = ar$ for some $r \in R$. But then $y = f(x) = f(ar) = f(a)f(r) \in (f(a))$. Thus, $I \subseteq (f(a))$. Now suppose that $f(a)s \in (f(a))$, where $s \in S$. Since f is an epimorphism, there is an $r \in R$ such that $f(r) = s$. Hence, $f(a)s = f(a)f(r) = f(ar)$ implies that $ar \in (a) = f^{-1}(I)$. Thus, $f(ar) \in I$, and so it follows that $f(a)s \in I$. Hence, $(f(a)) \subseteq I$, so we have shown that $I = (f(a))$, which shows that S is a Principal Ideal Domain. The answer to the second question is no.

9. First recall that in a Principal Ideal Domain D, an element is prime if and only if it is irreducible. Suppose that (p) is a maximal ideal of the Principal Ideal Domain D and that p is not prime. Then p is reducible, so there are elements $a, b \in D$ such that $p = ab$ and neither a nor b is a unit in D. Since $(p) \subseteq (a) \subseteq D$ and (p) is maximal, either $(p) = (a)$ or $(a) = D$. If $(a) = D$, then there is an element c in D such that $ac = e$. This implies that a is a unit, so $(a) \neq D$. Thus, it must be the case that $(p) = (a)$. Hence, $a \in (p)$, and so there is a $d \in D$ such that $a = pd$. But $p = ab$, so $a = abd$, which implies that $bd = e$. Hence, b is a unit, which is a contradiction. Therefore, if (p) is maximal, then p must be prime. Conversely, assume that p is prime and let's show that (p) is a maximal ideal of D. Suppose that (a) is an ideal of D such that $(p) \subseteq (a) \subseteq D$. Since $p \in (a)$, there is a $b \in D$ such that $p = ab$. Now p is irreducible, so a or b must be a unit. If a is a unit, it follows that $(a) = D$, and if b is a unit, then $(a) = (p)$. Therefore, when p is prime, (p) is a maximal ideal of D.

11. If ν is a constant function, then $\nu(a) = \nu(e)$ for all nonzero a in D. But by part (c) of Exercise 3, this implies that each nonzero a in D is a unit. Thus, D is a field.

Problem Set 7.3

1. 5 is irreducible in $\mathbf{Z}[\sqrt{-2}]$ but not in $\mathbf{Z}[i]$.

3. Use induction.

5. If $x = a + b\sqrt{-3} \in \mathbf{Z}[\sqrt{-3}]$, then $N(x) = a^2 + 3b^2$. Since x is a unit if and only if $N(x) = 1$, we see that x will be a unit if and only if $a^2 + 3b^2 = 1$. Since $a = \pm 1$ and $b = 0$ are the only solutions in $\mathbf{Z}$ of this equation, the only units in $\mathbf{Z}[\sqrt{-3}]$ are $x = 1$ and $x = -1$.

9. Since a Euclidean Domain must be a Unique Factorization Domain, if
    $\mathbf{Z}[\sqrt{-3}]$ is not a Unique Factorization Domain, then $\mathbf{Z}[\sqrt{-3}]$ cannot be a
    Euclidean Domain. We show that $\mathbf{Z}[\sqrt{-3}]$ is not a Unique Factorization
    Domain. Since $2 \cdot 2 = 4 = (1 + \sqrt{-3})(1 - \sqrt{-3})$, we will be finished if we
    can show that 2, $1 + \sqrt{-3}$, and $1 - \sqrt{-3}$ are irreducible in $\mathbf{Z}[\sqrt{-3}]$ and that
    2 is not an associate of either $1 + \sqrt{-3}$ or $1 - \sqrt{-3}$. If 2 is reducible, then
    $2 = (a + b\sqrt{-3})(c + d\sqrt{-3})$, where $a + b\sqrt{-3}, c + d\sqrt{-3} \in \mathbf{Z}[\sqrt{-3}]$. But
    then $4 = N(2) = N(a + b\sqrt{-3})N(c + d\sqrt{-3}) = (a^2 + 3b^2)(c^2 + 3d^2)$. Since
    4 can be factored as $1 \cdot 4$ or as $2 \cdot 2$, it follows that

$$1 = a^2 + 3b^2 \quad \text{and} \quad 4 = c^2 + 3d^2, \tag{1}$$
$$4 = a^2 + 3b^2 \quad \text{and} \quad 1 = c^2 + 3d^2, \quad \text{or} \tag{2}$$
$$2 = a^2 + 3b^2 \quad \text{and} \quad 2 = c^2 + 3d^2. \tag{3}$$

Equations (1) show that $N(a + b\sqrt{-3}) = 1$, so $a + b\sqrt{-3}$ is a unit in
$\mathbf{Z}[\sqrt{-3}]$. Similarly, equations (2) show that $c + d\sqrt{-3}$ is a unit. Since
equations (3) have no solutions in the integers, it follows that 2 is
irreducible in $\mathbf{Z}[\sqrt{-3}]$. Next, suppose that $1 + \sqrt{-3} =$
$(a + b\sqrt{-3})(c + d\sqrt{-3})$. Then

$$4 = N(1 + \sqrt{-3})$$
$$= N(a + b\sqrt{-3})N(c + d\sqrt{-3})$$
$$= (a^2 + 3b^2)(c^2 + 3d^2).$$

But this leads to a set of equations exactly like equations (1), (2), and (3), so
$1 + \sqrt{-3}$ is irreducible in $\mathbf{Z}[\sqrt{-3}]$. Similarly, $1 - \sqrt{-3}$ is irreducible in
$\mathbf{Z}[\sqrt{-3}]$. Finally, let's show that 2 is not an associate of $1 + \sqrt{-3}$. The fact
that 2 is not an associate of $1 - \sqrt{-3}$ has a similar proof. If 2 is an associate of
$1 + \sqrt{-3}$, then there is a unit $a + b\sqrt{-3}$ in $\mathbf{Z}[\sqrt{-3}]$ such that $2(a + b\sqrt{-3}) =$
$1 + \sqrt{-3}$. Hence, we see that $2a + 2b\sqrt{-3} = 1 + \sqrt{-3}$, and this in turn
implies that $2a = 1$ and $2b = 1$ for some integers a and b, a contradiction.
Thus, 2 is not an associate of $1 + \sqrt{-3}$.

11. (a) If $x = a + b\sqrt{2}$ and $y = c + d\sqrt{2}$, then $v(x) = |a^2 - 2b^2| \le$
    $|a^2 - 2b^2||c^2 - 2b^2| = v(x)v(y)$. The next step is to compute y^{-1} in
    $\mathbf{Q}[\sqrt{2}]$:

$$y^{-1} = \frac{1}{c + d\sqrt{2}} = \frac{c - d\sqrt{2}}{(c + d\sqrt{2})(c - d\sqrt{2})} = \frac{c}{c^2 - 2d^2} - \frac{d}{c^2 - 2d^2}\sqrt{2}.$$

Note that $c^2 - 2d^2 \neq 0$ since $\sqrt{2}$ is not a rational number. Since $xy^{-1} = c' + d'\sqrt{2} \in \mathbf{Q}[\sqrt{2}]$, choose integers m and n closest to c' and d', respectively. Then $|c' - m| \leq \frac{1}{2}$ and $|d' - n| \leq \frac{1}{2}$.

From this we see that

$$x = y(c' + d'\sqrt{2})$$
$$= y(c' - m + m) + y(d' - n + n)\sqrt{2}$$
$$= y(m + n\sqrt{2}) + y[(c' - m + d' - n)\sqrt{2}]$$
$$= yq + r,$$

where $q = m + n\sqrt{2}$ and $r = y[(c' - m) + (d' - n)\sqrt{2}]$. Since $x, y, q \in \mathbf{Z}[\sqrt{2}]$, we see that $r = x - yq \in \mathbf{Z}[\sqrt{2}]$. We will be finished if we can show that $r = 0$ or $v(r) < v(y)$. If $r \neq 0$, then $v(r) = v(y[(c' - m) + (d' - n)\sqrt{2}]) = v(y)v((c' - m) + (d' - n)\sqrt{2}) = v(y)|(c' - m)^2 - 2(d' - n)^2| \leq v(y)(c' - m)^2 \leq \frac{1}{4}v(y) < v(y)$. Hence, if $r \neq 0$, then $v(r) < v(y)$.

(b) The norm of each of these elements of $\mathbf{Z}[\sqrt{2}]$ is a prime integer, so each is irreducible in $\mathbf{Z}[\sqrt{2}]$.

(c) No, there can't be a contradiction. It must be the case that $2 + \sqrt{2}$ is an associate of $2 - \sqrt{2}$ or $5 + \sqrt{2}$, and a similar observation holds for $11 - 7\sqrt{2}$. Let's see if we can find a unit $u = a + b\sqrt{2}$ of $\mathbf{Z}[\sqrt{2}]$ such that $2 + \sqrt{2} = (a + b\sqrt{2})(2 - \sqrt{2})$. If such an equation holds, then $2 + \sqrt{2} = (2a - 2b) + (-a + 2b)\sqrt{2}$. Hence, it must be the case that $2a - 2b = 2$ and $-a + 2b = 1$, and these equations have a solution $a = 3$ and $b = 2$. Thus $u = 3 + 2\sqrt{2}$ will work if u is a unit. Since $N(u) = 1$, u is a unit, so we have shown that $2 + \sqrt{2}$ and $2 - \sqrt{2}$ are associates. In a like manner, it can be shown that $11 - 7\sqrt{2}$ and $5 + \sqrt{2}$ are associates. Thus, unique factorization is not contradicted.

Problem Set 8.1

1. (a) Is not a basis for $\mathbf{C}/\mathbf{R}$.   (c) Is a basis for $\mathbf{C}/\mathbf{R}$.

   (e) Is a basis for $\mathbf{Q}/\mathbf{Q}$.   (f) Is not a basis for $\mathbf{C}/\mathbf{R}$.

5. Suppose that $S = \{y_1, y_2, \ldots, y_k\}$ is a nonempty subset of B, and let $B' = \{z_1, z_2, \ldots, z_{n-k}\}$ be the remaining elements of B. That is, $B' = B \setminus S$. If S is linearly dependent, then there are $c_i \in F$, not all of which are zero, such that

$c_1 y_1 + c_2 y_2 + \cdots + c_k y_k = 0$. But then we have $c_1 y_1 + c_2 y_2 + \cdots + c_k y_k + 0z_1 + 0z_2 + \cdots + 0z_{n-k} = 0$, which is a linear combination of the elements of B, and not all of the coefficients are zero. This cannot be since B is a linearly independent set. Thus, S must be linearly independent. If $k < n$ and S were a basis for E / F, then each element of B' would be a linear combination of the elements of S, which would contradict the fact that B is linearly independent.

9. If $[E : F] = 1$, then E / F has a basis with a single element, say, $\{x\}$. Since $F \subseteq E$, we need to show that $E \subseteq F$. If $y \in F$, $y \neq 0$, then there is a $c \neq 0$ in F such that $y = cx$. But then $x = c^{-1} y \in F$. From this it follows easily that $E \subseteq F$ since every element of E is an element of F times $x \in F$ and this product is in F. Conversely, if $E = F$, then the identity e of F is a basis for F / F. Thus, $[E : F] = 1$.

11. If S spans E, then S is a basis for E / F, and we can let $B = S$. If S does not span E, there is a vector $x_1 \in E$ that is not in the span of S. Exercise 8 shows that $S_1 = S \cup \{x_1\}$ is linearly independent. If S_1 does not span E, then there is an element x_2 of E that is not in the span of S_1. Invoking Exercise 8 again shows that $S \cup \{x_1, x_2\} = S_1 \cup \{x_2\}$ is linearly independent. If $S \cup \{x_1, x_2\}$ does not span E, the process can be repeated. Since E / F is finite dimensional, this process must terminate with a finite set S_n that is a basis for E. Thus, we need only let $B = S_n$.

13. If K is an intermediate field, then Theorem 8.1.9 shows that $[E : K][K : F] = [E : F] = p$. Hence, either

$$[E : K] = 1 \quad \text{and} \quad [K : F] = p \qquad \text{or}$$
$$[E : K] = p \quad \text{and} \quad [K : F] = 1.$$

If $[E : K] = 1$, then Exercise 9 shows that $K = E$, and if $[K : F] = 1$, then Exercise 9 shows that $K = F$. Thus, F is the only proper intermediate field when $[E : F] = p$, with p a prime.

Problem Set 8.2

1. (a) $f(x) = x^2 - 4x + 13$ (c) $f(x) = x^4 - 4x^2 + 1$
    (d) $f(x) = x^4 - 2x^2 + 2$ (f) $f(x) = x^4 - 6x^2 + 11$

3. (a) $B = \{1, \sqrt{3}\}$ (c) $B = \{1, \sqrt[4]{2}, \sqrt[4]{4}, \sqrt[4]{8}\}$

5. If $x^2 - 5$ is reducible in $\mathbf{Q}(\sqrt{2})[x]$, then $x^2 - 5$ must have a root in $\mathbf{Q}(\sqrt{2})$. Hence, there must be an element $a + b\sqrt{2} \in \mathbf{Q}(\sqrt{2})$ such that

$(a + b\sqrt{2})^2 - 5 = 0$. Thus, we have $(a^2 + 2b^2 - 5) + 2ab\sqrt{2} = 0$, which implies that $a^2 + 2b^2 - 5 = 0$ and $ab = 0$. Since $\mathbf{Q}$ is an integral domain, $ab = 0$ implies $a = 0$ or $b = 0$. If $a = 0$, then $2b^2 = 5$, which implies that $\sqrt{5}$ is a rational number, which it is not. The same holds if $b = 0$. Thus, such an element $a + b\sqrt{2}$ cannot exist, so $x^2 - 5$ is irreducible in $\mathbf{Q}(\sqrt{2})[x]$.

7. The minimal polynomial of $\sqrt[3]{p}$ and of $\sqrt[3]{p}\omega$ is $f(x) = x^3 - p$, so Theorem 8.2.8 shows that $\mathbf{Q}(\sqrt[3]{p})$ and $\mathbf{Q}(\sqrt[3]{p}\omega)$ are isomorphic to $\mathbf{Q}[x] / (f(x))$. Consequently, $\mathbf{Q}(\sqrt[3]{p})$ and $\mathbf{Q}(\sqrt[3]{p}\omega)$ are isomorphic. However, $\mathbf{Q}(\sqrt[3]{p}) \neq \mathbf{Q}(\sqrt[3]{p}\omega)$ since $\sqrt[3]{p}\omega \notin \mathbf{Q}(\sqrt[3]{p})$.

9. (a) $2 + 5\sqrt{2}$ is a root of $f(x) = (x - 2 - 5\sqrt{2})(x - 2 + 5\sqrt{2}) = (x - 2)^2 - 50 = x^2 - 4x - 46 \in \mathbf{Q}[x]$ and so is algebraic over $\mathbf{Q}$. Note that $a + b\sqrt{2} \in \mathbf{Q}(\sqrt{2})$ implies that $(a - \frac{2}{5}b) + \frac{1}{5}b(2 + 5\sqrt{2})$ is in $\mathbf{Q}(2 + 5\sqrt{2})$, so $\mathbf{Q}(\sqrt{2}) \subseteq \mathbf{Q}(2 + 5\sqrt{2})$. Conversely, if $a + b(2 + 5\sqrt{2}) \in \mathbf{Q}(2 + 5\sqrt{2})$, then $(a + 2b) + (5b)\sqrt{2} \in \mathbf{Q}(\sqrt{2})$, so $\mathbf{Q}(2 + 5\sqrt{2}) \subseteq \mathbf{Q}(\sqrt{2})$.

   (b) Yes to both questions.

   (c) If $u + v(a + bc) \in F(a + bc)$, then $(u + va) + (vb)c \in F(c)$. Thus, $F(a + bc) \in F(c)$. If $u + vc \in F(c)$, then $(u - v\frac{a}{b}) + \frac{v}{b}(a + bc)$ is in $F(a + bc)$. Hence, $F(c) \subseteq F(a + bc)$, so $F(a + bc) = F(c)$. If $c \in F$, then $a + bc \in F(c)$ implies that $a + bc \in F$. Thus, $F(c) \subseteq F$. Since it is always the case that $F \subseteq F(c)$, $F = F(c)$ when $c \in F$.

11. It is sufficient to show that K is closed under addition and multiplication and that if $c \in K, c \neq 0$, then $c^{-1} \in K$. If $c_1, c_2 \in K$, then $F(c_1, c_2)$ is a field, so $c_1 + c_2, c_1c_2 \in F(c_1, c_2)$. Since $c_1c_2 \in K$, c_1 and c_2 are algebraic over F, and so every element of $F(c_1, c_2)$ is algebraic over F. Thus, $c_1 + c_2$ and c_1c_2 are algebraic over F, which shows that $c_1 + c_2$ and c_1c_2 are in K. If $c \in K$, $c \neq 0$, then every element of $F(c)$ is algebraic over F. Since c^{-1} is in $F(c)$, c^{-1} is algebraic over F, so $c^{-1} \in K$. Therefore, K is a subfield of E. If $a \in F$, then a is the root of $f(x) = x - a \in F[x]$, and so $a \in K$. Hence, $F \subseteq K \subseteq E$.

15. Since F is a field, $m(x)$ has at most two roots in some field extension of F. Part (c) of Exercise 9 shows that $F(c) = F(\sqrt{b^2 - 4d})$. We also see that

$$\left(\frac{-b+\sqrt{b^2-4d}}{2e}\right)^2 + b\left(\frac{-b+\sqrt{b^2-4d}}{2e}\right) + d$$

$$= \left(\frac{-b+\sqrt{b^2-4d}}{2e}\right)\left(\frac{-b+\sqrt{b^2-4d}}{2e}+b\right) + d$$

$$= \left(\frac{-b+\sqrt{b^2-4d}}{2e}\right)\left(\frac{b+\sqrt{b^2-4d}}{2e}\right) + d$$

$$= \frac{b^2-4d-b^2}{4e} + d = -d + d = 0,$$

where the calculations take place in $F(c)$. Hence, $\frac{-b+\sqrt{b^2-4d}}{2e}$ is a root of

$m(x)$. It can be shown in a like fashion that $\frac{-b-\sqrt{b^2-4d}}{2e}$ is also

a root of $m(x)$. Thus, $m(x)$ has two distinct roots in $F(c)$ if $b^2 - 4d$ is not a

perfect square in F. If char $F = 2$, then $2e = 0$, and we would be dividing by

0 in $F(c)$, which is impossible in a field. Hence, it must be the case that char

$F \neq 0$. The conclusion that if F is a field with char $F \neq 2$, then any field

extension E/F such that $[E : F] = 2$ must be of the form $F(\sqrt{a})$, where $a \in F$

is not a perfect square in F can be argued as follows. Since $[E : F] = 2$, the

field extension E/F has a basis with two elements, say, $B = \{e, c\}$. Hence, the

set $\{e, c, c^2\}$ is linearly dependent, so there are scalars $a_0, a_1, a_2 \in F$, not all of

which are zero, such that $a_0 e + a_1 c + a_2 c^2 = 0$. Note that $a_2 \neq 0$, for if so, it

would follow that $a_0 = 0$ and $a_1 = 0$ since B is a basis. If we multiply $a_0 e +$

$a_1 c + a_2 c^2 = 0$ through by a_2^{-1}, then we see that $a_2^{-1} a_0 e + a_2^{-1} a_1 c + c^2 = 0$,

so c is a root of a polynomial of the form $m(x) = x^2 + bx + d$, where $b = a_2^{-1} a_1$

and $d = a_2^{-1} a_0$. It follows that $m(x)$ is irreducible in $F[x]$ and has

roots $c = \frac{-b \pm \sqrt{b^2-4d}}{2e}$, so $m(x)$ is the minimal polynomial of c. If we let

$a = b^2 - 4d$, then $F(\sqrt{a}) = E$.

17. (a) $(1 + 2\sqrt{3})^{-1} = -\dfrac{1}{11} + \dfrac{2}{11}\sqrt{3}$

   (b) $(3 + 2\sqrt{2}i)^{-1} = \dfrac{3}{17} - \dfrac{2}{17}\sqrt{2}i$

   (c) $(1 + \sqrt[3]{2} + \sqrt[3]{4})^{-1} = \dfrac{6}{22} + \dfrac{4}{22}\sqrt[3]{2} - \dfrac{1}{22}\sqrt[3]{4}$

Problem Set 8.3

1. If $f(x) = x^2 - 7$ is reducible in $\mathbf{Q}(\sqrt{3})[x]$, then $f(x)$ has a root in $\mathbf{Q}(\sqrt{3})$. If $a + b\sqrt{3}$ is a root of $f(x)$, then $(a + b\sqrt{3})^2 - 7 = 0$. Hence, $a^2 + 2b\sqrt{3} + 3b^2 - 7 = 0$ or $(a^2 + 3b^2 - 7)1 + 2ab\sqrt{3} = 0$. Since $B = \{1, \sqrt{3}\}$ is a basis for $\mathbf{Q}(\sqrt{3})/\mathbf{Q}$, the last equation implies that $a^2 + 3b^2 - 7 = 0$ and that $ab = 0$. But $ab = 0$ implies $a = 0$ or $b = 0$. If $a = b = 0$, then $-7 = 0$, which is clearly impossible. If $a \neq 0$ but $b = 0$, then $a^2 = 7$. Now $\sqrt{7}$ is an irrational number, so there is no rational number a such that $a^2 = 7$. If $a = 0$ but $b \neq 0$, then this would imply that $\sqrt{\dfrac{7}{3}}$ is a rational number, which is not the case. Therefore, $f(x)$ cannot have a root in $\mathbf{Q}(\sqrt{3})$, so $f(x)$ is irreducible in $\mathbf{Q}(\sqrt{3})[x]$.

3. $[\mathbf{Q}(\sqrt{3}) : \mathbf{Q}] = 2$ and $B = \{1, \sqrt{3}\}$

7. $[\mathbf{Q}(\sqrt{3}, \sqrt{5}i) : \mathbf{Q}] = 4$.

9. The tower of field extensions is $\mathbf{Q} \subseteq \mathbf{Q}(\sqrt{2}) \subseteq \mathbf{Q}(\sqrt{2}, \sqrt{3}) \subseteq \mathbf{Q}(\sqrt{2}, \sqrt{3}, \sqrt{5})$. The minimal polynomial of $\sqrt{2}$ in $\mathbf{Q}[x]$ is $m(x) = x^2 - 2$, so $B = \{1, \sqrt{2}\}$ is a basis for $\mathbf{Q}(\sqrt{2})/\mathbf{Q}$. The minimal polynomial of $\sqrt{3}$ in $\mathbf{Q}(\sqrt{2})[x]$ is $m(x) = x^2 - 3$. Hence, $B' = \{1, \sqrt{3}\}$ is a basis for the field extension $\mathbf{Q}(\sqrt{2}, \sqrt{3})/\mathbf{Q}(\sqrt{2})$. The minimal polynomial of $\sqrt{5}$ in $\mathbf{Q}(\sqrt{2}, \sqrt{3})[x]$ is $m(x) = x^2 - 5$, so $B'' = \{1, \sqrt{5}\}$ is a basis for the field extension $\mathbf{Q}(\sqrt{2}, \sqrt{3}, \sqrt{5})/\mathbf{Q}(\sqrt{2}, \sqrt{3})$. Thus, a basis for $\mathbf{Q}(\sqrt{2}, \sqrt{3}, \sqrt{5})/\mathbf{Q}$ can be formed by taking all possible products of elements of
$$B = \{1, \sqrt{2}\}, \ B' = \{1, \sqrt{3}\}, \text{ and } B'' = \{1, \sqrt{5}\}.$$
Therefore, a basis for $\mathbf{Q}(\sqrt{2}, \sqrt{3}, \sqrt{5})/\mathbf{Q}$ is
$$\{1, \sqrt{2}, \sqrt{3}, \sqrt{5}, \sqrt{6}, \sqrt{10}, \sqrt{15}, \sqrt{30}\}.$$

11. If c_1, c_2 are algebraic over F, then $F(c_1, c_2)$ is a field, and every element of $F(c_1, c_2)$ is algebraic over F. Since $c_1, c_2 \in F(c_1, c_2)$ and since $F(c_1, c_2)$ is closed under addition, subtraction, and multiplication and has inverses, the result follows immediately.

Problem Set 8.4

1. The Rational Root Theorem 6.3.8 can be used to show that neither $f(x)$ nor $g(x)$ has a root in $\mathbf{Q}$, so because $f(x)$ and $g(x)$ both have degree 2, they must be irreducible in $\mathbf{Q}[x]$. $f(x)$ has $\pm\sqrt{5}$ for its roots, so $\mathbf{Q}(\sqrt{5})$ is the splitting field for $f(x)$. $g(x)$ has $1\pm\sqrt{5}$ for its roots, so $\mathbf{Q}(1 + \sqrt{5}) = \mathbf{Q}(\sqrt{5})$ is the splitting field for $g(x)$ as well. Note that $\mathbf{Q}(1 + \sqrt{5}) = \mathbf{Q}(\sqrt{5})$ by part (c) of Exercise 9 of Problem Set 8.2.

3. (a) $x^2 + [1]$ factors in $\mathbf{Z}_2[x]$ as $(x + [1])(x + [1])$. Thus, $\mathcal{S} = \mathbf{Z}_2$ is the splitting field for $f(x)$. $B = \{[1]\}$ is a basis for $\mathcal{S}$ and $[\mathcal{S} : \mathbf{Z}_2] = 1$.

7. Suppose that $\deg(f(x)) = 1$. Then $f(x) = ax + b$. If c_1 is the root of $f(x)$, then $f(c_1) = 0$, so $ac_1 + b = 0$, which gives $c_1 = -a^{-1}b \in F$. Hence, we see that $\mathcal{S}_1 = F(c_1) = F$, and so since $B = \{e\}$ is a basis for $\mathcal{S}_1 / F$, $[\mathcal{S}_1 : F] = 1 \leq 1!$. Next, make the induction hypothesis that if $f(x)$ is any polynomial in $F[x]$ with $\deg(f(x)) = k$, then $[\mathcal{S}_k : F] \leq k!$, where $\mathcal{S}_k$ is the splitting field of $f(x)$. Let $f(x)$ be a polynomial in $F[x]$ with $\deg(f(x)) = k + 1$. Suppose also that $c_1, c_2, \ldots, c_{k+1}$ are the roots of $f(x)$. Then $\mathcal{S}_{k+1} = F(c_1, c_2, \ldots, c_{k+1})$ is the splitting field for $f(x)$. Since $f(x) = [a(x - c_1) \cdots (x - c_k)](x - c_{k+1})$, then $g(x) = a(x - c_1) \cdots (x - c_k)$ is a polynomial with $\deg(g(x)) = k$, so by the induction hypothesis $[\mathcal{S}_k : F] \leq k!$, where $\mathcal{S}_k$ is the splitting field for $g(x)$. Now the splitting field for $f(x)$ is $\mathcal{S}_{k+1} = \mathcal{S}_k(c_{k+1})$ and since $f(c_{k+1}) = 0$, the degree of the minimal polynomial in $\mathcal{S}_k[x]$ for c_{k+1} is at most $k + 1$. (Recall that the minimal polynomial is a monic polynomial in $\mathcal{S}_k[x]$ of smallest degree that has c_{k+1} as a root.) Hence, $[\mathcal{S}_{k+1} : \mathcal{S}_k] \leq k + 1$. But $[\mathcal{S}_{k+1} : F] = [\mathcal{S}_{k+1} : \mathcal{S}_k][\mathcal{S}_k : F] \leq (k + 1)k! = (k + 1)!$, so the result follows by induction.

11. If $a, b \in F$, then there are positive integers s, t such that $a \in F_s$ and $b \in F_t$. Suppose that our notation has been chosen so that $s \leq t$. Then $F_s \subseteq F_t$, so $a, b \in F_t$. Define $a + b$ and ab to be the sum and product of a and b in F_t. Since $F_t \subseteq F$, F is closed under addition and multiplication, and these operations are well-defined since, in the tower, F_k is a subfield of F_{k+1} for each k. If $a \in F$, $a \neq 0$, then $a \in F_s$ for some positive integer s. But then

$a^{-1} \in F_s \subseteq F$, so F has inverses. All the other field properties hold since if not, there would exist a field F_s in the tower in which one or more of the field properties would fail. For example, suppose that $a, b, c \in F$ are such that $a(bc) \neq (ab)c$. If one goes far enough up in the tower, there is a field F_s for which $a, b, c \in F_s$ and $a(bc) \neq a(bc)$ in F_s, which is a contradiction.

Problem Set 8.5

3. First, we will show that $(a) \Rightarrow (c)$. Since we can construct an angle that measures α degrees, suppose that this is done as shown in the figure. Use a compass to measure the unit length from the point P on the terminal side of the angle. Label this point U. Construct a perpendicular through U to the initial side of the angle, and label this point A. Since the length of PU is 1, we see that $\sin \alpha$ is the length of AU. Thus, $\sin \alpha$ is constructible.

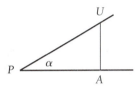

$(c) \Rightarrow (b)$. Now $\sin \alpha$ is constructible, so it follows that $1 - \sin^2\alpha$ is constructible since, by Theorem 8.5.2, the set of all constructible numbers is a field and so is closed under multiplication and subtraction. Since $1 - \sin^2\alpha \geq 0$, Lemma 8.5.7 shows that $\sqrt{1 - \sin^2\alpha}$ is constructible. But $\cos \alpha = \pm\sqrt{1 - \sin^2\alpha}$, and this shows that $\cos \alpha$ is constructible since $|\cos\alpha| = \sqrt{1 - \sin^2\alpha}$.

$(b) \Rightarrow (a)$. We show that if $\cos \alpha$ is constructible, then α is constructible. Construct a half-line, and label the end of the half-line P. Using a ruler and compass, construct a line segment of length $\cos \alpha$ that has one end at point P. Label the other end of this line segment A. Next, construct a perpendicular to the half-line at point A. Set the compass so that the space between its points represents the unit length. Place one end of the compass at point P, and strike the arc of a circle that intersects the perpendicular. Label this point of intersection B. Draw a half-line from P through B. It follows that the angle $\angle BPA$ measures α degrees.

9. To show that it is not possible to triple a cube, construct a cube with each side of length 1. If we can triple the volume of this cube, then we must be

able to construct a cube with sides of length s such that $s^3 = 3$. Thus, s must be a root of the polynomial $f(x) = x^3 - 3$. Since $f(x)$ has no rational roots, $f(x)$ is irreducible in $\mathbf{Q}[x]$, so s has a minimal polynomial in $\mathbf{Q}[x]$ of degree 3, and 3 is not a power of 2. (See Theorem 8.5.9.) Thus, s is not a constructible number, so it is impossible to construct a cube whose volume is 3. This clearly holds if 3 is replaced by any prime integer p.

Problem Set 9.1

3. First we show that if $m = 3uv$, $u^3 - v^3 = n$, and $\omega^3 = 1$, then $x = \omega u - \omega^2 v$ is a solution of $x^3 + mx = n$. This follows since
$$(\omega u - \omega^2 v)^3 + (3uv)(\omega u - \omega^2 v)$$

$$= \omega^3 u^3 - 3\omega^4 u^2 v + 3\omega^5 uv^2 - \omega^6 v^3$$
$$= u^3 - 3\omega u^2 v + 3\omega^2 uv^2 - v^3 + 3\omega u^2 v - 3\omega^2 uv^2 - v^3$$
$$= u^3 - v^3 = n.$$

The fact that under the same conditions $x = \omega^2 u - \omega v$ is also a solution follows in exactly the same manner.

5. (b) The product
$$\left(x - 3\sqrt[3]{1 + \sqrt{\tfrac{28}{27}}} + 3\sqrt[3]{-1 + \sqrt{\tfrac{28}{27}}}\right)\left(x - \omega\,3\sqrt[3]{1 + \sqrt{\tfrac{28}{27}}} + \omega^2\,3\sqrt[3]{-1 + \sqrt{\tfrac{28}{27}}}\right)$$
$$\left(x - \omega^2\,3\sqrt[3]{1 + \sqrt{\tfrac{28}{27}}} + \omega\,3\sqrt[3]{-1 + \sqrt{\tfrac{28}{27}}}\right),$$
where $\omega = -\tfrac{1}{2} + \tfrac{\sqrt{3}}{2}\,i$, gives
$$x^3 + 3\left(1 + \sqrt{\tfrac{28}{27}}\right)\left(-1 + \sqrt{\tfrac{28}{27}}\right)x - 2 = x^3 + x - 2.$$

(c) The product $(x - \sqrt{2})(x + \sqrt{2})(x - \sqrt{3})(x + \sqrt{3})$ gives
$(x^2 - 2)(x^2 - 3) = x^4 - 5x^2 + 6$.

Problem Set 9.2

3. H is a solvable group since, by Theorem 9.2.4, subgroups of solvable groups are solvable. Hence, by Theorem 9.2.8, there is a positive integer n such that $H^{[n]} = \{e\}$. Now suppose that $H \neq \{e\}$ and $H' = H$. Then $H^{[2]} = (H')' = H' = H \neq \{e\}$. It follows by induction that $H^{[n]} \neq \{e\}$ for each positive integer n, and this is a contradiction. Thus, if $H \neq \{e\}$, then $H' \neq H$.

Problem Set 9.3

1. (b) Note that $\mathrm{Gal}(E/F) \neq \varnothing$ since the identity mapping $\epsilon : E \longrightarrow E$ is in $\mathrm{Gal}(E/F)$. Therefore, to show that $\mathrm{Gal}(E/F)$ is a subgroup of $\mathrm{Aut}(E)$, it suffices, by Theorem 2.2.3, to show that $f \circ g^{-1} \in \mathrm{Gal}(E/F)$ whenever $f, g \in \mathrm{Gal}(E/F)$. If $x, y \in F$ and $g^{-1}(y) = x$, then $y = g \circ g^{-1}(y) = g(x) = x$, so $g^{-1}(x) = x$ for each $x \in F$. Hence, $f \circ g^{-1}(x) = f(g^{-1}(x)) = f(x) = x$ for each $x \in F$, and so $f \circ g^{-1}$ is in $\mathrm{Gal}(E/F)$.

5. Each of these polynomials can be factored as follows:

$$f(x) = (x - \sqrt{5})(x + \sqrt{5})$$

$$g(x) = (x - \sqrt{5})^2 (x + \sqrt{5})^2$$

$$h(x) = \left(x - \frac{5}{2} - \frac{3}{2}\sqrt{5}\right)\left(x - \frac{5}{2} + \frac{3}{2}\sqrt{5}\right)$$

If $\sqrt{5}$ is adjoined to $\mathbf{Q}$, then $\mathbf{Q}(\sqrt{5})$ contains the roots of $f(x)$, $g(x)$, and $h(x)$. Thus, $\mathcal{S} = \mathbf{Q}(\sqrt{5})$ is the splitting field for $f(x)$, $g(x)$, and $h(x)$. It follows that the Galois group of each of these polynomials is $\mathrm{Gal}(\mathcal{S}/\mathbf{Q}) = \{\epsilon, \alpha\}$, where $\epsilon(a + b\sqrt{5}) = a + b\sqrt{5}$ and $\alpha(a + b\sqrt{5}) = a - b\sqrt{5}$.

7. The roots of an irreducible quadratic polynomial $ax^2 + bx + c$ in $\mathbf{Q}[x]$ are given by the quadratic formula. Hence, $x = \dfrac{-b \pm \sqrt{b^2 - 4ac}}{2a}$, where $\sqrt{b^2 - 4ac} \notin \mathbf{Q}$. Hence, the splitting field of $ax^2 + bx + c$ is $\mathbf{Q}(\sqrt{d})$, where $d = b^2 - 4ac$. Since the minimal polynomial in $\mathbf{Q}[x]$ of $\sqrt{d}$ is $m(x) = ax^2 + bx + c$, $B = \{1, \sqrt{d}\}$ is a basis for the field extension $\mathbf{Q}(\sqrt{d})/\mathbf{Q}$. There are two automorphisms in $\mathrm{Gal}(\mathbf{Q}(\sqrt{d})/\mathbf{Q})$: the identity map $\epsilon(a + b\sqrt{d}) = a + b\sqrt{d}$ and the automorphism $\alpha(a + b\sqrt{d}) = a - b\sqrt{d}$. The multiplication table for $\mathrm{Gal}(\mathbf{Q}(\sqrt{d})/\mathbf{Q})$ and the addition table for $\mathbf{Z}_2$ are

$\cdot$	ϵ	α
ϵ	ϵ	α
α	α	ϵ

and

$+$	$[0]$	$[1]$
$[0]$	$[0]$	$[1]$
$[1]$	$[1]$	$[0]$

respectively. The one-to-one correspondence given by $\epsilon \leftrightarrow [0]$ and $\alpha \leftrightarrow [1]$ clearly preserves the operations. Hence, the result.

9. (a) The automorphisms ϵ, α, β, $\gamma \in \mathrm{Gal}(\boldsymbol{S}/\mathbf{Q})$ are given by

$$\epsilon : \begin{pmatrix} \sqrt{2} & -\sqrt{2} & \sqrt{3}i & -\sqrt{3}i \\ \sqrt{2} & -\sqrt{2} & \sqrt{3}i & -\sqrt{3}i \end{pmatrix}, \qquad \alpha : \begin{pmatrix} \sqrt{2} & -\sqrt{2} & \sqrt{3}i & -\sqrt{3}i \\ -\sqrt{2} & \sqrt{2} & \sqrt{3}i & -\sqrt{3}i \end{pmatrix},$$

$$\beta : \begin{pmatrix} \sqrt{2} & -\sqrt{2} & \sqrt{3}i & -\sqrt{3}i \\ \sqrt{2} & -\sqrt{2} & -\sqrt{3}i & \sqrt{3}i \end{pmatrix}, \qquad \gamma : \begin{pmatrix} \sqrt{2} & -\sqrt{2} & \sqrt{3}i & -\sqrt{3}i \\ -\sqrt{2} & \sqrt{2} & -\sqrt{3}i & \sqrt{3}i \end{pmatrix},$$

and the mapping ϕ is given by

$$\epsilon \longmapsto \begin{pmatrix} 1\ 2\ 3\ 4 \\ 1\ 2\ 3\ 4 \end{pmatrix} = (1), \qquad \alpha \longmapsto \begin{pmatrix} 1\ 2\ 3\ 4 \\ 2\ 1\ 3\ 4 \end{pmatrix} = (1,2),$$

$$\beta \longmapsto \begin{pmatrix} 1\ 2\ 3\ 4 \\ 1\ 2\ 4\ 3 \end{pmatrix} = (3,4), \qquad \gamma \longmapsto \begin{pmatrix} 1\ 2\ 3\ 4 \\ 2\ 1\ 4\ 3 \end{pmatrix} = (1,2)(3,4).$$

The map ϕ is clearly a well-defined injection, so we need only show that it is a group homomorphism. This can be accomplished by constructing a multiplication table for the groups $\mathrm{Gal}(\boldsymbol{S}/\mathbf{Q})$ and $G = \{(1), (1, 2), (3, 4), (1, 2)(3, 4)\}$ and showing that products of corresponding elements correspond. Inspection of the following multiplication tables shows that this is indeed the case.

$\cdot$	ϵ	α	β	γ
ϵ	ϵ	α	β	γ
α	α	ϵ	γ	β
β	β	γ	ϵ	α
γ	γ	β	α	ϵ

$\cdot$	(1)	(1,2)	(3,4)	(1,2)(3,4)
(1)	(1)	(1,2)	(3,4)	(1,2)(3,4)
(1,2)	(1,2)	(1)	(1,2)(3,4)	(3,4)
(3,4)	(3,4)	(1,2)(3,4)	(1)	(1,2)
(1,2)(3,4)	(1,2)(3,4)	(3,4)	(1,2)	(1)

13. If $\alpha \in H$, then $\alpha(0) = 0$, $\alpha(e) = e$, and thus $0, e \in F_H$. Therefore, not only is $F_H \neq \varnothing$ but F_H contains at least two elements. If $a, b \in F_H$ and $\alpha \in H$, then $\alpha(a - b) = \alpha(a) - \alpha(b) = a - b$. Thus, $a - b \in F_H$, which shows that F_H is an additive subgroup of the additive group of F. We also see that $\alpha(ab) = \alpha(a)\alpha(b) = ab$, so $ab \in F_H$. Hence, F_H is closed under multiplication. Therefore, it remains only to show that F_H has multiplicative inverses. If $a \in F_H$, $a \neq 0$, then a^{-1} exists in F. But for any ring homomorphism, $\alpha(a^{-1}) = \alpha(a)^{-1}$, and since $\alpha(a) = a$, we have $\alpha(a)^{-1} = a^{-1}$. Thus, $\alpha(a^{-1}) = a^{-1}$. The other field properties are inherited from F; that is, they must hold in F_H since $F_H \subseteq F$.

Problem Set 9.4

1. (c) $f(x) = (x^2 - 2)(x^2 + 3) = x^4 + x^2 - 6$ and $\text{Gal}(\mathcal{S}/\mathbf{Q}) = \{\epsilon, \alpha, \beta, \gamma\}$, where $\epsilon, \alpha, \beta,$ and γ are given by

$$\epsilon(a + b\sqrt{2} + c\sqrt{3}i + d\sqrt{6}i) = a + b\sqrt{2} + c\sqrt{3}i + d\sqrt{6}i,$$
$$\alpha(a + b\sqrt{2} + c\sqrt{3}i + d\sqrt{6}i) = a - b\sqrt{2} + c\sqrt{3}i - d\sqrt{6}i,$$
$$\beta(a + b\sqrt{2} + c\sqrt{3}i + d\sqrt{6}i) = a + b\sqrt{2} - c\sqrt{3}i - d\sqrt{6}i,$$
$$\gamma(a + b\sqrt{2} + c\sqrt{3}i + d\sqrt{6}i) = a - b\sqrt{2} - c\sqrt{3}i + d\sqrt{6}i.$$

The subgroups of $\text{Gal}(\mathcal{S}/\mathbf{Q})$ are $H_1 = \{\epsilon\}$, $H_2 = \{\epsilon, \alpha\}$, $H_3 = \{\epsilon, \beta\}$, $H_4 = \{\epsilon, \gamma\}$, and $H_5 = \{\epsilon, \alpha, \beta, \gamma\}$. The fixed fields of $H_1, H_2, H_3, H_4,$ and H_5 are $\mathcal{S}$, $\mathbf{Q}(\sqrt{3}\,i), \mathbf{Q}(\sqrt{2}), \mathbf{Q}(\sqrt{6}\,i)$ and $\mathbf{Q}$, respectively. The Galois correspondence is

$$H_1 \leftrightarrow \mathcal{S} \qquad\qquad H_4 \leftrightarrow \mathbf{Q}(\sqrt{6}i)$$
$$H_2 \leftrightarrow \mathbf{Q}(\sqrt{3}i) \qquad H_5 \leftrightarrow \mathbf{Q}.$$
$$H_3 \leftrightarrow \mathbf{Q}(\sqrt{2})$$

3. The splitting field for $f(x)$ is $\mathcal{S} = \mathbf{Q}(\sqrt[4]{2}, i) = \mathbf{Q}(\sqrt[4]{2}, \sqrt[4]{2}\,i)$ since $f(x)$ factors as $f(x) = (x - \sqrt[4]{2})(x + \sqrt[4]{2})(x - \sqrt[4]{2}\,i)(x + \sqrt[4]{2}\,i)$. The roots of $f(x)$ are $\pm\sqrt[4]{2}$ and $\pm\sqrt[4]{2}\,i$. The minimal polynomial in $\mathbf{Q}[x]$ for $\sqrt[4]{2}$ is $f(x)$, so $[\mathbf{Q}(\sqrt[4]{2}) : \mathbf{Q}] = 4$. The minimal polynomial in $\mathbf{Q}(\sqrt[4]{2})[x]$ for $\sqrt[4]{2}\,i$ is $m(x) = x^2 + \sqrt{2}$, and this gives $[\mathcal{S} : \mathbf{Q}(\sqrt[4]{2})] = 2$. Hence, $[\mathcal{S} : \mathbf{Q}] = [\mathcal{S} : \mathbf{Q}(\sqrt[4]{2})][\mathbf{Q}(\sqrt[4]{2}) : \mathbf{Q}] = 2 \cdot 4 = 8$, so $|\text{Gal}(\mathcal{S}/\mathbf{Q})| = 8$ since $|\text{Gal}(\mathcal{S}/\mathbf{Q})| = [\mathcal{S} : \mathbf{Q}]$.

7. Since $f(x) = (x^2 - 2)(x^2 - 3)$, the roots of $f(x)$ are $x = \pm\sqrt{2}$ and $x = \pm\sqrt{3}$. Hence, the splitting field of $f(x)$ is $\mathbf{Q}(\sqrt{2}, \sqrt{3})$. The minimal polynomial in $\mathbf{Q}[x]$ of $\sqrt{2}$ is $x^2 - 2$, so $[\mathbf{Q}(\sqrt{2}) : \mathbf{Q}] = 2$. The minimal polynomial in $\mathbf{Q}(\sqrt{2})[x]$ of $\sqrt{3}$ is $x^2 - 3$, so $[\mathcal{S} : \mathbf{Q}(\sqrt{2})] = 2$. Thus, $[\mathcal{S} : \mathbf{Q}] = [\mathcal{S} : \mathbf{Q}(\sqrt{2})] [\mathbf{Q}(\sqrt{2}) : \mathbf{Q}] = 2 \cdot 2 = 4$. Therefore, $|\text{Gal}(\mathcal{S}/\mathbf{Q})| = 4$, and these automorphisms can be determined by permuting the roots of $f(x)$ that have the same minimal polynomial. Hence, we have

$$\epsilon: \begin{cases} \sqrt{2} \longmapsto \sqrt{2} \\ -\sqrt{2} \longmapsto -\sqrt{2} \\ \sqrt{3} \longmapsto \sqrt{3} \\ -\sqrt{3} \longmapsto -\sqrt{3}, \end{cases} \quad \alpha: \begin{cases} \sqrt{2} \longmapsto -\sqrt{2} \\ -\sqrt{2} \longmapsto \sqrt{2} \\ \sqrt{3} \longmapsto \sqrt{3} \\ -\sqrt{3} \longmapsto -\sqrt{3}, \end{cases} \quad \beta: \begin{cases} \sqrt{2} \longmapsto \sqrt{2} \\ -\sqrt{2} \longmapsto -\sqrt{2} \\ \sqrt{3} \longmapsto -\sqrt{3} \\ -\sqrt{3} \longmapsto \sqrt{3}, \end{cases}$$

$$\gamma: \begin{cases} \sqrt{2} \longmapsto -\sqrt{2} \\ -\sqrt{2} \longmapsto \sqrt{2} \\ \sqrt{3} \longmapsto -\sqrt{3} \\ -\sqrt{3} \longmapsto \sqrt{3}, \end{cases}$$

so $\text{Gal}(\mathcal{S}/\mathbf{Q}) = \{\epsilon, \alpha, \beta, \gamma\}$. The subgroups of $\text{Gal}(\mathcal{S}/\mathbf{Q})$ are $H_1 = \{\epsilon\}$, $H_2 = \{\epsilon, \alpha\}$, $H_3 = \{\epsilon, \beta\}$, and $\text{Gal}(\mathcal{S}/\mathbf{Q})$. Both H_2 and H_3 are isomorphic to the additive group $\mathbf{Z}_2$ via the mappings given by

$$\phi_2 : H_2 \longrightarrow \mathbf{Z}_2 : \begin{cases} \epsilon \longmapsto [0] \\ \alpha \longmapsto [1] \end{cases} \quad \text{and} \quad \phi_3 : H_3 \longrightarrow \mathbf{Z}_2 : \begin{cases} \epsilon \longmapsto [0] \\ \beta \longmapsto [1] \end{cases}.$$

(See Exercise 7, Problem Set 9.3.) Since $\epsilon = \epsilon\epsilon$, $\alpha = \alpha\epsilon$, $\beta = \beta\epsilon$, $\gamma = \alpha\beta$, and $H_2 \cap H_3 = \{\epsilon\}$, we see that $\text{Gal}(\mathcal{S}/\mathbf{Q}) = H_2 \times H_2$. If follows easily that the mapping

$$\phi : \text{Gal}(\mathcal{S}/\mathbf{Q}) \longrightarrow \mathbf{Z}_2 \times \mathbf{Z}_2 : \begin{cases} \epsilon \longmapsto ([0], [0]) \\ \alpha \longmapsto ([1], [0]) \\ \beta \longmapsto ([0], [1]) \\ \gamma \longmapsto ([1], [1]) \end{cases}$$

is a group isomorphism.

Problem Set 10.1

1. (b) Since $e + (-e) = 0$, for any $v \in V$, it follows that $(e + (-e))v = 0v$. Hence, $ev + (-e)v = 0v$. But part (a) gives $0v = 0$, so we have $ev + (-e)v = 0$. Since $ev = v$, $v + (-e)v = 0$. Now add $-v$ to both sides of this equation, and the result follows.

3. (a) Is not a subspace.   (b) Is a subspace.   (c) Is not a subspace.

7. If $c_1(u + v) + c_2(v + w) + c_3(u + w) = 0$, then $(c_1 + c_3)u + (c_1 + c_2)v + (c_2 + c_3)w = 0$. Since the set of vectors $\{u, v, w\}$ is linearly independent, it must be the case that $c_1 + c_3 = 0$, $c_1 + c_2 = 0$, and $c_2 + c_3 = 0$. Subtracting $c_1 + c_3 = 0$ and $c_1 + c_2 = 0$ gives $c_2 - c_3 = 0$, and adding this to $c_2 + c_3 = 0$ produces $c_2 = 0$. Now $c_2 = 0$ and $c_1 + c_2 = 0$ tell us that $c_1 = 0$ and $c_1 = 0$ together with $c_1 + c_3 = 0$ yield $c_3 = 0$. Thus, $c_1(u + v) + c_2(v + w) + c_3(u + w) = 0$ implies that $c_1 = c_2 = c_3 = 0$, so the set of vectors $\{u + v, v + w, u + w\}$ is linearly independent when the set of vectors $\{u, v, w\}$ is linearly independent.

13. (a) If $c_{11}E_{11} + c_{12}E_{12} + c_{21}E_{21} + c_{22}E_{22} = \begin{pmatrix} 0 & 0 \\ 0 & 0 \end{pmatrix}$, then

$$c_{11}\begin{pmatrix} 1 & 0 \\ 0 & 0 \end{pmatrix} + c_{12}\begin{pmatrix} 0 & 1 \\ 0 & 0 \end{pmatrix} + c_{21}\begin{pmatrix} 0 & 0 \\ 1 & 0 \end{pmatrix} + c_{22}\begin{pmatrix} 0 & 0 \\ 0 & 1 \end{pmatrix}$$

$$= \begin{pmatrix} c_{11} & 0 \\ 0 & 0 \end{pmatrix} + \begin{pmatrix} 0 & c_{12} \\ 0 & 0 \end{pmatrix} +$$

$$\begin{pmatrix} 0 & 0 \\ c_{21} & 0 \end{pmatrix} + \begin{pmatrix} 0 & 0 \\ 0 & c_{22} \end{pmatrix} = \begin{pmatrix} c_{11} & c_{12} \\ c_{21} & c_{22} \end{pmatrix} = \begin{pmatrix} 0 & 0 \\ 0 & 0 \end{pmatrix}$$

implies that $c_{11} = c_{12} = c_{21} = c_{22} = 0$. Therefore, B is a linearly independent set of vectors in $M_2(\mathbf{Q})$. If $\begin{pmatrix} a & b \\ c & d \end{pmatrix} \in M_2(\mathbf{Q})$, then $\begin{pmatrix} a & b \\ c & d \end{pmatrix} = aE_{11} + bE_{12} + cE_{21} + dE_{22}$, and B spans $M_2(\mathbf{Q})$. Hence, B is a basis for $M_2(\mathbf{Q})$.

15. If S is not a minimal set of generators of V, then there is a proper subset S_1 of S such that $\langle S_1 \rangle = V$. If S_1 is a minimal set of generators of V, then S_1 is, by Theorem 10.1.15, a basis for V. If S_1 is not a minimal set of generators of V, then there is a proper subset S_2 of S_1 such that $\langle S_2 \rangle = V$. Continuing in this way, a descending chain $S \supset S_1 \supset S_2 \supset \cdots$ of sets of generators of V can be constructed. Since S is finite and since each set in the chain contains fewer elements than the set that immediately precedes it, this chain must terminate. If the chain terminates at S_k, then $S_k = S_{k+1} = S_{k+2} = \cdots$ is a

minimal set of generators of V and so is, by Theorem 10.1.15, a basis of V. Note that it may be the case that $S_k = \emptyset$, in which case $V = \{0\}$. Since S_k contains a finite number of elements, V is finite dimensional.

17. (a) Let $v = u_1 + u_2$ and $v = u'_1 + u'_2$. Then $u_1 + u_2 = u'_1 + u'_2$, and this implies that $u_1 - u'_1 = u'_2 - u_2 \in U_1 \cap U_2 = \{0\}$. Hence, we see that $u_1 = u'_1$ and $u_2 = u'_2$.

   (b) The vector space U_1 has a basis B_1, and we know from Theorem 10.1.16 that there is a basis B of V such that $B_1 \subseteq B$. Let $B_2 = B \setminus B_1$, and set $U_2 = \langle B_2 \rangle$. We know from Theorem 10.1.12 and Definition 10.1.13 that U_2 is a subspace of V. Since B_2 spans U_2, B_2 will be a basis for U_2 if B_2 is a linearly independent set of vectors. But this follows immediately since $B_2 \subseteq B$. Thus, B_2 is a basis for U_2. If v is a vector in V, then $v = c_1 v_1 + c_2 v_2 + \cdots + c_n v_n$. Suppose that our notation has been chosen so that $v_1, v_2, \ldots, v_k \in B_1$ and $v_{k+1}, v_{k+2}, \ldots, v_n \in B_2$. Then $u_1 = c_1 v_1 + c_2 v_2 + \cdots + c_k v_k \in U_1$ and $u_2 = c_{k+1} v_{k+1} + c_{k+2} v_{k+2} + \cdots + c_n v_n \in U_2$. Hence, $v = u_1 + u_2$ is in $U_1 + U_2$, so it follows that $V = U_1 + U_2$. Now suppose that $v \in U_1 \cap U_2$. Then there are vectors $v_1, v_2, \ldots, v_k \in B_1, v_{k+1}, v_{k+2}, \ldots, v_n \in B_2$, and ring elements $c_i, i = 1, 2, \ldots, n$ such that $v = c_1 v_1 + c_2 v_2 + \cdots + c_k v_k$ and $v = c_{k+1} v_{k+1} + c_{k+2} v_{k+2} + \cdots + c_n v_n$. It follows from this that $c_1 v_1 + c_2 v_2 + \cdots + c_k v_k - c_{k+1} v_{k+1} - c_{k+2} v_{k+2} - \cdots - c_n v_n = 0$. But the v_i's are vectors in B, which is a basis for V. Thus, the equation $c_1 v_1 + c_2 v_2 + \cdots + c_k v_k - c_{k+1} v_{k+1} - c_{k+2} v_{k+2} - \cdots - c_n v_n = 0$ tells us that $c_1 = c_2 = \cdots = c_n = 0$. Hence, $v = 0$, so $U_1 \cap U_2 = \{0\}$. Therefore, $V = U_1 \oplus U_2$.

Problem Set 10.2

3. (a) $\ker \phi = \{c(1,1,1) \mid c \in \mathbf{R}\}$ (c) $\ker \phi = \{(0,0,0)\}$

5. If $f(x) = a_0 + a_1 x + a_2 x^2 + a_3 x^3 + a_4 x^4 \in P_4(\mathbf{R})$, then relative to the basis B given in Exercise 4,

$$f(x) = (a_0 - a_1 + a_3) + (a_1 - a_3)(1 + x) + a_2 x^2 + a_3(x + x^3) + a_4 x^4.$$

Hence, the coefficient vector in $\mathbf{R}^5$ is given by

$$a = \zeta_{P_4(\mathbf{R})}(f(x)) = (a_0 - a_1 + a_3, a_1 - a_3, a_2, a_3, a_4).$$

Hence,

$$aA_D = (a_0 - a_1 + a_3, a_1 - a_3, a_2, a_3, a_4) \begin{pmatrix} 0 & 0 & 0 & 0 \\ 1 & 0 & 0 & 0 \\ 0 & 2 & 0 & 0 \\ 1 & 0 & 3 & 0 \\ 0 & -4 & 0 & 4 \end{pmatrix}$$

$$= (a_1, 2a_2 - 4a_4, 3a_3, 4a_4) \qquad \text{and}$$

$$\zeta_{P_3(\mathbf{R})}^{-1}((a_1, 2a_2 - 4a_4, 3a_3, 4a_4))$$

$$= a_1 1 + (2a_2 - 4a_4)x + 3a_3 x^2 + 4a_4(x + x^3)$$

$$= a_1 + 2a_2 x - 4a_4 x + 3a_3 x^2 + 4a_4 x + 4a_4 x^3$$

$$= a_1 + 2a_2 x + 3a_3 x^2 + 4a_4 x^3$$

$$= f'(x).$$

This suffices to show that the diagram is commutative.

7. Let U_λ be the set of all vectors $u \in U$ such that $\phi(u) = \lambda u$. Note that $U_\lambda \neq \emptyset$ since $\phi(0) = 0 = \lambda 0$. If u and u' are in U_λ and $c \in \mathbf{R}$, then $\phi(u + u') = \phi(u) + \phi(u') = \lambda u + \lambda u' = \lambda(u + u')$, so $u + u' \in U_\lambda$. Furthermore, $\phi(cu) = c\phi(u) = c(\lambda u) = (c\lambda)u = (\lambda c)u = \lambda(cu)$ shows that $cu \in U_\lambda$. Hence, U_λ is closed under addition and under scalar multiplication, and so U_λ is a subspace of U.

    If v is a nonzero vector in V and U' is the set of all vectors $u \in U$ such that $\phi(u) = v$, then U' is not a subspace of U. It may be the case that U' is empty, in which case U' is not a subspace of U. If $U' \neq \emptyset$, suppose that U' is a subspace of U. If $u \in U'$, then $\phi(2u) = 2\phi(u) = 2v$. But 2 is a scalar and U' is closed under scalar multiplication, so $2u \in U'$. Hence, it must be the case that $\phi(2u) = v$. Therefore, we have $2v = v$, which shows that $v = 0$, a contradiction. Thus, U' cannot be a subspace of U.

13. Suppose that $u, u' \in U$ are such that $u = c_1 u_1 + c_2 u_2 + \cdots + c_n u_n$ and $u' = c_1' u_1 + c_2' u_2 + \cdots + c_n' u_n$. Then $u + u' = (c_1 + c_1')u_1 + (c_2 + c_2')u_2 + \cdots + (c_n + c_n')u_n$. Hence, we have $\overline{\phi}(u + u') = (c_1 + c_1')\phi(u_1) + (c_2 + c_2')\phi(u_2) + \cdots + (c_n + c_n')\phi(u_n) = c_1\phi(u_1) + c_2\phi(u_2) + \cdots + c_n\phi(u_n) + c_1'\phi(u_1)) + c_2'\phi(u_2) + \cdots + c_n'\phi(u_n) = \overline{\phi}(u) + \overline{\phi}(u')$. If $c \in F$, then $cu = (cc_1)u_1 + (cc_2)u_2 + \cdots + (cc_n)u_n$, so $\overline{\phi}(cu) = (cc_1)\phi(u_1) + (cc_2)\phi(u_2) + \cdots + (cc_n)\phi(u_n) = c(c_1\phi(u_1) + c_2\phi(u_2) + \cdots + c_n\phi(u_n)) = c\overline{\phi}(u)$. Therefore, $\overline{\phi}$ is a linear transformation.

15. Since ϕ is a vector space isomorphism, ϕ is a bijection, and thus, by
Theorem 0.3.7, there is an inverse function $\phi^{-1} : V \to U$ such that $\phi \circ \phi^{-1}$
$= 1_V$ and $\phi^{-1} \circ \phi = 1_U$. If $v, v' \in V$, let $u, u' \in U$ be such that $\phi(u) = v$ and
$\phi(u') = v'$. Then $\phi^{-1}(v) = u$, $\phi^{-1}(v') = u'$, and $\phi(u + u') = \phi(u) +$
$\phi(u') = v + v'$. Hence, we see that $\phi^{-1}(v + v') = u + u' = \phi^{-1}(v) + \phi^{-1}(v')$.
Moreover, if $c \in F$, then, $\phi(cu) = c\phi(u) = cv$, so it follows that $\phi^{-1}(cv) =$
$cu = c\phi^{-1}(v)$. Therefore, $\phi^{-1} : V \to U$ is a linear transformation. Finally, if
$\phi^{-1}(v) = \phi^{-1}(v')$, then $\phi(\phi^{-1}(v)) = \phi(\phi^{-1}(v'))$, so $\phi \circ \phi^{-1}(v) =$
$\phi \circ \phi^{-1}(v')$. Hence, $v = v'$, so ϕ^{-1} is injective. Finally, if $u \in U$, then
$v = \phi(u) \in V$ is such that $\phi^{-1}(v) = \phi^{-1}(\phi(u)) = \phi^{-1} \circ \phi(u) = u$.
Consequently, ϕ^{-1} is injective and thus is a vector space isomorphism.

Problem Set 10.3

1. (a) $\dfrac{\sqrt{77}}{4}$ (b) $\dfrac{\sqrt{217}}{6}$

3. (a) Is not an inner product. (c) Is an inner product.

5. (a) $\begin{pmatrix} 1 & \frac{1}{2} \\ \frac{1}{2} & \frac{1}{3} \end{pmatrix}$

9. (a) $2\sqrt{\dfrac{14}{15}}$

(c) $d(f, g) = \| f - g \| = \sqrt{(f - g | f - g)} = [(f - g | f - g)]^{1/2}$, so the result

follows since $(f - g | f - g) = \displaystyle\int_a^b (f(x) - g(x))^2 dx$.

13. (a) $\cos \theta \cong 0.998768$ (b) $\theta = \cos^{-1}\left(\dfrac{23}{12} \dfrac{105}{239} \dfrac{15}{28} \right) \cong 63.2°$

Problem Set 11.1

1. M is not a left R-module under these operations.

3. Consider Example 2 of 11.1.2 and let $R = \mathbf{R}$. If $\begin{pmatrix} 1 & 0 \\ 1 & 0 \end{pmatrix}$, then

$\begin{pmatrix} 1 & 1 \\ 0 & 0 \end{pmatrix} \in M_2(\mathbf{R})$ and $\begin{pmatrix} 1 \\ 1 \end{pmatrix} \in \overline{\mathbf{R}^2}$. Since $\left(\begin{pmatrix} 1 & 0 \\ 1 & 0 \end{pmatrix} \begin{pmatrix} 1 & 1 \\ 0 & 0 \end{pmatrix} \right) \begin{pmatrix} 1 \\ 1 \end{pmatrix} =$

$\begin{pmatrix} 1 & 1 \\ 1 & 1 \end{pmatrix} \begin{pmatrix} 1 \\ 1 \end{pmatrix} = \begin{pmatrix} 2 \\ 2 \end{pmatrix}$ and $\left(\begin{pmatrix} 1 & 1 \\ 0 & 0 \end{pmatrix} \begin{pmatrix} 1 & 0 \\ 1 & 0 \end{pmatrix} \right) \begin{pmatrix} 1 \\ 1 \end{pmatrix} = \begin{pmatrix} 2 & 0 \\ 0 & 0 \end{pmatrix} \begin{pmatrix} 1 \\ 1 \end{pmatrix} = \begin{pmatrix} 2 \\ 0 \end{pmatrix}$, the

result follows.

5. Consider the ring $M_2(\mathbf{R})$ and the left $M_2(\mathbf{R})$-module $\overline{\mathbf{R}^2}$. Then $\begin{pmatrix} 1 & 0 \\ 0 & 0 \end{pmatrix}$ and

$\begin{pmatrix} 0 \\ 1 \end{pmatrix}$ are both nonzero, yet $\begin{pmatrix} 1 & 0 \\ 0 & 0 \end{pmatrix} \begin{pmatrix} 0 \\ 1 \end{pmatrix} = \begin{pmatrix} 0 \\ 0 \end{pmatrix}$. Let V be a vector space

over a field F, and suppose that $c \in F$, $c \neq 0$, and $x \in V$, $x \neq 0$, are such that $cx = 0$. Then $c \neq 0$ implies that c^{-1} exists in F. From this we see that $c^{-1}(cx) = (c^{-1}c)x = ex = x$. But $c^{-1}(cx) = c^{-1}0_V = 0_V$ by part 1 of Exercise 4. Hence, $x = 0$, and this is a contradiction.

9. Let $\{M_\alpha\}_{\alpha \in \Delta}$ be the family of submodules of M that contain $\{x_1, x_2, \ldots, x_n\}$. We know from Exercise 7 that $\cap_{\alpha \in \Delta} M_\alpha$ is a submodule of M, so it remains only to show that $\cap_{\alpha \in \Delta} M_\alpha = Rx_1 + Rx_2 + \cdots + Rx_n$. Since each $x_i \in M_\alpha$ for all $\alpha \in \Delta$, $Rx_i \subseteq M_\alpha$ for $i = 1, 2, \ldots, n$. Hence, $Rx_1 + Rx_2 + \cdots + Rx_n \subseteq M_\alpha$ for each $\alpha \in \Delta$, so $Rx_1 + Rx_2 + \cdots + Rx_n \subseteq \cap_{\alpha \in \Delta} M_\alpha$. To see the reverse containment, note that $Rx_1 + Rx_2 + \cdots + Rx_n$ is a submodule of M that contains the set $\{x_1, x_2, \ldots, x_n\}$. Hence, $Rx_1 + Rx_2 + \cdots + Rx_n$ is one of the M_α's, and so $\cap_{\alpha \in \Delta} M_\alpha \subseteq Rx_1 + Rx_2 + \cdots + Rx_n$.

15. Let $f(x) = \Sigma a_k x^k$ and $g(x) = \Sigma b_k x^k$ be polynomials in $R[x]$, and suppose that $a \in R$. Then $D[f(x) + g(x)] = D[\Sigma a_k x^k + \Sigma b_k x^k] = D[\Sigma(a_k + b_k)x^k] = \Sigma k(a_k + b_k)x^{k-1} = \Sigma k a_k x^{k-1} + \Sigma k b_k x^{k-1} = D[f(x)] + D[g(x)]$, and $D[af(x)] = D[a\Sigma a_k x^k] = D[\Sigma(aa_k)x^k] = \Sigma k(aa_k)x^{k-1} = a\Sigma k a_k x^{k-1} = aD[f(x)]$. This shows that D is a linear mapping from $R[x]$ to $R[x]$.

19. Since M is an additive abelian group, all that we need to show is that multiplication of elements of M by elements of R satisfies the requirements to be a module. Toward this end let $x, y \in M$ and $a, b \in R$. Then $(a + b) * x = f(a + b)x = (f(a) + f(b))x = f(a)x + f(b)x = a * x + b * x$, $a * (x + y) = f(a)(x + y) = f(a)x + f(a)y = a * x + b * y$, $(ab) * x = f(ab)x = (f(a)f(b))x = f(a)(f(b)x) = a * (b * x)$, and $e_R * x = f(e_R)x = e_s x = x$.

21. (a) First, note that $(0 : x) \neq \emptyset$ since $0 \in (0 : x)$. If a and b are in $(0 : x)$, then $(a + b)x = ax + bx = 0 + 0 = 0$, so $a + b$ is in $(0 : x)$. Next, let $c \in R$. Then $(ca)x = c(ax) = c0 = 0$, so ca is in $(0 : x)$. If we view R as a left R-module, this shows that $(0 : x)$ is a submodule of R. But the submodules of R viewed as a left R-module are the left ideals of R. Hence, $(0 : x)$ is a left ideal of R.

    (c) Consider the mapping $f : R \rightarrow Rx$ that is defined by $f(a) = ax$.

23. It was shown in part (a) of Exercise 12 that if $pr_1 : M_1 \times M_2 \rightarrow M_2$ is such that $pr_1((x,y)) \longmapsto x$, then pr_1 is an R-linear mapping. This mapping is an epimorphism since if $x \in M_1$, then for any $y \in M_2$, $(x, y) \in M_1 \times M_2$ and $pr_1((x, y)) = x$. Now $(x, y) \in \ker pr_1$ if and only if $pr_1((x, y)) = x = 0$, so it follows that $\ker pr_1 = \{0\} \times M_2$. Hence, Theorem 11.1.12 shows that $(M_1 \times M_2)/(\{0\} \times M_2)$ and M_1 are isomorphic left R-modules. The proof that $(M_1 \times M_2)/(M_1 \times \{0\})$ and M_2 are isomorphic is similar; just work with the second component of the ordered pairs rather than the first.

Problem Set 11.2

7. (a) Let $S^{-1}P = \{\frac{x}{s} \mid x \in P \text{ and } s \in S\}$. First, we need to show that $S^{-1}P$ is an ideal of $S^{-1}R$. If $\frac{x}{s}, \frac{y}{t} \in S^{-1}P$ and $\frac{a}{u} \in S^{-1}R$, then $\frac{x}{s} + \frac{y}{t} = \frac{tx + sy}{st}$ and $\frac{a}{u}\frac{x}{s} = \frac{ax}{us}$. Since $x, y \in P$, $tx + sy \in P$, so $\frac{tx + sy}{st} \in S^{-1}P$ since $st \in S$. Likewise, $ax \in P$, so $\frac{ax}{us} \in S^{-1}P$. Hence, $S^{-1}P$ is an ideal of $S^{-1}R$. We claim that $S^{-1}P$ is a maximal ideal of $S^{-1}R$. Suppose that I is an ideal of $S^{-1}R$ and $S^{-1}P \subseteq I \subseteq S^{-1}R$. If $S^{-1}P \neq I$, then there is an element $\frac{x}{s} \in I$ and $\frac{x}{s} \notin S^{-1}P$. Hence, we see that $x \notin P$, so $x \in S$. Therefore $\frac{s}{x} \in S^{-1}R$ and $\frac{e}{e} = \frac{sx}{sx} = \frac{s}{x}\frac{x}{s} \in I$. Therefore, $I = S^{-1}R$, and we have shown that $S^{-1}P$ is a maximal ideal of $S^{-1}R$. Finally, we need to show that $S^{-1}P$ is the only maximal ideal of $S^{-1}R$. If M is a maximal ideal of $S^{-1}R$, we claim that $M \subseteq S^{-1}P$. If M is not contained in $S^{-1}P$, there is an $\frac{x}{s} \in M$ such that $\frac{x}{s} \notin S^{-1}P$. But then $x \notin P$, and so $x \in S$. Thus, $\frac{s}{x} \in S^{-1}R$, so $\frac{e}{e} = \frac{s}{x}\frac{x}{s} \in M$. Therefore, $M = S^{-1}R$, which is a contradiction since a maximal ideal of $S^{-1}R$ must be, by definition, a proper ideal of $S^{-1}R$. Hence, $M \subseteq S^{-1}P$, so either $M = S^{-1}P$ or $S^{-1}P = S^{-1}R$. If $S^{-1}P = S^{-1}R$, we again have a contradiction to the fact that $S^{-1}P$ is a maximal ideal. Thus, $M = S^{-1}P$, and so $S^{-1}P$ is unique.

(b) Yes, $S^{-1}I$ is an ideal of $S^{-1}R$. The proof is exactly the same as the proof that $S^{-1}P$ is an ideal given in part (a). Just replace P by I.

9. The mapping $f: R \to Q_{cl}(R)$ is given by $f(x) = \frac{x}{e}$. If $x = y$, then $\frac{x}{e} = \frac{y}{e}$, so $f(x) = f(y)$. Thus, f is well-defined. Now suppose that $\frac{x}{s}$ and $\frac{y}{t}$ are arbitrary elements of $Q_{cl}(R)$. Then $f(x + y) = \frac{x+y}{e} = \frac{x}{e} + \frac{y}{e} = f(x) + f(y)$ and $f(xy) = \frac{xy}{e} = \frac{x}{e}\frac{y}{e} = f(x)f(y)$, and so f is a ring homomorphism. If $f(x) = f(y)$, then $\frac{x}{e} = \frac{y}{e}$, so there is an $s \in S$ such that $s(ex - ey) = 0$ or $s(x - y) = 0$. But S is the set of regular elements of R, and so s is not a zero-divisor. Therefore, $s(x - y) = 0$ implies that $x = y$, so $\frac{x}{e} = \frac{y}{e}$. This shows that f is injective.

11. (a) Yes. If $\frac{x}{s}, \frac{y}{t} \in S^{-1}M$ are such that $S^{-1}f\left(\frac{x}{s}\right) = S^{-1}f\left(\frac{y}{t}\right)$, then $\frac{f(x)}{s} = \frac{f(y)}{t}$, so there is a $u \in S$ such that $u(tf(x) - sf(y)) = 0$. Hence, $f(utx) = f(usy)$, so $utx = usy$ since f is injective. But $u(tx - sy) = 0$ implies that $\frac{x}{s} = \frac{y}{t}$. Therefore, $S^{-1}f$ is injective.

(b) Yes. If $\frac{y}{s} \in S^{-1}N$, then $y \in N$, and there is an $x \in M$ such that $f(x) = y$. Hence, $S^{-1}f\left(\frac{x}{s}\right) = \frac{f(x)}{s} = \frac{y}{s}$.

Problem Set 11.3

3. (a) Let $\{M_\alpha\}_{\alpha \in \Delta}$ be the family of submodules of M that contain S, and let's denote $\bigcap_{\alpha \in \Delta} M_\alpha$ by $\langle S \rangle$. We already know that $\langle S \rangle$ is a submodule of M (Theorem 11.1.5), and since $S \subseteq M_\alpha$ for each $\alpha \in \Delta$, $S \subseteq \langle S \rangle$. Hence, $\langle S \rangle$ is a submodule of M containing S. Now suppose that N is a sub-module of M such that $S \subseteq N$. Then N is a member of the family $\{M_\alpha\}_{\alpha \in \Delta}$, so $\langle S \rangle \subseteq N$. This shows that $\langle S \rangle$ is the smallest submodule of M that contains S.

(c) Since $S \subseteq Rx_1 + Rx_2 + \cdots + Rx_n$, $Rx_1 + Rx_2 + \cdots + Rx_n$ is a member of the family $\{M_\alpha\}_{\alpha \in \Delta}$ of all submodules of M that contain S. Hence, $\langle S \rangle \subseteq Rx_1 + Rx_2 + \cdots + Rx_n$. Since $S \subseteq \langle S \rangle$ and $\langle S \rangle$ is a submodule of M, $Rx_i = \{ax_i \mid a \in R\} \subseteq \langle S \rangle$ for $i = 1, 2, \ldots, n$. Thus, $Rx_1 + Rx_2 + \cdots + Rx_n \subseteq \langle S \rangle$, so $\langle S \rangle = Rx_1 + Rx_2 + \cdots + Rx_n$.

5. (a) $B = \{([1]_2, [1]_5)\}$ (c) $\mathbf{Z}_2 \times \mathbf{Z}_2 \times \mathbf{Z}_3$ is not a free $\mathbf{Z}_6$-module.

7. Suppose that N is free with basis $f(B)$. Let's show that f is an isomorphism. Suppose that $f(x) = 0$. Since B is a basis for M, there exist $a_1, a_2, \ldots, a_n \in R$ and $x_1, x_2, \ldots, x_n \in B$ such that $x = a_1 x_1 + a_2 x_2 + \cdots + a_n x_n$. Hence, $f(a_1 x_1 + a_2 x_2 + \cdots + a_n x_n) = 0$. From this it follows that $a_1 f(x_1) + a_2 f(x_2) + \cdots + a_n f(x_n) = 0$, so because $f(B)$ is a basis, it must be the case that $a_1 = a_2 = \cdots = a_n = 0$. Hence, $x = 0$, so $\ker f = \{0\}$. Thus, f is an injection. Next, note that if $y \in M$, then there are $a_1, a_2, \ldots, a_n \in R$ and $f(x_1), f(x_2), \ldots, f(x_n) \in f(B)$ such that $y = a_1 f(x_1) + a_2 f(x_2) + \cdots + a_n f(x_n)$. But then $a_1 x_1 + a_2 x_2 + \cdots + a_n x_n \in M$, and it follows easily that $f(a_1 x_1 + a_2 x_2 + \cdots + a_n x_n) = y$. Hence, f is surjective, and we have shown that f is an isomorphism.

    Conversely, suppose that f is an isomorphism. Let's show that $f(B)$ is a basis for N. If $y \in N$, then there is an $x \in M$ such that $f(x) = y$. If $a_1, a_2, \ldots, a_n \in R$ and $x_1, x_2, \ldots, x_n \in B$ are such that $x = a_1 x_1 + a_2 x_2 + \cdots + a_n x_n$, then $a_1 f(x_1) + a_2 f(x_2) + \cdots + a_n f(x_n) = f(a_1 x_1 + a_2 x_2 + \cdots + a_n x_n) = f(x) = y$, which shows that $f(B)$ spans N. Finally, if $a_1 f(x_1) + a_2 f(x_2) + \cdots + a_n f(x_n) = 0$, then $f(a_1 x_1 + a_2 x_2 + \cdots + a_n x_n) = 0$, so $a_1 x_1 + a_2 x_2 + \cdots + a_n x_n = 0$ since f is an injection. Thus, $a_1 = a_2 = \cdots = a_n = 0$, which shows that $f(B)$ is linearly independent. Hence, $f(B)$ is a basis for N.

Index

D

E

T

U